**Series Editor: Herbert C. Friedmann**
**The University of Chicago**

Volume

1 ENZYMES / *Herbert C. Friedmann*
2 PELLAGRA / *Kenneth J. Carpenter*
3 MOLECULAR PROCESSES IN VISION / *Edwin W. Abrahamson and Sanford Eugene Ostroy*

A BENCHMARK® Books Series

# ENZYMES

Edited by
**HERBERT C. FRIEDMANN**
**The University of Chicago**

Hutchinson Ross Publishing Company
Stroudsburg, Pennsylvania / Woods Hole, Massachusetts

Benchmark Papers in Biochemistry, Volume 1
Library of Congress Catalog Card Number: 79-22573
ISBN: 0-87933-367-7



83 82 81 1 2 3 4 5
Manufactured in the United States of America.

LIBRARY OF CONGRESS CATALOGING IN PUBLICATION DATA
Main entry under title:
Enzymes.
(Benchmark papers in biochemistry; 1)
Includes bibliographical references and indexes.
1. Enzymes—Addresses, essays, lectures. I. Friedmann, Herbert Claus, 1927- II. Series.
QP601.E524 574.1'925'08 79-22573
ISBN 0-87933-367-7

Distributed world wide by Academic Press,
a subsidiary of Harcourt Brace Jovanovich,
Publishers.

# CONTENTS

## PART II: CONSOLIDATIONS: DEVELOPMENTS OF TECHNIQUES, TERMS, AND THEORIES



## PART III: THE WORD ENZYME: THE END OF ONE ERA, THE BEGINNING OF ANOTHER

## PART IV: FIRST MECHANISTIC OBSERVATIONS: ENZYME-SUBSTRATE COMPLEX POSTULATED, STEREOSPECIFICITY, REVERSIBILITY, ENERGETICS, KINETICS, GRAPHICAL METHODS, AND pH

## PART V: FIRST CRYSTALLINE ENZYMES: PROTEINS WITH ENZYMATIC ACTIVITY



## PART VI: ENZYME SUBSTRATE INTERACTIONS: ACTIVE CENTER, TERNARY COMPLEX, SUBSTRATE ACTIVATION, ENZYME-SUBSTRATE COMPLEX, THREE-POINT ATTACHMENT, AND STEREOSPECIFIC HYDROGEN TRANSFER

## PART XII: ENZYME STRUCTURE: AMINO ACID SEQUENCE AND THREE-DIMENSIONAL STRUCTURE



## PART XIII: ENZYME SYNTHESIS: RIBONUCLEASE

# PREFACE

> We have no knowledge, that is, no general principles, drawn from the contemplation of particular facts, but what has been built up by pleasure, and exists in us by pleasure alone. The Man of Science, the Chemist and Mathematician, whatever difficulties and disgusts they may have had to struggle with, know and feel this.
>
> [Preface to the Third Edition of *Lyrical Ballads,* William Wordsworth, 1802]

> The happy characteristic that belongs to every classic, that which makes it classic and immortal, is the absolute harmony of the two forces, form and content. This concord is so absolute that a later reflective age will scarcely be able to separate, even for thought, the two constituent elements here so intimately united, without running the risk of entertaining or provoking a misunderstanding.
>
> [*Either/Or,* Søren Kierkegaard, 1843]

> The words and phrases used by men and women throughout the ages are the loveliest flowers of humanity...the whole past from the time when the word was coined is crystallized in it; it represents not only clear ideas, but endless ambiguities.
>
> [*A History of Science,* George Sarton, 1952]

In the year 8 of the French Republic, the Institut National des Sciences et Arts formulated the central problem of enzymology. It wanted to know the difference between what we now call *enzyme* and what we now call *substrate*: "What are the characteristics by which animal and vegetable substances which act as ferments can be distinguished from those which they are capable of fermenting?" (translation by Arthur Harden 1911). The person who submitted an acceptable answer was to be given a generous reward: a medal worth, in the brand-new metric system, one kilogram of gold. The late eighteenth century had been particularly full of such prize questions, the precursors of endowed awards that have become fashionable since. The institute was impatient; two years later, when it had received no satisfactory answers, it announced the prize again. Entrants were to submit their reports within two years. Citizen Louis Jacques Thénard was the first to accept the challenge (Thénard 1803) with an important study on fermentation that served as a basis for Liebig's views, but he was not given the valuable prize. A year later (1804), four years after the initial announcement, the prize was withdrawn. In the intervening 170-odd years, much water has flowed under the bridges of Paris. One can say that the insights gained during this long period constitute a variety of approaches to this one perceptive question of the institute, a question basic to an understanding not just of the nature and activity of "ferments" but to that of most areas of biology. These accomplishments would amaze the worthy members of the institute and would undoubtedly deserve their medal many times over. While we can provide answers to the original query from a variety of starting points, there are many aspects of the problem that puzzle us as much as they did the scientists of the year 1800. We are still confronted with a host of unanswered questions about the formation, detailed structure, function, and control of enzymes, all of which fall under the purview of the institute's question. These areas continue to be actively investigated. Present research is built on the edifice of curiosity, ingenuity, and perseverance that is the composite result of the efforts of earlier generations. The field attracted, long before our days of specialization, some of the greatest names of eighteenth- and nineteenth-century chemistry and biology, among them Reaumur, Spallanzani, Thénard, Schwann, Berzelius, Liebig, Berthelot, Pasteur, Eduard Buchner, Emil Fischer, not to mention those, many happily still active, of the present century.

We will meet these builders in this book, mainly as writers of scientific papers. A scientific paper may be considered a classic for any of a variety of reasons: It may impress the reader by the originality of its observations, by a felicitous turn of phrase, by the invention of a term that sums up an attitude or predicts a trend. It may appeal to the present-day reader because long-ignored or contested viewpoints have braved the test of time. It may, again, have had a special impact on the scientific community at the time that it was written. Scientific papers have a time-honored way of presenting discovery in the framework of antecedents, of showing the author's or authors' new ideas and observations against the background of accepted fact or established theory. Hence scientists, by consciously placing themselves in a given intellectual universe, necessarily write their own history. This history, however, is limited by the thoroughness with which the writer probes into past or contemporary discovery, by his judgment as to the important versus the trivial, by the background that he assumes of his readers, and often, indeed, by the oblivion into which his silence may, by design or by default, cast the achievements of competitors or of half-forgotten predecessors. A scientific classic, whether written yesterday or today, does not have to be completely correct: It may be mistaken in fact, in interpretation, in prediction, but it always manifests at least one outstanding attribute that compensates for weaknesses or omissions or errors that the wisdom of hindsight can so readily pillory with misplaced disdain. A great paper, above all, speaks to the reader with an immediacy, a presence, an aura of conviction, that the textbook writer, the reviewer, or the historian, all once removed, cannot and in fact do not strive to attain. In this manner a great paper, whether written two hundred years ago or today, allows the reader to look beyond its shortcomings, to temper criticism with respect. Mutual interests forge a bond of continuity between past and present, akin to the bond conjured up by a melody, a poem, a building, or a painting, whose first experience and expression are removed in time and environment from ours.

The reading of a scientific paper, however, more than the experience of a work of art, is at least as much intellectual as it is esthetic. Any paper, no matter how great, necessarily reflects the factual knowledge and the theoretical insight of its time. Hence a reader has to avoid the danger, on the one hand, of being carried away to read more into a paper than is actually written in it, and on the other hand, due to unaccustomed nomenclature or style, of finding less in it than it deserves. Between these two extremes there is the possible injustice of regarding a paper so exclusively from the viewpoint of later tacit assumptions and explicit knowledge that any insights, new at the time of writing, are taken for granted by the later reader.

The preceding remarks pertain to the single paper. A paper that is part of an anthology, however, must be read not only in the perspective of its period but also in that of the accompanying reports. History is a record of anticipations colored by hindsight, and an anthology is a record of accomplishments judged by later conventions. The anthology, as a selection, is characterized not only by what it includes but more perhaps by what it omits. It hence is both compressed as well as distorted history. The compression enables the excitement of perspective to be added to the excitement of the feeling of contemporaneousness discussed above. The distortion, however, partly counteracts the opportunities of compression by coloring the past with the prejudices of the present. These prejudices act on the anthology in exactly the same manner as on the individual paper, and they must hence be tempered by considerations similar to those required for the latter. The various editorial comments are designed to provide links, backgrounds, and clarification, and some biographical information about the writers who are no longer alive.

Many names and papers have had to be omitted because of the strictures of time, of space, and of the editor's ignorance. It is hoped that the papers selected here will, as representative and leading publications, give the reader a taste of high adventure, of deep frustration, and of the joy of discovery that follow each other as the true inspiration and reward of scientific research.

A word about organization: One way to arrange papers is by date of publication. This sequence has the advantage that a climate of opinion is common to a number of papers that hindsight would classify into separate subdisciplines, and the disadvantage that it may confuse larger trends. The other way to arrange papers is by division into topics justified by historical developments. For this volume a blend of these two approaches has been found the most appropriate. Thus early papers on activity and on ideas of structure that prepared the groundwork and that are more or less isolated in their incisiveness or experimental clarity are arranged chronologically, while later papers on chemistry, kinetics, and regulation are grouped by topics.

With one exception (Paper 5), the translations were made by the editor from the original German or French. In only two cases were contemporary translations into English available (Spallanzani 1789 and Pasteur 1879), and excerpts from these are reproduced directly. It was decided not to translate the first paper in the series, that by Reaumur (1752), so that the reader could enjoy the beauty of the original type. A summary of this paper is provided. The excerpts from Victor Henri's book (1903) and from Kurt G. Stern's translation (1932) into German of the modification of J. B. S. Haldane's 1930 book are given directly since they are readily understood. It is hoped that the challenge to maintain a balance between readability and reliability has been met in the translations and that, as far as possible, the highly variable individual styles and the personal idiosyncrasies of the various writers have been retained.

This volume would have been more difficult, if not indeed impossible, to compile without the use of the splendid collections of old books and journals of the University of Chicago Libraries and, in a few cases, of the Center for Research Libraries, located within a block of the university campus. All the papers reproduced or translated here, with two exceptions (Papers 18 and 29), and all the extensive secondary references, with three exceptions (the books by Hughes and by Browne quoted in Part I and Irvine's book, quoted in a footnote to Paper 9), were available at these libraries. Paper 18, which contains the first use of the word *enzyme*, by Willy Kühne in 1877, was obtained from the University of Illinois in Urbana; Paper 29, containing Victor Henri's 1903 derivation of what is usually called the Michaelis-Menten equation, is part of the collection of the John Crerar Library in Chicago. The two eighteenth-century books, by Griffith Hughes and by Patrick Browne, on the use in the West Indies of papaya extract as a meat tenderizer, were consulted in the remarkable library of the Field Museum of Natural History in Chicago. William Irvine's 1805 book, of which there appear to be only a few copies in the United States, was made available by the Boston Athenaeum.

I wish to express my gratitude to Dr. John Westley for his critical reading of my comments in some of their incarnations and for valuable suggestions; to Dr. Philip Handler for pointing out that in all editions after the first one of Arthur Harden's book *Alcoholic Fermentation*, it is mentioned why Pasteur, many years before Buchner, failed to obtain alcoholic fermentation with cell-free yeast extracts; and to Dr. Otto Hoffmann-Ostenhof, who suggested that I read the historical section of his book *Enzymologie,* which points to the mention of an enzyme in Homer, from where it was an easy step, via the footnotes of an edition of the *Iliad* by Walter Leaf, to find most of the various other references to enzymes in antiquity. Finally I wish to thank the many authors who generously provided reprints, often their very last ones, for photoreproduction.

HERBERT C. FRIEDMANN

# CONTENTS BY AUTHOR

# INTRODUCTION

In biological studies, perhaps more than in other scientific studies, it sooner or later becomes apparent that any division of subjects of interest is not only arbitrary and artificial but also restrictive and limiting. One has to face the paradox that the selection required for analysis constitutes at one and the same time a road to an understanding and to an ignoring of the whole. However, no matter what the objections be in theory, one can cite a remarkable pragmatic vindication of the historical process of ever more narrow specialization by abstraction from the whole: Here I do not have in mind the practical applications of scientific discoveries but rather a further paradox that to a large extent complements and answers the previous paradox, namely, that side by side with specialization, one constantly observes a coalescence of specialties. New discoveries, dug out as it were by the hard shovel of specializations, cry out for the abolition of the very specializations that brought about these discoveries in the first place. The coalescence of discoveries in various fields affords a pragmatic justification of the initial abstractions. The probability is vanishingly small that self-consistency is accidental and meaningless and that the abstractions dictated by convenience or by historical accident distort understanding to the point of arbitrariness.

Now biochemistry, which as its name shows, resulted directly from the need to bridge or to obliterate separate biological and chemical approaches to the understanding of various restricted questions, is in turn bridging and to a large extent obliterating the distinctions between a host of older disciplines such as botany, zoology, microbiology, genetics, toxicology, embryology, nutrition, general physiology, pharmacology, behavioral psychology, endocrinology, immunology, and pediatrics. If one includes technical advances that feed and are fed by these disciplines, one can extend the list to further areas such as physical, organic, analytical, and inorganic chemistry, metallurgy, and many branches of physics.

The field of enzymology is one of many that illustrates the preceding points. First of all, it is born out of practical and theoretical abstraction: It takes a whole cell, a collection of cells, or a whole tissue, and extracts and perhaps purifies from it one, often minute, component. It studies the chemical and functional properties of this component, and in so doing it builds bridges to other enzymes similarly studied, obtained from the same or from other cells, tissues, and organisms. Before long a number of generalizations stand out with glaring inevitability, for instance, that life is inconceivable without enzymes, that enzymes from different sources have certain definable similarities and play comparable roles in the overall function and control of chemical and biological processes in the living forms from which they were obtained, and that at a certain level of interest it is irrelevant whether the enzyme that one happens to study came from a plant, an animal, or a microorganism. A given enzyme may play a fundamental role in fetal development or in the expression of a particular genetic characteristic, it may be essential for certain transport processes, for the controlled capture or release of certain forms of energy, for the manifestation of certain inherited diseases, for the formation of active forms of certain vitamins, for the inactivation of certain poisons, for the activity of certain drugs, for the function of various hormones. The list goes on and on. Enzymes stand at the crossroads of biological studies not only because of their versatility—as a consequence of which they encompass all biological substances as substrates or as products—but also because they channel chemical processes in certain limited, fundamentally important directions, and because they are subject to regulatory constraints or do themselves exercise regulatory effects. Furthermore, the structure, versatility, and specificity of enzymes pose fascinating molecular problems that can fertilize and be informed by various branches of chemistry and that lead directly to questions about the origin and the evolution of life. Clearly any one enzyme is a selection or practical abstraction not only from its living environment but also from all the other enzymes with which it may there have been structurally or functionally associated. In addition, any one enzyme is a distinct chemical individual.

A perusal of the early literature on enzymes demonstrates that the field of enzymology has grown out of observations that addressed themselves primarily to three distinct questions: What do these agents do? What is their nature? What is their biological role? These questions are interrelated since the phenomenological description of the activity of any given enzyme must of course be understood both from the viewpoint of its chemical structure and of its biological function. It is hence most instructive to learn that progress in enzymology has been and is possible by emphasizing any one of the preceding three questions at the expense of the other two. Thus the very fact that enzymes are catalysts has made it possible for aspects of their activity such as their kinetic behavior and their specificity to be studied without the slightest inkling as to what these substances are. It is only fairly recently that kinetic analysis has begun to be correlated with mechanistic studies at a structural level and with aspects of biological behavior and control. Again the enormously tortuous groping toward an understanding of the molecular architecture of enzymes as individual chemical substances has to a large extent been carried out without immediate concern as to chemical behavior or biological function. Only recently again have the elegant results of x-ray structural studies in many instances made such correlations inevitable. Last, the study of biological function of enzymes, illustrated by the vast field of intermediary metabolism, has been forcefully helped by an understanding of kinetic relationships and of the chemical basis of enzymatic regulatory behavior.

This collection emphasizes the first two of the preceding questions. Although it would have been futile to limit the selections to "pure" enzymologic papers, the ever-present temptation to burgeon into the corporate fields of intermediary metabolism and of vitamin and coenzyme function had by and large to be resisted. Hence this collection does not contain the classical papers, say, by Lavoisier on respiration, by Gay-Lussac on fermentation, by Harden and Young on cozymase, by Thunberg and by Wieland on dehydrogenases, from Meyerhof's laboratory on hexokinase, from Warburg's laboratory on Atmungsferment, on triosephosphate dehydrogenase, and so on. It is hoped that these papers will be found in future Benchmark books.

Part I

# THE BEGINNINGS: ISOLATED OBSERVATIONS

# Editor's Comments on Papers 1 Through 8

1 **REAUMUR**
Excerpts from *Sur la Digestion des oiseaux, Second Mémoire: De la mannière dont elle se fait dans l'estomac des oiseaux de proie*

2 **SPALLANZANI**
Excerpts from *Dissertations Relative to the Natural History of Animals and Vegetables*

3 **PLANCHE**
*Note on the Adulteration of Jalap Resin and on Means to Recognize It, Etc.*

4 **PLANCHE**
Excerpts from *On the Substances Which Develop Blue Color in Guaiac Resin*

5 **KIRCHHOFF**
*Formation of Sugar in Cereal Grains Converted to Malt, and in Flour Steeped in Boiling Water*

6 **THÉNARD**
*New Observations on Oxygenated Water*

7 **LEUCHS**
*Action of Saliva on Starch*

8 **BEAUMONT**
Excerpts from *Experiments and Observations on the Gastric Juice and the Physiology of Digestion*

Many sciences such as astronomy, geology, or genetics have been preceded by what one may call a period of folklore or prehistory, during which valuable insights, undistorted by the exigencies of later abstractions, were obtained. Such prehistorical periods and pursuits differ from the later or historical ones in a number of respects: First of all, the originators of the various ideas or observations are by and large unknown; second, most of these insights are devoted to practical ends; third, valid fact is frequently distorted by what appears to us as a mythological superstructure; fourth, the insights, as a consequence of their practical ends, are not developed or changed over long periods of time; fifth, the exercise of these insights often partakes of elements of art rather than of objective definition; sixth, the original insights are not necessarily the ones that serve as the foundations of the later sciences: a conceptual and observational jump is needed so that the shackles of custom can be ignored in favor of the kind of abstractions that are essential for the development of a science.

Now enzymology also has a long prehistory. Much of this prehistory is centered around fermentative processes with intact microorganisms such as yeasts and acetic acid bacteria. The antiquity of mankind's use of alcoholic beverages (see Lichine 1971) need hardly be stressed. It suffices to mention that there are extensive records concerned with alcoholic beverages in ancient Babylonia (in the Codex of Hammurabi, about 2100 B.C.), Egypt (associated with Osiris and depicted in many inscriptions) Greece and Rome (Dionysos, Bacchus), India (an ancient drink, used perhaps before the soma of the Rig-Veda, was *madhu,* Sanskrit word for honey, related to the English word *mead*) and China (see Needham 1965). It is seen from this list that not only the consumption but also the production of alcohol was frequently associated with a mixture of practical ends and mythological or religious beliefs. A similar mixture of attributes was associated with the formation of the other product of yeast fermentation, carbon dioxide, in the ancient practice of bread-making.[1] Thus one need hardly be reminded of the ramified connotations in the Judeo-Christian tradition of expressions such as unleavened bread, one's daily bread, and so on.

The production and use of dilute acetic acid, known in an impure form as vinegar (French: *vinaigre,* sour wine) was also widely distributed in antiquity. Two points may be made: First, dilute acetic acid in this form was the strongest acid known in the ancient world, and in fact the words *acid* and *acetic* are related; second, vinegar was widely used in the preparation of foodstuffs and in medicine, but without the overt religious overtones that one finds in the case both of alcohol and of carbon dioxide whose effects are more dramatic.

In contrast to the above examples there are others from the

prehistory of enzymology that depend not on the activity of intact microorganisms but rather on the action of what we would now call enzyme extracts. These examples are centered around the important process of cheese manufacture and the practice of meat tenderizing. Right from the start these areas appear to be by and large free of religious adumbrations. In fact, a perusal of early descriptions, particularly of cheese making, reveals remarkable insights into the properties of enzymes. It will be instructive to cite some of these very early examples of the use of enzymes since they teach us that in spite of their accuracy they were singularly sterile in inducing what we would now call scientific approaches or explanations or disciplines. Moreover, these observations were of no use whatsoever to the development of the scientific study of enzymes that had its slow beginnings in the eighteenth century.

Cheese manufacture has the great appeal of bringing about the conversion of an important part of milk into a pleasant-tasting material that is more concentrated and above all more stable than milk. It may have been discovered accidentally upon storage of milk in a dead animal's stomach. The earliest recorded example does not, however, use rennet but rather an extract of the fig tree (now recognized as the source of the proteolytic enzyme, ficin). In this example, found near the end of the fifth book of Homer's Iliad, the rapidity of milk coagulation by such a juice is used as a simile for the speed with which an injury was cured:

> As the juice of the fig-tree curdles milk, and thickens it in a moment though it is liquid, even so instantly did Paeëon cure fierce Mars. (Butler 1898)

The fig was a staple component of ancient Greek food. Homer's use of this particular simile implies that this use of fig juice must have been very widely known.

We now traverse at least five hundred years. In the writings of Aristotle (384–322 B.C.) we find frequent references to the curdling of milk both by fig juice and by rennet (see Partington 1970). This great systematizer had vague inklings that the *in vitro* action of fig juice and the *in vivo* action of rennet are similar, although the passage from the *Parts of Animals* [de partibus animalium] where this idea is expressed includes wrong information:

> What goes by the name of Rennet is present in all animals which have a multiple stomach; the hare is the only animal with a single stomach which has it. . .The hare has rennet because it feeds on herbs with fig-like juice; and this juice can coagulate the milk in the stomach of sucklings. (Peck 1960)

In another work, *History of Animals* [Historia animalium], Aristotle tries to explain the basis of the action of rennet. He bases his idea on the then prevalent notion of the four elements:

> Rennet is a sort of milk; it is formed in the stomach of young animals while still being suckled. Rennet is thus milk which contains fire, which comes from the heat of the animal while the milk is undergoing concoction. (Peck 1965)

It would probably be going too far to claim that this is the earliest theory of enzyme action.

As we continue to follow the chronological sequence we find the following passage on cheese making in the *De re rustica,* or "On Agriculture," by Marcus Terentius Varro (116–27? B.C.), the most erudite person and the most prolific writer of his time, credited with the writing of about 620 books, of which only very few survive:

> In spring the milk for cheese making is drawn in the morning, while at other seasons the milking takes place toward midday; but the practice is not entirely uniform because of differences in locality and food. To two congii[2] of milk is added a bit of rennet the size of an olive, to make it coagulate; this is better when made from a hare or a kid than when made from a lamb. Others use, instead of rennet, the milk from the stem of a fig, and vinegar; they also curdle with various other substances. . . . "I should not be surprised, I remarked, if that is the reason that a fig tree was planted by shepherds near the shrine of the goddess Rumina; you know at that place sacrifice is offered with milk instead of with wine and sucklings. . . ." (Hooper and Ash 1934)

One cannot help giving a short quotation from the *Historia naturalis* of Pliny the Elder (c. 23–79), who died in the eruption of Vesuvius that destroyed Pompei and Herculanum and who covered an enormous range of material in this encyclopedic work, not always, as in this case, very accurately:

> There is also a juice in the body of trees, which must be looked upon as their blood. It is not the same in all trees—in figs it is a milkly substance, which has the property of curdling milk so as to produce cheese, in cherries it is gummy, in elms slimy, sticky and fat, in apples, vines and pears watery. The stickier this sap is, the longer the trees live. (Rackham 1945)

By far the most comprehensive, elegant, and accurate presentation of the enzymatic basis of cheese making is found in the *De re rustica* of Lucius Junius Moderatus Columella (flourished first century A.D.):

> It will be necessary too not to neglect the task of cheese-making, especially in distant parts of the country, where it is not convenient to take milk to the market in pails. . . .Cheese

should be made of pure milk which is as fresh as possible, for if it is left to stand or mixed with water, it quickly turns sour. It should usually be curdled with rennet obtained from a lamb or a kid, though it can also be coagulated with the flower of the wild thistle or the seeds of the safflower, and equally well with the liquid which flows from a fig-tree if you make an incision in the bark while it is still green. The best cheese, however, is that which contains only a very small quantity of any drug. The least amount of rennet that a pail of milk requires weighs a silver *denarius;* and there is no doubt that cheese which has been solidified by means of small shoots from a fig-tree has a very pleasant flavour. A pail when it has been filled with milk should always be kept at some degree of heat; it should not, however, be brought into contact with the flames, as some people think it proper to do, but should be put to stand not far from the fire, and, when the liquid has thickened, it should immediately be transferred to wicker vessels or baskets or moulds; for it is of the utmost importance that the whey should percolate as quickly as possible and become separated from the solid matter. (Forster and Heffner 1954)

In these writings then, five fundamental aspects of enzyme action have been clearly described: rapidity, effectiveness in small amounts, heat activation, heat denaturation, and wide distribution. In a rather later publication, the 1640 *Theatrum botanicum* of John Parkinson, yet another source of a so-called cheese rennet is described which goes back to Dioscorides (c.40–c.90), namely *Galerion* or *Galarion,* identified as the golden or yellow lady's bedstraw, *Galium verum* of the madder tribe (see "Notes and Queries," 1889; for further references, see Gow 1950).

Meat tenderizing constitutes a further application of the proteolytic activity of products of the fig tree that is already mentioned in antiquity.[3] As will become clear shortly, the relatively recent descriptions of this use of another plant protease are of immediate interest in the context of this book. It so happens that in the modern world meat tenderizing is indeed among the very first procedures with enzyme extracts to be described. Our examples, which come from the West Indies, use meat tenderizing with the fruit and other parts of the papaya or pawpaw tree, *Caryca papaya* (now known to contain the proteolytic enzyme papain). The first description is found in a book, *The Natural History of Barbados,* published in 1750 by the Reverend Mr. Griffith Hughes, A.M., F.R.S., Rector of St. Lucy's Parish in Barbados:

[The Popaw Tree] is generally of an undivided Trunk, and distinguished into the Male and Female; as well as the Fruit into the long *Mango Popo,* and the round Sort.... Both these Fruits, especially the round Sort, are ... when near ripe, boiled and eaten with any kind of flesh-meat, and esteemed wholesome, if they are cleansed of the milky corrosive juice they contain, and eaten but seldom. This juice is of so penetrating a Nature, that if this unripe Fruit, when unpeeled, is boiled with the toughest old salt Meat, it will soon make it soft and tender; and if Hogs are for any considerable Time fed with it, especially raw, it is said that it will wear off all the mucous slimy Matter, which covers the Inside of the Guts, and would in time, if not prevented by a Change of Food, intirely lacerate them.

I know of no physical Virtue in any Part of this Tree, unless that the milky Juice of the *Popo* is sometimes made use of to cure Ring-worms, and such cutaneous Eruptions.[4]

This book is provided with the most exquisite drawings of many of the plants described, including plates of the male and female papaya trees. It is rather ironic, in our context, that the name Barbados is derived from the bearded fig tree (*Ficus bengaliensis*) found there by the Portuguese.

The second passage is taken from a book, *The Civil and Natural History of Jamaica,* published in 1756, and in two later editions, both 1789, by Patrick Browne, M.D.[5] His description is rather more explicit than that of the Reverend Mr. Hughes (it is taken from the second edition):

The tree is full of an acrid milky juice, which is commonly said to cure the ring-worms; but how true this assertion may be, I am not able to determine... . Water impregnated with the milky juice of this tree, is thought to make all sorts of meat washed in it very tender; but eight or ten minutes steeping, it is said, will make it so soft, that it will drop in pieces from the spit before it is well roasted; or turn soon to rags in the boiling.

That this knowledge was certainly not subsequently forgotten is shown by a passage from an 1851 book *A Naturalist's Sojourn in Jamaica* by the eminent naturalist Philip Henry Gosse:

The Papaw, whose large fruit has the singular property of rendering tender the toughest meat with a few drops of its juice, and the Cocoa-nut which supplies meat and drink, are fine examples of tall and slender grace.

Schwimmer (1954) tells a papaya story from Hawaii, certainly a magnificent example of enzymatic folklore. While it is not exactly reliable as to enzyme specificity, this story pays telling tribute to the catalytic efficiency of enzymes:

Use of the leaves of the papaya tree for tenderizing meats has been known to the natives of Hawaii for several centuries. The story is told that once a native, when introduced to this seemingly occult phenomenon, reflected deeply and concluded that the leaves must contain some magical power which could

> digest all food. Being inordinately fond of rice, he reasoned that ingestion of copious quantities of the papaya leaf should increase his capacity for rice consumption indefinitely. Being a man of action as well as thought, our native hero amassed several bushels of papaya leaves, added them to several gallons of cooked rice, and then proceeded to partake of this feast in the late evening in the quiet of a grove of papaya trees.
>
> When his friends came to his scene of repast the next morning, all they could find was a pile of rice in which was intermingled a collection of bones. The rest of our hero had been digested by the enzymes of the leaves!

It is of very great interest that it was just around the time of the publications, in London, by Hughes and by Browne of the meat-tenderizing activity of papaya juice that Reaumur (1751) and Spallanzani (1780), both, like Hughes, Fellows of the Royal Society of London, carried out their studies on the gastric digestion of meat, studies that may be regarded as initiating the scientific era of enzymology. These two eminent scientists were impressed, as we shall see, by their own laboriously established observations that a chemical, not a physical, activity was responsible for the dissolution of meat in the stomach. They had no inkling that a process of chemical meat tenderizing had already been known in antiquity and on the other side of the Atlantic. They did not know, again, that a very similar process was at work in the age-old manufacture of cheese by the action on milk of another gastric enzyme, rennin, or of what was called vegetable rennet such as fig-tree sap. One wonders, indeed, in what way the development of the science of enzymology would have differed if these workers or the ones immediately following them had been acquainted with these old observations, and if they had been able to institute the necessary correlations. Here it is important to note that the scientific discovery of the chemical activity of gastric juice followed, as we shall see, upon the pursuit of a limited and specific question, namely, is gastric digestion in birds without gizzard analogous to the physical trituration observed in birds with gizzard? Furthermore, the correlation between the curdling of milk and the dissolution of meat could not have been made without prior answers to such limited questions as, what chemical substances are contained in milk and in meat?, and, what is the chemical nature of protein?, and what happens when proteins are denatured? The scientific study of papain was begun only in 1874, and the name *papain* was introduced only in 1879 (Wurtz and Bouchut 1879; see Hwang and Ivy 1951). Scientific studies and the name of ficin go back only to 1930 (Robbins 1930). In a similar fashion we shall see that the problem of the nature of alcoholic fermentation by yeast was approached by some of the protagonists in our book not by asking a general question such as, what does yeast do?, but rather by asking more particular questions which at first sight may seem trivial or irrelevant, such as, is yeast alive? and, is there a difference between a ferment and a yeast? and, can a ferment be extracted from yeast? Clearly hindsight is no substitute for the painstaking way by which much of science is eked out from poker-playing nature. The six points of difference between prehistorical and historical science enumerated above are but an expression of this built-in limitation to the gathering of scientific knowledge.

As this book develops we shall see that the topics which occupy the folklore of enzymology are most pertinent as forerunners of the scientific development of this subject, for in this development the study of proteolytic enzymes runs parallel, like a musical counterpoint, to studies on fermentation and on the hydrolysis of starch.

We now come to the historical or scientific study of enzymes. It often happens that in the development of a science, an enormous lag period, at time dotted with fertile but scientifically unexploited observations, is followed by a series of scientific studies that, initially infrequent and largely independent of each other, begin to gather momentum over a clearly defined and relatively short time span. The end of such an incubation period is signalled by a generalizing, unifying principle: A catchy term is invented that is both a summing up of the more or less scattered observations and a prognosticator of further discoveries, and by this baptismal act, a new discipline or subdiscipline has been sent on its way. In our case the prebaptismal period that encompasses the discovery of various reactions, later called *catalyzed,* both in the inorganic and in the biological fields, extends from about 1752 to about 1835. The former date marks the publication of two lengthy studies of digestion in birds in the Mémoires of the Paris Royal Academy by the 68-year-old René Antoine Ferchault, Sieur de Reaumur (1683–1757), and the latter date the coining of the word *catalysis* by Jöns Jacob Berzelius (1779–1848) in his fifteenth *Årsberättelser öfver Vetenskapernas Framsteg,* or *Annual Surveys of Progress in the Sciences.*

Reaumur[6] was probably the most prestigious member of the French Royal Academy in the first half of the eighteenth century; he had been its director twelve times, its subdirector nine times, and he was a member of the Royal Society of London, of the Academies of Science of Prussia, Russia, and Sweden, and of the Institute of Bologna. He was a metallurgist, worked on porcelain production (a crystalline ceramic that he discovered was used in the twentieth century to protect rocket nose cones from overheating), and one of the greatest naturalists of his or of any age. He wrote a six-volume work on insects with a famous section on bees. He was compared favorably to Charles Darwin by Darwin's great champion, Thomas Henry Huxley.[7] Reaumur is now remembered chiefly for the temperature scale (80° from the freezing to the boiling point of water), which was due to a misunderstanding of one of his articles and which bears his name.

The present study (Paper 1), which describes his experiments on digestion in birds of prey, probably contains the very first systematic observations of what we now call an enzyme. Frequently one initial paper or publication in what will develop into a new discipline manifests certain unmistakable qualities of experimental thoroughness and intellectual insight that set the tone for further developments. In our case this special quality can be found in Reaumur's highly entertaining article. The article preceding the present one deals with the physical trituration of food in the gizzard of various birds. Reaumur now is curious to know how birds without a gizzard digest their food; does the stomach take the place of the gizzard by physically grinding up the digested substances?[8] He bases his experiments on the falconer's knowledge that carnivorous birds throw up material they cannot digest. This is his starting point for the design of an indigestible cage (actually a small metal tube covered with wire gauze at the ends) containing a piece of meat that he lowers into the stomach of a buzzard. He finds that most of the meat has disappeared after 24 hours and that the residue is soft but not putrefied. He concludes: "This process hence can only be the work of a solvent whose existence has been well demonstrated." He further discovers specificity since plant materials are not digested. He asks, "What is the nature of this liquid which somehow acts on meats and on bones just as aqua regia acts on gold but which has little power on starchy substances as has aqua regia on silver? To which of the many solvents given to us by Chemistry can it be compared?" He decides to perform *in vitro* experiments, using sponges contained in his cage to take up some of the solvent from the bird's stomach. Unfortunately his buzzard dies at about this time, but he continues his work on a vulture given to him by a certain Commander Godeheu from Malta. By the use of litmus paper ("blue paper") he discovers that the liquid is acidic. He performs two of these *in vitro* experiments. However, he does not obtain enough gastric juice: He trickles the fluid from his sponges on a piece of meat the size of a hazel nut, but it does not suffice to cover the meat completely. He incubates the meat in an oven that he uses to hatch eggs ("one of these ovens was chosen to have the functions of the stomach"). He uses a similar piece of meat covered with water as a control. However, in spite of precautions taken to prevent evaporation, the meat had dried after 24 hours; the experiment was unsuccessful. The second experiment, with a much thinner piece of meat, also did not succeed. He concludes by speculating that digestion in animals with "membranous stomachs" such as dogs, pigs, fish, reptiles, several insects, should also be brought about by the action of a solvent. In support of this idea he describes one more experiment, using an old dog to which he force-feeds two small pieces of bone; after strangulation of the dog 26 hours later, the bones in its stomach had become smaller and flexible like horn.

Only excerpts from this long article, which is in very clear French, could be reproduced here (Paper 1).

The book by Abbé Lazzaro Spallanzani (1729–1799), *Dissertazioni Di Fisica Animale, e Vegetabile* (1780) appears to have been more influential than Reaumur's work almost thirty years earlier; at any rate his name is always quoted in textbooks of biochemistry, while references to Reaumur are much rarer. Spallanzani, Professor of Natural History at the University of Pavia for the last thirty years of his life, has been called "one of the world's greatest experimenters...A man of the highest intellect and indomitable perseverance" (Bulloch 1938, p. 398). He wrote on subjects whose variety is somewhat bewildering to the modern reader, ranging all the way from the origin of streams, the phenomenon of stone skipping on water, to regeneration in earthworms and many other small animals, to blood circulation, to reproduction, and spermatozoal form and behavior. He performed ingenious experiments to disprove the spontaneous generation of life, continuing along the lines of Francesco Redi [1626–1697(8)] and antedating Pasteur, who had a full-length portrait of Spallanzani in the drawing room of his apartment. His demonstration of the phenomenon of heat sterilization made food canning possible. (Here we have an early example of a far-reaching application of fundamental research.) "His last work, on respiration, was of monumental proportions and included the record of over 12,000 experiments" (Bulloch 1938, p. 398). He instituted the first marine zoological laboratory (in Portovenere). He was a member of many distinguished scientific societies, including the Royal Society of London. The world at large remembers his name in connection with that of the wizard in E. T. A. Hoffmann's *Tales*, written about ten years after his death and immortalized in Jacques Offenbach's opera (where his name is spelled with only one *l*).

In our context one is impressed by the detailed extension in the "Dissertazioni" of Reaumur's experiments: the subject of digestion comprising the first volume of this two-volume work is developed in six dissertations arranged in 264 sections. A French translation of most of this book appeared in 1783, and a translation into English (by Thomas Beddoes, the father of the poet Thomas Lovell Beddoes) in 1784. A second English edition, *Dissertations Relative to the Natural History of Animals and Vegetables*, was published already in 1789. Our excerpts (Paper 2) are taken from this edition. Most textbooks mention the Abbé's *in vivo* experiment on a hawk with the wire cage containing a piece of meat (actually, as we saw, due to Reaumur, whose work he quotes in his book)[9]. Even a cursory reading of Spallanzani's treatise shows, however, that he performed this kind of experiment on an impressive variety of animals (we include here part of a particularly entertaining section, on snakes). In addition, his *in vitro* observations on his own gastric juice showed that it could dissolve meat. He describes control experiments using water instead of gastric juice—a term first used in this treatise—and clearly recognizes the instability of the active principle over long storage times, as well as the requirement for warm temperatures during incubation.[10] In addition, we have probably the first observation in enzyme kinetics, viz. that "digestion of food is proportional to the quantity of gastric juice acting upon it" (Section LXIX). These observations were made before the phenomenon of catalysis had been recognized and before one understood the chemical nature of meat and hence the chemical changes that occur during its dissolution. It is significant that, in a part of the book not reproduced here, Spallanzani seeks to find analogies to the process of fermentation, but he rejects the notion that the dissolution is a fermentative process since it is not associated with the formation of bubbles!

Illustrative of the principle of the historical contiguity of independent scientific observations (usually described by the dictum "the time is/was ripe") is the circumstance that our next three enzyme papers were published within nine years of each other, thirty years after Spallanzani's work, and on widely different enzymes. In 1810 (Paper 3) Louis Antoine Planche (1776–1840), one of the editors of the *Bulletin de Pharmacie*, discovered that horseradish gives a blue coloration with guaiac extract. In 1820 (Paper 4) he listed 25 different roots that had the same effect on guaiac and a number of vegetable substances that had no effect. Planche has the further distinction of coining in the 1820 paper the first term, *cyanogen* ("blue-former"), for an enzyme, although $C_2N_2$ had already been described by Gay-Lussac. Three points may be made: First, one realizes how far we have come since Planche's time, for he thought that the disappearance of activity on heating might indicate the active principle to be volatile. An experiment showed this not to be so. Second, this substance from horseradish, much later called *peroxidase*, was destined to play an important role in the later history of enzymology, for exactly one hundred years after Planche, studies in the influential chemist Willstätter's laboratory led to the conclusion that enzymes cannot possibly be proteins. This conclusion was based principally on the remarkable successes achieved by Willstätter's group with the purification of peroxidase to a stage at which the available methods indicated absence of protein. Third, this enzyme, along with catalase, discovered by Thénard in 1819, was probably the only nonhydrolytic enzyme known until oxidative enzymes were beginning to be studied near the start of the twentieth century.

Chronologically what we now call *α-amylase* is the next enzyme discovered, in malt, by Gottlieb Sigismund Constantin Kirchhoff (1764–1833), royal apothecary in St. Petersburg (now Leningrad). His 1815 paper (Paper 5) is of interest for the following reasons: Kirchhoff had found some four years earlier that acid hydrolysis of starch leads to glucose and that the acid is not consumed in this process. It is through an interest in the biological formation of glucose ("sugar"), derived biologically from starch as a precursor of alcohol, that Kirchhoff discovered α-amylase. So we have, near the very beginning of the scientific study of our subject, the clear demonstration of a biological process that is analogous to a chemical one. The chemical basis of the reaction is very clearly expressed in the summary to this paper: "The formation of sugar in germinated grains is a chemical process, and not the result of vegetation."[11] The biological significance of this chemical change is clearly recognized since it is correlated with the physiological process of germination. Furthermore, Kirchhoff shows quite clearly that a given amount of malt gluten can convert more starch to "sugar" than corresponds to the amount of starch originally present in the grain. Kirchhoff belonged to a number of learned societies, including the Boston Academy of Sciences.

Louis Jacques Thénard (1777–1857), the next author in our series, interests us as the discoverer of hydrogen peroxide. He wrote a number of papers dealing with the decomposition of this substance to oxygen and water by various agents, including biological ones. A translation of the paper that deals with this latter topic is included here (Paper 6). This paper is of profound importance for several reasons, besides the fact of being the first that describes what we now know at catalase: Thénard is the first to study an enzymatic reaction quantitatively; he points out, as did Kirchhoff, the correspondence between a reaction brought about by inorganic and by animal substances; he recognizes that the capacity of the animal materials to bring about this change is, like that of the mineral substances, not changed at the end of the reaction. However, the most impressive statement in this elegant short paper occurs at the end and marks the intuition of genius: after describing what he considers to be the general force responsible for the hydrogen peroxide-decomposing ability of various biological substances (not just of fibrin, the material always mentioned in textbooks) and of various metals, Thénard asks the question: "Is it unreasonable to think, accordingly, that all animal and vegetable secretions are due to an analogous force?" His answer, "I do not think so," ushers in the field of enzymology. It is highly significant that the two biological discoveries that are discussed by Berzelius, in the section of Berzelius' *Yearbook* that introduces the word *catalyst* in the context of a "new natural force" (see Paper 11) are those of Kirchhoff and of Thénard. This editor cannot help feeling that Berzelius' emphasis on a catalytic *force,* and his assumption that thousands of catalytic processes occur in plants and animals, was suggested directly by a reading of Thénard's work. It is rather interesting in this connection to note that in 1820 Thénard published an extensive review (103 pages) of his studies on hydrogen peroxide, which, after giving a lengthy list of animal substances that decompose hydrogen peroxide, repeats, verbatim (p. 466) the preceding categorical question and answer. To put Thénard's review, published five years after Napoleon's defeat at Waterloo, into the perspective of contemporary science, it is instructive to note that this same 1820 volume of the *Mémoires* contains a paper by Biot on the polarization of light, and mathematical papers by Laplace and by Poisson.

Thénard was an outstanding chemist. Among other things he had a share in the discovery of the element boron (1808), and he found a method to prepare pure metallic potassium. To painters and porcelain makers he is known as the discoverer of Thénard's blue. He wrote an influential multivolume "Traité de chimie élémentaire, théoretique et pratique" which had six editions between 1813–1816 and 1833–1836. Berzelius, who met him in Paris in 1818, found that he and Gay-Lussac "lecture like angels, a pure pleasure" (quoted by Bodman 1952). Many years later this son of poor peasants became chancellor of the University of Paris.

Fifteen years after Kirchhoff's work on plant α-amylase, a completely independent report (Paper 7) on the ability of saliva to convert starch to sugar was published by Erhard Friedrich Leuchs (1800–1837). This is undoubtedly the first work on an enzyme from human sources after Spallanzani. We find here a very clear investigation of enzyme specificity and a clear realization, following folklore precedents, of the possible practical use of this discovery. This paper, translated here in its entirety, is most probably the shortest ever on the discovery of a new enzyme.

William Beaumont's (1785–1853) classic investigation of Alexis St. Martin's digestive processes is probably the first widely remembered physiological and biochemical work performed outside Europe.[12] It affords a fascinating contrast to the more ebullient styles of Reaumur and of Spallanzani, and it provides insight into eating habits a century and a half ago. This work was done after the eminent English chemist William Prout had in 1824 made the "startling discovery...that the principal acid in gastric juice was in fact hydrochloric acid... . The announcement...that such a strong mineral acid existed in the stomach, instead of the much milder lactic or acetic acids...caused a sensation among chemists and physiologists" (Coley 1973, p. 76).[13] On a superficial reading, Beaumont (who quotes Prout's result) does not appear to add much to Spallanzani, except that he carries out his incubations in acid rather than in water. At the very end of Beaumont's book, however, one comes to the twenty-fourth of fifty-one "Inferences": (Gastric Juice) contains free *Muriatic Acid* and some other active *chemical* principles" (italics in the original). It was Beaumont's work that led Johann Nepomuk Eberle a mere one year later (1834) to show that one's experiments do not have to be confined to secreted gastric juice, but that an acid *extract* of gastric mucosa dissolves coagulated egg white. This in turn provided the direct stimulus for Schwann's classic investigation, published only a year later, on the action of pepsin (Paper 10). The ethical aspects of Beaumont's work are described in an article (Numbers 1979) that also cites instances before Beaumont of fistulous patients used in the study of human digestion.

Although Beaumont expresses disappointment at not having received an analysis of the gastric juice that he had sent with some difficulty across the ocean to the great Berzelius, the present readers will be relieved to know that Berzelius did receive the bottle and that he refers to it and its precious contents in his chemistry textbook: "I received a small bottle of this gastric juice through Professor Silliman in New Haven. Before it reached Stockholm it had been en route for 5 months in the hot summer of 1834 and hence it was not to be considered unchanged and suitable for an exact analysis. It was a clear, yellowish liquid without the slightest smell, and gave a strong red color to litmus paper... . A part of this gastric juice, after being stored for 2 years, remained without any sign of putrefaction" (Berzelius 1840, editor's translation).

## NOTES

1. The formation of bubbles has given fermentation its name. The Latin noun for ferment, *fermentum,* is a contraction of *fervimentum,* related to the verb *ferveo* or *fervere,* to boil. Compare with the English adjective *fervent,* the verb *to effervesce,* and the phrase *to be in a ferment.*
2. About one-and-a-half gallons, or 5.5 l (cf. Davis 1965, p. 4).
3. Descriptions of the meat tenderizing activity associated with fig do not go back as far as those of cheese making. The earliest description appears to be that of Pliny. In his *Natural History,* referring to wild figs, he states: "Beef can be boiled soft with a great saving of fuel if the stalks be added to the water. An application of the unripe figs soften and disperse scrofulous sores and every kind of gathering; to a certain degree the leaves do the same" (see Jones 1961). Observations on the meat tenderizing activity associated with fig are also found in Plutarch (c.46–120) (see Clement and Hoffleit 1969), in the medical works of Galen (c.130–c.200), of Oribarius (fourth century) and in the writings of Michael (Constantine) Psellus (1018–1078) (see Bussemaker and Daremberg 1851).
4. Another very early use of the enzymatic activity of papaya and fig products in various parts of the world was in the treatment of intestinal worms and of dyspepsia (see Robbins 1930 for ficin and Hwang and Ivy 1951 for papain). Thus as pointed out by Hwang and Ivy, the anthelmintic properties of the milky juice of the papaya tree were first used in India by Hernandez in the seventeenth century. Again it is of interest that in what appears to be the first scientific study of the action of ficin, Robbins (1930) concluded that it was a proteolytic enzyme from observations that included its effect on the destruction of live ascaris worms. In what must be one of the most unusual proteolytic assays ever devised, Robbins determined the optimum pH for ficin activity by using the live worms as substrate. He based his assay on the wide use of *Leche de higueron,* the sap of *Ficus laurifolia* and of *Ficus glabrata,* "as a general anthelmintic by the natives of South America and Panama."

5. Patrick Browne (1720?–1790) corresponded with Linneaus. His herbarium of more than 1000 rare plants was bought by Linneaus for 8 guineas (see Beaglehole 1962).
6. His name was Reaumur and not, as commonly seen, Réaumur (Partington 1962).
7. "From the time of Aristotle to the present day I know of but one who has shown himself Mr. Darwin's equal in one field of research—and that is Réaumur" (Th. H. Huxley, quoted in L. Huxley 1901).
8. We tend to take this quite modern-sounding type of questioning for granted; however, this sort of approach was not always obvious. Thus, for example, Jean Fernel (1497–1558), a famous French physician, assumed "the existence of some occult power which performed the work of digestion" (Leicester 1974). Johannes Baptista van Helmont (1579–1644), the eminent iatrochemist, already regarded digestion in chemical terms, as a type of fermentation, but his ferments "were more spiritual essences than material substances" (Coley 1973, p. 12).
9. It must be noted that Spallanzani had yet another forerunner, a certain Dr. Edward Stevens of Edinburgh who published the results of his inaugural dissertation, *Experiments Concerning Digestion* in 1777 (for original reference see Partington 1962). His work is given in a short appendix to the English translation of Spallanzani's book (pages 375–391 of the 1789 edition) and makes fascinating reading. His human subject was "an Hussar, a man of weak understanding, who gained a miserable livelihood, by swallowing stones for the amusement of the common people, at the imminent hazard of his life. . . I gave the subject of my experiments a hollow silver sphere, divided into two cavities by a partition, and perforated on the surface with a great number of holes." One of numerous experiments is described as follows: He swallowed a sphere "containing in one partition some roasted turkey, and in the other some boiled salt herring. In forty-six hours it was voided, and nothing of the turkey or herring now appeared, both having been completely dissolved." He carried out ten such experiments. "It was my intention to make more experiments of this kind, but as the Hussar left Edinburgh soon afterwards, I was obliged to have recourse to dogs and ruminating animals." Stevens was actually the first to succeed with *in vitro* experiments on the action of gastric juice: "I divided a piece of putrid mutton into two parts, each of which was put into a separate phial, and to one, half an ounce of the recent gastric fluid of a dog was added, and to the other, which was designed as a term of comparison,

as much water. They were set in a cool place, and two days afterwards I examined them, when the latter emitted an intolerably putrid smell, and the other, though it had yet a bad odour, did not smell so disagreeably as the preceding, nor even so disagreeably as at first. Upon shaking the phial, the meat fell to pieces, but it was not quite dissolved. This perhaps happened because it was not exposed to sufficient heat." In another experiment incubation at 102° Fahrenheit caused conversion of his material "into a fluid by the solvent power of the gastric liquor." He summarized his conclusions as follows: "These experiments throw great light on digestion. They shew [sic], that it is not the effect of heat, trituration, putrefaction, or fermentation alone, but of a powerful solvent, secreted by the coats of the stomach, which converts the aliment into a fluid, resembling blood. . . . It is probable that every species of animal has its peculiar gastric liquor, capable of dissolving certain substances only. . ."

10. Spallanzani erred in regarding gastric juice as neutral (Section CCXLIV, not reproduced). He considered that a meal, for instance, of an "agreeable mixture" of "strawberries with sugar and white wine at dinner and supper" is the source of acidity observed when "the contents of [the] stomach rise almost into [the] mouth. . . . Every man must some time or other have been sensible of his meat and drink having turned sour" (Section CCXL).
11. Actually, William Irvine (1743–1787), lecturer in materia medica and chemistry at the University of Glasgow, had already reached the same conclusion some thirty years earlier. In an essay "On Fermentation" read in 1785 before the Glasgow Literary Society (!) he stressed that the conversion by germinating seeds of starch into sugar does not require "the powers of vegetable life." However, this essay did not reach a wider public in the eighteenth century. "Avenues of publication from Glasgow, at this time, were difficult of access" (Kent 1950). This essay, along with many others on various scientific topics, was only published posthumously by William Irvine's son William (Irvine 1805). The relevant parts are given as an insert for Paper 9 (p. **119**).
    At this point one must mention the closely related and more influential work of Giovanni Valentino Mattia Fabbroni (1752–1822), assistant director at the Museum of Physics and Natural Sciences in Florence. In 1787 Fabbroni had received a prize from the Academy of Florence for a work on fermentation which, he said, "is nothing else than a decomposition of one

substance by another. . . . The sugar-decomposing substance in vinous effervescence is of vegeto-animal nature. It is located in specialized utricles in grapes and in wheat. In crushing grapes, one mixes the glutenous substance with sugar, as if one were pouring acid and a carbonate into a vase. As soon as the two substances are in contact, the effervescence or fermentation starts, as in any chemical reaction." In 1798 Fabbroni had been one of the scientists from various countries who participated in the work of the Commission on Weights and Measures in Paris that soon thereafter introduced the metric system. Following his departure his memoir on fermentation and related topics was read before the Société Philomatique, and a detailed discussion was published by the influential chemist Fourcroy (1799). This work must have created quite a stir. In the view of Pasteur who quotes the above passage (Pasteur 1860, p. 364), "Fabbroni's assertion made the question of the nature of fermentation the order of the day" (translations by Lechevalier and Solotorovsky 1965). Pasteur pointed out that it was this 1799 publication on the nature of ferments, and some earlier work by Lavoisier on the nature of the fermented substrate, that suggested the query in 1800 by the French National Institute concerning the distinction between ferments and their substrates. Some confusion exists regarding the identity of the Fabbroni responsible for the important 1787 Florentine *Dell'arte di fare il vino*. This may very well have been Giovanni's brother Adamo (see Bulloch 1938, pp. 42–43).

12. Dr. William C. Rose in prefatory remarks to the facsimile reproduction of the inaugural dissertation *An Experimental Inquiry into the Principles of Nutrition and the Digestive Process* for the M.D. degree, submitted June 1803 to the Trustees and Medical Professors of the University of Pennsylvania, calls the author, John R. Young (1782–1804), the first American biochemist.
13. It took thirty years, however, until it was generally accepted that gastric juice contained no free acid besides hydrochloric acid (see Leicester 1974, pp. 164–165). Johannes Baptista van Helmont apparently knew hydrochloric acid as the gastric acid (Pagel 1956).

# 1

Reprinted in the original size from pages 701–705, 710–712, 715–717, 724–725, and 730–743 of *Acad. R. Sci. (Paris) de l'Année 1752, Mém. Math. Phys. tirê des Regist.*, ser. 2, **3**:701–752 (1761)

SECONDE SUITE DES

# MEMOIRES

DE

MATHEMATIQUE

ET

DE PHYSIQUE,

Tirés des Regiſtres de

L'ACADEMIE ROYALE DES SCIENCES,

DE L'ANNÉE M. D. CCLII.

NOUVELLE CENTURIE.

TOME TROISIEME.

A AMSTERDAM,
Chez J. SCHREUDER,
Et PIERRE MORTIER, le Jeune.
M. D. CCLXI.
*Avec Privilege de N. S. les Etats de Hollande & de Weſt-Friſe.*

## * *SUR LA DIGESTION DES OISEAUX.*

## SECOND MEMOIRE.

*De la manière dont elle ſe fait dans l'eſtomac des Oiſeaux de proie.*

Par Mr. DE REAUMUR.

LE Mémoire précédent nous a appris combien la tritura ion a de part à la digeſtion, dans les oiſeaux qui vivent de grains, dans ceux qui ont cet eſtomac ſi muſculeux appellé géſier, il nous a fait voir que ce géſier eſt chargé de faire la fonction de nos dents; quoiqu'il ne ſoit qu'une maſſe de chair, qu'un aſſemblage de fibres charnues, il broie plus facilement des corps très-durs que les meilleures dents ne le pourroient faire: nous avons vu que des coques de noiſettes & de noix, que les nôtres ne pourroient venir à bout de

*Ii 3* caſ-

**Editor's Note:* The column notes referred to by the asterisks within the text have been omitted. They cite the page numbers corresponding to the appearance of that material as it was originally published in *Acad. R. Sci. (Paris) de l'Année 1752, Hist. avec Mém. Math. Phys.*, 1756, pp. 461–495.

caffer, étoient brifées, les premières par le géfier du coq, & les unes & les autres par celui du dindon, dès qu'elles y étoient entrées; & que ces géfiers étoient capables de mettre en pièces des corps qui oppofoient des réfiftances bien fupérieures à celle que peuvent oppofer les plus dures de ces coques. Enfin, il a été bien démontré dans ce Mémoire, que fi les alimens n'étoient pas broyés dans le géfier, ils ne s'y digéreroient point; qu'il ne s'y trouvoit aucun diffolvant qui eût le pouvoir de les divifer; que leur divifion, pouffée au moins auffi loin que celle qui fe fait fous les meules de nos moulins à blé, étoit uniquement dûe à la force avec laquelle ce vifcère agit fur eux.

La trituration a-t-elle une auffi grande part à la digeftion qui fe fait dans les oifeaux dont les eftomacs font autrement conftruits, & à celle qui fe fait dans les animaux munis de dents? quelle part y a-t-elle? y en a-t-elle quelqu'une?

Les Phyficiens qui fe font le plus déclarés pour la trituration, * ayant cru qu'il étoit très-bien prouvé par la ftructure du géfier des oifeaux de différentes espèces, que la digeftion y étoit fon ouvrage, ont voulu qu'elle fût de même uniquement faite par fon moyen, dans les eftomacs des différentes fortes d'animaux, dans les eftomacs fimplement membraneux comme dans les plus charnus. Ceux au contraire qui n'ayant pas jugé les eftomacs membraneux capables de broyer, ont prétendu que la



digeftion y étoit opérée par un diffolvant, ont affuré qu'elle étoit de même dûe à un diffolvant dans les eftomacs les plus charnus. On cède trop volontiers au penchant qui porte à généralifer fes idées; il eft commode de s'épargner des difcuffions: d'ailleurs on s'y croit autorifé par des analogies qu'on étend fouvent trop loin en regardant les loix de la Nature comme plus uniformes qu'elles ne le font réellement. Nous avons pourtant par-tout des preuves, fi nous voulons y faire attention, que fon Auteur a voulu employer des moyens différens pour arriver à des fins femblables. Les oifeaux nous en fourniffent affez d'exemples: on fera furpris combien les formes de leurs becs diffèrent, fi on les étudie & fi on les compare entre elles; ces becs font cependant tous deftinés à prendre & à faire paffer dans l'intérieur du corps, des alimens quelquefois femblables. Combien l'Auteur par excellence femble-t-il s'être plu à mettre des variétés dans fes ouvrages! combien en a-t-il mis dans l'extérieur des animaux! Il n'en a pas fait entrer de moins confidérables dans la ftructure de leur intérieur: celles-ci ne paroiffent-elles pas prouver qu'il a voulu produire les mêmes effets par des moyens différens? Il a établi que la plupart des animaux, fans en excepter l'homme, les oifeaux, les quadrupèdes, devroient leur accroiffement & la durée de leur vie à une liqueur laiteufe, au chyle, qui eft préparé en partie dans l'eftomac; mais a-t-il voulu que cette

Ii 4

liqueur fût extraite des alimens dans tous les animaux, par des opérations semblables? nous avons au moins lieu d'en douter, puisqu'il y a employé des estomacs dont la conformation est différente. Quoiqu'il soit donc très-prouvé que la trituration est le grand agent de la digestion dans les gésiers, * nous n'en avons pas moins besoin de nous assurer par des expériences, si elle se fait par la même méchanique, ou par une méchanique différente dans les estomacs membraneux.

Les estomacs des oiseaux de proie sont les plus propres à nous donner des lumières sur cette question, ils sont de ceux qui ont le plus de rapport avec le nôtre. Il est pourtant vrai que le pouvoir de triturer sembleroit leur être plus nécessaire qu'à celui de l'homme: ces oiseaux voraces, avalent souvent de très-gros morceaux de viande que leur bec a arrachés; ils n'ont point de dents dont ils puissent se servir pour les diviser; le gésier supplée aux dents qui manquent à d'autres oiseaux, il en fait l'office. J'ai cru me devoir instruire de la manière dont se fait la digestion dans les oiseaux de proie, & que ce que nous en apprendrions ne seroit pas inutile pour nous donner des idées justes de la manière dont elle s'opère chez nous-mêmes. Pour peu qu'on se souvienne du Mémoire précédent, on prévoit que je me suis proposé de leur faire avaler bien des tubes différemment conditionnés. Quoiqu'on n'ait pas de basse-cours peuplées de ces oiseaux comme on en a qui le sont de poules, de dindons, de canards, &c. j'ai pensé avec plaisir que je pourrois multiplier sur eux les expériences à mon gré. On ne doit point être porté de compassion pour des oiseaux qui ôtent impitoyablement la vie à tous ceux à qui ils sont supérieurs en force, qui ne subsistent que de carnage; j'ai pourtant été content de voir que je tirerois d'eux autant d'éclaircissemens que j'en souhaiterois, sans devenir le vengeur des autres oiseaux, sans être obligé d'ôter la vie à un seul de ces meurtriers.

Pour peu qu'on ait lu quelque ouvrage de fauconnerie, on pourra prévoir que je ne devois pas me trouver dans la nécessité de tuer un oiseau de proie, pour examiner ce qui seroit arrivé au tube qui auroit passé vingt quatre heures dans son estomac: on se rappellera, & je me rappellai heureusement, que les oiseaux carnaciers rejettent par le bec les matières que leur estomac n'a pas pu digérer. Il leur est fort ordinaire d'y faire entrer par voracité des plumes de l'oiseau infortuné de * la chair seule duquel ils voudroient se rassasier: ces plumes, qui ne s'y digèrent point, ne sortent pas de l'intérieur du corps par la voie des excrémens, elles sont chassées par celle qui les y avoit conduites.

[*Editor's Note:* Material has been omitted at this point.]

Ces circonſtances, que je n'ai pas dû laiſſer ignorer, ne ſont pas ce qu'on eſt le plus curieux, & ce qu'il eſt le plus important de ſavoir. Le morceau de viande arrêté dans le tube par un fil avoit-il été digéré? voilà de quoi on demande à être inſtruit: en quel etat fut-il trouvé? il avoit été réduit * à moins du tiers, peut-être au quart de ſon premier volume & de ſon premier poids; ce qui en reſtoit étoit bien retenu par le fil, & couvert par une eſpèce de bouillie, venue probablement de celles de ſes parties qui avoient été diſſoutes. Après que la bouillie eut été enlevée, le reſte de chair qui fut mis à découvert, parut avoir à peu près ſon ancienne couleur, peut-être néanmoins étoit-elle un peu plus blanchâtre; mais cette chair avoit perdu de ſa conſiſtance; en la tirant doucement avec la pointe d'un canif en différens ſens, on la mettoit en charpie; ſon odeur n'étoit point celle de viande pourrie, elle en avoit pris une qui n'avoit rien de ſi deſagréable.

La conſidérable déperdition qu'avoit faite le morceau de chair, & l'eſpèce de bouillie dont étoit enveloppé ce qui en reſtoit, doivent, ce ſemble, convaincre les plus prévenus pour le ſyſtème de la trituration, qu'elle n'eſt pas l'agent principal de la digeſtion dans les oiſeaux de proie. Ce n'avoit pas été par des broiemens que les deux tiers ou les trois quarts du morceau de viande



viande aſſujetti par un fil dans le tube, avoient été enlevés: cet ouvrage n'avoit pu être que celui d'un diſſolvant qui avoit détaché peu à peu les petites parcelles qui formoient le ſédiment, l'eſpèce de bouillie, dont s'étoit trouvé recouvert ce qui n'avoit pas encore été diſſous. Un trop fort attachement au ſyſtème de la trituration ne laiſſera-t-il point néanmoins encore quelques doutes? ne fera-t-il point imaginer que le morceau de viande a pu être broyé dans le tube par les frottemens auxquels il a été expoſé? que ces frottemens ont pu être produits par cette force qui avoit mis, peut-être trop tard, un bouchon de duvet à un des bouts du tube? cette force, avant que le bouchon fût en place, n'a-t-elle point fait mouvoir continuellement dans le tube des matières ſolides qu'elle y avoit introduites, & qu'elle pouſſoit alternativement d'un bout vers l'autre? n'eſt-ce pas par de pareils frottemens répétés pendant près de vingt-quatre heures que le morceau de chair avoit été uſé, pour ainſi dire, en grande partie & réduit en bouillie?

Heureuſement qu'il étoit aiſé d'imaginer une expérience * qui apprit ce qu'on devoit penſer des difficultés précédentes, une expérience qui démontrât de la manière la plus rigoureuſe dont un fait de Phyſique peut être démontré, ſi la digeſtion eſt opérée dans les oiſeaux carnaciers par la ſeule action d'un diſſolvant, & par la fermentation qu'il fait naître. Une addition aſſez légère faite à notre tube, le rendra pro-

propre à faire cette expérience si décisive: plaçons dedans un morceau de viande qui n'occupe qu'une partie de sa longueur, & qu'il soit à égale distance de l'un & de l'autre de ses bouts; au lieu de laisser ceux-ci entièrement ouverts, donnons-leur à chacun un grillage qui bouche l'entrée à tout corps solide, & qui ne permette qu'à de la liqueur de pénétrer dans le tube. Il est de toute évidence que si le morceau de viande est réduit en bouillie & digéré dans ce tube où il est isolé, & seulement accessible à de la liqueur, ç'aura été par un dissolvant. Tout ce qu'il sembleroit y avoir à craindre, c'est que le dissolvant, s'il y en a un dans l'estomac, n'y fût pas en assez grande quantité pour fournir celle qu'il faudroit qui s'introduisît dans le tube pour agir avec succès contre le morceau de chair, dont les bouts seuls sont exposés à son action.

[Editor's Note: Material has been omitted at this point.]



Avec la lame du canif, je continuai à retirer du tube alternativement par l'un & par l'autre bout, la matière qui y étoit contenue; toute celle que j'en fis sortir jusqu'au quart de la longueur du tube de chaque côté, c'est-à-dire, celle qui se trouvoit dans deux étendues égales ensemble à la moitié de la longueur de ce tube, étoit parfaitement semblable; au jugement des yeux

yeux & du toucher, à celle que je viens de décrire: je mis ensuite à découvert une matiére un peu rougeâtre, & qui avoit plus de consistance; l'une & l'autre avoient fait partie du morceau de bœuf avant qu'il entrât dans le tube: la première, venue des fragmens qui en avoient été plutôt * détachés, étoit plus parfaitement dissoute, plus parfaitement digérée que la seconde: une couche de cette seconde matière enveloppoit ce qui étoit resté du morceau de bœuf sous une forme solide, & avec à peu près sa couleur naturelle. Après avoir lavé doucement ce reste de chair pour emporter la pâte qui l'enveloppoit, je jugeai qu'il n'étoit pas une huitième partie de celui qui étoit entré dans le tube, sa longueur n'étant que la moitié de celle de l'autre, & la différence entre les diamètres étoit au moins aussi grande.

Au reste, la bouillie fournie par les sept huitièmes du morceau de viande, & ce qui étoit resté de celle-ci sous une forme solide, n'avoient aucunement l'odeur d'une viande corrompue; ils n'en avoient pas cependant une agréable, elle ne tiroit pas sur l'aigre, elle n'étoit pas pénétrante, elle étoit plutôt fade.

Il est donc incontestablement prouvé par l'expérience précédente, que de la viande peut être digérée dans l'estomac des oiseaux carnaciers, non seulement sans y avoir été broyée, mais sans même y avoir souffert les plus légers frottemens: cette opération peut donc être uniquement l'ouvrage



vrage d'un dissolvant, dont l'existence est bien démontrée.

[*Editor's Note:* Material has been omitted at this point.]

Les oiseaux vraiment carnaciers, du nombre desquels étoit notre buse, ne se nourrissent que de la chair des autres oiseaux, de celle des quadrupèdes, de celle des reptiles, & quelques-uns de celle des poissons: la faim la plus pressante ne sauroit les contraindre à recourir au grain, de quelque espèce que ce soit; ils se laisseroient périr d'inanition auprès d'un tas de blé, & de même auprès des meilleurs fruits. Seroit-ce parce que le dissolvant de leur estomac n'a de prise que sur les chairs & les os, & qu'il ne peut rien sur les productions du règne végétal? ces oiseaux agissent-ils comme s'ils savoient qu'ils n'en mourroient pas moins de faim, quand ils rempliroient leur estomac de grain, parce qu'il y resteroit sans pouvoir y être digéré? La Nature donne aux animaux des leçons sûres, celles qu'ils ont besoin d'avoir, & ils ne manquent pas de les suivre. Il étoit donc à présumer, & il étoit curieux de s'en assurer, que ce dissolvant de l'estomac de la buse, si puissant contre les chairs



chairs & les os, ne pouvoit rien ou que très-peu sur les matières végétales: nous avons déjà rapporté une expérience, répetée bien des fois, qui paroît le prouver; les grilles de fil de nos tubes se sont soutenues sans qu'aucun de leurs brins ait été cassé; le dissolvant qui a réduit des os en gelée, n'a pas même affoibli ces fils, ils sont restés très-sains.

[*Editor's Note:* Material has been omitted at this point.]

Quelle eſt la nature de cette liqueur, qui agit en quelque ſorte ſur les chairs & ſur les os comme l'eau régale agit ſur l'or, mais qui n'a pas plus de pouvoir ſur les ſubſtances farineuſes que l'eau régale n'en a ſur l'argent? auquel de tant de diſſolvans que la Chymie nous fournit, peut-elle être comparée?

J'ai mis pluſieurs fois ſur ma langue de cette gelée, en laquelle elle avoit converti de la viande ou des os, & qui devoit en être pénétrée; je lui ai toujours trouvé de l'amertume, tantôt plus, & tantôt moins, mêlée avec un peu de ſalure. J'ai auſſi appliqué ſur ma langue des reſtes des os qui avoient été diſſous en partie; ces reſtes d'os devoient au moins être pénétrés de cette liqueur diſſolvante près de leur ſurface: je leur ai auſſi trouvé un goût amer, mêlé de ſalé, mais ce goût étoit plus foible que celui de la gelée.

Je ne dois pas laiſſer ignorer qu'ayant voulu ôter à un tube l'odeur puante qu'il avoit priſe dans une des expériences que j'ai rapportées, & la ſeule où une portion reſtante de la viande ſe fût pourrie, je le mis ſur des charbons ardens: bientôt il ſortit de ſon intérieur une flamme qui dura pendant plus d'une minute. Je ne me preſſerai pas de conclurre de ce fait, que la matière inflammable entre pour beaucoup dans le diſſolvant que nous devons ſouhaiter de connoître; il faut auparavant l'avoir ſoumis à bien d'autres examens, & pour cela en avoir des quantités ſur leſquelles on puiſſe opérer.



On auroit peut-être pas oſé ſe promettre d'avoir aſſez de cette liqueur qui agit dans un eſtomac avec tant d'efficacité ſur les alimens, pour la pouvoir mettre à toutes les * épreuves par leſquelles on fait paſſer celles dont on veut découvrir la compoſition: nos tubes néanmoins, qui nous ont valu tant de connoiſſances certaines ſur la manière dont ſe fait la digeſtion dans des eſtomacs différemment conſtruits, peuvent encore nous procurer aſſez de cette liqueur pour varier ſur elle des eſſais qui ont un objet auſſi utile que curieux. Je n'ai encore qu'ébauché ces eſſais, mais je ne doute pas qu'ils ne ſoient multipliés autant qu'ils méritent de l'être, par pluſieurs Savans qui ſeront frappés de l'importance de leur objet; il ne s'agit que de les faire penſer au moyen ſimple dont ils peuvent ſe ſervir pour ſe fournir d'une liqueur ſi intéreſſante: la quantité qui en entre dans un tube grillé, pendant les vingt-quatre heures qu'il ſéjourne dans l'eſtomac d'un oiſeau de proie, eſt conſidérable; elle ſuffit pour mouiller tout l'extérieur des morceaux de viande qui y ſont logés, & pour les pénétrer intimement.

Kk 6 Ce

Ce ne sont pas ces morceaux de viande qui attirent la liqueur dissolvante dans le tube; fût il vuide, elle ne s'y introduiroit pas en moindre quantité; il ne s'agiroit donc que de l'y retenir, & on l'y retiendra si le tube est rempli d'une matière qui s'en laisse imbiber, & qui ne soit pas capable d'émousser son activité, c'est-à-dire, qui ne lui soit pas dissoluble. Celle qui doit être employée ici par préférence ne se fait pas chercher, l'éponge se présente la première; elle n'est point de celles que mange un oiseau de proie, & ce que nous avons vu ci-dessus nous a appris à en conclurre qu'elle n'est donc point de celles que son estomac peut digérer: aussi n'ayant aucun doute sur le succès de l'expérience que j'allois tenter, je fis entrer plusieurs petits morceaux d'éponge dans le tube, je l'en remplis sans les y trop presser, il fut grillé, avalé par la buse & rendu à l'ordinaire. Lorsque les morceaux d'éponge y furent introduits, ils ne pesoient que treize grains: je les pesai dès que je les en eus retirés, alors ils en pesoient soixante-trois; ils s'étoient donc chargés de cinquante grains de liqueur, qu'il me fut aisé d'exprimer en grande partie dans un vase destiné à la recevoir.

* Cette expérience seule suffit pour faire voir qu'on pourroit se fournir aisément d'une quantité assez considérable d'une pareille liqueur: deux ou trois tubes garnis d'éponge, qu'on feroit prendre le même jour à la buse, en donneroient une quantité à peu près double ou triple, cent ou cent cinquante grains pesant; & si on nourrissoit deux ou trois de ces oiseaux, ce qui se feroit sans de fort grands frais pendant quelques semaines, on pourroit avoir chaque jour deux ou trois cens, ou même quatre cens cinquante grains de liqueur. Ceux qui, par leur goût pour cette chasse où les oiseaux de proie montrent combien ils sont supérieurs en vitesse de vol & en force aux oiseaux de tant d'autres genres, sont engagés à en nourrir chez eux plusieurs, pourroient nous procurer encore une plus grande quantité de cette liqueur dissolvante sans aucune dépense. S'ils craignoient que le tube de fer-blanc, dont les bords sont tranchans, ne blessât l'estomac de leurs oiseaux, ou le conduit par lequel ils y arrivent & en sortent, on pourroit substituer au tube de fer-blanc, qui cependant ne m'a jamais paru avoir fait aucun mal à ma buse, des tubes de plomb dont les bords seroient émoussés & doux. De pareils tubes ne nuiroient pas plus à l'oiseau de proie que les cures qu'on lui fait prendre pour assurer sa santé, & lui en tiendroient lieu. Mais si, peu satisfait de la quantité de la liqueur dissolvante qu'il est permis de tirer de l'estomac d'une buse ou d'un oiseau de proie qui lui est au plus égal en grandeur, on alloit la puiser dans celui des plus grands oiseaux de cette classe, dans celui d'un aigle ou dans celui d'un vautour, les éponges qu'on retireroit du grand tube, ou des grands tubes qui au-

Kk 7 roient

roient séjourné dans cet estomac, donneroient une quantité de liqueur qui suffiroit à beaucoup d'expériences. La mort de ma buse est arrivée avant que la suite de celles auxquelles je l'avois destinée ait été exécutée, & j'ai à me reprocher ma négligence à la remplacer par une autre, ou par quelque oiseau de proie d'une espèce différente de la sienne; mais je me promets de réparer cette faute, & de faire les expériences qui semblent être le plus à desirer: j'en * vais indiquer les principales, pour inviter à les tenter de leur côté, les Physiciens qui en trouveront des occasions commodes.

Lorsque je perdis ma buse, je n'avois encore tiré que deux fois de son estomac, au moyen des éponges, de cette liqueur qui y dissout les chairs & les os: celle que la pression de mes doigts obligea de sortir de quelques morceaux d'éponge qui en étoient imbibés, fut reçue dans un vase, elle ne ressembloit nullement par sa limpidité à la liqueur que différentes distillations nous donnent; loin d'avoir cette belle transparence, elle étoit un peu opaque & trouble, sa couleur étoit très-louche & d'un blanc un peu jaunâtre. Au reste, je suis incertain si sa couleur & sa transparence naturelles n'avoient pas été altérées, & c'est sur quoi de nouvelles expériences ne manqueront pas de nous instruire: dans les premières, je ne pris pas la précaution de bien laver les éponges: si des parties terreuses, ou de quelqu'autre nature, se sont trouvées dans leur intérieur, elles auront changé la couleur de la liqueur qui s'en sera chargée, & elles lui auront ôté de la transparence.

Indépendamment de ce que cette liqueur aura pu emporter des éponges, une autre cause peut l'avoir empêché d'être aussi pure qu'il eût été à desirer de l'avoir, la liqueur, avant que d'entrer dans le tube, aura trouvé dans son chemin les morceaux de viande qui étoient dans l'estomac, sur lesquels elle n'aura pas manqué d'agir au moins un peu; d'ailleurs, il est presque impossible que quelques portions de la viande digérée & réduite en bouillie, & même rendue plus liquide, ne se soient pas mêlées avec la liqueur: quoique celle qui avoit imbibé les morceaux d'éponge eût eu le pouvoir de dissoudre de la viande, elle ne devoit pas cependant être regardée comme pure, mais on peut parvenir à en avoir qui le soit, ou qui le soit au moins beaucoup plus que ne l'étoit celle dont nous parlons; il ne faut pour cela qu'avoir attention à ne faire avaler à l'oiseau de proie le tube garni d'éponge, que dans un temps où son estomac est vuide, & se donner de garde de lui faire prendre aucune nourriture pendant tout * celui où il gardera le tube dans son corps. Ce jeûne auquel nous voulons faire mettre cet oiseau, ne lui sera pas aussi difficile à soutenir qu'il le pourroit sembler: la Nature a mis ceux de ce genre en état d'en supporter de très-longs; leurs chasses ne sont pas tou-

toujours heureuses, il leur arrive souvent de passer des journées sans rien prendre, & par conséquent sans manger. Mr. le Commandeur Godcheu étant à Malte, reçut de Tripoli un vautour vivant, qu'il destina pour mes cabinets où il est actuellement, sur lequel il fit une expérience qui montre combien il est peu nécessaire aux oiseaux de son espèce, de prendre des alimens journellement. Pour le faire dessécher plus aisément après qu'il seroit mort, il crut le devoir rendre très-maigre, & il en prit le plus sûr moyen, il lui retrancha totalement toute nourriture: ce vautour résista pendant dix-sept jours à un jeûne si rude.

J'ai mis sur ma langue de cette liqueur dont les éponges s'étoient imbibées dans l'estomac de la buse; son goût m'a paru tenir plus du salé que de l'amer, quoiqu'au contraire la gelée d'os qui avoit été l'ouvrage d'une pareille liqueur, & les restes des os sur lesquels elle avoit agi, m'aient fait sentir, comme je l'ai déjà dit, un goût plus amer que salé.

Du papier bleu a été mouillé avec cette même liqueur, elle l'a rougi.

Une des premières expériences qui sembloient demander à être tentées avec cette liqueur, comme des plus curieuses, & des plus propres à nous démontrer qu'elle étoit incontestablement celle qui réduit les chairs & les os en bouillie, eût été de lui faire dissoudre des chairs dans un vase comme elle les dissout dans l'estomac de l'oiseau. De véritables digestions d'alimens, opérées dans un lieu si différent de celui où elles se font faites jusqu'à ce jour, eussent été un phénomène aussi singulier qu'intéressant. Quoique mes tentatives pour y parvenir ne m'aient point réussi, je ne laisserai pas de les rapporter; je n'en ai fait que deux, & elles demandent à être répétées avec des précautions que je n'ai pas prises: peut être ne saura-t-on toutes celles qu'elles exigent pour avoir un plein * succès, qu'après qu'elles auront été refaites un très-grand nombre de fois.



Je réservai la liqueur dont quelques-uns de mes morceaux d'éponge étoient imbibés, pour la faire travailler sur la viande, & voici l'usage que j'en fis. Dans un tube de fer-blanc un peu plus grand que celui dont il s'est agi jusqu'ici, & dont un bout étoit bouché par une plaque de même fer, qui y étoit soudée, je fis entrer un morceau de viande gros comme une noisette; sur ce morceau de viande, je fis tomber toute la liqueur que je pus exprimer des éponges en les pressant entre mes doigts; je fis ensuite rentrer dans le même tube les éponges, & cela dans la vue de diminuer au moins l'évaporation d'une liqueur qui étoit en trop petite quantité par rapport à la viande, car celle-ci n'en étoit pas entièrement couverte La liqueur, pour être assez active, a besoin d'être aidée de la chaleur qui règne dans l'estomac: celle des fours où je fais couver des œufs, n'est probablement inférieure à l'autre

tre que de peu de degrés: un de ces fours fut choisi pour faire une des fonctions de l'estomac; mais avant que d'y faire placer le tube, je le logeai dans un poudrier qui avoit un couvercle de papier ficelé autour de son bord, & cela encore dans la vuë de diminuer l'évaporation: dans ce même poudrier je mis un morceau de viande coupé à la même pièce de laquelle celui du tube avoit été pris, & à peu près d'égale grosseur; il devoit servir de terme de comparaison: pour l'y rendre plus propre, il avoit été trempé dans de l'eau ordinaire. Lorsque le poudrier qui contenoit le morceau de viande & le tube, entra dans le four, il y regnoit une chaleur de trente-trois degrés.

Le lendemain, c'est-à-dire, au bout de vingt-quatre heures, je retirai le poudrier du four, pour examiner l'état des matières qui lui avoient été confiées: les changemens qui s'étoient faits dans le tube, n'étoient pas ceux que j'avois cherché à occasionner; le morceau de viande qui étoit immédiatement dans le poudrier, s'étoit desséché, il n'avoit d'humide que la partie qui étoit immédiatement appliquée contre le verre; * il sentoit très-mauvais, il s'étoit corrompu: les éponges supérieures du tube s'étoient aussi desséchées, mais les inférieures étoient encore humides; la viande cependant qu'elles couvroient, n'étoit pas dissoute, elle étoit simplement ramollie, mais elle avoit pris une mauvaise odeur, quoique moins forte que celle de l'autre morceau.



Le peu de succès de l'expérience précédente, me parut devoir être attribué à la petite quantité de la liqueur par laquelle la viande avoit été attaquée: je refis provision de cette liqueur, en faisant séjourner dans l'estomac de la buse un tube rempli de fragmens d'éponge, qui ne pesoient ensemble que onze grains, & qui lorsqu'ils furent retirés du tube en pesoient cinquante-trois, ils avoient donc retenu quarante-trois grains de liqueur. Pour en faire un meilleur usage que de la première que j'avois employée, je pris un morceau de viande beaucoup plus mince; celui que je plaçai dans le fond du tube, fut entièrement couvert par la liqueur que les éponges fournirent; je les mouillai un peu avant que de les faire entrer dans le tube; enfin le fond du poudrier qui devoit recevoir le tube, fut couvert d'une couche d'eau épaisse de quelques lignes, & ce fut dans cette eau que fut plongé le morceau de viande qui devoit être comparé avec celui du tube. Ce poudrier passa, comme le premier, vingt-quatre heures dans le four à poulets; la viande qui étoit immédiatement dans l'eau s'y corrompit au point de répandre une odeur détestable; les éponges du tube ne se desséchèrent point, la viande cependant au dessus de laquelle elles étoient, ne fut point dissoute; elle s'étoit même un peu corrompue, mais très-peu en comparaison de celle qui avoit été tenue

tenue dans l'eau ordinaire; le diſſolvant avoit au moins empêché qu'elle ne ſe corrompît autant qu'elle auroit fait. Pour le mettre en état d'opérer avec plus de ſuccès, il reſte, comme je l'ai déjà dit beaucoup de tentatives à faire, ſoit par rapport à la quantité du diſſolvant, ſoit par rapport au degré de chaleur, ſoit par rapport à l'évaporation; peut-être que pour arrêter celle-ci ſuffiſamment, le tube demanderoit à être bouché * par les deux bouts: ne ſuffiroit-il pas qu'il le fût par un bouchon de liege? peut-être faut-il s'en prendre au défaut du degré de chaleur. C'eſt de quoi on s'inſtruira en faiſant entrer dans l'eſtomac de l'oiſeau de proie, le tube bouché par les deux bouts, dans lequel il y auroit un mince morceau de viande couvert de la liqueur puiſée dans l'eſtomac du même oiſeau: on verroit, lorſqu'il en ſeroit ſorti, ſi la liqueur dont la viande auroit été entourée, auroit ſuffi pour la diſſoudre. Peut être faudroit-il renouveller de temps en temps la liqueur du tube tenu dans le four à poulets.

Il n'eſt guère permis de ſoupçonner à cette liqueur une volatilité ſi grande, qu'elle perde tous les principes de ſon activité dès qu'elle eſt expoſée à l'air libre, ſon odeur, qui n'eſt nullement pénétrante, n'annonce rien de pareil; il convient cependant d'éprouver ſi elle ne s'altérera pas dans des bouteilles de verre où elle ſera renfermée avec ſoin : lorſqu'on ſera parvenu à en avoir des quantités un peu con-



ſidérables dans ces ſortes de bouteilles, on la mettra à toutes les épreuves auxquelles elle mérite d'être miſe.

Les conſéquences qu'on peut tirer de la comparaiſon des faits rapportés dans ce Mémoire, avec ceux qui l'ont été dans celui qui l'a précédé, ſe préſentent ſi naturellement, que je ne dois pas m'arrêter à les détailler ni à les développer. Il ne paroîtra pas qu'il y ait quelque lieu de douter que la digeſtion ne ſe faſſe dans tous les oiſeaux dont l'eſtomac eſt épais, muſculeux, en un mot un géſier, par la trituration, comme elle ſe fait dans les dindons, les poules & les canards, & qu'elle ne ſoit opérée par un diſſolvant dans tous les oiſeaux dont l'eſtomac eſt ſimplement membraneux, comme elle l'eſt dans celui de la buſe: il paroîtra très-vraiſemblable que le broiement & le diſſolvant y concourent dans les eſtomacs d'une ſtructure moyenne, je veux dire, tant dans ceux qui ſont mi-partis, comme celui du pic-vert, qui eſt géſier par un bout, & eſtomac membraneux par l'autre, que dans ceux qui ſont d'une conſiſtance & d'une épaiſſeur moyennes entre celles des géſiers & celles des eſtomacs ſimplement membraneux. * C'eſt ce qui pourra être vérifié par des expériences, à meſure qu'on aura des occaſions d'en faire ſur des oiſeaux qui ont en partage un eſtomac à qui l'une ou l'autre des deux dernières ſtructures eſt propre.

On ne ſe croira pas non plus en riſque de

de tirer des conséquences qui pourroient être démenties par les expériences, lorsqu'on jugera que de quelque classe que soient les animaux qui ont un estomac membraneux, la digestion est faite dans le leur par un dissolvant comme elle l'est dans l'estomac membraneux des oiseaux; que les alimens sont dissous par une liqueur convenable dans l'estomac de plusieurs quadrupèdes, par exemple, dans celui des chiens, dans celui des cochons, dans celui des chevaux; qu'ils le font de même dans celui au moins de la plupart des poissons, dans celui des reptiles, & dans celui de divers insectes: je dis de divers insectes, parce qu'il y en a des genres dont l'estomac a une structure qui paroît le rendre plus propre à broyer avec succès, comme celui de courtillières ou taupes grillons, que ne l'est le gésier des oiseaux. Qu'on ne pense pas cependant être absolument dispensé de faire des expériences sur l'estomac de quelques-uns au moins des animaux de chacune de ces différentes classes, elles ne seront pas inutiles; elles donneront le plus grand degré de certitude à un sentiment qui, tant qu'elles ne l'appuieront pas, ne sera qu'extrêmement vraisemblable.

Une seule expérience m'a paru suffire pour démontrer que la trituration n'a pas plus de part à la digestion dans l'estomac des chiens, que dans celui des oiseaux de proie: cette expérience fut faite sur une petite chienne dont l'âge m'étoit inconnu, & qui avoit un air vieillot, quoiqu'elle fût en chaleur. Pendant qu'on lui tenoit la gueule ouverte, je fis entrer forcément dans son gosier deux os qui furent bientôt conduits dans l'estomac, leur figure étoit à peu près cylindrique; ils avoient chacun sept lignes de long & un peu moins de deux lignes de diamètre; ils avoient été pris de la partie la plus compacte d'un gros os, & façonnés ensuite: c'étoient pour la chienne deux morceaux qui devoient être cause de sa mort, * elle fut étranglée après qu'ils eurent resté dans son estomac pendant vingt-six heures: sur le champ j'ouvris ce viscère pour les y chercher, ils y étoient encore, & ne furent pas difficiles à trouver, mais ils avoient perdu de leur volume; des lames longitudinales sembloient en avoir été enlevées. Je ne m'assurai point par des mesures, de la quantité qu'ils avoient perdue de leurs dimensions & de leur poids, je me contentai, par rapport à une expérience que je m'étois proposé de répéter, d'avoir eu une preuve nullement équivoque de l'existence d'un dissolvant des os dans l'estomac des chiens, non seulement par une perte visible qu'avoient fait ceux dont il étoit question, mais sur tout en ce que les restes de deux os si durs, & par conséquent si roides, avoient été rendus aussi flexibles que de la corne.



[*Editor's Note:* Material has been omitted at this point.]

# 2

# DISSERTAZIONI

DI FISICA ANIMALE, E VEGETABILE

DELL'

## ABATE SPALLANZANI

REGIO PROFESSORE DI STORIA NATURALE NELL' UNIVERSITA' DI PAVIA,

E SOPRANTENDENTE AL PUBBLICO MUSEO DELLA MEDESIMA;

SOCIO DELLE ACCADEMIE DI LONDRA, DI PRUSSIA, STOCKOLM, GOTTINGA, BOLOGNA, SIENA, DE' CURIOSI DELLA NATURA DI GERMANIA, E BERLINO, CORRISPONDENTE DELLA SOCIETA' REALE DELLE SCIENZE DI MONTPELLIER, EC.

*Aggiuntevi due Lettere relative ad esse Dissertazioni dal celebre Signor Bonnet di Ginevra scritte all' Autore.*

TOMO I.

IN MODENA. MDCCLXXX.

PRESSO LA SOCIETA' TIPOGRAFICA.

*Con licenza de' Superiori.*

# DISSERTATIONS

RELATIVE TO THE

# NATURAL HISTORY

OF

# ANIMALS AND VEGETABLES.

TRANSLATED FROM THE ITALIAN OF THE

## ABBÉ SPALLANZANI,

Royal Profeſſor of Natural Hiſtory in the Univerſity of PAVIA, Superintendent of the PUBLIC MUSEUM, and FELLOW of various learned SOCIETIES.

In TWO VOLUMES.

A NEW EDITION CORRECTED AND ENLARGED.

---

VOL I.

---

LONDON:

PRINTED FOR J. MURRAY, NO. 32, FLEET-STREET.

M DCC LXXXIX.

Reprinted from pages i–ii, 36–42, 55–58, 74–75, 140–142, 280–288, and 298–304 of *Dissertations Relative to the Natural History of Animals and Vegetables*, vol. 1, T. Beddoes, transl., London: J. Murray, 1789, 391pp.

# INTRODUCTION

IN the courſe of my public demonſtrations in the year 1777, I repeated in the preſence of my hearers thoſe celebrated experiments of the Academy of Cimento, which ſhew that the ſtomachs of fowls and ducks exert ſo aſtoniſhing a force as to reduce hollow globules of glaſs to powder in the ſpace of a few hours. Finding them perfectly exact, I conceived the deſign of extending them to ſome other individuals of that claſs of birds which have been termed birds with *muſcular ſtomachs* or *gizzards*. Such were the firſt lines of an undertaking, of which till that time I had never entertained the ſmalleſt idea, and which afterwards became more and more extenſive, as my curioſity concerning ſo fine and uſeful a ſubject as the important function of Digeſtion increaſed. Hence from animals with muſcular ſtomachs, I was induced to proceed to thoſe with intermediate, and from theſe again to animals with membranous ſtomachs (*a*). Thus I enjoyed the pleaſure of extending my reſearches to the principal claſſes of animals, not neglecting Man, the nobleſt and moſt intereſting of all. But theſe phyſiological reſearches laid me under the neceſſity of examining the moſt celebrated ſyſtems concerning Digeſtion, and of enquiring whether it is effected by trituration, by a ſolvent, by fermentation, or by an incipient putrefaction: or whether, according to the opinion of the great Boerhaave, it rather depends upon all theſe cauſes operating in conjunction. Thus I was obliged to enter anew upon a queſtion of very ancient date, and though diſcuſſed at great length by many phyſiologiſts, yet not in my opinion ſufficiently elucidated; ſince moſt writers have choſen to follow the deluſive invitation of theory and hypotheſis, rather than the unerring direction of deciſive experiments. The impartial and judicious reader, when he ſhall have peruſed the preſent eſſay, will be able to determine, whether what I aſſert, be true or falſe.

(*a*) The I, LVIII, and CIV paragraphs will explain what is meant by birds with muſcular, intermediate, and membranous ſtomachs.

[*Editor's Note:* Material has been omitted at this point.]

XLI. A ſubſtance not ſoluble by ſimple maceration, and at the ſame time ſofter than grain, upon which the gaſtric juices have no action (III, IV, V, VI, VII.), was wanting to clear up the doubt. Fleſh ſeemed to correſpond to this deſcription. Fleſh is digeſted by many birds with gizzards, which for the moſt part are both frugivorous and granivorous; I therefore filled four tubes with veal (*a*) bruiſed very ſmall in order to ſupply the want of trituration, and forced them into the ſtomach of a hen. They were taken out in twenty-four hours, and the fleſh was in the following ſtate: In the tube that came firſt to my hands it did not amount to above one-twentieth of its original bulk, in two others it had ſuffered nearly the ſame diminution; the only difference appeared in the fourth, which was not open at both extremities like the other three, but cloſed at one end with a circular plate of iron. The fleſh contiguous to the plate preſerved its red colour and na-

(*a*) Wherever I mention fleſh without an epithet. I mean raw fleſh.

tural

tural consistence, and did not seem at all wasted; but at the open end it was reduced to two thirds of the length of the tube, of which it had at first occupied the whole; the part that continued firm and red retained the true flavour of flesh; at the opposite end it had entirely lost that flavour, and the surface, to the depth of a full line, was besides reduced to a pulp, and had acquired a cineritious colour. The inconsiderable residuums in the other tubes were altered in the same manner.

The immediate consequences of these experiments are self-evident. The remarkable diminution of the flesh arose from its having been in great measure dissolved and digested; for all physiologists agree in considering the change of colour and taste, and the transmutation of the food to a pultaceous mass in the stomach, as the characteristic marks of digestion. The three tubes, of which the sides were perforated and the ends open, admitted the gastric liquor at every part. Hence a considerable waste of the flesh in them. The case was different in the tube closed at one extremity, and nothing can be more obvious than the reason; for as the liquor could only enter at one end, it could only there dissolve the flesh.

D 3 XLII. This

XLII. This experiment decifively proves, that the gaftric liquor was the caufe of digeftion in the prefent inftance; and it was eafy to forefee, that others upon the fame clafs of birds would be attended with the fame refult. Some tubes filled with flefh were next introduced into the gizzard of a very large turkey cock, but the lattice work at the open ends, though it confifted of iron, could ill withftand the action of fuch powerful mufcles. Upon examination feven hours afterwards, it was found feparated from the tubes, and coiled up in one mafs near the pylorus, in the midft of the pebbles and fcoriæ of the food, fome of which were jammed fo tight in the tubes, that there was difficulty in forcing them out with the point of a penknife. I could not perceive the fmalleft fragment of flefh amongft them, and remained in doubt whether it had been digefted, or expelled by thefe extraneous bodies. I refolved to fubmit this fpecies of bird to further experiments, but was obliged to abandon the tubes, and have recourfe to the hollow fpherules, of which I have fpoken above (VII). They were made thick and ftrong, with many fmall pores over the whole furface, in order to obviate two inconveniencies, the one left the receivers fhould be unable to refift the violent impulfes impulſes of the ſtomach, the other to prevent the matters compreſſed and agitated by the action of the muſcles from entering ſo readily into them. Two of theſe ſpheres were given to a turkey cock eleven months old, and in twenty-four hours were taken out of the gizzard. They contained at firſt about twenty-eight grains each of beef and veal bruiſed very ſmall. Upon opening them after the ſame interval as before, and weighing the fleſh, the beef was found to have loſt nine, and the veal thirteen grains. I muſt not however omit to remark, that they were both fully impregnated with gaſtric liquor, and conſequently would have weighed ſtill leſs if they had been free from it. The beef and veal, when touched with the point of a knife, ſeemed tenderer than in their natural ſtate, and reſembled a ſoft paſte rather than fleſh. They had the bitter taſte of the gaſtric juice with which they were impregnated, and the colour approached more to white than red. They were replaced in the ſphere, and kept twelve hours in the gizzard of another turkey-cock. Upon a freſh examination, the beef weighed only eight, and the veal only five grains. The gaſtric fluid had therefore produced a new ſolution, and this proceſs was entirely completed after the ſpheres, into

D 4 which

which the flesh was put for the third time, had continued five hours in the stomach of a third turkey cock.

XLIII. Flesh is digested by the gastric liquor of geese as well as of turkies. Eleven grains of beef, inclosed in a spherule, were entirely dissolved in two days in the gizzard of one of these large birds.

I will not describe three other results obtained, one from an hen, and the two others from two capons; since, with respect to the digestion of the flesh, they are exactly like those just mentioned.

All these experiments were made with flesh bruised very small; this condition is not indeed indispensably requisite, but it very much promotes digestion. The bruised flesh was always dissolved in two days, but when entire that process was not completed in four, and sometimes not even in five days. The reason of this difference is obvious. The more flesh is bruised, the larger surface does it acquire; and in proportion to the increase of surface, more points are exposed to the action of the gastric liquor, which will consequently sooner complete the solution.

XLIV. Before I proceed further and conclude the present dissertation, I must notice an experiment of Reaumur, which does not perfectly

perfectly agree with thofe juft related. The greateft part of his memoir is employed in fhewing the great force of the gizzard of gallinaceous fowls in triturating the food; in the remainder he endeavours to prove, that this vifcus contains no menftruum of fufficient efficacy to produce folution. In fupport of this propofition, befides the argument derived from barley continuing unaltered within the tubes, he adduces the following fact, which requires to be particularly related. It is well known, how greedily ducks devour, and how foon they digeft, flefh. In order therefore to obtain the information he wanted, Reaumur had recourfe to this bird. Having provided fix tubes, four of lead, and two of tin, he inclofed in the former bits of veal of the fize of a grain of barley, and in the latter fome confiderably larger. Thefe fix tubes he gave to a duck at different times; viz. a leaden one at ten o'clock in the morning, and another at eight in the evening; next day a third was given at fix in the morning, together with the two tin tubes; laftly, at nine the fame morning the animal was made to fwallow the laft leaden tube, and at ten was killed. Of the four leaden tubes, one was voided the preceding day at nine in the evening; it was that which had been taken

taken at ten in the morning; the other five remained in the gizzard, and the flesh was not only entire, but as firm as at first. Some of the pieces retained their red colour, three of them however had lost it. Of some of the tubes the whole capacity was no longer filled by the flesh; not that it had suffered any diminution, but because it was compressed by the stones and food, which had been admitted at the open ends of the tubes. From this experiment Reaumur infers, that no menstruum had acted on the flesh, since it was not either comminuted or dissolved. And though he does not affirm, that in the gallinaceous class digestion is the effect of trituration alone, he yet concludes, that the gizzard contains no solvent capable of decomposing and digesting the aliment.

[*Editor's Note:* Material has been omitted at this point.]

LVI. At whatever time the ſtomachs of theſe birds happen to be opened, they always contain a certain quantity of gaſtric liquor. But it is leſs abundant when they are full of food, (being in this caſe abſorbed by the food) than when they contain little or none. If we wiſh therefore to be provided with a large quantity of this liquor for experiments, it ſhould be taken from the empty ſtomach. Beſides, in this caſe it is purer than when mixed with the food. When examined in a ſtate of purity, its tranſparency, if we except a ſlight yellow tinge, is little inferior to that of water. It has likewiſe the fluidity, but not the inſipidity of water, being always a little bitter, as well as ſalt. I have found that the gizzards of turkeys and geeſe abound moſt in gaſtric juices, probably on account of

E 4 their

their ſuperior ſize. I was induced by the quantity they afforded to attempt an experiment, which if it ſucceeded, would ſtill further prove that trituration is only an aſſiſting or prediſpoſing, and not the efficient cauſe of digeſtion. It conſiſted in trying, whether theſe juices retain their ſolvent power out of the ſtomach. For this purpoſe, I took two tubes ſealed hermetically at one end, and at the other with wax: into one I put ſeveral bits of mutton, and into the other ſeveral bruiſed grains of wheat, and then filled them with the gaſtric liquor. In order that they might have that condition which in theſe animals precedes digeſtion, they had been macerated in the craw of a turkey cock. And as the warmth of the ſtomach is probably another condition neceſſary to the ſolution of food, I contrived to ſupply it by communicating to the tubes a degree of heat nearly equal, by fixing them under my axillæ. In this ſituation I kept them at different intervals for three days, at the expiration of which time I opened them. The tube with the grains of wheat was firſt examined; moſt of them now conſiſted of the bare huſk, the flour having been extracted, and forming a thick grey ſediment at the bottom of the tube. The fleſh in the other tube was in great meaſure

ſure diſſolved, (it did not exhale the leaſt putrid ſmell) and was incorporated with the gaſtric juice, which had become more turbid and denſe. What little remained had loſt its natural redneſs, and was now exceedingly tender. Upon putting it into another tube, and adding freſh gaſtric liquor, and replacing it under the axilla, the remainder was diſſolved in the courſe of a day.

I repeated theſe experiments with other grains of wheat bruiſed and macerated in the ſame manner, and likewiſe upon ſome fleſh of the ſame kind, but inſtead of gaſtric juice I employed common water. After the two tubes had remained three days under my axillæ, I found that the grains, where they were broken, were ſlightly excavated, which was occaſioned by an incipient ſolution of the pulpy ſubſtance. The fleſh had alſo undergone a ſlight ſuperficial ſolution, but internally it appeared fibrous, red, firm, and in ſhort, had all the characters of fleſh. It was alſo putrid; the wheat too had acquired ſome acidity, two circumſtances, neither of which took place in the grains and fleſh immerſed in the gaſtric liquor. Theſe facts are then irrefragable proofs that the gaſtric juice, even out of its natural ſituation, retains the power

power of diſſolving animal and vegetable ſubſtances in a degree far ſuperior to water.

LVII. The gaſtric juice which I employed was taken from a turkey. That of a gooſe produced ſimilar effects. I have further found, that in order to operate the ſolution of animal and vegetable ſubſtances, this juice ſhould be freſh. It loſes its efficacy, when it has been kept ſome time in veſſels, eſpecially if they ſhould happen to be open. It alſo becomes inefficacious after it has been uſed for one experiment. Laſtly, a conſiderable degree of heat, equal to the temperature of man or birds, muſt be applied; otherwiſe, the gaſtric juices are not more effectual in diſſolving fleſh and vegetables than common water. This artificial mode of digeſtion is well calculated to illuſtrate the ſubject I have undertaken to treat; but I ſhall have opportunities of ſpeaking of it at greater length in the ſubſequent diſſertations.

[*Editor's Note:* Material has been omitted at this point.]

LXIX. Young crows, as well as all other young birds, eat more than the adult; hence I ſuſpected their digeſtion to be quicker. Having a neſt of the grey ſpecies brought me in June, I made, among others, the experiment related in the laſt paragraph. The reſult was very ſatisfactory. A quarter of an ounce of beef, faſtened as before, to a thread, had ſcarce touched the ſtomach, when the ſolution began, and in forty-three minutes was completed; but an equal quantity diſtributed in ſeveral tubes, required four hours and a half to be diſſolved. Upon opening the ſtomachs of the two young birds, I immediately perceived the cauſe of this rapid ſolution; they contained half a ſpoonful of gaſtric fluid; a quantity ſeldom met with in the ſtomach of adult crows. As the neſtlings require more food, Nature has furniſhed them

them with the means of an eafier and more fpeedy digeftion.

It is fcarce neceffary to remark, that the experiments related in the LVth and following paragraphs, clearly evince this important truth, that the digeftion of food is proportional to the quantity of gaftric juice acting upon it. When this liquor comes in contact only with a few points, the decompofition is very flow and inconfiderable (LXVII); when freer accefs is allowed, the folution takes place more fpeedily, and is more confiderable (LXV, LXVI); it is very rapid, when every obftacle is removed, and the food is on all fides expofed to the action of the folvent liquor, (LXVIII, LXIX).

[*Editor's Note:* Material has been omitted at this point.]

CXXI. It is eafy to learn the nature of the food of water-fnakes, and we ought in con- fequence ſequence to provide it for our experiments. Among the antients Oliger Jacobeus, where he treats of frogs, and among the moderns Valiſneri will ſatisfy us, that theſe reptiles live chiefly upon frogs. Next to man water-ſnakes may be denominated their greateſt ſcourge. They particularly frequent the water of ditches, puddles, ponds, lakes, ſuch in ſhort as is frequented by frogs; and here they make an eaſy prey of them, notwithſtanding they mutually give each other notice when they perceive the ſnake at a diſtance, by a kind of whiſtle or outcy of diſtreſs, as I have often obſerved, at which all fly with the utmoſt precipitation: Dante was not acquainted with this circumſtance.

Come le rane innanzi l'inimica
Biſcia per l'acqua ſi dileguan tutte,
Finchè alla terra ciaſcuna s'abbica (*a*).

A fiſherman having brought me three very large and vigorous water-ſnakes, I gave each at the ſame time a tube encloſing a different part of a frog; one muſcle, the other liver, and the third ſpleen. The tubes were left three days and an half in the ſtomach. Upon forcing them out, I obſerved the ſame kind

(*a*) Infern. Cant. 9. Fol. 161, &c. As frogs ſcour along the water, at the approach of the water-ſnake, without ſtopping, till they have gained the dry ground.

of

of digeſtion that I had before ſeen in frogs (CV, CVI). The fleſh was beginning to be changed into an adheſive cineritious gluten; the interior parts were unaltered. The tubes were now introduced a ſecond time into the ſtomachs, and when they had continued there two days they were found empty; ſome of the adheſive matter ſtuck to the outſides of two of them.

[*Editor's Note:* Material has been omitted at this point.]

CCXV. In the CCIIId paragraph I remarked, that the chief experiments on man were reducible to two heads, thoſe which relate to the natural proceſs, as it may be obſerved by means of tubes and ſuch contrivances, and thoſe which relate to artificial digeſtion,

digeſtion, provided the gaſtric juices can be procured. Having treated the former of theſe diviſions as well as circumſtances would permit, it remained for me to make ſome enquiries relative to the ſecond. It was firſt neceſſary to deviſe a method of procuring the gaſtric fluid. The firſt idea that ſtruck me was to ſearch for it in dead bodies, but after examining ſeveral ſtomachs I was obliged to abandon this ſearch; for they were either without any fluid, or elſe what they contained was ſo turbid and ſo much adulterated with heterogeneous matters, that it would by no means ſuit my purpoſe. Nor were the little ſpunges, which had ſerved ſo well in animals, better adapted to the preſent occaſion. Two ſpunges would not ſupply me with a ſufficient quantity, and I could venture only to ſwallow two tubes at once, for fear of forming an obſtruction in my ſtomach. Beſides, the juice thus procured would have been very impure, on account of the heterogeneous matters that the tubes muſt neceſſarily have imbibed during their paſſage through the inteſtines.

There remained only to obtain it by exciting vomiting while the ſtomach was empty. To effect this, I choſe rather to tickle the fauces than drink warm water, as in this caſe the

the gaſtric fluid muſt have been diluted. In this manner therefore, before I took meat or drink, I procured in two mornings a quantity ſufficient for a few experiments, of which the reſult ſhall be related below. I could have wiſhed to have made a greater number, but the diſagreeable feelings occaſioned by the act of vomiting, the convulſions of my whole frame, and more eſpecially of my ſtomach, that continued for ſeveral hours after it, left upon my mind ſuch a repugnance for the operation, that I was abſolutely incapable of repeating it, notwithſtanding my earneſt deſire of procuring more gaſtric liquor.

CCXVI. I was therefore obliged to content myſelf with what theſe two vomits afforded me. The firſt time it amounted to an ounce and thirty-two grains. It was frothy at its being thrown up, and ſomewhat glutinous. After it had been at reſt a few hours and depoſited a ſmall ſediment, it was as limped as water. It was a little ſalt to the taſte, but not at all bitter. It did not, either when thrown on the fire or brought near a candle, ſhew any token of inflammability (*a*). It evaporated

(*a*) From this and the LXXXIſt, CXXIIId, CXLIXth, and CLXXXVth paragraphs we may collect, that the gaſtric juices both of man and animals are deſtitute of inflammability. I made

evaporated in the open air, and when I put fifty-two grains into a veſſel and ſet it on hot coals, it emitted a thick ſmoke. Another ſmall portion, weighing eighty-three grains, was put in a phial, which was cloſed with a ſtopple to prevent it from evaporating. It did not change colour or taſte, nor did it acquire any bad ſmell, notwithſtanding it was kept above a month in the hotteſt ſeaſon of the year. I thus employed about one half, the remainder was uſed for an attempt to obtain artificial digeſtion. It was put into a glaſs tube two inches long, ſealed hermetically at one end, and very narrow at the other; I then introduced a ſmall quantity of maſti-

made theſe experiments, becauſe Reaumur thought that that of his kite was inflammable, which quality Dr. Batigne imputes to the bile, a fluid conſiſting principally of oil (*premiere Reflexion ſur les Exper. de Reaumur*). But were this true, the gaſtric juices of moſt of my birds ought to have taken fire. As all mine are contrary to Reaumur's ſingle experiment, I ſhould ſuſpect, that what he obſerved was owing to accident. His experiment was the following: To take away the ſmell of putrid fleſh, which one of his tubes had acquired, he ſet it upon ſome burning coals, when immediately there iſſued a flame from the inſide that laſted above a minute (Seconde Mem.). But it is eaſy to perceive, that this might have been owing to ſome fat of the encloſed fleſh adhering to the tube. I am more confirmed in this ſuſpicion from having obſerved, that the gaſtric fluid of a kite, ſuch as Reaumur's, mentioned in a note to paragraph CLXXV, was not more diſpoſed to take fire than the other gaſtric juices which I examined.

cated

cated boiled beef, and ſtopping the tube with cotton, ſet it in a ſtove cloſe to a kitchen fire, where there was a conſiderable heat, though not perhaps exactly equal to the temperature of my ſtomach. By the ſide of this tube I placed another, containing the ſame quantity of fleſh immerſed in water. The appearances in both were the following: In twelve hours the fleſh in the former began to loſe its fibrous ſtructure, and in thirty-five it had ſo far loſt its conſiſtence, that when I attempted to lay hold of it, it ſlipped from between my fingers. But though to the naked eye it appeared to be reduced to a pultaceous maſs and to have loſt its fibrous texture, yet the microſcope rendered the fibres viſible; they were however reduced to a great degree of minuteneſs. After this ſemifluid ſhapeleſs maſs had continued two days longer in the gaſtric fluid, the ſolution did not ſeem to have made any further progreſs, and the reduced fibres were ſtill juſt as apparent. The fleſh did not emit the leaſt bad ſmell, while that immerſed in water became putrid in ſixteen hours, and grew worſe and worſe the two following days. It loſt in ſome meaſure its fibrous ſtructure, as always happens during putrefaction; but this appearance did not proceed

proceed ſo far as in the other portion, for the fibres were entire on the third day.

CCXVII. I vomited the ſecond time more gaſtric fluid, and was now enabled to examine it again as I had done before; and it appeared to poſſeſs exactly the ſame properties. In order to determine the influence of heat two tubes were filled with it, and ſome fleſh was immerſed as before (CCXVI). One of the tubes was placed in the ſtove, and the other left in the open air. In the former the fleſh was juſt as much diſſolved as in the preceding experiment; but in the latter the ſolution proceeded no farther than when water was employed (CCXVI). There was however no putrid ſmell, though the fleſh was left immerſed in the gaſtric fluid ſeven days.

Before I conclude this account, I muſt mention a circumſtance that happened the ſecond time I procured gaſtric liquor by vomiting. Four hours before I ſubmitted to this diſagreeable operation, I had ſwallowed two tubes filled with beef, one of which was thrown up; the fleſh was thoroughly ſoaked in the fluid of the ſtomach, and the ſurface was ſoft and gelatinous; it had moreover waſted from fifty-three to thirty-eight grains. This experiment proves, that there takes place a conſiderable degree of digeſtion in the ſtomach,

ſtomach, before the food paſſes into the inteſtines.

CCXVIII. We may now ſafely lay down ſome general conſequences concerning digeſtion in Man and animals. In the experiments on birds with muſcular ſtomachs, we have ſeen how trituration diſpoſes the food to be digeſted. Hence Nature has furniſhed that claſs with gaſtric muſcles of ſufficient power to effect this neceſſary preparation. But we have likewiſe ſeen how digeſtion, which conſiſts in the tranſmutation of the aliment into chyme, is the effect of the juices alone with which the ſtomach abounds (Diſſ. I).

We next proceeded to birds with intermediate ſtomachs, ſuch as crows and herons, and found, that in them digeſtion was owing to the gaſtric fluid alone (Diſſ. II).

We next conſidered animals with membranous ſtomachs, a claſs ſo numerous and various, that it comprehends almoſt every family of living creatures; it includes the inhabitants of ſalt and freſh water; amphibious animals, as the frog, the newt, and water-ſnake; reptiles, as the viper, the land-ſnake, and many others; quadrupeds, as the cat, the dog, the horſe, the ox; birds, as birds

birds of prey: to this catalogue Man himſelf is alſo to be added.

In ſeveral of theſe animals we have ſeen the neceſſity of previous trituration, as in the ruminating order and in Man; in them it is produced by the teeth, as in gallinaceous fowls by the muſcles of the ſtomach. But in others, as in the frog, the newt, ſerpents, and birds of prey, it has no ſhare in the proceſs of digeſtion. But in the latter, as well as the former caſes, we have ſeen how the food is diſſolved and digeſted by the gaſtric fluid (Diſſ. III, IV).

In every order of animals, Nature, always uniform in her operations, employs one principle for the performance of this vital function. Hence ſhe has ſo copiouſly furniſhed the œſophagus and ſtomach with glands, follicles, and other contrivances that anſwer the ſame end, whence continually flow the juices ſo neceſſary to the life of Man and animals. Theſe juices agree in many properties, but the difference of effect ſhews, that they differ in others. In the frog, the newt, ſcaly fiſhes, and other cold animals, the gaſtric fluid produces digeſtion in a temperature nearly equal to that of the atmoſphere. But the gaſtric fluid of hot animals is incapable of diſſolving the aliment in a degree of heat

heat lower than that of the animals themſelves. There is alſo a difference in celerity of action, and in efficacy. In celerity, becauſe the food in hot animals is digeſted in a few hours; whereas, in the oppoſite kind it requires ſeveral days and even weeks, particularly in ſerpents. In efficacy, becauſe the gaſtric juices of ſome animals, as the gallinaceous claſs, can only diſſolve bodies of a ſoft and yielding texture, and ſuch as have been previouſly triturated; while thoſe of others, as ſerpents, the heron, birds of prey, the dog, decompoſe ſubſtances of great tenacity, as ligament and tendon, and of conſiderable hardneſs, as the hardeſt bone. Man belongs to this diviſion; but his gaſtric fluid ſeems to have no action on the hardeſt kind of bones. Further, ſome ſpecies, as birds of prey, are incapable of digeſting vegetables. But Man, the dog, the cat, crows, &c. diſſolve the individuals of both kingdoms alike. In general theſe juices produce their effects out of the body, as the numerous inſtances of incipient digeſtion under this circumſtance, both with the gaſtric fluid of animals and Man abundantly ſhew.

[*Editor's Note:* Material has been omitted at this point.]

CCXXIII. I will conclude this differtation, with fome remarks on a problem clofely connected with refearches concerning the efficient caufe of digeftion. Mr. Hunter, one of the beft Englifh anatomifts of the prefent age, frequently found in the dead bodies which he opened that the great curvature of the ftomach was confiderably eroded, and fometimes entirely diffolved. In the former cafe, the edges of the wound were as foft as half-digefted food, and the contents of the ftomach had got into the cavity of the abdomen. He obferves, that fuch a wound could not have exifted in life, as it had no connection with the difeafe, and more frequently appeared in perfons persons who died violent deaths. In order to discover the cause of this phænomenon, he examined the stomachs of various animals, both immediately and some time after death. In several he observed the same appearance. Hence he thought he was enabled to assign the cause. He supposes the solution to be owing to a continuance of digestion after death, and that the gastric fluid is capable of dissolving the stomach when it has lost its vital principle. From this discovery he infers, that digestion neither depends on the action of the stomach nor on heat, but on the gastric juices, which he considers as the true menstruum of the food (*a*).

CCXXIV. When Mr. Hunter's short but sensible paper came to my hands, I was engaged in experiments on digestion. I had discovered the primary importance of the gastric fluid in this process, and that it acts out of the body; that is to say, in the dead body. I knew also, that after death this fluid issues from the coats of the stomach. From these data I had little difficulty in believing the fact related by the English anatomist, and adopting his explanation of it: nevertheless it was proper to repeat the experiment. Being un-

(*a*) Ph. Transf.

provided

provided with human ſubjects, I had recourſe to animals. Some were opened ſooner, and others later after death; but among the numbers I inſpected, not one had its great curvature diſſolved, or much eroded. I ſay, much eroded, becauſe I have often ſeen a little eroſion, eſpecially in different fiſhes, in which, when I had cleared the ſtomach of its contents, the internal coat was wanting. The injury was always confined to the inferior part of the ſtomach. If theſe facts are favourable to Mr. Hunter, a great number are againſt him. They do not however deſtroy his obſervations; mine are only negative, his are poſitive; and we know that a thouſand of the former do not deſtroy a ſingle one of the latter, provided it is well aſcertained. I have no reaſon to diſtruſt Mr. Hunter, for his paper has the air of ingenuouſneſs and candour which uſually accompanies truth.

CCXXV. The ill ſucceſs of my experiments did not induce me to abandon the idea of digeſtion after death, it only led me to conſider it in another point of view. If it be true, ſaid I to myſelf, that the gaſtric fluid exerts its action after death, it muſt produce ſome ſolution of the food. Let then an animal be fed and immediately killed, after ſome time

time let it be opened, and let us fee whether the food has been at all digefted. I determined to bring this obvious inference to the teft of experiment; I therefore kept a raven fafting feven hours in order to empty its ftomach, and then fet before it an hundred and fourteen grains of beef, which were immediately eaten, and muft have paffed into the ftomach, as this bird has no crop. I then killed it, and as it was winter, put it into a ftove, where it was left fix hours. Suppofing this to be a fufficient time for the gaftric fluid to exert its action, I opened the ftomach, and found the flefh in the following ftate. It was impregnated with gaftric fluid, and was become tender; the colour was changed to a pale red, and the furface had a bitter tafte, while the internal parts retained the tafte of flefh. After the gaftric fluid was wiped away, it weighed only fifty-two grains; it had therefore loft above half its weight in fix hours, or, what amounts to the fame thing, was above half digefted. The pylorus, and the duodenum for about an inch, were occupied by an afh-coloured mucus, which muft have been the diffolved part of the flefh.

At the fame time I gave another raven, that had in like manner been kept fafting feven hours,

hours, an equal quantity of flesh, and killed it in two hours and a quarter. My view was to observe the difference between what had lain six hours in the dead, and two and a quarter in the living stomach, and it was very great; for in this latter case the flesh was totally dissolved, except a little cellular substance, which I have found to be always longer in being digested than the muscular fibres; the mucus was the same as before, only in larger quantity, and occupied more of the duodenum. These two experiments compared together prove two things, first, that digestion continues after death; and secondly, that it is then far less considerable than in the living animal, though in the present instance the heat of the stove, which was about 100° (*a*), must have promoted it not a little. The heat of the living raven did not exceed 30° (*b*).

CCXXVI. Another dead raven was kept five hours in the same stove, after I had forced two dead lampreys, weighing together an hundred and twelve grains, down its throat. One lay in the œsophagus, the other had reached the stomach and was completely de-

(*a*) Two hundred fifty-seven deg. Fahr. Ther.

(*b*) One hundred nine and an half ditto.

composed,

composed, while the former was indeed entire, but soft and flaccid. This accident proves, that the gastric fluid is capable of producing a sensible degree of digestion at a time when the œsophageal juices are inert.

CCXXVII, These experiments were made in winter. I determined to repeat them the next summer, because then I could expose the dead animals to a greater heat. Accordingly in that season some bruised veal was given to two ravens, which were immediately killed, and left seven hours in a window exposed to the sun. We have already seen in several passages, the influence of heat in promoting artificial digestion (CXLII, CLXXXVI, CCI, CCXVII). Nor did it now appear less considerable. Each raven had eaten sixty-eight grains of flesh, of which there was not an atom left entire; it was all dissolved into the usual gelatinous pulp, and the greater part had passed through the pylorus.

These facts, I think, decisively prove, that animals, at least the species just mentioned, continue to digest after death. If we consider the matter rigorously, it will be proper to obviate a difficulty that may be started. However careful we are to kill the

the animal immediately after it has ſwallowed food, it is certain, that there will be a ſhort interval between the time the food gets into the ſtomach and the death of the animal, and that the gaſtric fluids act upon it during this interval. Moreover, after death they will act for ſome time juſt as in life, ſince the vital heat is not inſtantly exſtinguiſhed. The digeſtion therefore obſerved in dead animals may, if not entirely, at leaſt in part, be produced by the gaſtric fluid acting during life, and a ſhort time after death.

Nothing could be more eaſy than to aſcertain the juſtneſs of this ſuſpicion, ſince we have only to thruſt a little food into the ſtomach of a dead and cold animal, and obſerve the conſequence. I made the experiment upon a raven that had been dead an hour, and had now only the temperature of the atmoſphere. Forty-two grains of beef cut into pieces were forced into the ſtomach, which was opened after the bird had lain ſeven hours expoſed to the ſun. And here inſtead of pieces of ſolid fleſh, I found only the uſual pulpy maſs, partly in the ſtomach and partly in the duodenum. The ſolution was therefore effected by the gaſtric fluid, independently of the powers of life.

CCXXVIII.

[*Editor's Note:* Material has been omitted at this point.]

# 3

# NOTE ON THE ADULTERATION OF JALAP RESIN AND ON MEANS TO RECOGNIZE IT, ETC.

**Louis Antoine Planche**

*This article was translated expressly for this Benchmark volume by H. Friedmann, the University of Chicago, from "Note sur la sophistication de la résine de jalap et sur les moyens de la reconnaitre, etc." in* Bull. Pharm. (Paris) **2**:578–580 (1810).

For a long time the dangerous art of adulterating drugs has been attributed to the Egyptians, to the druggists of Frankfurt and of Marseilles. One may, however, conclude after reading the following article that the English are no less skilled than other people in this type of business.*

Various drugs, making up the cargo of a purchase from England and brought last year to the port of Dieppe, included Jalap resin, which was consigned to a Paris wholesale merchant. Shortly afterward, some commercial brokers offered this resin to various druggists of the city. Mr. A., among others who received this material, felt that it had been adulterated. He sent me a sample to be tested so that his mind could be put at ease about this substance for which I found the following properties:

It had a greenish color, darker than that of pure Jalap resin; this difference became more pronounced when one compared the two resins ground to a fine powder. A more agreeable and lightly aromatic fragrance, a sweetish flavor, a specific gravity higher than that of pure Jalap resin made me suppose that the substance I examined had been mixed with guaiac resin. Experiment has proved that my suspicions were well founded. I dissolved a lump of each resin in alcohol, and with each of these solutions moistened two pieces of very white linen, which I introduced one after the other into a bottle filled with nitric acid vapors, whose action on guaiac resin is, as is known, very rapid and very remarkable. No sooner did the linen impregnated with the suspect resin contact the nitric vapor than it took on a highly intense blue color. The other was not apparently changed at all by this chemical. I am not concerned with the amount of guaiac resin in the resinous material. The essential point was to know what the added substance was. Direct experiments with Jalap resin showed that 1/40 of guaiac resin mixed with the former and dissolved in alcohol can be detected by nitric acid vapor.

I have observed that fresh horseradish root also had the property

**Editor's Note:* This article was written during the height of the Napoleonic wars. Napoleon's defeat at Trafalgar in 1805, not long before this paper was written, may perhaps have helped to exacerbate certain Anglo-French animosities.

of imparting a blue color to tincture of guaiac resin. For this experiment, all that had to be done was to dip a piece of this root into a glass containing the resinous tincture. The liquid gradually becomes colored a beautiful blue resembling that of a dilute indigo solution in sulfuric acid. The colored principle is apparently very unstable or it has a marked tendency to combine with the ingredients of guaiac resin, for this blue color does not last longer than a quarter hour. After that the color progressively reverts to its initial state.

Sulfurous acid gas formed by the slow combustion of sulfur brings about a more or less similar phenomenon, and probably, following what we already know, many other different types of substances may bring about a similar color formation with guaiac resin; but it seems to me that nitric acid vapor is the most reliable reagent in all cases where one has to detect guaiac resin in some kind of mixture.

# 4

# ON THE SUBSTANCES WHICH DEVELOP BLUE COLOR IN GUAIAC RESIN

**Louis Antoine Planche**

*These excerpts were translated expressly for this Benchmark volume by H. Friedmann, the University of Chicago, from pages 17 and 23–25 of "Sur les substances qui développent la couleur bleue dans la résine de gaïac" in* Bull. Pharm. (Paris) **6**:16–25 (1820).

[*Editor's Note:* In the original, material precedes these excerpts.]

If the conversion of guaiac resin to a blue or green color is due only to the action of air and of light, why will guaiac resin, under conditions identical as to these two agents, turn green or blue when mixed with one substance, while not undergoing a change in its natural color when mixed with another?

The answer to this question is perhaps more difficult than appears at first; at least, this is how I look at it; it requires demanding experiments that may be undertaken only by a chemist more experienced than I am in this kind of research. But although I cannot pretend to answer this question completely, I may be allowed to clarify it. With this approach, I carried out the experiments that I am going to report.

[*Editor's Note:* Material has been omitted at this point.]

In general, the substances that color guaiac resin blue lose this property upon heating. Thus the root of *solanum tuberosum,* of *pastinaca sativa,* the fruit of *fagus castanea,* when boiled in closed vessels in their own juices act neither on the powdered resin nor on the mixture.

Mucilage of gum arabic prepared in the cold brings about a blue coloration as first observed by Mr. Boullay but does not produce any change when the gum is treated with boiling water.

[*Editor's Note:* Material has been omitted at this point.]

When considering the effect of heat on the substances containing the coloring principles of guaiac resin, I thought at first that this principle is volatile and that it may be possible, by heating one of the substances to different degrees, to coerce it into an apparatus arranged to receive liquid and gaseous products, just as for boiling water. The emulsion I chose for this experiment did not yield any satisfactory result. I noticed nothing special when a piece of linen saturated with guaiac tincture or impregnated with guaiac powder diluted with water was suspended in the hollow of the retort. The small amount of liquid that passed into the receiver, and the air of the vessels released by the heat and collected in a test-tube showed nothing remarkable on contact with guaiac resin.

From this moment I had to give up the idea that this principle was volatile. It hence seems to me more likely that this species of *cyanogen,* no matter what its nature, is absorbed by the substances that normally allow it to be active upon attaining a certain temperature, that it hence obeys other laws and forms new compounds, as shown by its coloring properties. Besides, this is a simple guess that I submit to the judgment of scholars who have already worked on the same subject and who, much more than I, are able to fathom it.

# 5

# FORMATION OF SUGAR IN CEREAL GRAINS CONVERTED TO MALT, AND IN FLOUR STEEPED IN BOILING WATER

**Gottlieb Sigismund Constantin Kirchhoff**

*Member of the Academy of Sciences of Petersburg (1)*

*This article was translated expressly for this Benchmark volume by H. Friedmann, the University of Chicago, from "Formation du sucre dans les graines céréales converties en malt, et dans la farine infusee dans l'eau bouillante" in* J. Pharm. Sci. Accessoires (Paris) **2**:250–258 (1816), the French translation of the original German paper *"Ueber die Zuckerbildung beim Malzen des Getreides, und beim Bebrühen seines Mehls mit kochendem Wasser" in* Schweigger's J. Chem. Phys. (Nuremberg) **14**:389–398 (1815).

Up to now the formation of the sweet substance that occurs in cereal grains during germination and in cereal flours steeped in warm water has not been explained in a satisfactory manner.

Cruikshank and Saussure, Jr., have been particularly concerned with this topic. Cruikshank thought his experiments led to the conclusion that during germination the grains absorb a certain quantity of oxygen that they remove from the atmosphere; that a part of this oxygen combines with the flour of the grain to form sugar; while the rest of this oxygen, combined with a part of the carbon of the grain, escapes in the form of carbon dioxide. Saussure repeated these experiments and proved that during germination the grains in fact lose carbon that is removed by oxygen acting on them; this oxygen, however, does not at all combine with these cereals but contributes only to the formation of carbon dioxide, which develops during germination (1). Yet he does not explain the formation of the sweet substance, for he merely states: "The formation of sugar which occurs with the help of oxygen gas in grains undergoing germination is a very remarkable phenomenon which I shall not explain" (1).

Thomson undertook to give an explanation. "It is seen," he says, "from these experiments (Saussure's) that the flour part is converted to sugar by losing carbon, and that because of this loss the proportions of the remaining hydrogen and oxygen are increased" (2).

It can be seen that with this explanation, Thomson attributes the formation of sugar to the loss of carbon, which the germinating grains undergo with oxygen gas. But if one considers that flour is not a simple substance, that it is composed, on the contrary, of many heterogeneous parts, all of which are probably not equally concerned with the formation of sugar, one will ask which of these parts has lost carbon (3); since Saussure's experiments do not give any answer to this question, Thomson's explanation is insufficient.

*We will recall to our readers that Kirchhoff revealed to us a few years ago how starch is converted to sugar by means of sulfuric acid; now he has just enriched the sciences by a discovery that is of no less interest for the general theory of fermentation; that is to say, to convert starch into sugar with the help of gluten. A. V. [*Editor's Note:* This is A. Vogel, one of the editors of the *Journal de Pharmacie.*]

No sugar is formed if starch and gluten stand separately under conditions that favor germination; neither does germination take place when starch and gluten are steeped in warm water separately.

It follows from these experiments that an agent other than oxygen participates when sugar is formed either during germination or while flour is steeped in warm water (4).

Although the circumstances surrounding the germination of grains and the steeping of flour in warm water are quite different, I think that in both cases the formation of sugar has the same cause, due to the interaction of starch and gluten.

In order to obtain some firmer ideas on this subject, I performed some experiments with preparations of starch and gluten. The gluten I used had been obtained from fresh wheat flour, was cut into small pieces with scissors, promptly dried at a moderate heat, and ground to powder. In order to work with pure starch, I used potato starch for my experiments.

The first experiment I performed already proved that starch and gluten interact; after some time the starch had lost its starchy consistency, and sugar had been formed.

Since the successful formation of sugar by the interaction of starch and gluten depends very much on the temperature, and since when working with small quantities one can easily miss the appropriate temperature, the experiment did not always succeed.

I was often successful with the following method: I first poured four parts of cold water into two parts of starch; then with continous stirring of the mixture, I added twenty parts of boiling water; a thick starch was formed; into this starch, still warm, I mixed a part of the pulverized gluten, and I incubated the mixture for 8 or 10 hours at a temperature of 40-60° Réaumur (5). After 1 or 2 hours, the starch slowly began to liquefy, so that after 8 to 10 hours and even after a shorter time interval it passed through filter paper. This filtered liquid, separated from its residue, is water-clear. It contains a little bit of acid, which still remains even after concentration to a syrupy consistency by boiling.

This syrup tastes as sweet as that obtained from malt; it has less aftertaste than the latter and is colored only feebly yellow. When acid yeast is added, it can undergo alcoholic fermentation. It is only partially soluble in alcohol, and the part dissolved in alcohol crystallizes after evaporation of the liquid; the crystals are white, very small, and indistinct.

The alcohol-insoluble residue dissolves almost completely in water. A gall-nut infusion forms no precipitate in this aqueous solution. This residue seems to be modified starch, which has not yet been completely converted into sugar.

The amount of syrup obtained is proportional to the amount of starch used. It appears that in the course of this operation, only very little gluten dissolves; for on the filter almost all the gluten used is recovered. During the digestion with starch it becomes acidic; it is probably by this means that it modifies the nature of the starch and brings about the formation of sugar. It is remarkable, however, that when one adds a few drops of concentrated sulfuric acid to the starch before admixing the gluten, no sugar is formed.

I determined whether the gluten remaining on the filter could convert a fresh portion of the starch into sugar; I found that it acted feebly and imperfectly.

These experiments convinced me that when flour is steeped in warm water, the formation of sugar is brought about by the action of gluten on starch.

I wanted to satisfy myself experimentally as to whether, as one might suppose by analogy, the same thing occurs during the germination of cereal grains.

With this aim I performed experiments with air-dried barley malt. Herewith a description of the experiments that gave me the most satisfying result.

I poured 4 parts of lukewarm water (30° Réaumur) into a mixture of 1 part pulverized malt and 2 parts starch; then with continuous shaking of the mixture I added 14 parts of boiling water; I covered the vessel and let it stay for 1 hour in a warm place. After this time the starch had been converted to a sweet liquid which, after filtration and concentration by evaporation, formed a syrup.

This experiment seemed to me to prove that the property of gluten to convert starch into sugar is singularly increased by germination and that malt-gluten can convert not only the starch contained in this malt into sugar but in addition a portion of added starch as well.

I washed the pulverized malt to obtain the starch. I digested this starch for 24 hours with a 3 percent solution of potash. After this time I decanted the yellow liquid, and I washed the starch with water until it did not show even the slightest trace of potash.

Starch treated in this manner had lost the property of forming sugar on infusion with warm water; it behaved like ordinary starch and formed starch paste with boiling water. The potash solution had removed the gluten that solubilizes it and that prepares it for conversion into sugar. From the potash solution the dissolved gluten is precipitated by acids. This experiment proves that Thomson is wrong when he states that the malt-starch has been modified (2). It follows that it is the gluten that brings about the formation of sugar in germinated cereal seeds and that the modification undergone by the germinated seeds depends on a modification in their gluten. During the germination the gluten changes the previously insoluble starch into sweet material in accord with the young plant's requirement for nourishment during its development.

The experiments I have just reported confirm the chemical principle that sugar is the substance that can undergo alcoholic fermentation, for when one steeps cereal seeds, or flour prepared from them, in warm water, then their starch is converted into sugar by the action on them of gluten. If one pays some attention to my experimental result, it is readily seen why starch and flour, which lack gluten, do not show the clearcut phenomena of alcoholic fermentation (6) when they are steeped in warm water and when yeast is added.

This explains again why plant substances that, like potato, contain starch but lack gluten, cannot undergo successful alcoholic fermentation (7) unless substances containing gluten such as flour or malt are added. It is easily understood how the addition of these substances favors alcoholic fermentation. It is the gluten of malt that predisposes potato starch steeped in warm water to be converted to sugar. Again it is gluten of malt added to flour that brings about the more rapid and perhaps also the more complete conversion of starch contained in flour into sugar than is possible for the gluten contained in this flour itself. Thus it has long been observed that it is advantageous to add malt to flour derived from cereal seeds that are to be used for alcoholic fermentation.

It has been claimed that malt used by itself yields no more brandy than a mixture of equal parts of malt and flour (8). It seems to me that a given weight of a mixture of malt and of flour should produce more alcohol than the same weight of malt, always assuming that the proportions of the components in the mixture be well chosen; for in this case the conversion of the starch of flour into sugar is partly determined by the gluten of the flour itself, and partly by the gluten of the malt.

Proust found 0.57 to 0.58 of starch in barley flour, all of whose grains had germinated completely; he assumes that this starch does not participate at all in alcoholic fermentation (9). Sugar of starch had not yet been discovered at the time that Proust performed this experiment. But very recently Hermbstaedt has maintained that starch occurs only accidentally in the mixture undergoing alcoholic fermentation and that it causes cloudiness in the beer (10). According to this chemist, starch dissolved in the malt infusion can be detected by adding gall-nut infusion to it; a precipitate is formed that is redissolved at a temperature of 60° Réaumur. This precipitate would be a combination of astringent principle and of sugar of starch. But since this did not agree with my ideas, I performed the following experiment to clear up my doubts. I prepared an infusion of 1 part malt and 2 parts starch. I poured a sufficient quantity of gall-nut infusion into 100 parts of this filtered decoction. I waited at least to obtain in precipitated form the starch used as such and that contained in the malt. But the weight of the dried precipitate was found to be but 2 percent of the malt and of the starch used. This statement of Hermbstaedt and of some other chemists is hence false (11). The astringent principle does not form a stable combination with the sugar of starch; the small amount of precipitate that I obtain is probably only gluten modified by germination and that, because of this modification, dissolves at a lower temperature.

Although starch constitutes the most abundant part of cereal flours and although the amount of brandy yielded by different kinds of flour is proportional to the amount of starch they contain, it could not be known for certain before sugar of starch was discovered that starch contributes to the formation of alcohol, all the more so since it was not possible to obtain alcoholic fermentation with this substance by itself. As soon as it was known that starch can be converted to sugar by the action of acid, it was very likely that one would not find starch to remain inert in the fermenting mixture. The experiments that I just reported prove that starch can be converted to sugar by the action of gluten as well. It can hence be stated that starch is the principal source of the constituents of alcohol formed by fermentation, that it is the main foundation of beer and of vinegar (obtained from cereals).

## CONCLUSIONS

1. Gluten is the substance that brings about the formation of sugar in germinated seeds and in flour steeped in warm water.
2. Starch has not undergone any change in germinated seeds, and only at a temperature above 40° Réaumur will gluten convert it to sugar.
3. Starch is the part of flour that serves as the main source for the formation of alcohol.
4. As a result of germination, gluten acquires the capacity to convert a larger quantity of starch into sugar than is contained in the seed.

5. The formation of sugar in germinated seeds is a chemical process and not a result of vegetation.
6. Starch is found in malt decoction in the sweetened state [The French translation, by quoting the original German phrase, "im versüssten Zustande," indicates that the meaning is not quite clear] and in this state gall-nut infusion does not act on it.

REFERENCES


1. Scherer, *Journal de Chimie,* Vol. IV, p. 81.
2. Thomson, *Système de Chimie.*
3. Ibid.
4. Ibid.
5. That is, 50-75°C. [*Editor's Note:* On this scale ice melts at 0°, water boils at 80°.]
6. Gehlen, *Journal de Chimie,* Vol. I, p. 668.
7. Ibid.
8. Gehlen, *Journal de Chimie,* Vol. I, p. 71.
9. Gehlen, *Journal de Chimie,* Vol. II, p. 377.
10. Hermbsteadt, *Principes chimiques pour l'art du brasseur* [Chemical Principles for the Brewer's Art].
11. Thomson.

# 6

# NEW OBSERVATIONS ON OXYGENATED WATER

**Louis Jacques Thénard**

*This article was translated expressly for this Benchmark volume by H. Friedmann, the University of Chicago, from "Nouvelles observations sur l'eau oxigenée" in* Ann. Chim. Phys. (Paris) **11**:85–87 (1819).

In the last observations about oxygenated water that I had the honor to present to the Academy, I tried to demonstrate that water saturated with oxygen contains exactly twice as much oxygen as pure water or, which comes to the same thing, that pure water can absorb up to 616 times its volume of this gas at a temperature of 0° and under a pressure of 0.76 m.[1] I have at the same time come to know the physical properties of this new liquid and the remarkable phenomena that arise from its contact with a certain number of mineral substances. I have since studied its action on almost all the other mineral substances and on almost all vegetable and animal substances. I will not here report all the results that I have obtained; I will quote only a single one that to me seems to deserve attention: that is, that several animal substances possess, in the same fashion as platinum, gold, silver, etc., the property of liberating oxygen from oxygenated water without undergoing any change, at least when the liquid is diluted with distilled water.

I have taken pure oxygenated water, and I have diluted it until it contained no more than 8 times its volume of oxygen; of this material, I introduced 22 measures into a tube filled with mercury, and then I introduced into it a small amount of perfectly white fibrin recently extracted from blood: immediately oxygen began to be released from the water; the mercury level in the tube decreased at a visible rate; at the end of 6 minutes the water was completely deoxygenated, for it did not effervesce any more upon addition of silver oxide. Having then measured the liberated gas, I found 176 measures, that is to say, as much as was contained in the liquid. Besides, this gas contained neither carbon dioxide nor nitrogen: it was pure oxygen. The same fibrin, repeatedly placed in contact with new liquid, behaved in exactly the same manner.

Urea, liquid or solid albumen, and gelatin did not liberate oxygen even from very highly oxygenated water; but lung tissue cut into thin slices and well washed, that of kidney and of spleen, discharge oxygen from the water at least as readily as fibrin. Skin and venous ducts also possess this property, but to a lesser extent.

1. In the issue of last March where these observations were reported, the number 850 was written instead of the number 616.

However, since fibrin, lung tissue, spleen, kidney, etc., like platinum, gold, silver, etc., possess the property of liberating oxygen from oxygenated water, it is very probable that all these effects are due to the same force. Is it unreasonable to think, accordingly, that an analogous force is the basis for all animal and vegetable secretions? I do not imagine so. One could thus understand how an organ, without absorbing anything, without yielding anything, can constantly act on a liquid and transform it into new products. This point of view, besides, agrees with some ideas that have been advanced recently and for which the evidence begins to become available [et qui deviennent en quelque sorte palpables] on the basis of the experiments presented in this note.

# 7

# ACTION OF SALIVA ON STARCH

**Erhard Friedrich Leuchs**

*This article was translated expressly for this Benchmark volume by H. Friedmann, the University of Chicago, from "Wirkung des Speichels auf Stärke" in* Ann. Phys. Chem. (Leipzig) **22**:623 (1831).

Tiedemann and Gmelin found that the gastric juice of a goose that had been fed to death with starch-flour contained sugar in addition to starch-flour. Since saliva is very similar to gastric juice, this observation suggested an experiment to me. When saliva was added to boiled starch, the mixture upon warming after a while soon turned thin and markedly sweet. Saliva appeared to act on unboiled starch only to the extent that the starch upon warming combined with water to form amidin [amylodextrin], and this probably also explains why all farinaceous foods become more easily digestible upon boiling or baking.

Egg white, glue, and the water-extractable part of hard-boiled egg do not change starch; also saliva does not sweeten gum arabic, and lactose is not changed either.

The Indians in some parts of South America have the custom of initially chewing the maize from which they prepare a kind of beer (*chicha*), a treatment that is supposed to improve it. The above readily explains this. In ancient Greece, the wet-nurses often chewed the food for the children in order to make it more digestible; it will be possible to use saliva and gastric juice of killed animals very successfully in cases of defective digestion.

# 8

EXPERIMENTS

AND

OBSERVATIONS

ON THE

GASTRIC JUICE

AND THE

PHYSIOLOGY OF DIGESTION.

---

BY WILLIAM BEAUMONT, M. D.

Surgeon in the U. S. Army.

---

PLATTSBURGH.

PRINTED BY F. P. ALLEN.

1833.

Reprinted from pages 81, 77–78, 83–84, 125–131, 170–171, 198–199, 227–228, and 278 of *Experiments and Observations on the Gastric Juice and the Physiology of Digestion*, Plattsburgh: F. P. Allen, 1833, 280pp.

[*Editor's Note:* In the introduction to this book, Beaumont gives a detailed clinical description of the wound the Canadian voyageur, Alexis St. Martin, received following the accidental "discharge of a musket, on the 6th of June, 1822," while working for the American Fur Company. St. Martin was "about eighteen years of age, of good constitution, robust and healthy." "The charge," Beaumont continues, "consisting of powder and duck shot, was received in the left side of the youth, he being at a distance of not more than one yard from the muzzle of the gun. The contents entered posteriorly, and in an oblique direction, forward and inward, literally blowing off integuments and muscles of the size of a man's hand, fracturing and carrying away the anterior half of the sixth rib, fracturing the fifth, lacerating the lower portion of the left lobe of the lungs, the diaphragm, and perforating the stomach." Under Beaumont's meticulous care, the patient made good progress: "For seventeen days, all that entered his stomach by the oesophagus, soon passed out through the wound" (p. 11). Soon thereafter his normal digestion returned. "By the 6th of June, 1823, one year from the time of the accident, the injured parts were all sound, and firmly cicatrized, with the exception of the aperture in the stomach and side" (p. 16). "In the spring of 1824 he had perfectly recovered his natural health and strength; the aperture remained; and the surrounding wound was firmly cicatrized to its edges... . In the month of May, 1825, I commenced my first series of gastric experiments with him, at Fort Mackinac, Michigan Territory" (p. 17). The remaining history of Alex St. Martin does not concern us here. He stayed with Beaumont with one four-year and one shorter interruption until March 1833. He died in 1880 at he ripe old age of 83 years, father of four children. In our excerpts we must ignore the physiological observations to be found in Beaumont's book, such as pioneering studies on gastric movements; we confine ourselves to portions of particular pertinence to the field of enzymology.]

At the instance of Professor SILLIMAN, I committed to the care of Mr. GAHN, Consul of his Swedish Majesty in New-York, a bottle, containing one pint, of gastric juice, to be transmitted by him to Professor BERZELIUS, of Stockholm, one of the most eminent chemists of the age, with a request that he would favour me with an analysis. Some unavoidable delay was experienced in forwarding the bottle; and no returns have yet been received. It is hoped, however, that they will arrive in time to be attached in an appendix to this volume.

[*Editor's Note:* Material has been omitted at this point.]

[*Editor's Note:* Material has been omitted at this point.]

That chymification is effected by the *solvent* action of the gastric juice, aided by the *motions* of the stomach, and the natural *warmth* of the system, not a doubt can remain in the mind of any candid person, who has had an opportunity to observe its effects on alimentary substances, or who has the liberality to credit the opinions of those who have had such opportunities.

It has been objected to this hypothesis, that the *sensible* properties of the gastric juice contradict the opinion of its active *solvent* effect. But we should recollect that many things which make very little impression on our external senses, produce most astonishing effects in other situations. The air which we breathe, by which we are surrounded, and which, to our external senses, is almost inappreciable, is one of the most powerful and destructive agents in nature—one portion of which is capable of combining with all grades of matter, either slowly and imperceptibly, as in the gradual decay of all substances, or rapidly, as in the combustion of wood, or even the hardest metals—and which by means, inexplicable to us, sustains in life and being the whole of animated nature.

[*Editor's Note:* Material has been omitted at this point.]

The discrepance of results in the reports of those who have had opportunities of examining the process of, and have made experiments on, *artificial digestion*, by the gastric juice, as well as in the chemical examination of this fluid, has been owing more to the difficulty of obtaining it pure, in sufficient quantity, and under proper circumstances, than to any real difference in its effects. Under the circumstances in which the following experiments were made, I flatter myself that these difficulties have been obviated; and if the inferences are incorrect, the blame must be attached to the experimenter. He can only say, that the experiments were made in good faith, and with a view to elicit facts.

I think I am warranted, from the result of all the experiments, in saying, that the gastric juice, so far from being " inert as water," as some authors assert, is the most general solvent in nature, of alimentary matter—even the hardest bone cannot withstand its action. It is capable, *even out of the stomach*, of effecting perfect digestion, with the aid of due and uniform degrees of heat, (100° Fahrenheit,) and gentle agitation, as will be seen in the following experiments.

The fact that alimentary matter is *transformed*, in the stomach, into chyme, is now pretty generally conceded. The peculiar process by which the change is effected, has been, by many, considered a problem in physiology. Without pretending to explain the exact *modus operandi* of the gastric fluid, yet I am impelled by the weight of evidence, afforded by the experiments, deductions and opinions of the ablest physiologists, but more by direct experiment, to conclude that the change effected by it on aliment is *purely chemical.* We must, I think, regard this fluid as a chemical agent, and its operation as a chemical action. It is certainly every way analogous to it; and I can see no more objection to accounting for the change effected on the food, on the supposition of a chemical process, than I do in accounting for the various and diversified modifications of matter, which are operated on in the same way. The decay of the dead body is a chemical operation, separating it into its elementary principles—and why not the solution of aliment in the stomach, and its ultimate assimilation into fibrine, gelatine and albumen? Matter, in a natural sense, is indestructible. It may be differently combined; and these combinations are chemical changes. It is well known that all organic bodies are composed of very few simple principles, or substances, modified by excess or diminution of some of their constituents.

[*Editor's Note:* Material has been omitted at this point.]

# EXPERIMENTS, &c.

## FIRST SERIES.

### *Experiment* 1.

*August* 1, 1825. At 12 o'clock, M., I introduced through the perforation, into the stomach, the following articles of diet, suspended by a silk string, and fastened at proper distances, so as to pass in without pain—viz. :—a piece of high seasoned *a la mode beef;* a piece of *raw, salted, fat pork ;* a piece of *raw, salted, lean beef;* a piece of *boiled, salted beef ;* a piece of *stale bread ;* and a bunch of *raw, sliced cabbage ;* each piece weighing about two drachms; the lad continuing his usual employment about the house.

At 1 o'clock, P. M., withdrew and examined them —found the *cabbage* and *bread* about half digested: the pieces of *meat* unchanged. Returned them into the stomach.

At 2 o'clock, P. M., withdrew them again—found the *cabbage, bread, pork,* and *boiled beef,* all cleanly digested,† and gone from the string; the other pieces of

† These Experiments are inserted here, as they were originally taken down in my Note Book, with very little alteration of phraseology, and none of the sense. Subsequent experiments have sometimes convinced me of errors in former ones. When this has been the case, I have generally made the corrections in the way of remarks, or observations, as in this experiment.

meat but very little affected. Returned them into the stomach again.

At 2 o'clock, P. M., examined again—found the *a la mode beef* partly digested: the *raw beef* was slightly macerated on the surface, but its general texture was firm and entire. The smell and taste of the fluids of the stomach were slightly rancid; and the boy complained of some pain and uneasiness at the breast. Returned them again.

The lad complaining of considerable distress and uneasiness at the stomach, general debility and lassitude, with some pain in his head, I withdrew the string, and found the remaining portions of aliment nearly in the same condition as when last examined; the fluid more rancid and sharp. The boy still complaining, I did not return them any more.

*August* 2. The distress at the stomach and pain in the head continuing, accompanied with costiveness, a depressed pulse, dry skin, coated tongue, and numerous white spots, or pustules, resembling coagulated lymph, spread over the inner surface of the stomach, I thought it advisable to give medicine; and, accordingly, dropped into the stomach, through the aperture, half a dozen *calomel pills*, four or five grains each; which, in about three hours, had a thorough cathartic effect, and removed all the foregoing symptoms, and the diseased appearance of the inner coat of the stomach. The effect of the medicine was the same as when administered in the usual way, by the mouth and œsophagus, except the nausea commonly occasioned by swallowing pills.

This experiment cannot be considered a fair test of the powers of the gastric juice. The cabbage, one of the articles which was, in this instance, most speedily dissolved, was cut into small, fibrous pieces, very thin, and necessarily exposed, on all its surfaces, to the action of the gastric juice.

The stale bread was porous, and, of course, admitted the juice into all its interstices; and probably fell from the string as soon as softened, and before it was completely dissolved. These circumstances will account for the more rapid disappearance of these substances, than of the pieces of meat, which were in entire solid pieces when put in. To account for the disappearance of the fat pork, it is only necessary to remark, that the fat of meat is always resolved into oil, by the warmth of the stomach, before it is digested. I have generally observed that when he has fed on fat meat or butter, the whole superior portion of the contents of the stomach, if examined a short time after eating, will be found covered with an oily pellicle. This fact may account for the disappearance of the pork from the string. I think, upon the whole, and subsequent experiments have confirmed the opinion, that fat meats are less easily digested than lean, when both have received the same advantages of comminution. Generally speaking, the looser the texture, and the more tender the fibre, of animal food, the easier it is of digestion.

This experiment is important, in a pathological point of view. It confirms the opinion, that undigested portions of food in the stomach produce all the phenomena of fever; and is calculated to warn us of the danger of all excesses, where that organ is concerned. It also admonishes us of the necessity of a perfect comminution of the articles of diet.

*Experiment* 2.

*Aug*. 7. At 11 o'clock, A. M., after having kept the lad fasting, for seventeen hours, I introduced the glass tube of a Thermometer (Fahrenheit's) through the perforation, into the stomach, nearly the whole length of the stem, to ascertain the natural warmth of the stomach. In fifteen minutes, or less, the mercury rose to 100°, and there remained stationary. This I determined by marking the height of the mercury on the glass, with ink, as it stood in the stomach, and then withdrawing it, and placing it on the gaduated scale again.

I now introduced a gum-elastic (caoutchouc) tube, and drew off one ounce of pure gastric liquor, unmixed with any other matter, except a small proportion of mucus, into a three ounce vial. I then took a solid piece of *boiled, recently salted beef*, weighing three drachms, and put it into the liquor in the vial; corked the vial tight, and placed it in a saucepan, filled with water, raised to the temperature of 100°, and kept at that point, on a nicely regulated sand bath. In *forty minutes* digestion had distinctly commenced over the surface of the meat. In *fifty minutes* the fluid had become quite opaque and cloudy; the external texture began to separate and become loose. In *sixty minutes*, chyme began to form.

At 1 o'clock, P. M., (digestion having progressed with the same regularity as in the last half hour,) the cellular texture seemed to be entirely destroyed, leaving the muscular fibres loose and unconnected, floating about in fine small shreds, very tender and soft.

At 3 o'clock, the muscular fibres had diminished one half, since last examination, at 1 o'clock.

At 5 o'clock, they were nearly all digested; a few fibres only remaining.

At 7 o'clock, the muscular texture was completely broken down; and only a few of the small fibres floating in the fluid.

At 9 o'clock, every part of the meat was completely digested.

The gastric juice, when taken from the stomach

was as clear and transparent as water. The mixture in the vial was now about the colour of whey. After standing at rest a few minutes, a fine sediment, of the colour of the meat, subsided to the bottom of the vial.

*Experiment* 3.

At the same time that I commenced the foregoing experiment, I suspended a piece of *beef*, exactly similar to that in the vial, (Ex. 2d) into the stomach, through the aperture.

At 12 o'clock, M., withdrew it, and found it about as much affected by digestion as that in the vial; there was little or no difference in their appearance. Returned it again.

At 1 o'clock, P. M., I drew out the string; but the meat was all completely digested, and gone.

The effect of the gastric juice on the piece of meat, suspended in the stomach, was exactly similar to that in the vial, only more rapid after the first half hour, and sooner completed. Digestion commenced on, and was confined to, the surface entirely, in both situations. Agitation accelerated the solution in the vial, by removing the coat that was digested on the surface; enveloping the remainder of the meat in the gastric fluid; and giving this fluid access to the undigested portions.

*Experiment* 4.

*Aug*. 8. At 9 o'clock, A. M., I drew off an ounce and a half of gastric juice, into a three ounce vial; suspended two pieces of *boiled chicken*, from the breast and back, into it, and placed it in the same situation and temperature as in the second experiment; observing the same regularity and minuteness.

Digestion commenced and progressed much the same, as in the second experiment, but rather slower; the *fowl* appearing to be more difficult of digestion than the *flesh*. The texture of the *chicken* being clo-

R

ser than that of the *beef*, the gastric juice appeared not to insinuate itself into the interstices of the muscular fibre, so readily as into the beef; but operated entirely upon the outer surface, dissolving it as a piece of gum arabic is dissolved in the mouth, until the last particle was digested.

The colour of the fluid, after digesting the chicken, was of a greyish white, and more resembled a milky fluid than whey, which was the colour of the chyme from the beef.

The contents of both vials, kept perfectly tight, remained free from any fœtor, acidity, or offensive smell or taste, from the time of the experiments, (7th and 8th August,) to the 6th of September; at which time, that containing the solution of *boiled beef*, became very offensive and putrid; while that containing the chyme from the *boiled chicken*, was perfectly bland and sweet. Both were kept in exactly similar situations.

It is, perhaps, unnecessary to make any comments on the result of the above experiments. Each one will make up his opinion from the facts. *These* demonstrate, at least, that the stomach secretes a fluid which possesses *solvent* properties. The change in the solid substances is effected too rapidly to be accounted for on the principle of either maceration or putrefaction. I shall be able to show, in some of the following experiments, that aliment undergoes the same changes in the stomach, as is effected in the mode here adopted.

The young man, who was the subject of these experiments, left me about this time, (September, 1825) and went to Canada, the place of his former residence. The experiments were consequently suspended.

# EXPERIMENTS, &C.

## SECOND SERIES.

FORT CRAWFORD, *Upper Mississippi*,
*June* 20th, 1829.

ALEXIS ST. MARTIN having returned from Canada, after an absence of nearly four years, with his stomach in the same, or very similar condition, as when he left me in September, 1825, I continued to prosecute the gastric experiments, which were commenced before he left me.

[*Editor's Note:* Material has been omitted at this point.]

## EXPERIMENTS, &c.

### THIRD SERIES.

WASHINGTON, D. C. 1832.

*Experiment* 1.

*Dec.* 4. At 2 o'ck, 30 ms., P. M.—Weather cloudy, damp and snowing—Th: 35°—Wind N. W. and brisk—the temperature under the tongue was 99°; in the stomach, 101°. Dined, at 3 o'clock, 30 mins., on *beef soup, meat* and *bread.* 4 o'clock, 15 mins.—took out a portion—particles of beef slightly macerated, and partially digested. 5 o'clock, 15 mins.—took out another portion—digestion more advanced—meat reduced to a pulp; particles of bread and oil floating on the top. Temperature of stomach, 100°. 6 o'clock, 45 mins.—digestion not completed—contents considerably diminished. 7 o'clock, 45 mins.—stomach empty—chyme all passed out.

*Experiment* 2.

*Dec.* 5. At 7 o'clock, A. M., temperature of the stomach, 100°; of the atmosphere, 30°.

At 1 o'clock, P. M.—temperature of stomach, 100°—atmosphere, 40°—he ate eleven *raw oysters,* and three *dry crackers;* and I suspended one *raw oyster* into the stomach, through the aperture, by a string. 1 o'clock, 30 mins.—examined—stomach full of fluids---digestion not much advanced. The oyster on the string appeared entire, though perhaps slightly affected on the surface. 2 o'clock—examined, and took out oyster--about one third digested, but retained its shape. 2 o'clock, 30 mins.---oyster gone from the string, except a small piece of the heart. Temperature of the stomach 101½°. Fluids less considerable. 4 o'clock, 15 mins.--stomach empty.

*Experiment* 3.

At 3 o'ck, 45 ms., P M., same day, he dined on *roast turkey*, *potatoes* and *bread*. 4 o'clock, 30 mins.---examined, and took out a portion. Turkey nearly all dissolved—vegetables half reduced. 5 o'clock, 15 mins.---took out another portion, almost completely chymified. 5 o'clock, 45 mins.---examined again—stomach nearly empty. 6 o'clock---some chyme yet remaining. 6 o'clock 15 mins.---stomach empty.

*Experiment* 4.

*Dec.* 6. At 8 o'clock, 30 mins., A. M., he breakfasted on *bread and butter*, and one pint of *coffee*. 9 o'clock, 45 mins.---examined—stomach full of fluids. 10 o'clock, 30 mins.---examined, and took out a portion, resembling thin gruel, in colour and consistence, with the oil of the butter floating on the top; a few small particles of the bread, and some mucus, falling to the bottom.---about two thirds digested. It had a sharp, acid taste. Temperature of the stomach, 100°---atmosphere, 38°. 11 o'clock, 30 mins., stomach empty.

*Experiment* 5.

At 4 o'clock, 30 mins., P. M., same day--he dined on *sausage* and *bread*; full meal. 5 o'clock, 30 mins. ---stomach full of fluids; digestion but very little advanced. 6 o'clock, 30 mins.--digestion considerably advanced: few distinct particles of sausage and bread to be seen entire. 7 o'clock, 30 mins., stomach empty.

[*Editor's Note:* Material has been omitted at this point.]

*Experiment* 50.

At 9 o'clock, 30 mins. A. M., same day, I took three vials, and put into each two drachms pure gastric juice, fresh from the healthy stomach. To one, I added one drachm of *albumen*—white of egg—to the second, half a drachm of the *yolk*—and to the third, another drachm of *albumen*. Put the two first, in axilla, and the other on the mantle-piece.

At 9 o'clock, P. M., the albumen in the warm gastric juice, in the axilla, had become quite opaque, with loose, light coloured sediment at the bottom. The albumen in the cold gastric juice remained unaltered. That containing the yolk, exhibited the appearance of a mere mixture of fine yellow coagulæ, resembling sulphur and milk, mixed together.

On the 12th, at 8 o'clock, P. M., both vials having been continued on the bath, or in the axilla, through the day, the difference observed last evening, between the cold and warm vials of albumen, was very little increased.

The yolk was considerably altered from a loose coagulæ, generally diffused through the gastric juice, to a fine compact body of coagulæ, rising upon the top of a perfectly clear, transparent fluid, free from a particle of sediment.

[*Editor's Note:* Material has been omitted at this point.]

[*Editor's Note:* Material has been omitted at this point.]

*Experiment* 104.

*Feb.* 14. At 9 o'clock, A. M., I took forty grains *masticated, broiled beef steak*, divided into two equal parts—put one into four drachms gastric juice, and the other, into four drachms of a mixture of dilute *muriatic* and *acetic* acids, reduced with water to the flavor of the gastric fluid, as nearly as practicable—three parts of the muriatic to one part of the acetic, Placed them together on the bath. At 6 o'clock, P, M., the meat in the gastric juice was all dissolved; that in the dilute acids, when filtered, left a residuum of nine grains, of a jelly-like consistence. The fluids, also, differed in appearance. That from the gastric juice was opaque, and of a lightish grey colour, depositing a brown sediment on standing. The other was also opaque, and of a reddish brown colour, but deposited no sediment.

This was an attempt to imitate the gastric juice. It was not satisfactory. Probably the gastric juice contains some principles inappreciable to the senses, or to chemical tests, besides the alkaline substances already discovered in it.

[*Editor's Note:* Material has been omitted at this point.]

I regret, exceedingly, that I have not been able to obtain returns from Professor BERZELIUS, to whom I transmitted, about seven months ago, a bottle of gastric juice for chemical examination. I could not, however, consistently with the expectations and wishes of my friends, further delay the publication of these experiments.

[*Editor's Note:* Material has been omitted at this point.]

Part II

# CONSOLIDATIONS: DEVELOPMENTS OF TECHNIQUES, TERMS, AND THEORIES

# Editor's Comments on Papers 9 Through 17

9 **PAYEN and PERSOZ**
Excerpts from *Memoir on Diastase, the Principal Products of Its Reactions, and Their Applications to Industrial Arts*

10 **SCHWANN**
Excerpts from *On the Nature of the Digestive Process*

11 **BERZELIUS**
Excerpts from *On a So Far Rarely Observed Force Which Is Probably Active in the Formation of Organic Substances*

12 **LIEBIG**
Excerpts from *On the Phenomena of Fermentation, Putrefaction, and Decay, and Their Causes*

13 **BERTHELOT**
*On the Glucosidic Fermentation of Cane Sugar*

14 **BRÜCKE**
Excerpts from *Contributions to the Theory of Digestion, Second Part*

15 **DANILEVSKY**
Excerpts from *On Specifically Acting Substances of Natural and Artificial Pancreatic Juice*

16 **TRAUBE**
Excerpts from *The Chemical Theory of the Activity of Ferments, and the Chemistry of Respiration. Answer to Statements of Mr. Hoppe-Seyler*

17 **KÜHNE**
Excerpts from *On the Digestion of Proteins by Pancreatic Juice. I.*

Paper 9, by Anselme Payen (1795–1871) and Jean François Persoz (1805–1868) on diastase, extends Kirchhoff's observations on starch hydrolysis by malt gluten. It is a classic in enzymology because it is the very first report of an enzyme purification and of a description of the properties of the enriched material, including loss of activity upon boiling. Ethanol was used to precipitate the active principle, and the biological implications of this work are, as in Kirchhoff's study, clearly recognized. The practical aspects of this research are stressed. This paper is also remarkable in that it is a very early example in enzymology of a work by more than one author. Another important facet of this paper is that it contains the first name for an enzyme that was accepted by others (unlike Planche's *cyanogen*), a name that was destined to play an important role in the development of enzymology: *diastase*. This name (derived from the Greek *diastasis*, "making a breach," to describe the physical effect of this enzyme on starch grains) was long used, especially in France, for what we now call enzymes. For example, the word was employed in this sense by Pierre Émile Duclaux[1] in his 1883 textbook *Chimie Biologique*, by Gabriel Bertrand in 1895 for describing the first oxidase ("Sur la laccase et sur le pouvoir oxydant de cette diastase," see comments to Paper 22), and as late as 1903 by Victor Henri in the title of his classic book on enzyme kinetics (see Paper 29). Another work, *ferment*, had different connotations, as will be seen later in this book.

Both Payen and Persoz were eminent chemists. To give a small sampling, Payen discovered inulin and pectin, introduced the use of charcoal as a decolorant, and found an assay method for morphine in opium, while Persoz, who had been an assistant to Thénard from 1826 to 1832, discovered dextrin and was the first to isolate cellulose from wood.

The word *pepsin* is first used in Theodor Schwann's classic 1836 paper (Paper 10). This study, following Johann Eberle's discovery two years earlier (see comments to Paper 8) uses the extraction of the active principle from gastric mucosa, a method that circumvented the tedious and cumbersome procedures for obtaining gastric juice that had limited the works of Reaumur, Stevens, Spallanzani, and Beaumont. The fact that acid was needed for peptic activity posed a special challenge concerning an understanding of its possible participation. It is fascinating to read the description of the very careful quantitative experiments that led Schwann by a logical whittling away of alternatives to the conclusion that acid was necessary but not sufficient and that it was not changed at the end of the reaction. The investigation into the role of acid led to what probably constitutes the very first mechanistic enzyme study. One must recognize that this work was done eighty years before Sørensen (Paper 33) recognized the importance of a fixed hydrogen ion concentration for enzyme activity. Schwann in this remarkable paper stresses the analogy of the action of unchanged acid in gastric digestion to the action of acid in the hydrolysis of starch, and recognizes that digestion by pepsin is yet another example of the type of activity represented by alcoholic and by acetic acid fermentation, and by the decomposition of hydrogen peroxide by gold oxide. In later sections of this paper, not translated here, further highly original work is described. Thus Schwann observed the irreversible inactivation of pepsin both by excess acid and excess alkali. More remarkable is the fact that we find here the first suggestion of a method for the purification of an enzyme that uses a sequence of different purification steps. This sequence is based on Schwann's discoveries that while in a precipitation step with potassium ferrocyanide under acid conditions, pepsin is not removed, the enzyme is precipitated both by lead acetate and by mercuric chloride, and can be redissolved upon treatment with hydrogen sulfide. This paper was published at exactly the time that Wöhler was translating Berzelius's work that introduced the word *catalysis*. Accordingly we find here the older term *contact action*. The expression *contact substance* had been introduced recently by Eilhard Mitscherlich (1834) (see below and Fruton 1972, p. 47).

The author of this paper is the same Theodor Schwann (1810–1882) who extended to animals his friend Mathias Jakob Schleiden's cell theory for plants. Among his many other accomplishments, Schwann coined the term *metabolism*, discovered the myelin sheath that covers peripheral axons (known as Schwann cells), observed the basic principle of embryology that the egg is one cell that develops into the complete organism, and was one of the investigators who showed yeast cells to be living organisms. He is sometimes called the father of cytology. His experimental procedure to disprove the spontaneous generation of life directly influenced Pasteur's much later work on this topic. Schwann performed the bulk of his brilliant and diversified research in his twenties (1834–1839) as an assistant to the eminent Johannes Müller (editor of the journal in which Paper 10 was published) at the Anatomical Museum in Berlin, and he accomplished much less in the remaining forty-three years of his life.

Our next author, Jöns Jacob Berzelius, as mentioned earlier in Part I, invented the term *catalysis*. He introduced this word in his 1835 *Årsberättelser*. In the third edition (1837) of his *Lärbok i*

*Kemien* [Textbook of chemistry], this section is repeated with some very slight changes. The present translation was made from his former student Friedrich Wöhler's translation into German of the Swedish textbook. The editor compared the textbook with Wöhler's translation of the slightly earlier *Annual Survey*, and the small differences between the two are indicated.[2]

Three points must be stressed: First, Berzelius's assumption of a special force that, as pointed out by Mittasch (1938) (cf. Färber 1937), has alchemical overtones, was made long before the ascent of chemical thermodynamics. Even the enunciation of the Law of Mass Action by the fellow Scandinavians Guldberg and Waage was still thirty years away. However, the Laws of Constant and Multiple Proportions in chemical reactions had been well established, as is clear, for instance, from Schwann's paper. What strikes one in reading Berzelius is the sense of mystery, generated by the impression that a catalyst can by its mere presence and in apparent contradiction to these laws, bring about a chemical change and not be affected by this change. If one realizes that catalysis remains intuitively a most impressive phenomenon, even after one has come to recognize activation energies and even after, more than a hundred years later, one has a better though still incomplete understanding of enzymatic rate accelerations (see Part IX and Luisi 1979), then it is not hard to see the excitement that this fairly recently discovered phenomenon must have generated in the minds of Berzelius and of his contemporaries. It is most impressive to see that Berzelius stresses the analogy to the action of heat, and that he asks two fundamental questions in addition to the question already posed by Thénard of the universality and diversity of catalytic processes in biological systems:[3] Is rate acceleration due to a catalyst as unselective as that due to heat? and, can different catalysts decompose a given substance in different ways? Berzelius was regarded by chemists such as Liebig[4] and Gerhardt[5] merely to have invented a new word for the "contact" terminology suggested two years earlier by his pupil and life-long friend Mitscherlich. A reading of Berzelius's text clearly indicates, however, that he did much more than replace one set of words by another. With an impressive inductive flair, he uses the term *catalysis* to drive home far more emphatically than had been done before what appeared to be the special nature of this phenomenon. The new term was accepted in the context of an "awakening of slumbering affinities" followed by a reorganization of chemical relationships so that "a larger electrochemical neutrality" results—ideas suggesting much later notions of free energy decrease.[6] Berzelius, in fact, had recognized that the "contact" terminology was not as descriptive as appeared at first sight, since all reactions require contacts between the reactants (see Mittasch and Theis 1932, p. 96). Second, in the words of Malcolm Dixon (1971): "It is important to realize that it was very largely the action of enzymes that gave rise to the idea of catalysis, not the converse as is often assumed." Along these lines it must also be stressed that Berzelius made no distinction between his catalytic "force" in animate and in inanimate nature since his notion of catalysis was based on observations with systems from both sources. This lack of a distinction contains the seed of the challenge to vitalistic thinking that was yet to sprout in various forms during the second half of the nineteenth century as we shall repeatedly see later in this book (Liebig, Berthelot, and Traube versus Pasteur; unorganized versus organized ferments; Kühne's invention of the word *enzyme* to perpetuate this distinction; the excitement caused by Buchner's demonstration of sucrose fermentation by cell-free yeast extracts). As Northrop (1961) pointed out in a perceptive essay, Berzelius was fifty years ahead of his time since it took this long—actually longer—for his ideas to become accepted. It seems clear, however, that Berzelius, just as Thénard and Kirchhoff before him, did not feel as challenged as some of their successors to stress the mechanistic implications of their ideas. Berzelius's thinking was somewhat more in accord with the temper of his times (his friend Wöhler had synthesized urea just a few years earlier) than it would have been thirty years later. Hence claims to prescience must be slightly tempered when accidents of historical circumstance are taken into account. This consideration in no way detracts from the value of Berzelius's generalization or from the effect of this generalization on the generation of physiologically inclined chemists that followed his own. Third, it should perhaps be stressed that any notions of the possible participation of an enzyme in enzymatic catalysis had to await the rate studies of O'Sullivan and Tompson (Paper 23), the stereochemical studies of Emil Fischer, still some sixty years away (Papers 24 and 25), and so on (see Part IV).[7]

At the time he coined the word *catalysis*, Berzelius was "the arbiter and dictator of the chemical world" (Harden 1911). His influence on the development of chemistry is so fundamental that one tends to forget the quite unusually broad scope of his contributions, made in addition to the five editions of his multivolume textbook, and his *Annual Surveys*, which he wrote for twenty-seven years until his death. A mere recitation of his contributions provides insight into the creativity of this astounding man: besides the word *catalysis*, he also coined the words *isomer, polymer, allotropy, am-*

*monium,* and *protein,*[8] and he first used the word *globulin* in its present-day context.[9] He discovered no less than three elements (cerium, selenium, and thorium) and isolated three more (silicon, zirconium, and titanium) in free form. He is considered to be the father of gravimetric analysis, he introduced into the laboratory the use of desiccator, filter paper, rubber tubing, water bath, and wash bottle, and he improved techniques in the use of the blowpipe. He determined the combining weights of 43 elements and analyzed some 2000 compounds with his own hands. He is responsible for the use of oxyen as the basis for atomic weights. The modern chemical symbols for the chemical elements are due to him (see Crosland 1962), and so is the electrochemical theory. Of particular interest to biochemists is the fact that he was the first to prepare pyruvic acid (by dry distillation of tartaric acid, hence his name *acidum pyruvicum* for the substance) and, a sign of his thoroughness, that he prepared no less than 20 different salts of pyruvic acid (Berzelius 1835). He was also the first to isolate lactic acid from muscle—it had been discovered in 1780 in sour milk by another Swede, Carl Wilhelm Scheele—and to show that it was optically active (see Jorpes 1966, p. 36). Berzelius's laboratory consisted "of two rooms without furnace, flues, water, or gas...[He] regularly worked 12 to 14 hours each day" (Partington 1964, p. 148).

Berzelius, who "was a member of eighty-five learned societies in other countries...was probably the last chemist capable of holding in his brain all contemporary chemical knowledge. In his scientific position, he was to some degree a parallel to Linné in the field of botany during the preceding century" (Bodman 1952). It is next to impossible, for instance, for a present-day reader to see how one man could have written the enormous textbook that presents methodology, inorganic, theoretical, organic, plant, and animal chemistry ("Tierchemie"), and quite a bit of physiology not only thoroughly but with an inescapable elegance.

Among all the papers selected for this work, Paper 12 by Johann Justus von Liebig (1803–1873) requires the most background information and commentary. Difficulties do not a classic unmake, and this paper is a classic for many reasons. It presents the very first theory of what one would now call *enzyme action*. However, this theory was exceedingly influential in its day primarily as an explanation of the processes of fermentation, putrefaction, and decay. This influence was due not only to the Baron von Liebig's enormous prestige as a leading chemist,[10] skillful polemicist, and renowned teacher but due also to the revolutionary ideas contained in this theory. Moreover, this is the theory that, about twenty years later, sparked the famous Pasteur-Liebig controversy on the nature of fermentation and that, modified in various details, was to occupy Liebig until the end of his life.

We take it for granted now that yeast is a living organism, but this was not known until the practically simultaneous work of three separate investigators (Cagniard-Latour, Kützing, and Schwann), just before Liebig's paper was written. In fact, this notion about the nature of yeast was scornfully rejected by Berzelius in his 1839 *Annual Survey*, and Liebig for many years rejected it as well. So one of the first difficulties with Liebig's paper—and we ignore here his discussion of putrefaction and decay—is a certain vagueness, common at the time, as to what was meant by the word *ferment*. The very notion of yeast as a living organism was considered to detract from the idea of fermentation as a chemical process. Liebig's contemporaries understood his rejection of the participation of living agents in fermentation not only as an affirmation of the commonly held view of this process, but as a veiled attack on the claim by the above three investigators that ferment was indeed the living yeast cell. In the words of Florkin (1972, p. 140): "Liebig, cock of the walk, who ruled over the whole fabric of the "biochemistry" of the time, had no positive argument to offer and he resorted to a most unpalatable and dishonest procedure (a lesson for those who ... believe that the practice of laboratory work develops the ethical sense of the scientist)." In 1839, shortly before the appearance of Liebig's paper on fermentation, putrefaction and decay, the good friends Wöhler and Liebig (see Delbrück and Schrohe 1904) had published in the very same international journal (which they edited with some others) an anonymous scathing attack on the notion that fermentative activity could possibly be due to a living cell (Wöhler and Liebig 1839). This attack was felt particularly strongly by the young and sensitive Theodor Schwann and spelled the virtual end of his remarkably creative career. It contributed to send him that very same year into exile from Germany to a professorship in Belgium (Florkin 1972, p. 141; 1975, p. 244). Even after the idea that yeast was alive had been accepted, it was not at all obvious that one should distinguish ferment in the sense of a living organism from ferment in the sense of an action due to a living organism. (We will return to this point, most important in the history of enzymology, when we come to discuss excerpts from Pasteur, Paper 20, as well as the reasons for Kühne's introduction, forty years after Liebig's paper, of the word *enzyme*, a word (Papers 18, 19) based on a widely prevalent, mistaken interpretation of the preceding second meaning of the work *ferment*.)

Liebig's theory consists of two parts, both of which have a chemical foundation and are based on chemical analogies: First, the initiating material, say, yeast, exhibits certain vibrations. As Dixon (1971, p. 20) comments: "The sugar (is) almost literally shaken to pieces." Second, it is not just the sugar molecule that is decomposed, but the yeast brings about its own decomposition as well by this initiating act, namely, the vibration.

A few points have to be made: First, Liebig here attempts to elaborate a theory of fermentation and related processes by analogy to the observation of energy transfer from one burning substance to another. Thus transformations due to putrefaction and due to heat are compared both from the point of view of chemical changes and of spontaneity. Second, the importance of this theory lies in the fact that it attempts to explain certain processes in chemical terms, particularly how a stable substance such as sugar is actively decomposed on contact with the yeast. One can say, in fact, that a refinement of Liebig's notion of molecular thrusts has a part in modern ideas of enzyme activity. Third, it may strike one as strange, in so far as this is regarded as the first theory of enzyme action, that as perhaps first pointed out by Arthur Harden (1911), this theory demands that the ferment itself is decomposed in the course of its activity. Hence this theory clearly does not regard the ferment as a catalyst.[11] Indeed, Liebig (1840) in at least one of his textbooks published just a year after this paper, discusses catalysis at great length, but one infers from the section on fermentation, which immediately follows the discussion of catalysis, that fermentation is not considered to be related to catalysis. The reason for this distinction is clearly pointed out by A. K. Balls (1942): In Liebig's theory the substances that caused fermentation and digestion acted not "as Berzelius maintained,... merely by their presence but instead participated most thoroughly in the reaction. Liebig realized that ferment action required not merely an exposition, but explanation. He was not satisfied to make a new word serve in lieu thereof. When Berzelius spoke of catalysis, Liebig in effect asked of Wöhler what that was except a new word (Volhard 1909)." The fourth point to make is that the notion of the decomposition of molecules by the transmission of violent internal motions is derived from ideas suggested 140 years before Liebig by the eminent Georg Ernst Stahl (1660–1734) of phlogistic fame in his book (1697) whose telling title is *Zymotechnica fundamentalis*. From this point of view, hence, Liebig's theory forms a direct link, via Stahl's ideas, to even more ancient notions of fermentation.

In retrospect, one can say that Liebig's ideas helped to expand

Berzelius's concept of a catalytic force into a general catalytic mechanism (see Kuznetsov 1966).

Paper 13, by Berthelot, is of great importance in the history of enzymology, for as the first on the extraction of a "ferment" from yeast cells, it provided an early contribution to the elimination of vitalistic thinking in biochemistry. Berthelot's ferment was not regarded as representative of all ferments; in fact, a distinction was soon made between extractable or unorganized ferments and nonextractable or organized ferments. The latter, in distinction to the former, were held to have intrinsic vital attributes.

A weakness inherent in this distinction was, of course, that it was based not on a positive criterion but simply on an experimental difficulty, the inability of laboratory workers to extract certain activities from living cells. Almost twenty years after the publication of the paper given here, between 1876 and 1879, Berthelot and Pasteur attacked each other in a series of exchanges in the *Comptes rendus* on the nature of alcoholic fermentation. Berthelot took a chemical view of the process and fortified his position by that apparently taken by the great physiologist Claude Bernard shortly before his death in 1878, while Pasteur took a strictly biological view. Berthelot (1876) maintained that one must distinguish between "the role of microscopic beings which secrete ferments and that of the ferments themselves" while Pasteur held to his view of the "organism-ferment."

The flavor of these exchanges is exemplified by a short example from Pasteur's pen (1878): "It is possible [for some one] to agree with me when, first, it is accepted that fermentations proper require as an absolute prerequisite the presence of microorganisms.... Will Mr. Berthelot or will he not contradict [this position]..., not with à priori points of view, but with serious facts? If yes, let our fellow member [i.e., of the Academy, undoubtedly intended sarcastically, ed.] have the kindness to say so; if no, there is nothing for us to discuss."[12]

It is most interesting to note that both participants in this controversy actually based their opposing views on very different interpretations of their own identically negative results. In the 1870s and even earlier, various scattered attempts had been made to prepare cell-free yeast extracts that could bring about alcoholic fermentation (see for example Kohler 1971, pp. 39–40; Fruton 1972, pp. 62–63). Pasteur's long-time associate Émile Roux in a remarkable lecture delivered a few years after Pasteur's death revealed that both Berthelot and Pasteur had also tried to obtain such active extracts, but failed (Roux 1898, pp. 836–837). Although for all practical purposes their views, as expressed in the above controversy, were in fact à priori, each thought that these views were backed by the results of their own experiments. It is almost certain, however, that neither knew of the other's work. Pasteur, for one, never published his results (Kohler 1971, p. 40), and it is clear that he knew of no such investigations by his opponent. How, then, could these two eminent scientists use their identical failures to support opposite views? Berthelot postulated that his failure to obtain active extracts was due to an inherent instability or self-consumption of the chemical constituents of the yeast in the act of extraction or assay, while Pasteur held failure to be simply a consequence of the fact that alcoholic fermentation was indissolubly correlated with the life of the yeast cell. These two different interpretations of similar data are an object-lesson in the facility be which the lure of negative results can feed the complacency of prior conviction. In the words of Pasteur's first biographer, the scientist Pierre Emile Duclaux: "Pasteur came out of [this discussion] more fixed in his ideas, and Berthelot, apparently, without having yielded any of his. This should lead us to distrust all discussions, even scientific ones... . A fact, even of the physical order, is nothing by itself. It becomes something only when it passes into the state of an intellectual fact, by traversing an intelligence the imprint of which it receives" (Duclaux 1896, p. 266).[13] Pasteur (1897, p. 54) later softened his stand (see Fruton 1972, pp. 61–62).

Why did the two, and others, fail? The answer was given many years later by another great experimentalist, Arthur Harden, who repeated Pasteur's experiments: "The nature of the yeast is of paramount importance. Thus while Munich (bottom) yeast usually gives a good result, a top yeast from a Paris brewery was found to yield extracts containing neither zymase nor its co-enzyme in whatever way the preparation was conducted. The existence of such yeasts is of great interest, and it was probably due to the unfortunate selection of such a yeast that Pasteur was unable to prepare active fermenting extracts..." (Harden 1914[14]). Marked differences between different yeasts in yielding active cell-free extracts had already been noted by Eduard Buchner (1897). (See discussion of Buchner's paper, Paper 21.)

It is in the light of this celebrated "querelle des fermentations" that Eduard Buchner's extraction from yeast cells, two years after Pasteur's death, of a fluid that converted sucrose to alcohol (Paper 21) was so revolutionary. The word *enzyme* ("in yeast"), coined by Kühne (Paper 18) roughly half-way between Berthelot's and Buchner's papers and with direct reference (Paper 19) to Berthelot's work, was intended to perpetuate this very distinction between extractable enzymes and unextractable ferments. It is often forgotten that our use of the word *enzyme* way beyond its original intent pays historical obeisance to a wrong theory and to a long-discarded view of vital processes (see Part III).

Pierre Eugene Marcellin Berthelot (1827–1907) was one of the most influential and versatile chemists in the second half of the nineteenth century. His work is "astounding in its volume, originality and importance. Jungfleisch lists some 1600 titles" (Partington 1964, p. 467). Among his many achievements are the first syntheses of methanol, formic acid, benzene and other compounds and, as a partial indication of his range of interests, the coining of the terms *exothermic* and *endothermic, principle of maximum work, partition coefficient, explosion wave,* and the words *saccharose* and *acetylene.* He was among the first to use the word *synthesis* to denote the production of organic compounds from their elements. He was one of the earliest in the nineteenth century to believe that a special vital force is not needed for the formation of biological substances and that there are no special chemical or physical laws to explain vital processes. His monumental book *Chimie Organique Fondèe sur la Synthèse* (1860) in which this attitude was stressed (see end of Paper 13) was highly influential in establishing the science of organic chemistry. In collaboration with L. P. de Saint-Gilles he was the first to derive a second-order rate equation, and this work directly initiated Guldberg and Waage's first publication on the law of mass action (see Partington 1964, p. 589). So Berthelot can be regarded as one of the progenitors of chemical kinetics as well. He rediscovered the action of microorganisms in nitrogen fixation by soils. He wrote classic and still extensively used treatises (see Multhauf 1966) on the history of alchemy and of chemistry, translations from Greek alchemical works into French, a classic biographical study of Lavoisier, and so on. His chemical analyses of ancient objects from Egypt and Mesopotamia laid the foundation of chemical archeology (see Partington 1964, p. 474). He helped to establish *La Grande Encyclopédie* to which he contributed many articles. As though all this were not enough, he served in the French government for short periods as minister of education and as foreign minister. As one of France's most eminent sons he is buried in the Panthéon in Paris, next to Victor Hugo and Sadi Carnot.

Paper 14, by Brücke, is important because it is the first to use adsorption methods for enzyme purification (on calcium phosphate and on cholesterol). In this paper different successive purification steps are used, an approach that had been suggested previously by Theodor Schwann in 1836 (Paper 10). The end of the part translated here indicates that Brücke was clearly aware of the fact that his approaches, although better than those used before him, did not lead to pepsin in a pure state. This accomplishment had to wait seventy years until the work of J. H. Northrop, although C. A. Pekelharing had come very close to this achievement more than thirty years before Northrop's work (see Northrop 1946, and comments on Papers 37 through 40 in Part V). The yields that Brücke obtained were very low (Northrop 1939, p. 21), and it is in fact possible that his purified preparation was not pepsin (see Kraut and Tria 1937). However, this does not detract from the originality of his approach. In our 1861 paper Brücke still leaves open the possibility that the lack of chemical reactions in his preparations may simply be due to high dilution, but in his physiology textbook of 1874 Brücke stated quite firmly that his purified material did not give protein reactions and that it hence could not be a protein (Brücke 1874). This conclusion, which perhaps had a bearing on Willstätter's influential views some fifty years later (Willstätter quotes Brücke in his papers), is discussed in Part V of this work.[15]

Ernst Wilhelm von Brücke (1819–1892) founded the influential Vienna physiology school. He was one of the most versatile physiologists of his time and counted among his students Willy Kühne and Sigmund Freud. The latter, who spent six years in his laboratory, considered him to be the most highly respected teacher and the most outstanding authority in physiology that he had ever met.[16] The range of Brücke's interests is indicated by some of his contributions: He discovered the ciliary muscle named after him, found that blood is not coagulated in uninjured vessels, laid the foundations for Helmholtz's invention of the ophthalmoscope, was a well-known microscopist, and carried out research in phonetics. Clark (1980) has the following to say about Brücke's laboratory: It "...was still extremely primitive when Freud joined it [that is long after the time Paper 14 was written]. The Institute was housed in the basement and ground floor of an ancient building...the only water available came from a well, from which it was drawn by a caretaker each morning. Here, in the most unsophisticated conditions, Brücke helped to lay the foundations for Vienna's future success as a medical center... ." We shall see that Kühne specially refers to Brücke when discussing his reasons (Paper 19) for inventing the term *enzyme*.

While Brücke used his adsorption techniques to achieve an extensive enzyme purification, Danilevsky (Paper 15) in Kühne's laboratory used this approach within a year of Brücke's paper to tackle a different and perhaps more demanding problem, namely, to determine whether various catalytic activities in a given tissue extract are due to one or several ferments. His starting material was pancreatin, a preparation from pancreas that was known to exhibit what we would now call hydrolytic activity on proteins, starch, and fats.[17] His paper is the first on preparative enzyme separation. By the use of selective adsorption on collodion, he was able to show that distinctive agents were responsible for the first two activities, and by selective denaturation he showed that the third activity is not associated with the former two. This paper hence is important not only by adding to our knowledge of the nature of the pancreatic secretion and by showing for the very first time that many problems in metabolism can be answered only with purified enzymes, but it is, in addition, a fundamental contribution to the problem of enzyme specificity.

Alexander Jakovlevich Danilevsky (1838–1923) was professor of physiological chemistry at the University of Khazan, and later at Kharkov. He spent many years working in various renowned laboratories such as those of Virchow, DuBois-Reymond, Brücke, Hoppe-Seyler, and Kühne. The editor was not able to determine whether Danilevsky's work in Kühne's laboratory was influenced directly by his experience in Brücke's laboratory or whether he went to Vienna after his stay in Berlin. (For biographical information see Bylankin 1950.)

Paper 16 (1877) by Moritz Traube is ahead of its time for two reasons: It champions the protein or protein-like nature of ferments, and it vigorously asserts the validity of chemical rather than vitalistic approaches to a study and understanding of living processes. In the latter respect Traube's position resembles that taken by Berthelot in his exchanges with Pasteur, which had just begun. Traube's ideas, which had already been expressed in his 1858 book *Theorie der Fermentwirkungen* [Theory of the activities of ferments] take issue with Liebig's theories of ferment activity, with Schwann's ideas,[18] and with Pasteur's assertions on the impossibility of dissociating the action of ferments of putrefaction (Fäulnissfermente) from the processes of the living cell. Traube's chemical approach was fortified by Frédéric Alphonse Musculus' discovery, about 1876, of a soluble ferment in ammoniacal urine that decomposes urea, later called *urease*. Traube's ideas appear not to have been as influential as one would expect from the vantage point of later insights, perhaps because of the greater renown of Pasteur, who attacked Traube's ideas along with those of Berthelot.

Among the outstanding scientists of his time, Traube (1826–

1894) was exceptional in not holding a position in a university. Following the death of a brother, he had to work in his birthplace of Ratibor (now Racibórz, Poland) in his father's liquor store, which he developed into a successful business. He performed his chemical and biochemical researches in his spare time in his own laboratory in the small attic of the store. In 1866 he moved his business to the bigger town of Breslau (now Wrocław, Poland) and continued his scientific work in the laboratories of various friends until he again set up his own laboratory. Twenty years later he resigned from his business to become a full-time scientist. His versatility is attested by the fact that his interest in the process of formation of living cells led him to produce (1864) certain artificial cells with model membranes. A study of these synthetic membranes led him to the discovery of the phenomenon of semipermeability (1866). Since it was in this year that Traube moved to Breslau, this work was probably at least in part still carried out in Ratibor. This is probably one of the few major scientific discoveries made by a wine merchant in the attic of his store. As pointed out by Jacques Loeb (1906, p. 13), Traube "was also the first to recognize that oxidations occur in the cells and not, as had been assumed before, in the lungs or the blood." (See also endnote 14, Part III.) Traube knew Karl Marx and kept him informed of the latest developments in the natural sciences (see Hoffmann-Ostenhof 1978).

The last paper in this section (Paper 17) is by Wilhelm Friedrich Kühne. Kirchhoff, Payen and Persoz, and Berzelius had pointed out the analogy between ferment-catalyzed and acid-catalyzed hydrolyses of a plant material such as starch. Schwann had used this analogy for pepsin. Kühne's 1867 paper elaborates this analogy for another animal enzyme; his experiments, moreover, go a step further since they demonstrate, in a part of the paper not translated here, intestinal protein digestion *in vivo*. It may be noted here that ten years later (see Paper 18) Kühne invented the word *trypsin* to denote the material studied in Paper 17 and that, in fact, he invented the word *enzyme* at the same time. Trypsin was the first substance to be called an enzyme. The other achievements and the personality of Kühne are discussed in the next section of this work.

## NOTES

1. Duclaux's use of this term is of great importance, for it is on page 124 of his textbook that the proposal is first made to use the suffix "ase" for all enzymes, and that this be preceded by

the name of the substrate. The -ase ending is hence derived directly from Payen and Persoz's paper.

2. Friedrich Wöhler (1800–1882) is well remembered *inter alia* as the discoverer of the elements aluminum (1827) and beryllium (1828) and as the synthesizer (also in 1828) of urea. A measure of his (and Berzelius's!) industry is given by the fact that the third and largest German edition of the *Textbook of Chemistry* comprised 6215 pages in ten volumes, all translated by Wöhler from Berzelius's original Swedish manuscript (Jorpes 1966, p. 95). In addition, he found time to translate Berzelius's *Annual Surveys* for over twenty years. He wrote three textbooks of his own (inorganic, organic, and analytical chemistry) and published 281 papers. He cultivated a life-long friendship with Berzelius and Liebig.
3. As pointed out by Walden (1949, p. 15), Mitscherlich in 1835, hence also before Berzelius, had stated quite clearly that processes as varied as digestion, milk coagulation in cheese manufacture, acid hydrolysis of starch to sugar, formation of alcohol and carbon dioxide from sugar, and oxidation of alcohol to acetic acid in the presence of platinum were all due to what he called contact substances.
4. See comments to Wöhler, quoted in the Editor's Comments on Paper 12.
5. "To call the phenomenon catalytic is not to explain it; it is nothing but the replacement of a common word by a Greek word." (Gerhardt 1856, quoted and translated by Fruton 1972, p. 48).
6. Although the notion of a special catalytic force soon came under attack, the general context of Berzelius's presentation promoted acceptance of the "catalytic" over the "contact" terminology. Berzelius's use of the word *force* was soon seen as not being intended to be interpreted too literally (see Mittasch and Theis 1932, pp. 104–177, for a scholarly discussion of this topic). However, even many years later one can still find references to "catalytic force." Thus as late as 1894 the eminent physical chemist Wilhelm Ostwald, a champion of the ideas of the great Josiah Willard Gibbs, found it necessary, in reviewing a paper that still insisted on the existence of such a force, to state, on thermodynamic grounds, "It is. . .erroneous to regard catalytic activity as resembling a force which brings about something that would not occur without the catalytic substance; it is even less valid to assume the latter to perform work. The understanding of the phenomenon might be helped

when I especially stress that the concept of chemical energy does not contain that of *time*. . . ." (Ostwald 1894).

7. However, Kutznetsov (1966) cites a number of early authors (1835–1852) who "found a way of subjecting catalysis to stoichiometric laws. . . . [Hence] nonstoichiometric participation of a catalyst was explained by stoichiometric equations describing the formation and splitting of an intermediate compound, which obeyed the laws of Proust and Dalton." Segal (1959, p. 17) states that since Berzelius "explicitly stated that catalysts exerted their action only by their presence and not by virtue of their affinity," he excluded the observations of intermediates that had already been made in 1806 by Désormes and Clément "in the form of brown oxides of nitrogen, in their studies of the catalysis by nitric acid of $H_2SO_3$ oxidation by molecular oxygen, and [who] proposed a mechanism for the reaction based on intermediate complexes." (See also Lemay 1949 and Fruton 1972, p. 83.)
8. In a letter dated 10 July 1838 to the Dutch chemist Gerardus Johannes Mulder (see Fruton 1972, p. 96, and references in Jorpes 1966, p. 104).
9. In a letter to Mulder dated 13 November 1839 (see Söderbaum 1916, pp. 128–130).
10. From the viewpoint of enzymology, it is of interest that Wöhler and Liebig (1837) published a paper on the discovery of "emulsin" ($\beta$-glucosidase) from bitter and sweet almonds. Almost sixty years later (see Paper 24) this type of enzyme figured prominently in Emil Fischer's classic research on enzyme mechanism, structure, and specificity.
11. From the quantitative data given in the paper (not translated) and from the overall treatment, it is not clear whether Liebig thought a decomposing sugar molecule, by virtue of the vibrations that it has received from the ferment, to bring about vibrations and decompositions in another sugar molecule, and so on.
12. This salvo was published (3 December 1878) near the middle of their exchange. Three more communications by each of the two protagonists were yet to come hard upon one another, with Pasteur having the last word (10 February 1879). For details see Velluz 1964, pages 223–227. For a masterful exposé of this exchange, see "Chemicus" (1974).
13. These passages give one a flavor of this magnificent biography, which has been called "one of the most impressive and per-

ceptive books ever written on the development of a scientist's thought" (Geison 1974).

14. This passage is found in the second edition (1914) and all subsequent editions (up to the fourth in 1938) but not in the first edition (1911) of Harden's classic book.
15. Brücke was not alone in reaching this kind of conclusion. Thus J. Cohnheim (1863) in Kühne's laboratory found that various purified amylases gave no protein reactions and concluded that they could not possibly be proteins.
16. Brücke must have been a formidable figure. Freud has the following to say about him in his *The Interpretation of Dreams* (1955): "It came to Brücke's ears that I sometimes reached the students' laboratory late. One morning he turned up punctually at the hour of opening and awaited my arrival. His words were brief and to the point. But it was not they that mattered. What overwhelmed me were the terrible blue eyes with which he looked at me by which I was reduced to nothing. . . .No one who can remember the great man's eyes, which retained their striking beauty even in his old age, and who has ever seen him in anger, will find it difficult to picture the young sinner's emotions."
17. Pancreatic lipase had been discovered by Claude Bernard (1849). Although it had been known earlier, the digestive action of pancreatic juice on proteins was first studied in some detail by Lucien Corvisart (1857, 1859) (cf. Paper 17), who is also remembered as the personal physician to the emperor Napoleon III. As to the conversion of starch to sugar by this secretion, this was thought by Danilevsky (in part of Paper 15 that is not translated here) to have been recognized "for ever" ("von jeher").
18. It is somewhat hard to see Schwann cast by Traube as an advocate of a vital force. Indeed, according to a detailed analysis by Watermann (1960), it seems clear that already in 1837 Schwann emphasized ferments to be chemical substances and that a ferment should be distinguished from the fungus in which it is contained. Watermann emphasizes that throughout his life Schwann was an opponent of vitalism.

# 9

# MEMOIR ON DIASTASE, THE PRINCIPAL PRODUCTS OF ITS REACTIONS, AND THEIR APPLICATIONS TO THE INDUSTRIAL ARTS

**Anselme Payen and Jean François Persoz**

*These excerpts were translated expressly for this Benchmark volume by H. Friedmann, the University of Chicago, from pages 73–76, 77–78, 88–90 and 91 of "Mémoire sur la Diastase, les principaux Produits de ses Réactions, et leurs applications aux arts industriels" in* Ann. Chim. Phys. (Paris) **53**:73–92 (1833).

Since the learned researches and the laborious work of Luwenhoeck, Saussure, Kirschoff, Vauquelin, of the English brewers, of Messrs. Dubrunfaut, Raspail, Guibourt, and Couverchel, etc., the physiological conformation of starch has been known; it was known that its globules consist of an envelope enclosing a *mucilaginous* substance; that when the temperature was increased somewhat, a part of the inner material oozed out; that water and sulfuric acid caused the teguments to break and to change the starch to sugar and even to *gum* if the reaction was carried out for a shorter time; that with the help of germinated barley, of water, and of heat (Dubrunfaut, Memoir read in April 1823), the starch was changed to sugar.

In 1785 Dr. Irvine showed that the sweet products of malt were increased when flour from uncooked grains was added (*Accum.*, transl. by Riffaut).

[*Editor's Note:* A few sentences from the 1785 essay by William Irvine, published in 1805 (see note 11 in the Editor's Comments in Part I), are given here since they show an uncanny feeling for a chemical basis of biological processes and since they have apparently never been directly quoted before:

> Most kinds of grain have little or no sweet matter in them in their natural state. When put into the earth, they gradually acquire a sweet taste, and retain this during the tender state of the young plant. . . .Many have thought, and still think, that the powers of vegetation only can produce this change in the farinaceous matter. . . .But whether we consider this alteration in the farinaceous matter as a distinct species of fermentation or not, certainly the powers of vegetable life are in no way concerned, or necessary to it. It is not during the growth only of the seed that this change can be effected; but a quantity of the sweet matter produced by the growth of the seed mixed with a quantity of the same seed ground into powder, and the whole mixed with a proper quantity of water, will all become sweet, and be changed into spirits in the same manner as if the whole had been previously altered by the vegetation of the seed. . . .Certain mixtures of farinaceous matter with sweet matter may be wholly

changed into saccharine matter, and the sweet matter with which it is mixed might be considered in the same light with yest in the vinous fermentation. But many things are wanted before this opinion can be supported by experiment.]

Did anything remain to be discovered along this beaten path? We thought so, all the while paying homage to the important work of our predecessors.

It appears to us, in fact, that no economic procedure has so far been worked out for extracting from starch the interior substance that Mr. Biot in recent work has shown to be characterized by a new optical phenomenon; it appears to us, again, that after several years of research, far from knowing the active principle that is developed by the process of germination, one had first ascribed its reactions to hordeine and then to some kine of *soluble gluten,* which we alone had established to be inert.

It appears to us, furthermore, that although the transformation of starch to sugar by this agent had been accepted, the liberation of *dextrin,* which today is becoming the source of numerous applications, had not been observed.

It appears to us yet again that the parameters of the phenomenon of starch saccharification in the presence of germinated barley had not been specified; that the voluminous material written on this subject by the English brewers and distillers and by our authors left a host of practical anomalies unforeseen and unexplained.

It appears to us, finally, that no one has been able to catch a glimpse of the consequences of the discovery of *diastase* for organic chemistry, for physiology, and for the industrial arts.

Since our announcement to the Academy of Sciences of a new way to prepare *dextrin* by directly separating and removing the envelopes of starch, we continued to persevere in our search for the principle that produces this curious reaction.

This substance, which we have just isolated, contains less and less nitrogen the purer it becomes, and in addition, it possesses the following properties: It is *solid, white, amorphous, insoluble* in alcohol, *soluble* in water and in dilute alcohol, its aqueous solution is neutral and without obvious taste, it is not precipitated by lead subacetate. When left by itself, it is altered more or less rapidly, depending on the room temperature, and it becomes acidic. When heated to 65-75° with starch, it shows the remarkable power of promptly detaching the envelopes from the modified inner substance, namely the *dextrin,* which readily dissolves in water while the teguments that are insoluble in this liquid rise to the surface or are precipitated, depending on the movements of the liquid. Because of this curious property of separating, we give the substance that possesses it the name of *diastase,* a name that expresses this fact precisely.

The operation, which is conveniently brought about, gives purer *dextrin* than had ever been prepared before. Also one finds again very clearly the large power of rotation that characterizes it and that is not obtained to an equal degree by any other procedure. Nevertheless, the solution of *diastase* gradually converts the dextrin to sugar, which is precipitated neither by baryta nor by lead subacetate. The temperature must be maintained at 65-75° during the contact, for upon heating the diastase solution to the boiling point, it loses the ability to act on starch and on dextrin.

[*Editor's Note:* Material has been omitted at this point.]

Cereals and potatoes before germination contain no diastase; it can be extracted from germinated barley by the procedure described below, and the yield can be increased by regular germination and provided that the plumule has grown to the same length as the grains.

After the mixture of water and of germinated barley has been macerated in cold water for a short time, it is filtered or, better, subjected to strong pressure followed by filtration of the solution. The clear liquid is heated in a waterbath to 70°. This temperature coagulates most of the nitrogenous material, which then has to be separated by a new filtration; the filtered liquid contains the active principle plus a little nitrogenous material, colored material, and an amount of sugar that depends on the extent of germination; to separate the latter, alcohol is added to the liquid until precipitation stops; diastase that is insoluble in alcohol is deposited in the form of flakes that can be collected and dried at a low temperature. It is essential in order not to alter it for the worse that, above all, heating of the moist material up to 90-100° be avoided; to obtain it even purer, it must be dissolved in water and precipitated anew by alcohol, and these solutions and precipitations can even be repeated twice. The diastase is again obtained free of nitrogenous material simply by several precipitations with alcohol, provided it is not coagulated by a temperature increase. After each precipitation less and less material dissolves, and the diastase becomes ever whiter and purer. The following is our most successful procedure: Freshly germinated barley is ground in a mortar and moistened with about half its weight of water. This mixture is subjected to strong pressure, and the liquid that runs out is mixed with sufficient alcohol to destroy its viscosity and to precipitate most of the nitrogenous material. The latter is separated by filtration. The filtered solution upon precipitation with alcohol yields impure diastase. It is purified by three alternating solutions in water and precipitations with excess alcohol.

The diastase solution, irrespective whether pure or containing sugar, separates the dextrin from all starches and starch-like substances, and thus makes a direct analysis of flours, of rice, of bread, and so on possible. When the extraction of this new principle has been done carefully, it has sufficient power for 1 part by weight to solubilize the inner material from 2000 parts of dry starch in warm water, and then to convert the dextrin into sugar. These reactions occur more readily and the first of these two is more rapid if a large excess of diastase is used. Thus if the dose is doubled to 1/1000 the solution of starch can be completed in 10 minutes.

[*Editor's Note:* Material has been omitted at this point.]

Crude dextrin, if certain teguments are excluded, in general consists of three substances: The first is insoluble in cold but soluble in warm water, colored by iodine, and identical to the inner substance of starch. The second is soluble in cold and in warm water and in weak alcohol and is not colored by iodine, just like gum. The third is a sugar soluble in water, in 35° alcohol [in the French, l'alcool à 35 degrés], not colored by iodine, fermentable, and so on.

Obviously the protracted action of diastase changes these three substances to the last two by completing the transformation of the first.

[*Editor's Note:* Material has been omitted at this point.]

The new principle, whether more or less pure, will above all be useful to assay starch, flour, bread, and sundry starchy substances. It is one of the most elegant processes of organic analysis.

When in solution, it is the agent for manufacturing commercial dextrin and sugar of dextrin. These operations have already been performed quite precisely and simplified as much as possible.

It is now possible to obtain starch teguments completely free of all the material colored by iodine, to obtain large amounts of this latter substance, or to convert it at will into two other substances, a sugar and a gum.

This agent helps to explain the passage of the products of starch through the sap; from this point of view it is seen that we have searched and found the active principle near the points where starch is consumed by the plants.

[*Editor's Note:* Material on the commercial applications of diastase and dextrin has been omitted at this point.]

A vast area of research remains to be pursued concerning the occurrence of diastase in the various parts of the plant [or: of the vegetable kingdom (l'organisation végétale)], its atomic weight, its elementary composition, its combinations, and the products of its rather special action on starch-containing plants.

[*Editor's Note:* The last paragraph, which acknowledges collaboration from industrial manufacturers, has been omitted here.]

# 10

# ON THE NATURE OF THE DIGESTIVE PROCESS

**Theodor Schwann**

*Assistant at the Anatomical Museum in Berlin*

*These excerpts were translated expressly for this Benchmark volume by H. Friedmann, the University of Chicago, from pages 90–92, 93–95, 96–101 and 136 of "Ueber das Wesen des Verdauungsprocesses" in* Arch. Anat. Physiol. Wiss. Med., pages 90–183 (1836).

Through the brilliant discovery of *Eberle*, we have learned to use mucous membranes treated with dilute acids as a material to bring about artificial digestion as efficiently as natural digestion. Through the investigations of Professor *Müller* as well as through those that I carried out in collaboration with him, this result was confirmed for protein, and it may be expected that *Eberle's* results will be confirmed also for the remaining substances. One can also regard *Eberle's* proof of the identity of the changes undergone by foodstuffs via artificial as well as via natural digestion as having been on the whole confirmed. However, in spite of the brilliant services of *Eberle*, his experiments have given us no information about the nature of the digestive processes. From his experiments we learn neither what the digesting principle is nor how it acts, whether it is merely a solvent like the normal chemical agents or whether it acts by means of contact. *Eberle* indeed regards the acid mucus as the active material, and since according to *Berzelius* many mucus types are insoluble in acids it would follow that an undissolved substance brings about digestion; thus it could also be assumed concerning the mode of action of the mucus that it is a contact action; in *Gerson's* essay, which contains the collaborative experiments of Professor *Müller* and myself, we have in fact regarded the matter in this fashion. However, at the end of the investigation, which I have carried out with Professor *Müller*, I observed that digestion was also brought about by the digestive fluid that had been prepared by treatment of the gastric mucosa with hydrochloric and acetic acid and filtration through linen. This preparation was only slightly turbid and formed only a trace of sediment. These experiments are reported in the preceding paper. Subsequently I filtered this liquid through paper and obtained a perfectly clear, yellowish fluid with undiminished digestive power. The digesting principle had hence been completely dissolved in it, and the view that it acts by contact hence had lost its basis, and thus *Eberle's* view that the mucus is the active principle was not actually contradicted but very strongly shaken. In this clear state, the digestive fluid was in fact very suited for a closer investigation.

Hence I first of all prepared large amounts of digestive fluid. The mucous membrane from the third and fourth stomach of an ox was prepared and cut into small pieces. The material weighed a few pounds.

Some water and a few ounces of hydrochloric acid were added, and the whole was digested twice for 24 hours at 32°R. [The Reaumur scale consisted of 80° between the freezing and boiling points of water. Hence 32°R equals 40°C.] This treatment dissolved most of the mucosa. A turbid fluid resulted, at the bottom of which undissolved mucus and small pieces of mucosa were lying and in which such material was also floating. The fluid was decanted and filtered first through linen, then through paper. In this manner, I obtained about 3/4 quart of an admittedly turbid yellowish solution that, however, even after standing for months, formed no sediment; even after repeated filtration, this liquid did not become clear. It contained 2.75 percent dry matter. It was called No. 1. Fresh water with acid was now poured onto the undissolved residue. Digestion and filtration were repeated. The filtrate that was obtained at this stage was quite clear, resembling in color a highly saturated urine. It measured 1/2 quart and was called No. 2. The residue was digested once more with water and acid and was then filtered again. This digestive fluid, No. 3, was also clear, not as intensely colored as the former one, but straw yellow. The amount of acid in all three fluids was about the same.

[*Editor's Note:* Material has been omitted at this point.]

The question concerning the nature of the digestive process can be divided first of all into two parts: What are the materials responsible for digestion, and, how do they act?

To answer the first question, one first has to consider free acid. In all experiments performed so far, both with artificial and with natural digestion, free acid was present. It does not, however, follow immediately that this is the essential active ingredient necessary for digestion. The acid could serve for the formation of some other essential digesting material that, once formed, brings about digestion by itself. In order to test this, I neutralized the digestive fluid with potassium carbonate and then incubated it with egg white. [The German word *Eiweiss* means both protein and egg white. Schwann almost certainly used coagulated egg white for the experiments reported in this paper.] The egg white, however, was not digested at all, nor changed in any way. When I now again added hydrochloric acid in appropriate amounts, the egg white was completely digested. Free acid is hence essential for activity during digestion.

However, the acid is not the only active agent. *Müller, Eberle, Beaumont,* and I have unsuccessfully carried out experiments to digest merely with dilute acids. After I had recognized that very much depends on the degree of dilution of the acid, I attempted the digestion of the egg white with hydrochloric acid diluted to the same extent as the digestive fluid, namely, 3.3 grain concentrated hydrochloric acid to 1/2 loth water. [A *loth* (modern spelling: "lot") corresponds to 1/2 ounce.] But even in this solution the egg white suffered no change in 24 hours at 32°R. Hence the following questions still remain: What is the second substance active besides acid, and, how do both act?

I first turn to answer the latter question, and first of all, how does the acid act? (1) Does it act as mere solvent of the second digesting substance; or (2) does it form a chemical compound with it, perhaps analogous to acid salts, and is it perhaps this compound that brings about the digestion; or (3) does the acid serve to dissolve the products that are formed from the digested substances; or (4) does it become decomposed during the digestion in order to become incorporated into the structure of the products of digestion; or finally (5) does it facilitate the decomposition of the digestible substances merely by contact and without being itself decomposed. It is these questions that are to be answered here.

In order to answer the first question, whether the acid serves merely as solvent of the second digesting principle, a part of the digestive fluid No. 3 was given so much potassium carbonate that more than half of the acid in it was neutralized so that the liquid was still clearly acidic and no turbidity was formed. Hence, although none of the second digesting principle could have been precipitated, digestion nevertheless did not proceed. The acid hence does not act as a mere solvent of the other digesting substance.

Hence the second view had to be tested, whether the acid perhaps forms a chemical compound with the other digesting principle in analogy to acid salts, where, although removal of a part of the acid indeed does not precipitate anything, the compound containing less acid nevertheless shows other activities. In this case the amount of acid must show a certain quantitative relationship to the amount of the other digesting substance. Here it is first of all important that a given quantity of digestive fluid must contain a certain amount of acid.

[*Editor's Note:* The very detailed description of quantitative experiments with varying amounts of hydrochloric acid added to extract No. 3 is omitted here. The following is Schwann's conclusion concerning this part of the investigation.]

It follows . . .that the required amount of acid does not depend on the amount of the second digestive principle; hence the above view cannot be the correct one either.

Hence the third hypothesis is also eliminated, namely, that the acid serves to dissolve products that are formed during digestion and that perhaps are soluble only in acid. It is true that during the digestion of egg white, a substance is formed that is still very similar to egg white and that is soluble only in acid. Here hence the amount of acid would have to depend on the amount of egg white or of this substance. However, an amount of digestive fluid that suffices in the undiluted state to dissolve a certain amount of egg white does not bring about digestion when it is diluted with water. Yet this substance is also soluble in very dilute acids, and the other digesting principle also does not lose its power upon rather strong dilution (with acidified water). The requisite amount of acid hence depends only on the amount of water, and the dissolution of the products formed during digestion can at the least not be the only purpose of the acid.

One can ask further, does the acid perhaps become incorporated into the structure of the products formed during digestion? In that case the amount of free acid must be changed during the digestion. In order to test this, one loth of digestive fluid No. 3 was digested with one dram of egg white in a closed vessel. After 24 hours everything except a very small residue had been dissolved. One loth of the original digestive fluid before digestion required 4.6 grain of potassium carbonate for saturation. Since it was diluted by the solution of one dram of egg white, it follows that if the acid content had not changed, then 1/2 loth after digestion would require 184 grain of potassium carbonate for saturation. Experimentally it was found that not quite 1.9 grain were required. The amount of free acid hence remains unchanged during the digestion of the egg white.

Hence only the last assumption remains, that during digestion the acid participates in the decomposition of the organic substances by its presence, without itself being changed, just as in the case of the change of starch to sugar by boiling with dilute acids. A secondary purpose may indeed be the dissolution of such products of digestion that are soluble only in acid.

We now proceed to the second question: How does the other essential digesting principle, required in addition to acid, act? Chemistry knows two ways by which one substance can bring about the decomposition of another: Either it or one of its constituents combines with the substance to be dissolved or with one of its constituents, or the first substance brings about the spontaneous decomposition of the other by means of contact with it, but without forming a new compound with it. The former takes place in the usual chemical processes. Of the latter type we have in inorganic Nature the decomposition of hydrogen peroxide by gold oxide and other examples. Examples from organic Nature that belong here are alcoholic and acetic acid fermentation. In the case of the usual chemical processes, the amount of decomposing and decomposed substance always shows a constant ratio, and one never sees a very considerable disproportion in the quantity of the two substances. On the other hand, in the case of actions by contact the quantity of the substance that initiates the decomposition has much less influence, and a relatively rather small quantity suffices to decompose a large quantity of another substance. This is one criterion of an action by contact. Second and more fundamental, neither the substance that brings about the decomposition nor one of its constituents combines with the products of the decomposed substance. This criterion is shown most clearly when the decomposing substance remains completely unchanged, for example, in the conversion of alcohol to acetic acid by the platinum preparation discovered by *E. Davy*. When, however, the decomposing substance is itself decomposed, then it is necessary if one wishes to regard the process as action by contact to show that the products of decomposition do not combine with one another. Thus when hydrogen peroxide is decomposed by gold oxide, oxygen, gold, and water, which do not combine with each other, are formed. In the case of alcoholic and acetic acid fermentation, we have only the first criterion, the small quantity of ferment that brings about the fermentation of a large amount of material. The second criterion, however, is still problematical, for during the fermentation the ferment loses its ability to cause further fermentation. The ferment is hence decomposed as a consequence of the fermentation, and since one does not know what happens to its products--they might possibly combine with the products of the fermenting substance--it is still unclear whether fermentation should be regarded as an action by contact. One may, however, state in favor of this view that one can at least bring about acetic acid fermentation by a true action by contact, namely, through the action of that platinum preparation.

We now have the question: Does digestion belong to the usual chemical processes in the course of which substances can be dissolved, or must it be classified with the actions by contact?

We begin with the first characteristic, namely, how much of the digesting principle is needed to digest a given amount of protein.

The first test tube contained 1/2 loth of normal digestive fluid No. 1, in a second tube 9.6 grain of the same digestive fluid was mixed with 1/2 loth of acidified water. This test tube hence contained only 8 percent of digestive fluid or, since this contained 2.75 percent solids, the diluted digestive fluid contained only 0.22 percent solids. In a third test tube 4.8 grain digestive fluid were mixed with 120 grain of acidified water. This hence contained 4 percent of pure digestive fluid or 0.11 percent solids. The fourth test tube contained in the same quantity of acidified water only 1 percent digestive fluid No. 1 or 0.055 percent solids. The fifth contained only 1/2 percent digestive fluid or 0.0275 percent solid material; the sixth one, finally, contained only 0.3 grain digestive fluid with 1/2 loth of acidified water, hence only 1/4 percent digestive fluid or 0.0137 percent solids. Next to these, finally, was placed a test tube that contained equally diluted hydrochloric acid alone. Into all test tubes were placed a few thin slices of egg white weighing a few grains. After 12 hours the egg white had completely dissolved in the test tubes containing 8 and 4 percent digestive fluid. Closest to this was the egg white in the tube containing 2 percent digestive fluid. Here only the thickest parts of the egg white remained undissolved. In the test tube that contained only 1 percent digestive fluid, the egg white was exceedingly soft and translucent, but the shape of the pieces was still recognizable. The protein in the undiluted digestive fluid behaved approximately like the one in this test tube and had been digested much less than in the former tube. In the test tube that contained only 1/2 percent digestive fluid, an initial digestion was also seen; the protein had become somewhat transparent and soft, and even in the test tube with 1/4 percent digestive fluid, a rather noticeable change had still taken place, and the egg white in it differed markedly from that in the test tube containing only dilute acid where the egg white was unchanged. As to time, the normal digestive fluid hence did not digest more rapidly than acidified water containing only 1 percent of it. Mixtures that contained 8 or 4 percent of it actually digested more rapidly than the undiluted digestive fluid. A marked activity was still seen even for a dilution containing only 1/4 percent digestive fluid or 0.0137 percent solid digesting material. The question now arose as to how the digesting force of the diluted and undiluted digestive fluid is related to the amount of egg white. Hence 4.8 grain of digestive fluid No. 1 were mixed with 1 dram of coagulated triturated egg white (weighed moist) and 1/2 loth of acidified water. The same amount of egg white was added to 1/2 loth of undiluted digestive fluid. After 24 hours the egg white in both had dissolved but for some few residues, and actually somewhat less remained in the diluted digestive fluid than in the undiluted one. Hence 4.8 grain digestive fluid or 0.11 dry digesting substance had dissolved 60 grain moist egg white that contains about 10 grain solid material; that is, 1 part had brought about the decomposition of about 100 parts, a ratio that occurs only in actions by contact and in fermentations. In the case of alcoholic fermentation, according to *Thénard,* 1 1/2 part of ferment brings about the decomposition of 100 parts of sugar. In this respect the digestion of egg white hence corresponds to actions by contact and to fermentation.

[*Editor's Note:* Material has been omitted at this point.]

Fibrin and muscle . . . are digested in the same manner as coagulated egg white, namely, by free acid in conjunction with another substance that is active even in minimal amounts. Since the latter hence actually brings about the digestion of the most important animal nutrients, one could justifiably give it the name *pepsin* [emphasis added]. This substance acts initially even on uncoagulated casein, milk, and so on, in the same way as it does during digestion, by precipitating these. It is hence, in general, active on substances related to animal protein.

[*Editor's Note:* The rest of the article is omitted.]

# 11

# ON A SO FAR ONLY RARELY OBSERVED FORCE WHICH IS PROBABLY ACTIVE IN THE FORMATION OF ORGANIC SUBSTANCES

**Jöns Jacob Berzelius**

*These excerpts were translated expressly for this Benchmark volume by H. Friedmann, the University of Chicago, from pages 19–22 and 23–25 of "Ueber eine, bei der Bildung organischer Verbindungen wirksame, bis jetzt wenig bemerkte Kraft" in* Lehrbuch der Chemie, *3rd ed., vol. 6, F. Wöhler, transl., Dresden and Leipzig: Arnoldische Buchhandlung, 1837, 658 pp.*

Investigations over the last few years have made us familiar with a force, active in both inorganic and organic Nature, which differs from the forces that were known earlier and whose history, which so far has been developed only slightly, I will sketch here briefly.

New compounds are formed in inorganic Nature by the mutual interaction of several substances. This follows from the fact that tendencies for combinations that seek to satisfy themselves are expressed, since, on the one hand, substances with strong affinities form compounds with one another while, on the other hand, the residual weaker ones also combine. Up to 1800 one did not suspect that the expression of the degree of affinity could be effected by factors other than heat and at times light. At that point the influence of electricity was discovered. It was soon found that electrical and chemical relationships are one and the same thing and that the affinities that are selected merely result from the more pronounced reciprocal electrical relationships, which are increased by heat and by light. At that time the formation of new compounds hence could only be explained in terms of an encounter of substances whose electrical relationships could be neutralized better by transposition of their constituents. When we turned with the experience that we had gained from inorganic Nature toward the study of the chemical processes that occur in living Nature, we found that in their organs substances are produced with highly differing compositions, for which the raw material in general is a single fluid or solution that is carried around in the ducts at varying slow speeds. This was especially clear in the case of the animals; here one sees an uninterrupted uptake of blood by the ducts and, without addition of any other fluid that could bring about double decompositions in them, a discharge of milk, bile, urine, and so on from their openings. It was obvious that something occurred here for which inorganic Nature had not yet provided a key. Now *Kirchhof* discovered that when starch is dissolved at a certain temperature in dilute acids, it is converted first into gum and then into grape sugar. It was very much in line with our approach to the study of such changes to determine what the acid had taken up from the starch, so that the remainder could combine to form sugar; however, nothing gaseous escaped, nothing was found combined with the acid, all the acid originally present could be removed again with bases, and in the liquid only sugar was found, rather more by weight than corresponded to the amount of starch used. The matter remained for us as mysterious as a secretion in organic Nature. Then *Thénard* discovered a liquid in which only a weak force held the constituents in combination with each other, I mean the superoxide of hydrogen. Under the influence of acids the combination of the constituents remained undisturbed; under the influence of alkalis the tendency of the constituents to separate was stimulated anew, and a type of slow fermentation occurred, during which oxygen gas escaped and water remained. However, not only substances that were soluble in this liquid effected this decomposition; even solids, organic as well as inorganic, brought this about, especially manganese dioxide, silver, platinum, gold, and among the organic substances, blood fibrin. The substance that here caused the decomposition of the constituents did this not by being incorporated into the new compounds; it remained unchanged and hence acted by means of an inherent force whose nature is still unknown, although its existence has been observed in this manner.

Shortly before *Thénard's* discovery, *Humphry Davy* observed a phenomenon whose relationship with the preceding one had not been immediately recognized. He found that platinum, heated up to a certain temperature, had the property, upon contact with a mixture of atmospheric air and alcohol or ether vapors, to support a combustion of the latter but that gold and silver did not possess this property. Not long afterwards, his relative, *Edmund Davy,* discovered a platinum preparation, later found to consist of metallic platinum in a high degree of dispersion, which possessed the property of glowing at normal air temperatures upon being moistened with alcohol, because of the combustion of the latter, or, upon dilution with water, to oxidize it to acetic acid. Now came the discovery that, as it were, crowned the preceding ones, namely, *Döbereiner's* discovery of the ability of platinum sponge to ignite hydrogen gas that is allowed to escape into the air. Shortly afterwards, this discovery was pursued further in a collaborative investigation by *Dulong* and *Thénard*. It followed from this that many simple and complex substances possess this property, but to such varying degrees that while it is seen even much below the freezing point [of water] in the case of platinum, iridium, and other companions of platinum, a higher temperature is required for gold, a still higher one for silver, while a temperature of at least 300° is required for glass. Thus this property was not an isolated, exceptional behavior but proved to be a more general one, exhibited by substances to varying extents. It now became possible to attempt to find applications for this phenomenon. We have found, for instance, that the conversion of sugar to carbon dioxide and alcohol, which occurs in fermentation through the influence of an insoluble substance known by the name of *ferment* and which can be replaced, although less actively, by animal fibrin, by coagulated plant protein, cheese, and similar substances, could not be explained by a chemical reaction between sugar and ferment resembling double decomposition. However, when compared with phenomena known in inorganic Nature, the preceding phenomenon most closely resembled the decomposition of hydrogen peroxide under the influence of platinum, silver, or fibrin; it was hence very natural to imagine an analogous activity in the case of the ferment. However, we did not remember any case comparable to the action of alkalis on hydrogen peroxide, that is, where this inexplicable influence could be exerted by a dissolved substance on another substance contained in the same solution. The formation of sugar from starch under the influence of sulfuric acid was not yet recognized as such an example.

We owe this recognition to *Mitscherlich's* ingenious experiments concerning the formation of ether.

[*Editor's Note:* A short passage on ether formation has been omitted at this point.]

The conclusions at which *Mitscherlich* arrived by the method of preparing ether had not been derived by anyone before him. He showed that at this [high] temperature, sulfuric acid acts on alcohol with a force of the same type as is exerted by alkalis on hydrogen peroxide; . . .this led him, in turn, to the conclusion that the action of sulfuric acid on starch during the conversion of the latter into sugar is of the same type.

It has hence been established that many substances, simple as well as complex, solid as well as dissolved, possess the property of acting on complex substances by means of an influence very different from ordinary chemical affinity. They accomplish this by bringing about a reordering of the components of the substance into other relationships without the necessary participation of their own components, although this may occasionally be the case.

We have here a new force, belonging both to inorganic and to organic Nature, for bringing about chemical activity. This force is almost certainly more widely distributed than had been thought up to now, and its nature is still hidden. When I call it a new force, I do not at all, however, intend to explain it as a capacity independent of the electrochemical relationships of matter; on the contrary, I can only assume it to be a special way by which these are expressed. As long, however, as the manner in which these forces are connected to one another remains hidden, it will help our researches to regard this new force for the time being as a distinct force, just as discussions about it are simplified when we have a special name for it. I hence, using a derivation well known in chemistry, will name it the *catalytic force* of the substances, and I will name decomposition by this force *catalysis,* just as we understand by the word analysis the separation of the component parts of substances based on ordinary chemical affinity. The catalytic force seems in fact to be shown in the capacity that substances have, by their mere presence and not by their own affinity, to awaken affinities that are slumbering at a given temperature. As a consequence of this force the elements in a complex substance become organized in other relationships that result in a larger electrochemical neutrality. In this context catalytic substances behave on the whole in the same way as heat, and one can ask the question whether a difference in the amount of catalytic force may bring about the same difference in catalytic products in different substances, as is often the case with heat or with different temperatures? One can hence ask further whether substances with different catalytic properties yield various catalytic products from a given compound? It cannot as yet be decided whether these questions should be answered by yes or no. Another question is whether substances with catalytic force exert it on many compounds or whether, as still appears to be the case at the present time, they catalyze some substances without acting on others? The answers to these and to other questions must be left to future research. For the present it suffices to have pointed out the existence of the catalytic force with a sufficient number of examples.

If we now turn with this idea to the chemical processes in living Nature, then we see things in a completely new light. [In the *Annual*

*Survey,* Vol. 15 (submitted to the Swedish Academy of Sciences, 31 March 1835) but not in the *Textbook,* there are some interesting sentences here that run as follows: When Nature, for instance, has placed diastase in the eyes of the potato (*Annual Survey,* 1830, p. 283) but not in the tuber or on its projecting sprouts, then we see how catalytic force converts insoluble starch into gum and into sugar, and we realize that the environment of the eyes becomes a secretory organ for the soluble substances from which the sap for the growing shoot will be formed. From this, however, it does not follow that this catalytic process is going to be the only one in the vegetable kingdom.] When we do this, we will find solid reasons to assume that in living plants and animals, thousands of catalytic processes occur among tissues and fluids, generating the multitude of differing chemical compositions whose formation from the common raw material, the plant sap or the blood, we could never understand in terms of an acceptable cause. In the future we will perhaps discover this cause to be the catalytic force of the organic tissue of which the organs of the living body are composed. [The *Annual Survey* ends at this point and the remainder of this paragraph is found only in the *Textbook.*] Even after life has been extinguished, this force may still continue to be active. We find that an animal or a plant after its death undergoes a spontaneous decomposition by change of the elements into other substances that are usually simpler and that tend more and more to approach inorganic compounds in their composition. It seems to have been the purpose of the Originator of Nature thus to remove traces of what has lived, so that sufficient material and space will not be lacking for whatever receives life in its place. The changes that occur here can be summarized as three processes: namely, fermentation of wine, acid fermentation, and putrefaction, all of which will be discussed further below in more detail from the point of view of the products formed.

[*Editor's Note:* The section on catalysis ends at this point.]

# 12

# ON THE PHENOMENA OF FERMENTATION, PUTREFACTION, AND DECAY, AND THEIR CAUSES*

**Justus Liebig**

*These excerpts were translated expressly for this Benchmark volume by H. Friedmann, the University of Chicago, from pages 250, 262, 268, 274, 277, 281, and 286–287 of "Ueber die Erscheinungen der Gährung, Fäulniss und Verwesung und ihre Ursachen" in Ann.* Pharm. (Heidelberg) **30**:250–287 (1839).

In the following I shall try to develop a few principles with which one must from the present viewpoint of organic chemistry regard the spontaneous phenomena, brought about by unknown causes, that accompany the decompositions and changes of organic substances.

If we first of all consider the components of organic substances, then it is known that *carbon, hydrogen, nitrogen,* and *oxygen* are attracted toward one another to highly different degrees.

We know that in the combustion of a substance composed of carbon and oxygen, it is the hydrogen that is oxidized first and that the carbon is oxidized only when more oxygen is present than is needed by the hydrogen to form water.

When the amount of oxygen admitted is no more than corresponds to the amount that combines with the hydrogen, then the carbon is precipitated as lamp black.

If a compound of nitrogen and carbon is burned in air or in oxygen, then only the carbon combines with it, the nitrogen separates as gas, and only in rare cases do we get an oxidation stage of the nitrogen.

[*Editor's Note:* Material has been omitted at this point.]

I wish now to direct the attention of scientists to a so far unobserved cause that brings about the changes and the phenomena of decomposition that are in general denoted by the terms *decay, putrefaction, fermentation,* and *rotting.*

*This cause is the ability possessed by a substance that undergoes a decomposition or a combination, that is, a chemical action, either to bring about the same activity in another substance with which it is in contact, or to prepare it to be subject to the same change that it itself undergoes.*

This type of activity may best be represented by a burning substance (that is, a substance in action) that can cause the same activity in other substances on being brought close to them.

The capacity possessed by a substance that is undergoing decomposition or combination, to evoke previously absent affinities in another substance, or to increase the relationship of its constituents to such a degree that compounds are formed that were not formed before, is peculiar, it represents a special manifestation of affinity, whose activity is equivalent to a peculiar force. In the case of the burning body, this cause is the elevated temperature.

*Reprint from *Organic Chemistry* of J. L., published by Winter's Bookstore.

The wide distribution of this cause can be established by means of innumerable examples, but it will suffice when I here give just some of these.

*Platinum,* for example, does not possess the ability to decompose nitric acid and to dissolve in it. Platinum alloyed with *silver* readily dissolves in nitric acid. Silver transfers to platinum an ability that platinum does not possess by itself. Both metals take up oxygen by decomposing nitric acid.

[*Editor's Note:* Material has been omitted at this point.]

In this context the products of putrefaction are absolutely similar to the products formed by the influence of heat on organic substances.

The only difference between dry distillation and putrefaction is that the influence of the volatility or heat stability of the products on their own formation is absent in the case of putrefaction, so that both processes can be completely compared with one another only in case all the carbon participates in the change. Putrefaction can actually be regarded as a dry distillation that proceeds in water and at a somewhat higher than normal temperature . . . .

[*Editor's Note:* Material has been omitted at this point.]

The ferment is no special causative agent but brings about putrefaction and fermentation only as a result of the change that it itself undergoes.

*The ferment is a substance that undergoes putrefaction and decay.* It converts the oxygen of the surrounding air into carbon dioxide, and it furthermore forms carbon dioxide out of its own substance (*Colin*). Under water, it continues to produce carbon dioxide and after a few days evil-smelling gases (*Thénard*). Finally, it is changed into a substance similar to old cheese (*Proust*). Its ability to bring about fermentation disappears with the completion of this putrefaction.

[*Editor's Note:* Material has been omitted at this point.]

Sugar thus contains neither finished carbon dioxide nor alcohol or ether, nor any one of the numerous other products that are formed by the action of foreign agents. Its behavior characterizes it as a complex organic atom, its decomposition into alcohol and carbon dioxide occurs as a consequence of a rearrangement of its constituents.

Hence in the course of the fermentation of sugar, as follows from the amounts of the products obtained, the components of the ferment do not participate at all; we deal with a cleavage of a complex atom, resembling in its result the change that heat brings about in organic atoms; there is a difference, however, in that it occurs in water and that the constituents of water combine with one of the products.

[*Editor's Note:* Material has been omitted at this point.]

In the preceding presentation, I have stressed the cases in which two substances are changed side by side without the constituents of one of the substances participating in the formation of products produced by the decomposition of the other.

[*Editor's Note:* Material has been omitted at this point.]

Certain scientists were misled to explain the ferment as living organic beings, as plants or as animals, that incorporate the components of sugar for their development and that then expel these as excrements in the form of alcohol and carbon dioxide; in this manner they explain sugar decomposition and the increase in the amount of the added ferment during beer fermentation.

This view is self-contradictory. . . .

[*Editor's Note:* Material has been omitted at this point. Here Liebig discusses what he calls the microscopic animals that develop on putrefying material. He proceeds as follows.]

Many scientists regard the chemical process of putrefaction as a mere consequence of the generation of these animals. . . . This view does not hold when one realizes that the animals die along with the disappearance of the putrefying substance and that after their death an agent must be present in order to bring about the destruction of the components of the muscles and organs of their organism to solid and gaseous new products. This cause is then indeed a chemical process after all.

Organic chemistry offers two contrasting general phenomena in the behavior of its compounds.

1. *Substances with new and changed properties are formed when the constituents of several atoms of a simpler compound are decomposed and combined to form an atom of a higher order.*

2. *The assembled atoms of a higher order dissociate into two or more less complex atoms of a lower order when the equilibrium in the attraction of their constituents is removed.*

This disturbance is brought about by

a. heat,
b. contact with a different substance,
c. the influence of a substance that is undergoing a change.

# 13

# ON THE GLUCOSIDIC FERMENTATION OF CANE SUGAR

**Pierre Eugène Marcellin Berthelot**

*This article was translated expressly for this Benchmark volume by H. Friedmann, the University of Chicago, from "Sur la fermentation glucosique du sucre de canne" in* Acad. Sci. (Paris) C. R. **50**:980–984 (1860).

1. One of the most remarkable changes that brewer's yeast produces in cane sugar is to convert it to invert sugar. Actually, Dubrunfaut's research taught us thirty years ago that cane sugar on treatment with yeast changes first of all to an uncrystallizable sugar; Persoz's research has shown that the rotation of this sugar has a sign opposite to that of the starting sugar. Hence the name *invert sugar.*

What is the exact nature [caractère] of this phenomenon of inversion? Is it due to a special action of yeast, necessitated by the circumstance that cane sugar cannot be fermented directly? Or, rather, is the inversion of cane sugar the result of some secondary, chemical, influence not directly dependent on the action of the ferment? All these points are still uncertain. In this connection, let us quote the words of Pasteur in his recent study on alcoholic fermentation:[1]

> All that has been written on this subject. . .lacks solid proof. As far as I am concerned, I think that the formation of "invert" sugar is related simply to the continuous formation of succinic acid, so that it is only an adventitious phenomenon. . . . In other words, I do not think that yeast globules have any special power to transform cane sugar into "invert" sugar. But since succinic acid is continuously being produced by alcoholic fermentation, its effect on sugar must be the same as that generally brought about by the action of acids.

Since I found myself in the position of having to explain and to sum up the principal phenomena of fermentations, I was led to resume the study of the preceding questions. These questions are not without interest; for it is necessary to know whether yeast exhibits several successive activities on cane sugar or whether it exhibits only a single activity; whether it represents several ferments that can bring about numerous results; finally, whether some of the effects that it brings about are the same as those developed by contact with dilute acids.

[1]See also Regnault, *Cours de Chimie,* Vol. IV: Article on Alcoholic Fermentation, 1860.

2. I have performed three types of experiments. In one of these I determined whether succinic acid, under conditions identical to those used during fermentation, really can invert cane sugar. In the second type I brought about alcoholic fermentation while maintaining the liquid alkaline; under these conditions all acidic influence was excluded. Since these experiments proved that the inverting action is in fact due to the beer yeast, I was led to isolate the actual ferment that produces the inversion of cane sugar, and I studied its activity by itself.

Herewith the summary of my observations:

3. *Comparison of the effect of brewer's yeast and of succinic acid on cane sugar.* I take 200 g of candied sugar and dissolve it in water to a total volume of 1000 cc. This liquid rotates light by +29.2° in a tube 200 mm long. I divide it into two equal parts. To one I add 0.8 g of succinic acid, which is more than the yield from the same amount of sugar in an ordinary alcoholic fermentation. To the other portion I add 10 g of well-pressed brewer's yeast.

After 16 hours at a temperature between 15 and 20° the solution that had been mixed with yeast is undergoing full fermentation. It strongly reduces alkaline copper tartrate; it is inverted and rotates light by -9° (to the left). On the other hand, the solution mixed with succinic acid reduces only to a negligible extent; it rotates by +28.9° (to the right).

There is no doubt about the difference between these two results. Hence the inversion that follows the action of yeast cannot be attributed to succinic acid.

Even more decisive results are obtained when the influence of succinic acid is completely suppressed.

4. *Inversion of cane sugar by brewer's yeast in an alkaline fluid.* I have taken 200 g of candied sugar and 20 g of sodium bicarbonate, and I dissolve these in cold water to a total volume of 1 l. The rotation of light is the same as above: +29.2°. To this fluid I add 20 g of well-pressed yeast, and I leave the whole set-up at a temperature between 15 and 20°. Alcoholic fermentation develops, although a little bit more slowly than in the preceding case.

After 16 hours the rotation has dropped to +9°, and the liquid strongly reduces alkaline copper tartrate.

At the end of 40 hours alcoholic fermentation continues while the liquid remains alkaline; the sugar is inverted, and the rotation is -7°. The next day it is -8°; the liquid continues to be alkaline. Half of it after distillation yields 7 g of alcohol. The rest continues to undergo alcoholic fermentation.

These facts prove that the inversion of cane sugar is due to an activity inherent in beer yeast and independent of the acidity of the fluids. Hence one would like to know whether the inverting action of yeast is due to yeast as a whole or rather due to some principle contained in the tissues of this plant [végétal]. Hence the new experiments.

5. *Glucosidic ferment.* I wished first of all to determine what effect the soluble parts of yeast had on sugar. I extracted these by macerating the yeast in the cold. I diluted the yeast, which had first been exposed to pressure [préalablement exprimée] in twice its weight of water, digested it for several hours, and then filtered it. The resulting liquid contained 1.5 percent of soluble matter. But upon contact with its own volume of a 1/5-concentrated solution of sugar containing 1/50 part of bicarbonate, it inverted the sugar in the same way as yeast itself, and without acidifying the liquid. A very rapid

way to show its activity is by means of alkaline copper tartrate.

Under these conditions the process stops with the inversion of the sugar, without its subjection to alcoholic fermentation and without the immediate development of organized beings [sans donner lieu au developpement immédiat d'êtres organisés].

Yeast extract hence contains a specific [particulier] ferment, soluble in water and able to convert cane sugar to invert sugar.

This ferment can be studied in more detail. It is enough to mix the extract of yeast obtained in the cold with an equal volume of alcohol. A precipitate of white flakes that collects at the bottom of the vessel can be seen. This is decanted and washed with alcohol; then the flakes are dried at room temperature. A yellowish and corneous mass is obtained having approximately one-fifth the weight of the soluble material contained in the extract. This mass is made up of a specific [particulier] nitrogenous principle comparable to diastase and to pancreatin, coagulable by heat and by nitric acid. Once isolated, it can be redissolved in water, reprecipitated by alcohol, and so on. But these repeated treatments slightly weaken its specific activity [activité spécifique]. In its original condition 1 part suffices to invert 50 to 100 parts of cane sugar.

Let us finally state that this ferment appears to be reproduced at the expense of yeast since it is secreted by it. In fact, no matter how often the yeast was washed on a filter or by decantation, no matter how much water it had been in contact with, as long as the yeast had not deteriorated [n'a pas été altérée], it sufficed to digest it subsequently with a small amount of water to be able to see that the glucosidic ferment appeared in this water. This explains why washed yeast, upon contact with a solution of sugar, does not delay its inversion.

6. The facts that have just been shown throw a new light on the nature of brewer's yeast and on the nature of the phenomena that it brings about. Actually, they prove that yeast does not constitute a unique and definite ferment.

It is known from the researches of Cagniard de Latour and above all from those of Pasteur that brewer's yeast consists of a mycodermic plant [végétal]. On the basis of new experiments that I have just reported, I think that this plant [végétal] acts on sugar not because of a physiological activity, but simply by means of the ferments that it has the property of secreting, in the same way as germinated barley secretes diastase, almonds secrete emulsin, the pancreas of an animal secretes pancreatin, and the stomach of the same animal secretes pepsin. Among the secreted ferments, those that are soluble can be isolated and purified up to a certain point if carefully defined principles are followed. I have just shown that the same is true for the glucosidic ferment, which is one of those that beer yeast contains. On the other hand, the insoluble ferments remain entangled in the organized tissues and cannot be separated from them.

In short, in the cases enumerated above that refer to soluble ferments, it is seen clearly that the living being is not the ferment; but it gives rise to it. Also once the soluble ferments are produced, they act independently of any further vital act; this activity shows no necessary correlation with any physiological phenomenon. I insist on these words in order not to leave any ambiguity about my way of regarding the action of soluble ferments. It is evident, besides, that each of these ferments can be formed preferentially, if not exclusively, by this or that given plant or animal: This organized being produces and multiplies the ferment corresponding to it to the same extent and in the same fashion by which it produces and multiplies all the other chemically defined immediate principles of which it is made. Hence the success of the very important experiments of Pasteur on the sowing [l'ensemencement] of ferments or rather, in my opinion, of organized beings that secrete the true ferments.

If a deeper study leads to an extension of the viewpoint that I propose and to its firm application to the insoluble ferments, then all fermentations will be found to lead back to one and the same general concept, and it will be possible to study them under the same heading as the activities due to contact with acids ["et elles pourront être définitivement assimilées aux actions de présence provoquées par le contact des acides. . ."] and chemical agents properly so called. [Berthelot uses almost the same words in his influential textbook of organic chemistry (Berthelot 1860), but he adds a few phrases and continues, significantly: "To banish life from all explanations relating to organic chemistry (relatives à la chimie organique), that is the aim of our studies. Only thus will we succeed in building a science that is complete and that can exist by itself, that is to say, one that can be used efficiently (c'est à dire telle qu'elle doit être pour concourir efficacement) to understand physiological changes and their artificial reproduction. . . ."]

# 14

# CONTRIBUTIONS TO THE THEORY OF DIGESTION, SECOND PART

**Ernst Wilhelm Brücke**

*These excerpts were translated expressly for this Benchmark volume by H. Friedmann, the University of Chicago, from pages 602–603, 606, 607, and 608 of "Beiträge zur Lehre von der Verdauung, zweite Abteilung" in* Kais. Akad. Wiss. Wien, Math.-Naturwiss. Cl. Sitzungsber.* **43**, Pt. 2, 601–623 (1861).

[*Editor's Note:* In the original, material precedes these excerpts.]

The separated glandular tissue of two pig stomachs was digested with dilute tribasic phosphoric acid at 38°C until it disintegrated. The digestive fluid thus obtained was saturated with lime water until blue litmus paper was just colored violet, the calcium phosphate was filtered off on a cloth and pressed dry in a screw press. By this procedure almost all pepsin is precipitated along with the calcium phosphate. If the latter is again dissolved in some hydrochloric acid, a liquid is obtained that even at very high dilutions still digests vigorously, while the liquid from which calcium phosphate has been removed by filtration shows only slight traces of digestive action upon acidification. However, when warmed with nitric acid and then saturated with ammonia, the latter liquid becomes colored an intense yellow, while the above vigorously digesting fluid gives much less color by the same treatment. This latter liquid, hence, associated with a disproportionately higher pepsin content, clearly contains less foreign substances than the product originally obtained by self-digestion of the gastric mucosa.

[*Editor's Note:* Material has been omitted at this point.]

Now I used a different procedure. I saturated a mixture consisting of 4 parts ethanol (using alcohol with 94 percent by volume strength) and 1 part of ether with cholesterol in the cold. Then I prepared a pepsin solution using phosphoric acid and lime water as described above. After a second precipitation with lime water, followed by pressing, I dissolved the calcium phosphate and the pepsin that adhered to it in dilute hydrochloric acid and filtered into a large bottle. Into this bottle I then placed a long funnel that reached to the bottom, and through this I poured the cholesterol solution in small portions. The cholesterol that separated accumulated on the surface in the form of a white slime. When this had become about 1 in thick, the funnel was removed and the bottle was closed and shaken continuously in order to bind as much pepsin as possible to the finely dispersed cholesterol. Then I filtered

*[*Editor's Note:* Reports of the Sessions of the Mathematical and Scientific Section of the Royal Academy of Sciences, Vienna.]

and washed, initially with water acidified with acetic acid, then with pure water. The washing was continued until the wash water gave neither a turbidity with silver solution nor an acidic reaction. Now I poured the moist cholesterol into a powder glass [Pulverglas] and covered it with ether that had previously been shaken with distilled water in order to remove any alcohol from it.

I used ether to dissolve the cholesterol, while the adhering water separated from it and formed a turbid layer at the bottom of the glass. The ether was decanted, replaced with fresh ether, again shaken, decanted, and so on. After this had been repeated several times, I allowed the vessel to stand open until the last thin ether layer, which could not be decanted without loss, had evaporated. Then I filtered. A small amount of a slimy substance remained on the filter paper. The perfectly neutral water-clear filtrate showed a vigorous activity upon acidification; it not only very rapidly dissolved a fibrin flake that had been thrown in it, but a single drop added to 5 cc of hydrochloric acid, of an acidity degree of 1 (1 g ClH/l) contributed so much digestive capacity that it dissolved an added fibrin flake in about an hour.

[*Editor's Note:* Material has been omitted at this point.]

Now this liquid did not show a series of reactions that several authors have described as belonging to pepsin solutions.

[*Editor's Note:* Material has been omitted at this point.]

One can raise the objection against these experiments that since it is not known how little pepsin suffices for activity it is also not known whether so little was perhaps present in the investigated liquid that it could not be detected by chemical reagents in spite of its activity.

[*Editor's Note:* Material has been omitted at this point.]

For the time being I have here discontinued the experiments concerning the chemical nature of pepsin. The thread that had been followed up to now had torn in my hand; in order to find a new thread, I would have had to prepare pure pepsin, and this I could not consider offhand since I had no hope of being able to crystallize pepsin or to convert it into a crystalline compound. I hence at first studied other points that can be investigated in more detail even without pure preparations of pepsin.

[*Editor's Note:* The remainder of this article has been omitted.]

# 15

# ON SPECIFICALLY ACTING SUBSTANCES OF NATURAL AND ARTIFICIAL PANCREATIC JUICE*

**Alexander Jakovlevich Danilevsky**
*Medical Practitioner from Kharkov*

*These excerpts were translated expressly for this Benchmark volume by H. Friedmann, the University of Chicago, from pages 279, 282, 285, 289–291, 295, 300–303, and 304–306 of "Ueber specifisch wirkende Körper des natürlichen und künstlichen pancreatischen Saftes" in* Virchow's Arch. Pathol. Anat. Physiol. Klin. Med. **25**: 279–307 (1862).

It is the purpose of this paper to show that both in natural and in artificial pancreatic juice, every specific physiological activity is associated with special substances and that some of these substances can, without losing their properties, be prepared in a more or less pure state.

[*Editor's Note:* The next part of this article deals with the history of investigations on the nature and activity of pancreatic juice, with a criticism of the term *pancreatin* as implying only one active substance, and with the difficulties of obtaining the natural pancreatic secretion in contrast to an "artificial" pancreatic juice or infusion. It has been omitted in this translation.]

From what has been stated it follows that three principal physiological reactions have been more or less unanimously ascribed to natural and to artificial pancreatic juice: (a) the conversion of starch to sugar (and further to lactic and butyric acid); (b) the emulsification and decomposition of neutral fats; and (c) the digestion of coagulated proteins.

It is peculiar that three so very different activities should be due to a single secretion; it is even more peculiar that all these activities should be ascribed to a single ferment, "pancreatin."

[*Editor's Note:* A section describing partial separation of protease and amylase using cholesterol treatment, based on Brücke's method of pepsin purification, is omitted here. A more successful method using collodion as a precipitant is described next.]

After repeated attempts, I was able to bring the method to the point where it yields a specific substance from artificial pancreatic

*My investigations have been carried out during the winter semester in the Chemical Laboratory of the Pathological Institute in Berlin. It is my duty to express publicly my warm thanks to Dr. Kühne for his friendliness shown to me by word and deed. [The rest of this footnote, apologizing for the incomplete nature of this investigation, has been omitted here.]

juice, which only dissolves fibrin and not at all starch and neutral fats.

[*Editor's Note:* Material containing preparative descriptions has been omitted here.]

After the physiological reactions of the artificial pancreatic juice have been tested, one can proceed to isolate the substance acting on fibrin. As stated, I use for this end the precipitation of the substance with collodion. Since success depends mainly on the method of work-up and on the suitable properties of the precipitate, I take the liberty of describing the technical part in some detail.

The juice is poured into a bottle to occupy 1/3 of the total volume. Then without stirring one pours into the liquid almost 1/4 of its volume of thick commercial collodion. The bottle is tightly stoppered and is now shaken with a strong and sudden movement. This procedure must be continued for some minutes. During this process the collodion is precipitated as a semitransparent viscous and sticky mass that on standing only gradually flows together toward the top. Shaking must be repeated a few more times. Then the mixture is poured into a wide measuring cylinder that is kept *continuously* stirred with a glass rod to allow most of the ether to escape. As the ether gradually escapes, the precipitate, which is at first voluminous and semitransparent, becomes more compact and white. Because of the steady stirring with a glass rod, the precipitate cannot change into a big lump, but rather is divided into small, more or less round white grains. When the precipitate has assumed this form, one can decrease the shaking, and allow the liquid to remain standing undisturbed; but preferably the liquid is immediately poured onto a patch of linen and allowed to run through. After the ether and alcohol have been removed, either by spontaneous evaporation, or preferably *in vacuo,* the filtrate must once again be treated with collodion as above and filtered through the same patch. The filtrate (I) is stored. The precipitate on the patch is repeatedly washed with 60-70 percent alcohol over another vessel, in order to remove the original liquid completely. This is an important point. The well-washed precipitate is squeezed, along with the patch, between two layers of blotting paper, distributed with a spatula, and exposed to the air in a clean area of the room.

After a few hours or by the next day, the precipitate is completely dry, and if the preceding procedure was carried out correctly, it is present as a light powder. It is better to use somewhat more collodion rather than too little, but there is a certain limit to both extremes. With much collodion it is difficult to obtain a good fine precipitate; too little collodion cannot take up enough of the required substance. It is best that the ratio of the quantity of fluid to that of the collodion be 3:1.

Thus we now have a solid precipitate and a liquid. On the basis of my earlier experiments, I hoped to be able to obtain the substance that acts on fibrin from the former, and the substance that acts on starch from the latter. Let us first turn to the precipitate.

Ether containing a small amount of absolute alcohol is added to the dry mass in a tall narrow glass cylinder that is then stoppered.

The precipitate eventually dissolves after it is shaken and allowed to swell. Thereupon one obtains a very turbid solution of collodion in which the required substance, which is insoluble in alcohol and ether, is suspended. The liquid is allowed to stand undisturbed for one or two days. Under these circumstances a yellow precipitate is formed at the bottom of the vessel. The supernatant liquid, which is still quite turbid, is poured off, diluted with a new portion of ether, distributed into two similar cylinders and again allowed to stand undisturbed for a few days. The yellowish sediment is also shaken with fresh ether and allowed to settle; the supernatant ether is again removed; the sediment is washed out once more with fresh ether, and so on. Each sediment must be treated in this manner until all collodion is removed from it. The last, still turbid, collodion can be removed by filtration through Swedish paper; the required substance remains on the filter. The ether from the last washing of the sediment is removed *in vacuo*.

The residual yellow mass consists of the required substance and some protein. The former is readily dissolved in cold distilled water; the latter has been so firmly coagulated by the lengthy action of alcohol and ether that in most cases none of it dissolves in water. After the requisite extraction of the precipitate with pure water and filtration, a yellowish solution is obtained that contains the specific substance of the artificial pancreatic juice that dissolves fibrin. The solution is neutral, and at a fairly high concentration, it gives, one can say, no xanthoproteic reaction with nitric acid. It is precipitated by dilute acetic or hydrochloric acid, but not by alkalis.

[*Editor's Note:* Material on the enzymatic properties of this preparation follows here. It has been omitted in this translation.]

Hence by the use of collodion it has been possible to prepare a chemically more or less pure substance from artificial pancreatic juice that most decidedly manifests *one* specific physiological reaction, that is, under suitable conditions it digests *only* fibrin.

[*Editor's Note:* Sections on the variation in chemical purity from one preparation to the next, on the mode of assay with fibrin fibers, on the incubation time for assay, on ignorance concerning the products of the reaction compared to those obtained with pepsin, and on the fact that a preparation from pancreatic secretion obtained by means of a fistula from a big dog showed identical behavior, have been omitted here.]

I now proceed to the investigation of the second specific substance of natural and of artificial pancreatic juice, which vigorously converts starch to sugar and which is not precipitated by collodion.*

After addition of collodion to the artificial pancreatic juice and removal of the resulting precipitate by filtration, a yellow, clear liquid (I) was left that showed an alkaline reaction and contained a part of the ether and alcohol derived from the collodion used. In spite of the presence of these two liquids, the filtrate (I) very vigorously converts starch to sugar. In order to separate the specific substance that acts on starch, I propose the following procedure,

*The circumstance that one specific substance is precipitated by collodion, the other not, already shows that the two substances differ not only physiologically but also physically and perhaps chemically. How the collodion takes up the substance is completely unclear. [The rest of this footnote, which deals with analogous adsorptive properties of pepsin and of salivary amylase, has been omitted here.]

based on several experiments.

The filtrate (I) is evaporated with the help of an air pump. Soon after the ether and alcohol have been removed, a fine turbidity is formed, as well as a precipitate of collodion that had been kept in solution by the ether. Then one filters. The filtrate shows, even before the last filtration, a wonderful phenomenon: It contains two proteins in solution that differ from one another since one of these starts to coagulate already at 37°C and appears completely in the form of flakes at 44°C, while the other coagulates like a normal protein, at 72°C only. In order to determine the significance of the first substance, I tested the filtrate (I) for its physiological reactions before and after removal of this substance, and I convinced myself that this substance does not contribute anything to these. Hence by warming the liquid to 43-44°C it is possible to remove a large quantity of protein. . . .This second specific substance is active in neutral, alkaline, and acidic solutions; in the latter, however, the activity is considerably decreased.

[*Editor's Note:* Material on preparative procedures and on repetition of the purification with pancreatic secretion has been omitted at this point.]

The third specific reaction that has been assigned to pancreatic juice remains to be discussed. This concerns the decomposition of neutral fats by the juice. I have no wide experience concerning this property of the natural juice, but I can note the following about this property of the artificial juice.

This property of the artificial juice depends entirely on the way it is prepared. When obtained by the above method that I use to precipitate with collodion, then, as already noted above, this juice has completely lost this property. Nevertheless, it is excellent for converting starch to sugar, and it very rapidly digests fibrin. Now the question is, on what does the property of the juice to decompose fats depend? Is it perhaps due to the presence of a third specific substance? I believe that I can answer this question in the affirmative on the basis of the following observation.

When the pancreas is removed from an animal (surrounded by very little fat, hence from a dog or rabbit) in the fifth to seventh hour of digestion, cut into small pieces as rapidly as possible, digested for a short time with some water and filtered through paper, then a clear liquid is obtained showing an alkaline or at least a neutral reaction and possessing all three specific physiological reactions, that is, it changes starch into sugar, it decomposes neutral fats, and it digests fibrin. As soon, however, as the pancreas is surrounded and infiltrated by much fat (hence derived from pig or ox) or when the preparation of the artificial juice takes a long time, then the infusion shows a more or less strongly acidic reaction due to liberated fatty acids. If one part of such a milky-white infusion is allowed to stand undisturbed at room temperature, then one observes very readily how the acidic reaction increases rather rapidly. Hence the fats are still decomposed by the components of the gland. If, however, already at the beginning of the fat decomposition--that is, when the acid reaction is not yet very strong--one supersaturates the other part of the milky-white infusion with magnesium hydroxide and follows by thorough stirring, standing, and filtration, then one obtains a pure, clear, alkaline filtrate that converts starch to sugar, dissolves fibrin but cannot decompose fats. Hence, clearly this decomposition of neutral fats cannot depend on substances that act on starch and on fibrin, but rather a third specific substance must occur in the artificial juice that mediates the decomposition of the neutral fats and which, in this procedure, magnesium hydroxide either destroys or precipitates. This unfortunately is all that I can now cite for the existence of a third specific substance.

[*Editor's Note:* The summary has been omitted here.]

# 16

# THE CHEMICAL THEORY OF THE ACTIVITY OF FERMENTS, AND THE CHEMISTRY OF RESPIRATION. ANSWER TO STATEMENTS OF MR. HOPPE-SEYLER[1]

**Moritz Traube**

*This excerpt was translated expressly for this Benchmark volume by H. Friedmann, the University of Chicago, from pages 1984–1985 and 1992 of "Die chemische Theorie der Fermentwirkungen und der Chemismus der Respiration. Antwort auf die Aeusserungen des Hrn. Hoppe-Seyler" in* Dtsch. Chem. Ges. Ber. **10**:1984–1992 (1877).

The main content of my theory of the activity of ferments published in 1858 is the following:

1. The ferments are not, as assumed by *Liebig*, substances in the process of decomposition that can convey their chemical movement to otherwise passive substances, but rather they are chemical substances similar to proteins that, although they can up to now not be prepared pure, undoubtedly possess, just as all other substances, a definite chemical composition, and that bring about changes in other substances by the expression of certain chemical affinities.[2]

2. The hypothesis of *Schwann* (later adopted by *Pasteur*), according to which fermentations are to be regarded as effects of the vital forces of lower organisms, is unsatisfactory.[3]

Chemistry may in fact be able to explain physiological processes, but physiology cannot explain chemical processes.

The reverse of *Schwann's* hypothesis is correct:

> Ferments are causes of the most important vital-chemical processes and not only, indeed, in the lower but in the higher organisms as well.

In this sense a correct chemical theory of the activity of ferments constitutes an essential foundation for biological-chemical research in general.

[*Editor's Note:* The main parts of this fascinating paper that deal with oxidative processes and that still make rewarding reading have been omitted for lack of space, and only the following statement repeated from above, near the end of the paper, is translated here.]

. . .It will then be found that it is impossible to reach an understanding of the chemical process of muscular activity without the aid of the chemical theory of the activity of ferments. This theory has, in fact, already been verified here in the form that it should be: *as an essential foundation for biological-chemical research in general.*

[*Editor's Note:* The final brief paragraph of this paper, which attacks

Hoppe-Seyler, has been omitted here.]

1. These Berichte, X, 693.

2. M. *Traube*, Theory of Ferment Actions. Berlin, published by F. Dümmler, pp. 15-21 and 57-61.

3. *Pasteur's* excellent investigations have established that the ferments of putrefaction are not, as I myself at the time assumed, products of a spontaneous decomposition of the proteins, but are always produced in microscopic organisms. The mode of action of the ferments themselves, however, does not for that reason cease to be a purely chemical one. This is proven by the discovery of *Musculus* who extracts the ferment of urea fermentation from the organisms that produce it and hence is the first to have established that the operation of this ferment is tied to no living activity. Likewise my more recent experiments (these Berichte, VIII, p. 1399, and X, p. 512) have directly contradicted *Pasteur's* assertion that the alcoholic fermentation of sugar is tied to the respiration of yeast.

# 17

# ON THE DIGESTION OF PROTEINS BY PANCREATIC JUICE. I.

**Willy Kühne**

*These excerpts were translated expressly for this Benchmark volume by H. Friedmann, the University of Chicago, from pages 130–131, 154, and 172 of "Ueber die Verdauung der Eiweissstoffe durch den Pankreassaft. I"* in Virchow's Arch. Pathol. Anat. Physiol. Klin. Med. **39**: 130–174 (1867).

I have long known of the digestive action of pancreatic juice on proteins both from seeing *Corvisart's* experiments myself ["durch eigene Anschauung der Versuche *Corvisart's*." Kühne must have seen Corvisart at work since he spent the years 1858-1860 in Paris, and it was just during this time that Corvisart's work on this topic was published (see note 17 in the Editor's Comments to Part II).] and in the past several years from many experiments that I have occasionally performed myself. It hence seemed to me desirable to investigate this important part of the field of digestion in more detail. Although much is known about gastric digestion, very little is known about the further fate of proteins in the small intestine so that each new systematic investigation in this area must almost certainly lead to the discovery of physiologically important facts.

Unlike the digestion of proteins by gastric juice, that by pancreatic juice is not generally recognized. It was at first completely ignored by all those researching the processes of digestion, and it was only incidentally suspected by *Cl. Bernard*; after being emphatically asserted by *Corvisart*, it has become again so much forgotten in France itself that, for example, *Ch. Robin* in his very detailed, just published work (*Leçons sur les humeurs normales et morbides du corps de l'homme*. Paris: Bailliere et fils, 1867) did not even have a single syllable about *Corvisart's* teachings.

[*Editor's Note:* Material has been omitted at this point.]

It cannot be denied that the decomposition of fibrin by pancreatic juice is very similar to the already known artificial ways in which proteins can be cleaved; I am thinking of their decomposition by melting potash, by boiling sulfuric acid, and by so-called putrefaction. Hence the time has come and, on the basis of our new experiments, is in fact overripe, to get to the root of the question of the existence of an actual pancreatic digestion that decides the fate of the proteins in the small intestine, or whether we are merely led astray by quite incidental putrefactive phenomena.

[*Editor's Note:* Extensive material has been omitted at this point, including an experiment investigating the digestion of fibrin by ligated intestine. The paper ends as follows.]

It has often been doubted whether decompositions of the constituents of food brought about by apparently violent chemical means might find their likeness in the gradual course of animal metabolism. Now it is a fact that protein decomposes in the small intestine as though it would be boiled for hours with sulfuric acid.

# Part III

# THE WORD ENZYME: THE END OF ONE ERA, THE BEGINNING OF ANOTHER

# Editor's Comments on Papers 18 Through 22



> It's in words that the magic is—Abracadabra, Open Sesame, and the rest—but the magic words in one story aren't magical in the next. The real magic is to understand which words work, and when, and for what; the trick is to learn the trick.
>
> [*Chimera*, John Barth, 1972]

> Convictions are more dangerous enemies of truth than lies.
>
> [*Human, All-too-Human*, Friedrich Wilhelm Nietzsche, 1878]

This section contains papers that deal with the coining of the term *enzyme* by Kühne; with Pasteur's insistence that fermentative and living processes cannot be separated; and with Buchner's discovery of cell-free fermentation, a discovery that placed a new perspective on the term *enzyme* and that above all caused many of the ideas associated with Pasteur's name to be very thoroughly modified. In addition, a paper by Bertrand has been included. This paper, which dates from the same year as Buchner's, is on laccase, an oxidase, that is a completely new type of enzyme.

Kühne used the word *enzyme* for the first time at the 4 February 1876 session of the Scientific and Medical Society (Naturhistorisch-Medicinischer Verein) of Heidelberg.[1] The first page of Kühne's presentation and a translation are given here (Paper 18).[2] In this t 'k Kühne introduced the word *trypsin* as well, and it is most probable that he here also used the word *zymogen* for the first time.

The reasons for the introduction of the word *enzyme* are spelled out in much more detail in a paper from the 1877 issue of the house journal of Kühne's institute, the pertinent parts of which have been translated here (Paper 19). Kühne points out clearly that although he derived the word *enzyme* from *in yeast* (and he apologizes for not finding a more pleasant-sounding word),[3] it should be extended as a unifying term beyond Berthelot's yeast invertin to the unorganized (*ungeformte*) ferments of higher organisms as well. Thus this new word should help to stress the similarities between multicellular and unicellular organisms. But Kühne had additional reasons for introducing a new term. He states quite clearly that the unorganized ferments studied up to that time without exception bring about only hydrolytic and not reductive or oxidative changes. It is within the context of this experimental limitation that Kühne finds just one term, *ferment*, to have outlived its usefulness, since different types of processes and the associated different extractabilities must be due to quite different types of causative agents. Furthermore, the use of this particular term, *ferment*, happens to have been beset with additional confusions, since it was applied both to processes and to organisms. Kühne hence wished to limit the new term, *enzyme*, to agents that are both (a) extractable and (b) hydrolytic, a correlation that applied for a further twenty years until Eduard Buchner's work in 1897.

With the introduction of the word *enzyme*, Kühne sought to remove the multifarious types of vagueness that he regarded to be inherent in the use of the word *ferment*. He did so by trying to codify a distinction between the so-called unorganized (*ungeformte*) or extractable or hydrolytic ferments, and the so-called organized (*geformte*) or unextractable and as it seemed qualitatively different ferments that were thought to be associated inextricably with intracellular chemical reactions. It is rather ironic to realize that the older term *ferment*, which did not imply such a distinction, is hence much closer than the later word *enzyme* to subsequent unifying ideas of biocatalytic agents as substances, chemically and biologically related, that happen to catalyze different types of reactions. In the words of R. E. Kohler, Jr.: "When the term 'enzyme' lost the precision of Kühne's definition, it acquired the scope of Liebig's 'ferment'" (Kohler 1973, p. 193).

The word *enzyme* was for a long time not accepted universally. Felix Hoppe-Seyler, dean of German physiological chemists, whose views Kühne had cited in his paper (Paper 19) as one reason for the introduction of the new term, reacted strongly and swiftly. In a paper published shortly after Kühne's he reaffirmed his position, similar to that of Liebig, Traube and Berthelot before him, that "fermentations are chemical processes that must have chemical causes." It is interesting to realize that Hoppe-Seyler's opposition to the new term was in fact based on a view of living processes which eventually became commonplace following Buchner's demonstration of cell-free fermentation some twenty years later (Paper 21). Hoppe-Seyler attacked Kühne in the sarcastic tone that we find so commonly in the nineteenth century: "Recently Kühne found it necessary to oppose my distinction [between ferment as chemical substance and ferment as organism which produces that chemical substance], but since he gives absolutely no reason worth noting in favor of his position I do not consider it necessary to say anything in reply. The new word, enzyme, can be added to the large number of new names that Kühne has invented, all of which denote substances that are still completely unknown" (Hoppe-Seyler 1878, editor's translation). As we will see from the titles to the papers of Part IV, some authors around the turn of the century still preferred the word ferment to the word enzyme. This preference actually continued much later. Carl Oppenheimer's exhaustive treatise was called *Die Fermente und ihre Wirkungen* (The ferments and their activities) through many editions until the last supplement to be published (1939). Harden and Young, in their classic studies on alcoholic fermentation, referred to the alcoholic ferment of yeast juice in titles to papers between 1904 and 1911 (see Harden 1914, pp. 144–145). Warburg resolutely held on to the term "Ferment" (1938, 1946), and in a classic book published in 1949, he still referred to the glycolytic enzymes as "Gährungsfermente" (awkardly translatable as "Fermentative Ferments").

No problem has been associated with the gender of *enzyme* in English or of *das Enzym* in German. In the French language, however, where every noun must be either masculine or feminine, there were difficulties. Here the word *enzyme* had gradually replaced the word *diastase* (see Editor's Comments to Part II).

*L'enzyme* was born feminine, but drifted towards the masculine. Plantefol (1968) in a most learned and lengthy paper argued for a return to the feminine gender, and in fact the Académie Française ruled as recently as 5 February 1970 in favor of the feminine (see Fruton 1972, p. 74).

In the introduction to a *1976 Supplement of FEBS Letters* (Vol. 62, 4 February 1976) commemorating the centenary of the first use of the word *enzyme* by Kühne, H. Guttfreund pointed out that the inventor[4] of this word was given the name Willy, but that he changed it to Wilhelm Friedrich. In case the reader should draw any hasty conclusions, it must be mentioned that Kühne (1837–1900) in the words of his long-time collaborator R. H. Chittenden (1930) "was as large and broad mentally as he was physically, kindhearted to a high degree, an ideal teacher..."[5] Kühne received excellent training. At the University of Göttingen, where he graduated in 1856, he learned chemistry from Friedrich Wöhler and anatomy from Jacob Henle. He later studied under some of the leading physiologists of his time: Emil H. Du Bois-Reymond (Berlin), Claude Bernard (Paris), E. W. von Brücke and Karl F. W. Ludwig (Vienna). After working with Rudolf Virchow in Berlin, he spent a short time as professor of physiology in Amsterdam. In this capacity he moved to Heidelberg in 1871 where he succeeded the great Hermann von Helmholtz. He stayed in Heidelberg for the rest of his life. His apartment was on the top floor of the physiological institute. From 1883 until his death he edited the prestigious *Zeitschrift für Biologie* with Carl Voit.

Kühne performed research work on the physiology of muscle (he discovered and named *myosin*, a name first used in his textbook of physiological chemistry of 1868), on nerve conduction, on vision (fundamental work on rhodopsin, he introduced the term *visual purple*, see Hubbard 1976), and on the chemistry and enzymology of digestion. Much of this latter work was done with Chittenden between 1883 and 1890. This collaboration is of interest for a number of reasons. In some of this work the rather new technique of protein fractionation with ammonium sulfate was extensively used for the first time.[6] Although the fractionations obtained with this and with other salts "yielded little fruitful knowledge" (Fruton 1972, p. 114) it was in the course of this work that Kühne introduced the use of what he called the "Schlauchdialysor" or dialysis tubing (Kühne and Chittenden 1883, p. 165; see Hofmeister 1900, p. 3879). It is remarkable that Kühne and Chittenden did not collaborate in the same laboratory. Chittenden, who had twice been at Kühne's laboratory in Heidelberg, did his part of the work at New Haven where he was in charge at Yale University of the first physiological chemistry laboratory in the United States. In the words of Chittenden (1930, p. 38): "The experimental work done by Kühne and Chittenden during the eight years was carried on in the two laboratories at Heidelberg and New Haven, all discussions relating to methods and the significance of the findings being by correspondence." The resulting papers were published in Germany, and some were likewise published in English in the United States (see Chittenden 1930, p. 39; Leicester 1974, p. 262). Collaboration under the conditions of trans-oceanic mail service almost a hundred years ago points to a somewhat more leisurely pace of research than one would be comfortable with nowadays. "Such collaboration between an American and a European scientist was almost unique at that period" (Vickery 1971, p. 257).

The editor has felt it appropriate to give some biographical background on some of the early workers in the field of enzymology. This is not necessary in the case of Louis Pasteur (1822–1895) who "belongs in the history of science to a small group of great men" (Partington 1964, p. 749). He combined chemical and biological insights with rigorously established experimental facts, enormous courage (as in the famous first inoculations against rabies), versatility, and a skillful and ruthless polemical ardor. It is in the latter manifestation that we have already encountered Pasteur in the note on Berthelot (Part II); but it seemed highly rewarding to show some more excerpts from one of his writings in the present work (Paper 20). A few pages are given from the 1879 English translation of Pasteur's famous and influential book, *Studies on Fermentation, The Diseases of Beer, Their Causes and the Means of Preventing Them*. The topic selected here again concerns Pasteur's insistence that fermentation is necessarily and inextricably linked to the life of the yeast cell. He attacks Liebig's ideas with ruthless diligence so that the reader would in fact never guess that not only this translation but even the original (1876) was published after Liebig's death in 1873. Pasteur's polemics were strictly professional and not personal since, for example, when Berthelot wrote a very friendly letter to Pasteur in 1879 on the occasion of the marriage of Pasteur's daughter, Pasteur in turn replied in the most cordial terms: "The scientific discussions in which I have engaged have never suggested the least bitterness toward my adversaries..." (see Velluz 1964, pp. 128–129; "Chemicus" 1974).

Eduard Buchner's paper (1897) (Paper 21) on cell-free fermentation by an extract prepared from yeast cells is a watershed in the history of enzymology and of biology in general. This discovery eventually helped to shatter the then still prevalent ideas associated with terms such as *vitalism, protoplasm, organized* and *unorganized ferments* and with the names of Liebig and particularly of Pasteur (who had died two years before). Indeed Buchner himself, in a superb example of the conflict between the force of tradition and the insistency of experimental evidence, stated: "We all grew up in the atmosphere of Pasteur's views...I hence understandably was very skeptical when I...obtained experimental facts that appeared to indicate cell-free fermentation. Publication hence followed only after...experiments repeated two months later led to exactly the same results as the first" (Buchner 1898, p. 568, editor's translation). Buchner's results were attacked from various sides, but within a few years they were almost universally accepted (see Kohler 1972; Fruton 1972, p. 86). It is most interesting that the first public acceptance of Buchner's work came already in 1897 from none other than Pierre Émile Duclaux, Pasteur's first biographer (see endnote 13, Part II), his immediate successor as director of the Pasteur Institute in Paris, and founder as well as editor of the influential *Annales de l'Institut Pasteur*. One would have expected the new findings to be opposed by a disciple of Pasteur, since they could be readily interpreted to conflict with Pasteur's views. However, Duclaux's respect for experimental observations, similar to Pasteur's, prevailed, but he did not feel that Buchner's findings contradicted Pasteur's position. He wrote in the *Annales* in one of several 1897 papers championing Buchner's work (see Kohler 1972, p. 338): "Some scientists fully accept this discovery and even derive extreme conclusions from it by pretending that it overturns Pasteur's doctrine on fermentations. Pasteur's doctrine will be overturned the day that alcoholic fermentation will be achieved purely chemically and without any vital activity. But as long as yeast is needed to produce alcoholic diastase Pasteur's theory can express what the master himself would have said: Here is yet another vital activity [action vitale] which is manifested by a chemical mechanism" (Duclaux 1897, p. 348, editor's translation). This position resembles that taken already in 1858 by Moritz Traube (Paper 16). A year later another former close Pasteur associate, Émile Roux, made a statement similar to Duclaux's, but the emphasis is rather more chemical: "Certainly, the decomposition of sugar by alcoholase is a purely chemical reaction, but the formation of the enzyme is an act associated with life, and since it is not yet possible to make alcoholase without a living cell it follows that alcoholic fermentation remains correlated with the life of the yeast" (Roux 1898, p. 839, editor's translation). Buchner's own position vis-à-vis Pasteur's views was very modest: "The famous Frenchman's theory consists of a physiological part: *fermentation is life without oxygen*, that is, fermentative organisms obtain their reservoir of energy by the process of fermentation while the other living beings obtain it by means of respiration; and the theory consists of a fermentative-chemical half: *no fermentation without organisms*. The first statement is not at all changed by the discovery of zymase; the second statement requires only one modification: no fermentation without zymase, which is formed in organisms" (Buchner and Rapp 1898, p. 211, editor's translation). Perhaps Bernal (1954, ch. 9, sec. 5) put it most succinctly: "In the end both von Liebig and Pasteur were right. Fermentation is brought about by a ferment, but that ferment can only be elaborated by a living organism." However, it is important to stress that the views of Duclaux and of Roux were not at all shared by all scientists. Many still argued that yeast extract contained "living protein" or "bits of protoplasm" (see Fruton 1972, p. 86). Kohler makes the important point that "Buchner's 'proof' of the chemical view did not seem so unambiguous at first. Biologists' reactions to zymase correlate very closely with their previous disciplinary commitments" (Kohler 1975, p. 295). "Initially, at least, zymase was less a determinant of opinion than a touchstone of pre-existing opinion...The primary effect of the zymase debate...was on those who were already inclined to the new view" (Kohler 1972, pp. 351–352). We have here an example of the futility of argument over the dictates of opposing prejudice that is forcibly reminiscent of Duclaux's comment on the Berthelot-Pasteur dispute on the same topic, long before the new evidence was available (see Editor's Comments on Paper 13). In the long run time, not argument, prevailed over fashion. "Conceptual change came about not by wholesale conversions of individuals but by a process akin to natural selection, whereby the composite character of a population changes" (Kohler 1975, p. 295). The zymase debate occupied Buchner until his last papers. As an example of its persistence in much more recent times Richard Willstätter, whom we will meet again as an influential gadfly (mainly in Part V), maintained in his very last paper (Willstätter and Rohdewald 1940): "It must be concluded that Buchner's press juice and macerated yeast react with sugar in a fashion which differs from that of living yeast." He elaborated on this in his celebrated autobiography, published posthumously: "Some fermentation potential can be isolated [from yeast], but I consider it different from the fermentation effect of the living yeast cell" (Willstätter 1949, p. 63, English translation 1965, p. 66). Willstätter's views were by no means unique. Thus in 1940, again, F. F. Nord, who was to be the distinguished editor from 1941 to 1971 of the annual *Advances in Enzymology* stated in a detailed review on the mechanism of alcoholic fermentation: "It is not conclusive if, from the enzymatic behavior of structurally destroyed systems, which can be more or less complete, forceful conclusions as to the qualitative actions of the parent systems within the living cell are drawn" (Nord 1940).[7]

Buchner's paper (Paper 21) is very clearly written and easy to follow, but no reasons are given in it for the preparation of yeast extract, for the development of the method used, and for the addition of sucrose. Eduard Buchner (1860–1917), who had earlier and repeatedly attempted to open yeast cells, decidedly did not set out to prepare yeast extract to see whether it would ferment sucrose. In the present context the impetus for the preparation of intracellular protein extracts from microorganisms came from the immunologist Hans Buchner (1850–1902), Eduard's brother. Hans Buchner's interest was derived from an immunochemical theory involving bactericidal "alexines." It was stimulated by Robert Koch's preparation, in 1890, of tuberculin, by work on immunization against anthrax (C. E. Chamberland and Émile Roux) and against tetanus and diphtheria (Emil von Behring and Shibasaburo Kitasato), by von Behring's ideas on "antitoxins," and by Carl v. Nägeli's earlier theory of inflammation. In 1890 Hans Buchner had in fact extracted proteins from pneumococci with dilute alkali. Sand-grinding was being used in many laboratories at about this time to prepare extracts for various purposes and with varying success from yeast, fungi, and bacteria. Robert Kohler (1971, p. 56), in his superb review "The Background to Eduard Buchner's Discovery of Cell-Free Fermentation" uses the phrase "this spate of sand-grinding." Hans Buchner, the immunology researcher in Munich, had often consulted his chemist brother Eduard about the preparation of yeast extracts. By 1896 Eduard Buchner was at the University of Tübingen. Ironically, the successful method (sand-grinding, followed by admixture of kieselguhr and application of hydraulic pressure[8]) described by Eduard Buchner in his historical paper was worked out in Hans Buchner's laboratory by a certain Dr. Martin Hahn (1865–1934) who found, in addition, that these extracts coagulated on standing and that added antiseptics did not prevent this coagulation. Ironically, again, Martin Hahn and Hans Buchner discussed these results and decided to try other preservatives, among them 40 percent sucrose. At this very time (end of 1896 summer), Martin Hahn left for his vacation. Eduard Buchner in turn arrived in Munich a few days later from Tübingen for *his* autumn vacation, which he intended to spend with work on yeast press juice. Such a juice, with added 40 percent sucrose, was already standing in the laboratory when he arrived! Eduard, who was familiar with the phenomenon of fermentation, noticed a steady stream of bubbles generated in this liquid. Hans and Eduard Buchner studied this phenomenon together, and after Eduard's return to Tübingen, further experiments were carried out in Hans Buchner's laboratory by Rudolf Rapp. Hans Buchner reported on this work at a talk in Munich in March 1897. A few weeks later Eduard Buchner's first paper on the subject (Paper 21) appeared. The activity was ascribed to a single substance to which the name *zymase*[9] was given.

It is most ironic that while Pasteur, Berthelot, and several others (see Roux 1898, pp. 836–837; Buchner 1907, p. 107; Harden 1911, pp. 14–15; Kohler 1971, pp. 37–39; Fruton 1972, p. 62) deliberately looked for cell-free alcoholic fermentation with yeast extracts many years before Buchner and failed, Eduard Buchner literally stumbled onto it. Buchner stated in an elegant lecture (Buchner 1898) that in October 1896 the experimental facts "fell into my hands" (fielen mir in die Hände). It turns out that there was an even greater element of luck associated with Buchner's discovery than we had mentioned earlier. In his second publication on the subject of alcoholic fermentation without yeast cells (Buchner 1897), not long after his preliminary note (Paper 21), Buchner already pointed out that not all yeasts yielded active extracts. He had obtained good fermentation with press juice from compressed Munich brewers' bottom yeast (untergährige Presshefe), an inexpensive brewery waste product. However, a more costly compressed yeast obtained elsewhere and preferred by bakers (Getreidepresshefe) yielded an essentially inactive press juice.[10] It was soon found that another brewers' yeast was also much less suitable than the first, but even the first yeast gave less active extracts if it had been stored in the cold for as little as three days (Buchner and Rapp 1897). (As mentioned in the Editor's Comments on Paper 13, Arthur Harden, about fifteen years after Buchner, obtained inactive extracts with top yeast from a Paris brewery in experiments designed to explain Pasteur's negative results.) Nature regards the researcher's path to success or to failure with indifference. In this way, again, the development of science is constrained by an elemental and ruthless pragmatism. From the viewpoint of the progress of science the question as to who is the greater scientist, the visionary who fails because of an unsuspected contingency, or the accidental discoverer, is unimportant, since it is the discovery that matters and not the process by which it was obtained. Adolf von Baeyer commented not very charitably upon his former student's discovery of zymase: "This will bring him fame, even though he has no chemical talent" (quoted by Willstätter 1949, p. 63, see Fruton 1972, p. 86). Buchner was lucky, and modern biochemistry was born. We will see an example of a near success, which would have had a similarly profound effect on scientific developments, in the circumstance, recognized by J. H. Northrop many years later (see the Editor's Comments on Papers 37 through 40), that C. A. Pekelharing had almost stumbled upon crystalline pepsin, again near the end of the nineteenth century. In this case luck was thwarted, and some thirty years had to pass before enzyme crystallizations began to be accomplished (Part V).

Hans Buchner's immunochemical theory, which provided the immediate impetus for the preparation of yeast extract in his laboratory, turned out to be a "red herring." However, "it led to the discovery of how the black box of the cell could be opened up. The Buchner-Hahn method of getting press juice revolutionized the study of intracellular enzymes. Finally, Buchner's theory. . .was responsibility for the discovery of zymase, the most important precedent for the belief that all vital functions of the cells were due to the activity of enzymes. This new belief was soon to become the keystone of the new science of biochemistry" (Kohler 1971, p. 61).

A careful reading of Eduard Buchner's paper shows that it was strongly influenced by the protoplasmic theory of vital functions. Thus zymase was considered to be a special kind of enzyme since, contrary to yeast invertase, it was not precipitated in an active form by ethanol. Buchner felt (Paper 21, p. 124), however, that zymase acted outside the cell, like a proper enzyme (in Kuhne's sense).[11]

A classical book on zymase was written by the Buchner brothers and M. Hahn (1903), and papers on cell-free fermentation were published by Buchner and various co-workers until his death. The impact of Buchner's work can be gauged from the fact that he was awarded the 1907 Nobel Prize in chemistry "for his biochemical researches and his discovery of cell-free fermentation." Hans Buchner had died of cancer in 1902.

Eduard Buchner was "a forthright, good-natured and outgoing Bavarian" (Kohler 1971, p. 53). A devoted mountaineer, he climbed a few hundred mountains. An example of his joie de vivre is provided by the wonderful story that he all but spent the night before his wedding (1900) in the Tübingen police station for singing a loud, comic song when riding home through the quiet Tu-

after an initial period as an officer at the front in World War I and recall to teaching at the University of Würzburg, he volunteered at age 57 for a second round of active duty. He was wounded in action in Rumania and died two days later. His obituary notice (Harries 1917) carries a photograph of Captain Buchner, the Nobel Laureate and usherer of the new science of biochemistry, in military uniform with a group of his officers.

Toward the end of the nineteenth century the genie is out of the bottle: The number of papers on enzymes increases exponentially with time. An interesting yardstick is provided by the growth of Carl Oppenheimer's monumental treatise *Die Fermente und ihre Wirkungen* (The ferments and their activities). The first edition (1900) has 309 text pages and no less than 1280 references; the second edition (already in 1903!) has swollen to 375 text pages and 1694 references. The fifth edition consists of four huge volumes; of these the first two (1925–1926), with 1871 much larger pages than the 1900 and 1903 editions and 95 pages of authors' names, carry the earlier subject areas, while two later volumes (1929) of 1930 pages deal with methods and technology. A 1939 supplement has 1707 pages plus 128 pages of bibliography. The scientists' fascination with enzymes or, alternatively, the enzymes' superb capacity to generate scientific papers, is brought home even more strongly when it is recognized that the considerable early literature, say, to 1903, deals with a very limited number of enzymes. As Malcolm Dixon (1971, p. 16) has pointed out, even by 1920 only about a dozen enzymes were known,[13] and until about that time just one enzyme, $\beta$-D-fructofuranosidase (or invertase, invertin, saccharase, $\beta$-*h*-fructosidase) played an outstanding role (see also Myrbäck 1960 and endnote 1 of the Editor's Comments in Part IV).

We end this section with one of several papers by Bertrand (Paper 22) on laccase, a far less glamorous enzyme than Buchner's "zymase," and one that has received rather less attention than invertase, but one that is of great historical significance: Bertrand realized that this recently discovered "soluble ferment" constituted a rather new type of biological catalyst, and he introduced the word *oxidase* to describe it (Bertrand 1895).[14] In addition, his work with this enzyme led to the important recognition of trace elements and to the quite novel idea of *coferments*. Bertrand discovered this enzyme in the course of an investigation of the process by which the colorless latex of the Vietnamese lacquer tree (shades of the French empire?) becomes jet-black and hard. The paper reproduced here, from the same year as Buchner's, was bingen streets after his wedding-eve party (Polterabend).[12] In 1917,

selected because in it the term *coferment* is used for the first time. His identification of manganese as the metal in laccase, although it led to highly influential new ideas in enzymology and in pathology, was wrong since copper rather than manganese is the metallic constituent of this enzyme. It is important that in earlier papers Bertrand stressed the marked analogy of certain manifestations of laccase activity, in terms of oxygen uptake and carbon dioxide evolution, to respiration in plants (Bertrand 1895), and in plants and animals (Bertrand 1897): "This first example of a diastatic [i.e., enzymatic] reaction with gas exchange is very remarkable. This phenomenon is rather like artificial respiration... . I think this hypothesis is not farfetched" (Bertrand 1895, p. 269, translation by R. E. Kohler, Jr., 1973, p. 189).

Gabriel Émile Bertrand (1867–1962) who developed widely used methods for the microanalysis of metals, is the first person in this volume who, undoubtedly because of his longevity, reaches into rather recent times. (He was honorary president, at age eighty-five, of the Second International Congress of Biochemistry, Paris, 1952). He succeeded Pierre Émile Duclaux as professor of biochemistry in the Faculté des Sciences at Paris, a position he held for twenty-nine years until his retirement in 1937.

## NOTES

1. This society had been founded in 1856; its first president (1858) was Helmholtz.
2. This report must have been printed in a hurry. In a reprint of this paper shown in the 1976 enzyme centenary supplement of *FEBS Letters,* the name J. v. Meyer at the bottom of the page has been changed in ink to J. Müller.
3. As is pointed out by Dixon and Webb (1979, p. 2), Kühne's intent was to emphasize the "en" part of the word "enzyme;" he wished "to denote something which was *in* yeast, in contrast to the yeast itself."
4. It is almost certain that the word *Enzym* was coined by Kühne, but as pointed out by Fruton (1978), this word was already used in the Middle Ages for "the leavened bread with which the Eucharist is celebrated in the Greek Church (Oxford English Dictionary, Supplement, Vol. 1, p. 957)."
5. Chittenden's account (1930, pp. 27–32) of his first meeting with Kühne is most entertaining and throws a great deal of light on Kühne's personality. Chittenden had gone to Europe from New Haven to work in the laboratory of Felix Hoppe-Seyler at

Strassburg. "In 1878 any American student desirous of making progress in physiological chemistry had no recourse other than going to Germany for the knowledge and experience needed to help him on his way." However, for various reasons he did not like the laboratory and he did not like the city. He decided to proceed, unannounced, to Kühne's laboratory in Heidelberg. "Kühne's reception was a gracious one, but it was not difficult to see that the young American, with his imperfect command of German, and with his lack of the customary credentials presented a case quite out of the ordinary, and the outcome seemed dubious. It was explained that all necessary credentials would be forthcoming from America as soon as possible, but that did not seem to interest Kühne particularly. He appeared more interested in the visiting card he held in his hand, and much to the writer's surprise he said, 'Are you the Chittenden who published in Liebig's *Annalen* a year or two ago an article on glycogen and glycocoll?' I was to learn later that Kühne possessed a wonderful memory; anything he had read he rarely forgot...Going into his library, he came back with the volume of the *Annalen* containing the article...The atmosphere was completely changed, and my spirits rose accordingly, reaching a still higher level when Kühne remarked that he would find a place for me in the laboratory at once...At the Heidelberg laboratory, numbers were small, hence there was ample opportunity for Kühne to exercise his influence over us all. He himself carried on his own experimental work nearby, from which we all derived some profit..."

6. The technique of ammonium sulfate precipitation had first been described by Camille Jean Marie Méhu (1878) and was applied specifically to the salting out of proteins by Adriaan Heynsius (1884). Much of the spade work on its use in the physiological institute at Heidelberg was done by Richard Neumeister. Although Neumeister published a paper on this topic only in 1888, Kühne and Chittenden acknowledge his work already in 1886 and use this technique in a second paper from the same year (Kühne and Chittenden 1886a, p. 410, and 1886b).
7. F. F. Nord here quotes Hofmeister (1901, p. 8) for the analogy of the difficulties one faces in trying to understand the working of a watch from a chemical, not just a mechanical, analysis of its shattered parts. In fact Hofmeister stresses in this quite prophetic speech that this analogy is quite misleading when applied to preparations obtained from living cells. He states that

chemical analysis of different tissue constituents has provided a plethora of important findings, and that it turned out to have been a bit premature to assume that the destruction of the living cell completely destroys its vital functions. He proceeds to stress that it is indeed only by such destruction that it has been found possible to establish the presence in cells of agents, such as enzymes, that are active during life. This short lecture is an elegant vindication of the value of the analytical approach to the unravelling of vital processes and remains rewarding reading to this day. Concerning the importance of this lecture, see Kohler (1972, p. 346, and 1973, p. 185).

8. As mentioned elsewhere (Editor's Comments on Papers 13 and 21), the type of yeast used was also, as Buchner soon discovered, of supreme importance.
9. Old ideas die hard. In the sixth (!) enlarged and revised English edition (1960) of Fritz Feigl's standard treatise *Spot Tests in Organic Analysis,* zymase is included (p. 633) in a list of "Individual Compounds" for which an identifying test is given (p. 477).
10. Some forty years after Buchner's original observations, Fritz Lipmann (1938) showed that bakers' yeast actually gives highly active extracts, provided that one uses as starting material not the fresh yeast, but rather yeast that had been air-dried under certain specified conditions. The simple method of air-drying for obtaining highly active yeast extracts had been introduced by Lebedew already in 1911 (Lebedeff 1911; Lebedew 1911). It is rather incredible that for almost thirty years after that no one appears to have tried to use this method with bakers' yeast, which as Lipmann states is actually easier to obtain than brewers' yeast. There are certain enzymatic differences in regard to the response of Lebedew yeast extracts from bakers' and from brewers' yeast toward inorganic phosphate (Lipmann 1938) and these are probably related to their relative content in an enzyme that breaks down adenosine triphosphate, but the discussion of this matter, based on later work by others, would go beyond the scope of this book.
11. The editor was alerted to these facts through an article by Robert E. Kohler (1972). For further insights into the Buchners' and others' views of the discovery of zymase, see this article.
12. Kohler's statement that Buchner was actually kept for the night in the police station is due to a misreading of the corresponding passage in the obituary notice by Harries: "Darauf sang

Buchner...mit so lauter Stimme in den schweigenden Tübinger Gassen, dass nicht viel gefehlt hätte, und er musste die Nacht auf der Polizeiwache verbringen" (Harries 1917, p. 1846).

13. Joseph Fruton (1972, p. 73), on the basis of several turn-of-the-century authors, gives the number of soluble enzymes known by 1900 as about two dozen. Fifty years later the number and types of known enzymes had grown so prolifically, and terminology had become so confusing that—a sign of the times—an International Commission on Enzymes was established in 1955 under the auspices of the International Union of Biochemistry, whose main task was to classify and in many instances to rename the enzymes. The resulting *1961 Report of the Enzyme Commission* included 712 enzymes, ingeniously classified into just six different reaction types or classes and various sub-classes. Each enzyme received an Enzyme Commission number and a recommended name. The work of compilation and classification continues. By 1975 the list of enzymes had grown to over 1900. The latest available list, completed in 1978 (*Enzyme Nomenclature 1978*) contains more than 2100 enzymes. The number of enzymes added per year between 1965 and 1978 was less than in the previous periods, but the list will doubtlessly continue to grow for many years to come. A good example of the rapid growth of known enzymes is provided by the type II restriction endonucleases, whose value as reagents in genetic research "has prompted the most massive hunt for additional examples in the history of enzymology" (Kornberg 1980). The 1979 edition of the 1978 *Enzyme Nomenclature* list cites some 45 of these enzymes. Kornberg in his book, completed in the middle of 1979, mentions that "nearly 200 of these enzymes are now known."
14. Bertrand is usually given credit for the discovery of this type of enzyme (see, for example, Kohler 1973, p. 188). Actually, three years earlier Alfred Jaquet, working in Strassburg, had demonstrated a heat-sensitive oxidative activity (conversion of salicylaldehyde to salicyclic acid) in tissue extracts (Jaquet 1892). Earlier still, Hikorokuro Yoshida (1883) in a study on urushi, "the material for the well-known Japanese lacquer varnish," had come to the following conclusion: "The phenomenon of the drying of urushi juice is due to the oxidation of urushic acid, $C_{14}H_{18}O_2$, into oxy-urushic acid, $C_{14}H_{18}O_3$, which takes place by the aid of diastase in the presence of oxygen and moisture." The predecessor of all these and many later workers is Moritz Traube (see Editor's Comments on

Paper 16) who "proposed, during 1858–1886, a theory of biological oxidations based on the activation of molecular oxygen by intracellular enzymes" (Fruton 1972, p. 291). However, in the words of Marjory Stephenson (1949, Introduction) referring specifically to Paper 16, "it was purely speculative and not founded on experimentally acquired knowledge. Hence Traube's contribution...was forty or fifty years before its time; when that time had elapsed, Traube's views were forgotten. It is always thus with guesses unsupported by experimental evidence."

# 18

Reprinted from page 190 of *Naturhist.-Med. Ver Heidelberg Verh.*, n.s., **1**:190–193 (1877)

## Ueber das Verhalten verschiedener organisirter und sog. ungeformter Fermente.

**W. Kühne**

### Sitzung am 4. Februar 1876

Hr. W. Kühne berichtet über das Verhalten verschiedener organisirter und sog. ungeformter Fermente. Um Missverständnissen vorzubeugen und lästige Umschreibungen zu vermeiden schlägt Vortragender vor, die ungeformten oder nicht organisirten Fermente, deren Wirkung ohne Anwesenheit von Organismen und ausserhalb derselben erfolgen kann, als *Enzyme* zu bezeichnen. — Genauer untersucht wurde besonders das Eiweiss verdauende Enzym des Pankreas, für welches, da es zugleich Spaltung der Albuminkörper veranlasst, der Name *Trypsin* gewählt wurde. Das Trypsin vom Vortr. zuerst dargestellt und zwar frei von durch dasselbe noch verdaulichen und zersetzbaren Eiweissstoffen, verdaut nur in alkalischer, neutraler, oder sehr schwach sauer reagirender Lösung.

[*Editor's Note:* Material has been omitted at this point. An English translation of the above text follows.]

# 18

# ON THE BEHAVIOR OF VARIOUS ORGANIZED AND SO-CALLED UNORGANIZED FERMENTS

**Willy Kühne**

*This excerpt was translated expressly for this Benchmark volume by H. Friedmann, the University of Chicago, from page 190 of "Ueber das Verhalten verschiedener organisirter und sog. ungeformter Fermente" in* Naturhist.-Med. Ver. Heidelberg Verh., Neue Folge,*1:190–193 (1877).

Session of 4 February 1876

Mr. *W. Kühne* reports on the behavior of various organized and so-called unorganized ferments. In order to prevent misunderstandings and in order to avoid tedious circumlocutions, the lecturer proposes to denote the unformed or unorganized ferments, whose activity can take place without the presence of organisms and outside of these, as *enzymes*. A rather detailed investigation was instituted on the protein-digesting enzyme of the pancreas for which, since it also cleaves albumins, the name *trypsin* was chosen. Trypsin, first prepared by the speaker and actually free of proteins that could still be digested and decomposed by it, digests only in alkaline, neutral, or very weakly acidic solutions.

[*Editor's Note:* The rest of this report deals with the preparation of trypsin using precipitation with salicyclic acid and with some of its properties. Both the word *enzyme* and the word *trypsin* are used as a matter of fact and quite repeatedly. The paper immediately following, pages 194-198 also by Kühne, entitled "About Trypsin (Enzyme of the Pancreas)," is the first to use the word *enzyme* as well as *trypsin* in a title and, in addition, presumably for the first time uses the word *zymogen*. It is not translated here.]

*[*Discussions of the Natural History and Medical Society at Heidelberg, New Series.*]

# 19

# EXPERIENCES AND REMARKS ON ENZYMES AND FERMENTS

**Willy Kühne**

*These excerpts were translated expressly for this Benchmark volume by H. Friedmann, the University of Chicago, from pages 293–296 of "Erfahrungen und Bemerkungen über Enzyme und Fermente" in* Physiol. Inst. Univ. Heidelberg Unters. **1**:291–324 (1877).

[*Editor's Note:* In the original, material precedes these excerpts.]

Recently. . .the need for a distinction between the two classes of phenomena [hydrolytic and fermentative] appears to have been accepted and already expressed by the names *organized* and *unorganized ferments* [*geformte* und *ungeformte Fermente*]. However, this naming still recognizes these two types of processes to be related to each other, and it distinguishes these processes only from the point of view of the agent that brings them about.

It is known that the latter designations [that is, organized and unorganized ferments] have not been generally accepted, for, on the one hand, it was explained that chemical substances such as ptyalin [salivary amylase], pepsin, and so on, cannot be called "ferments" since that name has already been given to yeast cells and other *organisms* (*Brücke*), while on the other hand it was said that yeast cells could not be a ferment and could not be called thus since then all organisms, including man, would be included (*Hoppe-Seyler*). Without wishing to examine further why this name is so offensive to the opposing sides, the existence of this antagonism prompted me first of all to propose a new name; I took the liberty of giving the name *enzymes* to some of the better-known substances that many call unorganized ferments. This name made no claim to any particular hypothesis; it stated only that something occurs in the zyme [yeast] that has an activity of one kind or another which can be regarded as fermentative. However, by not limiting the expression to the invertin of yeast, it implied that more complicated organisms from which the enzymes pepsin, trypsin, and so on can be obtained do not differ as fundamentally from the unicellular organisms as, for example, *Hoppe-Seyler* seems to think. There was yet a second reason that I looked for a new name (and I would have liked to find a more pleasant-sounding one for our language). As is known, the chemical processes that one so far knew to be brought about by the so-called unorganized ferments are, without exception, hydrolytic ones. On the other hand, this type of reaction is brought about only by a few of the organized ferments. Indeed, one can prove that the latter bring about reductions and oxidations and in many cases yield a number of products whose formation cannot be in any way explained chemically by mere *cleavage and decomposition of the fermenting material;* the formation of these products is, indeed, as inexplicable as if one would wish to derive all compounds formed by higher animal and plant organisms simply by cleavages of compounds contained in their food. Thus, it was indeed proper to get rid of our name and to replace it with a new one when the time had come to ask whether it was reasonable to use the old name for denoting different types of processes as though they shared common features, while this old name already recognized different causative agents.

[*Editor's Note:* Material has been omitted at this point.]

As is known, *Berthelot* in 1860 began to assign all processes related to fermentation to *one* class of chemicals, namely, the unorganized ferments. He did this in so far as he prepared cell-free aqueous extracts from yeast and proved that they inverted sugar. Since all alcoholic fermentation of dextrorotatory sugars starts with inversion, a part of the whole act of fermentation, namely, the beginning, had in fact been imitated independently and outside of the organism; thus in this instance a chemical experiment using an agent of the organism had been substituted for the physiological one. Since this has been accomplished, seventeen years have passed. During this time one did in fact get to know and to isolate invertin better, but without being able to obtain a second similar new substance out of yeast. It is not alcoholic fermentation but rather something preceding it that has been, as it were, removed from the organism and imitated outside of it, and although in the meantime many new methods have been found to kill yeast without destroying the invertin, and many good agents have been found to bring it into solution, to precipitate and to redissolve it, in all this work the presumed other substance responsible for alcohol and $CO_2$ formation was not encountered.

Clearly the discovery of invertin justifies some hope and seems to have already been followed by *Musculus'* discovery of the enzyme that decomposes urea. Hence one may hope for the day when there will be as many enzymes available as the number of fermentations known up to that time. It only remains to ask whether in the meantime it is most suitable to build hypotheses on hopes and to designate these hypotheses as theories, whether one should try to extract enzymes from organisms with all possible solvents or whether one should for the time being not attempt this and rather try to unravel all individual processes within the life of the cell. It seems to me that one should undertake the two latter approaches but that the first of the three should be used as little as possible. Above all, one should not be content if nothing should work.

[*Editor's Note:* The rest of this paper has been omitted.]

# 20

# STUDIES ON FERMENTATION

## THE DISEASES OF BEER

## THEIR CAUSES, AND THE MEANS OF PREVENTING THEM

BY

L. PASTEUR

MEMBER OF INSTITUTE OF FRANCE, THE ROYAL SOCIETY OF LONDON, ETC.

*A TRANSLATION, MADE WITH THE AUTHOR'S SANCTION, OF "ÉTUDES SUR LA BIÈRE," WITH NOTES, INDEX, AND ORIGINAL ILLUSTRATIONS*

BY

FRANK FAULKNER

AUTHOR OF "THE ART OF BREWING," ETC.

AND

D. CONSTABLE ROBB, B.A.

LATE SCHOLAR OF WORCESTER COLL., OXFORD

London

MACMILLAN & CO

1879

Reprinted from pages 143, 144, 196–197, 235–236, 266–267, 270, 321–323, and 326–328 of *Studies on Fermentation: The Diseases of Beer, Their Causes, and the Means of Preventing Them*, F. Faulkner and D. C. Robb, transl., London: Macmillan, 1879, 418 pp.

# CHAPTER V.

## THE ALCOHOLIC FERMENTS.

### § I.—ON THE ORIGIN OF FERMENT.

AMONGST the productions that appear spontaneously, or, we should rather say, without direct impregnation, in organic liquids exposed to contact with the air, there is one that more particularly claims our study. It is that one which, by reason of its active energy as an agent of decomposition, has been distinguished and utilized from the earliest times, and is considered as the type of ferments in general; we mean the ferment of wine, beer, and more generally, of all fermented beverages.

Yeast is that viscous sort of deposit which takes place in the vats or barrels of must or wort that is undergoing fermentation. This kind of ferment presents for consideration a physical fact of the most extraordinary character. Take a morsel of the substance and put it in sweetened water, in must, or in dough, which always contains a little sugar; after a time, the length of which varies, a few minutes often sufficing, we see these liquids or the dough rise, so to speak. This inflation of the mass, which is due to a liberation of carbonic acid gas, may cause it to overflow the vessels containing it, if their capacity is not considerably greater than the volume of the matters fermenting.

[*Editor's Note:* Material has been omitted at this point.]

As for the nature of yeast, the microscope has taught us what it is. That marvellous instrument, although still in its infancy, enabled Leuwenhoeck, towards the close of the 17th century, to discover that yeast is composed of a mass of cells. In 1835 Cagnard-Latour and Schwann took up Leuwenhoeck's observations, and by employing a more perfect microscope, discovered that these same cells vegetate and multiply by a process of gemmation. Since then the physical and chemical phenomena already mentioned, such as the raising of the mass, the liberation of carbonic acid gas, and the formation of alcohol, have been announced as acts probably connected with the living processes of a little cellular plant, and subsequent researches have confirmed these views.

In introducing a quantity of yeast into a saccharine wort, it must be borne in mind that we are sowing a multitude of minute living cells, representing so many centres of life, capable of vegetating with extraordinary rapidity in a medium adapted to their nutrition. This phenomenon can occur at any temperature betweeen zero and 55° C. (131° F.), although a temperature between 15° C. and 30° C. (59° F. and 86° F.) is the most favourable to its occurrence.

[*Editor's Note:* Material has been omitted at this point. The footnote below refers to the following sentence, which appears on the portion of page 196 that has been omitted here, "Lastly, the beer possesses a flavour and delicacy which cause it to be held in higher esteem by consumers than beers produced by means of other ferments.*" It is included as an amusing contrast to the comment made by the Frenchman Planche at the beginning of Paper 3.]

A ferment is a combination of cells, the individuals of which must differ more or less from each other. Each of these cells

* [On this point again Dr. Graham expresses some dissent (" Nature," loc. cit.): " Here surely M. Pasteur must be thinking rather of the inferior products of the surface fermentation in France and Germany, than of those of England and Scotland."—D. C. R.]

has certain generic and specific peculiarities which it shares with the neighbouring cells; but over and above this, certain peculiar characteristics which distinguish it, and which it is capable of transmitting to succeeding generations. If, therefore, we could manage with some species of ferment to isolate the different cells that compose it, and could cultivate each of these separately, we should obtain as many specimens of ferments, which would, probably, be distinct from one another, inasmuch as each of them would inherit the individual peculiarities of the cell from which it originated. Our endeavours are directed to realizing this result practically, by first thoroughly drying a ferment and reducing it to fine powder. We have seen (Chap. III. § 6) that this mode of experiment is practicable, that in a powder composed of yeast and plaster the ferment preserves its faculty of reproduction for a very long time.

[*Editor's Note:* Material has been omitted at this point.]

# CHAPTER VI.

## THE PHYSIOLOGICAL THEORY OF FERMENTATION.

### § I.—On the Relations existing between Oxygen and Yeast.

The object of all science is a continuous reduction of the number of unexplained phenomena. It is observed, for instance, that fleshy fruits are not liable to fermentation so long as their epidermis remains uninjured. On the other hand, they ferment very readily when they are piled up in heaps, more or less open, and immersed in their saccharine juice. The mass becomes heated and swells; carbonic acid gas is disengaged, and the sugar disappears and is replaced by alcohol. Now, as to the question of the origin of these spontaneous phenomena, so remarkable in character as well as usefulness for man's service, modern knowledge has taught us that fermentation is the consequence of a development of vegetable cells, the germs of which do not exist in the saccharine juices within fruits; that many varieties of these cellular plants exist, each giving rise to its own particular fermentation. The principal products of these various fermentations, although resembling each other in their nature, differ in their relative proportions and in the accessory substances that accompany them, a fact which alone is sufficient to account for wide differences in the quality and commercial value of alcoholic beverages.

Now that the discovery of ferments and their living nature, and our knowledge of their origin, may have solved the mystery of the spontaneous appearance of fermentations in natural saccharine juices, we may ask whether we must still regard the reactions that occur in these fermentations as phenomena inexplicable by the ordinary laws of chemistry. We can readily see that fermentations occupy a special place in the series of chemical and biological phenomena. What gives to fermentations certain exceptional characters, of which we are only now beginning to suspect the causes, is the mode of life in the minute plants designated under the generic name of *ferments*, a mode of life which is essentially different from that in other vegetables, and from which result phenomena equally exceptional throughout the whole range of the chemistry of living beings.

[*Editor's Note:* Material has been omitted at this point.]

### § II.—Fermentation in Saccharine Fruits Immersed in Carbonic Acid Gas.

The theory which we have, step by step, evolved, on the subject of the causes of the chemical phenomena of fermentation, may claim a character of simplicity and generality that is well worthy of attention. Fermentation is no longer one of those isolated and mysterious phenomena which do not admit of explanation. It is the consequence of a peculiar vital process of nutrition which occurs under certain conditions, differing from those which characterize the life of all ordinary beings, animal or vegetable, but by which the latter may be affected, more or less, in a way which brings them, to some extent, within the class of ferments, properly so called. We can even conceive that the fermentative character may belong to every organized form, to every animal or vegetable cell, on the sole condition that the chemico-vital acts of assimilation and excretion must be capable of taking place in that cell for a brief period, longer or shorter it may be, without the necessity for recourse to supplies of atmospheric oxygen ; in other words, the cell must be able to derive its needful heat from the decomposition of some body which yields a surplus of heat in the process.

[*Editor's Note:* Material has been omitted at this point.]

In short, fermentation is a very general phenomenon. It is life without air, or life without free oxygen, or, more generally still, it is the result of a chemical process accomplished on a fermentable substance, *i.e.* a substance capable of producing heat by its decomposition, in which process the entire heat used up is derived from a part of the heat that the decomposition of the fermentable substance sets free. The class of fermentations, properly so called, is, however, restricted by the small number of substances capable of decomposing with the production of heat, and at the same time of serving for the nourishment of lower forms of life, when deprived of the presence and action of air. This, again, is a consequence of our theory, which is well worthy of notice.

[*Editor's Note:* Material has been omitted at this point.]

It is true that there are circumstances under which yeast brings about modifications in different substances. Doebereiner and Mitscherlich, more especially, have shown that yeast imparts to water a soluble material, which liquefies cane-sugar and produces inversion in it by causing it to take up the elements of water, just as diastase behaves to starch or emulsin to amygdalin.

M. Berthelot also has shown that this substance may be isolated by precipitating it with alcohol, in the same way as diastase is precipitated from its solutions.* These are remark-

* DOEBEREINER, *Journal de Chimie de Schweigger*, vol. xii. p. 129, and *Journal de Pharmacie*, vol. i. p. 342.

MITSCHERLICH, *Monatsberichte d. Kön. Preuss. Akad. d. Wissen. zu Berlin*,

able facts, which are, however, at present but vaguely connected with the alcoholic fermentation of sugar by means of yeast. The researches in which we have proved the existence of special forms of living ferments in many fermentations, which one might have supposed to have been produced by simple contact action, had established beyond doubt the existence of profound differences between those fermentations, which we have dis-

and *Rapports annuels de Berzelius*, Paris, 1843, 3rd year. On the occasion of a communication on the inversion of cane sugar, by H. Rose, published in 1840, M. Mitscherlich observed: "The inversion of cane sugar in alcoholic fermentation is not due to the globules of yeast, but to a soluble matter in the water with which they mix. The liquid obtained by straining off the ferment on a filter paper, possesses the property of converting cane sugar into uncrystallizable sugar."

BERTHELOT, *Comptes rendus de l'Académie*. Meeting of May 28th, 1860. M. Berthelot confirms the preceding experiment of Mitscherlich, and proves, moreover, that the soluble matter of which that author speaks may be precipitated with alcohol without losing its invertive power.

M. Béchamp has applied Mitscherlich's observation, concerning the soluble fermentative part of yeast, to fungoid growths, and has made the interesting discovery that fungoid growths, like yeast, yield to water a substance that inverts sugar. When the production of fungoid growths is prevented by means of an antiseptic the inversion of sugar does not take place.

We may here say a few words respecting M. Béchamp's claim to priority of discovery. It is a well-known fact that we were the first to demonstrate that living ferments might be completely developed, if their germs were placed in pure water, together with sugar, ammonia, and phosphates. Relying on this established fact, that moulds are capable of development in sweetened water, in which, according to M. Béchamp, they invert the sugar, our author asserts that he has proved that, "living organized ferments may originate in media which contain no albuminous substances." (See *Comptes rendus*, vol. lxxv. p. 1519.) To be logical, M. Béchamp might say that he has proved that certain moulds originate in pure sweetened water, without nitrogen or phosphates or other mineral elements, for such a deduction might very well be drawn from his work, in which we do not find the least expression of astonishment at the possibility of moulds developing in pure water, containing nothing but sugar without other mineral or organic principles.

M. Béchamp's first Note on the inversion of sugar was published in 1855. In it we find nothing relating to the influence of moulds. His second, in which that influence is noticed, was published in January,

tinguished as fermentations proper, and the phenomena connected with soluble substances. The more we advance, the more clearly we are able to detect these differences. M. Dumas has insisted on the fact that the ferments of fermentation proper multiply and reproduce themselves in the process, whilst the others are destroyed.* Still more recently M. Müntz has shown that chloroform prevents fermentations proper, but does not interfere with the action of diastase (*Comptes rendus*, 1875.) M. Bouchardat had already established the fact that "hydrocyanic acid, salts of mercury, ether, alcohol, creosote, and the oils of turpentine, lemon, cloves, and mustard destroy or check alcoholic fermentation, whilst in no way interfering with the glucoside fermentations (*Annales de Chimie et de Physique*, 3rd series, t. xiv., 1845.) We may add, in praise of M. Bouchardat's sagacity, that that skilful observer has always considered these results as a proof that alcoholic fermentation is dependent on the life of the yeast-cell, and that a distinction should be made between the two orders of fermentation.

[*Editor's Note:* Material has been omitted at this point.]

1858, that is, subsequently to our work on lactic fermentation, which appeared in November, 1857. In that work we established, for the first time, that the lactic ferment is a living organized being, that albuminous substances have no share in the production of fermentation, and that they only serve as the food of the ferment. M. Béchamp's Note was even subsequent to our first work on alcoholic fermentation, which appeared on December 21st, 1857. It is since the appearance of these two works of ours that the preponderating influence of the life of microscopic organisms, in the phenomena of fermentation, has been better understood. Immediately after their appearance M. Béchamp, who, from 1855, had made no observation on the action of fungoid growths on sugar, although he had remarked their presence, modified his former conclusions. (*Comptes rendus*, January 4th, 1858.)

* "There are two classes of ferments; the first, of which the yeast of beer may be taken as the type, perpetuate and renew themselves if they can find in the liquid in which they produce fermentation food enough for their wants; the second, of which diastase is the type, always sacrifice themselves in the exercise of their activity." (DUMAS, *Comptes rendus de l'Académie*, t. lxxv. p. 277, 1872.)

Liebig, who, as well as M. Fremy, was compelled to renounce his original opinions concerning the nature of ferments, devised the following obscure theory (Memoir by Liebig, 1870, already cited) :—

"There seems to be no doubt as to the part which the vegetable organism plays in the phenomenon of fermentation. It is through it alone that an albuminous substance and sugar are enabled to unite and form this particular combination, this unstable form under which alone, as a component part of the mycoderm, they manifest an action on sugar. Should the mycoderm cease to grow, the bond which unites the constituent parts of the cellular contents is loosened, and it is through the motion produced therein that the cells of yeast bring about a disarrangement or separation of the elements of the sugar into other organic molecules."

One might easily believe that the translator for the *Annales* has made some mistake, so great is the obscurity of this passage.

Whether we take this new form of the theory or the old one, neither can be reconciled at all with the development of yeast and fermentation in a saccharine mineral medium, for in the latter experiment fermentation is correlative to the life of the ferment and to its nutrition, a constant change going on between the ferment and its food-matters, since all the carbon assimilated by the ferment is derived from sugar, its nitrogen from ammonia, and phosphorus from the phosphates in solution. And even all said, what purpose can be served by the gratuitous hypothesis of contact-action or communicated motion ? The experiment of which we are speaking is thus a fundamental one; indeed, it is its possibility that constitutes the most effective point in the controversy. No doubt Liebig might say, "but it is the motion of life and of nutrition which constitutes your experiment, and this is the communicated motion that my theory requires." Curiously enough, Liebig does endeavour, as a matter of fact, to say this, but he does so timidly and incidentally: "From a chemical point of view, which point of view I would not willingly abandon, a *vital action* is a phenomenon of motion, and, in this double sense of *life* M. Pasteur's theory agrees with my own, and is not in contradiction with it (page 6)." This is true. Elsewhere Liebig says :—

"It is possible that the only correlation between the physiological act and the phenomenon of fermentation is the production, in the living cell, of the substance which, by some special property analogous to that by which emulsin exerts a decomposing action on salicin and amygdalin, may bring about the decomposition of sugar into other organic molecules; the physiological act, in this view, would be necessary for the production of this substance, but would have nothing else to do with the fermentation (page 10)." To this, again, we have no objection to raise.

Liebig, however, does not dwell upon these considerations, which he merely notices in passing, because he is well aware that, as far as the defence of his theory is concerned, they would be mere evasions. If he had insisted on them, or based his opposition solely upon them, our answer would have been simply this: "If you admit with us that fermentation *is* correlated with the life and nutrition of the ferment, we agree upon the principal point. So agreeing, let us examine, if you will, the actual cause of fermentation;—this is a second question, quite distinct from the first. Science is built up of successive solutions given to questions of ever-increasing subtlety, approaching nearer and nearer towards the very essence of phenomena. If we proceed to discuss together the question of how living, organized beings act in decomposing fermentable substances, we will be found to fall out once more on your hypothesis of communicated motion, since, according to our ideas, the actual cause of fermentation is to be sought, in most cases, in the fact of life without air, which is the characteristic of many ferments."

Let us briefly see what Liebig thinks of the experiment in which fermentation is produced by the impregnation of a saccharine mineral medium, a result so greatly at variance with his mode of viewing the question.* After deep consideration he pronounces this experiment to be inexact, and the result ill-founded. Liebig, however, was not one to reject a fact without grave reasons for his doing so, or with the sole object of evading a troublesome discussion. "I have repeated this experiment," he says, "a great number of times, with the greatest possible care, and have obtained the same results as M. Pasteur, excepting as regards the formation and increase of the ferment." It was, however, the formation and increase of the ferment that constituted the point of the experiment. Our discussion was, therefore, distinctly limited to this: Liebig denied that the ferment was capable of development in a saccharine mineral medium, whilst we asserted that this development did actually take place, and was comparatively easy to prove. In 1871 we replied to M. Liebig before the Paris Academy of Sciences in a Note, in which we offered to prepare in a mineral medium, in the presence of a commission to be chosen for the purpose, as great a weight of ferment as Liebig could reasonably demand.† We were bolder than we should, perhaps, have been in 1860; the reason was that our knowledge of the subject had been strengthened by ten years of renewed research. Liebig did not accept our proposal, nor did he even reply to our Note. Up to the time of his death, which took place on April 18th, 1873, he wrote nothing more on the subject.

[*Editor's Note:* Material has been omitted at this point.]

* See our Memoir of 1860 (*Annales de Chimie et de Physique*, vol. lviii. p. 61, and following, and especially pp. 69 and 70, where the details of the experiment will be found.

† PASTEUR, *Comptes rendus de l'Académie des Sciences*, vol. lxxiii. p. 1419, 1871.

# 21

# ALCOHOLIC FERMENTATION WITHOUT YEAST CELLS

**Eduard Buchner**

*This article was translated expressly for this Benchmark volume by H. Friedmann, the University of Chicago, from "Alkoholische Gährung ohne Hefezellen (Vorläufige Mittheilung)" in* Dtsch. Chem. Ges. Ber. **30**:117–124 (1897). *(Corrections given on pages 335 and 1110 of this journal have been incorporated.)*

*Received 11 January*

Preliminary Note

So far it has not been possible to effect a separation of fermentative activity from living yeast cells; in the following a procedure that solves this problem is described.

One thousand grams of brewer's yeast,[1] purified for the preparation of compressed yeast but not yet containing added potato starch, is carefully mixed with the same weight of quartz sand[2] and 250 g of Kieselguhr and is then ground up until the mass has become moist and pliable. Now 100 g of water are added to the paste and, wrapped in cheesecloth, it is gradually subjected to a pressure of 400-500 atm: 350 cc of press juice are obtained. The residual cake is once more ground up, sieved, and 100 g of water are added; when again subjected to the same pressure in the hydraulic press, it yields a further 150 cc of press juice. From 1 kg of yeast, one hence obtains 500 cc press juice, which contain almost 300 cc of substances derived from the cell content. In order to remove a trace of turbidity, the press juice is finally shaken with 4 g of Kieselguhr and filtered through a paper filter. In this process repeated filtration of the first aliquots is required.

The press juice obtained in this way is a clear, slightly opalescing yellow liquid with a pleasant yeast smell. The specific gravity was once found to be 1.0416 (17°C). A large amount of coagulum separates on boiling, so that the liquid almost completely solidifies. The formation of insoluble flakes starts already at 35-40°; even below this temperature one notices gas bubbles to rise which can be shown to be carbon dioxide that hence saturates the liquid.[3] The press juice contains over 10 percent of dry weight. A press juice prepared by an earlier, inferior procedure contained 6.7 percent dry weight, 1.15 percent ash and, judging from the nitrogen content, 3.7 percent protein.

The most interesting property of the press juice is its capacity to bring about fermentation of carbohydrates. When it is mixed with the same volume of a concentrated solution of cane sugar, one finds that, already after 1/4 to 1 hour a regular evolution of carbon dioxide takes place that lasts for days. Glucose, fructose, and maltose behave in the same manner. However, no fermentation occurs in mixtures of the press juice with saturated solutions of lactose and mannitol, just as, indeed, these substances are not fermented by living brewer's yeast cells either. Mixtures of press juice and sugar solution that have been fermenting for several days and that have been placed in the ice box gradually become turbid, although no organisms can be found microscopically. On the other hand, upon 700-fold magnification, one can see quite many protein coaguli whose separation was probably brought about by the acids formed in the course of the fermentation. Saturation of the mixture of press juice and saccharose solution with chloroform does not prevent the fermentation but causes an early slight separation of protein. Similarly, filtration of the press juice through a sterilized *Berkefeldt-Kieselguhr* filter, which undoubtedly retains all yeast cells, does not destroy the fermentative capacity; fermentation of the mixture of the absolutely clear filtrate and of sterilized cane sugar solution begins, although with some delay, after about one day, even at ice-box temperature. When a parchment paper tubing filled with press juice hangs in 37 percent cane sugar solution, then after a few hours the surface of the tubing becomes covered with innumerable tiny gas bubbles. Vigorous gas development could, of course, also be observed on the inside because of diffusion of sugar solution. Further experiments are needed to decide whether the carrier of the fermentative activity can actually dialyze through the parchment paper, as appears to be the case.

[*Editor's Note:* This observation was later retracted, see Buchner and Rapp, 1898.]

In time the fermentative capacity of the press juice gradually becomes lost. Press juice that had been stored five days in ice water in a half-filled bottle proved to be inactive toward saccharose. It is remarkable, on the other hand, that press juice that contains added cane sugar, that is, press juice with fermentative activity, does retain this activity in the ice box for at least two weeks. In this connection one should probably first of all consider the favorable effect of the carbon dioxide formed during the fermentation in keeping atmospheric oxygen away; it is also possible, however, that the sugar, which is readily assimilated, contributes to the maintenance of the agent.

As to the nature of the active substance in the press juice, so far only a few experiments have been performed. When the press juice is heated to 40-50° first carbon dioxide develops, and then gradually coagulated protein separates; after one hour the material was filtered off, and the filtrate was repeatedly refiltered. The clear filtrate still possessed weak fermentative power toward cane sugar in one experiment, but not in another; hence the active substance either seems to lose its activity already at this remarkably low temperature, or it coagulates and precipitates. Furthermore, 20 cc of the expressed material were added to three volumes of absolute alcohol, the precipitate was filtered by suction and dried *in vacuo* over sulfuric acid; 2 g of dry substance were obtained, only a very small part of which redissolved upon digestion with 10 cc of water. The filtrate from this had no fermentative activity toward cane sugar. These experiments have to be repeated; in addition, a special attempt will be made to isolate the active principle by means of ammonium sulfate.

As to the *theory of fermentation,* the following conclusions can be drawn up to now. First of all, it has been proved that for the fermentative process to take place, it is not necessary to have an apparatus as complicated as the yeast cell. Rather a dissolved substance, doubtlessly a protein, is to be considered as the carrier of the fermentative activity of the press juice. This carrier will be called *zymase*.

The view that a special type of protein substance derived from yeast cells brings about fermentation had already been expressed in 1858 by *M. Traube* as *enzyme*, or *ferment*, *theory* and was later defended particularly by *F. Hoppe-Seyler*. The separation of such an enzyme from yeast cells had not, however, been accomplished up to now.

It is still questionable whether zymase may be added directly to the list of enzymes that have already been known for awhile. As has already been stressed by *C. v. Nägeli*,[4] important differences exist between fermentative activity and the activity of the ordinary enzymes. The latter are merely hydrolases that can be imitated by the simplest chemical means. Although *A. v. Baeyer*[5] has made it easier to understand alcoholic fermentation from the point of view of the chemical process by the use of analogies to relatively simple principles, the decomposition of sugar into alcohol and carbon dioxide is still one of the more complicated reactions. In this process carbon bonds are broken with a completeness [Vollständigkeit, referring to thoroughness or high yields] that is not so far possible in any other way. In addition, a significant difference exists in the effect of heat.[6]

Invertin can be extracted with water from yeast cells that had been killed by dry heat (1 hour to 150°); upon precipitation with alcohol, it can be isolated as a powder readily soluble in water. It is not possible in this manner to obtain the substance that causes fermentation. It is unlikely still to be present in yeast cells that have been heated to such a high temperature; alcohol precipitation converts it, if the preceding experiment permits one to reach a conclusion, into a water-insoluble modification. Hence it will scarcely be wrong to assume that zymase is one of the genuine proteins and that it is much more closely related than invertin to the living protoplasm of yeast cells.

Similar views have been expressed by the French bacteriologist *Miquel* with respect to urase [this rather than urease is the term used here], the enzyme excreted by the bacteria that carry out the so-called urea fermentation. He designates this enzyme directly as protoplasm since it does not need the protection of the cell wall, acts outside the cell wall, and on the whole differs only in this manner from the enzyme of the cell contents.[7] In addition the experiences of *E. Fischer* and *P. Lindner*[8] relative to the effect of the yeast fungus *Monilia candida* on cane sugar must be mentioned here. This fungus ferments saccharose; however, neither *Ch. E. Hansen* nor the above authors succeeded to extract with water from fresh or from dried yeast an invertin-like enzyme that would carry out the preceding cleavage into glucose and fructose. The experiment proceeded quite differently when *Fischer* and *Lindner* used fresh *Monilia* yeast in which at the beginning a part of the cells was opened up by means of careful grinding with glass powder. Now the inverting activity was obvious. In this case, however, the inverting agent seems to be not a stable water-soluble enzyme but a constituent of the living protoplasm.

The fermentation of sugar by zymase can occur inside the yeast cells;[9] it is more likely, however, that the yeast cells transfer this protein into the sugar solution where it brings about the fermentation.[10] The events that occur during alcoholic fermentation hence may perhaps be regarded as physiological only insofar as the living cells excrete the zymase. *Nägeli*[11] and *O. Löw* have shown that already after fifteen hours at 30° considerable quantities of protein, coagulable by boiling, diffuse out of yeast cells in a nutrient solution that is initially weakly alkaline (by means of $K_3PO_4$) and that later turns neutral. Actually, in addition, as shown in the above experiment, zymase can apparently go through parchment paper.

Fermentation Experiments

| *No.* | *Press Juice (cc)* | *Carbohydrate Solution (cc)* | | *Total Sugar Content (%)* | *Experimental Temperature* | *Remarks* |
|---|---|---|---|---|---|---|
| 1 | 30 | Saccharose | 30 | 37 | Ice-box | After 1 hour distinct gas development, not yet complete after 14 days. The froth layer eventually is 1 cm high. |
| 2 | 50 | " | 50 | 37 | " | Strong gas development and froth layer. The initially clear solution becomes opaque after 3 days without a precipitate. |
| 3 | 150 | " | 150 | 37 | " | The froth layer is 3/4 cm high after 3 days. |
| 4 | 20 | " | 20 | 37 | " | Gas development becomes visible after 2 hours and is not complete after 14 days; the initially clear solution at the end shows only minimal turbidity; froth layer 1 1/2 cm high. |
| 5 | 30 | " | 30 | 37 | " | Gas development begins after 1 day and is not yet complete after 1 week; during this time solution is still completely clear. |
| 6 | 20 | " | 20 | 37 | Room temp. | After 1 hour vigorous gas development; even after 2 weeks still slight formation of gas bubbles with only minimal turbidity. |
| 7 | 20 | " | 20 | 37 | 40° | After 2 hours already 10 cm high froth layer; after 1 day strong coagulum separation; gas formation is finished. |
| 8 | 30 | " | 30 | 12 | Ice-box | After 6 days still strong gas formation; in addition, turbidity consisting of very fine coagulum. |
| 9 | 5 | Maltose | 5 | 33 | " | After 1 hour start of gas development that still lasts after 12 days. |

Fermentation Experiments (con't)

| *No.* | *Press Juice (cc)* | *Carbohydrate Solution (cc)* | | *Total Sugar Content (%)* | *Experimental Temperature* | *Remarks* |
|---|---|---|---|---|---|---|
| 10 | 10 | Maltose | 5 | 26 | Ice-box | Gas development is extraordinarily strong already after 3 hours. |
| 11 | 10 | Glucose | 10 | 33 | " | Only after 20 hours strong gas development, which, however, still lasts after 12 days; froth layer 3/4 cm high. |
| 12 | 10 | " | 5 | 26 | " | Already after 1/2 hour rather strong gas development that lasts 12 days; the solution becomes turbid and deposits some precipitate (see below). |
| 13 | 10 | Fructose | 10 | 37 | " | Gas development is very strong already after 1/4 hour and is still vigorous after 3 days; the solution remains clear. |
| 14 | 10 | " | 5 | 25 | " | The frothy layer is already considerable after 15 minutes and measures 1 cm after 3 days. |
| 15 | 10 | Lactose | 10 | Sat.soln. | Room temp. | No gas development, not even after 6 days. |
| 16 | 10 | Mannitol | 10 | " | " | As for lactose. |

NOTES

In experiment 1, the escaping gas was passed into lime water 4 hours after it began to develop and was identified as *carbon dioxide*. In experiments 2 and 3, the *alcohol* formed by the fermentation after 3 days was identified: in experiment 2, 1.5 g, and in 3, 3.3 g ethyl alcohol were present. In this calculation, it is necessary to deduct the amounts that still adhered to the yeast derived from beer fabrication. In experiment 2 the yeast was washed 4 times with 5-liter aliquots of water before preparation of the press juice; then the alcohol was determined in 2/3 of the whole, the rest was worked up to obtain the press-juice. The results showed that the yeast used contained at the most 0.3 g alcohol. In experiment 3 commercial brewer's yeast purified for the manufacture of compressed yeast to which no starch had been added as yet was worked up directly; the alcohol content of the yeast needed to prepare 150 cc of press juice was calculated from the analysis to be 1.2 g. Hence in experiment 2, 1.2 g and in experiment 3, 2.1 g alcohol were formed by fermentation. In all cases the alcohol was identified by the iodoform reaction and finally potash was used for salting out of the aqueous solution. The precipitate obtained in experiment 3 distilled completely between 79 and 81° (734 mm); the distillate was colorless, combustible, and possessed the odor of ethyl alcohol.

*Microscopic investigation* was performed on experiments 2 and 3 after they had proceeded for 3 days. Furthermore, this was done for the slight sediment in experiment 8 after 6 days and in experiment 12 after 12 days of fermentation; in all cases, no organisms but merely coagulated protein could be found as the cause of the more or less strong turbidity. In addition, for experiment 3 upon interruption after a total of 3 days, 6 *plate cultures* were started. One cc of liquid was inoculated in each of 3 tubes with liquefied beer spice gelatin [Bierwürzegelatine] and 1 cc in each of 3 tubes of liquefied meat water peptone gelatin [Fleischwasserpeptongelatine]. After 6 days one of the former plates showed 11 colonies, the two others had remained sterile; the three peptone gelatin plates showed uniformly 50-100 colonies each and had become liquefied. In view of the large volumes inoculated in these experiments, the results prove that the fermentative activity was not due to microorganisms; this conclusion, besides, almost follows from the rapid appearance of the fermentative phenomena.

Finally, in experiments 4 and 5 the *press juice was filtered by suction through sterilized Berkefeldt-Kieselguhr filters.* In experiment 5, furthermore, the cane sugar solution had been sterilized in the autoclave, and the two liquids were mixed with complete aseptic precautions.

It has been found that the method described above for preparing the press juice is also suitable to obtain the contents of bacterial cells, and the pertinent experiments, particularly also with pathogenic bacteria, are proceeding at the Hygienic Institute in Munich.

* * * * *

1. Any water adhering to its surface has been removed to the point that no more water is obtained at a pressure of 25 atm.

2. Powdered glass is less suitable since it acts as a weak alkali.

3. The plant physiologists may decide whether this carbon dioxide is possibly derived from oxidation processes connected with respiration.

4. *Theorie der Gährung* (Theory of Fermentation), Munich, 1879, p. 15.

5. These Berichte 3, 73.

6. *A. Bouffard* has recently again determined the heat developed by budding yeast during alcoholic fermentation. *Compt. rend.* 121:357.

7. In this connection one has to note, however, that the so-called urea fermentation, the decomposition of urea to ammonia and carbon dioxide, in chemical respects differs extraordinarily from actual fermentation processes and hence many regard it not to be a fermentation. It is a simple hydrolysis that can be obtained already with water at 120°.

8. These Berichte 28:3037.

9. This appears possible because of the diosmotic circumstances. Compare *v. Nägeli* l.c., p. 39.

10. In this way one can probably explain the experiments of *J. de Rey-Pailhade* (*Compt. rend.* 118:201) who from fresh brewer's yeast by addition of some glucose prepared a weakly alcoholic (22 percent) extract.

After removal of microorganisms by means of filtration through a sterile *Arsonval* candle, this sugar-containing extract spontaneously developed carbon dioxide in the absence of oxygen.

11. l.c., p. 94. The experiments were repeated with the same success; but it was found that they take place not only in saccharose but also in lactose solutions. The diffusion processes hence are not linked to the fermentative activity, as was assumed by these authors.

# 22

# ON THE INTERVENTION OF MANGANESE IN THE OXIDATIONS BROUGHT ABOUT BY LACCASE

**Gabriel Émile Bertrand**

*This article was translated expressly for this Benchmark volume by H. Friedmann, the University of Chicago, from "Sur l'intervention du manganèse dans les oxydations provoquees par la laccase" in* Acad. Sci. (Paris) C. R. **124**: 1032–1035 (1897).

As I have already stated,[1] laccase, the soluble oxidizing ferment obtained from the lac tree, yields an ash upon incineration that is relatively rich in manganese oxide.

By combining colorimetry with Hoppe-Seyler's reaction, which consists in converting manganese to permanganic acid by the use of lead dioxide and nitric acid, I found that a gram of laccase extracted from annamite lac [Annam was a former kingdom in what is now a part of Vietnam. French rule started in 1883.] contained:

| | |
|---|---|
| Water of hydration . . . | 0.072 |
| Ash . . . . . . . . . . . | 0.046 |
| Manganese . . . . . . . | 0.00117 |

that is to say, the proportion of manganese is roughly 2.5 percent of the weight of the ash.

I then carried out a fractional precipitation with alcohol on an aqueous solution of this laccase and I thus obtained two new samples of ferment, one of which was more and the other less active than the original laccase. Now, on comparing all these samples, I observed that their oxidizing activity varied in proportion to their manganese content.

Thus the volume of oxygen taken up in 1 1/2 hours by 50 cc of a 2-percent solution of hydroquinone in the presence of 0.2 g of the product, assumed to be dry, was:

| | |
|---|---|
| With sample No. 1 . . . . | 19.1 cc |
| With sample No. 2 . . . . | 15.5 cc |
| With sample No. 3 . . . . | 10.6 cc |

while the corresponding amounts of manganese were

| No. 1 | No. 2 | No. 3 |
|---|---|---|
| 0.159% | 0.126% | 0.098% |

Was this a simple coincidence, or, on the other hand, was the activity of the soluble ferment due to the presence of manganese? It appeared to me important to establish this.

With this aim I first tried to eliminate all manganese from the laccase that I had. But since this problem doubtlessly was too finicky, I have not yet been able to resolve it satisfactorily. In certain cases the separation of the metal was too incomplete (and we will see later how far this can be pushed); in the others the reagent simultaneously modified the highly alterable organic material.

Happily I had another method. It is known that laccase, or at least substances very similar to it, can be found in most green plants.[2] I hence extracted it from a series of different species by a slight variation in the operating procedure. Probably because of a special composition of the cellular sap, I was able to obtain from Lucern a material with very little manganese. Under these conditions it was weakly active, but it regained its activity upon addition of a minimal amount of a manganese salt. This product was prepared as follows:

Several kilograms of ordinary Lucern (*Medicago sativa L.*) harvested at the beginning of the flowering season were immediately ground in a mortar and subjected to pressure. The juice, saturated with chloroform, was allowed to coagulate in the dark. After 24 hours the juice was filtered and then 2 1/2 volumes of alcohol were added; the precipitate was drained free of liquid and taken up in a little bit of water. On filtration a clear pale yellow liquid was obtained. Excess alcohol (about 5 volumes) separated nearly white flakes, easily collected and rapidly dried *in vacuo*.

This sample of laccase extracted from Lucern then showed:

| | |
|---|---|
| Water of hydration (determined at $+110^{\circ}$) | 12.4% |
| Organic material | 42.4% |
| Ashes | 45.2% |

and a rather small proportion of manganese, less than 1/50,000.

On dissolving it, using 0.1 g in 50 cc of a solution of hydroquinone, no red color is observed even after 24 hours of continuous shaking in contact with air. On the other hand, if 1 mg of manganese is added to the same solution (for example, as sulfate), the first crystals of quinhydrone can be seen after about 2 hours, clear evidence of oxidation.[3]

The experiment can, besides, be performed quantitatively using the method that I already described[4] and measuring the absorbed oxygen. Then, with the proportions of material given above and uniform agitation for 6 hours at the normal temperature (about $15^{\circ}$), one finds:

| | |
|---|---|
| 1. With manganese alone (control experiment) | 0.3 cc |
| 2. With laccase alone (derived from Lucern) | 0.2 cc |
| 3. With laccase and added manganese | 6.3 cc |

I must add that it did not appear possible to replace manganese to any useful extent by any other metal, not even by iron. I tried the following, always using 1 mg of metal in the form of sulfate: iron, aluminum, cerium, zinc, copper, calcium, magnesium, and potassium.

In no case did the volume of absorbed oxygen exceed a few tenths cubic centimeters. This faint oxidation may, again, be attributed to the very small amount of manganese that remained in the laccase used for the experiment.

These facts stress the physiological importance of manganese and define the role that it plays in plants. The circumstance that manganese is present only in small amounts in living beings actually increases the significance of this conclusion: It directs attention to a whole series of substances that so far were passed off as being of secondary importance because, just like manganese, they were present only in small amounts, and their physiological importance was ignored; examples are: zinc, as illustrated by the experiments of Raulin, boron, which the researches of Pasarini[5] and above all of Jay[6] have demonstrated to be very generally present in plants, and so on.

But still other consequences follow from these facts. I have shown with Mallèvre that pectase cannot transform pectin when it has been completely freed of calcium that accompanies it in cellular saps; that in trying to separate this soluble ferment from carrot juice using alcohol, one obtains an almost inert product not only because the alcohol had changed a part of the organic material but, in addition, because it had been separated from the mineral principle without which it is powerless; addition of a trace of a soluble calcium salt to the pectase solution sufficed to render it active.[7] Today I arrive at analogous results with laccase. This, hence, is a new idea that has been confirmed and that has to be extended. From now on it is necessary, when studying soluble ferments, to pay attention not only to the organic and highly alterable material that has up to the present been exclusively associated with the idea of a soluble ferment. One must, in addition, pay attention to those substances that one may call *coferments* (sometimes, as in the present case, minerals, sometimes perhaps organic), which together with the other material form the truly active system.

In a later communication I will report new facts that, together with the preceding ones, will allow us to interpret the chemical constitution of oxidases.[8]

1. *Comptes rendus* 122:1132 (1896).

2. G. Bertrand, *Comptes rendus* 121:166 (1895).

3. *Comptes rendus* 120:266 (1895).

4. Ibid.

5. *Staz. Sper. Agrar.* 20:471; 21:20 and 565 (1893).

6. *Comptes rendus* 121:896 (1895).

7. *Comptes rendus* 119:1012 (1894) and 120:110 (1895).

8. Work from the Chemical Laboratory of the Museum.

# Part IV

# FIRST MECHANISTIC OBSERVATIONS: ENZYME-SUBSTRATE COMPLEX POSTULATED, STEREOSPECIFICITY, REVERSIBILITY, ENERGETICS, KINETICS, GRAPHICAL METHODS, AND pH

# Editor's Comments on Papers 23 Through 33

Even a cursory reading of the papers in this section reveals that their approaches and conclusions could have been obtained without, and in fact for all practical purposes did not require, the knowledge that occupied us in Part III regarding enzymes versus ferments, or the existence of cell-free fermentation. The studies in the present section, some of which in fact pre-date Buchner's historical paper, have been grouped together because they show a physicochemical emphasis and because they tend to address themselves to the first of three enzymological questions posed at the beginning of this work: What do enzymes do? This question, in its various ramifications, could be studied as an abstraction from the sum total of enzymological experience precisely because of the generality of its application, irrespective of the detailed mechanism, nature, or function of the particular enzyme being studied. In fact, as in the classical papers of Emil Fischer included here, conclusions extending to areas of enzyme structure and function could often be drawn from various studies on enzyme mechanisms. Most of the papers in this section deal with invertase, which, as we saw earlier, served for a great many early enzyme studies, and with related hydrolases.

Parts of the long paper by Cornelius O'Sullivan (1842–1907) and Frederick William Tompson (1859–1930) on yeast invertase are reproduced here (Paper 23) not as an early study on enzyme kinetics (as pointed out by A. J. Brown [Paper 28] and others, they were wrong in concluding that the course of inversion was unimolecular), but because this is one of the very first papers that led to the conclusion that an enzyme-substrate complex is formed. The evidence for this conclusion is based on the observation that invertase is more heat stable in the presence than in the absence of cane sugar, its substrate. Actually a quite direct demonstration for the formation of an enzyme-substrate complex had been given ten years earlier by the eminent French chemist Adolphe Wurtz who, as mentioned in the Editor's Comments to Part I, with Bouchut (1879) had carried out the first scientific studies on papain and introduced this name for the enzyme. In 1880 he demonstrated that papain is completely removed from solution by fibrin and that the enzyme cannot be detached from the insoluble material by thorough washing (Wurtz 1880). He concluded that "papain acts by becoming attached to fibrin, and that the insoluble product [is] perhaps a combination of fibrin and papain." He cited the analogy of sulfuric acid, which catalyzes hydrolytic reactions via the "short-lived formation of combinations which are ceaselessly made and unmade."

Emil Fischer's 1894 paper (Paper 24) compares the activities of two different glucosidases, invertin[1] and emulsin,[1] on a variety of natural and synthetic sugars or sugar derivatives. This study was suggested by earlier observations in Fischer's laboratory that showed—in analogy to Pasteur's studies (Pasteur 1858) on various stereoisomers of tartaric acid—that intact yeast distinguishes between stereoisomeric hexoses. In Paper 24 that basis for this type of distinction is sought by means of *in vitro* enzyme studies. The fundamental observation is made that substrates for invertin are not substrates for emulsin and vice versa. On the basis of these simple experiments, Fischer arrives at far-reaching conclusions about (1) enzyme mechanisms, (2) enzyme structure, (3) the analogy between *in vitro* and *in vivo* enzymatic activities, and (4) the use of enzymes for structural studies. Perhaps even more impressive as a masterful deductive conclusion is the most celebrated part of this classical paper, the famous lock-and-key analogy to explain interactions between enzymes and substrates. It should be noted that this picture, deduced from the stereochemical experiments described, is with admirable restraint limited to them. Furthermore, this deduction from the observed stereospecificity was at one and the same time helped by and found consistent with the view that enzymes, as proteins or as substances very similar to proteins, are undoubtedly themselves asymmetric molecules. It must be realized that by 1894 not only the phenomenon of asymmetry and the notion of biological stereospecificity, both based on Pasteur's work, were very firmly established, but the subsequently enunciated van't Hoff and Le Bel theory of the tetrahedral direction of the carbon valences was already twenty years old.

Our second Fischer paper (1898/1899) (Paper 25) was published in a physiological chemistry journal shortly after Buchner's demonstration of cell-free fermentation, and it addresses itself more directly than the earlier one to the biological implications of Fischer's preceding enzyme studies. Fischer states near the beginning of this paper that his studies on fermentation from the viewpoint of stereochemistry are "distributed in a number of separate papers published in the Berichte [Reports] of the German Chemical Society, 1894 and 1895, and are often buried in the predominantly chemical material so that it appears justified to summarize them here... ." Only a few excerpts from this long paper are translated here. From these it will be seen that this paper is more than a mere elaboration of earlier work. Thus the phenomenon of the stereospecificity of enzymes is invoked to explain asymmetric synthesis in living forms. In addition it is recognized that enzyme purification and structure determination are needed to prove the assumption that enzymes are proteins. It is rather ironic to realize in retrospect that it was precisely the successful use of new methods of enzyme purification, primarily by the school of another eminent organic chemist, Richard Willstätter (see Papers 34 and 35), that temporarily delayed the general acceptance of the idea that enzymes are indeed proteins. Many years, again, were to separate the first crystallizations of enzymes from the determination of their structures.

(Hermann) Emil Fischer (1852–1919) was a giant among giants in organic chemistry at a time when this science was reaping the fruits of the first half-century or so of flowering since Wöhler's synthesis of urea. Fischer's analytic and synthetic work extends over an enormous range of biologically fundamental compounds: carbohydrates, purines, amino acids (of which he discovered valine, proline, and hydroxyproline), proteins, and neutral fats. At the same time there is a breadth of vision in his work that extends beyond the confines of what a later, more specialized age has associated with the stereotype of the organic chemist. It has been aptly stated, "The aim of all his investigations was to apply the methods of organic chemistry to the synthesis and processes of substances of living matter" (Farber 1972). In his celebrated Faraday lecture, delivered in 1907 and still fascinating to read, Fischer reminded his audience that organic chemistry had close connections to biology during its initial development in the days of Liebig and Dumas. In the intervening period organic chemistry had to separate from biology in order to elaborate its own theories and a "powerful armoury of analytical and synthetical weapons." The time had come, in Fischer's opinion, that this alliance should be re-established "both to its [organic chemistry's] own honour and to the advantage of biology" (Fischer 1907). Fischer received the Nobel Prize in chemistry in 1902.

The study by the young medical student Arthur Croft Hill (1863–1947) (Paper 26) is of historical importance since it is the very first to demonstrate the enzymatic synthesis of a molecule (maltose, actually iso-maltose) by an enzyme. We know now that the biosynthesis of this substance by this reaction is of no physiological importance, but these observations were in accord with Wilhelm Ostwald's view that catalysts speed up both directions of reversible reactions, and they hence confirmed the notion that enzymes are true catalysts. Second, this work led to the conclusion that enzymatic versatility extends to biosynthetic processes that hence do not require the participation of a special vital force. As pointed out by Kohler (1973), "Croft Hill's success had an immediate impact." Furthermore, Jacques Loeb (quoted by Kohler, 1973) later wrote, "It is no exaggeration to say that Hill's paper entirely changed the conceptions of the physiology of metabolism" (Loeb 1906, p. 9).

Hill's paper is quoted in van't Hoff's address of the same year (1898) (Paper 27) that discusses, *inter alia,* the thermodynamic constraints imposed upon catalyzed as well as on uncatalyzed reactions. Jacobus Hendricus van't Hoff (1852–1911) was professor at the universities of Amsterdam, Leipzig, and Berlin. He was one of the founders of physical chemistry as an independent discipline and made fundamental contributions to stereochemistry, reaction kinetics, the thermodynamics of chemical reactions, the influence of temperature on reaction rates and so on. He was the cofounder, with Wilhelm Ostwald, of the *Zeitschrift für physikalische Chemie* (1887) and was the first Nobel Laureate in chemistry (1901).

The study on yeast invertase by A. J. Brown (Paper 28) for the first time deduces the formation of an enzyme-substrate complex not from heat-stability data (O'Sullivan and Tompson) or from direct binding studies (Wurtz), but from careful rate measurements that indicated the initial rate of cane-sugar hydrolysis, except for very dilute solutions, to be independent of the concentration of cane sugar used. It was concluded that mass action is concealed in the case of this substrate-independent rate by a time factor associated with molecular combination.

Victor Henri (Paper 29) was the first to derive a hyperbolic rate equation for a single-substrate enzymatic reaction, equivalent to the equation used at the present time. He noted that the validity of a given rate equation does not necessarily prove the validity of the postulated mechanism. He showed that his rate equation predicts the catalyzed rate to be proportional to the substrate concentration at low, and independent of it at high substrate concentrations. An elegant analysis of his assumptions has been provided by Harold L. Segal who notes that Henri's contribution consists, in addition to the derivation of the rate equation, in the clear exposition of the general process used to derive such equations "under any circumstances in enzyme experiments, a process which is at the heart of all kinetic studies" (Segal 1959). It may be noted, again (see the Editor's Comments to Part II), that Henri uses the word *diastase* in the title and text of his book rather than *ferment* or *enzyme.*

Victor Henri (1872–1940) had a French father and a Russian mother. He began his scientific career in the fields of psychology and physiology. While still in his twenties he published numerous articles and books in these fields, some with Alfred Binet, the inventor of the I. Q. test. In 1897 he received a doctorate from the University of Göttingen for a thesis on tactile sensations. Already at the age of 23 his name appears on the title page of the first volume of the yearly publication *L'Année Psychologique*, whose chief editor was Binet. Henri's association with this respected publication continued for ten years until 1904, not long after the appearance of yet another dissertation, this time in physical chemistry. The break with Binet must have been rather sudden, for although volume 10 of *L'Année Psychologique* announced an article with a metaphysical title by Binet and Henri to appear in volume 11, it never did, and Henri's name disappeared from the title page with this same volume.

Henri quickly rose to become the "undisputed leader of French physical chemists" (v. Halban 1941). Already in 1914 he was elected to be president of the French physicochemical society. In 1953 in conjunction with its fiftieth anniversary, this society held a special celebration commemorating Henri's achievements in fields including, for example, various aspects of spectroscopy and chemical kinetics (see *Commémoration de Victor Henri*, 1953). His kinetic studies on hemolysis led to the proposal of "single-hit" theories of toxicity, and his photochemical studies resulted in the proposal and proof of chain reactions.[2]

The 1903 dissertation is none other than the classical study in enzyme kinetics from which our excerpt is taken.[3] This topic comes from a transition period in Henri's crowded career that required both a biological and a physicochemical emphasis. Since this dissertation, which was published as a book, is not readily available and since its elegant, concise, and remarkably modern-sounding preface has never been quoted or translated before, most of the preface is given here as a tribute to a clear and incisive thinker:

> When a theory tries to explain a group of phenomena by invoking the existence of absolutely new forces or forms of energy, two different cases are possible: these new factors, whose existence is assumed, can either be analyzed experimentally and hence can permit a deeper understanding of the mechanism of the phenomena studied, or else these factors cannot be investigated experimentally. In the first case the theory necessarily leads to the discovery of new facts, it will give rise to a whole series of experimental inquiries and hence, whether true or false, it will contribute to the development of science. In the second case, however, the theory will lead to purely speculative answers which will readily become dogmatic

> and which in that case can put an end to experimental inquiries.
>
> In the course of study of the general phenomena of the life of organisms two groups of theories have been advanced: one of these considers that vital manifestations are due to physico-chemical activities, the other group renounces this reduction and admits the existence of new forms of energy subsumed under the name of "vital." Since experimentation is based on the methods and the ideas of chemistry and physics, the vitalist theories renounce the possibility of performing experiments with this vital force; they are a sort of brake which stops experimental, i.e. scientific inquiry, and they transport discussion from the region of experiment to the field of speculation. These theories are hence detrimental in practice, since the usefulness of any theory is measured by the number and the importance of new facts which it helps to discover.
>
> Among the phenomena that occur in the living organism, the processes of nutrition, of absorption, of secretion and of the defense of the organism's cells, in sum the whole metabolism of the living being, can be regarded to be almost completely reducible to diastatic activities; actually as the number of experimental investigations increases, the number of activities formerly considered intimately linked to the vitality of the cells themselves decreases; the investigations of Buchner, E. Fischer and others yield both striking and suggestive examples. Hence interest in the study of the laws of diastatic activities becomes greater and greater all the time.
>
> The explanation of these diastatic activities has given rise to many discussions and theories which can again be divided into two groups: one of these does not allow diastatic activities to be classified along with the activities of general chemistry, it assigns them to a special group with its own laws; the other group, however, tries to lead diastatic activities back to the laws of general chemistry, and in this fashion these theories make some contributions to the theories which explain vital phenomena by means of physico-chemical activities.
>
> The present work has as its object the general laws of diastatic activities. Its aim is to study the activities of diastases by the use of the methods and results of physical chemistry... .

Henri's rate equation for an enzyme reaction, as an expression of a saturation phenomenon, was derived by Briggs and Haldane via another approach. The publication by Briggs and Haldane (Paper 30) on the steady-state derivation of the single-substrate enzyme saturation curve is remarkable for its lucidity and brevity. It is this treatment, which makes no assumptions about the equilibrium conditions and about the relative magnitudes of the various rate constants, that is used in most elementary textbooks and that is learned by almost every student of an introductory biochemistry course. It should be noted that the concept of the steady state had already been introduced twelve years earlier (in the same year as the equilibrium treatment of Michaelis and Menten [1913] by Bodenstein (1913) in a lengthy paper on the theory of photochemical reaction velocities.

John Burdon Sanderson Haldane (1892–1964), son of the eminent physiologist John Scott Haldane (1860–1936), was one of the most versatile scientists in the first sixty-odd years of this century. There is no space to list all the accomplishments of this remarkable man. He was a physiologist, biochemist, and geneticist, a great popularizer of science, and he originated, independently of Oparin in Russia, ideas on the origin of life. Seven years before his death from cancer, he left his native Britain to live in India where, after an unsatisfactory stay at the Indian Statistical Office in Calcutta, he set up a laboratory of genetics and biometry in Bhubaneswar, the state capital of Orissa. When his cancer was diagnosed in London shortly after he had attended a conference on the origin of life in Florida, he wrote a poem in a typical vein, "Cancer's a Funny Thing" that starts:

> I wish I had the voice of Homer
> To sing of rectal carcinoma.

Six of Haldane's classical papers on a mathematical theory of natural and artificial selection, along with related papers by Ronald Fisher and Sewall Wright, are included in the Benchmark volume *Evolutionary Genetics* (Jameson 1977).

The conversion of the Briggs-Haldane enzyme rate equation from hyperbola to straight line by the simple expedient of plotting the reciprocal of the initial velocity against the reciprocal of the substrate concentration is usually associated with the names of Lineweaver and Burk (1934). However, the consequence of the double-reciprocal plot, and of other expressions of the rate equation in linear forms, were first published in Kurt G. Stern's German translation (1932) of J. B. S. Haldane's book *Enzymes* (English edition, 1930). The derivation is clearly attributed to Barnet Woolf. Haldane's short note from *Nature* indicates that Woolf (who is represented in this work by Papers 43 and 44) did not publish this material due to illness.

Two aspects of enzyme studies, namely, that the activity of most enzymes is strongly affected by changes in the hydrogen ion concentration and that buffers are used to keep this hydrogen ion concentration constant, are now commonplace. The effect of changes in acidity on enzyme activity had, in fact, been recognized long before Sørensen's classic 1909 paper (see Paper 33). Furthermore, phosphate buffers had already been invented (Fernbach and Hubert 1900) precisely for the purpose of controlling the acidity of a protease from malt. Sørensen's work consolidated these scattered observations. His work, performed long after Arrhenius's demonstration of the partial dissociation of weak electrolytes, firmly established the importance of distinguishing in enzyme studies between the total or titrable acidity and the actual hydrogen ion concentration, which Sørensen expressed by means of the pH scale. Sørensen's great paper deals with different ways of measuring the latter (both electrometrically and with numerous dyes whose use he introduced), with the invention and standardization of a number of new buffers, and with the effect of changes in the hydrogen ion concentration on the activity of various enzymes. Indeed, one may say that Sørensen's invention of the pH scale was dictated by two factors: a strongly perceived need to replace the cumbersome numbers that express the actual hydrogen ion concentration encountered under most physiological conditions with a convenient scale, and the recognition that a way had to be found to drive home the importance of hydrogen ion concentrations in enzyme studies. Enzymologists will be gratified to know that the notions both of buffers and of pH were dictated as a consequence of studies on enzymes. Sørensen's paper, published simultaneously in three languages, is most remarkable for elegance, meticulous attention to detail, and reliability. A greatly expanded version was published some three years later (Sørensen 1912).

Søren Peter Lauritz Sørensen (1868–1939) received most of his training in inorganic chemistry. He turned to biochemistry, mainly amino acid and protein studies, after being appointed in 1901 to succeed Johan Gustav Kjeldahl (1849–1900) as director of the chemistry department of the Carlsberg Laboratory in Copenhagen. He kept this position until the year before his death. The phosphate salts $KH_2PO_4$ and especially $Na_2HPO_4 2H_2O$ are named after him. His formol titration method was long used for the quantitation of protein amino groups. He and his wife Margrete investigated topics such as lipoproteins and complexes between carbon monoxide and hemoglobin. They carefully studied crystalline egg albumin (1917). His demonstration by osmotic pressure measurements in 1917 that this purified protein had a molecular weight of 34,000 had a revolutionary effect on protein chemistry, since such a value was far higher than any that had up to that time been expected for any protein.

Although Sørensen's classic work is of signal importance in any study of the development of biochemistry, it must be stressed that work published in numerous papers from the laboratory of Leonor Michaelis in Berlin was highly influential in furthering the notion of the importance of the influence of hydrogen ion concentration on enzyme activity. "Michaelis and his collaborators made pH control a routine characteristic of all serious enzyme studies" (Cornish-Bowden 1979). In fact, Michaelis and Davidsohn (1911a), in a paper published shortly after Sørensen's, stressed that they had studied the effect of hydrogen ion concentration on invertin independently of Sørensen, but that they did not publish their data because of the appearance of Sørensen's paper.[4] Michaelis, in characteristic fashion, immediately proceeded to extend the theoretical basis of the meaning of the pH dependence of enzyme reactions (see Michaelis 1958).

## NOTES

1. It must be pointed out that in this study on stereospecificity Fischer uses the term *invertin* for an α-glycosidase extracted from *dried* yeast that acts on maltose (4-*O*-α-D-glucopyranosyl-D-glucopyranose). This is not the classical β-glycosidase, namely β-D-fructofuranosidase that acts on sucrose and that was extracted from *fresh* yeast by Berthelot (Paper 13). It is this latter enzyme, and not the former, that played an exemplary role in the development of enzymology (as mentioned near the end of the Editor's Comments to Part III) and that in our book is the subject of the studies, in addition to Berthelot's, by O'Sullivan and Tompson (Paper 23), Brown (Paper 28), Henri (Paper 29) and Sørensen (in an untranslated portion of Paper 33). In the present paper (at the end of the section on invertin) Fischer in fact expresses doubts about the identity of his "invertin" preparation with commercially available invertin, and he discusses possible reasons for their specificity differences. In two subsequent papers (Fischer 1894, 1895) Fischer proposes to restrict the term invertin to the enzme that hydrolyzes cane sugar rather than maltose or synthetic methyl α-glucoside, and he suggests the name maltase (first used by Bourquelot) for the maltose-hydrolyzing enzyme. The invertin of Paper 24 is hence the same as present-day yeast maltase. The source of the confusion is due not only to the fact that yeast contains both an α-glycosidase and a β-glycosidase, but also to the circumstance that sucrose (1-*O*-α-D-glucopyranosyl-β-D-fructofuranoside) is both an α-glucoside and a β-fructoside, so that optical inversion can be brought about by cleavage of either the α- or the β-glycosidic bond. As recently as 1950 Neuberg

and Mandl in a classic review article applied the term invertase to both $\alpha$-glycosidase and $\beta$-glycosidase, although they stated that the term *yeast invertase* should refer only to the $\beta$-glycosidase (Neuberg and Mandl 1950).

As to the term *emulsin,* in Fischer's paper this refers to the type of glycosidase, discovered by Wohler and Liebig (see endnote 10, Part II) in studies on the hydrolysis of amygdalin, which attacks cert B-glucosidic bonds. This term is no longer in use. It is now known that the classic emulsin from almonds was actually a mixture of enzymes of different stereospecificities (see Pigman 1944; Gottschalk 1950).

A word of explanation is needed about the kefir grains that Fischer used to hydrolyze lactose. These grains resemble miniature cauliflowers. They consist of casein, lactose-fermenting *Torula* yeasts and various bacteria including a strain of *Lactobacillus brevis,* formerly called *Lactobacillus caucasicus,* and glycogen-containing rod-shaped kefir bacilli. At least some of these microorganisms probably exhibit a symbiotic relationship. These grains change milk to the refreshing and nutritive material known as kefir by bringing about simultaneous alcoholic and lactic acid fermentation. Kefir is used chiefly in southern Russia. Kefir production is peculiar in that the inoculum consists of these grains that increase in size during the fermentation and can be strained out, dried, stored over long periods, and used again (see Webb and Johnson 1965). A complicated material such as kefir grain is not an ideal choice as an enzyme source, as Fischer recognized. However, the identity of the various yeasts that he would have preferred to use but that were not available to him is uncertain.

To appreciate Paper 24 it suffices to see that Emil Fischer had the ingenuity to recognize two glycosidases that differed in their stereospecificities for the types of glycosidic bond at the anomeric carbon, and that his far-reaching and elegant observations and interpretations were based on this fact.

2. Henri has been credited (Stent and Calendar 1978) with the discovery in 1914 of UV-induced bacterial mutagenesis, some 13 years before H. J. Muller's discovery of the mutagenic effect of X-rays on fruit flies. Actually, in 1911 Victor Henri and his wife had already received a Prix Bellion for various studies on the action of UV light on various microorganisms and for the effect of UV on cancerous tissues. However, while Victor Henri proceeded among other UV studies to initiate the use of UV absorption spectra to follow chemical reactions (Henri and Landau

1914), it was his wife who first demonstrated in two remarkable papers that anthrax bacteria undergo mutations after exposure to nonlethal doses of UV light (Mme. Henri 1914*a*, *b*). There is one joint paper by the Henris on this subject from the same year (Henri and Henri 1914). World War I interrupted these studies and the subject was not subsequently pursued by them.

3. Henri's enzyme studies received immediate recognition. Already in 1903 he received a Prix Montyon and a Berthelot medal for this work, and two years later he received a Prix Philipeaux for his studies on enzymes and on the hemolysis of erythrocytes.
4. It must be mentioned that the term buffer, used by Sørensen and translated from the French *tampon* suggested by Fernbach and Hubert (see Fruton 1972, p. 143), was not immediately accepted universally. Thus Michaelis and Davidsohn use the word "regulator" rather than "buffer" in their papers on invertin (1911*a*) and on trypsin (1911*b*). In his influential monograph on hydrogen ion concentration, Michaelis (1914) already uses both terms, and a few years later he has dropped the term "regulator" (Michaelis and Rothstein 1920).

# 23

Reprinted from pages 835–836, 838–839, 918–919, 921, and 926–929 of *Chem. Soc. J. (Trans.)* **57**:834–931 (1890)

# INVERTASE: A CONTRIBUTION TO THE HISTORY OF AN ENZYME OR UNORGANISED FERMENT

**C. O'Sullivan, F. R. S., and F. W. Tompson**

[*Editor's Note:* The table of contents has been omitted.]

## Part I.—General and Introductory.

### *Objects of the Work.*

The group of bodies known as unorganised ferments or enzymes are, as far as the products of their action are concerned, well defined; but, notwithstanding much useful and valuable work, we know very little of their mode of action and far less of their chemical constitution. They are a highly interesting and important class of substances, and are named unorganised ferments in contradistinction to the living or organised ferments, because they possess a life function without life. The function or functions of the organised ferments are highly complicated, while the function of the unorganised ferments is more or less a simple breaking down of some organic compound. They may be properly called *the transforming* ferments, for their function is a transforming or breaking down of one substance into two or more others. They may be looked upon as the reagents of life, and they are all products of it. Their function is a life function. Is there anything in this which can be distinguished from ordinary chemical action? If so, what?

Roberts, in his admirable little work on "The Digestive Ferments," holds that the living organism by which they are produced imparts to them a certain definite amount of vital force. This force is stored up in the ferment until it comes in contact with the particular substance upon which it is intended to act, and then it is used in altering it. This it continues to do until the whole of the vital force is exhausted. When this happens the ferment is still left—but it is dead.

Other observers think that the action of the enzyme is an ordinary chemical reaction, the ferment being either destroyed in the process, or remaining uninjured as in catalytic reactions.

To throw some light on these points was one of the objects of this work; the other was to arrive at some knowledge of the chemical constitution of the active substance. The generally received idea is that they are the products of some change in albuminoïds, but Hartley has shown that some of them at least do not act upon the spectrum in the same way as albuminoïds; and the published analyses represent them as containing far less nitrogen than that generally yielded by those bodies.

Of the many members of this class of compounds there are few that can be easily prepared in any quantity, and very few the function of which can be followed with moderate ease and sufficient accuracy. Our choice of material fell upon *invertase*, or, as it is often called, invertin. We prefer the former name, because we consider the ending (*ase*) should be the distinguishing affix of the names of this group. The reasons that we chose invertase were that we succeeded in discovering a method by which it is to be obtained in abundance, and because its function can be followed with ease and precision. We hope that the light thrown upon the mode of action and chemical constitution of one member of the group will be a key to a wider knowledge of the others.

[*Editor's Note:* Material has been omitted at this point.]

*Mode of Procedure.*

We have already stated that the two great questions we wished to solve by this work were, first, the nature of the inverting power or function of invertase; and, secondly, the chemical composition of the active agent. The function of invertase is to transform sucrose according to the equation—

$$\underset{\text{Sucrose.}}{C_{12}H_{22}O_{11}} + H_2O = \underset{\text{Dextrose.}}{C_6H_{12}O_6} + \underset{\text{Lævulose.}}{C_6H_{12}O_6}.$$

Our first object was to determine if by the aid of what we could learn from the above-mentioned publications, and by any further experiments that might suggest themselves to us, we could find some quantitative expression for this function. We have so far no other factor which acts as a criterion of the purity of the substance, and no means of estimating it. We merely know that invertase is the only known substance which transforms cane-sugar, as described above, in the cold. We have no characteristic to guide us but this. Hence, when we have found that a certain product possessed this action, we have considered that invertase was present, whilst, on the contrary, if it did not invert cane-sugar under suitable conditions, we always considered that invertase was absent. From this it will be seen that we discarded the idea of such a substance as *dead* invertase. If a solution which was formerly known to contain invertase was found, under suitable circumstances, to have no power of inverting sucrose, we concluded that invertase was no longer present, and that the chemical substance, invertase, had been in some manner changed into another compound, or compounds.

It was clear at starting that we had to establish a method of estimating a *function* and not a quantity, or rather we had to find a quantitative expression for a function.

For this purpose it was necessary to make a choice between two theories for a working hypothesis. If Roberts is right, it is obviously of no use for our purpose to estimate the strength of such a variable quantity as the unspent vital force. If, on the contrary, the chemical theory is correct, it is without doubt possible not only to prove the presence or absence of invertase, but also its relative amount.

Once the laws which govern the reaction properly understood, it would, by estimating the intensity of the function, or, in other words, the extent or speed of the reaction, be easy to calculate *relatively* the amount of invertase present, just as we might estimate the amount of free sulphuric acid by its action on a boiling solution of sucrose under standard conditions. And invertase has one great advantage over sulphuric acid, in that there is no other known substance that acts in the same manner, and therefore we are not misled by the presence of other inverting agents.

This necessitated a complete study of the action of invertase on sucrose.

[*Editor's Note:* Material has been omitted at this point.]

## PART V.—THEORETICAL.

### *Theory of Inversion by Invertase.*

When we look back over the whole of the facts we have adduced with regard to the action of invertase on cane-sugar, we are forced to admit that there is not much which throws light on the actual chemistry of the reaction. We have, however, shown that hydrolysis by invertase is a simple chemical reaction, differing in no important particular from the ordinary chemical reactions.

From the time curve of the reaction it is easy to make the deduction that the invertase itself is not injured or destroyed by the act of inversion.

We have shown that the time curve is the one that expresses a chemical change "of which no condition varies excepting the diminution of the changing substance." It therefore follows that the activity of the invertase does not vary during the reaction.

The same argument shows that the products of inversion have no influence on the pace of the reaction.

A necessary corollary of the fact that invertase is not injured by inversion is that there is no limit to its inverting power. This we have practically shown to be the case.

In these respects the action of invertase on cane-sugar agrees very closely with that of boiling acids. We have little doubt that this reaction would follow the same curve as the one under consideration, though we have seen no figures bearing on the point.

In Table XXXIV we have shown that invertase when in the presence of cane-sugar will stand without injury a temperature fully 25° higher than in its absence. This is a very striking fact, and, as far as we can see, there is only one explanation of it, namely, the invertase enters into combination with the sugar. It is very difficult to see how it can otherwise be explained. Further, this combination between the invertase and the sugar must last until the compound molecule comes in contact with another molecule of cane-sugar, otherwise towards the end of the reaction, when cane-sugar is getting scarce, the invertase would be destroyed by the heat in the intervals between the combinations. It might be thought that our experiment on the molecular weight of cane-sugar undergoing inversion would throw some light on this point, but it does not throw much, because we were obliged by the conditions of experiment, to make allowance for a constant error in the observations, and this constant error would include the error due to the combination between the invertase and sugar if it existed. It may be held, however, to show that invertase does not combine with many molecules of sugar at the same time. We are not able to give further proof of the combination, which we believe to exist, between invertase and sugar, but it is obvious if our idea is right that the sugar must be invert-sugar, not sucrose, otherwise if much invertase was used the optical activity of the solution at the end of the reaction would not show complete inversion.

[*Editor's Note:* Material has been omitted at this point.]

In conclusion we are able to state, from a consideration of the above results, that the invertase simply acts the part of a carrier of water to the sucrose, and that in doing so a loose combination is formed between the invertase and the inverted sucrose, and that the invertase is again freed from this combination in its former active state.

[*Editor's Note:* Material has been omitted at this point.]

## PART VI.—SUMMARY.

The following is a short summary of the principal results of our work, arranged as much as possible in the same order as has been observed in the body of the paper:—

1. The rate of inversion of cane-sugar by means of invertase may always be expressed by a definite time-curve; this curve is practically that given by Harcourt as being the one expressing a chemical change "of which no condition varies, excepting the diminution of the changing substance." Whatever the conditions may be under which inversion is taking place, as long as these conditions remain unchanged, this curve is adhered to. There are, however, some slight, but apparently constant, deviations from the theoretical curve.

2. When the degree of acidity is that most favourable for the action of invertase, the rapidity of the action is in proportion to the amount of invertase present.

3. The most favourable concentration of the sugar solution at a temperature of 54° is about 20 per cent. Below that, there is a rapid decline in the speed of inversion. Greater concentrations are only slightly less favourable until about 40 grams per 100 c.c. is reached. In saturated solutions inversion only proceeds with extreme slowness.

4. The speed of inversion increases rapidly with the temperature until 55—60° is reached. At 65°, the invertase is slowly destroyed,

and at 75°, it is immediately destroyed. At the lower temperatures, the speed of the action increases with rise of temperature in accordance with Harcourt's law, the rate being about doubled for 10° rise, but above 30° the increase is not nearly as rapid. Elevated temperatures have no permanent effect on the activity of invertase, so long as they are not sufficiently high to destroy it.

5. The caustic alkalis, even in very small proportions, are instantly and irretrievably destructive of invertase.

6. Minute quantities of sulphuric acid are exceedingly favourable to the action, but a slight increase of acidity beyond the most favourable point is very detrimental. The most favourable amount of acid increases to some extent with the proportion of invertase, and decreases with rise of temperature, but we have not been able to discover on what it depends. We find that in studying the action of invertase, it is of the utmost importance that the most favourable amount of acid should be employed, otherwise correct results cannot be obtained. At a temperature of 60°, the action is almost stopped, unless exactly the right amount of acid is used, whilst if this factor is properly adjusted, inversion proceeds at (probably) the maximum speed.

7. The influence of alcohol varies in direct proportion with the amount present. 5 per cent. of alcohol decreases the speed of the action by about one half.

8. The dextrose formed by the action of invertase is initially in the birotary state, and, therefore, the optical activity of a solution undergoing inversion is no guide to the amount of inversion that has taken place.

9. If a caustic alkali be added to a solution undergoing inversion, and the optical activity be allowed sufficient time to become constant, it is a true indicator of the amount of inversion that had taken place at the moment of adding the alkali.

10. A sample of invertase which had induced inversion of 100,000 times its own weight of cane-sugar was still active; and we have shown that invertase itself is not injured or destroyed by its action on cane-sugar. Under these circumstances there is evidently no limit to the amount of sugar which can be hydrolysed by a given amount of invertase.

11. The inversion of cane-sugar by means of invertase is a simple chemical change, differing in no important way from those which inorganic substances undergo.

12. The products of inversion have no influence on the rate of the action.

13. A solution of invertase will withstand a temperature 25° higher in the presence of cane-sugar than in its absence. From this fact we

are of opinion that when invertase hydrolyses cane-sugar, combination takes place between the two substances, and the invertase remains in combination with the invert-sugar. The combination breaks up in the presence of molecules of cane-sugar.

14. A means of estimating the activity of a material containing invertase has been devised; the result is recorded by means of the time factor, $\pm 0 = x$ min. In this equation, $\pm 0 =$ a certain definite amount of work, and $x =$ the time necessary to perform it. The expression means that the given inverting material takes $x$ minutes to invert a standard amount of cane-sugar under standard conditions. The number $x$ varies in inverse proportion to the actual amount of invertase contained in the material or materials under examination.

15. If sound brewer's yeast be pressed, and then kept at the ordinary temperature for a month or two, it does not undergo putrefaction, but changes into a heavy, yellow liquid. The product possesses no power of fermentation, but an apparent increase takes place in the inverting power.

16. From such liquefied yeast, it is easy to filter off a clear solution of high hydrolytic power; and we have shown that all the invertase of the yeast is in this solution. We have named it "yeast liquor," and it has a specific gravity of about 1080. It will remain for a long time unaltered, excepting that the colour darkens. If exposed to the air it may slowly become covered with mould.

17. If spirit be added to yeast liquor until the mixture contains 47 per cent. of alcohol, the whole of the invertase separates with only a slight loss of power. This precipitated invertase may be washed with spirit of the same strength and then the residue either dehydrated with strong alcohol and dried *in vacuo*, or else it may be extracted by means of 10 to 20 per cent. alcohol and then filtered. The filtrate contains the invertase. On one occasion the extent of the loss involved by this process was determined and it was found that all the invertase of the yeast liquor was present in the filtrate except 12·3 per cent.

18. We have not been successful in further purifying invertase preparations carefully made in this manner, as the slightest attempt at purification destroys the invertase. We have succeeded in preparing invertase almost free from ash.

19. The inverting power of pressed English yeast varies from $\pm 0 = 1000'$ to $\pm 0 = \pm 3000'$; this is equivalent to about $\frac{1}{3}$ of these numbers on the dry solid matter of the yeast.

20. The inverting power of the most active invertase preparation made was $\pm 0 = 25{\cdot}1'$ on the dry solid matter. We believe that pure invertase would approximately have $\pm 0 = 22{\cdot}5'$.

21. The dry solid matter of yeast contains from 2 to 6 per cent. of invertase. 5·8 per cent. of invertase was separated from one sample of yeast.

22. During the preparation of invertase from yeast liquor an albuminoïd is obtained, which is not redissolved by water. This is termed yeast albuminoïd, and it possesses all the characteristics of an albuminoïd.

[*Editor's Note:* Material has been omitted at this point.]

# 24

# INFLUENCE OF CONFIGURATION ON THE ACTIVITY OF ENZYMES

**Emil Fischer**

*First Berlin University Laboratory*

*This article was translated expressly for this Benchmark volume by H. Friedmann, the University of Chicago, from "Einfluss der Configuration auf die Wirkung der Enzyme" in* Dtsch. Chem. Ges. Ber. **27**:2985–2993 (1894).

The difference in the behavior of the stereoisomeric hexoses toward yeast has led *Thierfelder* and me to the hypothesis that the active chemical agents of the yeast cell can attack only those sugars whose configuration is related to that of these agents.[1]

This stereochemical view of the process of fermentation was bound to become more plausible when similar differences could also be demonstrated in the case of the ferments that one can separate from the organism, that is to say, the so-called enzymes.

I have now succeeded in demonstrating this unequivocally, using first of all two glucoside-cleaving enzymes, invertin and emulsin. This became possible via the synthetic glucosides that can be prepared in large numbers from different sugars and alcohols by the procedure that I discovered.[2] For comparative purposes, however, several natural products of the aromatic series and also some polysaccharides that I regard as the *glucosides of the sugars themselves* were included in this investigation. The result of this investigation may be summarized by stating that the action of the two enzymes is strikingly dependent on the configuration of the glucoside molecule.

## EXPERIMENTS WITH INVERTIN

The enzyme may, as is known, be extracted from brewer's yeast with water. Apparently alcohol precipitates it from this solution in an unaltered form. For reasons to be given later, I decided not to isolate it. The following experiments have, rather, been carried out directly with a clear solution, obtained by filtration of a 15-hour digest consisting of 1 part of air-dried brewer's yeast (*Saccharomyces cerevisiae* type *Frohberg*, pure culture) and 15 parts of water at 30-35°.

*Alcohol glucosides*. As I already reported earlier, the glucoside formula, which my experiments showed to be very likely, allows one to predict the existence of two stereoisomers that differ only in the arrangement of the glucosido group at the asymmetric carbon. For the derivatives of the hexoses, these would have the following configurations [note that in 1894 it was not yet known that hexoses are in the pyranose rather than in the furanose form]:

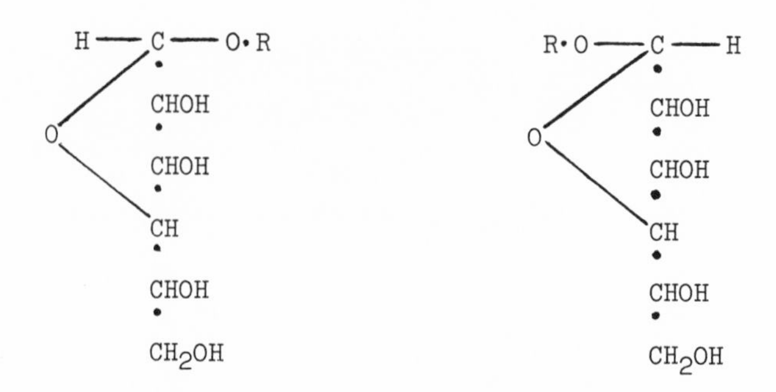


I have isolated one of the two compounds of grape sugar with methyl alcohol and described it as methylglucoside. The second compound is formed simultaneously, as I quite correctly assumed, and is found in the syrupy mother liquor.[3] However, *Alberda van Ekenstein*[4] got ahead of me with its isolation. A sample of the crystals that he kindly gave to me enabled me to obtain this substance in crystalline form as well.

I propose that the new isomer be distinguished as β-compound from the older α-methylglucoside and that for the time being the same manner of designation be used for all isomers of the same type. As I have already reported earlier, α-methylglucoside is cleaved by invertin. After an amount of enzyme ten times larger than the above had acted on it for 20 hours at 30-35°, about one half had been converted to grape sugar.

Under the same conditions, the β-methylglucoside remained completely unchanged. This is particularly remarkable since, according to the observation of *v. Ekenstein*, it is much more rapidly hydrolyzed by dilute acids than the isomer.

The crystallized *ethyl glucoside*[5] behaves toward invertin like the α-methyl compound and hence evidently belongs to the α-series.

*Benzyl-* and *glycerolglucoside* have up to now not been obtained in crystalline form. The amorphous products are most probably a mixture of α- and β-compound. Actually they are attacked by invertin but are split less completely than the pure α-glucosides.

All remaining alcohol glucosides known up to now that can be derived from other types of sugars such as methyl- and ethyl-galactoside, methyl-, ethyl-, and benzyl-arabinoside, methyl-, and ethyl-rhamnoside, are not attacked at all by the enzyme solution. The five former have been crystallized, and one could think that they represent the β-forms; however, I regard this as unlikely since they were prepared in exactly the same fashion as the α-methylglucoside. In any case it is desirable to look for the isomers here as well and to test their behavior toward the enzyme.

It seemed to me of particular interest, finally, to report on the investigation of a derivative of L-glucose. I have hence prepared its methyl compound, which I wish to call *methyl-L-glucoside*, in the same way as the α-methyl derivative of grape sugar.

Unfortunately, pure crystalline L-glucose could not be used for this experiment because it is too difficult to obtain. Since the slightly impure sugar was used, the isolation of the crystalline glucosides has failed so far. However, a syrup was obtained that undoubtedly contains the glucosides and, judging from the preparative method used, it has mainly the α-compound. Since, furthermore, the preparation was quite free of sugar, it was easy to investigate the action of the enzyme solution. *It was completely negative.*

*Polysaccharides.* As is well known, the enzyme is named from its action on *cane sugar.* However, it was assumed up to now that *maltose* is not cleaved by it. The experiment showed me, however, that *the opposite is correct* and that hydrolysis here proceeds as rapidly as, and more completely than, for the alcohol-glucosides. The grape sugar that is formed in this manner can be identified very easily and with complete certainty by means of phenylhydrazine, even in the presence of unchanged maltose. In order to avoid any misunderstandings, I wish to note that commercial solid invertin, in contrast to the freshly prepared solution, does not cleave maltose. [We might note that this observation was due to specificity differences, not stability differences, since conventional invertin is a β-glycosidase. See Fischer's comments below.]

The conversion of maltose to grape sugar has already been observed in the case of the action of pancreatic juice, of small intestine,[6] of Koji extracts,[7] and of the so-called glucase that was found in maize by *Cuisinier* and was later examined in more detail by *Geduld, Lintner,* and *Morris.* I cannot say whether any of these enzymes is related to that of yeast. In my opinion the preceding observation indicates that maltose is not, as was assumed so far, directly fermented by yeast, but is first of all converted to hexose, similarly as in the case of cane sugar.

*Lactose* is quite stable toward the yeast enzyme. After a 24-hour digestion of 1 part lactose with 10 parts of the preceding enzyme solution at 30-35°, the liquid, on the basis of the very sensitive osazone test, contained neither grape sugar nor galactose. I cannot explain how *Dastre*[8] could get the opposite result.

I regard the difference between lactose and maltose in their behavior toward invertin to be once again a consequence of their differing configuration. If one makes, that is to say, the very probable assumption that both possess the same glucosidic type of structure, then one of them would be the glucoside and the other the galactoside of grape sugar, corresponding to the following formulae:

*Maltose*

$CH_2OH \cdot CHOH \cdot CH \cdot CHOH \cdot CHOH \cdot CH \cdot O \cdot CH_2(CHOH)_4 \cdot COH$

(the two CH groups joined through O)

Glucose residue — Glucose residue

*Lactose*

$CH_2OH \cdot CH \cdot OH \cdot CH \cdot CHOH \cdot CHOH \cdot CH \cdot OCH_2(CHOH)_4 \cdot COH$

(the two CH groups joined through O)

Galactose residue — Glucose residue

They would hence be related to one another like methylglucoside to methylgalactoside, of which only the former is cleaved by invertin.

*Inulin* and *starch* in the form of starch paste are not changed by invertin solution.[9]

*Aromatic Glucosides.* Invertin solution does not act on salicin, coniferin, phloridzin, and on the phenolglucoside artificially prepared by *Michael*;[10] however, it readily splits off grape sugar from amygdalin. Since in the latter case, however, neither benzaldehyde nor hydrocyanic acid are formed, the process is obviously quite different from the one that occurs in the case of the action of emulsin. It appears, rather, that invertin removes only half of the two molecules of grape sugar that are probably contained in amygdalin in the form of maltose. [The two glucoses are linked differently in the two compounds.] This observation indicates that a type of glucosidic cleavage takes place for which so far no analogous case is known, and I intend to study it further.

The following has still to be noted concerning the treatment of invertin solution in practice. In the absence of antiseptic substances or when it is not kept very cold, it begins to putrefy already after a few days, becoming turbid and assuming an evil odor. For the hydrolytic experiments it is possible to avoid this danger, which is increased at higher temperature, by added chloroform, which does not inhibit the activity of the enzyme. A different behavior is shown by phenol which immediately produces a precipitate in the solution; even the addition of 1 percent phenol to the fresh enzyme solution markedly decreased the hydrolysis, and 2½ percent completely stopped it. In all cases, however, it is advisable to use the freshly prepared enzyme solution. Alternatively one can actually use an aqueous suspension of yeast directly, provided that the living activity of the cells and the fermentative action connected with this have been stopped, as well known, by addition of chloroform.

All experiments described were carried out with *Saccharomyces cerevisiae*, type *Frohberg*. I have, however, convinced myself that type *Saaz* acts in the same manner on the alcoholglucosides as well as on maltose. On the other hand, it can be expected that the Saccharomyces types which do not ferment maltose, such as *S. exiguus*, *Ludwigii*, or *apiculatus*, correspondingly do not form an enzyme that splits glucoside. I can already now assert the correctness of this assumption for lactose yeast.

Commercially available solid invertin behaves differently from the freshly prepared enzyme solution, for a preparation obtained from *E. Merck* in Darmstadt still showed a rather strong activity toward cane sugar but did not change the α-methylglycoside and maltose. I hope soon to be able to determine experimentally whether this lower activity is due to the type of yeast used or to the method of isolation, or whether, besides the known invertin, yeast contains a new, more strongly hydrolyzing enzyme.

## EXPERIMENTS WITH EMULSIN

For these, a preparation obtained by *E. Merck*, which acted strongly on amygdalin, was used. One part of the enzyme was incubated for 15-20 hours at 30-35° with 2 parts of the investigated glucoside and 20 parts of water.

Emulsin and invertin are identical insofar as the former attacks only the glucosides of grape sugar but causes no change in the galactosides, arabinosides, and rhamnosides mentioned above, and in methyl-L-glucoside.

On the other hand, there is a sharp difference in the two enzymes toward the α- and β-methylglucoside; for while the first is attacked by invertin, the second is attacked exclusively by emulsin. A quantitative experiment showed that, under the preceding conditions, 90 percent of the β-methylglucoside was converted by emulsin to grape sugar, while under the same conditions the α-compound showed no detectable hydrolysis.

It is true that the glucosides of glycerol and of benzyl alcohol are partially cleaved not only by invertin but also by emulsin. But these amorphous substances obviously are, as has already to be concluded from the method of preparation and as was already stressed, mixtures of the α- and β-compounds.

It is known that emulsin hydrolyzes many naturally occurring aromatic glucosides such as salicin, coniferin, arbutin, etc., or the artificially prepared phenolglucoside. Since the same products are not attacked by invertin--obviously the phenols that could be formed in this process would not, because of the high dilution, harm the latter--it can be assumed that these glucosides belong to the β-series (analogs of β-methylglucoside).

Under the preceding conditions, maltose[11] and cane sugar are not cleaved by emulsin in detectable amounts. The action of emulsin on lactose seems not to have been tested as yet. To my surprise I found that here hydrolysis takes place readily. A solution of 2 g of lactose in 20 cc water after addition of 0.5 g emulsin was warmed for 22 hours at 35°. The osazone test was used to determine the cleavage products. For this purpose a small amount of sodium acetate was added to the solution; after a few minutes of heating in a water bath, the amorphous precipitate was filtered off, and after addition of 2 g phenylhydrazine and 0.7 g pure acetic acid, the mixture was heated in the water bath for 1 1/4 hours. The precipitated osazones, which already formed at the elevated temperature, were filtered after cooling and freed of lactosazone by boiling with much water. The residue weighed 0.6 g. On recrystallization from dilute alcohol, the substance melted at 203-205° and showed the composition of the phenylhexosazones (*N* found 15.7; *N* calculated 15.6). The preparation undoubtedly was hexosazone, and, in fact, it was a mixture of glucosazone and galactosazone.

In the case of maltose and lactose, a similar contrast is seen between invertin and emulsin as in the case of the α- and β-glucosides, in spite of the fact that the difference in the configuration of the two sugars extends even further.

## ENZYME OF KEFIR GRAINS

After the action of invertin on maltose had been observed, so that the direct fermentatility of this sugar had become rather doubtful, it was assumed by analogy that lactose yeast synthesizes an enzyme that cleaves lactose and that thus renders its fermentation possible. Unfortunately, I did not have enough lactose yeast available to carry out the experiment in the same way as was done with brewer's yeast. Hence I first of all used kefir grains. The result was as expected. The *clear filtrate* of the aqueous extract of the grains, obtained by filtration through a *Pukall* clay filter, decomposed large amounts of lactose into its components at 30-35° in the course of 20 hours. The presence of these compounds was shown as before by the osazone test. Under the same conditions maltose remained unchanged. I shall, as soon as possible, repeat the experiment with pure lactose yeast, and

I will also attempt the isolation of the enzyme.

*Beyerinck* already maintained[12] that such a lactose-cleaving agent is present in *Saccharomyces kefir* and in *Saccharomyces tyrocola* (cheese yeast). However, there are serious doubts about the reliability of the results he obtained in his attempts to prove, by the use of luminescent bacteria, that the so-called lactase acts on lactose. In fact, *Beyerinck's* opinion has been very emphatically contested by *Schnurmans Stekhoven.*[13] The latter concludes, on the contrary, that the enzyme of kefir yeast decomposes cane sugar and raffinose but not lactose. The question as to who is correct may be easily resolved when the experiment is performed with pure kefir yeast in the way described above for kefir grains. I will do this experiment as soon as I have a sufficient quantity of the yeast available.

Furthermore, I intend to use a few related enzymes such as glucase, ptyalin, myrosin, and the pancreatic ferments for comparative purposes, and I wish in addition to extend the experiments to the rarer polysaccharides such as isomaltose, turanose, melibiose, melitriose, trehalose, melezitose, the artificial dextrins, and so on. Undoubtedly still more such contrasts will then be found such as the ones now known to exist between invertin and emulsin.

But the observations already available suffice to prove that in principle the enzymes are as selective as yeast and other microorganisms with regard to the configuration of the compounds that they attack. The analogy of the two phenomena in this respect appears to be so complete that they can be assumed to have the same cause, and so I again take up the hypothesis of *Thierfelder* and myself, mentioned earlier. As is known, invertin and emulsin have many similarities to proteins and, like these substances, they undoubtedly possess an asymmetrically built molecule. One may hence explain the restricted activity of these enzymes on the glucosides by assuming that these two types of molecule can approach one another in a fashion that leads to the initiation of the chemical process only when they have a similar geometrical structure. To use a metaphor I wish to say that in order to be able to act chemically on one another, enzyme and glucoside must fit together like lock and key. This view has undoubtedly become more probable and more valuable for stereochemical research following the transfer of the phenomenon itself from biological to purely chemical territory. This view constitutes an extension of the theory of asymmetry but is not directly derived from it; for in my opinion only new factual observations could have convinced anyone that the geometrical structure of the molecule exerts, even in the case of mirror images, such a large influence on the interaction of chemical affinities [literally, play of chemical affinities (Spiel der chemischen Affinitäten)]. This conclusion could certainly not have been reached on the basis of the experimental knowledge available up to now, which was limited to the knowledge that the salts formed from two asymmetric components may be distinguished by solubility and melting point. I am just as convinced that the phenomenon of interactions between asymmetric substances, established so far only for the complicated enzymes, will soon also be found for simpler asymmetric agents, as I am that enzymes can be used to determine what the configuration of asymmetric substances is.

The experience that molecular geometry restricts the activity of enzymes to such a large extent should also be of some use for physiological research. [It should be noted that Emil Fischer's reference to physiological research was made before the term *biochemistry* had been coined (cf. Kohler 1973).] For physiological research, however, it appears to me even more important to have proved that, as to molecular asymmetry, the formerly common assumption of a difference between the chemical activity of the living cell and the action of chemical agents does not hold. Hence the analogy between "living and nonliving ferments," emphasized so often especially by *Berzelius*, *Liebig*, and others, has to a not unimportant degree been restored.

In the execution of the preceding experiments, I have enjoyed the eager and able help of Dr. *Paul Rehländer*. Further, I am highly indebted to Dr. *H. Thierfelder*, Prof. *M. Delbrück*, and Dr. *O. Reinke* for the gifts of pure yeasts.

1. These *Berichte* 27:2036.

2. These *Berichte* 26:2400.

3. These *Berichte* 26:2404.

4. *Recueil d. trav. chim. d. Pays-Bas.* 13:183.

5. E. Fischer and L. Beensch, these *Berichte* 27:2479.

6. *Ann. d. Chem.* 204:228.

7. *Zeitschr. f. physiolog. Chem.* 14:297.

8. *Compt. rend.* 96:932.

9. *Kiliani* (*Ann. d. Chem.* 205:189) already observed that inulin is not changed by invertin isolated by the method of *Barth*.

10. *Michael*, *Compt. rend.* 89:355.

11. Compare *v. Mering*, *Zeitschr. für physiolog. Chemie* 5:190.

12. "Lactase, a New Enzyme" (*Centralbl. für Bacteriologie und Parasitenkunde* 6:44).

13. *Koch's Jahresbericht über Gährungsorganismen* [Koch's Yearly Report on Fermentative Organisms] 1891, 136.

# 25

# THE SIGNIFICANCE OF STEREOCHEMISTRY FOR PHYSIOLOGY

**Emil Fischer**

*These excerpts were translated expressly for this Benchmark volume by H. Friedmann, the University of Chicago, from pages 81, 82–83, 84, and 87 of "Bedeutung der Stereochemie für die Physiologie" in* Hoppe-Seyler's Z. Physiol. Chem. **26**:60–87 (1898/1899).

[*Editor's Note:* In the original, material precedes these excerpts.]

THEORETICAL CONSIDERATIONS

Nobody who has been closely occupied with this topic can doubt that the enzymes are incomparably more special reagents than the simpler compounds of inorganic and organic chemistry such as acids, bases, condensing agents, and so on. It is precisely their narrowly limited activity that determines the usefulness of these substances for the organism and for experimental research.

[*Editor's Note:* Material has been omitted at this point.]

The basis for these phenomena most probably lies in the asymmetric structure of the enzyme molecule. For even though these substances are not yet known in the pure state, their similarity to proteins is so great and their development from the latter so probable that they themselves undoubtedly have to be regarded as optically active and hence asymmetric molecular structures. This led to the hypothesis that there must be a similarity in molecular configuration between the enzymes and the object of their attack if a reaction is to occur.[1] In order to make this idea more attractive, I have used the picture of lock and key.[2] Far be it from me to intend to place this hypothesis alongside the extensively developed theories of our science. I gladly admit that it will be possible to test this hypothesis thoroughly only when we will be able to isolate the enzymes in a pure state and to investigate their configuration.

[*Editor's Note:* Material has been omitted at this point.]

If one proceeds to extend stereochemical considerations to the chemical processes that occur in the more highly developed organisms, then one arrives at the concept that, in general, the configuration of a molecule is as important as its structure for the changes brought about by active proteins, as they undoubtedly occur in the protoplasm. Hence it is no longer surprising when among two stereoisomeric substances, one is found to act strongly on our sensory organs such as taste or smell or on the central nervous system, while the other is quite indifferent or brings about only a much weaker reaction.

[*Editor's Note:* Material has been omitted at this point. The paper ends as follows.]

In addition, the present insights allow a new and plausible approach concerning the assimilation of carbon dioxide which, as is known, is converted in plant tissues to active sugars. This new approach follows from the fact that observations among the sugars have shown that when optically active substances participate, even artificial synthesis proceeds asymmetrically. This state of affairs holds for assimilation, since the transformation of carbon dioxide to sugar apparently proceeds with the participation of the optically active substances of the chloroplast. This observation, which I have presented in detail earlier, has successfully removed the main apparent contrast between the artificial and natural synthesis of asymmetric carbon compounds.[3]

1. Fischer and Thierfelder, *Ber. d. d. chem. Ges.* 27:2036.

2. *Ber. d. d. chem. Ges.* 27:2992.

3. *Ber. d. d. chem. Ges.* 27:3230.

# 26

Reprinted from pages 634–637, 649, and 651–658 of *Chem. Soc. J. (Trans.)*
73:634–658 (1898)

## *Reversible Zymohydrolysis*

By Arthur Croft Hill, B.A.

### Introductory.

This investigation was undertaken with the object of determining if hydrolysis through the agency of enzymes is, like other chemical reactions, reversible; if, in fact, under appropriate conditions, the corresponding synthesis can be obtained.

It was necessary to choose a very simple case of zymolytic action in which the exact stage of conversion could at any time be accurately determined, and then to ascertain if the reaction were retarded by the products formed and did not reach completion, as only such a reaction could be considered suitable for the purpose. In the case of cane-sugar and invertase, it had been found by O'Sullivan and Tompson (Trans., 1890, 57, 834) that the reaction tended to total conversion, and that the time curve gave no evidence of hindering by the products of inversion; but viewed in the light of other chemical reactions, it seemed possible that a hindering influence existed but had been small enough to escape detection. The hydrolysis of maltose differs from that of cane-sugar in that, whilst the latter is split into two dissimilar molecules, the former yields two similar molecules, and it seemed reasonable to expect that, in the hydrolysis of maltose, a hindering influence might be more marked and more easily observed. This reaction was therefore chosen, and the maltase of yeast selected as the enzyme.

Part I of this paper deals with the manner of extraction and with certain properties of the enzyme; Part II, with the analytical methods by which the amount of change at any stages of a hydrolysis has been estimated; Part III is taken up with an account of experiments which show a distinct hindering action by glucose, an action which increases with the concentration of the solutions; Part IV, with experiments showing a reverse change of glucose to maltose in concentrated solutions. It deals also with the equilibrium points, or limits of conversion, for several different concentrations.

## Part I. Extraction of the Enzyme.

The existence in yeast of this enzyme which converts maltose into glucose, was suspected by Bourquelot (*Journ. de l'Anatomie et de la Physiologie*, 1886, pp. 180, 200); he was unable, however, to get an active extract of it, using chloroform as an antiseptic. Lintner (*Zeits. d. Ges. Brauwesen*, 1892, 106), Fischer (*Ber.*, 1894, **27**, 1113), Röhmann (*Ber.*, 1894, **27**, 3251), and Lintner and Kröber (*Ber.*, 1895, **28**, 1050) have extracted it from dry yeast. Morris (Proc., 1895, **46**) called attention to the fact that fresh yeast in the presence of chloroform did not attack maltose; but it seems clear from Lintner and Kröber's (*Ber.*, 1895, **28**, 1050) observations, and from observations made in the course of this research, that chloroform slowly destroys the enzyme, whilst Fischer (*Ber.*, 1895, **28**, 1430) has found that fresh yeast hydrolyses maltose when, not chloroform, but thymol or toluene is used as an antiseptic.

Nevertheless, dried yeast hydrolyses much more actively than fresh yeast, for the enzyme does not leave the cells until they have been thoroughly dried. Even from dried yeast, Fischer (*Ber.*, 1894, **27**, 1479) found that only a feebly active extract was obtained, so that in most of his experiments he preferred using the dried yeast itself with an antiseptic to using the extract. Lintner and Kröber's extract was more active, but they do not enter into details as to the circumstances in the preparation and extraction of the yeast which give the best results, neither do they discuss the stability of their extracts. Röhmann mentions that his yeast was heated for 1 hour at 105° to 110°, according to Barth's method, and then extracted with thymol water. He does not state how active the extract was, nor whether it kept well.

The duration of heating and the range of temperature are important; although the enzyme is not destroyed by heating the yeast at 120° for a short time, or by heating above 105° for 6 hours, there is a decided disadvantage in such heating, for the extract filters badly, becoming distinctly more turbid when kept and losing much of its activity. The result appears to be due to rupture of the cell walls consequent on treating the highly exsiccated and damaged cells with water, whereby a most unstable proteid escapes and is gradually precipitated. A similar progressive precipitation was observed by E. Buchner (*Ber.*, 1897, **30**, 117) in his expressed yeast juice. The difficulty does not occur if the heating be limited in the way which will be presently described. By using very dilute alkali for the extraction instead of distilled water, the activity of the extract is usually much increased, the effect of the alkali being to neutralise any acidity of the yeast which would prevent the enzyme from going into solution. The digesting mixture must be neutral, not alkaline, as free alkali at once destroys the enzyme.

The following method has given the best results. Good, pressed, bottom fermentation yeast * after being pounded in a mortar with distilled water, and washed three times by decantation, is collected on a filter provided with a cover, and then spread on porous tiles and dried in a vacuum over sulphuric acid. After two days or so in the exsiccator, the yeast is dry and brittle; it is then powdered in a mortar and sifted through fine muslin, being thus obtained as a bright, yellowish-white powder. The heating is performed by suspending the powdered yeast on a double layer of fine muslin over the mouth of a glass jar, which is then placed in an oven previously heated to 40°, and the temperature is allowed to rise during successive quarters of an hour to 60°, 70°, 90°, 100°; it is then maintained at 100° for a quarter of an hour, when the jar is taken out of the bath and put in the exsiccator until cool. The yeast is weighed in a closed vessel, pounded in a mortar with ten times its weight of 0·1 per cent. sodium hydroxide solution, put into a flask, toluene added, and the whole allowed to digest at the temperature of the room for three days. The extract is then filtered, first through paper and afterwards through a Pasteur-Chamberland filter into sterilised apparatus.

As thus obtained, it is a pale yellow, clear liquid of neutral or faintly acid reaction, and yielding from 0·516 to 0·52 per cent. of ash, and from 2·45 to 3·14 per cent. of total solids. It remains bright, and has shown activity, although in diminished degree, after the lapse of some months. The test of activity is that 1 c.c. of the fresh extract acting on 20 c.c. of a 2 per cent. solution of maltose at 30°, will hydrolyse about 20 per cent. of the sugar in 40 minutes, and over 90 per cent. when left from one evening to the next morning.

If the extract, as first obtained, is feebly acid, when mixed with the sugar solution the mixture is not acid enough to react with litmus; it may be said, therefore, that the conversion proceeds in neutral solution. The addition of acid to the extract causes precipitation with loss of activity; addition of alkali, sufficient to make it alkaline to litmus, also destroys the activity. An attempt was made to see if

* The yeast was obtained from one of the chief breweries in Bern (Aebersold's, Schauplatzgasse).

the addition of a very minute quantity of alkali or of acid would have an appreciable effect on the rate of hydrolysis. Two flasks, containing each a similar mixture of enzyme and maltose solution, and differing only in that in one sodium hydroxide had been added to the extent of 1 part in 12,000, were put in the bath at 30°, and the hydrolysis was estimated after 45 minutes; in either case, it was about 20 per cent. The contents of both flasks were neutral to litmus.

Although alcohol, as pointed out by Fischer (*Ber.*, 1894, **27**, 1113) and by Röhmann (*Ber.*, 1894, **27**, 3251), so readily destroys the enzyme in the presence of water, it has no action when both it and the enzyme are dry. Some yeast powder, previously dried at 100°, was left for 3 days under alcohol dried over baryta; it was afterwards found to be quite active on maltose. A similar result was obtained with the dried and powdered extract.

Extracts prepared in the presence of chloroform were but feebly active. Further, the observation of Lintner and Kröber (*loc. cit.*), that chloroform retards the progress of the hydrolysis, has been confirmed.

[*Editor's Note:* Material has been omitted at this point.]

## PART IV. THE REVERSE ACTION.

It has been shown in Part III that the presence of glucose hinders the hydrolytic decomposition of maltose by maltase, and it will have been noticed that the effect is more marked in those experiments in which the sugar solutions are more concentrated. The following experiments show that this appearance of a hindering effect is due to a reverse action by which glucose undergoes change with synthetic production of maltose. It will be seen that in presence of the enzyme solutions of glucose of sufficient concentration there is increase in optical activity and decrease in cupric-reducing power, the magnitudes of these changes having to one another the relation required on the assumption that glucose is converted to maltose. No change takes place in control flasks, so that the presence of the active enzyme is as

[*Editor's Note:* The table appearing as page 650 with details on Experiment IV has been omitted.]

essential for the reverse action as for the ordinary hydrolysis; neither does a solution of the enzyme in absence of sugar develop of itself any such changes.

When the time allowed for the progress of the reaction is much extended, in order to find the limit of the conversion, the same equilibrium point is approached whether maltose is being hydrolysed to glucose or glucose synthesised to maltose, provided always that the concentration of total sugar is the same in each case.

The position of the equilibrium point is a function of the concentration, moving with increase of the latter in the direction of more maltose and less glucose; for example, a solution of maltose of 4 per cent. concentration or less does not arrive at equilibrium point until nearly the whole is converted into glucose, whilst with solutions of 40 per cent. concentration there is some considerable amount of maltose present when close to the equilibrium point.

No attempt has been made to separate the synthesised maltose from the excess of glucose with which it is mixed; but the osazone was separated from its mixture with glucosazone and identified, and it is hoped that the separation of the sugar itself will shortly be effected.

In the first experiment on the effect of adding a solution of the enzyme to a solution of glucose, a 20 per cent. concentration was used. Here the change was very small and the progress of the reaction slow, for in nineteen days the specific rotation had only increased by 1°, corresponding with a back conversion of 1·2 per cent. of the total sugar. The actual change in the observed rotation was 5′, or ten times the probable error in reading, and no change could be detected in the control flask; both these results were confirmed by several separate estimations. In the next experiment, a 40 per cent. concentration was used and a back conversion of 7·5 per cent. was noted, the value given by the copper method agreeing with the polarimetric to the nearest whole number. In Experiment VIII, of which details are appended, a freshly-prepared enzyme solution was used with a glucose concentration of 40 per cent. and a sufficient time was allowed for the reaction to be nearly completed. As much as 14·5 per cent. back conversion occurred in one flask and 15·5 per cent. in another in which the action was more rapid from the presence of a larger proportion of the enzyme; and it is probable, from this and from Experiment IX, in which the equilibrium point was approached from the other side, that at this concentration there is equilibrium when 16 per cent. of the whole sugar is present as maltose (hydrate).

EXPERIMENT VIII (See Fig. 5).

*Sugar concentration*, 40 per cent. (approx.).
*Temperature of bath*, 30°.

Flask A. ......... { 9·81 grams of glucose.
5 c.c. of enzyme solution.
Water to 25 c.c.
Toluene. }

Flask B. ......... (control) { 3·98 grams of glucose
2 c.c. of enzyme solution
Water to 10 c.c. } boiled.
Toluene.

Flask C. ......... { 19·65 grams of glucose.
15 c.c. of enzyme solution.
Water to 50 c.c.
Toluene. }

FIG. 5.

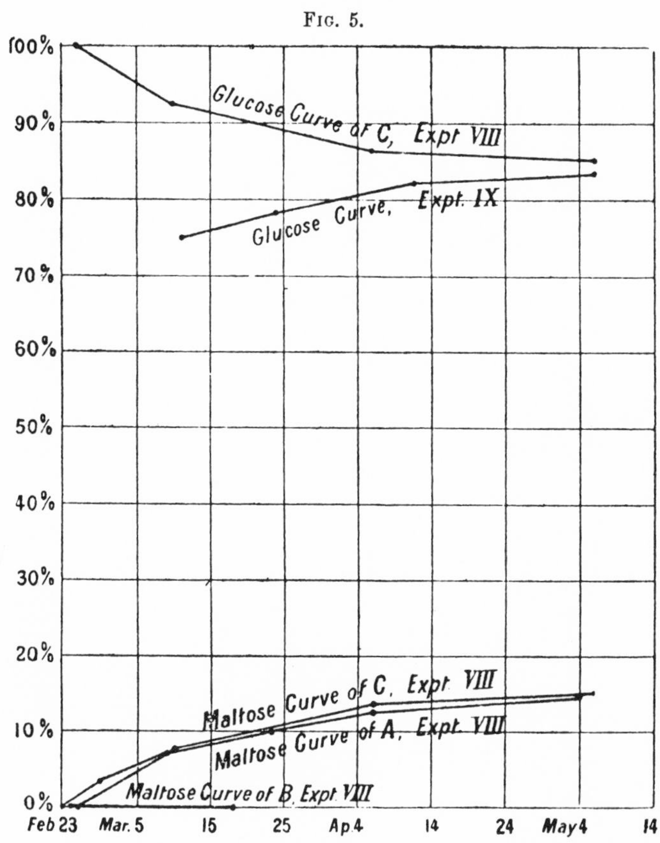

EXPERIMENT VIII.

| Date. | Flask. | Weight of the 1 c.c. withdrawn. | Weight of sugar in 20 c.c. for polarimeter in 200 c.c. for titration. | Observed rotation. | $[\alpha]_D$ | Percentage of glucose converted into maltose. | Titration value. | Titration of 0·2 per cent. standard glucose. | *R.* | Percentage of glucose converted into maltose. | Mean of the two methods. |
|---|---|---|---|---|---|---|---|---|---|---|---|
| | | grams. | | | | | | | | | |
| February 23. | A | 1·148 | 0·3925 | 2·06° | 52·5° | 0 | {4·8, 4·75} | 4·85 | 100·5 | 0 | 0 |
| February 28. | A | 1·147 | 0·392 | 2·16° | 55° | 3 | 4·7 | 4·55 | 97·7 | 3·5 | 3·25 |
| March 9. | A | 1·150 | 0·393 | 2·29° | 58·3° | 7·4 | 4·95 | 4·65 | 95·8 | 6·8 | 7 |
| March 23. | A | 1·148 | 0·3925 | 2·36° | 60·3° | 10 | 5·1 | 4·7 | 94 | 10 | 10 |
| April 6. | A | 1·152 | 0·394 | 2·47° | 62·7° | 13 | 5·05 | 4·6 | 92·5 | 12 | 12·5 |
| May 4. | A | 1·159 | 0·396 | 2·52° | 63·6° | 14 | 5·3 | 4·75 | 90·6 | 15 | 14·5 |
| February 24. | B | 1·148 | 0·398 | 2·08° | 52·3° | 0 | 4·7 | 4·7 | 100·5 | 0 | 0 |
| March 17. | B | 1·146 | 0·3975 | 2·075° | 52·2° | 0 | {4·4, 4·45} | 4·4 | 99·9 | 0 | 0 |
| February 25. | C | 1·1475 | 0·393 | 2·06° | 52·4° | 0 | 4·8 | 4·75 | 100·8 | 0 | 0 |
| March 10. | C | 1·148 | 0·393 | 2·315° | 59° | 8 | 5·05 | 4·75 | 95·7 | 7 | 7·5 |
| April 6. | C | 1·148 | 0·393 | 2·46° | 62·6° | 13 | 5·15 | 4·6 | 91 | 14 | 13·5 |
| May 4. | C | 1·151 | 0·394 | 2·52° | 64° | 14·5 | 5·4 | 4·8 | 90·4 | 15·5 | 15 |

For the estimations, 1 c.c. was withdrawn and weighed, and after adding 1 c.c. of 0·4 per cent. sodium hydroxide solution was diluted to 20 c.c., and examined by the polarimeter. Ten c.c. taken from the polarimeter tube were diluted to 100 c.c. and examined by the copper method.

The lower curves in Figure 5 show the percentage of maltose formed in the three flasks of Experiment VIII.

In order to find the point of equilibrium for a 40 per cent. concentration, Experiment IX was so arranged that the equilibrium point should be approached from the side opposite to that in Experiment VIII. Instead, however, of starting with all maltose, a mixture of 75 parts glucose and 25 parts maltose hydrate was used, that is, a concentration of 30 per cent. glucose and 10 per cent. maltose hydrate, since it had been found that a 40 per cent. concentration of maltose caused a precipitate in the enzyme solution, whilst a somewhat higher concentration of glucose did not. The upper curves in Figure 5 show the course of hydrolysis in Experiment IX, compared with the back conversion in flask C, Experiment VIII. For the purpose of comparison, the glucose curve of the latter is given, the maltose curve being at the lower part of the same figure.

EXPERIMENT IX (See Fig. 5, p. 652).

*Sugar concentration*, 40 per cent. (approx.).
*Temperature of bath*, 30°.
0·98 gram of maltose hydrate.
2·95 grams of glucose.
2 c.c. of enzyme solution.
Water to 10 c.c.

*At start* { 75 per cent. of total sugar = glucose.
25 " " " = maltose hydrate.

| Date. | Weight of the 1 c.c. withdrawn. | Weight of sugar in 20 c.c. for polarimeter or in 200 c.c. for titration. | Observed rotation. | $[\alpha]_D$. | Percentage of glucose in total sugar. |
|---|---|---|---|---|---|
| March 11 | 1·154 grams | 0·394 gram | 2·84° | 72·1° | 75 |
| March 24 | 1·156 grams | 0·395 gram | 2·76° | 69·9° | 78 |
| April 12 | 1·147 grams | 0·392 gram | 2·61° | 66·6° | 82 |
| May 6 | 1·153 grams | 0·394 gram | 2·56° | 65° | 84 } mean 83·25 |
| May 6. By copper method, $R = 89$; percentage of glucose in total sugar = 82·5 | | | | | |

### *The Equilibrium Points.*

While the limit reached in Experiment IX was 83·25 parts of glucose to 16·75 parts of maltose hydrate, in Experiment VIII it was 15 parts of maltose hydrate to 85 parts of glucose; and, since in these experiments the equilibrium point was approached from opposite sides, its position must lie somewhere between these points, and not far from 84 parts of glucose to 16 parts of maltose hydrate. The flasks A and B in Experiment IV, with concentrations of 20 per cent. and 10 per cent. respectively, were allowed to remain in the bath at 30° for some months after the period mapped in Figure 4, until all action appeared to have stopped; the limits then reached were 90·5 per cent. for the 20 per cent. concentration, and 94·5 per cent. of glucose for the 10 per cent. concentration. In another experiment with 4 per cent. concentration, 98 per cent. of the maltose hydrate was converted into glucose, and with 2 per cent. concentration 99 per cent. has been converted. These results may be tabulated thus.

| Sugar concentration. | Percentage of maltose hydrate hydrolysed to glucose. |
|---|---|
| 40 per cent. | About 84 |
| 20 ,, | At least 90·5 |
| 10 ,, | ,, 94·5 |
| 4 ,, | ,, 98 |
| 2 ,, | ,, 99 |

In dilute solutions, therefore, hydrolysis is practically complete; with a concentration of 20 per cent., a reverse action can be detected; and with a concentration of 40 per cent., it is well marked.

### *The Osazone.*

In each of the experiments made to test the reverse action, a part of the contents of the flask was treated with phenylhydrazine, and a more soluble osazone separated from the excess of glucosazone by extraction and recrystallising; the osazone had, in each case, the crystalline character of maltosazone. Examination of the control flasks gave a negative result. In Experiment VIII, a more extended examination of this osazone was made. It was prepared as follows: a portion of the contents of the flask was diluted with twice its volume of water, a few drops of acetic acid added, and the whole boiled on the water bath until a clear filtrate could be obtained. It was then treated on the water bath with phenylhydrazine hydrochloride and sodium acetate, the volume being such that the total sugar was in a concentration of about 10 grams in 150 to 170 c.c. The mixture quickly became almost solid from separation of glucosazone, which was three times filtered off on a hot funnel. The solution, having been at the boiling point for 1½ hours in all, was then allowed to cool in the water bath, and the crystals, which separated on cooling, were extracted with boiling water, and the soluble part recrystallised, this process being repeated; crystalline plates resembling those of ordinary maltosazone were finally obtained. There was, however, more tendency for these plates to be deposited around a central axis than was observed in an osazone prepared from pure maltose, but not differing in this respect from a maltosazone prepared from a trial mixture of pure maltose and glucose, 1 part of the former to 9 of the latter, to which also some boiled enzyme solution had been added in the proportion usual in the preceding experiments (see Fig. 6).

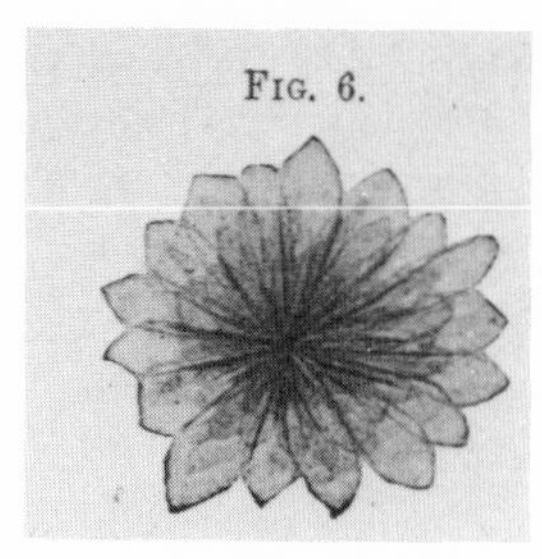

FIG. 6.

The osazone prepared from Experiment VIII, when the proportion of synthesised maltose had reached 10 per cent., melted at 174—176°, and another preparation, perhaps further purified, at 178—180°. The osazone made from the test mixture, containing ordinary maltose, melted at 173°, and that from Experiment VIII, when the proportion of synthesised maltose had reached 14 to 15 per cent. of the total sugar, at 186°. All these osazones showed signs of change below these temperatures, although in each case the bulk melted suddenly at the temperature given. When it is considered how greatly the melting point of maltosazone is depressed by a small amount of impurity, 186° is not lower than might be expected in the circumstances under which the substance was obtained. The melting point of maltosazone first given by Fischer was 193°, and afterwards 206°, whilst the melting point of isomaltosazone is about 150°.

Analysis of the osazone of the synthesised sugar by combustion with the aid of a Sprengel pump, gave the following results.

| Osazone. | $CO_2$ at 0° and 760 mm. | $N_2$ at 0° and 760 mm. | Percentage of carbon. | Percentage of nitrogen. | Ratio C : N. |
|---|---|---|---|---|---|
| 0·0325 gram | 33·3 c.c. | 2·7 c.c. | 55·3 | 10·46 | 5·29 |
| 0·035 gram | 35·9 c.c. | 3·1 c.c. | 55·3 | 11·1 | 4·98 |
| Calculated for maltosazone ..... .................. | | | 55·4 | 10 77 | 5·14 |
| Calculated for glucosazone.......................... | | | 60·4 | 15·65 | 3·86 |

It is, therefore, the osazone of a sugar of the $C_{12}H_{22}O_{11}$ class. The possibility of its being isomaltosazone is excluded by its solubility in water at 100°, its melting point, and its crystal form, in each of which particulars it differs widely from isomaltosazone, and resembles ordinary maltosazone.

*Summary of Evidence for the Reverse Action.*

1. The hydrolysis of maltose by the enzyme is hindered by the presence of glucose. It is incomplete, and both these effects are more marked the more concentrated the solution.

2. When a concentrated solution of glucose is treated with the enzyme, the optical activity increases *pari passu* with a decrease in reducing power, the amounts of these changes having that mutual relation which is required on the assumption that glucose disappears and maltose is formed.

3. The same equilibrium point is approached whether one starts with a solution of maltose or with a solution of glucose of the same concentration.

4. When the presence of synthesised maltose has been tested by the use of phenylhydrazine, an osazone has been obtained having the percentage composition and the ordinary physical characteristics of maltosazone.

A reverse action in the case of the zymolysis of maltose suggests that other enzyme actions are also reversible. In the majority of cases it is very doubtful whether such reverse action would admit of demonstration, for even in the present case the sugar solution must be of high concentration, and in the cases where the substances are of high molecular weight and where the products of the hydrolysis are heterogeneous, the concentration would probably have to be far higher than could be attained.

As to whether the reverse action occurs in living cells, it is worth while pointing out that it is by no means necessary for a high concentration to be realised ; it is only necessary for the synthesised product to be removed, perhaps by further synthesis through the agency of another enzyme, the process being repeated until an insoluble substance is formed. To take a concrete example : in the animal body glucose is present, also an enzyme capable of hydrolysing maltose, but no maltose is found. The hydrolysis of maltose takes place extra-cellularly in the alimentary canal ; it is possible that the reverse process takes place intra-cellularly, the maltose being further dealt with as fast as it is formed, so that its concentration is kept practically nil. Many applications of this idea in physiology and pathology suggest themselves but will be reserved for future investigation.

Part of the expenses of this research were defrayed by a grant from the Government Grants Committee of the Royal Society. I desire to record my thanks to this Committee and also to the Committee of the Davy-Faraday Laboratory of the Royal Institution for allowing me the use of a laboratory and of the greater part of my apparatus. The earlier part of the work was done in the Medizinisch-Chemisch. Institut, Bern, by the kind permission of the late Professor Drechsel.

DAVY-FARADAY LABORATORY,
ROYAL INSTITUTION, LONDON.

# 27

# CONCERNING THE INCREASING IMPORTANCE OF INORGANIC CHEMISTRY

## Address given at the 70th Meeting of the Society of German Scientists and Physicians in Düsseldorf

**Jacobus Hendricus van't Hoff**

*These excerpts were translated expressly for this Benchmark volume by H. Friedmann, the University of Chicago, from pages 1, 8, 9–10 and 12–13 of "Über die zunehmende Bedeutung der anorganischen Chemie" in* Z. Anorg. Chem. **18**:1–13 (1898).

Highly Honored Assembly!

It was certainly a lucky strike that a certain scientist[1] already 200 years ago established the main divisions of the field of chemistry on the basis of a correlation with the source of the respective substances; he separated substances occurring in organic nature, in the domain of living things, from those of the lifeless area of minerals, and hence distinguished organic substances from inorganic ones.

This division had right from the beginning an inner scientific justification as well since inorganic chemistry had the relatively simple task of explaining the chemical changes in dead matter, while organic chemistry was given the far more complicated problem of the processes in the living organism.

[*Editor's Note:* Material has been omitted at this point.]

We have first of all to mention the fundamental problem of affinity. Thermodynamics cannot base manifestations of affinity on the action of atoms on one another, but it rather pursues the interaction of affinities by measuring the external manifestations of this interaction: We learn from thermodynamics that affinity is not measured by the velocity or by the heat of a reaction but rather by the work that the reaction can yield maximally.

[*Editor's Note:* Material has been omitted at this point.]

Concepts of great consequence have been obtained. We have an unobjectionable principle for predicting reactions:

*A change can occur only if it can yield a positive amount of work; if this amount of work is negative, then the change can occur only in the reverse direction; if it is zero, then it can occur neither in one nor in the other direction.*

This work and hence the possibility for a reaction to occur may, however, be calculated if the equation for a given reaction is known, provided that the work obtainable for the formation of each of the participating substances from its elements has been once and for all determined and expressed, for example, in calories. This "work of formation" leads by simple addition and subtraction, just as when one calculates heat development, to "work of transformation" whose sign determines whether the transformation is possible. However, such a work of formation depends not only on the temperature but also on the given state (dissolved or undissolved, solvent and concentration).

[*Editor's Note:* Material has been omitted at this point.]

Second, we have obtained a fundamental rule for those reactions that proceed only in part because the reverse reaction also occurs. One then obtains a condition of so-called chemical equilibrium, as in the case of the combination of iodine and hydrogen and in that of ether formation, which, as is known, occur only in part. It is decisively important that in such cases changes in concentration occur during and because of the reaction, which in turn bring about a change, a decrease, in the work of transformation. These concentration changes eventually result in a decrease of this work to zero. Under these conditions the velocity of the reaction gradually decreases and eventually also becomes zero. Such a concentration change does not occur when, for example, copper nitrate and calcium nitrate combine to form a double salt; this reaction hence occurs either not at all or it goes to completion. By contrast, during the combination of iodine and hydrogen, the increasing concentration of the hydrogen iodide formed is equivalent to a gradually increasing opposing force that gradually brings the reaction to a halt.

With this consideration, however, we have obtained a further principle of wide applicability for predicting reactions. The point at which a reaction ceases may be calculated from the work of transformation.

[*Editor's Note:* Material has been omitted at this point.]

A final remark may be permitted. . . .In the field of organic chemistry the lessons of chemical equilibrium can and have actually been applied already; however, because of the enormous riches of forms and because of sluggish reactions, it is not easy to select a suitable material. Hence it is perhaps worthwhile to refer on this occasion to the most peculiar actions of ferments or enzymes that, if the latest investigations are confirmed, are superbly suited for the types of application mentioned above. On the one hand, *Fischer*[2] found that under the influence of ferments organic transformations are channeled into quite definite paths, a circumstance that completely excludes complications due to an abundance of products. On the other hand, it appears from the most recent investigations of *Tammann*,[3] *Duclaux*[4] and particularly of *Hill*[5] that equilibrium phenomena occur here. *Tammann* already observed that amygdalin is only partially cleaved under the influence of emulsin and that this cleavage proceeds further after removal of the cleavage products. If he had, conversely, added the cleavage products, he would probably have succeeded in synthesizing amygdalin. *Duclaux* used transformation equations that also point to the occurrence of an equilibrium, and *Hill* appears to have succeeded in synthesizing maltose from glucose by means of a yeast ferment. It follows from theoretical considerations that, in case a ferment is not changed during its action, a state of equilibrium and not a complete conversion must be attained. Hence it also follows that the reaction must be able to proceed in the opposite direction as well. One can justifiably ask whether (using equilibrium theory) formation of sugar from carbon dioxide and alcohol takes place under the influence of zymase if one exceeds the final pressure of carbon dioxide, and whether trypsin[6] may not under certain conditions given by equilibrium theory be able to form protein from the cleavage products that it itself forms.

[*Editor's Note:* The end of this address has been omitted here.]

1. Leméry, *Cours de Chimie*, 1675.

2. See also *Ber. deutsch. chem. Ges.* 27:2992.

3. *Zeitschr. phys. Chem.* 18:426.

4. *Bull. de l'Institut Pasteur*, 1898.

5. *Trans. Journ. Chem. Soc.* 1898:634.

6. Kossel, *Zeitschr. physiol. Chem.* 1898, 165.

# 28

Reprinted from *Chem. Soc. J. (Trans.)* **81**:373–388 (1902)

## *Enzyme Action*

By Adrian J. Brown.

### *Introduction.*

In a paper on the fermentative functions of yeast (Trans., 1892, 61, 380), the author described some experiments which showed that the character of the action of fermentation differed in a very marked manner from the character of the action usually attributed to enzyme change.

The author's experiments indicated that during fermentative change a constant amount of yeast decomposes an approximately *constant weight* of sugar in unit time in solutions of varying concentration, and that the velocity of fermentative action is therefore represented graphically by a *straight line*. On the other hand, the character of the action usually attributed to an enzyme is that a constant amount of the ferment changes in unit time a *constant fraction* of the reacting substance present, and that the velocity of its action is represented by the *logarithmic curve* of mass action.

At the time the author's work (*loc. cit.*) was published, the fermentative power of the yeast cell was considered to be a life function inseparable from the cell, and there appeared to be nothing specially remarkable in the observation that fermentation, a life function, differed in the velocity of its action from enzyme action. But since the more recent work of Buchner has demonstrated that the phenomenon of fermentation is caused by enzyme action, the question assumed another aspect. If fermentation is now regarded as an enzyme action, then, either the velocity of its action must be regarded as differing essentially from that which is usually attributed to other enzymes, or the experimental evidence on which the assumed difference rests must be regarded as misleading.

It was with the intention of investigating this question that the author commenced the work described in the following paper.

### *Velocity of the Action of Fermentation.*

If the view is adopted as a working hypothesis that the supposed difference in velocity of the actions of fermentation and ordinary enzyme change does not exist, but that it is due to some misconception, it is evident misconception may have arisen concerning either the velocity of fermentation or that of ordinary enzyme change, and consequently a re-examination of the experiments by which both velocities have been determined appeared desirable.

As the author is responsible for the experiments by which the velocity of fermentation has been determined, he commenced his investigations by repeating them. It does not appear necessary, however, to give the results of this work, for the experiments were similar to those described in the earlier paper (*loc. cit.*), and the results fully confirmed the conclusion that fermentative action does not proceed in accordance with the law of mass action.

As the general character of the action of fermentation appeared to be thus established, the author proceeded to examine the experimental evidence from which the conclusion is drawn that the velocity of enzyme action accords with the law of mass action.

### *Velocity of Enzyme Action.*

The generally accepted view regarding the velocity of enzyme action is based on the researches of Cornelius O'Sullivan and F. W. Tompson on the action of invertase on cane sugar (Trans., 1890, **57**, 865). These authors demonstrated the velocity of the action of invertase in the following manner.

Invertase was caused to act in solutions of cane sugar, and during the progress of the actions the quantities of sugar inverted during a succession of time intervals were determined. By this means, observations were obtained from which time curves were constructed which represented graphically the velocity of the action of inversion. When these curves were compared with the curve representing simple mass action, a very close agreement in shape was observed, which appeared to indicate that they were of the same order, and from this close agreement in shape, C. O'Sullivan and Tompson concluded that the action of invertase instanced the operation of the law of mass action.

This conclusion has also received confirmation from the researches of James O'Sullivan on the power of inversion of living yeast cells (Trans., 1892, **61**, 926), the experiments of this author indicating that the velocity of action of the living cell is the same as that of the extracted invertase used by C. O'Sullivan and Tompson in their experiments.

The evidence referred to is, so far as the author is aware, all that has been brought forward to support the conclusion that the velocity of enzyme action indicates the operation of a simple mass action.

Hitherto, no doubt, the want of additional evidence has not been felt, owing to C. O'Sullivan and Tompson's experiments appearing conclusive so far as invertase is concerned, and also to the fact that the conclusion these authors arrived at with regard to the character of enzyme action is one which there is every reason to anticipate.

But when the character of the action of fermentation, now very generally recognised as an enzyme action, was found to differ essentially from that attributed to invertase both in the free state and within the living yeast cell,* it raised doubt in the author's mind regarding the accuracy of C. O'Sullivan and Tompson's conclusion. Moreover, the author found that he was not alone in his distrust, for it has already been pointed out by Duclaux, in a criticism on C. O'Sullivan and Tompson's work (*Ann. Inst. Pasteur*, 1898, **12**, 96), that the logarithmic curve representing the action of invertase, on which C. O'Sullivan and Tompson founded their conclusion, may be shaped by other causes than the supposed action of mass. For Duclaux maintains that such a curve represents, not only a decrease in a changing substance, but also, and equally well, an increase in the products of change, and it is possible these products of change may act as the influence shaping the curve and not the influence of mass action. No experimental evidence sustaining this point is, however, brought forward by Duclaux.

As C. O'Sullivan and Tompson's conclusion rests entirely on the shape of the curve representing the action of invertase, the author considered it advisable first to repeat the experiments from which the curve was derived. Conditions of experiment similar to those used by C. O'Sullivan and Tompson were employed, but the invertase used was prepared in a different manner from the enzyme with which these authors experimented.

C. O'Sullivan and Tompson worked with invertase obtained from an extract of auto-digested yeast by precipitation with alcohol, and in so doing encountered the difficulty that the action of invertase prepared in this manner was very irregular unless it was associated with a small quantity of sulphuric acid. Moreover, the amount of acid required to reach the point described by these authors as "the most favourable condition of acidity," at which point it was necessary to work, varied in every experiment in a most remarkable manner.

It appeared very desirable to avoid this complication when repeating C. O'Sullivan and Tompson's experiments, so the author employed in his experiments an extract of invertase prepared from dried yeast by digestion with water. An extract of invertase prepared in this manner was quite suitable for the purpose of the experiments, and the risk of modifying the activity of the invertase by precipitation was avoided. That this method of obtaining a preparation of invertase suitable for experiment was preferable to that employed by C. O'Sullivan and Tompson was evidenced by the invertase being free from the irregularities of action associated with the precipitated in-

* Presumably, invertase within the wall of the living cell is in the same position as zymase with regard to its action as an enzyme.

vertase used by these authors, and, in consequence, it could be employed without the complicating addition of sulphuric acid.

The author's experiments, like those of C. O'Sullivan and Tompson, consisted in the addition of a suitable amount of invertase solution to a solution of cane sugar, and in determining, by means of a polarimeter, the fractions of the sugar inverted during successive intervals of time.

Instead, however, of expressing the velocity of the inversion change by means of a curve, the author preferred to make use of the value $k$, derived from the expression $\frac{1}{\theta}\log\frac{1}{1-x}$. This well-recognised means of expression, usually adopted now to demonstrate such changes as those of a mass action, has the advantage of avoiding certain difficulties which attend the comparison of calculated and experimental curves.

The results of two series of experiments determining the velocity of the action of invertase are given in Tables I and II:

TABLE I.—*Velocity of inversion change in* 9·48 *per cent. solution of cane sugar.* 500 *c.c. of solution of sugar and* 25 *c.c. of invertase solution used. Temp.* 30°.

| Duration of time interval in minutes. $\theta$. | Fraction of sugar inverted in $\theta$. $x$. | $k=\frac{1}{\theta}\log\frac{1}{1-x}$. |
|---|---|---|
| 30 | 0·265 | 0·00445 |
| 64 | 0·509 | 0·00483 |
| 120 | 0·794 | 0·00571 |
| 180 | 0·945 | 0·00698 |
| 240 | 0·983 | 0·00737 |
| 300 | 1·003 | |

During the course of a change proceeding as a simple mass action, it is well known that the value $k$, determined for any point of the action, is a constant. But in the experiments described in the above tables it will be noticed that the value $k$ increases in both experiments as inversion proceeds, until the value at the termination of the experiments is about 70 per cent. higher than at the beginning.*

Now, these results do not support the view that the action of inversion instances a mass action, as C. O'Sullivan and Tompson believed, for they differ very materially from the results these authors obtained. But in order to emphasise more distinctly the difference between the

* It will be noticed that there is no indication of "reversion" in these inversion experiments. An increase in the value of $k$ denotes an increasing velocity; reversion would lead to a decreasing velocity.

TABLE II.—*Velocity of inversion change in* 19·28 *per cent. solution of cane sugar.* 500 *c.c. of sugar solution and* 25 *c.c. of invertase solution used. Temp.* 30°.

| Duration of time interval in minutes. $\theta$. | Fraction of sugar inverted in $\theta$. $x$. | $k=\frac{1}{\theta}\log\frac{1}{1-x}$. |
|---|---|---|
| 30 | 0·130 | 0·00201 |
| 64 | 0·256 | 0·00201 |
| 120 | 0·454 | 0·00219 |
| 180 | 0·619 | 0·00232 |
| 240 | 0·738 | 0·00242 |
| 300 | 0·831 | 0·00257 |
| 360 | 0·890 | 0·00265 |
| 420 | 0·935 | 0·00283 |
| 480 | 0·961 | 0·00293 |
| 540 | 0·983 | 0·00327 |
| 581 | 0·990 | 0·00344 |

character of the action of inversion and that of a mass action, the results of an experiment involving mass action are given in Table III for the purpose of comparison. In this experiment, the author employed sulphuric acid to invert cane sugar, and thus obtained results from a typical mass action, which are directly comparable with those effected by invertase.

TABLE III.—*Velocity of inversion change of cane sugar by acid.* 600 *c.c. of a* 20 *per cent. solution of cane sugar, and* 35 *c.c. of normal* $H_2SO_4$ *used. Temp.* 48°.

| Duration of time interval in minutes. $\theta$. | Fraction of sugar inverted in $\theta$. $x$. | $k=\frac{1}{\theta}\log\frac{1}{1-x}$. |
|---|---|---|
| 30 | 0·165 | 0·00261 |
| 61 | 0·317 | 0·00271 |
| 90 | 0·433 | 0·00274 |
| 120 | 0·532 | 0·00275 |
| 150 | 0·617 | 0·00278 |
| 180 | 0·688 | 0·00281 |
| 243 | 0·785 | 0·00275 |
| 302 | 0 856 | 0·00279 |
| 362 | 0·902 | 0·00278 |
| | | Mean result 0·00275 |

The above experiments show very clearly, when a true mass action is followed under experimental conditions similar to those used when determining the character of the action of invertase, that a very different result is obtained. The small and irregular variation in the value of $k$* is very different from the regular and well marked increase in the value of $k$ observed in the experiments with invertase given in Tables I and II, and from this there can remain but little doubt that the order of progression of inversion differs essentially from that of a mass action. At present, the author has not attempted to determine any expression for the order of progression of inversion under the conditions of his experiment, because, for the immediate purpose of his investigation, it is only necessary to show that inversion does not proceed as a mass action.†

Although the author's experiments throw the greatest doubt on the accuracy of the conclusion that inversion evidences mass action, they cast very little light on the true character of the action of invertase, and in order to obtain more knowledge, it became necessary to adopt some method of experiment different from that which has already been described.

* The variations in value of $k$ are no doubt due to experimental error. Very slight changes in temperature have a marked influence on the velocity of inversion change by acid.

† Since writing the above, a communication from Victor Henri has been published (*Compt. rend.*, 1901, **133**, 891) on the velocity of inversion change. This author arrives at the conclusion that the action proceeds in accordance with the expression $2k_1 = \frac{1}{\theta}\log\frac{1+x}{1-x}$

On applying this purely mathematical expression of velocity to the inversion experiments described in Table II, the following results have been obtained:

TABLE II.—*Recalculated.*

| Duration of time interval in minutes. $\theta$. | Fraction of sugar inverted in $\theta$. $x$. | $2k_1 = \frac{1}{\theta}\log\frac{1+x}{1-x}$. |
|---|---|---|
| 30 | 0·130 | 0·00376 |
| 64 | 0·256 | 0·00355 |
| 120 | 0·454 | 0·00356 |
| 180 | 0·619 | 0·00346 |
| 240 | 0·738 | 0·00343 |
| 300 | 0·831 | 0·00353 |
| 360 | 0·890 | 0·00343 |
| 420 | 0·935 | 0·00351 |
| 480 | 0·961 | 0·00354 |
| 540 | 0·983 | 0·00383 |
| 581 | 0·990 | 0·00395 |

It will be observed from the remarkably close agreement of the numbers in the third column, representing the values $2k$, that the author's experiments indicate a very similar velocity for inversion change to that which is represented by Henri's expression. This is of interest as further assisting to establish the fact that the progress of an inversion change is not ordered in conformity with the law of mass action.

Such a method exists in causing a constant amount of invertase to act on varying amounts of cane sugar in constant volume of solution for a constant brief interval of time. Under these conditions, the variable in the actions is the amount of reacting substance (cane sugar) present, and in a simple mass action under these conditions, the amount of reacting substance changed in unit time is a *constant fraction* of the reacting substance present.*

In Table IV, the results of a series of five experiments are given in which a constant amount of invertase has acted on varying amounts of sugar under the conditions just named :

TABLE IV.—*Inversion changes by a constant amount of invertase acting in constant volume of cane sugar solutions of varying concentrations, in constant time.* 1 *c.c. of invertase solution added to* 100 *c.c. of cane sugar solution in each experiment. Temp.* 28°.

| No. of experiment. | Grams cane sugar per 100 c.c. | Grams cane sugar inverted in 60 minutes. | Fraction of cane sugar inverted in 60 minutes $=x$. | $k=\frac{1}{\theta}\log\frac{1}{1-x}$. |
|---|---|---|---|---|
| 1 | 4·89 | 1·230 | 0·252 | 0·00210 |
| 2 | 9·85 | 1·355 | 0·138 | 0·00107 |
| 3 | 19·91 | 1·355 | 0·068 | 0·00051 |
| 4 | 29·96 | 1·235 | 0·041 | 0·00031 |
| 5 | 40·02 | 1·076 | 0·027 | 0·00020 |

When the law of mass action was evidenced by Ostwald's experiments on methyl acetate (*loc. cit.*) under conditions similar to those employed in the above experiments, he found that a *constant fraction* of the methyl acetate present in each solution was hydrolysed in unit time, and therefore, if the action of invertase is an instance of simple mass action, a *constant fraction* of the cane sugar present in each of the above experiments should be inverted. But it will be noticed that instead of a constant fraction, a constant (or approximately constant) *weight* of the cane sugar is inverted. The fraction inverted diminishes in inverse proportion to the amount of cane sugar present in the experiments, and, as a consequence, the value $k$, which is constant in a true mass action, varies to a very large extent.

These experiments,† therefore, confirm the conclusion derived from the experiments given in Tables I and II, that the influence of

* For experimental confirmation of this necessary consequence of mass action, see Ostwald on the hydrolysis of methyl acetate by hydrochloric acid ("Outlines of General Chemistry," p. 353).

† Duclaux (*loc. cit.*) quotes some experiments with invertase derived from *Aspergillus niger* which fully confirm the author's experiments described in Table IV.

mass action does not rule in inversion change. Moreover, when the velocity of the action of inversion determined in the manner described in Table IV is examined, it will be noticed that the action is similar in character to that of fermentation, referred to at the commencement of this paper. During alcoholic fermentation, a constant amount of yeast decomposes in unit time an approximately *constant weight* of sugar in equal volumes of solution containing varying amounts of sugar. Invertase is now found to invert approximately constant quantities of cane sugar under similar conditions. Therefore the supposed difference in character of the two actions of fermentation and inversion which led the author to commence the investigation described in this paper does not exist, for the action of both, if expressed graphically, is represented approximately by a straight line. So far, therefore, the first object of the investigation is attained.

Although experiments carried out in the manner just described show that the general character of the action of invertase resembles that of fermentation, they do not explain the apparent paradox that when the action of invertase is studied during a series of consecutive changes in a single solution, the velocity of the action is then represented, not by a straight line, but by a curve, showing that there is a decrease in the *absolute amount* of sugar inverted during each time interval (see the experiments in Tables I and II and foot-note to p. 378 which indicate that, although the curve of the action is not so pronounced as the logarithmic curve of mass action, it is still very marked).

Apparently there are two causes which may lead to the production of such a curve during the continued action of invertase in a solution of cane sugar. Either it may be due to a natural weakening of the invertase by continued work,* or it may be due, as Duclaux has suggested (*loc. cit.*), to the action of invertase being influenced prejudicially by the accumulation of its own products of inversion. From what is known regarding the very large amount of cane sugar which is capable of being hydrolysed by a very small amount of invertase, the former cause appeared to be far less probable than the latter, so the author turned his attention to the investigation of the possible retarding influence of inversion products on the action of invertase.

### *Action of Inversion Products on the Velocity of Inversion Change.*

The method of experiment adopted by the author was to observe the action of a constant amount of invertase during a brief interval of time in equal volumes of solutions containing a constant amount of

* Some interesting experiments of Victor Henri (*loc. cit.*) indicate that sustained work does not weaken the action of invertase.

cane sugar *and varying amounts of invert sugar.* The invert sugar used in the experiments was prepared by the action of invertase on a concentrated solution of cane sugar until complete inversion was obtained, the invertase being then destroyed by raising the temperature of the solution to 90°.

The following table gives particulars of a series of experiments in which different amounts of this solution of invert sugar were mixed with a constant amount of cane sugar solution, the total volumes of the solutions in the different experiments being made constant:

TABLE V.—*Influence of invert sugar on the action of invertase. Volume of each experiment,* 100 *c.c.* 1 *c.c. of invertase solution used in each experiment. Time of change,* 80 *minutes. Temp.* 30°.

| No. of experiment. | Grams cane sugar present in 100 c.c. | Grams invert sugar present in 100 c.c. | Grams cane sugar inverted in 80 minutes. |
|---|---|---|---|
| 1 | 4·06 | none | 2·27 |
| 2 | 4·06 | 1·47 | 2·21 |
| 3 | 4·06 | 5·39 | 1·99 |
| 4 | 4·06 | 11·38 | 1·66 |
| 5 | 4·06 | 17·87 | 1·25 |

In these experiments, if the presence of invert sugar exerted no influence on the action of invertase, the quantities of cane sugar inverted in constant time would be constant, for the same quantities of cane sugar and invertase were present in all the experiments. But an examination of the table shows that the amount of cane sugar inverted decreased as the quantity of added invert sugar increased, until, in the last experiment (No. 5), the quantity of cane sugar inverted has been reduced to nearly one-half in the presence of 17·87 grams of invert sugar.

The series of experiments indicate, therefore, that the presence of invert sugar has diminished the activity of invertase, and that the arresting influence has increased as the amount of invert sugar has increased. But it is possible that the arresting influence of invert sugar may be due, not to the presence of the sugar as such, but to the increased viscosity of the solution containing it, for, owing to the manner in which the experiments were conducted, the total amount of sugars in the different solutions is an increasing one from the first to the last experiment. In order to investigate this question, a series of experiments was conducted in a similar manner to those described in Table V, excepting that lactose was used in the place of invert sugar. Lactose is a sugar which is not changed by the action of invertase, but its solutions possess a viscosity almost identical with that of solutions of invert sugar of similar concentrations. Therefore in a series of experiments with lactose in place of invert sugar, the factor of increasing viscosity is introduced apart from any special influence possessed by invert sugar alone.

The results of a series of experiments with lactose are given below:

TABLE VI.—*Influence of lactose on the action of invertase. Volume in each experiment,* 100 *c.c.* 1 *c.c. of invertase solution used. Time of change,* 60 *minutes. Temp.* 28°.

| No. of experiment. | Grams cane sugar present in 100 c.c. | Grams lactose present in 100 c.c. | Grams cane sugar inverted in 60 minutes. |
|---|---|---|---|
| 1 | 7·0 | none | 2·072 |
| 2 | 7·0 | 5·0 | 2·052 |
| 3 | 7·0 | 10·0 | 2·052 |
| 4 | 7 0 | 20·0 | 1·893 |

The results given in this table show that the influence on the action of invertase of the viscosity (or any other property) of the lactose used in the experiments is comparatively insignificant. In experiments 2 and 3, the retarding influence of 5 per cent. and 10 per cent. of lactose lies almost within the limits of experimental error, and in 4, in which the large amount of 20 per cent. lactose is present, the reduction in the amount of cane sugar inverted is only 9 per cent. On the other hand, it has already been shown (Table V, No. 4) that 17·8 per cent. of invert sugar under similar conditions reduced the amount of cane sugar inverted to the extent of 45 per cent. The major part of this reduction, therefore, is not due to viscosity, but must be occasioned by the arresting influence of invert sugar as such.

When the arresting influence of invert sugar on the action of invertase is thus established, there is then no difficulty in explaining the apparent paradox that the true action of invertase, which is indicated graphically by a straight line, is expressed by a curve when the action is determined for a series of progressive changes in one solution. Under the latter conditions, as the action of inversion proceeds, the products of inversion accumulate, and these consequently exert an increasing retarding influence on the action of inversion, and thus compel the action to follow the course of a curve.

*The Inversion Functions of Living Yeast Cells.*

So far, when discussing the action of invertase, the author has referred more especially to C. O'Sullivan and Tompson's experiments and conclusion regarding the velocity of its action. It now remains to discuss J. O'Sullivan's experiments, alluded to at the commencement of this paper as supporting C. O'Sullivan and Tompson's conclusion.

It will be remembered that J. O'Sullivan, when studying the velocity of the inversion change produced by living yeast in solutions of cane sugar (*loc. cit.*), found that the value $k$ derived from the expression $\frac{1}{\theta}\log\frac{1}{1-x}$, was constant, or nearly so, during the progression of the changes, and from this he concluded that the velocity of change followed the law of mass action.

There is no doubt that J. O'Sullivan's determinations—like those of C. O'Sullivan and Tompson—indicate a velocity approximating to that of mass action, when the progress of an inversion change is followed in one solution; but J. O'Sullivan has overlooked the fact—rendered evident by his own determinations—that, although the velocity in each separate change approximately follows the law, the value $k$ found for comparable experiments in which varying amounts of sugar have been used, shows that there is no conformity with mass action, but, on the contrary, indicates that a *constant amount* of sugar is inverted—an action similar to that which has been shown for free invertase.

For instance, in J. O'Sullivan's paper four comparable experiments are described, in which equal amounts of yeast were used in equal volumes of solution during equal intervals of time, the only variable being the quantity of cane sugar present in the solutions. The results of these experiments are given in the table on p. 384.

It will be noticed, on examining this table, that J. O'Sullivan has determined the progression of inversion in each of the four solutions at three time intervals, and the values of $k$ for the changes in each separate solution are fairly constant; but the values $k$ should also be constant for all four of the solutions if the velocity of change follows the law of mass action, because the solutions only differ in containing varying quantities of sugar. On the contrary, however, the value $k$ varies inversely as the amount of sugar present, in a similar manner to the value $k$ in the author's experiments with invertase, given in Table IV.

A similar conclusion may also be derived from the numbers in the column in Table VII showing the fractions of cane sugar inverted during the experiments. If the first numbers in each series are

TABLE VII.—*Velocity of inversion by living yeast cells* (*J. O'Sullivan*).

| Grams cane sugar per 100 c.c. | Grams of yeast used. | Time of action in minutes. $\theta$. | Fraction of sugar inverted. $x$. | $k=\frac{1}{\theta}\log\frac{1}{1-x}$. |
|---|---|---|---|---|
| 5 | 0·5 | 30 | 0·1636 | 0·0025 |
| | | 60 | 0·3164 | 0·0027 |
| | | 120 | 0·5442 | 0·0028 |
| 10 | 0·5 | 30 | 0·1042 | 0·0016 |
| | | 60 | 0·1544 | 0·0012 |
| | | 120 | 0·2780 | 0·0012 |
| 20 | 0·5 | 30 | 0·0627 | 0·0009 |
| | | 60 | 0·0850 | 0·0006 |
| | | 120 | 0·1467 | 0·0006 |
| 30 | 0·5 | 30 | 0·0366 | 0·0005 |
| | | 60 | 0·0495 | 0·0003 |
| | | 120 | 0·0862 | 0·0003 |

compared, it will be noticed that the fractions inverted are, approximately, in inverse proportion to the amounts of cane sugar present in the solutions—or, in other words, the actual *quantity* of sugar inverted is the same for all the experiments.

Thus J. O'Sullivan's experiments show that the velocity of action of the inversion function of yeast falls into line with the action of free invertase, and the action of fermentation,* previously demonstrated by the author.

### *Time and Molecular Change.*

It was stated at the commencement of this paper that the author's object was to examine, and, if possible, bring together, certain conclusions regarding the nature of enzyme action which seemed to be contradictory. Experimental evidence appeared to show that on the one hand the action of invertase, both in the free state and confined within the living cell, followed the law of mass action; and, on the other hand, that the action of the enzyme of alcoholic fermentation followed a different law. The author has now shown that these supposed differences in character of action do not exist, and that the actions of both inversion and fermentation follow approximately

* It is interesting to note the agreement in character of action of the inversion and fermentation functions of the *living yeast cell*, as it tends to strengthen the conclusion that fermentation is a true enzyme action.

the same order of progression—an order which is not that of mass action.

But this conclusion, that the actions of the two enzymes exhibit an exceptional order of progression differing from that of mass action, introduces a question which requires explanation.

It appears impossible to believe that enzyme change, however produced, is independent of mass action. According to our present conception of matter and its mechanics, such an idea appears to be inconceivable. Therefore, in looking for some explanation of the exceptional character of the actions of inversion and fermentation, the author concludes that the influence of mass in these actions, as they have been studied so far, must be limited or concealed by some other influence.

If such an influence is looked for, consideration shows that it may be due to the existence of a time factor in certain forms of complex molecular change.

When the law of mass action regulating simple chemical change has been confirmed by direct experiment, the reactions investigated have been changes such as the hydrolysis of methyl acetate by hydrochloric acid (Ostwald, *loc. cit.*) and the inversion of cane sugar by acids. In such experiments, the molecular change following collision of the reacting molecules takes place with extreme rapidity and the existence of a time factor is not in evidence in experimental determinations of the velocity of change. But it is quite conceivable, with regard to such a change as that of enzyme action, that the time elapsing during molecular union and transformation may be sufficiently prolonged to influence the general course of the action.

There is reason to believe that during inversion of cane sugar by invertase the sugar combines with the enzyme previous to inversion. C. O'Sullivan and Tompson (*loc. cit.*) have shown that the activity of invertase in the presence of cane sugar survives a temperature which completely destroys it if cane sugar is not present, and regard this as indicating the existence of a combination of the enzyme and sugar molecules. Wurtz (*Compt. rend.*, 1880, 91, 787) has also shown that papain appears to form an insoluble compound with fibrin previous to its hydrolysis. Moreover, the more recent conception of E. Fischer with regard to enzyme configuration and action, also implies some form of combination of enzyme and reacting substance.

Let it be assumed, therefore, that one molecule of an enzyme combines with one molecule of a reacting substance, and that the compound molecule exists for a brief interval of time during the further actions which end in disruption and change. Let it be assumed also that the interval of time during which the compound molecule of enzyme and reacting substance exists is 1/100 of a time unit.

Then it follows that a molecule of the enzyme may assist in effecting 100 completed molecular changes in unit time, but that this is the limit to its power of change.

Again, let it be assumed that the number of molecular collisions between the active and reacting molecules which lead to their combination bears some proportion to the number of possible completed molecular changes in unit time. Let the number of collisions be 20, then there may be 20 complete molecular changes; if 40, there may be 40 changes. In fact, the action of the mass law is observed, for other conditions being equal, the average number of molecular collisions must depend on the number of molecules, or mass, of the matter present.

But now assume that the mass of reacting substance is increased, so that the number of molecular collisions in unit time exceeds 100; let it be 150, 1000, or any other number larger than 100. Then, although the number of molecular collisions may exceed 100 by a number following the law of mass action, 100 molecular changes cannot be exceeded, for the compound enzyme and sugar molecule is only capable of effecting 100 complete changes in unit time.

It follows, therefore, that if, in a series of changes like the imaginary ones described, a constant amount of enzyme is in the presence of varying quantities of a reacting substance, and in all cases the quantity of reacting substance present ensures a greater number of molecular collisions in unit time than the possible number of molecular changes, then a *constant weight* of substance may be changed in unit time in all the actions.

When invertase acts in solutions of cane sugar of varying concentrations, an approximately *constant weight* of sugar is inverted in unit time, and the yeast cell, under similar conditions, ferments an approximately *constant weight* of sugar; it appears, therefore, that the exceptional character of these changes may be satisfactorily accounted for by the theory advanced.

Experimental evidence may also be brought forward in support of this theory.

In Table IV, the results of the author's experiments show that approximately constant quantities of cane sugar are inverted in unit time in solutions varying in concentration from 5 to 40 per cent. If the results of these experiments are looked at in the light of the author's theory, the number of molecular collisions in unit time in each experiment must have equalled, or exceeded, the possible number of changes by the compound molecule of enzyme and sugar. But this has happened in solutions containing 5 per cent. and upwards of cane sugar. It must, however, be possible to make solutions of varying quantities of cane sugar so dilute that the number of molecular collisions taking place in unit time between the sugar molecules and a constant number of invertase molecules will fall below the possible number of changes. Then, if the author's theory be correct, the progress of inversion in a series of these dilute solutions of cane sugar of different concentrations will exhibit an action in accordance with the law of mass action, for the time interval of change no longer restricts its effect.

It seemed very possible when commencing the attempt to demonstrate this experimentally that it might prove that the solutions of cane sugar required for the purpose were too dilute to use for experimental purposes. But when the attempt was made, it was found that the necessary dilutions are within the limit of experiment, as the results given in the following table (VIII) show :

TABLE VIII.—*Velocity of action of invertase in very dilute solutions of cane sugar.* 100 *c.c. of cane sugar solution and* 1 *c.c. of diluted invertase solution employed for each experiment. Time of change,* 60 *minutes. Temp.* 31°.

| No. of experiment. | Grams cane sugar per 100 c.c. | Grams cane sugar inverted in 60 minutes. | $k=\frac{1}{\theta}\log\frac{1}{1-x}$. |
|---|---|---|---|
| 1 | 2·0 | 0·308 | 0·00132 |
| 2 | 1·0 | 0·249 | 0·00219 |
| 3 | 0·5 | 0·129 | 0·00239 |
| 4 | 0·25 | 0·060 | 0·00228 |

The results given in the above table furnish very strong evidence in support of the view that in the dilute solutions of cane sugar employed the number of contacts of the sugar molecules with the invertase molecules in unit time have been reduced to a less number than the possible number of molecular changes. In experiment No. 1, in which a concentration of 2 grams of sugar per 100 c.c. has been used, the dilution appears to have been hardly sufficient to reach the desired point. In Nos. 2, 3, and 4, however, the quantities of sugar inverted in unit time are no longer constant quantities—as was found in the experiments with concentrations of 5 per cent. and upwards, given in Table IV., and decrease in direct proportion with the concentrations.

Moreover, the value $k$ in these experiments is a constant number.

These observations indicate a change in accordance with mass action, which, according to the author's theory, should be evidenced in solutions of sufficient dilution. There is, therefore, reason to believe from the results of the above experiments that the exceptional action of inversion in all but very dilute solutions of cane sugar is due to a time factor accompanying molecular combination and change which limits the influence of mass action.

It has been shown in this paper that the action of alcoholic fermentation follows approximately the same order of progression as that of inversion, and the work of Kastle and Loevenhart (*Amer. Chem. J.*, 1900, **24**, 491) shows that the action of lipase progresses in the same manner; it therefore appears probable that both these enzyme actions are regulated, like inversion, by a time factor accompanying complex molecular change.

It will be noticed that the author's theory demands, not only the formation of a molecular compound of enzyme and reacting substance, but the existence of this molecular compound for an interval of time previous to final disruption and change. Various speculations regarding the conditions ruling such an effect suggest themselves, but the author does not at present attempt to discuss this question.

THE BRITISH SCHOOL OF MALTING AND BREWING,
UNIVERSITY OF BIRMINGHAM.

# LOIS GÉNÉRALES

DE

# L'ACTION DES DIASTASES

PAR

VICTOR HENRI

DOCTEUR EN PHILOSOPHIE (UNIVERSITÉ DE GÖTTINGUE)
DOCTEUR ÈS-SCIENCES
PRÉPARATEUR DE PHYSIOLOGIE A LA SORBONNE

PARIS
LIBRAIRIE SCIENTIFIQUE A. HERMANN
ÉDITEUR, LIBRAIRE DE S. M. LE ROI DE SUÈDE ET DE NORVÈGE
6 et 12, rue de la Sorbonne, 6 et 12

1903

Reprinted from pages 85–93 of *Lois Générales de l'Action des Diastases*, Paris: Librairie Scientifique A. Hermann, 1903, 129 pp.

# THÉORIE DE L'ACTION DE L'INVERTINE

**Victor Henri**

**25. Théorie de l'action de l'invertine.** — J'ai repris l'étude de l'action de l'invertine sur le saccharose en prenant comme point de départ un certain nombre d'hypothèses qui peuvent être justifiées par des expériences.

On sait que *Emil Fischer* (1) a montré qu'entre l'action d'une diastase et la constitution stéréochimique d'un corps sur lequel elle agit existe une relation très étroite, de sorte que l'on peut prévoir d'avance si un ferment donné pourra hydrolyser un corps déterminé ou bien s'il restera sans action sur ce corps. Et, inversement, lorsqu'on trouve qu'un certain corps est hydrolysé par un ferment, on peut affirmer que la constitution chimique de ce corps a une forme déterminée.

Ce résultat capital a été confirmé depuis par plusieurs auteurs. C'est ainsi que nous savons, d'après les recherches de E. Fischer, que l'invertine agit sur les sucres qui donnent le lévulose par hydrolyse (saccharose, raffinose,

(1) E. Fischer, *Zeit. f. physiol. Chem.*, 46.

gentiobiose) ; que l'émulsine agit sur les sucres et les glycosides qui donnent du galactose et ainsi de suite.

Fischer a conclu de ses recherches sur les diastases que le ferment formait avec le corps à transformer une combinaison chimique intermédiaire, laquelle se décomposait ensuite en régénérant le ferment primitif.

L'étude de la vitesse d'inversion du saccharose, et surtout l'étude de l'influence de la concentration du sucre, montre que cette inversion suit des lois complexes qui rappellent celles que nous avons rencontrées en étudiant les réactions catalytiques, dans lesquelles il y avait formation de combinaisons intermédiaires. Il est donc naturel de chercher à expliquer l'action de l'invertine sur le saccharose en supposant qu'il y a formation de combinaisons intermédiaires.

Mais, d'autre part, nous avons vu que le sucre interverti exercait également une action sur la vitesse de la réaction, donc pour expliquer cette action nous ferons l'hypothèse de l'existence d'une combinaison intermédiaire entre le ferment et le sucre interverti. On pourrait même dire que cette combinaison intermédiaire a lieu entre le ferment et le lévulose, puisque nous avons vu dans le § 20 que l'action retardatrice du sucre interverti était due presque uniquement au lévulose. De cette façon l'action des produits de la réaction se trouve rapprochée du résultat de E. Fischer, d'après lequel l'invertine hydrolyse les sucres qui donnent du lévulose par dédoublement.

Dans la classification générale des actions catalytiques, nous avons vu que l'on pouvait soit admettre la formation de combinaisons intermédiaires complètes, soit, au contraire, supposer la formation de combinaisons incomplètes avec existence d'un certain état d'équilibre entre le catalysateur et le corps à transformer.

L'hypothèse de combinaisons complètes conduisait à une équation de la vitesse ayant pour expression $\frac{dx}{dt} = K$, la courbe serait donc dans ce cas une ligne droite.

Ces considérations théoriques nous conduisent donc à supposer que les combinaisons intermédiaires entre le ferment et le saccharose, ainsi que entre le ferment et le sucre interverti, sont incomplètes et donnent lieu à des états d'équibre.

Enfin, si nous admettons que ces combinaisons sont incomplètes, il en résulte qu'une certaine partie du ferment restera non combinée ; deux suppositions se présentent donc à l'esprit : 1°. On pourra supposer que la partie de ferment non combinée agit sur le saccharose et le transforme. 2° On pourra, au contraire, supposer que c'est la combinaison entre le ferment et le saccharose qui se décompose et produit ainsi l'hydrolyse du saccharose.

Quelle devrait être la loi d'action du ferment s'il obéissait aux différentes hypothèses que nous venons d'indiquer ? Telle est la question qu'il s'agit d'examiner maintenant.

Supposons que nous ayons une solution contenant la quantité $a$ de saccharose et que nous ajoutions la quantité $\Phi$ de ferment : au bout d'un temps égal à $t$ minutes nous voyons qu'il y a dans la solution $a - x$ saccharose et $x$ sucre interverti.

D'après les hypothèses indiquées plus haut, nous supposons que le ferment est réparti entre le saccharose et le sucre interverti et qu'une partie du ferment reste non combinée. Soit donc X la quantité de ferment qui n'est pas combinée, $z$ la quantité qui est combinée au moment $t$ avec le saccharose et $y$ la quantité combinée avec le sucre interverti. Nous supposons, en plus, que ces combinaisons présentent des états d'équilibre, c'est-à-dire qu'elles suivent la loi de l'action des masses.

Ecrivons donc les équations qui expriment les conditions d'équilibre. Nous avons pour l'équilibre entre le ferment et le saccharose dont la quantité est $a - x$ :

$$X.(a - x) = \frac{1}{m} z. , \tag{1}$$

pour l'équilibre entre le ferment et le sucre interverti :

$$X.x = \frac{1}{n} . y, \tag{2}$$

$m$ et $n$ sont les deux constantes d'équilibre ; enfin écrivons que la quantité totale de ferment combiné et libre est égal à $\Phi$ :

$$\Phi = X + y + z. \tag{3}$$

Ces trois équations permettent de calculer la quantité $x$ de ferment non combiné et la quantité $z$ de la combinaison entre le saccharose et le ferment. Nous obtenons en effet :

$$X = \frac{\Phi}{1 + m(a - x) + nx} \tag{4}$$

et

$$Z = \frac{m.\Phi.(a - x)}{1 + m(a - x) + nx}. \tag{5}$$

Ceci étant, examinons quelle devra être la vitesse d'inversion au moment $t$, suivant que nous admettons que c'est le ferment libre qui agit sur le saccharose ou bien que c'est la combinaison $z$ qui se décompose.

1[re] *hypothèse.* — La partie libre du ferment agit sur le saccharose ; dans ce cas la vitesse est proportionnelle à la quantité de ce ferment libre et à la quantité de saccharose, elle sera donc proportionnelle à X et à $a - x$, par conséquent au produit de ces deux valeurs. Donc :

$$\frac{dx}{dt} = \text{K.X.}(a - x),$$

ou en remplaçant X par sa valeur (4) on obtient :

$$\frac{dx}{dt} = \frac{\text{K.}\Phi.(a - x)}{1 + m(a - x) + nx}. \tag{6}$$

2[e] *hypothèse.* — C'est la combinaison entre le ferment et le saccharose qui se décompose, la vitesse de cette décomposition sera donc proportionnelle à la quantité de cette combinaison intermédiaire, c'est-à-dire à $z$ ; on aura : $\frac{dx}{dt} = \text{K}.z$ ou en remplaçant $z$ par sa valeur (5) on obtient :

$$\frac{dx}{dt} = \frac{\text{K}.m.\Phi.(a - x)}{1 + m(a - x) + nx}. \tag{7}$$

Ces deux expressions de la vitesse de la réaction (6) et (7), sont identiques entre elles, par conséquent, quelle que soit l'hypothèse que l'on fasse, la loi suivant laquelle se produira l'inversion du saccharose sera la même. Ce résultat présente un intérêt au point de vue de la chimie générale, mais je ne m'y arrête pas ici.

Donc la vitesse de la réaction a pour expression :

$$\text{(I)} \qquad \frac{dx}{dt} = \frac{K_3(a - x)}{1 + m(a - x) + nx}$$

ou $K_3$ est une constante proportionnelle à la quantité de ferment ; $m$ et $n$ sont deux constantes caractéristiques qui pourront varier avec le milieu et avec la température, mais qui, une fois déterminées, devront donner pour $K_3$ la meme valeur, quelle que soit la concentration de saccharose ou de sucre interverti.

Si, au début de la réaction, nous avions un mélange de $a_1$ saccharose et de $i$ sucre interverti, la vitesse au moment $t$ aurait pour expression :

$$\text{(II)} \qquad \frac{dx}{dt} = \frac{K_3(a_1 - x)}{1 + m(a_1 - x) + n(x + i)}.$$

On peut discuter la forme générale de l'action de l'invertine en se servant des équations (I) ou (II).

Considerons le cas où au début nous ayons seulement du saccharose et étudions la vitesse du début ; cette vitesse s'obtient évidemment en posant dans l'expression (I) $x$ egal à zéro. Donc

$$\text{Vitesse initiale} = \frac{K_3 a}{1 + ma}.$$

On voit que lorsque la concentration en saccharose est faible, c'est-à-dire si $a$ est petit, $ma$ disparaît er face de l'unité, et alors la vitesse initiale est proportionnelle à la concentration de saccharose. Mais à mesure que $a$ augmente la vitesse croît d'abord rapidement puis lentement, de sorte que lorsque $ma$ sera grand par rapport à l'unité, la vitesse initiale sera égale environ à $\frac{K_3a}{ma}$ c'est-à-dire à $\frac{K_3}{m}$, elle sera constante quelle que soit la concentration du saccharose.

Donc, pour les solutions diluées la vitesse d'inversion est influencée par la quantité de saccharose et pour les solutions plus concentrées elle en est presque indépendante. Graphiquement la relation entre la concentration de saccharose et la vitesse initiale est représentée par une hyperbole passant par l'origine et ayant une asymptote parallèle à l'axe des $x$ à une distance égale à $K_3$.

C'est bien la forme générale que l'on obtient expérimentalement, ainsi que nous l'avons vu plus haut.

Supposons maintenant que nous ayons au début une quantité $a_1$ de saccharose et $i$ de sucre interverti ; la vitesse au début sera alors :

$$\text{Vitesse initiale} = \frac{K_3a_1}{1 + ma_1 + ni}.$$

Cette expression nous donne une représentation complète de l'action du sucre interverti. En effet, pour la même quantité de saccharose $a_1$ la vitesse sera d'autant plus faible que $i$ sera plus grand (exemples § 19).

2° Pour la même quantité de sucre interverti $i$ le ralentissement produit par ce sucre sera d'autant plus intense que $a_1$ sera plus faible : une même quantité de sucre interverti ralentit plus la vitesse d'inversion d'une solution faible que d'une solution concentrée en saccharose (exemples § 19).

Calculons maintenant la formule qui donnera la constante $K_3$. Intégrons l'équation (I) que nous pouvons encore écrire sous la forme :

$$dx.\left[\frac{1+na}{a-x}+m-n\right]=K_3dt;$$

nous obtenons :

$$-(1+na)\ln(a-x)+(m-n)x=K_3t+\text{constante};$$

la constante est déterminée par la condition que $t=0$ pour $x=0$, on obtient :

$$-(1+na)\ln a=\text{constante},$$

donc la relation devient :

$$(1+na)\ln\frac{a}{a-x}+(m-n)x=K_3t,$$

ou encore :

$$\text{(III)}\quad K_3=\frac{a}{t}\left[(m-n)\frac{x}{a}+n\ln\frac{a}{a-x}\right]+\frac{1}{t}\ln\frac{a}{a-x}.$$

On obtient de même dans le cas où au début on a une quantité $a_1$ de saccharose et $i$ de sucre interverti l'expression suivante :

$$\text{(IV)}\quad K_3=\frac{a_1}{t}\left[(m-n)\frac{x}{a_1}+n\left(1+\frac{i}{a_1}\right)\ln\frac{a_1}{a_1-x}\right]+\frac{1}{t}\ln\frac{a_1}{a_1-x}.$$

On voit que l'expression de $K_3$ se compose de la somme de deux facteurs dont l'un dépend de la valeur *absolue* de $a$, tandis que l'autre ne dépend que du *rapport* $\frac{x}{a}$. C'est cette forme complexe qui exprime l'influence exercée par la concentration de saccharose sur la vitesse de la réaction.

[*Editor's Note:* The rest of this chapter, which consists mainly of experimental measurements, has been omitted here.]

**ERRATUM**

The footnote on page 85 should cite volume 26 instead of 16.

# 30


Reprinted from *Biochem. J.* **19**:338–339 (1925)

# A NOTE ON THE KINETICS OF ENZYME ACTION.

By GEORGE EDWARD BRIGGS
AND JOHN BURDON SANDERSON HALDANE.

*(From the Botanical and Biochemical Laboratories, Cambridge.)*

THE equation of Michaelis and Menten [1913] has been applied with success by Kuhn [1924] and others to numerous cases of enzyme action. It is therefore desirable to examine its theoretical basis. Consider the irreversible reaction $A \rightarrow B$, unimolecular as regards $A$, and catalysed by an enzyme. Suppose one molecule of $A$ to combine reversibly with one of enzyme, the compound then changing irreversibly into free enzyme and $B$, where $B$ may represent several molecules. We may represent this as:

$$\underset{(a-x)\,(e-p)}{A+E} \rightleftharpoons \underset{p}{AE} \rightarrow \underset{x}{B} + E.$$

Now let $a$ be the initial concentration of $A$. $e$ the total concentration of enzyme, $x$ the concentration of $B$ produced after time $t$, and $p$ the concentration of enzyme combined with substrate at time $t$. We suppose $e$ and $p$ to be negligibly small compared with $a$ and $x$. Then by the laws of mass action

$$\frac{dp}{dt} = k_1 (a-x)(e-p) - k_2 p - k_3 p,$$

where $k_1$, $k_2$, $k_3$ are the velocity constants of the reactions

$$A + E \rightarrow AE, \quad AE \rightarrow A + E, \quad \text{and} \quad AE \rightarrow B + E,$$

respectively. Now since $p$ is always negligible compared with $x$ and $a - x$, its rate of change must, except during the first instant of the reaction, be negligible compared with theirs.

For during the remainder of the reaction $p$ diminishes from a value not exceeding $e$ to zero, whilst $x$ increases from zero to $a$. Thus the average value of $-\frac{dp}{dt} \div \frac{dx}{dt}$ is less than $\frac{e}{a}$. And provided $\frac{e}{a}$ is small it is clear that if the amount of combined enzyme decreased for a measurable time at a rate comparable with that of its substrate the reaction would come to an end. To take a concrete example Kuhn [1924] calculates that a yeast saccharase molecule at 15° and $p_H$ 4·6 can invert 100 or more molecules of sucrose per second. Even if the enzyme concentration is so unusually large that the inversion of a strong sucrose solution is half completed in 10 minutes, $\frac{a}{e}$ cannot be less than 120,000, and if $-\frac{dp}{dt}$ attained 1 % of the value of $\frac{dx}{dt}$ for 1 second the reaction would stop owing to all the enzyme being set free. (Actually it may be shown that

$$\frac{dp}{dt} = \frac{-k_3(k_2+k_3)e(a-x)}{k_1\left(a-x+\frac{k_2+k_3}{k_1}\right)^3}.)$$

Hence $$k_1(a-x)(e-p) - k_2p - k_3p = 0.$$

$$\therefore p = \frac{k_1e(a-x)}{k_2+k_3+k_1(a-x)}$$

$$= \frac{e(a-x)}{a-x+\frac{k_2+k_3}{k_1}}$$

$$\therefore \frac{dx}{dt} = k_3p = \frac{k_3e(a-x)}{a-x+\frac{k_2+k_3}{k_1}}.$$

This is Michaelis and Menten's [1913] equation, $(k_2+k_3)/k_1$ representing their constant $K_s$. They assume that the reaction

$$A + E \rightleftharpoons AE$$

is always practically in equilibrium, and $K_s$ its equilibrium constant, *i.e.* that $k_3$ is negligible in comparison with $k_2$. Van Slyke and Cullen [1914] on the other hand assumed that the first stage of the reaction was irreversible, *i.e.* $k_2 = 0$, and arrived at the same equation. It is clear, however, that data as to the course of a reaction can give no indication of the ratio of $k_2$ and $k_3$, though when the velocity of the observed reaction, and hence $k_3$, is very large, the upper limit to $k_2$ deducible from the kinetic theory may possibly prove to be of the same order of magnitude.

It may be remarked that with this modification of their theory, Michaelis and Menten's analysis of the effects of the products of the reaction, or other substances which combine with the enzyme, still holds good.

REFERENCES.

Kuhn (1924). Oppenheimer's "Die Fermente und ihre Wirkungen," 185 et seq.
Michaelis and Menten (1913). *Biochem. Z.* **49**, 333.
Van Slyke and Cullen (1914). *J. Biol. Chem.* **19**, 141.

# 31



Reprinted from pages 119–120 of *Allgemeine Chemie der Enzyme*, Dresden and Leipzig: Verlag Theodor Steinkopf, 1932, 367 pp.

## GRAPHISCHE METHODEN

**J. B. S. Haldane and K. G. Stern**

Die folgenden graphischen Methoden, die wir im wesentlichen *Woolf* verdanken, haben sich bei der Ermittelung der *Michaelis*-Konstante von Enzymen als wertvoll erwiesen. Für eine gegebene Zahl von Werten verschiedener Substratkonzentrationen x und dazugehöriger Reaktionsgeschwindigkeiten v haben wir die zur Erfüllung der Gleichung $v = \frac{Vx}{K+x}$ erforderlichen, passenden Werte für V und K zu ermitteln. Die vorstehende Gleichung kann auf dreierlei Weise geschrieben werden:

a) $\frac{V}{v} = 1 + \frac{K}{x}$; Schnittpunkte auf den Koordinaten: $\frac{1}{V}, \frac{-1}{K}$

b) $v + K\frac{v}{x} = V$; „ „ „ „ $V, \frac{V}{K}$

c) $\frac{x}{v} + K = Vx$; „ „ „ „ $\frac{K}{V}, -K.$

In jedem Falle erhalten wir, wenn wir die beobachteten Werte für $\frac{1}{v}$ und $\frac{1}{x}$, bzw. v und $\frac{v}{x}$ oder $\frac{x}{v}$ und x gegeneinander auftragen, gerade Linien, wenn in dem untersuchten Falle die *Michaelis*-Beziehung Gültigkeit besitzt. Die Werte für K und V können aus den Abschnitten ermittelt werden, die von dieser Geraden auf den Koordinatenachsen abgeteilt werden. Allgemein betrachtet besitzen die Werte für x einen weiten Spielraum (z. B. unterscheiden sie sich in Fig. 13 um das 29fache voneinander). Die Werte für v können nur innerhalb eines engeren Bereiches verschieden sein und die Werte für $\frac{v}{x}$ und $\frac{x}{v}$ können sich nur um noch kleinere Beträge unterscheiden. Wenn wir gemäß Gleichung a) verfahren, so liegen die hohen Substratkonzentrationen entsprechenden Punkte dicht beieinander. Wenn wir hingegen die Gleichung c) benutzen, ist das gleiche bei den Punkten der Fall, die niedrigen Substratkonzentrationen entsprechen. Die Methode b) ist demnach im allgemeinen dann am bequemsten anzuwenden, wenn nur eine Reihe von Beobachtungen zu verarbeiten ist. Hierbei stellen die Abschnitte auf den Koordinatenachsen die Werte für V und $\frac{V}{K}$ dar. Wenn anderseits mehrere Versuchsserien vorhanden sind, bei denen V, K oder beide Größen nicht übereinstimmen, so erweist sich oft die Methode a) als zweckmäßiger. Beim Vorliegen einer kompetitiven Hemmung wird eine Linienserie erhalten, die durch den gleichen Punkt geht $\left(\frac{1}{v} = \frac{1}{V}\right)$, während die Linienschar bei nicht-kompetitiver Hemmung durch den Punkt $\frac{1}{x} = \frac{-1}{K}$ geht.

Wo eine kompetitive Hemmung vorliegt, liefert die Anwendung der Methode c) eine Schar paralleler Linien.

# 32



Reprinted from *Nature* **179**:832 (1957)

# GRAPHICAL METHODS IN ENZYME CHEMISTRY

**J. B. S. Haldane**

*University College, London*

IN 1932, Dr. Kurt Stern published a German translation of my book "Enzymes"[1], with numerous additions to the English text. On pp. 119–120, I described some graphical methods, stating that they were due to my friend Dr. Barnett Woolf. Michaelis's equation may be written $v = \frac{Vx}{x + K}$, where $x$ is the substrate concentration at any moment, $v$ the velocity with which the substrate is being destroyed, $V$ the velocity when the enzyme is saturated, and $K$ the Michaelis constant. Woolf pointed out that linear graphs are obtained when $v$ is plotted against $vx^{-1}$, $v^{-1}$ against $x^{-1}$, or $v^{-1}x$ against $x$, the first plot being most convenient unless inhibition is being studied. But competitive inhibition gives a pencil of lines through the point $(0, V^{-1})$, while non-competitive inhibition gives a pencil through the point $(-K^{-1}, 0)$, when $v^{-1}$ is plotted against $x^{-1}$.

These methods were afterwards rediscovered by others, the first of these rediscoveries being by Lineweaver and Burk[2]. Dr. Woolf was undergoing a serious illness as the result of an accident, and never published the results which he had shown me in draft form. Recently, Hofstee[3] has credited me with the invention of these methods without mentioning Woolf's name. Squabbles as to priority are undignified; but if anyone is to be credited with these methods, the credit belongs entirely to Woolf, and not to myself, unless indeed he and I had overlooked still earlier work.

[1] Haldane, J. B. S., and Stern, K., "Allgemeine Chemie der Enzyme" (Steinkopf, Leipzig and Berlin, 1932).

[2] Lineweaver, H., and Burk, D., *J. Amer. Chem. Soc.*, **56**, 658 (1934).

[3] Hofstee, B. H. J., *Enzymologia*, **17**, 273 (1956).

# 33

# ENZYME STUDIES, SECOND REPORT: ON THE MEASUREMENT AND SIGNIFICANCE OF THE HYDROGEN ION CONCENTRATION IN ENZYMATIC PROCESSES[1]

**Søren Peter Lauritz Sørensen**

*The Carlsberg Laboratory, Copenhagen*

*These excerpts were translated expressly for this Benchmark volume by H. Friedmann, the University of Chicago, from pages 131–134, 186–190, and 300–304 of "Enzymstudien. II. Mitteilung. Über die Messung und die Bedeutung der Wasserstoffionenkonzentration bei enzymatischen Prozessen" in* Biochem. Z. **21**:131–304 (1909).

[*Editor's Note:* This classic paper, published in Danish, French, and German, is more than 150 pages long. The whole paper cannot be translated for this volume; some, and by no means all, of the highlights are given here. The "Short Survey" with which the paper concludes is given in its entirety to indicate the scope of this study. The present translation was made from the German version.]

## INTRODUCTION

### 1. Degree of Acidity: Hydrogen Ion Concentration

It is a well-known fact that the velocity with which an enzymatic cleavage proceeds depends, among other things, on the degree of acidity or alkalinity of the solution used. Usually--indeed, it may be said always--the degree of acidity or alkalinity for enzymatic processes is calculated and reported on the basis of the total amount of added acid or base, and it is not at all common to pay attention to the degree of dissociation of the acid or base used; even more rarely does one consider the capacity of the respective solution to bind acid or base.

There is no doubt, however, that the views on the nature of solutions, based on *Arrhenius'* theory of electrolytic dissociation, have to be applied to these types of cases as well. When, for example, a digestion by pepsin occurs in a hydrochloric acid solution that is 0.1 normal relative to its total hydrochloric acid content, then one must not ignore the fact that 0.1 N hydrochloric acid is not completely dissociated, and that the "true degree of acidity," which is rationally to be designated as the *hydrogen ion concentration*, is hence a bit smaller than 0.1 N. At the same time one has to consider whether the solution contains salts--e.g., phosphates--with which hydrochloric acid can react, or other substances that can influence the hydrogen ion concentration; above all, it must not be forgotten that the substrate, in this special case a suitable protein, binds acid. Since the hydrogen ion concentration of the solution depends only on the amount of free dissociated acid present and since obviously the amount of acid bound to protein depends on the nature and amount of protein, it follows directly that two solutions, one of which, for example, contains 1 g and the other 5 g protein in 100 cc of 0.1 N hydrochloric acid and that hence, following the usual convention, have the same degree of acidity, exhibit a widely differing "true degree of acidity," a widely differing hydrogen ion concentration.

These statements concerning the degree of acidity and the hydrogen ion concentration of a digestion by pepsin apply to all enzymatic processes. I have chosen the example of cleavage by pepsin since this process proceeds best at a higher hydrogen ion concentration so that in this case the situation permits a clearer perspective and is also more or less familiar. However, in the case of enzymatic processes that proceed best at a weakly acidic, at a neutral or an alkaline reaction, a quite similar approach must also be used. It will be clear from the examples in the last sections of this treatise [not translated here] that the magnitude of the hydrogen ion concentration is of as fundamental an importance for the other enzymatic cleavages as for digestion by pepsin; the difference lies only in the order of magnitude of the hydrogen ion concentration with which one has to deal in the various cases.

## 2. The Magnitude of the Hydrogen Ion Concentration: The Hydrogen Ion Exponent

When the concentrations of the hydrogen ions, of the hydroxyl ions, and of water in an aqueous solution are denoted by $C_{H^{\cdot}}$, $C_{OH'}$, and $C_{H_2O}$ respectively, then, as is known, because of the law of chemical mass action the following equation will hold:

$$\frac{C_{H^{\cdot}} \times C_{OH'}}{C_{H_2O}} = \text{constant.}$$

Since $C_{H_2O}$ is to be regarded as constant for any reasonably dilute solution, it follows that the product

$$C_{H^{\cdot}} \times C_{OH'} = \text{constant as well.}$$

This product, which is usually, as also in this treatise, called the *dissociation constant of water*, is equal to $0.64 \times 10^{-14}$ at 18°; in a series of measurements that were carried out in this laboratory, we have found the mean value of $0.72 \times 10^{-14}$ or written differently, $10^{-14.14}$. This magnitude makes it possible, as is readily seen, to calculate the hydrogen ion concentration of an aqueous solution when the concentration of the hydroxyl ions is known, and vice versa. Since the value of the dissociation constant of water is, of course, subject to errors and since furthermore the concentration of the hydrogen ions can usually be determined more accurately and more conveniently than that of the hydroxyl ions, it is more rational as proposed by *H. Friedenthal*,[2] always as far as possible to determine the hydrogen ion concentration of a solution and to use this for calculations, even when the solution has an alkaline reaction. This procedure is hence also used here so that, for example, one denotes a solution whose normality relative to hydrogen ion is 0.01 by 0.01 N, or, omitting the normality designation, for short by $10^{-2}$. In the same way, the hydrogen ion concentration of a solution that, for example, is 0.01 normal relative to hydroxyl ions is given by $10^{-12.14}$ since $10^{-12.14} \times 10^{-2} = 10^{-14.14}$. Absolutely pure water and truly neutral solutions will, when one uses the same terminology, have a hydrogen ion concentration of $10^{-7.07}$ since $10^{-7.07} \times 10^{-7.07} = 10^{-14.14}$.

The magnitude of the hydrogen ion concentration accordingly is expressed by the normality factor of the respective solution relative to the hydrogen ions, and this factor is written *in the form of a negative power of 10.* While referring the reader to a subsequent section, I wish to indicate here only that I use the name "hydrogen ion exponent" and the symbol $p_{H^{\cdot}}$ for the numerical value of the exponent of this power. In the three preceding examples, the $p_{H^{\cdot}}$ hence becomes 2, 12.14, and 7.07, respectively.

[*Editor's Note:* Further material, the "subsequent section" referred to above, and the discussion referred to below, are not included in the present translation. This part mentions, among other things, that many years were spent to investigate about a hundred colorimetric hydrogen ion indicators, many of which were prepared in Sørensen's laboratory, and that one must be careful not to assume a given indicator that works with a protein-free solution to give reliable hydrogen ion concentration measurements in the presence of protein.]

It may be permitted to make a few remarks about the circumstances that determined my choice of the standard solutions that have just been discussed.

In a work, done in this laboratory by *Fr. Weis* and published[3] in 1902, dealing with the proteolytic enzymes of malt, the large role played in proteolysis by the phosphate content was already pointed out. *Weis*, in agreement with the data of *A. Fernbach* and *L. Hubert*,[4] found the proteolytic capacity of a malt extract to be increased upon addition of acid up to the point at which the secondary phosphate present is completely converted into primary phospate; upon addition of further acid, the proteolytic capacity, however, decreases again. On the other hand, upon addition of bases proteolysis is inhibited the more, the greater the amount added; the inhibition is shown most markedly, however, when more base is added than is required to convert the available primary phosphate into secondary phosphate.

A series of experiments on the proteolytic enzyme of yeast carried out by the engineer *Fr. Petersen* and the chemist *P. R. Sollied* in this laboratory in the years 1902 and 1903 had also shown that the amount of primary and secondary phosphates present in the experimental fluids exerted a marked effect on the speed of proteolysis.

Since it was known furthermore from the investigations of *Fernbach*[5] that the solubilizing activity of diastase toward starch is also dependent on the phosphates present, in that the primary phosphates accelerated the process, while the secondary ones inhibited it, the idea suggested itself that one was here dealing with a general effect of the preceding salt mixture.

Since it is scarcely thinkable that primary or secondary salts as such can influence enzymatic processes in either direction, one hence must look elsewhere to explain the importance of these salts for enzymatic cleavage, and the obvious thing is to consider the hydrogen ion concentration of the experimental fluid for which the ratio of primary to secondary phosphates is of course decisive. If a limited amount of acid or base is added to such a phosphate mixture, the result is only that a part of the secondary salt is converted to the primary salt or vice versa; the hydrogen ion content of the solution is, however, not pushed beyond the concentration that corresponds to that of the primary or secondary phosphate. Such a mixture of phosphates hence affords a natural protection against excessively precipitous changes in the ion concentration. It acts--using an image employed in this connection by *Fernbach* and *Hubert*[6]--like a *buffer*.[7]

Mixtures of normal carbonates and bicarbonate, as also bicarbonates with an excess of carbon dioxide, can act as buffers and certainly often function as such. It was mentioned earlier . . . that a 0.05 M solution of normal sodium carbonate has a hydrogen ion concentration that corresponds to the hydrogen ion exponent of 11.39. A 0.1 M solution of sodium bicarbonate, however, has the hydrogen ion exponent of 8.40, and when this solution is saturated with carbon dioxide, its $p_{H^{\cdot}}$ sinks to 6.8 to 6.9. It follows from this as shown by a glance at the main Table of Curves that a sodium carbonate solution that contains carbon dioxide will be able to serve as buffer for a concentration range similar to that covered by phosphate mixtures but lying somewhat more to the alkaline side.

There is, however, one enzymatic process--digestion by pepsin--for which the optimal hydrogen ion concentration lies completely outside the range mentioned here. In this process, hence, neither phosphates nor carbonates can be used as regulators; the substrate itself, that is, the protein and its cleavage products, both of which bind acid and base, here serves as buffer.

If we now wish to prepare standard solutions with hydrogen ion concentrations similar to those that play a role in enzymatic processes, then it is obvious that we should consider using substances employed as buffers in Nature. The first standard solutions that we used hence were phosphate mixtures and glycine mixtures, the latter since glycine is the simplest and most readily available representative of proteins and of their degradation products. The Table of Curves shows at a first glance how beautifully the curves corresponding to these mixtures supplement each other: The phosphate curve governs exactly that region in which the glycine curve is absolutely useless. It can scarcely be doubted that the remaining degradation products of proteins, as well as the proteins themselves, will behave in essentially the same way as glycine. Neutral solutions of these substances hence are displaced far from the point of neutrality ($p_{H^{\cdot}}$ = 7.07) when even small amounts of acid or base are added, unless the solutions contain phosphates or carbonates that can serve as buffers.

A closer view of the Table of Curves soon teaches, however, that it is indeed possible, as already mentioned, to use phosphate and glycine mixtures as standard solutions to cover the whole concentration range from 0.1 N HCl to 0.1 N NaOH; the shape of the curves shows, however, that there are two concentration segments, namely, at the end of the phosphate curve and at the beginning of the glycine curve (with HCl and NaOH, respectively) where the need for other standard solutions is apparent. Since this requirement, as shown by the Table of Curves, is particularly pressing in one segment (with $p_{H^{\cdot}}$ = about 4) and since this segment is of quite special interest for many enzyme cleavages . . . it soon became necessary to supplement the phosphate and glycine mixtures with new standard solutions. Among the solutions that we investigated from this point of view, we selected the citrate and borate mixtures mentioned earlier, and these have proved themselves to be very good from all points of view.

It has to be noted, in addition, that *Friedenthal* and *Salm* in their papers already mentioned [Sørensen cites numerous papers by these workers (and papers by W. Salessky and B. Fels) and mentions that their

Table of Curves. (*Reprinted from Biochem. Z.* **21**:131–304, 1909.)

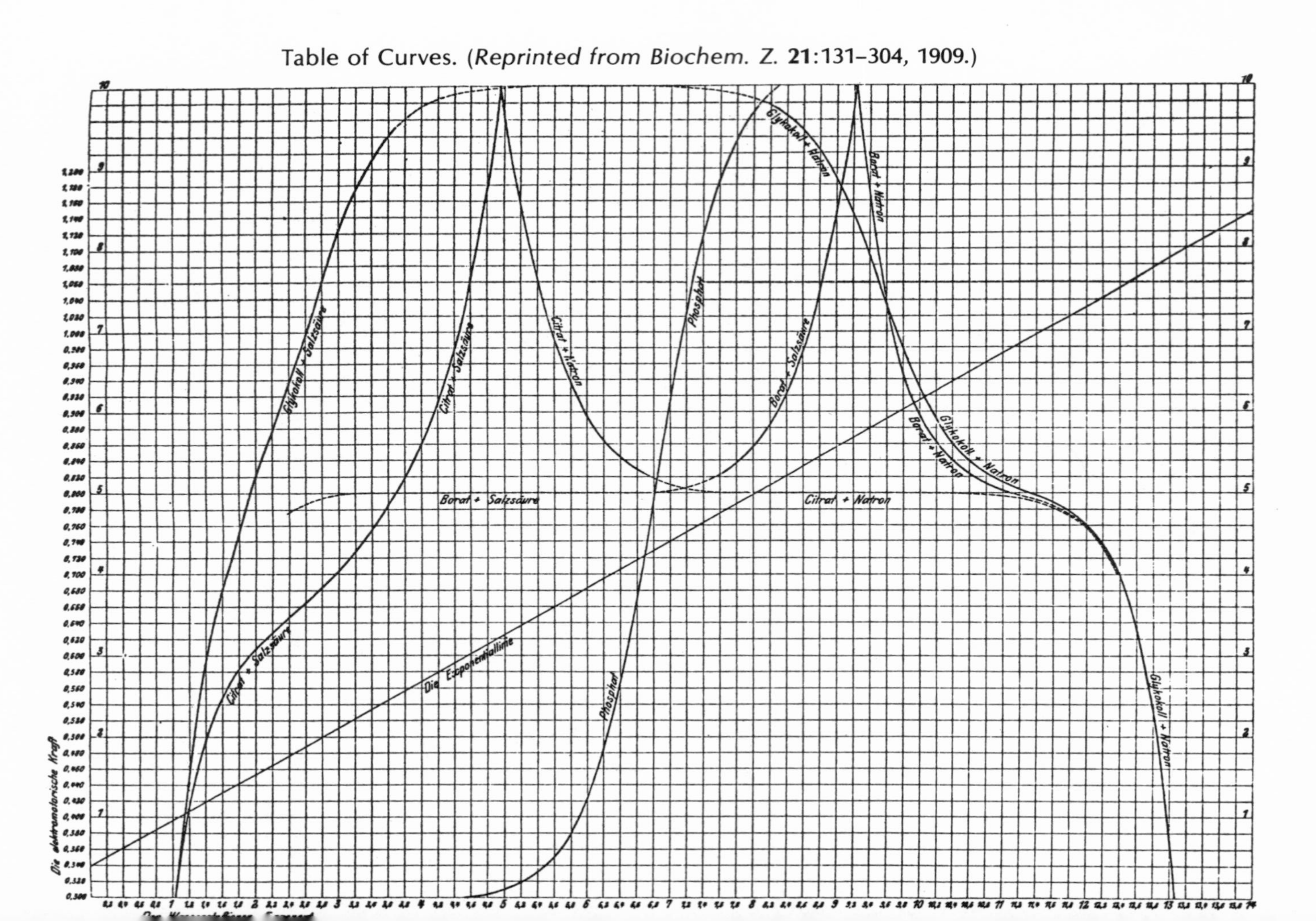

beautiful investigations give them the honor to have been the first to use the colorimetric method for determining hydrogen ion concentrations.], following the proposal of *Szilys*, used phosphate mixtures as standard solutions. Naturally, I do not wish to dispute the priority of these excellent researchers on this point. On the other hand, I did not wish to omit to report the main train of ideas which led to my choice of the standard solutions.

[*Editor's Note:* Extensive material that includes detailed experiments on the classic yeast invertase from fresh brewer's yeast, and somewhat less detailed studies on catalase and on pepsin, and the first ever pH-activity curves (for invertase and for catalase), has been omitted at this point. The following survey ends the paper.]

## SHORT SURVEY

1. One has to distinguish clearly between *degree of acidity* and *hydrogen ion concentration*; only the latter plays a role in enzymatic cleavages.

2. If the normality factor of a solution relative to hydrogen ions is expressed by the term $10^{-p}$, then it is proposed to use the name *hydrogen ion exponent*, represented by $p_{H^\bullet}$ for the numerical value of the power exponent.

3. In enzymatic cleavages the *hydrogen ion concentration of the experimental fluid plays a role similar to the experimental temperature.* By the *hydrogen ion concentration curve* of an enzyme, one understands the curve that is obtained when the substrate amounts that are cleaved under the given experimental conditions during a time unit are plotted as ordinates, while the hydrogen ion exponents of the experimental fluids are plotted as abscissae. The hydrogen ion concentration curves have a course similar to the temperature curves.

4. For the measurement of the hydrogen ion concentration of a fluid, methods that result in a change of the hydrogen ion concentration during the measurement are all useless; the *usual acidimetric* and *alkalimetric titration methods* hence are always unsuitable for the present purposes. Usually the "*catalytic*" methods of measurement are also unsuitable for determining the hydrogen ion concentration in enzymatic cleavages, and in the rare case where these methods can be used, this may be done only if conditions are controlled critically and strictly.

5. The following two complementary methods are to be recommended for the measurement of the hydrogen ion concentration in enzymatic cleavages: *the exact but cumbersome electrometric and the less exact but in practice exceedingly simple colorimetric method.*

6. *The electrometric method*

a. If $\pi$ denotes the electromotive force at 18° of an element that consists on the one hand of a mercury-calomel electrode in a 0.1 N potassium chloride solution, and on the other hand of a platinum-hydrogen electrode in an electrode fluid whose hydrogen ion exponent is $p_{H^\bullet}$, then we obtain the following equation:

$$\pi = 0.3377 + 0.0577 \times p_{H^\bullet}$$

*as an expression for the interdependence of the electromotive force of the element and of the hydrogen ion concentration of the electrode fluid* at 18°; for $p_{H^\bullet} = 0$, we get $\pi = 0.3377$.

b. In the main Table of Curves that accompanies this treatise, it is seen that the straight line that constitutes the graphic expression of this equation is denoted by the name "*exponential line*." By means of this straight line, it is possible purely graphically, that is, without any calculation, to convert a measured value of $\pi$ into the corresponding value of $p_H$., and vice versa.

c. Using the electrometric measurement of the hydrogen ion concentration of dilute sodium hydroxide solutions, the *dissociation constant of water* at 18° has been found equal to $0.72 \times 10^{-4} = 10^{-14.14}$.

d. By dissolving in water a series of simple substances whose purity can be readily controlled and that are commercially available in a guaranteed pure state (*A. F. Kahlbaum*, Berlin), it is possible to prepare *standard solutions*. When these are mixed in given proportions, one can obtain solutions of any hydrogen ion concentration from about $10^{-1}$ to about $10^{-13}$. These concentrations are, however, always accurately determined beforehand by the electrometric method. From the main *Table of Curves that accompanies this treatise, the hydrogen ion exponent of each of these mixtures can conveniently be read off graphically*.

When selecting the compounds used in the standard solutions, preference was given to substances that in the living organism provide a natural protection against sudden changes in the hydrogen ion concentration. It is repeatedly pointed out how important the presence of such "*buffers*" is for enzymatic processes.

e. In a number of special cases, quoted in this treatise, the *electrometric measurement* of such fluids, which can occur in enzymatic processes, *offers difficulties*.

7. *The colorimetric method*

a. *The procedure for colorimetric measurement* when one wishes to use mixtures of the above-mentioned standard solutions for *purposes of comparison* is described.

b. Numerous sources of error of the method are given. The most important of these is *the tendency of proteins and of their least degraded cleavage products to combine with the indicators*, whereby in certain cases the colorimetric method becomes difficult or even impossible. Because of these sources of error, it was necessary to test the suitability of each separate indicator under different conditions. Thus the useless ones could be discarded, and the accuracy of the useful ones could be determined. *For these control measurements, the electrometric measurement of the hydrogen ion concentration was used as the basic method.*

c. *The indicators investigated* here, some of which had not been known before, are divided into five groups depending mainly on the position of the inflection point. For the most important and best indicators of each group, a suitable selection is given of examples of the control measurements that were carried out with them.

d. Some of the most frequently used indicators, for example, *Congo red*, are to be regarded as quite useless to carry out measurements of the type mentioned here.

e. *On the basis of all control measurements carried out, 20 indicators are recommended as suitable.* Since these, however, are not equally useful under all circumstances, it is necessary, when choosing the best suitable indicator for a particular case, to consider a series of facts that have been carefully stressed in the treatise.

8. *The significance of hydrogen ion concentration in the cases of cleavage by invertin, catalase, and pepsin is shown by means of*

*examples.*

a. It is shown *that the optimal hydrogen ion concentration for cleavage by invertin is almost constant when the other experimental parameters are fixed, that is, it is independent of the type and amount of invertin and independent of the type of acid used in the experiments.* Under the experimental conditions chosen, the optimal hydrogen ion concentration corresponds to a $p_H$. = 4.4 to 4.6.

b. By means of a series of experiments on invertin, it is shown *that it is necessary to include a consideration of the hydrogen ion concentration when carrying out studies on reaction kinetics* with enzymes, and it is stressed that the widely diverging views about the reaction kinetics of cleavage by invertin expressed by previous researchers may at least in part be due to the circumstance that up to now insufficient attention has been given to hydrogen ion concentration.

c. Attention is drawn both to the importance of self-destruction of enzymes in studies of the type mentioned here, and to the related *interdependence of the incubation time, the experimental temperature, and the hydrogen ion concentration of the experimental solution.* In the case of invertin the optimal hydrogen ion concentration is shifted a bit toward the alkaline side with increasing incubation time.

d. At an experimental temperature of 0°, the optimal hydrogen ion concentration of *cleavage by catalase* is very close to neutrality, but with increasing incubation time, it seems to be shifted a little toward the acid side.

e. At the experimental temperature of 37°, the optimal hydrogen ion concentration for *cleavage by pepsin* is clearly time dependent; for short incubation times the optimal point corresponds to a hydrogen ion exponent somewhat smaller than 2, but with increasing incubation time, this is shifted toward the acidic side.


1. Published simultaneously in French in *Compt. rend. des travaux du Lab. de Carlsberg* 8:1-174 (1909). [Also in Danish in *Meddelelser fra Carlsberg Laboratoriet* 8:1-153 (1909-1910).]

2. *Zeitschr. f. Elektrochem.* 10:114, 1904.

3. *Compt. rend. des travaux du Lab. de Carlsberg* 5:211, and *Zeitschr. f. d. ges. Brauwesen* 26:558, 1903.

4. *Compt. rend.* 131:293, 1900.

5. *Ann. de la Brasserie et de la Distillerie* 5 Sept., 10 and 25 Oct., 1899; *Wochenschr. für Brauerei* 17:34, 1900. Compare also the later papers by *L. Maquenne* and *Eug. Roux* about the saccharification of starch by means of diastase (review: *Bull. Soc. Chim. Paris* [3] 35:I-XV, 1906, and *Ann. Chim. et Phys.* [8] 9:179, 1906).

6. L.c., p. 295.

7. Cf. also R. Höber, *Beiträge z. chem. Physiol. u. Pathol.* 3:528, 1903, and also these papers from more recent years: G. Bertrand, *Bull. Soc. Chim. de France* [4] 1:1130, 1907; L. J. Henderson, *Amer. J. of Physiol.* 15:257; 21:427--L. J. Henderson and O. F. Black, ibid. 18:250;

21:420--G. W. Hall, ibid. 18:283; the five last papers are quoted on the basis of reports in *Chem. Centralbl.* 1906 1:1031; 1908, 1:1190 and 1197; 1908, 2:335 and 335 [Printing error in original].

Tabelle III. Glykokollmischungen.

| Zusammensetzung der Mischung | Die elektromotorische Kraft $\pi$ | Der Wasserstoffionenexponent $p_H$. | Zusammensetzung der Mischung | Die elektromotorische Kraft $\pi$ | Der Wasserstoffionenexponent $p_H$. |
|---|---|---|---|---|---|
| 10 ccm Glykokoll | ca. 0,6900 | ca. 6,106 | 10 ccm Glykokoll | ca. 0,6900 | ca. 6,106 |
| 9,9 ccm Glykokoll + 0,1 ccm HCl | 0,5922 | 4,411 | 9,9 ccm Glykokoll + 0,1 ccm NaOH | 0,7883 | 7,809 |
| 9,75 ccm Glykokoll + 0,25 ccm HCl | 0,5680 | 3,991 | 9,75 ccm Glykokoll + 0,25 ccm NaOH | 0,8130 | 8,237 |
| 9,5 ccm Glykokoll + 0,5 ccm HCl | 0,5500 | 3,679 | 9,5 ccm Glykokoll + 0,5 ccm NaOH | 0,8325 | 8,575 |
| 9 ccm Glykokoll + 1 ccm HCl | 0,5305 | 3,341 | 9 ccm Glykokoll + 1 ccm NaOH | 0,8529 | 8,929 |
| 8 ccm Glykokoll + 2 ccm HCl | 0,5063 | 2,922 | 8 ccm Glykokoll + 2 ccm NaOH | 0,8780 | 9,364 |
| 7 ccm Glykokoll + 3 ccm HCl | 0,4881 | 2,607 | 7 ccm Glykokoll + 3 ccm NaOH | 0,8982 | 9,714 |
| 6 ccm Glykokoll + 4 ccm HCl | 0,4692 | 2,279 | 6 ccm Glykokoll + 4 ccm NaOH | 0,9228 | 10,140 |
| 5 ccm Glykokoll + 5 ccm HCl | 0,4492 | 1,932 | 5,5 ccm Glykokoll + 4,5 ccm NaOH | 0,9425 | 10,482 |
| 4 ccm Glykokoll + 6 ccm HCl | 0,4326 | 1,645 | 5,1 ccm Glykokoll + 4,9 ccm NaOH | 0,9763 | 11,067 |
| 3 ccm Glykokoll + 7 ccm HCl | 0,4196 | 1,419 | 5 ccm Glykokoll + 5 ccm NaOH | 0,9900 | 11,305 |
| 2 ccm Glykokoll + 8 ccm HCl | 0,4099 | 1,251 | 4,9 ccm Glykokoll + 5,1 ccm NaOH | 1,0050 | 11,565 |
| 1 ccm Glykokoll + 9 ccm HCl | 0,4038 | 1,146 | 4,5 ccm Glykokoll + 5,5 ccm NaOH | 1,0356 | 12,095 |
| 10 ccm HCl | 0,3976 | 1,038 | 4 ccm Glykokoll + 6 ccm NaOH | 1,0531 | 12,390 |
| | | | 3 ccm Glykokoll + 7 ccm NaOH | 1,0690 | 12,674 |
| | | | 2 ccm Glykokoll + 8 ccm NaOH | 1,0795 | 12,856 |
| | | | 1 ccm Glykokoll + 9 ccm NaOH | 1,0862 | 12,972 |
| | | | 10 ccm NaOH | 1,0916 | 13,066 |

Für „9 ccm Glykokoll + 1 ccm NaOH“ haben wir gefunden: 0,8525 — 0,8526 — 0,8529 — 0,8525 — 0,8532 — 0,8529 — 0,8531 — 0,8528 — 0,8531 — 0,8533;

Mittel: 0,8529.

(*Reprinted from page 174 of Biochem. Z.* **21**:131–304, 1909.)

**Tabelle IV. Phosphatmischungen.**

| Zusammensetzung der Mischung | Die elektromotorische Kraft $\pi$ | Der Wasserstoffionenexponent $p_H$. | Zusammensetzung der Mischung | Die elektromotorische Kraft $\pi$ | Der Wasserstoffionenexponent $p_H$. |
|---|---|---|---|---|---|
| 10 ccm sek. Phosphat . . . . . | 0,8167 | 8,302 | 4 ccm sek. Phosphat + 6 ccm prim. Phosphat | 0,7210 | 6,643 |
| 9,9 ccm sek. Phosphat + 0,1 ccm prim. Phosphat | 0,8092 | 8,171 | 3 ccm sek. Phosphat + 7 ccm prim. Phosphat | 0,7109 | 6,468 |
| 9,75 ccm sek. Phosphat + 0,25 ccm prim. Phosphat | 0,8015 | 8,038 | 2 ccm sek. Phosphat + 8 ccm prim. Phosphat | 0,6977 | 6,239 |
| 9,5 ccm sek. Phosphat + 0,5 ccm prim. Phosphat | 0,7914 | 7,863 | 1 ccm sek. Phosphat + 9 ccm prim. Phosphat | 0,6787 | 5,910 |
| 9 ccm sek. Phosphat + 1 ccm prim. Phosphat | 0,7790 | 7,648 | 0,5 ccm sek. Phosphat + 9,5 ccm prim. Phosphat | 0,6608 | 5,600 |
| 8 ccm sek. Phosphat + 2 ccm prim. Phosphat | 0,7616 | 7,347 | 0,25 ccm sek. Phosphat + 9,75 ccm prim. Phosphat | 0,6438 | 5,305 |
| 7 ccm sek. Phosphat + 3 ccm prim. Phosphat | 0,7500 | 7,146 | 0,1 ccm sek. Phosphat + 9,9 ccm prim. Phosphat | 0,6248 | 4,976 |
| 6 ccm sek. Phosphat + 4 ccm prim. Phosphat | 0,7402 | 6,976 | 10 ccm prim. Phosphat . . . . . | 0,5990 | 4,529 |
| 5 ccm sek. Phosphat + 5 ccm prim. Phosphat | 0,7308 | 6,813 | | | |

(*Reprinted from page 175 of Biochem. Z.* **21**:131–304, 1909.)

Tabelle V. Citratenmischungen.

| Zusammensetzung der Mischung | Die elektromotorische Kraft $\pi$ | Der Wasserstoffionenexponent $p_H$. | Zusammensetzung der Mischung | Die elektromotorische Kraft $\pi$ | Der Wasserstoffionenexponent $p$ . |
|---|---|---|---|---|---|
| 10 ccm Citr. . . . | 0,6238 | 4,958 | 10 ccm Citr. . . | 0,6238 | 4,958 |
| 9,5 ccm Citr. + 0.5 ccm HCl . | 0,6197 | 4,887 | 9,5 ccm Citr. + 0,5 ccm NaOH | 0,6275 | 5,023 |
| 9 ccm Citr. + 1 ccm HCl . . | 0,6164 | 4,830 | 9 ccm Citr. + 1 ccm NaOH . | 0,6325 | 5,109 |
| 8 ccm Citr. + 2 ccm HCl . . | 0,6061 | 4,652 | 8 ccm Citr. + 2 ccm NaOH . | 0,6443 | 5,314 |
| 7 ccm Citr. + 3 ccm HCl . . | 0,5943 | 4,447 | 7 ccm Citr. + 3 ccm NaOH . | 0,6590 | 5,568 |
| 6 ccm Citr. + 4 ccm HCl . . | 0,5776 | 4,158 | 6 ccm Citr. + 4 ccm NaOH . | 0,6821 | 5,969 |
| 5,5 ccm Citr. + 4,5 ccm HCl . | 0,5655 | 3,948 | 5,5 ccm Citr. + 4,5 ccm NaOH | 0,7030 | 6,331 |
| 5 ccm Citr. + 5 ccm HCl . . | 0,5507 | 3,692 | 5,25 ccm Citr. + 4,75 ccm NaOH | 0,7230 | 6,678 |
| 4,75 ccm Citr. + 5,25 ccm HCl . | 0,5413 | 3,529 | 5 ccm Citr. + 5 ccm NaOH . | 0,8600—0,9200 | 9,052—10,092 |
| 4,5 ccm Citr. + 5,5 ccm HCl . | 0,5318 | 3,364 | 4,5 ccm Citr. + 5,5 ccm NaOH | 1,0343 | 12,073 |
| 4 ccm Citr. + 6 ccm HCl . . | 0,5092 | 2,972 | 4 ccm Citr. + 6 ccm NaOH . | 1,0511 | 12,364 |
| 3,33 ccm Citr. + 6,67 ccm HCl . | 0,4689 | 2,274 | | | |
| 3 ccm Citr. + 7 ccm HCl . . | 0,4488 | 1,925 | | | |
| 2 ccm Citr. + 8 ccm HCl . . | 0,4195 | 1,418 | | | |
| 1 ccm Citr. + 9 ccm HCl . . | 0,4054 | 1,173 | | | |
| 10 ccm HCl . . | 0,3976 | 1,038 | | | |

*(Reprinted from page 176 of Biochem. Z.* **21***:131–304, 1909.)*

Tabelle VI. Boratenmischungen.

| Zusammensetzung der Mischung | Die elektromotorische Kraft $\pi$ | Der Wasserstoffionenexponent $p_H$. | Zusammensetzung der Mischung | Die elektromotorische Kraft $\pi$ | Der Wasserstoffionenexponent $p_H$. |
|---|---|---|---|---|---|
| 10 ccm Borat . . | 0,8709 | 9,241 | 10 ccm Borat . . | 0,8709 | 9,241 |
| 9,5 ccm Borat + 0,5 ccm HCl . | 0,8667 | 9,168 | 9 ccm Borat + 1 ccm NaOH . | 0,8778 | 9,360 |
| 9 ccm Borat + 1 ccm HCl . . | 0,8620 | 9,087 | 8 ccm Borat + 2 ccm NaOH . | 0,8860 | 9,503 |
| 8,5 ccm Borat + 1,5 ccm HCl . | 0,8574 | 9,007 | 7 ccm Borat + 3 ccm NaOH . | 0,8960 | 9,676 |
| 8 ccm Borat + 2 ccm HCl . . | 0,8517 | 8,908 | 6 ccm Borat + 4 ccm NaOH . | 0,9132 | 9,974 |
| 7,5 ccm Borat + 2,5 ccm HCl . | 0,8454 | 8,799 | 5 ccm Borat + 5 ccm NaOH . | 0,9768 | 11,076 |
| 7 ccm Borat + 3 ccm HCl . . | 0,8384 | 8,678 | 4 ccm Borat + 6 ccm NaOH . | 1,0518 | 12,376 |
| 6,5 ccm Borat + 3,5 ccm HCl . | 0,8285 | 8,506 | | | |
| 6 ccm Borat + 4 ccm HCl . . | 0,8160 | 8,289 | | | |
| 5,75 ccm Borat + 4,25 ccm HCl . | 0,8072 | 8,137 | | | |
| 5,5 ccm Borat + 4,5 ccm HCl . | 0,7958 | 7,939 | | | |
| 5,25 ccm Borat + 4,75 ccm HCl . | 0,7774 | 7,621 | | | |
| 5 ccm Borat + 5 ccm HCl . . | 0,7155 | 6,548 | | | |
| 4,75 ccm Borat + 5,25 ccm HCl . | 0,4745 | 2,371 | | | |

*(Reprinted from page 177 of Biochem. Z.* **21**:131–304, 1909.)

Part V

# FIRST CRYSTALLINE ENZYMES: PROTEINS WITH ENZYMATIC ACTIVITY

# Editor's Comments on Papers 34 Through 40



One would expect that it should have been fairly obvious that enzymes have a knowable chemical structure once it had been established by Buchner that ferments in Pasteur's sense are really no different from enzymes in Kühne's sense. One would have expected Buchner's work to have stripped the agents of chemical change in biological systems of inscrutable vital attributes. However, this was not so. Much of the inscrutability of the intact living cell had, in a reiteration of older attitudes, been shifted to the catalytically active cellular extract. As Kohler (1975, pp. 290–291) points out, the vaguely chemical, vaguely vitalistic concept of *protoplasm,* and particularly the concept of *colloid,* became prevalent in the early years of the twentieth century. Willstätter had hoped that it would be possible to understand enzymes as "a straightforward problem of organic chemistry, thus by-passing the whole colloidal mess" (Kohler 1975, p. 291). It is against this background that Willstätter's insistence that enzymes should be approached as chemically definable substances must be recognized as an advance over the chemically vague notion of "colloid" which was so prevalent in the thinking of his time.

Short excerpts from two of Richard Martin Willstätter's (1872–1942) many enzyme papers, both of these from lectures, are included in this volume since their attitude and conclusions, just as in the case of papers by Liebig, Pasteur, and Kühne, had such a strong impact on contemporary scientists that they must be regarded as classics (Papers 34 and 35). Willstätter developed Brücke's and Danilevsky's approaches to enzyme purification by the use of absorbents. He succeeded with his powerful and novel adsorption methods to purify enzymes (one of the best-remembered adsorbents that he introduced is alumina $C\gamma$) to the stage where their enzymatic activity was still strong but their reaction for protein by the available methods was negative. This success led him to the conclusion that enzymes cannot be proteins.

Willstätter was a powerful force in the 1920s and 1930s. He had succeeded his teacher Adolf von Baeyer, who himself had succeeded Liebig, to head the prestigious Chemical Laboratory of the Bavarian Academy of Sciences in Munich. For his pioneer researches in plant pigments, especially chlorophyll, he had received the Nobel Prize in chemistry in 1915. Because of his superb and varied achievements in organic chemistry, his views on enzymes were taken very seriously indeed. The tenor of his papers and speeches is not vindictive—much in contrast to the vitriolic writings of Liebig and Pasteur—and he stated in his 1933 Gibbs medal address in Chicago, where he was still talking of "the controversial protein nature of enzymes," that what matters is to gain new knowledge, not who is right. His autobiography *Aus meinem Leben* (From My Life)—the very same title had been used by Emil Fischer for his autobiography written some twenty years earlier—is a very moving book that shows him as a man of high personal principles. For an appreciation of Willstatters personality, environment, and work, see Nachmansohn (1979).

It is mainly because of Willstätter's views that John Batcheller Sumner's crystallization of urease in 1926 (Paper 36), and particularly John Howard Northrop's crystallization of pepsin in 1930 (Paper 37), rapidly followed by that of other digestive proteases by Northrop and Moses Kunitz (Papers 38, 39, and 40), had a strong emotional impact. This impact would have been much less if it had been generally agreed—as already proposed in 1858 by Moritz Traube and assumed by Emil Fischer in the 1890s (Papers 24 and 25) as well as by others, but disputed for example by Brücke (1874)—that enzymes are indeed proteins.[1]

J. B. Sumner (1887–1955) was interested in urease from his student days, for his Ph.D. thesis (1914) was called *The Formation of Urea in the Animal Body*. It was while he was engaged in his lengthy work on the purification of urease that he crystallized for the first time two other proteins of the jack bean (*Canavalia ensiformis*), concanavanin A and B (Sumner 1919). The term concanavanin had been used before (Jones and Johns 1916–1917), but the designations A and B are due to Sumner. His crystallization of urease while still an assistant professor at Cornell University is an example of dogged persistence leading to controversial success, for Willstätter's dictum that proteins are merely colloidal carriers of adsorbed low molecular weight enzymes with unknown structure was so influential that Sumner's views were not immediately accepted, in spite of the publication of more papers to support them. Very soon after Sumner's discovery, Willstätter in fact personally carried the controversy to Cornell University with a series of lectures (Willstätter 1927). It was only because of Northrop's crystallization of pepsin soon thereafter, and the further crystallizations of digestive proteases by Northrop and Kunitz, that Sumner's conclusions came to be accepted. So great was the eventual impact of these discoveries that Sumner, Northrop, and Wendell M. Stanley (who crystallized tobacco mosaic virus) received a Nobel Prize in chemistry in 1946, Sumner receiving one half, Northrop and Stanley the other half.

It is rather ironic that jack bean urease, whose crystallization initiated the view that enzymes are proteins, is actually not a simple protein. Some fifty years after Sumner's work it was shown that this enzyme contains nickel (Dixon et al 1980, and earlier references therein). The metal is essential for enzymatic activity. The fact that Sumner did not know this enzyme to contain a metal was fortunate since it resulted in the attendance by him and by others to the proteinous nature of this and other enzymes. The story of urease is an object lesson on the vagaries of scientific progress and on the pragmatic value of partial ignorance.

Sumner's scientific achievements are rendered all the more

remarkable by the fact that at the age of seventeen he lost his left arm due to a shooting accident; in addition, he had been left-handed. An example of the stultifying, if temporary, victory of convention over imagination is provided by the following episode from Sumner's life: "In 1921, when his research was still in its early stages, he had been granted an American-Belgian fellowship and decided to go to Brussels to work with Jean Effront, who had written several books on enzymes. The plan fell through, however, because Effront thought Sumner's idea of isolating urease was ridiculous" (Nobel Lectures 1964).

Papers 37–40 from the laboratory of J. H. Northrop (1891– ) and M. Kunitz (1887–1978) are remarkable for their thoroughness, elegance, original insights, and cogency. It is most impressive that all this work, representing the crystallization of pepsin, trypsin, trypsinogen, chymotrypsin, chymotrypsinogen, and various other proteins was accomplished in the span of but a few years and that by and large just two techniques, the use of ammonium sulfate (occasionally magnesium sulfate) along with pH changes, sufficed. One of these papers along contains methods for preparing and the photographs of, four crystalline proteins (Paper 40). Unfortunately, only a selection of a few of these classical papers can be given here.[2] This work, which is in direct line with the pioneering experiments of Reaumur, Spallanzani, Schwann, Beaumont, Brucke, Danilevsky, and Kuhne, was summarized in a masterly fashion in Northrop's book *Crystalline Enzymes* (1939) (second edition, with Kunitz and Herriott, 1948).
ments are in order. The conclusion that enzymes are proteins is seen to be somewhat tentative in the first paper (on crystalline pepsin), stronger at the beginning of the first paper on crystalline trypsin, but quite emphatically asserted in the discussion to the second trypsin paper. By the time the paper on crystalline chymotrypsin and chymotrypsinogen was written (Paper 39), this point was not argued any more, and the transformation of chymotrypsinogen to chymotrypsin is treated in terms of protein chemistry. The most impressive suggestion is made that this conversion is associated with an intramolecular rearrangement. (The experimental pages that detail the evidence for this conclusion are not reproduced here.) It should be noted that the words *chymotrypsin* and *chymotrypsinogen* are first used in this paper.

The background to Northrop's work was provided by a remarkable paper by the eminent Dutch biochemist Cornelis Adrianus Pekelharing from just before the turn of the century

(Pekelharing 1896/97; see also Pekelharing 1902; Ringer 1915). Pekelharing had used a method for the purification of pepsin quite different from that of Brücke. It did not employ adsorbents. Pekelharing concluded that the protein that he had purified was indeed pepsin.[3] Northrop, who had been stimulated by his association with Jacques Loeb at the Rockefeller Institute "to consider the protein nature of enzymes" (Northrop 1961), repeated Pekelharing's experiments about 1920 (see Northrop 1946) but made no advances then. It was Sumner's crystallization of urease in 1926 that encouraged him to repeat the pepsin preparation yet again, but on a much larger scale. In his opinion the protein that he crystallized "is probably the same as that isolated by Ringer[4] and Pekelharing 30 years before" (Northrop 1961). Furthermore, "had Pekelharing filtered his pepsin preparation and then dissolved it in a very small amount of water, instead of centrifuging the precipitate and dissolving it in a large amount of water, I am quite sure he would have crystallized the enzyme nearly fifty years ago" (Northrop 1946, p. 127). This clearly is a lesson that shows how a few experimental details can change the course of scientific history. Northrop (1961) notes that years of quantitative experiments by him and by many others were needed before the notion that the protein was actually pepsin and not a "carrier" of pepsin was accepted.[5]

The impact of the work of Sumner, Northrop, and Kunitz on the progress of enzymology is comparable to that of Eduard Buchner on cell-free fermentation and to Emil Fischer's work on enzyme mechanisms, some thirty to forty years earlier.

Among other enzymes that were crystallized early, it is necessary to note carboxypeptidase (Anson 1935), the "Old Yellow Enzyme" (NADH dehydrogenase) (Theorell 1934, 1935), catalase (Sumner and Dounce 1937), and alcohol dehydrogenase (Negelein and Wulff 1937*a*, 1937*b*). The last three of these did not catalyze hydrolytic reactions; among these alcohol dehydrogenase was the only one without an attached nonprotein group.

By 1948, that is, twenty-two years after Sumner's crystallization of urease, no less than 39 enzymes of various types had been crystallized (listed in Northrop et al. 1948), while a little less than twenty years after that an influential treatise on enzymes (Dixon and Webb 1964, pp. 793–808) carried an "atlas" containing the photographs of no less than 134 crystalline enzyme preparations. (The 1979 edition [Dixon and Webb 1979, pp. 665–680] shows 192 photographs.) By now, with the availability of a greater choice of methods for enzyme purification (see Part XI) and with standardized approaches to enzyme crystallization (see Jakoby 1971), the crystallization of an enzyme, although still an elegant achievement, has lost much of its glamor. In fact, more stringent criteria than crystallization are required before a crystalline protein is accepted to be homogeneous (a fact already stated by Kossel in 1901, see Edsall 1962), so that in enzyme purification one now often dispenses with crystallization as the final step.

## NOTES

1. Toward the end of the nineteenth century Eduard Buchner in addition to Emil Fischer felt that enzymes are proteins. Thus Buchner (1897, p. 1111) mentions that loss in the fermentative activity of cell-free yeast extracts on standing is probably due to the known presence in such extracts of "peptic enzymes," while later (Buchner 1898, p. 574) he uses the term "proteolytic enzymes" in the same context. However, many workers at this time were uncertain about the chemical nature of enzymes (see Fruton 1972, pp. 75–77), and Willstätter's work after World War I appeared for a while to have dealt the death blow to the protein theory of enzyme structure.
2. The papers on the crystallization of pepsinogen (Herriott and Northrop 1936 and Herriott 1938) are not included here.
3. Pekelharing's argument was clear; one of his protein preparations (starting with material obtained, of all places, from Armoux [possibly Armour?] in Chicago) still showed enzymatic activity on dilution of the protein to the stage where no reaction for protein was given. Conversely, such dilute solutions could be concentrated, with retention of enzyme activity, to the stage where protein characteristics could again be observed. Pekelharing's wise conclusion bears repeating: "Das Ausbleiben der Eiweissreactionen in gut wirkenden Pepsinlösungen beweist also nicht, dass Pepsin nicht ein Eiweissstoff ist" ("The absence of reactions for protein in highly active pepsin solutions hence does not prove that pepsin is not a protein"). (See also the Editor's Comments on Paper 14 and endnote 15, Part II.)
4. This is W. E. Ringer, not the Sidney Ringer of Ringer's solution fame.
5. As late as 1937 a paper was published (Kraut and Tria 1937) that cast doubts on the protein nature of pepsin. In this paper the products obtained by the purification methods of Northrop and of Brücke are compared.

# 34

# ON THE ISOLATION OF ENZYMES

**Richard Martin Willstätter**

*These excerpts were translated expressly for this Benchmark volume by H. Friedmann, the University of Chicago, from pages 3606 and 3611 of "Über Isolierung von Enzymen" in* Dtsch. Chem. Ges. Ber. **55**:3601–3623 (1922), *by permission of the publisher, Deutsche Chemische Gesellschaft.*

(Review Lecture, given under the auspices of the German Chemical Society on the occasion of the Centenary of the Society of German Scientists and Physicians in Leipzig, 20 September 1922.)

[*Editor's Note:* In the original, material precedes these excerpts.]

The phenomena of enzyme specificity, the observations concerning decomposition and stabilization of purified invertin solutions, and especially the influences of dispersion on enzymatic activity--which were found with other enzymes--lead to the view that the molecule of an enzyme consists of a colloidal carrier and an active group that functions purely chemically. The naturally occurring substances that accompany invertin and that have nothing to do with the enzyme associate with the colloidal carrier, and this addition is without effect on the hydrolysis of cane sugar that occurs on the active group.

[*Editor's Note:* Material has been omitted at this point.]

The increase in enzymatic purity is easily followed in the initial stages of purification by the disappearance of the sensitive reactions for proteins, purines, carbohydrates, and so on.

[*Editor's Note:* The rest of this paper, which deals with Willstätter's innovative and influential adsorption methods for enzyme purification, has not been translated.]

# 35

# ON PROGRESS IN ENZYME ISOLATION

**Richard Martin Willstätter**

*These excerpts were translated expressly for this Benchmark volume by H. Friedmann, the University of Chicago, from pages 1 and 12 of "Über Fortschritte in der Enzyme-Isolierung" in* Dtsch. Chem. Ges. Ber. **59**:1–12 (1926)*, by permission of the publisher, Deutsche Chemische Gesellschaft.*

(Delivered at the Session of 16th November 1925)

According to experiences gained over the last years, the basis of enzymatic activities is to be found not in special conditions by which suitable substances are dispersed but rather in certain organic compounds of unknown constitution. It is not yet possible to get to know the enzymes in the pure state, and it appears too early to initiate investigations concerning their formulae and structures. It has only been possible to increase the degree of purity of enzymes to such an extent that the first questions regarding their composition can be decided, that certain assumptions about their constitution can be excluded. The result is that the enzymes cannot be included among the proteins or carbohydrates, that in fact they do not belong to the known large groups of the more complicated organic compounds.

[*Editor's Note:* Material has been omitted at this point. We proceed to the end of the paper.]

If we attempt to sketch a picture of the chemical nature of an enzyme, then we see that the "molecule of an enzyme consists of a colloidal carrier and an active group that functions purely chemically." We know that every single chemically defined colloidal extraneous substance [Begleitstoff] can, on the basis of the experimental evidence, be removed with retention of the specific enzyme activity. We also know that enzyme affinities are not constant. Hence it is probable that the colloidal carriers are variable. A given colloidal carrier hence seems to be dispensable when another suitable one is available for the enzyme. The enzyme can change its aggregates [vermag seine Aggregate zu wechseln]. It has not yet been possible completely to separate, with retention of activity, the chemically active group, that is, the enzyme molecule proper, from the protecting colloids.

# 36


Reprinted from *J. Biol. Chem.* **69**:435–441 (1926)

## THE ISOLATION AND CRYSTALLIZATION OF THE ENZYME UREASE.

### PRELIMINARY PAPER.

By JAMES B. SUMNER.

(*From the Department of Physiology and Biochemistry, Cornell University Medical College, Ithaca.*)

(Received for publication, June 2, 1926.)

After work both by myself and in collaboration with Dr. V. A. Graham and Dr. C. V. Noback that extends over a period of a little less than 9 years, I discovered on the 29th of April a means of obtaining from the jack bean a new protein which crystallizes beautifully and whose solutions possess to an extraordinary degree the ability to decompose urea into ammonium carbonate. The protein crystals, which are shown in Fig. 1, have been examined through the kindness of Dr. A. C. Gill, who reports them to be sharply crystallized, colorless octahedra, belonging by this definition to the isometric system. They show no double refraction and are from 4 to 5$\mu$ in diameter.

While the most active solutions of urease prepared in this laboratory by Sumner, Graham, and Noback[1] and by Sumner and Graham[2] possessed an activity of about 30,000 units per gm. of protein present, the octahedra, after washing away the mother liquor, have an activity of 100,000 units per gm. of dry material. In other words, 1 gm. of the material will produce 100,000 mg. of ammonia nitrogen from a urea-phosphate solution in 5 minutes at 20°C. At this temperature the material requires 1.4 seconds to decompose its own weight of urea.

The crystals, when freshly formed, dissolve fairly rapidly in distilled water, giving a water-clear solution after centrifuging from the slight amount of insoluble matter that is present. The solution coagulates upon heating and gives strongly the biuret,

[1] Sumner, J. B., Graham, V. A., and Noback, C. V., *Proc. Soc. Exp. Biol. and Med.*, 1924, xxi, 551.

xanthoproteic, Millon, Hopkins and Cole, ninhydrin, and unoxidized sulfur tests. The phenol reagent of Folin and Denis gives a strong color, while the uric acid reagent gives none. The material can be entirely precipitated by saturating with ammonium sulfate. The Molisch test is negative and Bial's test is negative also. The absence of pentose carbohydrate, as shown by Bial's

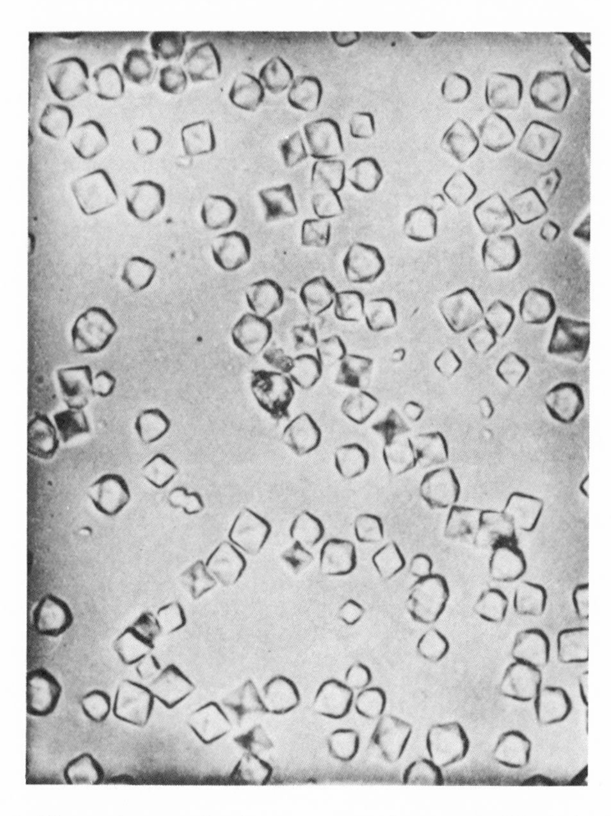

FIG. 1. Photomicrograph of urease crystals magnified 728 diameters.

test is especially pleasing as we have experienced a great deal of trouble in the past in freeing jack bean proteins from this substance.

The octahedral crystals, when freshly prepared, are very soluble in dilute alkali or dilute ammonia, and are either dissolved or coagulated by dilute mineral and organic acids, depending upon the concentration of acid. Even so weakly acid a substance as primary potassium phosphate is capable of causing an irreversible coagulation. Although the crystals dissolve in distilled water I am inclined to regard the material as globulin inasmuch as a precipitate is formed when carbon dioxide is passed into its solution and this precipitate immediately redissolves upon the addition of a drop of neutral phosphate solution.

Owing to the fact that I have not had large enough amounts of the material to work with I am unable to give accurate figures for its nitrogen content at the present time, but this can be stated to be not far from 17 per cent, as shown by micro-Kjeldahl determinations made on several preparations. The content of ash is certainly low, so low that a considerable amount of material will have to be used to obtain this figure. Determination of the enzyme activity of the crystals has been somewhat interfered with, owing to the fact that dilute solutions of the crystals produce less ammonia from urea than one would calculate from results obtained from more concentrated solutions. If this effect is real, rather than apparent, it may be due to the instability of the enzyme at great dilutions. When in concentrated solution the activity is not lost very rapidly, provided the material is kept in the ice chest.

When old the crystals are entirely insoluble in distilled water, salt solutions, and dilute ammonia. In this condition the enzyme activity is almost nil. I have made several attempts to purify the fresh crystals by a second crystallization but have never succeeded in obtaining more than traces of crystals and these have been insoluble in water and inactive.

It may be worth noting that practically all of our previous ideas concerning the nature of urease appear to be confirmed by the discovery of the octahedral crystals and by study of their properties. I undertook the task of isolating urease in the fall of 1917 with the idea that it might be found to be a crystallizable globulin, in which case the proof of its isolation would be greatly simplified. Other reasons for choosing urease were that the quantitative estimation of urease is both rapid and accurate, that urease can be reasonably expected to be an individual enzyme, rather than a mixture of enzymes, and that the jack bean appears to contain a very large amount of urease, if it is permissible to draw a parallelism between the urease content of the jack bean and the amounts of other enzymes found in other plant and animal materials.

In previous work in collaboration with Graham and Noback[1,2] and in unpublished work of my own it has been found that urease is very completely precipitated, together with the jack bean globulins, by cooling its 35 per cent alcoholic solution to −5 to −10°C., provided the reaction is sufficiently acid. We have found that urease can be precipitated by neutral lead acetate and neutralized cadmium chloride and that most of the urease can be re-extracted by decomposing the precipitate with potassium oxalate; that urease can be precipitated by tannic acid without very much inactivation and that urease can be rendered insoluble, with loss of a part of its activity, by the action of dilute alcohol or very dilute acid.

Although the literature contains numerous references to a coenzyme of urease, I believe that no specific coenzyme exists. My evidence rests upon the fact that the loss of activity that occurs when the octahedral crystals are separated from a jack bean extract is almost exactly equal to the activity obtained when these crystals are washed with dilute acetone and then dissolved in water. If anything could separate an enzyme from its coenzyme crystallization might be expected to do so. The proteins in impure urease solutions doubtless exert a protective action as buffers and both proteins and polysaccharides may exert protective colloidal action.

I present below a list of reasons why I believe the octahedral crystals to be identical with the enzyme urease.

1. The fact that the crystals can be seen by the microscope to be practically uncontaminated by any other material.

2. The great activity of solutions of the crystals.

3. The fact that solvents which do not dissolve the crystals extract little or no urease and that to obtain solutions of urease one must dissolve the crystals.

4. The fact that the other crystallizable jack bean globulins, concanavalin A and B, carry with them very little urease when they are formed from solutions that are comparatively rich in urease.

[2] Sumner, J. B., and Graham, V. A., *Proc. Soc. Exp. Biol. and Med.*, 1925, xxii, 504.

5. The unique crystalline habit of the octahedra and their ready denaturation by acid.

6. The fact that the crystals are purely protein in so far as can be determined by chemical tests, combined with evidence from previous work to the effect that urease behaves like a protein in its reactions towards heavy metals, alkaloid reagents, alcohol, and acids.

7. The fact that the crystals are nearly free from ash and the fact that we have previously prepared solutions of urease that contained neither iron, manganese, nor phosphorus.

The method which I have used to obtain the crystals is extremely simple. It consists in extracting finely powdered, fat-free jack bean meal with 31.6 per cent acetone and allowing the material to filter by gravity in an ice chest. After standing overnight the filtrate is centrifuged and the precipitate of crystalline urease is stirred with cold 31.6 per cent acetone and centrifuged again. The crystals can be now dissolved in distilled water and centrifuged free from insoluble and inactive matter that has passed through the filter during the filtration. Of the urease extracted from the meal as much as 47 per cent may be present in the crystals. If one uses coarsely ground jack bean meal that has not been freed from fat the crystals are still obtained, but in traces only. I have carried out the process described above about fifteen times since first discovering the crystals and have always had success. The method is described in the experimental part of this paper in detail. It is probable that not all of the precautions that I have taken are necessary, but I think it best to give a description of the method exactly as it has been used by me.

*Preparation of Urease Crystals.*

Distill commercial acetone from a mixture of fused calcium chloride and a little soda-lime to remove water and acids. Place 158 cc. of the distillate in a 500 cc. graduate. Dilute with distilled water to the 500 cc. mark, mix, and cool to 22°C. Dilute exactly to mark and pour upon 100 gm. of "Arlco" jack bean meal which has been placed in a beaker. Stir for 3 or 4 minutes to break up the lumps. Now pour the material upon a 28 cm. Schleicher and Schüll filter, number 595. It is best to pour the mixture on the filter where there are four thicknesses and to add more only when the paper has become moistened higher up, as in preparing the Folin-Wu blood filtrate, but even if this is done the filtrate may come through not quite clear at first and in this case the first 50 cc. should be refiltered. When all, or nearly all of the material is on the filter place in an ice chest at 2–2.5°C. and allow to remain overnight. The next morning centrifuge off the crystals that are present in the filtrate, using cold 50 cc. centrifuge tubes and cold holders for the tubes. (I have centrifuged for 7 minutes at a moderate speed and then cooled the tubes and holders in the ice chest for 10 minutes before centrifuging more of the filtrate.)

When all of the crystals have been centrifuged off drain the tubes against clean filter papers to remove the last drops of the mother liquor. The crystals can now be stirred up with 5 to 10 cc. of ice-cold 31.6 per cent acetone and centrifuged again and the tubes again drained against filter papers. The crystals are now dissolved in from 15 to 40 cc. of distilled water at room temperature by stirring, and the solution is centrifuged free from insoluble matter. For examination under the microscope it is best not to separate the crystals from all of the original mother liquor as they are insoluble in this.

The Arlco jack bean meal referred to above is an extremely finely ground, fat-free meal obtained from The Arlington Chemical Company, Yonkers, N. Y.

*Estimation of Urease Action.*

Make a solution containing 3 per cent of urea, 5.4 per cent of $K_2HPO_4$, and 4.25 per cent of $KH_2PO_4$; preserve with toluene and place in a bath of 20.0°C. Pipette 1 cc. of the diluted enzyme solution into a test-tube that contains no adsorbed mercury and allow to remain in bath to reach its temperature. Now rapidly add to the enzyme 1 cc. of the urea-phosphate solution and mix. Allow the action to proceed for 5 minutes. At the end of this interval blow into the solution 1 cc. of normal hydrochloric acid to inactivate the enzyme and mix immediately. Wash the material into a 100 cc. volumetric flask, dilute, Nesslerize, dilute to mark, mix, and compare with a standard containing 1 mg. of ammonium nitrogen that has been prepared at the same time. If the color of the unknown is much lighter or darker than that of the standard the test must be repeated, using a more concentrated, or a more dilute enzyme solution, as the case may be.

This procedure can be used only with solutions of urease that are relatively pure. For example, urease prepared by extracting jack bean meal with 30 per cent alcohol gives results by this method that are about 3 per cent too low on account of the interfering action of the protein that is present. The protein can be removed by precipitation with potassiomercuric iodide containing hydrochloric acid and an aliquot of the filtrate can be Nesslerized in cases of this sort.[3]

### *Determination of Dry Weight of Urease in Solutions.*

As the addition of absolute alcohol to solutions of the crystals did not precipitate all of the protein, which was probably on account of the lack of electrolytes, I was obliged to pipette the solution into a weighed glass dish and evaporate to dryness in a boiling water oven. The temperature of the oven was about 94°C. The dish and contents were then dried to constant weight in a desiccator over sulfuric acid.

### *Determination of Total Nitrogen.*

As I have not yet prepared the material on a large scale the macro-Kjeldahl method was not used. The material was digested in a large Kjeldahl flask with a small amount of sulfuric acid and copper sulfate. After digestion the material was washed into a large glass tube and the ammonia aerated off and Nesslerized. It was found impossible to digest the material in a glass tube, such as is ordinarily used, as it foamed very badly.

## CONCLUSION.

A new crystallizable globulin has been isolated from the jack bean, *Canavalia ensiformis*. From the reasons given elsewhere in this paper I am compelled to believe that this globulin is identical with the enzyme urease.

I wish to thank Professor A. C. Gill for examining the crystals and Professor S. H. Gage and Professor B. F. Kingsbury for photographing them.

[3] Sumner, J. B., *J. Biol. Chem.*, 1919, xxxviii, 59.

# 37


Reprinted from *J. Gen. Physiol.* **13**:739–766 (1930)

# CRYSTALLINE PEPSIN

## I. Isolation and Tests of Purity

By JOHN H. NORTHROP

(*From the Laboratories of The Rockefeller Institute for Medical Research, Princeton, N. J.*)

(Accepted for publication, May 9, 1930)

### I

### INTRODUCTION

Enzymes are in many respects connecting links between living and inanimate matter since their action is analogous to inorganic catalysts, although the enzymes themselves are found only in living organisms. As catalysts they increase the rate of one or more specific reactions and so act as directive agents for the reactions occurring in the organism. This directive property is undoubtedly essential for the existence of living cells. As a consequence of these properties the study of enzymes has been of interest to both chemists and biologists and has resulted in a great increase in the knowledge of their mode of action. The results of attempts to isolate the enzymes in pure form, however, have been singularly unsuccessful. There seems to be no convincing evidence that any enzyme has been obtained in the pure state; and only one, the urease described by Sumner (1), has been previously obtained in crystalline form. A number of methods have been found which allow the activity of an enzyme preparation to be increased almost indefinitely; at the same time, however, the preparation becomes more unstable and eventually the activity becomes lost.

In practically all the work the assumption has been made either explicitly or otherwise that the activity was a measure of the purity of the preparation and that any increase in activity was due to an increase in purity. This is not necessarily true. If the enzymes are analogous to inorganic catalysts then it is quite possible that the activity depends on the physical arrangement of the molecules or atoms (2). Evidence for this relation between the physical state and the activity was found by Fodor (3) in the case of the proteolytic enzymes of yeast, and by Kuhn and Wasserman (4) in the case of hemin. It is possible, on the other hand, that enzymes in general are of the type of hemoglobin (which might be considered an enzyme), and that they consist of an active group combined with an inert group. It might be possible under certain conditions to attach many more active groups to the inert group and so increase the activity above that of the original compound. Either of the above ideas would account for the well-known fact that crude preparations are much more stable than purified material and that the rate of inactivation of enzyme solutions practically always shows evidence of a mixture of stable and unstable forms (5, 6).

There is some reason to think, therefore, that enzymes exist in a more stable form for either physical or chemical reasons, and in view of the uniformly negative results which have been obtained in attempting to isolate the most active preparations it seemed advisable in attempting the isolation of pepsin to study the more stable as well as the most active fraction.

## II

## PRELIMINARY EXPERIMENTS

A number of methods have been proposed for the purification of pepsin, such as precipitation with safranin (7), etc., fractionation by various adsorbents, and precipitation by dialysis from acid solution (Pekelharing (8)). These and a number of other methods were tried and more or less active preparations obtained. The results with Pekelharing's method seemed the most encouraging, however, since the loss of activity was less and there was some indication that a constant activity was reached. This result has been reported by Pekelharing and also by Fenger, Andrew and Ralston (9) using a simiiar method. It was found, however, that the dialysis could be dispensed with and the process made more rapid and efficient by solution with alkali and subsequent precipitation with acid, after a preliminary precipitation with half saturated $MgSO_4$ or $(NH_4)_2SO_4$. The amorphous material so obtained contains about half the activity present in the original material and is 3 to 6 times as active as measured by the liquefaction of gelatin and about 5 times as active as measured by the digestion of casein or by the rennet action on milk.[1] Repetition of this procedure gave products of increasing activity as measured by the liquefaction of gelatin, and apparently this activity could be increased indefinitely. Several samples were obtained which were 100 times as active as the original preparation. They were also more unstable, so that each succeeding precipitation was accompanied by a larger and larger percentage loss until finally no more active material remained. This has been the fate of all previous attempts to isolate the most active fraction of a number of enzymes. When the activity of the various fractions was determined by the rate of hydrolysis of casein or by the rennet action on milk, however, it was found that the activity increased until it reached about 5 times that of the crude preparation

---

[1] *Determination of Activity.*—The rate of digestion of proteins by pepsin may be followed by determining the increase in carboxyl groups or amino nitrogen, or by the decrease in protein nitrogen or by the changes in viscosity. The increase in carboxyl groups probably represents the best measure of the progress of the reaction and has been used as the basis of the activity units used in this paper.

The most widely used unit of enzyme activity in general is that suggested by Euler (16) and is equal to the velocity constant of the reaction divided by the grams of enzyme. Theoretically this is undoubtedly correct since the activity of a catalyst can be expressed only as a velocity constant. Many enzymes do not give a velocity constant independent of the substrate concentration but instead the velocity constant decreases as the substrate concentration increases. This is especially true of pepsin and in addition the reaction does not follow any simple reaction rate. The end point of the reaction, which must be known in order to calculate the velocity constant, is also very difficult to determine. It is not practical therefore to use the velocity constant as a measure of activity of pepsin.

The activity of pepsin may conveniently be defined as the milliequivalents of carboxyl groups liberated per mole or gram of enzyme per minute at 35.5°C., optimum pH, and 5 per cent substrate concentration. Theoretically it would be better to use that substrate concentration at which the rate of reaction is a maximum, but experimentally this is difficult, owing to the high viscosity of such concentrated solutions. The enzyme concentration used should be in the range in which the activity is proportional to the enzyme concentration. In order to determine the activity in this way the rate of hydrolysis of casein, gelatin, edestin, and denatured egg albumin at pH 2.0 to 2.5, was determined by means of the increase in formol titration (19).

The following abbreviations are used in this paper:

PU = proteolytic units = milliequivalents carboxyl groups per minute.
$[PU]_{gm.}$ = " " per gram.
$[PU]_{gm.}^{cas.F}$ = " " " " as determined by cas. F. method, etc.

and then remained constant instead of increasing as did the gelatin liquefying power. This was the result reported by Pekelharing and also by Fenger, Andrew and Ralston. This material appeared to be protein, as previous workers had found, and was reasonably stable. Efforts were therefore made to isolate this protein in crystalline form.

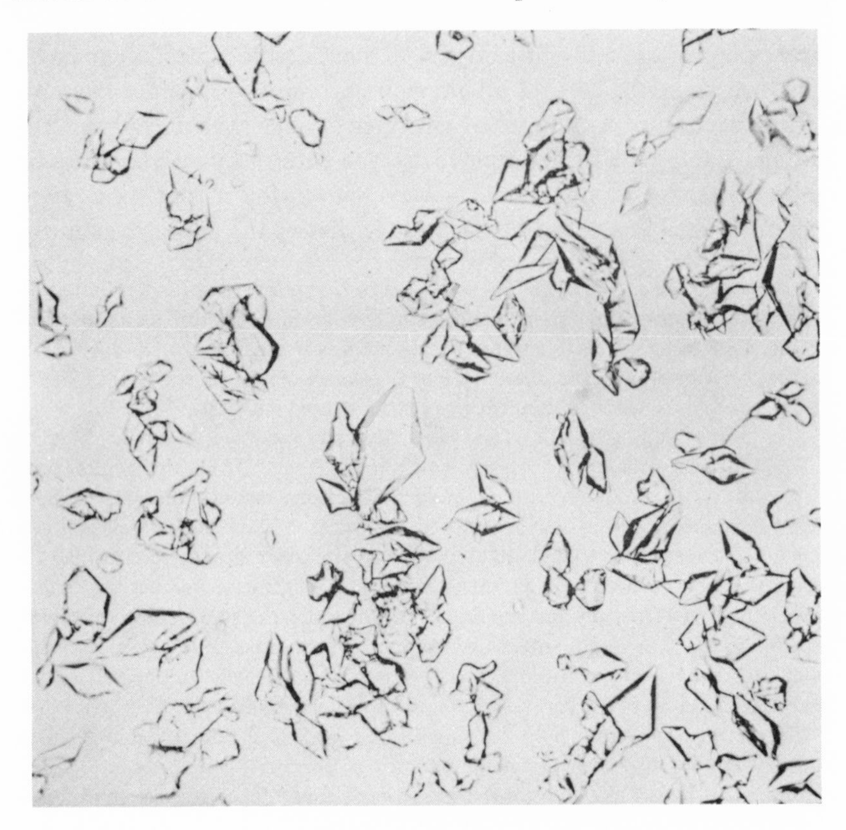

FIG. 1. Crystalline pepsin

## III

### *Isolation of the Crystalline Enzyme*

It was noticed that the precipitate which formed in the dialyzing sac when the procedure of Pekelharing was followed appeared in more or less granular form and filtered rather easily, as though it were on the verge of crystallization. This precipitate dissolved on warming the suspension and it was eventually found that it could be induced to crystallize by warming to 45°C., filtering, and allowing the filtrate to cool slowly. The crystals so obtained were regular hexahedra and showed a tendency to grow in clusters, especially when appearing from more acid solutions. They are remarkably similar to the urease crystals pictured by Sumner and differ only in that they have a hexagonal base while the urease has an octagonal base. On one occasion a few crystals with truncated pyramids were obtained. They had the same activity and optical activity as the usual form. The crystals showed positive double refraction and were optically active in solution. They possessed proteolytic activity, when dissolved, equivalent to 5 times that of the U. S. P. 1 to 10,000 pepsin as measured by hydrolysis of casein, and 2.5 times as measured by the liquefaction of gelatin.

*Improved Method for the Preparation of the Crystals.*—The isolation of the crystals in bulk by the above method was difficult owing to the dialysis. It was found that this could be avoided and the purification carried out as outlined above for the amorphous preparations except that the acid precipitate was dissolved at 45° C. in concentrated solution. On inoculation, this solution set to a solid paste of crystals. The following is an outline of the method as finally developed.

## IV

## *General Properties and Analysis of the Crystals*

The material prepared in this way has the general properties of a protein. It is coagulated by heat, precipitated by saturation of the solution with $MgSO_4$ or $(NH_4)_2SO_4$ and gives a strongly positive xanthoproteic test. The Millon test is negative. The crude material contains a large amount of yellowish pigment which is removed with difficulty. It may be largely removed by reprecipitation with $MgSO_4$ and becomes less as the material is recrystallized. All the preparations however, give a slightly yellowish solution when dissolved although the dry crystals are pure white after several crystallizations. There is no relation between the activity and the color. Analysis of the material gave the results shown in Table II.

The crystals are difficult to dissolve after drying and are best kept under saturated $MgSO_4$ at 5°C. They are instantly inactivated by alkali in solution and lose activity slowly in acid solutions. The

TABLE I

*Preparation of Crystalline Pepsin*

| Procedure | Activity per gm. dry weight Method | | | |
|---|---|---|---|---|
| | Gel. V. | Cas. S. | Rennet | E. A. |
| 500 gm. Parke, Davis pepsin U. S. P. 1:10,000 dissolved in 500 ml. $H_2O$ and 500 ml. 1 Normal $H_2SO_4$ added. 1,000 ml. saturated $MgSO_4$ added with stirring. Solution filtered through fluted paper (S. and S. No. 1450½) and then with suction. Filtrate discarded. | 0.075 | 2.5 | 2.5 | 4.5 |
| Precipitate 1. Wash twice with equal volume ⅔ saturated $MgSO_4$, filter with suction. Filtrate discarded. | | | | |
| Precipitate 2. Stir with water to thick paste and M/2 NaOH run in until complete solution. (Great care must be taken to avoid local excess of NaOH. pH never more than 5.0.) M/2 $H_2SO_4$ added with stirring until heavy precipitate forms, (pH about 3.0), 3 to 6 hrs. at 8°C., filter with suction. Filtrate discarded. | .15 | 7.5 | 7.5 | 7.0 |
| Precipitate 3. Stir with $H_2O$ to thick paste at 45°C., M/2 NaOH added carefully until precipitate dissolves, (filter if cloudy and discard precipitate). Beaker containing filtrate placed in a vessel containing about 4 liters of $H_2O$ at 45°C., inoculated and allowed to cool slowly, cooling should require 3 to 4 hrs. and heavy crystalline precipitate should form at about 30 to 35°C. Solution kept at 20°C. for 24 hours. Thick crystalline paste, filter with suction. | .15 | 10.0 | 10.0 | 8.0 |
| Precipitate 4. Wash with small amount of cold $H_2O$ and then with ½ saturated $MgSO_4$ and store under saturated $MgSO_4$ at 5°C. | .17 | 14.0 | [14.0] | 9.0 |
| Filtrate. M/2 $H_2SO_4$ added to pH 3.0, amorphous precipitate filtered off and treat as Precipitate 3. | | | | |

*Recrystallization*

Method 1. Crystalline paste filtered with suction on large funnel so as to form a thin layer of crystals and washed 3 times with cold M/500 HCl. Filter cake stirred to a paste with ½ its weight of water, the suspension warmed to 45°C. and M/2 NaOH run in slowly with constant stirring until the precipitate dissolves ($pH < 5.0$). M/2 $H_2SO_4$ is then run in until the solution is faintly turbid, a few crystals added and the solution allowed to cool slowly as before. A heavy crop of crystals should separate in about 24 hrs. The suspension is then warmed to 45°C. again and more $H_2SO_2$ added until the pH of the suspension is about 3.0. It is then allowed to cool slowly again, and filtered after 24 hrs. The crystals may be washed with M/500 HCl until free of $SO_4$.

Ammonium sulfate may be used in place of $MgSO_4$. Sodium acetate may be used in place of sodium hydroxide.

Method 2. Crystals dissolved with NaOH and treated as described for Precipitate 2.

inactivated material is digested by the remaining active material and a large amount of tyrosin crystallizes out. This process also occurs slowly in the ice box so that the crystals on standing become mixed with nonprotein material that is not precipitated by salt nor by heat. The crystals can be freed from this soluble material by thorough washing with $H_2O$ or by recrystallization. When freshly prepared in this way 98 to 99 per cent of the nitrogen is precipitated from solution by heating rapidly to boiling at pH 3 with sulfuric acid and $Na_2SO_4$, by saturation with $MgSO_4$ or $(NH_4)_2SO_4$, by the addition of alkali

TABLE II

*Analysis. Dried at 60° in vacuo for 24 Hrs.*

| Method | Dumas | Kjeldahl | Van Slyke | | | | | | | Precipitation with $BaCl_2$ |
|---|---|---|---|---|---|---|---|---|---|---|
| | Total N | Total N | Amino N | C | H | Cl | S | P | Ash | $SO_4^{=}$ |
| Per cent dry weight | 15.5<br>15.3 | 15.15<br>15.30 | 0.80 | 52.3<br>52.6 | 6.66<br>6.64<br>6.70 | 0.23<br>.20 | 0.88<br>.82 | 0.078 | 0.40<br>.55 | 0 |

TABLE III

*Activity and Composition of Various Preparations of Pepsin, Crystallized Once*

| Preparation | 1 | 2 | 3–5 | 5–10 | 10–12 | 12–20 | 22 |
|---|---|---|---|---|---|---|---|
| $[PU]_{gm.}^{gel. V.}$ | 0.17 | 0.16 | 0.18 | 0.154 | 0.16 | 0.14 | 0.17 |
| Per cent N | 15.2 | 15.1 | 15.1 | 15.3 | 15.2 | 15.0 | 15.1 |
| $[\alpha]_D^{22°}$ pH 4.5 | −70 | −72 | −74 | −73 | −74 | −71 | −70 |

and subsequent neutralization, or by heating with 10 per cent trichloracetic acid.

*Constant Activity of Various Preparations.*—About 2 kg. of the crystals have been prepared from six different lots of the commercial preparation during the course of this work and have all had the same percentage of nitrogen and the same activity within the experimental error. A summary of these properties of the various preparations is given in Table III. The preparations in some cases were combined at the final crystallization.

The relative activity with various substrates and various methods is shown in Table IV. The figures are the average of 6 to 10 determinations with different enzyme preparations. All crystalline preparations tested yielded the same result with these methods, but amorphous preparations frequently gave much higher activity as measured by gelatin hydrolysis and lower activity as measured by egg albumin,

TABLE IV

*Activity of Various Enzyme Preparations*

| Enzyme | Substrate | Method | Activity. Milliequivalents per gm. per min. = $[PU]_{gm.}$ | Milliequivalents per millimole or equivalents per mole per min. = $[PU]_{mole}$ |
|---|---|---|---|---|
| Crystalline pepsin | Casein | Formol | 14 ± 0.5 | 500 |
| | | Solution | [14] | |
| | | Rennet action | [14] | |
| | Gelatin | Formol | .17 ± 1 | 6 |
| | | Viscosity | [.17] | |
| | Edestin | Formol | 28 ± .8 | 1,000 |
| | Egg albumin | Formol | 9 ± .5 | 320 |
| U.S.P. pepsin 1:10,000 | Casein | Formol | 2.5 | |
| | | Solution | [2.5] | |
| | | Rennet action | [2.5] | |
| | Gelatin | Formol | .075 | |
| | | Viscosity | [.075] | |
| | Egg albumin | Formol | 4.5 | |

while the activity as measured by casein hydrolysis remained constant. On recrystallization of these abnormal amorphous precipitates the activity always returned to the characteristic value for the crystalline material, shown in the table.

The activity compared to other enzymes is less than that reported by Sumner for urease, on a weight basis, and much less than for several other enzymes, but comparison should be made on the basis of activity per mole rather than per gram of enzyme.

V

## *Evidence Concerning the Purity of the Crystalline Material*

*Constant Activity and Composition on Repeated Crystallization.*—The work reported above shows that a crystalline protein having proteolytic activity may be obtained in quantity from commercial preparations of pepsin. The question of interest is whether this proteolytic activity is a property of the protein molecule or of another molecular species associated with the protein. The usual criterion for the purity of a substance is constant composition and properties after repeated recrystallization. Unfortunately in the case of proteins the

TABLE V

*Summary of Recrystallized Pepsin*

| Crystallization No. and color | Quantity of crystals | $[PU]_{gm.}^{gel. V.}$ | $[PU]_{gm.}^{cas. S.}$ | N | P | $[\alpha]_D^{22°}$ pH 5.0 | Formol per gm. |
|---|---|---|---|---|---|---|---|
| | *gm.* | | | *per cent* | *per cent* | | *ml. 0.1 NaOH* |
| 1. Dark brown | 50 | 0.18 | 14.0 | 14.85 | 0.078 | −70 | 14.0 |
| 2. Brown | 22 | .17 | 14.5 | 15.0 | .080 | −70 | 12.5 |
| 3. Yellowish | 15 | .15 | 15.0 | 15.14 | .075 | −68 | 12.5 |
| 4. Slightly yellow fine powder | 9 | .16 | 14.5 | 15.16 | .076 | −67.6 | 11.0 |
| 5. Nearly white | 5 | .16 | 14.8 | 15.13 | .077 | −68 | 11.0 |
| 6. " " | 3 | .14 | 14.5 | | .073 | −72 | 11.0 |
| 7. " " | .5 | .17 | 15.0 | 15.15 | .080 | −72 | 12.0 |
| Original preparation | | .075 | 2.5 | | | | |

melting point, which is the most sensitive test of purity, cannot be used since proteins decompose before melting. Of the other properties the proteolytic activity, optical rotation, and percentage of nitrogen and of phosphorus were considered the most significant and these properties were determined on preparations obtained from a series of seven successive crystallizations. A summary of the results of this experiment is shown in Table V. There is no perceptible drift in any of the properties determined and therefore no indication that the material can ever be separated into fractions by crystallization under these conditions. This shows that the composition of the crystals is independent of the concentration and quantity of the solution from which they are formed.

Several systems could give this result (10). (1) A mixture of two or more substances present in amounts which are proportional to their solubilities. (2) A solid solution of such composition as to have a minimum solubility. This would correspond to a constant boiling point mixture of miscible liquids. (3) A solid solution of two or more substances having nearly the same solubility. (4) A pure substance.

The various possibilities noted above all depend on some definite relation between the composition of the mixture and the solubilities of the hypothetical components. If this relation were changed, the composition would be changed (except in the case of allotropic modifications). The possibility of such a relation would be greatly reduced if the recrystallization were carried out in a number of different solvents in which the solubility was different. It is quite possible that a mixture of constant composition might be obtained from one solvent just as constant boiling mixtures are frequently found, but it would be expected that the composition of the mixture would vary with different solvents just as the composition of constant boiling mixtures varies with the pressure. If it could be shown therefore that the material retained its constant proteolytic power and composition after recrystallization from a number of different solvents, it would be excellent proof that it was a pure substance. Unfortunately, however, it is not possible to crystallize the material from a series of solvents, but the same result may be obtained by studying the solubility in several solvents.

*Results of the Solubility Determinations.*—The values for the solubility obtained with different amounts of precipitate in various solvents are shown graphically[2] in Figs. 2 to 6. Since the solubility is found to be independent of the amount of precipitate this procedure does not affect the shape of the curve. The soluble nitrogen per milliliter of solution, the proteolytic activity per milliliter of solution as determined by the digestion of casein or gelatin ($[PU]_{ml.}$) and the optical rotation of the solution in a 1 dm. tube are plotted against the total nitrogen content per milliliter of the suspension. The values corresponding to solutions obtained from the undersaturated side are represented by solid dots and those corresponding to solutions obtained from the supersaturated side

[2] In order to save space the scale representing the total amount of nitrogen per milliliter of suspension has been changed to a logarithmic one in some of the curves after the amount has become large as compared to the amount in solution.

are represented by circles. The solid lines drawn for the curves representing the pepsin per milliliter and the optical activity were calculated on the assumption that the ratio of activity or optical activity to nitrogen was the same in the solution as in the original crystals. That

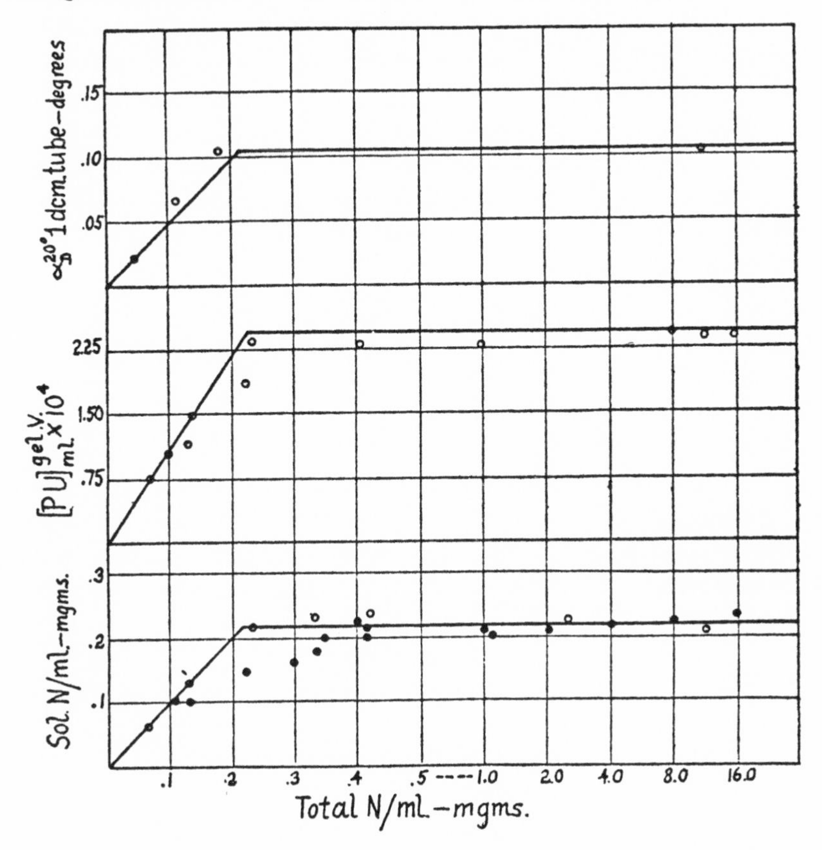


FIG. 2. Solubility of crystalline pepsin in 0.05 acetate and 0.5 saturated magnesium sulfate at 8°C.

is, 1.15 proteolytic units per gram of nitrogen as measured by the liquefaction of gelatin, 92 proteolytic units per gram of nitrogen as measured by the digestion of casein and a specific rotation per gram of nitrogen of −460° as measured at 22 to 24°C. with the sodium D line.

The results show: (1) that the solubility is independent of the quantity of solid present, (2) that the value for the solubility is an equilibrium value since it is the same whether obtained from supersaturated or undersaturated solutions, and (3) the specific proteolytic activity and the specific optical activity are the same for the material present in the solution as for the original material. The material behaves in regard to these determinations as a pure substance and there is no evidence of

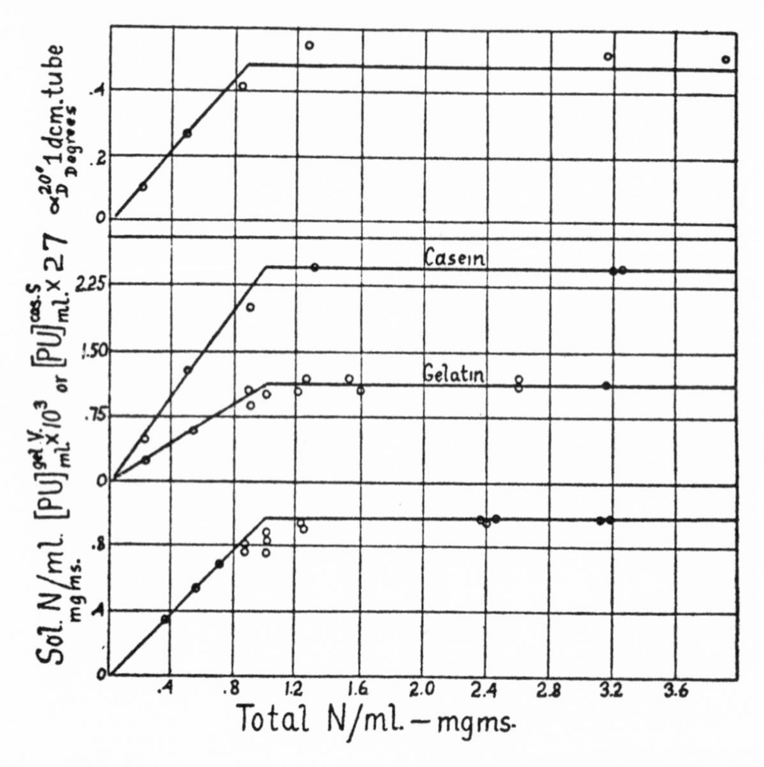


FIG. 3. Solubility of amorphous pepsin in 0.05 acetate and 0.5 saturated magnesium sulfate at 20°C.

mixture. They show conclusively that the activity cannot be ascribed to the presence of a minute amount of a very highly active material associated with the protein, unless it be further assumed either that this active material has about the same solubility as the protein or that it forms solid solutions with the protein. If the active material were soluble and *mixed* with the protein the activity per milliliter of solution would continue to increase as more and more solid material was added. While if it were present as a relatively insoluble material the activity per milliliter of the supernatant solution would become constant before the solubility of the protein had become constant. The figures show, however, that the proteolytic activity of the liquid becomes constant at the same point at which the concentration of the protein in the solution becomes constant. The only mixture which would behave in this

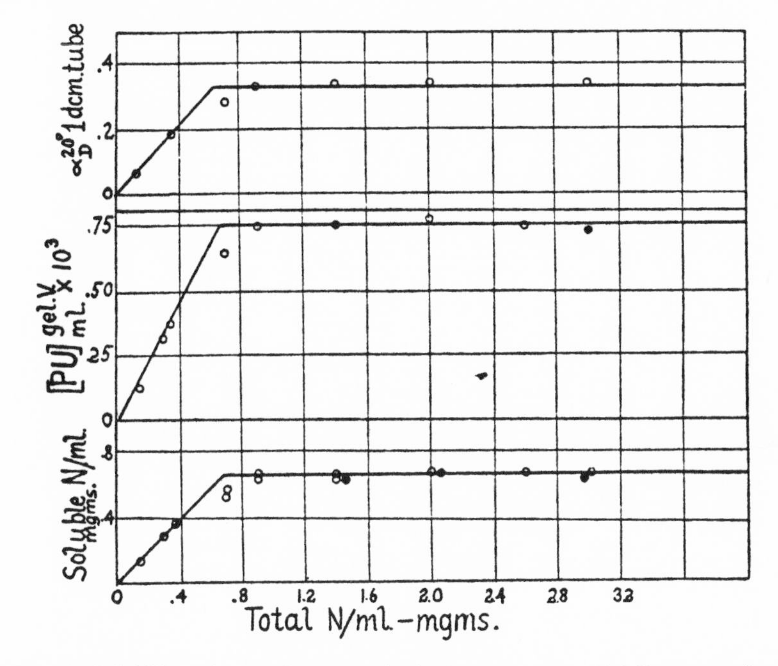


FIG. 4. Solubility of amorphous pepsin in 0.01 acetate and 0.5 saturated magnesium sulfate at 20°C.

way would be one in which the relative amounts of active material and protein were in almost exact proportion to the relative solubilities of the two substances. This might conceivably be the case in one solvent but the possibility that this relation would hold in several solvents appears so small as to be negligible.

There remains only the possibility that the active material forms a solid solution with the protein and is either present in small amount or has about the same solubility as the protein. If it were present in small amount and dissolved in the protein, its presence would not affect the total solubility curve. The ratio of the active material to the protein in the precipitate at the point where the precipitate first

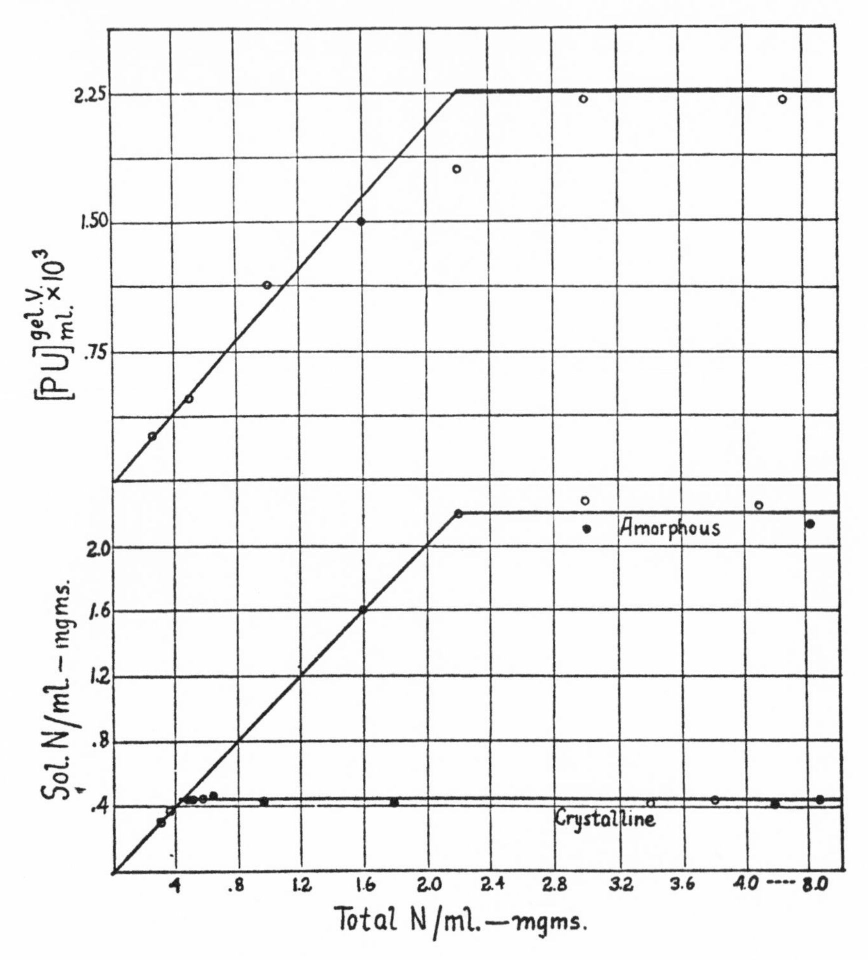


FIG. 5. Solubility of crystalline and amorphous pepsin in 0.055 acetate and 0.444 saturated magnesium sulfate at 20°C.

appears would differ, however, from the ratio in the original material. The ratio in the solution in that part of the curve where there is a large amount of solid present would also differ from the ratio in the original material. The figures show that this latter condition does not occur. The activity per gram of nitrogen in the solution in equilibrium with a

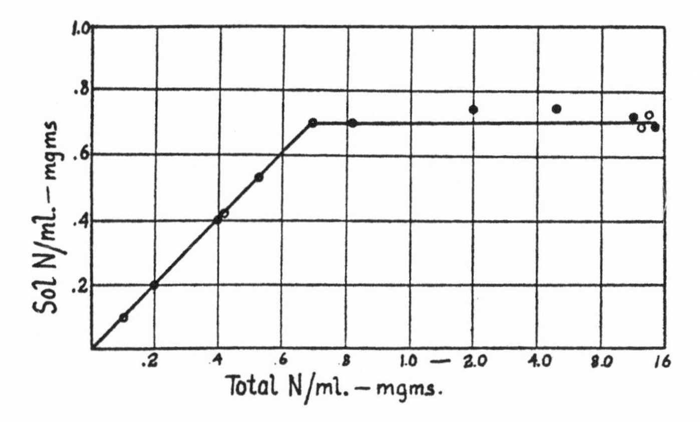


FIG. 6. Solubility of crystalline pepsin in 0.522 M sodium sulfate and 2.5 × $10^{-5}$ N sulfuric acid at 20°C.

TABLE VI

$$\textit{Ratio}\ \frac{\textit{Activity}}{\textit{Nitrogen}} = [\text{PU}]\ \frac{\text{gel. V.}}{\text{gm. N}}$$

| Experiment | Pepsin | Solvent | In precipitate when nearly all soluble | In solution when large excess of solid present | In original crystals |
|---|---|---|---|---|---|
| 26 | Crystals | 0.05 acetate pH 4.65, 0.50 sat. $MgSO_4$ | 0.75<br>1.10 | 1.00<br>1.09 | 1.11 ± 0.05 |
| 23 | " | " " " " " " " | 1.2 | 1.11 | |
| 22 | Amorphous | " " " " " " " | 1.20 | 1.16 | |
| 21 | " | 0.01 " " " " " " | 1.09 | 1.02<br>1.13 | |
| 20 | " | " " " " " " " | 1.11 | 1.21 | |

large excess of precipitate is the same within the experimental error as the ratio in the original material. The activity-nitrogen ratio of the first precipitate to appear is also the same in the experiments with amorphous and crystalline material as is shown in Table VI. If old preparations which have not been freed from the decomposition products are used, the activity-nitrogen ratio at this point will be low. An indication of this is shown in the first figure in the table and in some of the figures.

The material is therefore either a pure substance or a solid solution the components of which have nearly the same solubility in all the solvents tried.

A summary of the solubilities in the various solvents is given in Table VII.

TABLE VII

*Solubility of Pepsin*

| Preparation | Solvent | Temperature | Solubility N/ml. |
|---|---|---|---|
| | | | *mg.* |
| Crystals........... | 0.522 M $Na_2SO_4$, 2.5 × $10^{-5}$ N $H_2SO_4$ | 20° | 0.70 |
| " ........... | 0.444 M $Na_2SO_4$, 0.0556 acetate buffer pH 4.65 | 20° | .43 |
| Amorphous......... | " " " " " " " | 20° | 2.2 |
| Crystals........... | 0.05 acetate pH 4.65, 0.50 sat. $MgSO_4$ | 8° | .22 |
| Amorphous......... | " " " " " " " | 20° | 1.00 |
| " ......... | 0.01 " " " " " " | 20° | .65 |

*Inactivation.*—It was pointed out by Pekelharing that the temperature coefficient for the inactivation of enzymes and especially of pepsin was extremely high and agreed with that for the denaturization of proteins. No other reaction known has such an enormous temperature coefficient and the agreement of the two values suggests that the enzymes are proteins. Pekelharing also showed that loss of activity in strong acid solution was accompanied by the appearance of denatured, insoluble protein. It can be shown that inactivation either by heat or by alkali is quantitatively proportional to the denaturization of the protein.

*Inactivation by Alkali.*—It is well known that pepsin is very sensitive to alkali. Goulding, Wasteneys and Borsook (13) showed that this inactivation could be separated into two reactions, one of which was instantaneous; and further that if the amount of this inactivation were plotted against the pH a titration curve with a pK of 6.85 was obtained. These experiments were repeated and the amount of inactivation compared with the amount of denatured protein found.

*Experimental.*—A series of 1 per cent solutions of the pepsin was made and adjusted to about pH 5. Increasing amounts of alkali were then added to these solutions so as to give a series of solutions varying in pH from about 5.5 to 9. An amount of acid equivalent to the amount of alkali in the tube, was added immediately after the addition of the alkali and the activity of the resulting solution determined after neutralization. 5 ml. portions were added to 5 ml. of a solution of half-saturated sodium sulfate and M/500 sulfuric acid. The unchanged pepsin is soluble under these conditions (to about 1 per cent) but the denatured protein is insoluble. The denatured protein therefore precipitates and the unchanged protein nitrogen is determined from the filtrate. In this way the percentage inactivation and percentage denaturization may be determined. The denatured protein is rapidly digested by the active enzyme so these solutions must be precipitated and filtered as soon as possible.

Control experiments with known mixtures of denatured and active pepsin show that very little activity is carried down with this precipitate if the activity is determined by hydrolysis of casein. If the activity is determined by hydrolysis of gelatin, however, the precipitate of denatured protein may be quite active; and this result is evidently due to the abnormal activity observed throughout in relation to gelatin hydrolysis. This complication may be avoided by determining the activity of the solution before precipitation.

The results of such an experiment are shown in Fig. 7. The inactivation of the pepsin is quantitatively parallel with the denaturization of the protein since the percentage of activity left is the same as the

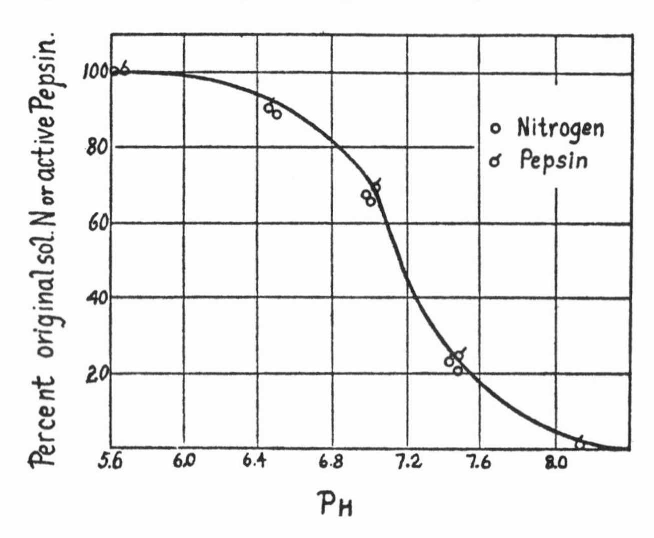


Fig. 7. Percentage inactivation and percentage denaturization of pepsin at various pH at 20°C.

percentage of soluble nitrogen remaining. The curve is very similar to that of Wasteneys and Borsook but is a little more to the alkaline side.

*Inactivation by Heat.*—Pekelharing noted that when pepsin solutions were allowed to stand in a range of acidity where the pepsin is active, the protein was digested and at the same time the activity of the solution decreased. He concluded from this that the pepsin digested itself. It can easily be shown, as was mentioned above, that the denatured protein is rapidly digested and it seems probable that the effect noticed by Pekelharing was the result of denaturization of the protein followed by hydrolysis. It is then unnecessary to suppose that the enzyme digests itself. Pepsin solutions when heated slowly or when maintained at a temperature at which the inactivation proceeds slowly, do not contain denatured, insoluble protein since the latter is hydrolyzed as rapidly as it is formed. Under these conditions inactivation of pepsin is accompanied by the appearance of soluble nitrogen instead of insoluble; and it is difficult to make a sharp separation of the original pepsin from the products of hydrolysis. The result of an experiment which shows the digestion of denatured pepsin by active pepsin is given in Table VIII in which mixtures of various amounts of active and denatured pepsin were made up at pH 3 and left at 35°C. for 24 hours.

TABLE VIII

*Digestion of Inactivated Pepsin pH 3.0, 35°C.*

Pepsin dissolved in HCl, pH 3.0 = Solution A
50 ml. A boiled, cooled = Solution D
A and D mixed in various proportions,
2 ml. sample of mixture taken and
2 ml. M/1 $Na_2SO_4$ added, filtered,
N/ml. determined in filtrate.

| A<br>D | 20<br>0 | 10<br>10 | 5<br>15 | 0<br>20 |
|---|---|---|---|---|
| Mg. soluble N in filtrate after: | | | | |
| 0 hrs. | .30 | .11 | .07 | 0 |
| 20 " | .30 | .30 | .28 | .01 |

The denatured pepsin is precipitated by M/1 $Na_2SO_4$ at pH 3, while the active pepsin is soluble. The table shows that the inactive and insoluble pepsin is dissolved by the active pepsin.

It may be shown, however, by means of solubility experiments, that the loss of inactivation in the solution is exactly paralleled by the disappearance of the original protein. If a saturated solution of pepsin in M/300 hydrochloric acid is placed at 65°C. the activity is rapidly lost so that only about half the original activity remains after 15 minutes. No precipitate occurs but the formol titration increases. If the solution is now cooled and stirred again with a large quantity of crystalline pepsin more of the crystals must go into solution to replace the material which has been changed. If the inactivation of the enzyme is accompanied by some chemical change in the protein then the amount of material which will dissolve must be proportional to the percentage loss in activity. That is, if half the activity is lost the increase in solubility must be equal to half the original solubility (assuming the solubility to be unaffected by the presence of the inactivated enzyme). On the other hand, the activity per milliliter of the solution after being stirred with the solid should return to its original value. The result of an experiment designed to test this assumption is shown in Table IX. The solution was heated for 15 minutes at 65°C. and in that time the activity per milliliter had been reduced to 50 per cent of its original value, whereas the formol titration

TABLE IX

*Inactivated Pepsin and Solubility. 65°C.*

| | $[PU]^{\text{gel. V.}}_{\text{ml.}} \times 10^3$ | N/ml. | Formol/ml. |
|---|---|---|---|
| | | *mg.* | *ml. 0.01 NaOH* |
| M/300 HCl, stirred 1 hr. with pepsin, 20°C. Centrifuge | 0.315 | 0.30 | 0.22 |
| Supernatant heated at 65°C. for 15 min. | .160 | .30 | .33 |
| Pepsin crystals added and stir 1 hr. Centrifuge | | | |
| Observed | .325 | .45 | .45 |
| Calculated | .315 | .455 | .44 |

had increased 50 per cent and the total N per milliliter was unchanged since no precipitate appeared. The solution was then cooled and stirred at 20°C. with more crystals. The suspension was then filtered and the supernatant analyzed. The activity per milliliter is now the same as that of the original solution while the nitrogen per milliliter is 50 per cent higher than the original value. The formol titration has been increased by 50 per cent of its original value also. The results show that the loss of activity is exactly paralleled by some chemical change in the protein molecule.

*The Rate of Inactivation at 65°C.*—It has frequently been found that the rate of inactivation of enzymes does not follow the course of a simple monomolecular reaction but becomes slower as the reaction proceeds. It was found by the writer (6) that this peculiarity was connected with the purity of the enzyme solution and that the rate of inactivation of more highly purified solutions agreed with the monomolecular reaction rate, while those solutions containing inhibiting substances did not. On this basis the rate of inactivation of the crystalline material would be expected to be monomolecular. The result of an experiment in which the rate of inactivation of a 0.06 per cent solution of pepsin in 0.001 M hydrochloric acid (pH 3) at 65°C. is shown in

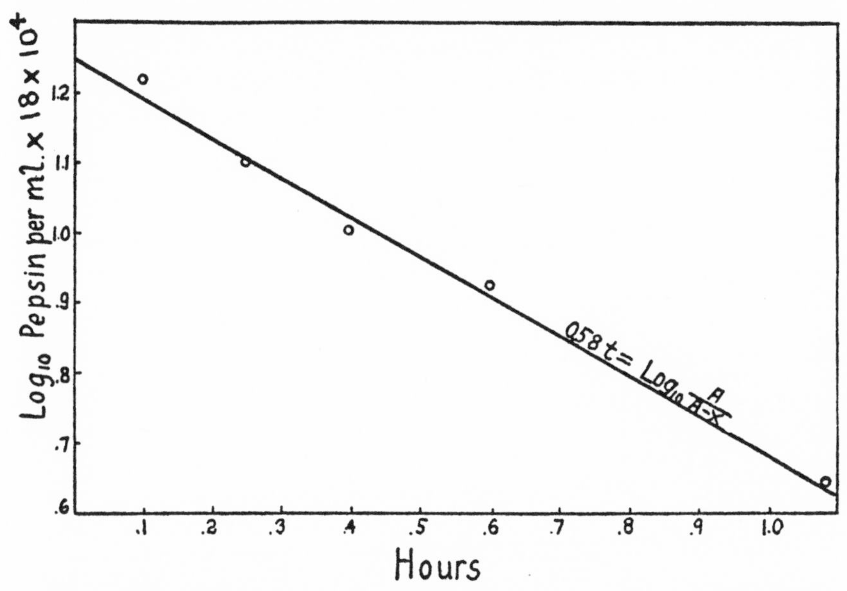


FIG. 8. Rate of inactivation of 0.06 per cent pepsin in 0.001 M hydrochloric acid at 65°C.

Fig. 8. The reaction follows the theoretical monomolecular course quite closely.

*Identity of the Diffusion Coefficient of the Protein and of the Active Material.*—The preceding experiments have shown that the activity cannot be separated from the protein molecule either by solubility measurements or inactivation experiments. It might be assumed, however, that the crystalline substance was analogous to a double salt and that the active material and the protein were in equilibrium with each other but were separate in solution. If this were the case it would be expected that the rate of diffusion of the two would be different. The diffusion coefficients were therefore determined by the method previously described (14). The experiments were carried out at 8°C. and the amount of material diffusing was determined by analyzing for nitrogen and also by activity determinations. The results of this experiment are shown in Table X. The diffusion coefficient was found to be the same when determined by either nitrogen or by activity. The radius (and presumably the weight) of the molecule containing the nitrogen and of the active molecule must therefore be the same, so that there is no indication that the protein molecule is different from the molecule responsible for the activity even in solution. The diffusion coefficient is found to be 0.048 ± 0.0005 cm.$^2$ per day which, according

TABLE X

*Diffusion Coefficient, Radius of Molecule and Molecular Weight*

| | | | |
|---|---|---|---|
| 8°C. | 1 per cent solution in M/1 acetate buffer pH 4.6, viscosity of buffer = 0.017. | | |
| D | from activity determination | = | .047 ± .0005 cm.$^2$/day |
| | from N determination | = | .0468 ± .0005 cm.$^2$/day |
| | | = | 5.44 × 10$^{-7}$ cm.$^2$/sec. |
| $r$ | = 2.22 × 10$^{-7}$ cm. | | |
| | = 2.22 mμ | | |
| $M$ | = 37,000 ± 1,000 | | |

to Einstein's formula, gives a molecular weight of 36,000. This figure agrees very well with the molecular weight of 37,000 as determined by osmotic pressure measurements. The agreement between these two figures shows that the molecules must be nearly all the same size since the molecular weight gives a mean value while the diffusion experiment as carried out gives a minimum value. The agreement also shows that the molecule is not hydrated to any great extent, since the value used for the specific gravity in calculating the molecular weight from the diffusion coefficient is the specific gravity of dry protein; and the value for the molecular weight so obtained agrees with that found by osmotic pressure.[3]

[3] Preliminary experiments (15) gave larger diffusion coefficients corresponding to a molecular weight of about 10,000, but this result was found to be due to an error in calibrating the diffusion cell.

*Antipepsin.*—If the enzyme is a protein an antibody should be formed on injection into an animal. The production of such antienzymes has been reported in the literature but the results are uncertain, as Euler has pointed out (16), because of lack of information in regard to the activity of the preparation used. In order to see if any antibody could be produced and if its production were associated with the activity of the preparation, two rabbits were immunized, one with an active preparation of the enzyme and one with a preparation of the same concentration which had been inactivated by alkali. 5 ml. injections of a 1 per cent solution of the active or inactive material were made intra-

TABLE XI

*Antiserum*

1 ml. 1 per cent pepsin solution in M/5 4.6 acetate buffer, diluted with buffer, + 0.15 ml. serum.

| Pepsin dilution | | 1 | 1 | 1/2 | 1/4 | 1/8 | 1/16 | 1/32 | 1/64 | 1/128 | 1/256 | 1/512 | 1/1,000 | 1/2,000 | 0 |
|---|---|---|---|---|---|---|---|---|---|---|---|---|---|---|---|
| Pepsin | Serum | 0 serum | .15 ml. serum | | | | | | | | | | | | |
| Active | Active | – | ++ | ++ | ++ | ++ | ++ | + | + | + | + | + | + | + | – |
| " | Denatured | – | ++ | ++ | ++ | ++ | + | – | – | – | – | – | – | – | – |
| " | Normal | – | – | – | – | – | – | – | – | | | | | | |
| Denatured | Active | – | ++ | ++ | + | + | + | – | – | – | – | – | – | – | – |
| " | Denatured | – | ++ | ++ | ++ | + | – | – | – | – | – | – | – | – | – |
| " | Normal | – | – | – | – | – | – | – | – | – | – | – | – | – | – |

peritoneally at intervals of about 1 week. Four injections were made; the rabbits were bled 2 weeks after the final injection and the serum obtained in the usual way. These sera gave a precipitin reaction with the pepsin solution, as shown in Table XI. Normal serum gave no precipitate with either the active or the inactive pepsin preparation. The serum obtained from the rabbit injected with active pepsin gave a positive precipitin reaction in a dilution of about 1/2,000 with the active pepsin solution and to about 1/16 with the inactive solution. The serum from the rabbit immunized with denatured pepsin gave a precipitin test with the denatured preparation in a dilution of about 1/8 per cent solution and with the active preparation to about the same extent. Both sera gave precipitin reactions with a 1 per cent solution of pepsin inactivated by boiling in a dilution of about 1/16. As might be expected, therefore, the sera were not strictly specific but gave cross precipitations with both active and denatured pepsin. The inhibiting effect of the sera on the activity was tested by mixing increasing amounts of the antiserum with a small amount of active pepsin and determining the rate of digestion of casein or gelatin with this mixture. The result of this experiment is shown in Table XII.

TABLE XII

*Effect of Antiserum on Digestion of Casein or Gelatin by Pepsin*

Serum diluted as noted + M/5 4.6 acetate buffer. Pepsin in 4.6 acetate buffer. 1 ml. pepsin solution + 1 ml. serum dilution.

*Gelatin*

| | Final dilution of serum | | | | | |
|---|---|---|---|---|---|---|
| Serum | 1/4 | 1/8 | 1/16 | 1/32 | 1/64 | 0 |
| | $[PU]_{ml.}^{gel. V.}$ after 3 hrs. × $10^3$ | | | | | |
| Normal | 0.075 | 0.10 | 0.11 | 0.12 | 0.11 | 0.12 |
| Antidenatured | .060 | .10 | .09 | .10 | .11 | .12 |
| Antiactive | .045 | .075 | .10 | .10 | .11 | .12 |

*Casein*

| Serum | 0 | Normal | Anti-denatured | Antiactive |
|---|---|---|---|---|
| Soluble N in filtrate, mg. | 2.2 | 2.0 | 1.87 | 1.87 |
| $\simeq [PU]_{ml.}^{cas. S.}$ | .015 | .008 | .007 | .007 |

Both the serum prepared by the injection of the active pepsin and that prepared by the injection of the denatured pepsin inhibit the action of the enzyme to about the same extent when tested by the solution of casein (cas. S. method). The normal serum inactivates somewhat less although there is no very great difference. It is difficult to compare the action of the various sera quantitatively from this experiment. The effect may be seen more clearly by determining the rate of hydrolysis of gelatin (gel. V. method) with a small amount of pepsin to which has been added increasing amounts of the various sera. The result of such an experiment is also shown in Table XII. The active serum inhibits quite strongly in a dilution of 1:4, the denatured serum less so and the normal serum still less. This inhibiting effect is evidently very weak compared with that of many antitoxins, although the precipitin reaction is about what would be expected. This may be due to the fact that the pepsin injected must be almost instantly inactivated since, as was shown above, the pepsin is instantly more than half inactivated at a pH of 7. It is probable therefore that the active pepsin is in the circulation of the animal for only a very short time.

The writer is indebted to Dr. F. S. Jones and Dr. A. P. Krueger for advice and assistance in connection with these experiments.

## VI

### *Conclusions as to the Purity of the Preparation*

The preceding experiments have shown that no evidence for the existence of a mixture of active and inactive material in the crystals could be obtained by recrystallization, solubility determinations in a series of solvents, inactivation by either heat or alkali, or by the rate of diffusion. It is reasonable to conclude therefore that the material is either a pure substance or a solid solution of two very closely related substances. If it is a solid solution of two or more substances it must be further assumed that these substances have about the same degree of solubility in the various solvents used, as well as the same diffusion coefficient and rate of inactivation or denaturization by heat. It must also be assumed that both substances are changed by alkali at the same rate and to the same extent. This could hardly be true with the possible exception of two closely related proteins. It is conceivable that two proteins might be indistinguishable by any of the tests applied in this work. But in this case it would follow that the enzyme itself was a protein and this, after all, is the main point. It does not necessarily follow even if the material represents the pure enzyme that it is the most active preparation that can be obtained nor that it is the only compound which has proteolytic activity. There is some evidence that the activity of the preparation may depend on its physical state as is known to be the case with the catalytic activity of colloidal metals. It is possible, on the other hand, that hemoglobin is the type structure for the enzymes and that they consist of an active group combined with a protein as suggested by Pekelharing. The active group may be too unstable to exist alone, but it is quite conceivable that a series of compounds may exist containing varying numbers of active groups combined with the protein, and that the activity of the compound would depend on the number of these active groups. This hypothetical complex would not differ much from that assumed by Willstätter (18) and his coworkers, except that it supposes a definite chemical compound with the protective group in place of an adsorption complex. It is of course possible that both types of complex may be formed under suitable conditions. The reactivation of enzymes as reported in the literature also suggests their protein nature since the conditions for this reactivation are similar to those found by Anson and Mirsky (20) to be suitable for the formation of native from denatured protein. The fact that the crystalline urease prepared by Sumner is also a protein and that the temperature coefficient for the rate of inactivation of enzymes in general is that characteristic for the denaturization of proteins, suggests that the protein fraction in the purification of enzymes be given special attention even though it may not be the most active fraction.

The crystalline pepsin resembles the amorphous preparations obtained previously by Pekelharing, Ringer, Fenger, Andrew and Ralston and other workers. It is probable that the preparations obtained by these workers were nearly pure pepsin. Both Pekelharing and Ringer (17) obtained preparations free from phosphorus so that there may be several proteolytically active forms.

Pekelharing showed that the same protein material could be obtained from gastric juice as from autolyzed gastric mucosa so that it is probable that the crystalline material could also be readily prepared from gastric juice. It seems fair to conclude therefore that the crystalline protein described in this paper is identical with the enzyme pepsin as secreted by the animal.

VII

*Proteolysis Is a Homogeneous Reaction*

Sørensen has shown that a solution of egg albumin is a one-phase system and the solubility measurements described in this paper show that a solution of pepsin consists also of only one phase. The reaction between dissolved egg albumin and dissolved pepsin therefore is a case of homogeneous, rather than heterogeneous catalysis.

SUMMARY

A method is described for the preparation of a crystalline protein from commercial pepsin preparations which has powerful peptic activity. The composition, optical activity, and proteolytic activity of this protein remain constant through seven successive crystallizations. No evidence for the presence of a mixture or of a solid solution is found in a study of the solubility of the protein in a series of different salt solutions, nor from the diffusion coefficient or from the rate of inactivation. These results indicate that the material is a pure substance or possibly a solid solution of two or more substances having nearly the same solubility in all the various solvents studied. It seems reasonable to conclude from these experiments that the possibility of a mixture must be limited to a mixture of proteins, so that the conclusion seems justified that pepsin itself is a protein.

BIBLIOGRAPHY

1. Sumner, J. B., *J. Biol. Chem.*, 1926, **69**, 435.
2. *Cf.* Taylor, H. S., Treatise on physical chemistry, Van Nostrand, New York, 1925, **2**, 956.
3. Fodor, A., Grundlagen der dispersoidchemie Steinkopff, Dresden, 1925.
4. Kuhn, R., and Wasserman, A., *Ber.*, 1928, **61 B**, 1550.
5. Vernon, H. M., *J. Physiol.*, 1904, **30**, 330.
6. Northrop, J. H., *J. Gen. Physiol.*, 1922, **4**, 261.
7. Marston, H. R., *Biochem. J.*, 1923, **17**, 851.
8. Pekelharing, C. A., *Z. Physiol. Chem.*, 1896, **22**, 233.
9. Fenger, F., Andrew, R. H., and Ralston, A. W., *J. Biol. Chem.*, 1928, **80**, 187.
10. Northrop, J. H., *J. Gen. Physiol.*, 1930, **13**, 781.
11. Anson, M. L., Unpublished results.
12. *Cf.* Taylor, H. S., Colloid Symposium Monograph, 1926, **4**, 19.
13. Goulding, A. M., Wasteneys, H., and Borsook, H., *J. Gen. Physiol.*, 1926–27, **10**, 451.

14. Northrop, J. H., and Anson, M. L., *J. Gen. Physiol.*, 1929, **12**, 543.
15. Northrop, J. H., *Science*, 1929, **69**, 580.
16. Euler, H., Chemie der enzyme, Bergmann, München, 3rd edition, 1925, Part 1, 390.
17. Ringer, W. A., *Z. physik. Chem.*, 1915, **95**, 195.
18. Willstätter, R., *Naturwissenschaften*, 1927, **15**, 585.
19. Northrop, J. H., *J. Gen. Physiol.*, 1926, **9**, 767.
20. Anson, M. L., and Mirsky, A. E., *J. Gen. Physiol.*, 1929, **13**, 121.

## ERRATUM

On page 743, line 4 of the first paragraph of part IV should read: "The Molisch test is negative."

# 38A


Reprinted from pages 267–280 and 289–294 of *J. Gen. Physiol.* **16**:267–294 (1932)

## CRYSTALLINE TRYPSIN

### I. Isolation and Tests of Purity

By JOHN H. NORTHROP and M. KUNITZ

(*From the Laboratories of The Rockefeller Institute for Medical Research, Princeton, N. J.*)

(Accepted for publication, June 22, 1932)

It has been known since the time of Corvissart and Kühne that pancreatic juice possesses the property of digesting proteins. Kühne assumed that this property was due to the presence in the juice of an unorganized ferment or enzyme which he called trypsin. Subsequent work by Fisher and Abderhalden, Cohnheim, Bayliss, Vernon, Schaffer and Terroine, Abderhalden, Willstätter, and Waldschmidt-Leitz and their collaborators has added greatly to our knowledge of the enzymatic properties of pancreatic juice (1). The kinetics of these reactions have also been partially worked out. According to Waldschmidt-Leitz the activated pancreatic juice contains at least five proteolytic enzymes—trypsinogen, trypsin-kinase, carboxy-peptidase and amino-peptidase, which hydrolyze polypeptides, and erepsin. There is no positive evidence[1] as yet that any of these enzymes has ever been obtained in pure form or separated completely from each other and it is probable that the separation can be carried still further, as already indicated by Abderhalden (3). The enzyme described in this paper differs essentially from any previously described. In the meantime the existence of these separate enzymes has been assumed purely from the behavior of certain solutions, as Linderström-Lang has pointed out (4).

In contrast to the marked advances in knowledge of the properties

[1] In one sense this will always be the case since it is not possible to furnish positive proof of the purity of a substance but only negative proof consisting of the fact that no evidence of mixtures can be found under conditions which would be expected to show such evidence. For a discussion of the difficulty of proving the existence of, or defining a "pure substance" see Lunn and Senior (2).

of the enzymes of the pancreatic juices, as shown by their catalytic effect in various reactions, little or no knowledge has been gained as to the chemical nature of these enzymes. The early workers, Kühne, Mays, Hammarsten, and Michaelis, considered them to be associated with the nucleic acid fraction; but Levene (5) showed that they were not nucleic acids themselves since hydrolysis of the nucleic acids did not destroy the activity. Willstätter and his collaborators (6) concluded that the enzymes belong to an unknown class of chemical compounds associated with proteins or other substances of high molecular weight and a somewhat similar point of view is expressed by Fodor (7). This conclusion, however, rests only on the purely negative evidence that no pure substance, *i.e.* one having constant properties, has been obtained which showed enzymatic activity. The known chemical properties of enzymes, on the other hand, such as the temperature coefficient of inactivation, effect of acid and alkali on stability, and the reactivation sometimes observed, as well as the ease with which they are adsorbed on colloidal particles, are those of proteins. In addition, crystalline proteins of constant composition and activity have been isolated by Sumner (8) in the case of urease, by Northrop (9) in the case of pepsin, and by Caldwell, Booher, and Sherman (10) in the case of amylase. The existing positive evidence, therefore, indicates the protein nature of enzymes and the conditions and methods used in the present work were those known to be favorable for the isolation of proteins; *i.e.*, concentrated solutions in concentrated neutral salt and low temperatures. These methods have led to the isolation of a crystalline protein having constant physical and chemical properties, including constant proteolytic activity. This enzyme attacks only proteins, and pepsin-peptone, so far as we have determined. Since the protease of pancreas has always been called trypsin the present enzyme will be referred to as trypsin. It differs markedly from the trypsin-kinase of Waldschmidt-Leitz in that it does not carry the hydrolysis of proteins nearly so far as does trypsin-kinase.

### *Methods of Determining Activity*

The activity of the various preparations was determined by the change in viscosity of gelatin and casein, the increase in formol titration of gelatin and casein, and the formation of non-protein nitrogen in casein solutions, etc. The following definitions and abbreviations are used in the paper.

$[\text{T. U.}]^{\text{4g} V}_{\text{mg. N}}$—per cent change in viscosity per minute per mg. trypsin nitrogen contained in 5.2 ml. 2.5 per cent gelatin, pH 4.0.

$[\text{T. U.}]^{\left\{\substack{\text{Cas. } F \\ \text{Gel. } F}\right.}_{\text{mg. N}}$—milliequivalents carboxyl groups liberated per minute, per mg. trypsin nitrogen contained in 6.0 ml. 4 per cent $\left\{\begin{array}{l}\text{casein} \\ \text{gelatin, pH 7.6.}\end{array}\right.$

$[\text{T. U.}]^{\text{Cas. } S}_{\text{mg. N}}$—milliequivalents nitrogen soluble in 5 per cent trichloracetic acid formed per minute per mg. trypsin nitrogen contained in 6.0 ml. 4 per cent casein, pH 7.6

$[\text{T. U.}]^{\text{Ren.}}_{\text{mg. N}}$—per cent increase in viscosity per minute per mg. trypsin nitrogen contained in 5.2 ml. of standard milk solution.

$[\text{T. U.}]^{\text{Clot.}}_{\text{mg. N}}$—ml. magnesium sulfate plasma clotted by 1 mg. trypsin nitrogen in 18 hours, 6°C.

$[\alpha]^{D}_{\text{mg. P. N.}}$—degrees optical rotation in a 1 dm. tube of a solution in $\frac{1}{4}$ saturated ammonium sulfate, M/10 pH 4.0 acetate, containing 1 mg. protein nitrogen per ml.; sodium $D$ line, at 20°C.

The methods used are described in the experimental part of the paper (11).

The determinations were confined in all cases except the rennet action, to the initial slope of the curves in which region the specific activity obtained is independent of the concentration of enzyme used.

### *Preliminary Method of Fractionation*

The first attempts at purification of the enzyme were similar to the experiments of Michaelis and Davidsohn (12) and consisted in a study of the precipitate obtained from crude trypsin extracts at about pH 3.0. The crude material used was Fairchild's trypsin which is prepared from beef pancreas.[2] If an aqueous extract of the dry powder is titrated to about pH 3.0 with acid a precipitate forms, as Michaelis found. Most of the activity is found in this precipitate. Upon repeated solution and reprecipitation much of the activity is lost and the specific activity of the precipitate becomes less and less. A number of other methods of fractionation were tried but it was found that

[2] The writers are indebted to Mr. Benjamin Fairchild of Fairchild Bros. and Foster for this information.

the precipitate obtained with strong ammonium sulfate was the only one which could be dissolved and reprecipitated indefinitely without loss of activity either as a whole or in regard to the precipitate. Systematic fractionation of an extract of the commercial preparation in $\frac{1}{4}$ saturated ammonium sulfate was then undertaken with various concentrations of ammonium sulfate. As the ammonium sulfate concentration is increased a series of precipitates is obtained which become more and more active. The most active precipitate appeared at about 0.6 saturated ammonium sulfate and further fractionation of this precipitate did not change its activity appreciably. The precipitate was a protein and gave some indications of crystallizing so that a large number of experiments were done in an attempt to crystallize it. It was found eventually that crystals could be obtained by adding saturated ammonium sulfate very cautiously to a 5 per cent solution of the precipitate (1 part filter cake dissolved in five times its weight of solution) in $\frac{1}{4}$ saturated ammonium sulfate made up in M/10 acetate buffer, pH 4.0, temperature 25–30°C. The first precipitate which appears under these conditions is usually amorphous. If this is filtered off and the filtrate allowed to stand at 25–30°C., small crystals of cubic form begin to appear in the solution and increase rapidly in amount. Good crystals are obtained only with slow crystallization. Otherwise the crystals are not well formed and usually appear spherical. The process is favored by stirring. The crystallization may be hastened by continued addition of ammonium sulfate (with stirring) and eventually practically all the protein may be obtained in crystalline form. This is the method of preparation already reported (13). Material obtained in this way is about ten times as active as the original commercial product on a total dry weight basis and about three times as active per milligram soluble protein nitrogen.

Repeated crystallization does not change the specific activity to any extent but it was noted that the first small amount of precipitate formed was always slightly less active than the succeeding crops of crystals. The results of such an experiment are shown in Table I.

It was very difficult to decide whether this result was due to actual fractionation or to loss in activity during the experiment. In order to test the purity of the preparation in another way a series of solubility experiments was done in $\frac{3}{4}$ saturated magnesium sulfate. The crystalline precipitate was stirred for 10 to 15 minutes with a mixture made up of 75 ml. saturated magnesium sulfate and 25 ml. M/10 pH 4.0 acetate. The suspension was then filtered and the precipitate again

TABLE I

100 gm. poorly crystalline cake dissolved in 600 ml. ¼ saturated ammonium sulfate in M/10 acetate buffer, pH 4.0. Saturated ammonium sulfate added with stirring to faint turbidity. More saturated ammonium sulfate run in very slowly. Crystalline precipitate formed and filtered off from time to time. Saturated ammonium sulfate added to filtrate until no further crystals are obtained. The precipitates were then combined, dissolved in 6 volumes ¼ saturated ammonium sulfate, pH 4.0 and the crystallization of Fraction b repeated four times. Samples of each fraction were analyzed for protein nitrogen, optical activity, and proteolytic activity by gelatin viscosity method.

| Crystallization No. | Weight of cake | Fraction | Precipitate | $[T. U.]^{4gV}_{mg. N}$ | $[\alpha]^{D}_{mg. N}$ |
|---|---|---|---|---|---|
| | *gm.* | | *gm.* | | |
| 1 | 100 | 1 a | 11 | 40 | |
| | | 1 b | 17 | 55 | 0.33 |
| | | 1 c | 15 | 57 | |
| | | 1 d | 11 | 57 | |
| | | Mother liquor | | 42 | |
| 2 | 39 | 2 a | 1.3 | 42 | |
| (1a + 1b + | | 2 b | 26.5 | 47 | 0.33 |
| 1c + 1d) | | Mother liquor | | 36 | |
| 3 | 26 | 3 a | 2.5 | 43 | |
| | | 3 b | 11.5 | 47 | 0.37 |
| | | Mother liquor | | 50 | |
| 4 | 11 | 4 a | 1 | | |
| | | 4 b | 7.5 | 45 | 0.39 |
| | | Mother liquor | | 40 | |
| 5 | 7 | 5 a | 0.5 | 42 | |
| | | 5 b | 3.6 | 45 | 0.36 |
| | | Mother liquor | | | |

stirred with the same solvent. This was continued until the precipitate had nearly all gone into solution and the filtrates and residue then analyzed for nitrogen and activity. The results of the experiments are shown in Table II.

The solubility decreases rapidly with successive extractions while the specific activity of the protein in solution also decreases to some extent. The final residue, however, has about the same activity as the original material. These results show definitely that the crystalline material prepared by this method is not a pure protein since the solubility depends on the quantity of precipitate. It is similar to the type of result found by Sörensen (22) with other proteins and indicates that the substance is probably a solid solution. There was some loss in activity during this experiment but hardly sufficient to account for the observed results.

TABLE II

*Solubility in Magnesium Sulfate, pH 4.0 Acetate =*
*{75 Ml. Saturated Magnesium Sulfate*
*{25 Ml. M/10 pH 4.0 Acetate*

20 gm. crystalline filter cake stirred for 10–15 minutes with 75 ml. solvent. Filtered and filtrate analyzed for total nitrogen and activity. Extraction repeated eight times.

| Extract No. | 1 | 3 | 5 | 7 | 9 | Residue from 9th extract |
|---|---|---|---|---|---|---|
| N/ml. | 1.92 | 0.94 | 0.80 | 0.49 | 0.29 | |
| $[T. U.]^{4gV}_{mg. P. N.}$ | 60 | 36 | 54 | 40 | 47 | 60 |

## *Effect of Heating in Acid Solution*

It was noted when a solution of the material was boiled in dilute hydrochloric acid and then cooled that a form of protein appeared which precipitated on the addition of magnesium sulfate although there was not very much loss in activity. This remarkable stability in dilute acid is a characteristic property of trypsin and has been noted by Mellanby and Wooley (14). It was thought at first that the precipitate formed on adding a solution, which had been heated and then cooled, to salt solutions was probably denatured trypsin protein and an experiment was made in order to determine whether the loss in activity was proportional to the quantity of this denatured protein formed. A solution of the material in N/10 hydrochloric acid was heated to 95°C. and the total activity of the solution determined as well as the quantity of denatured protein present. The results of an experiment of this kind are shown in Table III.

The table shows that the total protein nitrogen remains constant while the activity per milliliter of solution decreases. The specific activity per milligram total protein nitrogen therefore decreases with time. However, if instead of total protein nitrogen the concentration of protein which does not precipitate with magnesium sulfate is considered, the experiment shows that the specific activity, referred to this soluble protein, increases nearly 100 per cent in the first few min-

TABLE III

*Loss in Activity and Formation of Denatured Protein in Dilute Acid Solution at 95°C.*

2 gm. of crystalline filter cake dissolved in 50 ml. of M/10 hydrochloric acid and kept at 95°C. 1 ml. samples taken and added to 4 ml. M/10 pH 4.0 acetate buffer and this solution analyzed for protein nitrogen and total activity. Another set of 1 ml. samples taken, cooled, and added to 4 ml. cold M/10 magnesium sulfate in 0.5 normal sulfuric acid. The suspension centrifuged and supernatant liquid analyzed for protein nitrogen and activity.

| Time | Analysis from samples in acetate | | | Analysis of filtrate from $MgSO_4$ samples | | |
|---|---|---|---|---|---|---|
| | Protein N | $[T.U.]^{4gV}_{ml.}$ | $[T.U.]^{4gV}_{mg. P.N}$ | Protein N | $[T.U.]^{4gV}_{ml.}$ | $[T.U.]^{4gV}_{mg. P.N.}$ |
| *hrs.* | *mg.* | | | *mg.* | | |
| 0 | 1.6 | 100 | 62 | 1.6 | 100 | 62 |
| 0.1 | 1.57 | 80 | 51 | 0.58 | 63 | 110 |
| 0.2 | 1.5 | 64 | 43 | 0.56 | 54 | 97 |
| 0.4 | 1.7 | 43 | 25 | 0.52 | 27 | 52 |
| 0.8 | 1.7 | 18.5 | 11 | 0.45 | 7.4 | 16 |

utes, remains nearly constant for a while, and then decreases. The experiment shows conclusively that the material is not a pure substance but contains a protein which is denatured by heat and which has little or no activity, and another protein of high activity which either is not denatured by heat or which reverts to the native condition on cooling.[3] This behavior is so unusual that a detailed study of

[3] This result is very similar to that obtained recently by Waldschmidt-Leitz and Steigerwaldt (15) in following the digestion of urease with trypsin. It proves only that the material is not a pure protein but does not prove that no active protein is present.

the reaction was made and is reported in another paper (16). It turns out that the active protein is denatured when heated but reverts to the native condition very rapidly on cooling. At the same time the activity returns.

The experiment just described, however, also indicates a very efficient method of further fractionation. A study of the fractionation was therefore undertaken again. The raw material was either an aqueous extract of Fairchild's or other commercial trypsin, or pancreatic juice obtained from frozen pancreas. The frozen mass was sliced, spread on racks and allowed to thaw, and the liquid which drained out collected.

### *Final Method of Isolation*

The method of fractionation finally worked out consists essentially of a preliminary precipitation of juice obtained in this way with strong acid which removes most of the inert proteins, as has previously been reported by Fodor and by Schönfeld-Reiner (17). The filtrate from this precipitate is then fractionated with ammonium sulfate and dissolved in dilute acid, heated to 80°C., and again fractionated with ammonium sulfate. By this method a protein was obtained which had about twice the specific activity of the original crystalline material but which crystallized under the same conditions and in a similar crystalline form (*cf.* Fig. 1). Up to the present it has not been found possible to increase the specific activity any further. Whenever any of the protein is destroyed or removed there is a corresponding loss in activity. The details of the preparation are shown in Table IV. In this experiment the heating and fractionation were repeated four times. Ordinarily the fraction obtained after the first heating (No. 5 in the table) is used as the final material. This fraction contains nearly one-half of the original proteolytic activity. Most of the loss occurs when the solution is heated. This is due to the fact that there are other proteins present in the solution, since, as is described in the paper on heat inactivation (16), the purified material may be heated and cooled again indefinitely without any loss in activity. The specific activity is about six times that of the original crude extract on the basis of protein nitrogen and considerably more than six times as active on the basis of total dry weight.

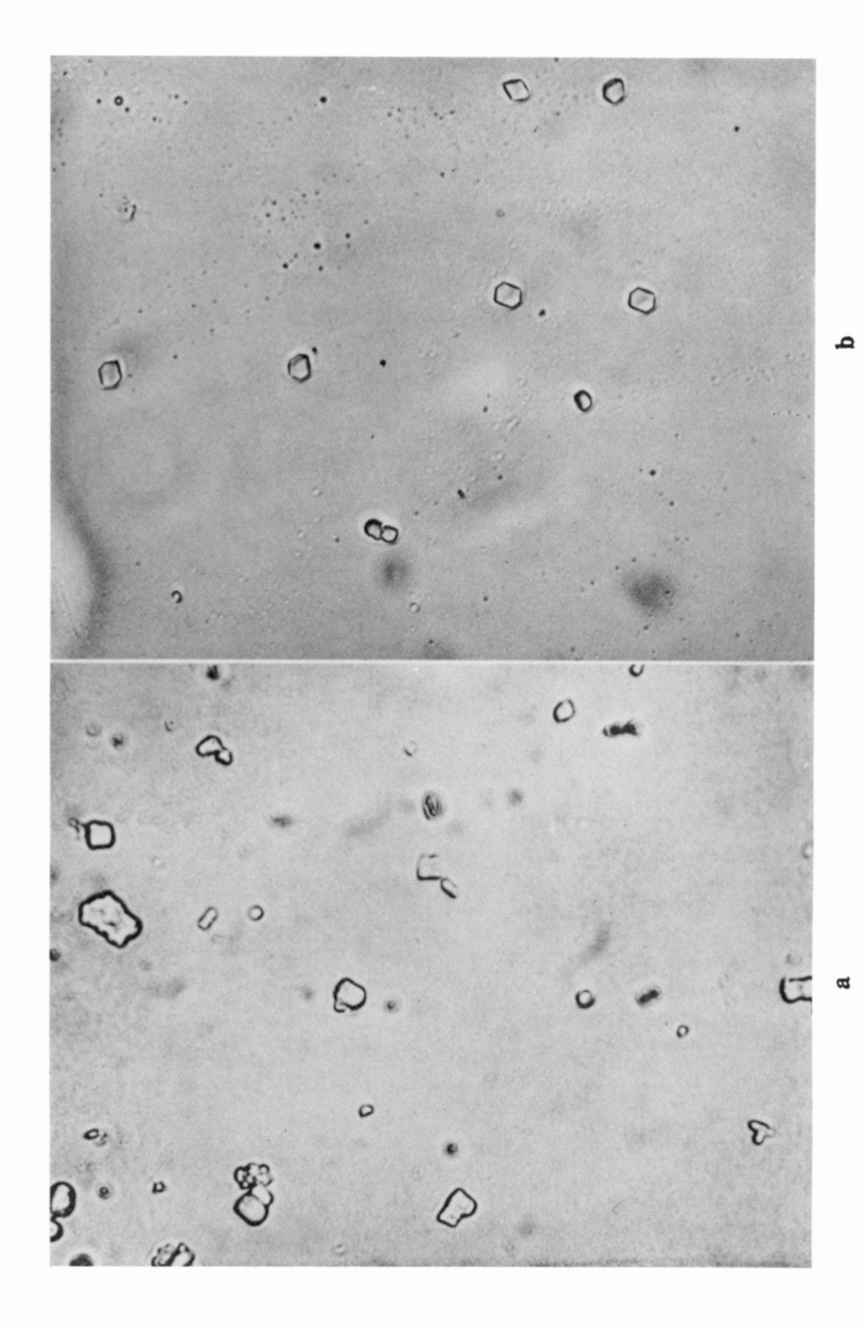


FIG. 1, *a* and *b*. (*a*) Trypsin crystals immediately after crystallization. (*b*) Trypsin crystals after standing in contact with the mother liquor for 1 or 2 months.

TABLE IV

*Preparation of Crystalline Trypsin*

| | Fraction No. | Weight | $[T. U.]^{4gV}_{mg. P.N.}$ | Total $[T. U.]^{4gV}$ |
|---|---|---|---|---|
| | | *gm.* | | |
| 150 kg. frozen pancreas cut in slices and allowed to thaw overnight at about 5°C. and the expressed fluid collected | 1 | 12,000 | 18 | $12 \times 10^6$ |
| Diluted with 1 volume water and 22 ml. concentrated hydrochloric acid added per liter (final concentration HCl 0.25 molar). Solution filtered in cold room through fluted paper (S. and S. No. 1450 1/2). Filtrate brought to 0.4 saturated ammonium sulfate by the addition of solid ammonium sulfate (250 gm. per liter) and refiltered through fluted paper. Filtrate brought to 0.7 saturated ammonium sulfate (250 gm. ammonium sulfate per liter) and filtered with suction. Precipitate | 3 | 520 | 37 | $9 \times 10^5$ |
| 500 gm. Precipitate 3 dissolved in 12 liters N/20 hydrochloric acid, heated* rapidly to 90°C. and cooled to 20°C., brought to 0.4 saturated ammonium sulfate, and filtered. Filtrate brought to 0.7 saturated ammonium sulfate and filtered with suction. Precipitate | 5 | 120 | 95 | $5 \times 10^6$ |
| This completes the usual method of preparation. The succeeding fractionations are reported to show that no further change in activity occurs | | | | |
| Precipitate 5 dissolved in 25 volumes M/20 pH 4 acetate buffer, heated rapidly to 90°C., cooled to 20°C., brought to 0.4 saturated ammonium sulfate, and filtered. Brought to 0.7 saturated ammonium sulfate and filtered with suction. Precipitate | 7 | 70 | 100 | $4 \times 10^5$ |
| No. 7 dissolved in 3 volumes ¼ saturated ammonium sulfate pH 4 brought to 0.4 saturated ammonium sulfate, filtered. Filtrate brought to 0.7 saturated ammonium sulfate, filtered with suction. Precipitate | 8 | 50 | 110 | $3 \times 10^5$ |

TABLE IV—*Concluded*

| | Fraction No. | Weight | $[T. U.]^{4gV}_{mg. P.N.}$ | Total $[T. U.]^{4gV}$ |
|---|---|---|---|---|
| | | *gm.* | | |
| No. 8 dissolved in 25 volumes N/20 hydrochloric acid, brought to 0.4 saturated ammonium sulfate, and filtered. Filtrate brought to 0.7 ammonium sulfate, filtered with suction. Precipitate | 9 | 35 | 105 | $2 \times 10^5$ |
| No. 9 dissolved in 3 volumes ¼ saturated ammonium sulfate, brought to 0.4 saturated ammonium sulfate by the addition of saturated ammonium sulfate, clear solution. Saturated ammonium sulfate added slowly until slight turbidity. Solution filtered, filtrate inoculated with crystals, and saturated ammonium sulfate added very slowly. Crystalline precipitate forms. Precipitate | 10 | 20 | 110 | $1 \times 10^5$ |

* The heating and cooling of large quantities of solution may be most conveniently done by running the solution through a glass coil immersed in boiling water and then through a coil in cold water. The rate of flow is regulated so that the solution leaves the hot coil at a temperature of 80–85°C. The coil used in these experiments was of 5 mm. (inside diameter) thin walled tubing, about 2 m. long.

It is possible to prepare crystalline material having the maximum activity by fractionation with ammonium sulfate alone and without heating. The process is laborious and the yield poor so that, for practical purposes, the method is not satisfactory. The material obtained in this way, however, is identical with that obtained after heating. There is no reason to suppose, therefore, that the properties of the enzyme are changed by the heating.

The various fractions were tested for tryptic activity by gelatin and casein viscosity methods at pH 4.0, by the formol titration of gelatin and casein, and by the production of soluble nitrogen from casein. They were also tested for their blood-clotting power (18), for ability to clot milk, and for amylase, lipase, and erepsin. The optical activity at pH 4.0 and the total increase in formol titration with casein and gelatin in the presence of excess enzyme were also determined. The results of these determinations are shown in Table V.

The proteolytic activity, as measured by any of the methods and

TABLE V

*Analysis of Fractions*

| Fraction No. | 1 | 3 | 5 | 7 | 8 | 9 | 10 | Glycerin extract of acetone dried pig pancreas | Crystalline pepsin |
|---|---|---|---|---|---|---|---|---|---|
| $[T. U.]^{4gV}_{mg. P. N.}$ | 18 | 37 | 100 | 100 | 110 | 105 | 110 | 20 | 13 |
| $[T. U.]^{Gel. F}_{mg. P. N.}$ | 0.05 | 0.11 | 0.32 | 0.24 | 0.34 | 0.33 | 0.29 | | 0.001 |
| $[T. U.]^{Cas. V}_{mg. P. N.}$ | 160 | 330 | | | | | 870 | | |
| $[T. U.]^{Cas. F}_{mg. P. N.}$ | 0.053 | 0.084 | 0.19<br>0.13 | 0.15 | 0.16<br>0.12 | 0.23 | 0.18 | 0.04 | 0.20 |
| $[T. U.]^{Cas. S}_{mg. P. N.}$ | 0.61 | 1.1 | 2.4<br>1.9 | 1.7 | 2.1 | 2.2 | 2.4 | 0.50 | 0.44 |
| $[T. U.]^{Clot.}_{mg. P. N.}$ | | 810 | | 1500 | | | 1500 | | |
| $[\alpha]^{D}_{mg. N}$ (1 dm. tube) 25°C. | | −0.458 | | −0.28 | −0.29 | | −0.26 | | |
| Amylase—mg. P. N. for positive test, *mg.* | <0.001 | >2 | | >2 | >2 | >2 | >2 | | |

| | | | | | | | | | |
|---|---|---|---|---|---|---|---|---|---|
| $[T. U.]^{Ren.}_{mg. P. N.}$ | 1000 | 4000 | 400 | 300 | 160 | 100 | 95 | | 280,000 |
| Lipase—$\frac{ml.\ N/10}{1\ mg.\ N}$ | 0.6 | 0.4 | | | | | | | |
| Erepsin (glycyl-glycine) | + | ± | | – | – | – | – | | |
| Maximum increase in formol titration—ml. N/50 NaOH per 5 ml. protein solution: casein | >32 | >15 | | 9.1 | 9.0 | 8.9 | 9.0 | | 17.0 |
| Maximum increase in formol titration—ml. N/50 NaOH per 5 ml. protein solution: gelatin | >25 | >10 | | 6.9 | 7.0 | 7.0 | 7.0 | | 7.0 |

also the blood-clotting property increases rapidly up to Fraction 5 and then remains constant. The percentage increase in activity between Fraction 1 and Fraction 5 is slightly more as determined by the gelatin viscosity method than by the other methods, which may be due either to the presence of more than one enzyme, one of which is removed during the course of purification, or to the fact that the other methods of measuring the activity measure secondary reactions due to peptone-splitting enzymes while the viscosity method measures only the change in the protein. The results as a whole indicate that the final preparation has constant activity as measured by any of the preceding methods and that no change in this activity occurs during the repeated fractionations between Fraction 5 and Fraction 10. The optical activity per milligram of nitrogen also reaches a constant value. The amylase activity is expressed as the number of milligrams protein nitrogen required to give a positive test under the conditions used by Willstätter. Fraction 1 has powerful amylolytic activity but no positive test for amylase can be obtained in succeeding fractions. The lipase activity, which is expressed as ml. N/10 alkali per milligram nitrogen under the conditions described in the experimental part (11) disappears in Fraction 5 as does the ability to digest dipeptides. The rennet action decreases rapidly as fractionation proceeds but even the last fractions have a weak effect on the clotting of milk which seems to remain constant. There is no doubt that there is originally present another enzyme with a much more powerful milk-clotting power than the final trypsin and it is possible that the small amount of rennet action noted with the final fractions is due to a trace of this enzyme carried through the preparation.

[*Editor's Note*: Material has been omitted at this point.]

### *Relation of the Proteolytic Activity to the Protein*

*Heat Inactivation.*—The experiments described show that under the conditions studied so far the material behaves like a pure substance, or

TABLE XI

*Decrease in Activity and Protein Concentration on Heating Trypsin Solutions in N/10 Hydrochloric Acid*

2 gm. crystalline filter cake dissolved in 40 ml. N/10 hydrochloric acid and solution heated to 95°C. by immersion in boiling water. 1 ml. samples taken, cooled to 0°C. for 2 minutes, and added to 4 ml. 0.7 saturated magnesium sulfate in N/1 sulfuric acid. Precipitate formed. Centrifuged and protein nitrogen and activity determined on supernatant liquid.

| Time at 95°C., *hrs*. | 0 | 0.10 | 0.20 | 0.30 | 0.40 |
|---|---|---|---|---|---|
| $[T. U.]^{4gV}_{ml.}$ | 118.0 | 32.8 | 22.4 | 7.0 | 0.94 |
| Protein nitrogen per ml., *mg*. | 1.50 | 0.315 | 0.245 | 0.245 | 0.105 |
| $[T. U.]^{4gV}_{mg. P. N.}$ | 80 | 104 | 92 | 29 | 9.0 |

possibly a solid solution. They indicate, therefore, that the proteolytic power of the preparation is a property of the protein molecule. This conclusion may be tested in a number of other ways. If the proteolytic activity is a property of the protein molecule then any chemical change in the protein should be reflected by a change in activity. When a solution of the protein is heated for a short time, the protein is denatured; *i.e.*, it is changed to a form which is precipitated by low concentrations of salt. At the same time the activity is lost. When the solution is allowed to cool, however, the protein is no longer precipitated by salt but reverts to its original native condition; at the same time its original activity is regained. This reaction has been studied in detail and reported in another paper (16). If a solution of the protein is heated for a longer time, however, it does not return to its original soluble condition on cooling nor does the activity return. There is evidently a second non-reversible reaction which changes the denatured protein into another form. If the rate of formation of this permanently denatured protein be compared with the loss in activity of the solution, it is found that the decrease in activity is proportional to the decrease in native protein. The results of such an experiment are shown in Table XI. This experiment is carried out in the same way as that already reported with the first crystalline material obtained (Table III). In that case the specific activity increased rapidly at first, showing that the original material undoubtedly contained an inactive protein. In the present experiment there is also a slight increase in the specific activity at first. The specific activity then remains practically constant until only 20 per cent of the original total activity is left; it then decreases rapidly. Evidently on long heating there is formed in the solution a compound which does not precipitate with salt solution but does precipitate with trichloracetic acid. Such compounds are always found in the acid hydrolysis of proteins and it is not surprising that they are found under these conditions. The material used in this experiment originally had a specific activity of slightly more than 100. On standing the activity decreased somewhat, evidently with formation of a protein which is rapidly and permanently denatured by heat, so that the first few minutes heating results in a slight increase in the specific activity to its original value. This behavior has been noted consistently and indicates that on standing at room temperature the active protein becomes transformed into an inactive one which has lost the remarkable property of reverting to the native condition immediately on cooling after being heated. This experiment shows that when the protein is denatured by prolonged heating there is a corresponding loss in activity and therefore agrees with the result expected if the activity were a property of the protein molecule.

*Pepsin Digestion.*—It is known that pepsin digestion destroys the tryptic activity of a solution (Long and Johnson; Long and Hull (23)). If the proteolytic activity were a property of the native protein molecule it would be expected that the loss in activity during pepsin digestion would be proportional to the loss of native protein. If, on the other hand, the proteolytic activity were due to some molecular species accompanying the protein, it would be expected that the protein concentration would decrease more rapidly than the activity since, so far as is known, pepsin acts only on proteins. This result would also be expected if the material were a mixture of an inactive with an active protein, since the rate of hydrolysis of different proteins with pepsin is highly specific. This would result in a change in the specific activity when calculated on the basis of total protein present. The same result would be obtained if any portion of the protein molecule

TABLE XII

*Decrease in Activity and Protein Concentration in Trypsin Solutions Digested by Pepsin*

1 gm. crystalline trypsin cake dissolved in 30 ml. M/100 hydrochloric acid at 35.5°C. 0.5 ml. crystalline pepsin solution containing 0.28 mg. protein nitrogen added. Protein nitrogen determined with 2.5 and 18 per cent trichloracetic acid. Activity determined by gelatin viscosity method. The activity due to pepsin was negligible.

| Time at 35°C., *hrs.* | 0 | 1 | 2 | 4 | 8 | 24 | 48 |
|---|---|---|---|---|---|---|---|
| $[T. U.]^{4gV}_{ml.}$ | 125 | 99 | 74 | 56 | 40 | 20 | 10.3 |
| Protein nitrogen per ml., *mg.* | | | | | | | |
| 18 per cent $CCl_3COOH$ | 1.49 | 1.24 | 1.1 | 0.96 | 0.79 | 0.51 | 0.40 |
| 2.5 " " $CCl_3COOH$ | 1.21 | 0.93 | 0.62 | 0.51 | 0.40 | 0.23 | 0.20 |
| $[T. U.]^{4gV}_{mg. P. N.}$ | | | | | | | |
| 18 per cent $CCL_3COOH$ | 84 | 80 | 67 | 58 | 51 | 39 | 26 |
| 2.5 " " $CCL_3COOH$ | 103 | 107 | 119 | 110 | 100 | 87 | 52 |

retained any appreciable activity. In this case the specific activity of the total protein left in solution would increase since the activity due to the fragment of the protein molecule would be added to that due to the unchanged protein itself. The result of an experiment in which crystalline trypsin was digested with a very small quantity of crystalline pepsin is shown in Table XII. The total protein present in solution was determined by precipitation with 2.5 per cent and also with 18 per cent trichloracetic acid. No quantitative method exists by which the native protein alone can be determined but the nitrogen precipitated by 2.5 per cent trichloracetic acid is probably not far from the correct amount. 18 per cent trichloracetic acid precipitates considerably more nitrogen than 2.5 per cent and if any of the higher split products of the protein retained any activity, it might be expected that the specific activity calculated from nitrogen precipitated with 18 per cent trichloracetic acid would be more constant than that obtained when 2.5 per cent trichloracetic acid is used. The results of the experiment, however, show that the specific activity calculated on the basis of the nitrogen precipitated with 2.5 per cent trichloracetic acid remains constant until more than 70 per cent of the total activity is lost, while the specific activity calculated on the basis of 18 per cent trichloracetic acid decreases rapidly. Apparently, therefore, as soon as the protein is split so that it no longer precipitates with 2.5 per cent trichloracetic acid all its activity is lost and even the fragments still large enough to precipitate with 18 per cent trichloracetic acid possess no activity. The results indicate that the proteolytic activity is a property of the entire molecule as is the case with the (reversible) oxygen combining power of hemoglobin. It was found that this experiment was a very sensitive test for the purity of the preparation used, as is the effect of heating in acid. Preparations which have become partially inactivated show at first a rapid increase in the specific activity under these conditions due to the fact that the inactivated, denatured protein is more rapidly hydrolyzed than the active, native protein. With such preparations, therefore the initial specific activity is low but increases rapidly until it reaches a figure of about 100 and then remains constant.

### *Alkali Inactivation*

When a solution of trypsin is allowed to stand in slightly alkaline solution the protein decomposes rapidly (16) so that these conditions furnish another method of comparing changes in the protein and changes in the activity. The results of an experiment under these conditions are shown in Table XIII. They are similar to those obtained with pepsin digestion in that the decrease in activity is very nearly proportional to the decrease in protein nitrogen as determined by 2.5 per cent trichloracetic acid, and is greater than the loss in protein nitrogen as determined by 18 per cent trichloracetic acid. They indicate again that the proteolytic property is lost so soon as any change occurs in the original native protein molecule.

The hydrolysis of the protein under these conditions may be ascribed to several possible mechanisms. It may be simply a case of the usual alkali hydrolysis of proteins, or it may be considered that the protein digests itself, and in either case it might be supposed that the formation of denatured protein was an intermediate step in the reaction so that the experiment as a whole might be considered simply as a variation of the result already obtained (16) to the effect that the activity is lost when the protein is denatured.

TABLE XIII

*Decrease in Activity and Protein Nitrogen Concentration of Trypsin Solutions in Sodium Bicarbonate at 35.5°C.*

0.50 gm. crystalline trypsin filter cake dissolved in 15 ml. N/100 sodium bicarbonate at 35.5°C. Protein nitrogen determined by precipitation with 2.5 or 18 per cent trichloracetic acid. Activity determined by viscosity pH 4.0 gelatin.

| | | | |
|---|---|---|---|
| Time at 35.5°C., *hrs* | 0 | 0.5 | 1.0 |
| Protein nitrogen per ml. solution { 18 per cent $CCl_3COOH$), *mg* | 1.0 | 0.66 | 0.54 |
| Protein nitrogen per ml. solution { 2.5 " " $CCl_3COOH$), *mg* | 0.85 | 0.46 | 0.40 |
| $[T. U.]^{4gV}_{ml.}$ | 116 | 52 | 34 |
| $[T. U.]^{4gV}_{mg. P. N.}$ { 18 per cent $CCl_3COOH$ | 116 | 79 | 62 |
| $[T. U.]^{4gV}_{mg. P. N.}$ { 2.5 " " $CCl_3COOH$ | (136)[1] | 113 | 84 |

[1] This figure is based on one experimental value and is probably an experimental error.

REFERENCES

1. General reviews of the literature are given in: Waldschmidt-Leitz, E., Die Enzyme, Braunschweig, Friedr. Vieweg and Sohn Akt.-Ges., 1926. Loewenthal, E., in Oppenheimer, C., Fermente und ihre Wirkungen, Leipsic, George Thien, 5th edition, 1926, **2**. Grassman, W., *Ergebn. Enzymforsch.*, 1932, **1**. Euler, H., Chemie der Enzyme, Munich, J. F. Bergmann, 2nd and 3rd edition, 1927, pt. 2. Northrop, J. H., *Harvey Lectures*, 1925–26, **21**, 36.
2. Lunn, A. C., and Senior, J. K., *J. Phys. Chem.*, 1929, **33**, 1027.
3. Abderhalden, E., and Schwab, E., *Fermentforschung*, 1931, **12**, 559.
4. Linderström-Lang, K., *Compt.-rend. trav. Lab. Carlsberg*, 1931, **19**, No. 3.
5. Levene, P. A., *Am. J. Physiol.*, 1901, **5**, 298.
6. Willstätter, R., *Naturwissenschaften*, 1927, **15**, 585.
7. Fodor, A., *Ergebn. Enzymforsch.*, 1932, **1**.
8. Sumner, J. B., *J. Biol. Chem.*, 1926, **69**, 435; 1926, **70**, 97. *Ergebn. Enzymforsch.*, 1932, **1**, 295.

9. Northrop, J. H., *J. Gen. Physiol.*, 1930, **13**, 739. *Ergebn. Enzymforsch.*, 1932, **1**, 302.
10. Caldwell, M. L., Booher, L. E., and Sherman, H. C., *Science*, 1931, **74**, 37.
11. Northrop, J. H., and Kunitz, M., *J. Gen. Physiol.*, 1932, **16**, 313.
12. Michaelis, L., and Davidsohn, H., *Biochem. Z.*, 1911, **30**, 481.
13. Northrop, J. H., and Kunitz, M., *Science*, 1931, **73**, 262.
14. Mellanby, J., and Wooley, B. J., *J. Physiol.*, 1913, **47**, 339.
15. Waldschmidt-Leitz, E., and Steigerwaldt, F., *J. Physiol. Chem.*, 1932, **206**, 133.
16. Northrop, J. H., *J. Gen. Physiol.*, 1932, **16**, 323.
17. Schönfeld-Reiner, R., *Fermentforschung*, 1930, **12**, 67.
18. Waldschmidt-Leitz, E., Stadtler, P., and Steigerwaldt, F., *Naturwissenschaften*, 1928, **16**, 1027.
19. Willstätter, R., and Waldschmidt-Leitz, E., quoted in Rona, P., Praktikum der physiologischen Chemie, Berlin, Julius Springer, 2nd edition, 1931, pt. 1, 278.
20. Northrop, J. H., *J. Gen. Physiol.*, 1931, **15**, 29.
21. Northrop, J. H., and Kunitz, M., *J. Gen. Physiol.*, 1930, **13**, 781.
22. Sörensen, S. P. L., Proteins, New York, Fleischmann Co., 1925. *Compt.-rend. trav. Lab. Carlsberg*, 1923, **15**, No. 11.
23. Long, J. H., and Johnson, W. A., *J. Am. Chem. Soc.*, 1913, **35**, 1188. Long, J. H., and Hull, M., *J. Am. Chem. Soc.*, 1916, **38**, 1620.
24. Waldschmidt-Leitz, E., and Purr, A., *Ber. chem. Ges.*, 1929, **62**, 2217.
25. Abderhalden, E., and von Ehrenwall, E., *Fermentforschung*, 1931, **12**, 411; 1931, **13**, 47.

# 38B


Reprinted from pages 295 and 303–311 of *J. Gen. Physiol.* **16**:295–311 (1932)

## CRYSTALLINE TRYPSIN

### II. General Properties

By JOHN H. NORTHROP and M. KUNITZ

(*From the Laboratories of The Rockefeller Institute for Medical Research, Princeton, N. J.*)

(Accepted for publication, June 22, 1932)

The isolation of a crystalline protein having constant properties, including powerful proteolytic activity, was described in the preceding paper (1). Experiments were also described which showed that destruction of the protein by any method tried resulted in a corresponding loss in proteolytic activity. The present paper describes some of the general properties of the preparation.

[*Editor's Note:* Material has been omitted at this point.]

*General Protein Tests*

A 0.5 per cent solution of the material gave positive tests with Biuret, Millon, xanthoproteic, and Folin's tyrosine reagent (12).

In 0.0005 per cent solution, which is approximately the concentration used for activity determinations, all the protein tests were negative. It is therefore perfectly possible to have active solutions or preparations of the enzyme which give no protein reaction, simply because they are too dilute.

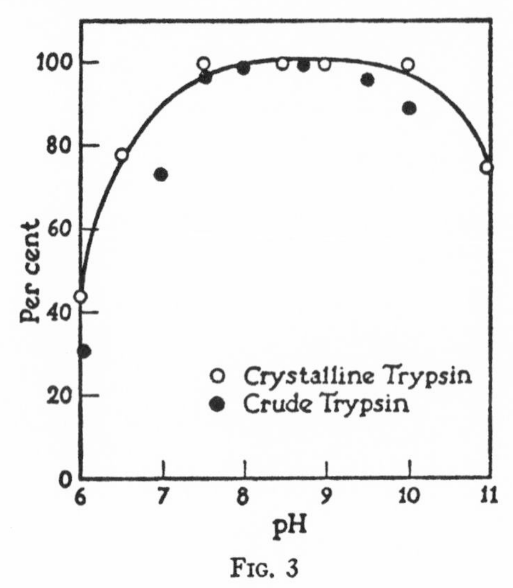


FIG. 3

*Number of Carboxyl Groups*

0.137 gm. of the protein dissolved in 5 ml. water (pH 8.0) required 7.0 ml. N/50 alkali for formol titration. This corresponds to 0.001 equivalents of amino (or carboxyl) groups per gm., or 34 per mole.

*Molecular Weight*

The molecular weight of the protein was determined by osmotic pressure measurements in $\frac{1}{4}$ saturated ammonium sulfate and pH 4.0 acetate buffer. The experiments were done with different concentrations of the protein and the pressure was read some time after it remained constant. The results of the experiment are shown in Table IV. The molecular weight is about 34,000.

### *Determination of the Diffusion Coefficient*[1]

The determination was carried out as described by Northrop and Anson (13) and the quantity of material diffusing determined both by total nitrogen and by proteolytic activity. In the case of Cell 2 less

TABLE IV

*Osmotic Pressure*

20 gm. crystalline trypsin filter cake dissolved in ¼ saturated ammonium sulfate pH 4.0 acetate buffer. Solution put in rocking osmometer in same solvent at 5°C. Osmotic pressure determined after 24 hours. Solution analyzed for protein nitrogen.

| | | | | | | |
|---|---|---|---|---|---|---|
| Mg. protein per ml. $= \frac{\text{mg. N}}{\text{ml.}} \times 6.5$ | 72 | 71 | 50 | 49 | 22.5 | 18 |
| Pressure, mm. Hg $= P$ | 39 | 38 | 22.0 | 21.5 | 11 | 9 |
| Molecular weight $\frac{278 \times 760 \times 22.4}{273\ P} \times$ gm. protein/l | 32,000 | 32,300 | 39,300 | 39,500 | 34,600 | 34,600 |

TABLE V

*Diffusion Constant of Crystalline Trypsin*

| | | | | |
|---|---|---|---|---|
| Cell No. | 2 | 4 | 2 | 4 |
| Cell constant (hemoglobin standard) | 0.054 | 0.0315 | 0.054 | 0.0315 |
| Solvent | 0.5 saturated magnesium sulfate N/10 acetate buffer pH 4.0 | | 0.5 saturated magnesium sulfate sulfuric acid pH 3.0 | |
| $\eta$ solvent 5°C. erg sec. cm.$^{-3}$ | 0.0303 | 0.0303 | 0.0362 | 0.0362 |
| Sp.g. solvent 5°C. gm. cm.$^{-3}$. | 1.115 | 1.115 | 1.145 | 1.145 |
| Average $D$ cm.$^2$ day$^{-1}$ | | | | |
| nitrogen | 0.0207 ±0.001 | | 0.0187 ±0.001 | |
| activity | 0.0218 ±0.001 | | 0.020 ±0.002 | |
| Radius trypsin molecules | 2.72 × 10$^{-7}$ cm. | | 2.5 × 10$^{-7}$ cm. | |

than 10 per cent of the dissolved protein diffused out during the course of the experiment, while with Cell 4 about 25 per cent diffused out. The results of the experiment are shown in Table V. The value for

[1] These experiments were carried out by Dr. Henry W. Scherp.

the diffusion coefficient is the same within the experimental error for both cells and for both methods of determination.

This is direct experimental evidence that the molecules responsible for the proteolytic activity are the same size as those which contain the protein nitrogen and that the enzyme molecule cannot be separated from the protein molecule by fractional diffusion. The average value of the diffusion coefficient is $0.020 \pm 0.001$ cm.[2] per day, corresponding to a molecular radius of about $2.6 \times 10^{-7}$ cm.

The volume of 1 mole of (hydrated) protein is therefore $(2.6 \times 10^{-7})^3 \times 4/3\pi \times 6.06 \times 10^{23} = 44{,}700$ cm.[3] From the osmotic pressure measurements the molecular weight of the anhydrous protein is 34,000. If the specific volume of anhydrous protein is assumed to be 0.75 (the value generally found for protein), this corresponds to a molecular volume of (anhydrous) protein of 26,200 ml. Therefore, each mole of protein carried with it about 18,500 gm. of water, which corresponds to about 0.53 gm. of water per gm. of anhydrous protein, and a molecular weight for the hydrated protein of 53,500.

### *Hydration from Viscosity Measurements*

Hydration of trypsin was determined by measuring the viscosity of various concentrations of crystalline trypsin in 0.5 saturated magnesium sulfate pH 4.0 at 5°C. The results showed that the trypsin is hydrated to the extent of 0.5 gm. of water per gm. of dry trypsin and is independent of the concentration of trypsin in the range of 4.0 to 0.8 gm. of protein per 100 ml. of solution. The hydration value was calculated from the specific viscosity by means of Kunitz's formula (14). The specific gravity of the anhydrous protein was assumed to be 1.33. The value obtained in this way agrees very closely with that calculated above from the diffusion and osmotic pressure measurement.

### *Isoelectric Point*

If a solution of the crystalline trypsin is added to boiling M/10 phosphate buffer of various pH, a precipitate is formed between pH 7.0 and 8.0. This precipitate undoubtedly consists of denatured protein and the relation of this zone of precipitation of the denatured protein to the isoelectric point of the native protein is therefore somewhat doubtful. Cataphoresis measurement showed that the isoelectric point of collodion particles suspended in dilute solutions of trypsin in different buffers is about pH 6.0 in M/50 acetate, about pH 7.0 in M/50 phosphate, and between pH 5.0 and 6.0 in the presence of M/50 ammonium sulfate. These results indicate that the isoelectric point is in the neighborhood of pH 7.0 but its exact position can only be determined by transport measurements. Apparently, however, it is more towards the acid side than that found for crude trypsin preparations by means of the distribution of the enzyme in gelatin particles (Northrop (16)) which indicated an isoelectric point at about pH 10.0. Willstätter (17) also concluded from the behavior with various adsorbents that the enzyme with which he was working had an isoelectric point quite

TABLE VI

*Inactivation of Trypsin Solutions at 30°C. and Various pH*

About 5 per cent solution of crystalline trypsin made up in ¼ saturated ammonium sulfate and increasing amounts of sulfuric acid. pH measured and activity determined after various time intervals at 35°C.

| Final concentration $H_2SO_4$ | N/4 | N/8 | N/16 | N/32 | N/64 | N/128 | N/256 | N/512 | N/1024 |
|---|---|---|---|---|---|---|---|---|---|
| pH (electrometric) | 1.5 | 1.8 | 2.16 | 2.42 | 2.64 | 2.81 | 2.95 | 3.0 | 3.12 |
| Time at 30°C. | Per cent of original activity after increasing time at 30°C. | | | | | | | | |
| *hrs.* | | | | | | | | | |
| 46 | 81 | 100 | 95 | 80 | 60 | 50 | 48 | 44 | 31 |
| 70 | 67 | 76 | 74 | 62 | 39 | 35 | 30 | 26 | 26 |

far on the alkaline side. The enzyme described in this paper, however, is undoubtedly different from that with which Willstätter's experiments were done.

## *The Effect of pH on the Stability of Trypsin*

The rate of inactivation of a solution of crystalline enzyme in ¼ saturated ammonium sulfate at 30°C. was determined. The activity determinations were made by the gelatin viscosity method. The results of the experiment are shown in Table VI. There is a rather sharp maximum for stability at about pH 2 under these conditions.

The results in general are quite different from those obtained by the writer (26) and by Pace (27) with crude trypsin preparations since the crude preparations were found to be most stable at about pH 6.0.

## DISCUSSION

The experiments described in these papers show that a crystalline protein may be isolated from pancreas which has constant physical and chemical properties including intense proteolytic activity. The protein has been studied under a variety of conditions which would be expected to show evidence of mixtures without causing any demonstrable change in its characteristic properties. If the material were other than a protein these experiments would justify the statement that it was a pure substance. Since it is a protein, however, it is quite possible that the material may be a solid solution, as in the case of proteins such solid solutions frequently exist and are extremely difficult to fractionate into their components. The problem is rendered unusually difficult in this case by the extremely unstable nature of the protein. It seems unlikely, however, that the material contains any non-protein molecular species. The constant composition under various conditions of fractionation precludes the possibility of an adsorption compound since it is characteristic of these compounds that their composition varies with the conditions of precipitation.

Even though the crystalline material is a mixture or solid solution and not a pure substance, there seems good reason to believe that the proteolytic activity and the protein properties are attributes of the same molecule. This conclusion is confirmed by a number of experiments in which it was found that any change in the protein properties caused a corresponding decrease in the activity of the solution. Denaturation of the protein by heat, hydrolysis by acid or pepsin or alkali all cause the concentration of native protein in the solution to decrease and this decrease is accompanied by a corresponding decrease in activity. In addition, the denatured, inactive, protein formed by heating the solution reverts to the native condition when the solution is cooled and at the same time the normal specific activity returns (18). In order to account for these results on the assumption that the activity is due to the presence of some non-protein molecule, it is necessary to assume that this hypothetical molecule cannot exist in the absence of the protein and also that it regains its activity under the same conditions as cause the denatured protein to return to the native form. In the absence of positive proof for the existence of such a hypothetical molecule these assumptions seem unlikely. So far as the writer is aware, there is no positive proof of the existence of such molecules and the assumption that they exist rests merely on the negative fact that most of the attempts to prepare pure substances, *i.e.* those with constant properties including enzymatic activity, have been unsuccessful. On the other hand it is, of course, impossible to disprove the existence of such molecules. Since nothing is known of the properties of these hypothetical active molecules it would be perfectly logical to assume that they are proteins themselves, especially since the general properties of enzymes such as inactivation by heat, adsorption on surfaces, and destruction by strong acid or alkali are in general those of proteins.

Active enzyme preparations have been obtained which contain very small amounts of protein; on the other hand extremely active preparations of urease, pepsin, and trypsin, and amylase have been obtained which are pure, or nearly pure proteins. If it be assumed that the activity of these protein preparations is due to the presence of some minute amount of a non-protein molecule, it is equally reasonable to assume that the activity of the non-protein preparation is due to the presence of a minute amount of protein.

The fact that in other cases the enzymatic activity may vary independently of the total protein content of the preparation proves only that some of the protein present is inactive but not that all of it is inactive.

Numerous experiments have been reported in the literature in which solutions of pepsin and other enzymes have been found to give negative protein tests although they are active. These experiments are also inconclusive since the activity test is far more delicate than the chemical test for proteins. For instance, a solution of crystalline trypsin or pepsin containing less than 1/1,000,000 of a gram of protein nitrogen per milliliter has an accurately measurable effect on the digestion of casein, while solutions of pepsin containing less than 1/10,000,000 of a gram of nitrogen have a very powerful effect on the coagulation of milk. Such solutions give negative results with protein tests but the dry material from which the solutions are made is practically pure protein. The minimum concentrations of these enzymes which can be detected are at least ten times less than the concentrations mentioned above and are of the same order of magnitude as the concentration of respiratory ferment in yeast as calculated by Warburg and Kubowitz (19).

It appears to the writers, therefore, that the assumption that enzymes are proteins is in the best accord with the facts up to the present time. Since these proteins possess characteristic enzymatic activity, in addition to the usual properties of proteins, they must possess some characteristic chemical structure which may or may not be an amino acid complex. The problem is the same as in the case of insulin (20). In general, most properties of molecules cannot be considered quantitatively as the sum of the properties of the various groups or atoms of which they are made, but must be considered as properties of the whole molecule. Thus, the optical activity, color, strength of acid groups, etc., of any one molecule depends qualitatively on the presence of a certain group or groups, but quantitatively, the property is affected by any change in the molecule. For instance, to possess optical activity a molecule must contain an asymmetric atom but its specific optical activity will change with any change in the molecule and it is impossible to isolate a group from the compound possessing the optical activity of the whole molecule. The same is true of the color of dyes to a more marked degree.

Hemoglobin presents perhaps the best example. This substance has the general properties of a protein but in addition possesses the remarkable property of combining reversibly with oxygen. It acts as a catalyst in certain oxidation reactions and might, therefore, be considered an enzyme. The property of combining reversibly with oxygen is assumed to be due to the presence of the iron-pyrrol group but denaturation of the protein, a reaction common to all proteins, destroys its power of combining reversibly with oxygen although the denatured protein still contains the prosthetic group.

Krebs (21) has shown that heme itself is a very poor catalyst but when combined with certain nitrogenous groups it forms hemochromogens, some of which are very effective catalysts. Thus the catalytic properties of hemoglobin and these related compounds are all due to the presence of the heme group, but this group when isolated has little or no catalytic activity and the catalytic power of the various heme compounds depends upon the substance with which the heme is combined. It is quite possible that the same general condition applies to other enzymes and that there are an indefinite number of closely related enzymes depending upon the compound with which the characteristic group is combined. This point of view does not differ very much from that developed by Willstätter (22) and his collaborators except that it regards the various active compounds as definite chemical individuals rather than as adsorption complexes of varying composition.

At the present time, however, there is no direct evidence of the existence of any peculiar prosthetic group not found in other proteins and it is quite possible that their activity depends on some peculiar arrangement of the amino acids, as Jensen (20) has suggested in the case of insulin.

### *Mechanism of the Catalytic Effect*

Sörensen (23) has shown that protein solutions in the presence of the solid phase are in true equilibrium and that the system as a whole is a two phase one as defined by the phase rule. The protein solution, therefore, consists of one phase. The solubility experiments with pepsin (24) and trypsin give the same result. These results show that the catalytic reactions caused by pepsin and trypsin in protein solutions are homogeneous rather than heterogeneous.

## SUMMARY

A method is described for isolating a crystalline protein of high tryptic activity from beef pancreas. The protein has constant proteolytic activity and optical activity under various conditions and no indication of further fractionation could be obtained. The loss in activity corresponds to the decrease in native protein when the protein is denatured by heat, digested by pepsin, or hydrolyzed in dilute alkali.

The enzyme digests casein, gelatin, edestin, and denatured hemoglobin, but not native hemoglobin. It accelerates the coagulation of blood but has little effect on the clotting of milk. It digests peptone prepared by the action of pepsin on casein, edestin or gelatin.

The extent of the digestion of gelatin caused by this enzyme is the same as that caused by crystalline pepsin and is approximately equivalent to tripling the number of carboxyl groups present in the solution.

The activity of the preparation is not increased by enterokinase.

The molecular weight by osmotic pressure measure is about 34,000.

The diffusion coefficient in $\frac{1}{2}$ saturated magnesium sulfate at 6°C. is $0.020 \pm 0.001$ cm.$^2$ per day, corresponding to a molecular radius of $2.6 \times 10^{-7}$ cm.

The isoelectric point is probably between pH 7.0 and pH 8.0.

The optimum pH for the digestion of casein is from 8.0–9.0.

The optimum stability is at pH 1.8.

REFERENCES


1. Northrop, J. H., and Kunitz, M., *J. Gen. Physiol.*, 1932, **16**, 267.
2. Willstätter, R., Waldschmidt-Leitz, E., Dunaiturria, S., and Künstner, G., quoted by Rona, P., Praktikum der physiologischen Chemie, Berlin, Julius Springer, 2nd edition, 1931, pt. 1, 311.
3. Schönfeld-Reiner, R., *Fermentforschung*, 1930, **12**, 67.
4. Northrop, J. H., *J. Gen. Physiol.*, 1932, **16**, 339.
5. Abderhalden, E., and Schwab, E., *Fermentforschung*, 1929, **11**, 92.
6. Waldschmidt-Leitz, E., and Simons, E., *Z. physiol. Chem.*, 1926, **156**, 99.
7. Abderhalden, E., *Fermentforschung*, 1931, **12**, 411; 1931, **13**, 47.
8. Northrop, J. H., and Kunitz, M., *J. Gen. Physiol.*, 1932, **16**, 314.
9. Quoted by Rona, P., Praktikum der physiologischen Chemie, Berlin, Julius Springer, 2nd edition, 1931, pt. 1, 278, 318.
10. Waldschmidt-Leitz, E., *Z. physiol. Chem.*, 1923–24. **132**, 181.
11. Northrop, J. H., *J. Gen. Physiol.*, 1922, **5**, 263.
12. Folin, O., and Ciocalteu, V., *J. Biol. Chem.*, 1927, **73**, 627.
13. Northrop, J. H., and Anson, M. L., *J. Gen. Physiol.*, 1929, **12**, 543.
14. Kunitz, M., *J. Gen. Physiol.*, 1926, **9**, 715.
15. Northrop, J. H., and Kunitz, M., *J. Gen. Physiol.*, 1925, **7**, 729.
16. Northrop, J. H., *J. Gen. Physiol.*, 1924, **6**, 337.
17. Willstätter, R., and Waldschmidt-Leitz, E., *Z. physiol. Chem.*, 1922–23, **125**, 139.
18. Northrop, J. H., *J. Gen. Physiol.*, 1932, **16**, 323.
19. Warburg, O., and Kubowitz, F., *Biochem. Z.*, 1928, **203**, 95.
20. Jensen, H., *Science*, 1932, **75**, 614.
21. Krebs, H. A., *Biochem. Z.*, 1928, **193**, 347.
22. Willstätter, R., *Naturwissenschaften*, 1927, **15**, 585.
23. Sörensen, S. P. L., Proteins, New York, Fleischmann Co., 1925.
24. Northrop, J. H., *J. Gen. Physiol.*, 1930, **13**, 739.
25. Levene, P. A., Simms, H. S., and Pfaltz, M. H., *J. Biol. Chem.*, 1924, **31**, 445.
26. Northrop, J. H., *J. Gen. Physiol.*, 1922, **4**, 261.
27. Pace, J., *Biochem. J.*, 1930, **24**, 605.

# 39


Reprinted from pages 433–440, 445–449, 450, 455, and 457–458 of *J. Gen. Physiol.* **18**:433–458 (1935)

# CRYSTALLINE CHYMO-TRYPSIN AND CHYMO-TRYPSINOGEN

## I. Isolation, Crystallization, and General Properties of a New Proteolytic Enzyme and Its Precursor*

By M. KUNITZ and JOHN H. NORTHROP

(*From the Laboratories of The Rockefeller Institute for Medical Research, Princeton, N. J.*)

(Received for publication, May 28, 1934)

Kühne (1) and Heidenhain (2) showed that extracts of fresh pancreas or freshly secreted pancreatic juice have no proteolytic activity. The preparations become active when mixed with the enterokinase of the small intestine, as found by Schepowalnikow (3) or when the pancreas is allowed to stand in slightly acid solution. The mechanism of this activation has been the subject of controversy for many years (4). Pavlov, Bayliss, Zunz, Wohlgemuth, Vernon, Delezenne, and others found the activation reaction to be catalytic and considered enterokinase to be an enzyme. Hamburger and Hekma, Dastre and Stassano and Waldschmidt-Leitz found the reaction to be stoichiometric and considered that the enterokinase formed an addition compound with the inactive zymogen. Vernon (5) found that activation could be caused by trypsin as well as by enterokinase but this was denied by Bayliss and Starling (6). The contradictory nature of the numerous experimental results indicates that there is more than one proteolytic enzyme in pancreatic extracts. Vernon showed (7) that the activity, as determined by the clotting of milk, could be partially separated from the proteolytic activity, as determined by protein hydrolysis, and concluded that there were at least two enzymes. He also showed that one of these was more stable than the other and that activation was caused by the less stable one.

The crystalline trypsin previously reported by the writers (8) was obtained from pancreas which had been allowed to activate spontaneously. The present experiments were carried out with fresh inactive pancreatic extracts in order to obtain the inactive form of the enzyme and to study the kinetics of activation.

* Kunitz, M., and Northrop, J. H., *Science*, 1933, **78**, 558.

In the course of these attempts to isolate the inactive precursor of crystalline trypsin a crystalline inactive protein was isolated from inactive (cattle) pancreatic extracts. This protein was called chymo-trypsinogen. It cannot be activated by enterokinase but is changed into an active proteolytic enzyme by crystalline trypsin. The new proteolytic enzyme formed in this way was also crystallized and was called chymo-trypsin. It differs from chymo-trypsinogen in crystalline form, optical activity, and number of amino groups, and it is more soluble and less stable. The molecular weight and molecular radius are about the same as the corresponding values for chymo-trypsinogen.

The new enzyme differs from the crystalline trypsin previously described in that it clots milk but does not clot blood and has a weaker action on protamines. It resembles crystalline trypsin in that it digests denatured proteins in slightly alkaline solution.[1]

The results agree, in general, with Vernon's experiments since they show that there are at least two proteolytic enzymes present in activated pancreatic extract, trypsin and chymo-trypsin. Fresh, inactive pancreatic extracts contain at least two zymogens, chymo-trypsinogen and trypsinogen. Enterokinase transforms trypsinogen into trypsin and this in turn transforms chymo-trypsinogen into chymo-trypsin.

Qualitatively this mechanism accounts for the peculiarly shaped curves frequently observed for the activation of crude pancreatic extract by enterokinase (5). The activation of chymo-trypsinogen by trypsin is a simple catalytic monomolecular reaction but when this is superimposed upon the primary activation of trypsinogen by enterokinase the combined result yields a complicated asymmetrical curve. The transformation of chymo-trypsinogen into chymo-trypsin is accompanied by a change in optical activity and a slight increase in amino nitrogen. There is no detectable non-protein nitrogen fraction formed nor is there any significant change in molecular weight. The reaction, therefore, is probably an internal rearrangement, possibly due to the splitting of a ring. It is possible, however, that a small part of the molecule containing no nitrogen is split off although there is no evidence for this at present.

[1] Waldschmidt-Leitz and Akabori (*Z. physiol. Chem.*, 1934, **228**, 224) have recently shown that pancreatic "proteinase" probably represents a mixture of trypsin and chymo-trypsin.

Both the active and inactive form of the enzyme may be recrystallized repeatedly without change of properties. Denaturation or hydrolysis of the protein results in a corresponding loss in activity. There is reason, therefore, to believe that the preparations represent pure proteins and that the proteolytic activity is a property of the protein molecule.

## EXPERIMENTAL RESULTS

### *Isolation and Crystallization of Chymo-Trypsinogen*

The material used in these experiments was cattle pancreas removed from the animal immediately after slaughter and immersed in cold N/4 sulfuric acid. Acid prevents spontaneous activation and removes practically all the potentially active material from the pancreas while most of the inert protein is precipitated. The acid extract obtained in this way (when brought to pH 7.0–8.0) is rapidly activated by enterokinase but cannot be activated by small amounts of trypsin. A protein was crystallized from this extract which could be activated either by enterokinase or trypsin; it was called chymo-trypsinogen. After repeated crystallization, however, the protein could not be activated by enterokinase but only by trypsin. The explanation of these apparently contradictory results is that the crude extract and the once crystallized protein contain trypsinogen and also some substance which inactivates small amounts of trypsin. When small amounts of trypsin are added to such preparations, therefore, no activation of chymo-trypsinogen occurs since the trypsin added is inactivated, but when enterokinase is added sufficient active trypsin is formed from the trypsinogen to overcome the inhibiting action of the solution and so activate the chymo-trypsinogen. The same result can be obtained by adding enough trypsin even in the presence of the inhibitor. The effect of repeated crystallization is simply to remove the last traces of these impurities, and the experiment is a good example of the efficiency of recrystallization as a method of purification.

The mechanism outlined above was confirmed by mixing pure crystalline chymo-trypsinogen with the mother liquor from the first crystallization and adding trypsin or enterokinase. The results of such an experiment are shown in Table I. As stated above, the recrystallized chymo-trypsinogen is activated only by trypsin while the mother liquor from the first crystallization, or the chymo-trypsinogen when mixed with the mother liquor, cannot be activated by small amounts of trypsin but can be activated by enterokinase. The inhibiting ef-

TABLE I

*Effect of Addition of Mother Liquor from Chymo-Trypsinogen Crystallization on the Activation of Chymo-Trypsinogen by Enterokinase and Trypsin*

| Quantity of material activated in 10 ml. M/25 phosphate buffer pH 7.6 | Chymo-trypsinogen 0.4 mg. protein nitrogen | | Mother liquor 0.16 mg. protein nitrogen | | Chymo-trypsinogen 0.4 mg. protein nitrogen + mother liquor 0.16 mg. protein nitrogen | |
|---|---|---|---|---|---|---|
| Activating agent | Trypsin | Enterokinase (20) | Trypsin | Enterokinase | Trypsin | Enterokinase |
| Quantity in 10 ml. | 0.002 mg. protein nitrogen | 2 ml. | 0.002 mg. protein nitrogen | 2 ml. | 0.002 mg. protein nitrogen | 2 ml. |
| $[T.U.]^{Hemoglobin}_{ml.}$ after 69 hrs. at 5°C. | 0.002 | <0.0001 | <0.0001 | 0.0015 | <0.0001 | 0.004 |

TABLE II

*Inhibiting Effect of Mother Liquor on Digestion of Hemoglobin by Chymo-Trypsin and Trypsin*

1 ml. { Crystalline trypsin solution containing 0.03 mg. protein nitrogen or Crystalline chymo-trypsin containing 0.07 mg. protein nitrogen } added to 4 ml. M/20 phosphate buffer pH 7.6 containing increasing amounts trypsinogen-mother liquor; activity of solution determined by hemoglobin method.

| Protein nitrogen in mother liquor, added, *mg.* | 0 | 0.04 | 0.08 |
|---|---|---|---|
| $[T. U.]^{Hb}_{ml.} \times 10^{-4}$ crystalline trypsin | 7.2 | 6.3 | 5.6 |
| $[T. U.]^{Hb}_{ml.} \times 10^{-4}$ chymo-trypsin | 8.8 | 8.8 | 8.8 |

fect of the mother liquor may be demonstrated directly by determining the effect upon the digestion of hemoglobin. The results of such an experiment are shown in Table II. The mother liquor inhibits the digestion of hemoglobin by crystalline trypsin but not the digestion of hemoglobin by chymo-trypsin.

### *Isolation of Chymo-Trypsinogen*

The method finally adopted for the preparation of chymo-trypsinogen is as follows.

The pancreas is removed from cattle immediately after slaughter and immersed in cold ¼ normal sulfuric acid. Fat and connective tissue are removed and the pancreas minced in a meat grinder, suspended in 2 volumes of ice cold N/4 sulfuric acid, and the suspension allowed to stand in the cold room at 5°C. overnight. It is then strained through gauze on a large Büchner funnel and the precipitate resuspended in an equal volume of N/4 sulfuric acid and refiltered. The combined filtrates and washings are brought to 0.4 saturated ammonium sulfate by the addition of solid ammonium sulfate and the suspension filtered through soft fluted paper (S. and S. No. 1450 ½) in the cold room. The filtrate is brought to 0.7 saturated ammonium sulfate and the suspension allowed to settle in the cold room for 48 hours. The supernatant fluid is decanted and the suspension filtered with suction. The filter cake is dissolved in 3 volumes of water[2] and 2 volumes saturated ammonium sulfate added. The suspension is filtered and the precipitate discarded. The filtrate is brought to 0.7 saturated ammonium sulfate by the addition of solid ammonium sulfate or an equal volume of saturated ammonium sulfate. The suspension is filtered with suction. The filter cake is dissolved in 1.5 volumes water and brought to ¼ saturated ammonium sulfate by the addition of saturated ammonium sulfate solution. The solution is adjusted to pH 5.0 (brick red color with methyl red on test plate) by the addition of 5 N sodium hydroxide. About 1.5 ml. per 100 ml. of solution is required. The solution is allowed to stand for 2 days at room temperature (about 20°C.). A heavy crop of crystals gradually forms. They are filtered with suction. The isolation of the chymo-trypsinogen is practically complete in one crystallization.

[2]The volume of the semi-dry filter cake is usually determined by weight. The specific volume of the filter cake is assumed, for convenience, to be equal to one. The expression "the filter cake is dissolved in *n* volumes of solvent" as used in the text, means that 1 gm. of filter cake is dissolved in *n* ml. of solvent.

### *Recrystallization*

The crystalline filter cake is suspended in 3 volumes of water and 5 N sulfuric acid added from a burette with stirring until the precipitate is dissolved. The solution is brought to $\frac{1}{4}$ saturated ammonium sulfate by the addition of 1 volume of saturated ammonium sulfate. An equivalent amount of 5 N sodium hydroxide is then added with stirring and the solution inoculated and allowed to stand at 20°C. Crystallization should be practically complete in an hour.

If the crystals are to be used for the preparation of active chymo-trypsin the crystallization should be repeated seven or eight times as otherwise difficulty is encountered in crystallizing the active enzyme.

TABLE III

*Fractional Crystallization of Chymo-Trypsinogen*

| Times crystallized | Optical activity in M/10 acetic acid, 25°C. $[\alpha]^{D}_{\text{mg. protein nitrogen}}$ | Specific activity after activation, 18 hrs., 6°C. in pH 8.0 M/50 phosphate buffer solution + 0.0007 mg. crystalline trypsin/ml. | | |
|---|---|---|---|---|
| | | Hemoglobin $[\text{T. U.}]^{\text{Hb}}_{\text{mg. protein nitrogen}}$ | Gelatin viscosity $[\text{T. U.}]^{\text{Gel. V}}_{\text{mg. protein nitrogen}}$ | Casein formol $[\text{T. U.}]^{\text{Cas. F}}_{\text{mg. protein nitrogen}}$ |
| 1 | 0.480 | 0.036 | 13.5 | 0.069 |
| 3 | 0.475 | 0.037 | 15.9 | 0.094 |
| 5 | 0.473 | 0.037 | 13.5 | 0.073 |
| 8 | 0.480 | 0.040 | | 0.070 |
| 10 | 0.482 | 0.038 | 13.9 | 0.056 |

The above outline describes the preparation from fresh inactive pancreas. The preparations vary somewhat and occasionally the first crystals retain a brown coloring matter. This coloring matter may be removed by the addition of 2 volumes of saturated ammonium sulfate to the acid solution of the crystals. An amorphous precipitate is formed which carries down with it the coloring matter and which may be removed by filtration. Most of the enzyme may be recovered from the precipitate by washing the precipitate on the filter paper with N/100 sulfuric acid.

The material may be isolated from frozen fresh inactive pancreas although this procedure is more troublesome. The gland must be frozen rapidly and immediately after removal. Ordinary commercial frozen pancreas is active and cannot be used. Crystals prepared from frozen pancreas frequently contain small amounts of foreign inert protein. This may be removed as follows: The crystals are dissolved in 3 volumes of water and sulfuric acid added, as described for re-crystallization, and the solution neutralized by the addition of an equivalent

FIG. 1. Chymo-trypsinogen.

amount of sodium hydroxide. A gelatinous precipitate appears and is filtered off. The filtrate is acidified again with a few drops of 5 N sulfuric acid, 1 volume saturated ammonium sulfate added, and the solution brought to pH 5.0 with sodium hydroxide and inoculated.

The preparation is conveniently carried out with about 10 fresh cattle pancreas. About 15 gm. of once crystallized filter cake is usually obtained from 10 pancreas.

The properties of the crystalline chymo-trypsinogen are constant through at least ten fractional recrystallizations as shown in Table III. The crystals of chymo-trypsinogen are shown in Fig. 1.

[*Editor's Note*: Material has been omitted at this point.]

### *Isolation and Crystallization of Chymo-Trypsin*

The final method adopted for the preparation of chymo-trypsin from chymo-trypsinogen is as follows.

The chymo-trypsinogen should be recrystallized eight times. 10 gm. of crystalline chymo-trypsinogen filter cake is suspended in 30 ml. water and dissolved by the addition of a few drops of 5 N sulfuric acid. 10 ml. M/2 pH 7.6 phosphate buffer is added and a quantity of molar sodium hydroxide equivalent to the acid is also added. About 0.5 mg. crystalline trypsin is added and the solution left at about 5°C. for 48 hours. Any active trypsin preparation (of equivalent activity) may be used instead of the crystalline trypsin. After 48 hours the solution is brought to pH 4.0 by the addition of about 5 ml. N/1 sulfuric acid, 25 gm. solid ammonium sulfate is added, and the precipitate filtered with suction.

### *Crystallization*

The filter cake is dissolved in 0.75 volumes N/100 sulfuric acid and filtered if the solution is not clear. The clear filtrate is inoculated and allowed to stand at 20°C. for 24 hours. About 5 gm. of crystalline filter cake should form.

### *Recrystallization*

The crystalline filter cake is dissolved in 1.5 volumes N/100 sulfuric acid; 1 volume of saturated ammonium sulfate is then added cautiously until crystallization commences. The solution is allowed to stand at room temperature and practically complete crystallization should take place.

A further crop of crystals may be obtained by precipitating the mother liquors with saturated ammonium sulfate and treating the precipitate obtained in this way as described under crystallization.

The optical activity and specific enzymatic activity of the chymo-trypsin remain constant through at least three fractional crystallizations as shown in Table VI. The chymo-trypsin crystals are shown in Fig. 5.

### *Change in Activity with Decrease in Native Protein*

When the chymo-trypsin protein is denatured in M/10 hydrochloric acid the decrease in activity is proportional to the decrease in native

FIG. 5. Chymo-trypsin.

protein concentration, as shown in Table VII. The per cent loss in activity under these conditions is the same when measured either by digestion of hemoglobin or by rennet activity (Table VIII). This indicates that the hemoglobin digestion and rennet action are due to

TABLE VI

*Fractional Crystallization of Chymo-Trypsin*

| No. of times crystallized | Optical activity M/10 acetic acid, 25°C. $[\alpha]^{D}_{mg.\ protein\ nitrogen}$ | Specific activity/mg. protein nitrogen | | | |
|---|---|---|---|---|---|
| | | Hemoglobin $[T.\ U.]^{Hb}_{mg.\ protein\ nitrogen}$ | Gelatin V. $[T.\ U.]^{Gel.\ V.}_{mg.\ protein\ nitrogen}$ | Casein S. $[T.\ U.]^{Cas.\ S.}_{mg.\ protein\ nitrogen}$ | Casein F. $[T.\ U.]^{Cas.\ F.}_{mg.\ protein\ nitrogen}$ |
| 1 | 0.386 | 0.039 | 11.3 | 0.98 | 0.077 |
| 2 | 0.416 | 0.037 | 12.0 | 1.05 | 0.079 |
| 3 | 0.380 | 0.038 | 10.7 | 1.01 | 0.073 |

TABLE VII

*Changes in Activity and Native Protein of Chymo-Trypsin Solutions in N/10 Hydrochloric Acid, 20°C.*

10 ml. chymo-trypsin solution (0.8 mg. protein nitrogen/ml.) + 10 ml. N/5 hydrochloric acid, 20°C. 2 ml. samples taken and added to 2.0 ml. N/10 sodium hydroxide, solution (No. 2). Activity by hemoglobin method.

Native protein nitrogen: 2 ml. (No. 2) + 2 ml. 2 M sodium chloride in M/200 hydrochloric acid. Precipitate = denatured protein; filter. Protein in filtrate determined by turbidity method.

| Time at 20°C., *hrs.* | 0 | 1.3 | 4 | 7 | 16 |
|---|---|---|---|---|---|
| $[T.\ U.]^{Hb}_{ml.}$ | 0.015 | 0.011 | 0.081 | 0.0052 | 0.0016 |
| Native protein N/ml., *mg.* | 0.390 | 0.256 | 0.172 | 0.144 | 0.031 |
| $[T.\ U.]^{Hb}_{mg.\ protein\ nitrogen}$ | 0.038 | 0.043 | 0.047 | 0.036 | 0.052 |

TABLE VIII

*Inactivation of Chymo-Trypsin Solution in M/10 Hydrochloric Acid 25°C. Measured by Hemoglobin Method and Rennet Action*

2 ml. chymo-trypsin solution (1 mg. protein nitrogen/ml.) + 8 ml. M/8 hydrochloric acid, 25°C. Analyzed for activity by hemoglobin and rennet methods.

| Time at 25°C., *min.* | | 0 | 20 | 60 | 240 |
|---|---|---|---|---|---|
| Per cent activity by | Hemoglobin | [100] | 78 | 55 | 33 |
| | Rennet | [100] | 79 | 60 | 30 |

the same molecule and confirms the results of the activation experiment described in Table V.

If the chymo-trypsin is heated to 100°C. in M/400 hydrochloric acid it is very rapidly and completely inactivated with the formation of denatured protein as shown by the fact that the protein is com-

TABLE IX

*Reversible and Irreversible Inactivation of Chymo-Trypsin at 100°C.*

10 ml. chymo-trypsin solution (0.33 mg. protein nitrogen/ml.) in M/400 hydrochloric acid; immersed in boiling water (No. 1).

Activity and native protein in hot solution: 1 ml. (No. 1) + 4 ml. hot M/400 hydrochloric acid, 2 ml. of this solution + 2 ml. 2 M sodium chloride (20°C.) filter, activity and protein nitrogen determined on filtrate.

Activity and native protein after reversal by cooling: 1 ml. (No. 1) + 4 ml. cold M/400 hydrochloric acid, 10 minutes 20°C. 2 ml. + 2 ml. 2 M sodium chloride. Activity and protein nitrogen determined on filtrate.

| Time at 100°C., *min* | Not heated | 1 | 5 | 15 | 30 |
|---|---|---|---|---|---|
| **Activity and native protein in hot solution** | | | | | |
| Activity $[T. U.]^{Hb}_{ml.}$ | 0.017 | 0 | 0 | | |
| Native protein nitrogen/ml., *mg* | 0.33 | 0 | 0 | | |
| Per cent total inactivation | 0 | 100 | 100 | | |
| **Activity and native protein after reversal by cooling** | | | | | |
| $[T. U.]^{Hb}_{ml.}$ | 0.017 | | 0.014 | 0.0084 | 0.0054 |
| $[T. U.]^{Rennet}_{ml.}$ | 1.90 | | 1.7 | 1.0 | 0.54 |
| Protein nitrogen/ml., *mg* | 0.33 | | 0.28 | 0.22 | 0.15 |
| $[T. U.]^{Hb}_{mg.\ protein\ nitrogen}$ | 0.052 | | 0.05 | 0.038 | 0.036 |
| $[T. U.]^{Rennet}_{mg.\ protein\ nitrogen}$ | 5.8 | | 6.1 | 4.6 | 3.6 |

pletely precipitated when the hot solution is poured into an equal volume of 2 M sodium chloride, and by the fact that the filtrate from the salt precipitate is completely inactive. However, if the heated solution is cooled and allowed to stand at 20°C. the solution recovers its original activity and the protein, like unheated chymo-trypsin is soluble in M/1 sodium chloride. Thus, the denaturation and inactivation of chymo-trypsin by heat is completely reversible as has already been shown to be true in the case of trypsin (9). If the native protein is assumed to be merely a carrier for an hypothetical "active group" it is necessary to assume that the active group becomes inactive when the protein is denatured and then becomes active again when the protein reverts to the native condition. On longer heating this reversibly inactivated and denatured form gradually changes to an irreversibly inactivated and denatured form which does not become active and salt-soluble again on cooling and standing at 20°C. (Table IX).

If chymo-trypsin solutions are allowed to stand at pH 9.0 and 37°C. there is a loss in protein nitrogen paralleled by the loss in activity

TABLE X

*Decrease in Protein Nitrogen and Activity in Chymo-Trypsin Solutions pH 9.0, 37°C.*

20 ml. chymo-trypsin solution (0.48 mg. protein nitrogen/ml.) + 20 ml. M/10 borate buffer pH 9.0, activity and protein nitrogen determined.

| Time at 37°C., *hrs.* | 0 | 1 | 2 | 19 |
|---|---|---|---|---|
| $[T. U.]^{Hb}_{ml.} \times 10^{-4}$ | 1.0 | 0.84 | 0.60 | 0.22 |
| Protein nitrogen/ml., *mg.* | 0.23 | 0.19 | 0.13 | 0.050 |
| $[T. U.]^{Hb}_{mg.\ protein\ nitrogen}$ | 0.043 | 0.044 | 0.046 | 0.044 |

(Table X). This reaction is probably analogous to the inactivation of trypsin in alkali (10) and is due to the formation and subsequent hydrolysis of denatured protein.

The connection between the protein and the activity may also be tested by pepsin digestion. The hydrolysis of the protein by pepsin is accompanied by a corresponding decrease in activity (Table XI).

[*Editor's Note*: Material has been omitted at this point.]

TABLE XI

*Decrease in Activity and Protein Nitrogen during Digestion of Chymo-Trypsin by Pepsin pH 2.0, 35°C.*

25 ml. crystalline chymo-trypsin solution in M/100 hydrochloric acid (0.1 mg. protein nitrogen/ml.) + 1 ml. crystalline pepsin solution (0.1 mg. protein nitrogen). Analyzed for protein nitrogen and activity by hemoglobin method.

| Time at 35°C., *hrs* | 0 | 0.5 | 1.0 | 2.0 |
|---|---|---|---|---|
| $[T. U.]^{Hb}_{ml.} \times 10^{-3}$ | 4.3 | 0.89 | 0.66 | 0.53 |
| Protein nitrogen/ml., *mg* | 0.09 | 0.025 | 0.015 | 0.011 |
| $[T. U.]^{Hb}_{mg.\ protein\ nitrogen}$ | 0.048 | 0.036 | 0.044 | 0.048 |

[*Editor's Note:* Material has been omitted at this point.]

TABLE XV

*Summary of the Properties of Chymo-Trypsinogen, Chymo-Trypsin, and Crystalline Trypsin*

| | | Chymo-trypsinogen | Chymo-trypsin | Trypsin |
|---|---|---|---|---|
| Crystalline form | | Long, square prisms | Rhombo-hedrons | Short prisms |
| Elementary analysis per cent dry weight | Carbon | 50.6 | 50.0 | 50.0 |
| | Hydrogen | 7.0 | 7.06 | 7.1 |
| | Nitrogen | 15.8 | 15.5 | 15.0 |
| | Chlorine | 0.17 | 0.16 | 2.85 |
| | Sulfur | 1.9 | 1.85 | 1.1 |
| | Phosphorus | 0 | 0 | 0 |
| | Ash | 0.1 | 0.12 | 1.0 |
| Amino nitrogen as per cent total nitrogen | By formol | 4.7 | 6.0 | 9.3 |
| | By Van Slyke | 4.75 | 6.0 | |
| Tyrosine + tryptophane equivalent milli-equivalents/mg. total nitrogen | | $2.5 \times 10^{-3}$ | $2.7 \times 10^{-3}$ | $3 \times 10^{-3}$ |
| Optical activity, 25°C. | | In M/10 acetic acid | | pH 4.0 in 0.25 sat. ammonium sulfate |
| [α]D line, per mg. nitrogen | | −0.48 | −0.40 | −0.27 |
| Solubility in distilled water | | Slight | Very soluble | Very soluble |
| | | In M/2 $K_2SO_4$, M/10 acetate pH 4.0 | | In 0.5 sat. $MgSO_4$ |
| Diffusion coefficient, 6°C. cm.$^2$/day | By nitrogen | 0.039 | 0.037 | 0.023 |
| | By hemoglobin | | 0.039 | |
| | By rennet | | 0.037 | |
| Molecular volume from diffusion coefficient, cm.$^3$/mole | | 52,000 | 52,000 | 65,000 |
| Molecular weight from osmotic pressure | | 36,000 (32,000) | 41,000 | 36,500 |
| Hydration, gm. water/gm. protein, from osmotic pressure and diffusion coefficient | | 0.7 | 0.5 | 0.8 |
| By viscosity | | 0 | 0.1 | 0.6 |
| Isoelectric point from cataphoresis of collodion particles | | 5.0 | 5.4 | 7–8 |
| Specific activity [T. U.] per mg. protein nitrogen | Substrate | | | |
| | Hemoglobin | $<1 \times 10^{-5}$ | 0.04 | 0.17 |
| | Casein, sol. | $<1 \times 10^{-5}$ | 1.0 | 2.4 |
| | Casein, F. | <0.01 | 0.08 | 0.18 |
| | Gelatin V. | <0.1 | 12.0 | 100 |
| | Rennet | <0.01 | 8.5 | <0.1 |
| | Clot blood | <2.0 | <2.0 | 1500 |
| | Sturin F. | $<2 \times 10^{-4}$ | 0.018 | 0.63 |
| pH optimum for digestion casein | | | 8–9 | 8–9 |
| Total digestion casein, ml. M/50 sodium hydroxide/5 ml. 5 per cent casein | | | 17 | 9–11 |

[*Editor's Note*: Material has been omitted at this point.]


SUMMARY

A new crystalline protein, chymo-trypsinogen, has been isolated from acid extracts of fresh cattle pancreas. This protein is not an enzyme but is transformed by minute amounts of trypsin into an active proteolytic enzyme called chymo-trypsin. The chymo-trypsin has also been obtained in crystalline form.

The chymo-trypsinogen cannot be activated by enterokinase, pepsin, inactive trypsin, or calcium chloride. There is an extremely slow spontaneous activation upon standing in solution.

The activation of chymo-trypsinogen by trypsin follows the course of a monomolecular reaction the velocity constant of which is proportional to the trypsin concentration and independent of the chymo-trypsinogen concentration. The rate of activation is a maximum at pH 7.0–8.0. Activation is accompanied by an increase of six primary amino groups per mole but no split products could be found, indicating that the activation consists in an intramolecular rearrangement. There is a slight change in optical activity but no change in molecular weight.

The physical and chemical properties of both proteins are constant through a series of fractional crystallizations.

The activity of chymo-trypsin decreases in proportion to the destruction of the native protein by pepsin digestion or denaturation by heat or acid.

Chymo-trypsin has powerful milk-clotting power but does not clot blood plasma and differs qualitatively in this respect from the crystalline trypsin previously reported. It hydrolyzes sturin, casein, gelatin, and hemoglobin more slowly than does crystalline trypsin but the hydrolysis of casein is carried much further. The hydrolysis takes place at different linkages from those attacked by trypsin. The optimum pH for the digestion of casein is about 8.0–9.0. It does not hydrolyze any of a series of dipeptides or polypeptides tested.

Several chemical and physical properties of both proteins have been determined.

REFERENCES

1. Kühne, W., *Virchows Arch. path. Anat.*, 1867, **39**, **130**.
2. Heidenhain, R., *Arch. ges. Physiol.*, 1874, **10**, 557.
3. Schepowalnikow, N. P., *Jahresber. Fortschr. Tierchem.*, 1900, **29**, 378.
4. For review of the literature on this subject see Loewenthal, E., in Oppenheimer, C., Fermente und ihre Wirkungen, Leipsic, Thieme, 1926, **2**, 916.
5. Vernon, H. M., *J. Physiol.*, 1901, **27**, 269; 1913–14, **47**, 325.
6. Bayliss, W. M., and Starling, E. H., *J. Physiol.*, 1904, **30**, 61.
7. Vernon, H. M., *J. Physiol.*, 1902, **28**, 448.
8. Northrop, J. H., and Kunitz, M., *J. Gen. Physiol.*, 1932, **16**, 267.
9. Northrop, J. H., *J. Gen. Physiol.*, 1932, **16**, 323. Anson, M. L., and Mirsky, A. E., *J. Gen. Physiol.*, 1934, **17**, 393.
10. Kunitz, M., and Northrop, J. H., *J. Gen. Physiol.*, 1934, **17**, 591.
11. Waldschmidt-Leitz, E., *Z. physiol. Chem.*, 1933, **222**, 148.
12. Waldschmidt-Leitz, E., and Kollmann, T., *Z. physiol. Chem.*, 1927, **166**, 262.
13. Northrop, J. H., and Anson, M. L., *J. Gen. Physiol.*, 1929, **12**, 543. The cells were standardized against 0.23 M NaCl at 5°C., using Ohölm's value of 0.74 $cm.^2$/day for the diffusion coefficient.
14. Willstätter, R., Waldschmidt-Leitz, E., and Hesse, A. R., *Z. physiol. Chem.*, 1923, **126**, 155.
15. Willstätter, R., and Memmen, F., *Z. physiol. Chem.*, 1923, **129**, 1.
16. Rona, P., Praktikum der physiologischen Chemie, Berlin, Julius Springer, 2nd edition, 1931, pt. **1**, 389.
17. Northrop, J. H., and Kunitz, M., *J. Gen. Physiol.*, 1932, **16**, 313.
18. Anson, M. L., and Mirsky, A. E., *J. Gen. Physiol.*, 1933, **17**, 151.
19. Kunitz, M., *J. Gen. Physiol.*, 1935, **18**, 459.
20. Waldschmidt-Leitz, E., *cf.* Rona, P., (16), p. 318.
21. Northrop, J. H., and Kunitz, M., *J. Gen. Physiol.*, 1925, **7**, 729. The cell used is manufactured by A. H. Thomas and Co., Philadelphia. Loeb, J., Proteins and the theory of colloidal behavior, The international chemical series, New York, McGraw-Hill Book Co., Inc., 2nd revised edition, 1924, 323.

Reprinted from *J. Gen. Physiol.* **19**:991–1007 (1936)

# ISOLATION FROM BEEF PANCREAS OF CRYSTALLINE TRYPSINOGEN, TRYPSIN, A TRYPSIN INHIBITOR, AND AN INHIBITOR-TRYPSIN COMPOUND

BY M. KUNITZ AND JOHN H. NORTHROP

(*From the Laboratories of The Rockefeller Institute for Medical Research, Princeton, N. J.*)

(Accepted for publication, November 15, 1935)

The isolation of trypsin from active pancreatic extract (1) and of chymo-trypsinogen and chymo-trypsin (2)[1] from fresh inactive pancreas has been described in previous papers. Crystalline trypsinogen has also been obtained (3) from inactive cattle pancreas and transformed into active trypsin which may then be crystallized much more readily than by the earlier method. During the course of this work a polypeptide, which has a powerful inhibiting effect on trypsin, as well as a compound of this substance with trypsin, was also obtained in crystalline form (4). The present paper contains detailed descriptions of methods of preparing these substances and a brief description of their properties.

## GENERAL PROPERTIES

### *Trypsinogen*

Trypsinogen is obtained as small triangular prisms (Fig. 1). When these crystals are dissolved in neutral solution the trypsinogen is rapidly transformed into active trypsin and it has, therefore, been impossible so far to recrystallize trypsinogen. The original crystalliza-tion occurs without activation owing to the presence of the inhibitor and if inhibitor is added to a solution of trypsinogen recrystallization may be carried out without activation. Numerous attempts have been made to recrystallize inhibitor-free trypsinogen under conditions

[1] It has recently been found that poorly formed needle-shaped protein crystals, which have about the same activity as the usual chymo-trypsin crystals, may be obtained from the mother liquor of the chymo-trypsin crystallization. The properties of these new crystals are now being investigated.

which would not at the same time cause activation, but so far without success. Analyses and properties of this substance are therefore somewhat uncertain.

The transformation of trypsinogen into trypsin is accelerated by the addition of trypsin or enterokinase or concentrated solutions of magnesium sulfate or ammonium sulfate (5). The addition of inhibitor retards activation by all three methods and a large quantity of inhibitor will completely prevent activation. A solution of trypsino-

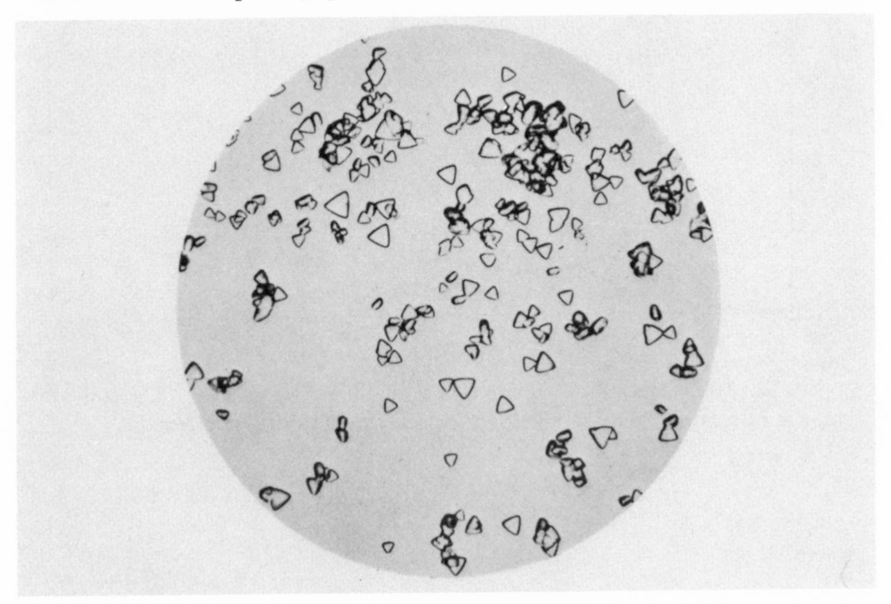

FIG. 1. Trypsinogen crystals.

gen to which inhibitor has been added behaves, therefore, just as the crude trypsinogen solution previously described (2). The fact that activation is accelerated by the addition of trypsin indicates that activation is autocatalytic and this is borne out by the kinetics of the reaction as shown in Fig. 2. Under these conditions the reaction follows quite closely that of a simple autocatalytic reaction. The rate of activation depends on the pH and is maximum at pH 7.0–8.0. It follows from this that if trypsinogen could be prepared completely free from active trypsin it would remain inactive. Owing to the extremely minute amounts of trypsin required to activate, however,

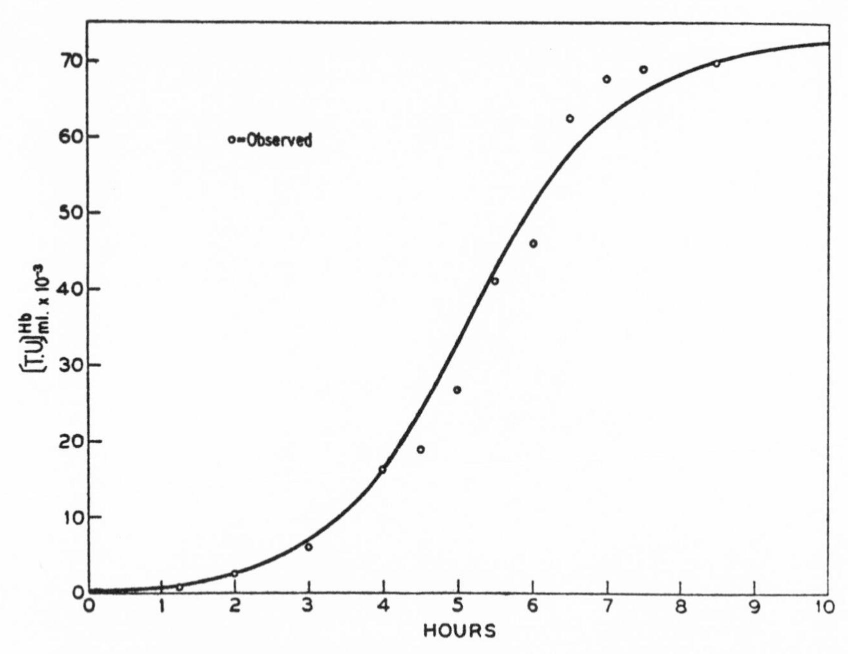


FIG. 2. Autocatalytic activation of crystalline trypsinogen. The smooth curve is calculated from the equation for a simple autocatalytic reaction

$$KA_e t = 2.3 \log \frac{A(A_e - A_0)}{A_0(A_e - A)}$$

where

$$K = 14.6$$

$$A_e = 0.072 \text{ (T. U.)}^{Hb}$$

$$A_0 = 0.0003 \text{ (T. U.)}^{Hb}$$

Crystalline trypsinogen—10 gm. crystalline filter cake dissolved in 200 ml. N/400 hydrochloric acid, 200 ml. 5 per cent trichloracetic acid added. Left at 20°C. for ½ hour. Filtered. Precipitate dissolved in 25 times its weight of N/50 hydrochloric acid. Ammonium sulfate added to 0.4 saturated ammonium sulfate. Filtered. Filter cake dissolved in 3 times its weight of N/200 hydrochloric acid and dialyzed against N/200 hydrochloric acid at 6°C. 2 ml. dialyzed solution plus 2 ml. M/1 pH 5.0 acetate buffer plus 16 ml. water at 8°C. 1 ml. samples removed at intervals and added to 9 ml. M/75 hydrochloric acid. Activity determined by the hemoglobin method.

it has so far been impossible to obtain such preparations. Activation by the addition of enterokinase or concentrated magnesium or ammonium sulfate is also autocatalytic in type (5). The mechanism whereby the activation is accelerated by kinase and neutral salts is still uncertain.

## *Trypsin*

Crystalline trypsin obtained from inactive pancreatic extract (Fig. 3) has, as a rule, slightly lower activity than that obtained from

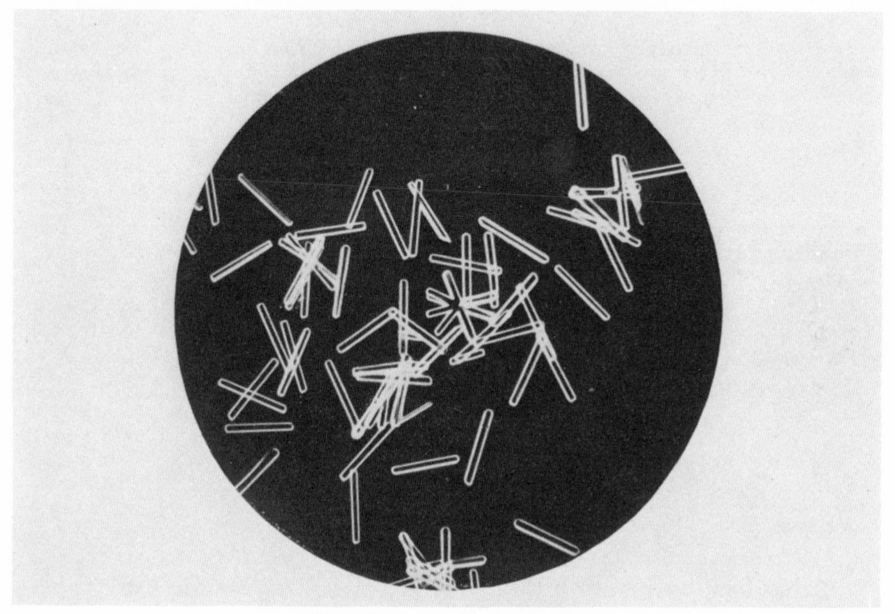

FIG. 3. Crystalline trypsin.

active extracts. This is due partly to the presence of small amounts of inactivated trypsinogen and partly to the presence of some inhibitor. The activity may be increased to the maximum value by repeated crystallization or more rapidly by precipitation with trichloracetic acid. Either process results in trypsin having the same activity as that prepared from active pancreatic extract (see Table I). The composition, molecular weight, specific activity, etc. on various substrates are the same as previously described for trypsin from active pancreatic extract. These values are summarized in Table II.

The trypsin described above appears to be identical with the "proteinase" recently described by Waldschmidt-Leitz and Akabori (6). Proteinase was originally described (6) as having a lower activity on clupean than crystalline trypsin but according to Holter and the writers (7) both enzymes hydrolyze clupean at the same rate and to the same extent. This result has also been recently obtained by Professor Waldschmidt-Leitz (personal communication from Professor Waldschmidt-Leitz).

TABLE I

*Activity of Various Preparations of Crystalline Trypsin*

| Preparation | [T. U.]$_{mg. N}$ determined by | | |
|---|---|---|---|
| | Gelatin viscosity pH 4.0 | Hemoglobin | Casein formol |
| No. 50 prepared from crystalline trypsinogen—1 × crystallized | 77 | 0.10 | 0.15 |
| 3 × " | 73 | 0.12 | 0.15 |
| 5 × " | 75 | 0.12 | 0.14 |
| 8 × " | 90 | 0.14 | 0.15 |
| 3 × crystallized, precipitated + trichloracetic acid (See Par. I-5) | | 0.16 | |
| Crystals prepared from active pancreatic juice—1 × crystallized | 98 | 0.17 | |
| 3 × " | 120 | 0.15 | |
| 5 × " | 90<br>100 | 0.16 | 0.16 |

### *Inhibitor*

It has long been known that pancreatic extracts contain some substance which inhibits trypsin and this was clearly brought out by Willstätter and Rhodewald (8). It has been suggested by Dyckerhoff, Miehler, and Tadsen (9) that trypsin exists in fresh pancreas as a compound with the inhibitor so that the activation of this inactive compound consists merely in the removal of the inhibitor and that no special inactive form of trypsin; *i.e.*, trypsinogen, exists. The actual isolation of trypsinogen, as described in the present paper, shows that this explanation is incomplete. The inhibitor, however, does play a very important part in regulating the activation of trypsinogen and there is no doubt that in partly activated pancreatic extracts more or less active trypsin occurs in the form of an inactive compound with

TABLE II

*Analysis and General Properties*

| | Tryp-sinogen | Trypsin | Inhibitor-trypsin compound | | Inhibitor |
|---|---|---|---|---|---|
| | | | Natural | Synthetic | |
| Elementary analysis per cent dry weight, calculated ash-free* — C | 50.1 | 50.2 | 48.0 | | 38.6 |
| H | 6.9 | 6.6 | 6.9 | | 7.6 |
| N | 15.3 | 16.13 | 15.4 | | 11.25 |
| S | 1.1 | 1.1 | | | |
| P | — | — | — | — | — |
| Protein nitrogen as per cent of total nitrogen | | 93 | 76 | 73 | 0 |
| Amino nitrogen as per cent total nitrogen, formol | | | | | 6.5 |
| Van Slyke | | | | | 6.5 |
| Amino nitrogen as per cent total nitrogen after acid hydrolysis | | | | | 80–90 |
| Tyrosine + tryptophane: milli-equivalents tyrosine per mg. total nitrogen | | $4.3 \times 10^{-3}$ | $3.7 \times 10^{-3}$ | $3.5 \times 10^{-3}$ | $2.1 \times 10^{-3}$ |
| Optical rotation $[\alpha]^D$ per mg. nitrogen | | −0.26° | −0.33° | −0.33° | −0.65° |
| | | M/10 pH 5.5 acetate | | | In distilled water |
| Molecular weight by osmotic pressure in 0.25 saturated magnesium sulfate 6°C. | | 36,500 | 40,000 | | 6,000 |
| Specific activity [T. U.] per mg. protein nitrogen† — *Substrate* | | | | | |
| Hemoglobin | — | 0.17 | | | |
| Casein S | — | 2.4 | | | |
| Casein F | — | 0.16 | | | |
| Gelatin V | — | 100. | | | |
| Rennet | — | <0.1 | | | |
| Clot blood | — | 1500. | | | |
| Sturin F | — | 0.6 | | | |
| Clupein F | — | 0.8 | | | |
| Clupean F | — | + | | | |
| Total digestion casein pH 8.0 ml. M/50 sodium hydroxide per 5 ml. 5 per cent casein | | 9–11 | | | |

* The trypsinogen and trypsin were prepared for analysis by precipitation with trichloracetic acid followed by washing with trichloracetic acid, distilled water, and alcohol and then dried with ether. The crystals of inhibitor-trypsin compound were washed with distilled water and alcohol and dried with ether. The inhibitor was dried in the form of filter cake with saturated magnesium sulfate. The analytical results are therefore of somewhat uncertain significance since they were not obtained upon the unchanged active preparations.

The elementary analyses were carried out by Dr. Elek in Dr. P. A. Levene's laboratory.

† Activity determinations were carried out as previously described (12). The sturin formol and clupein formol figures were obtained in exactly the same way as those for casein formol activity except that 5 cc. 5 per cent clupein or sturin solutions were used.

inhibitor. The inhibitor (Fig. 4) has the general properties of a polypeptide. It gives a faint biuret test and is precipitated by saturated magnesium sulfate or 0.7 saturated ammonium sulfate, but is not precipitated nor changed by 2.5 per cent trichloracetic acid, either hot or cold, nor by boiling. It diffuses slowly through a collodion membrane and has a molecular weight, by osmotic pressure, of about 6,000. The carbon and nitrogen content (Table II) appear to be lower than usual for proteins. The amino nitrogen content is

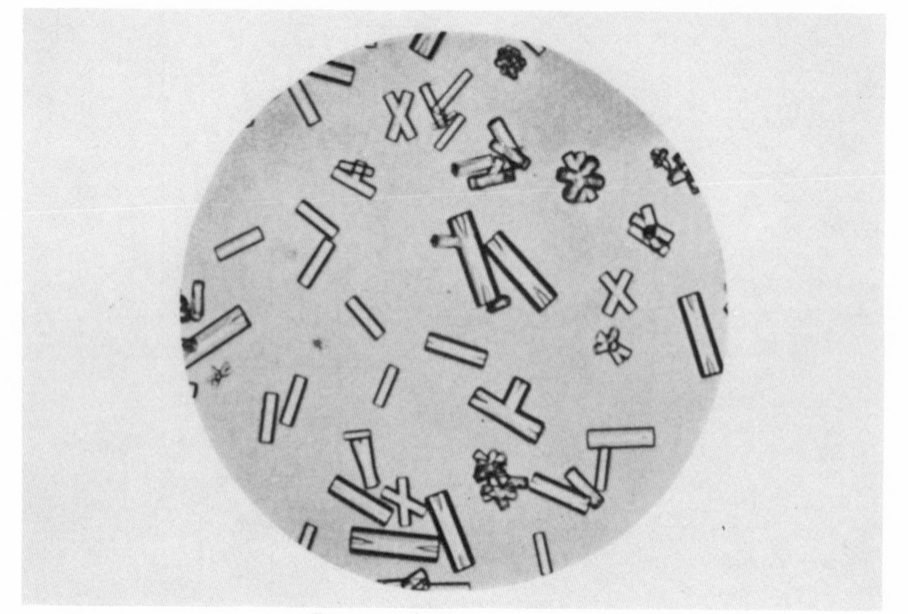

FIG. 4. Inhibitor crystals.

low but after acid hydrolysis amounts to 80–90 per cent of the total nitrogen so there is reason to believe that the substance is made up largely of amino acids.

*Reaction between Inhibitor and Trypsin*

When a solution of the inhibitor is mixed with a solution of trypsin of equal molecular strength at pH 7.0 the activity of the mixture decreases rapidly with time and after about ½ hour at 6°C. it is completely inactive as measured by the digestion of hemoglobin. If the solution is allowed to stand in the pH range of 7.0–3.0 it remains inactive upon addition to hemoglobin digestive mixture (10), but if titrated to pH 1.0 before addition to hemoglobin the activity rapidly reappears and in about ½ hour will have completely returned. The cycle may be repeated indefinitely. The inhibitor evidently reacts with trypsin to form an addition compound which dissociates in acid

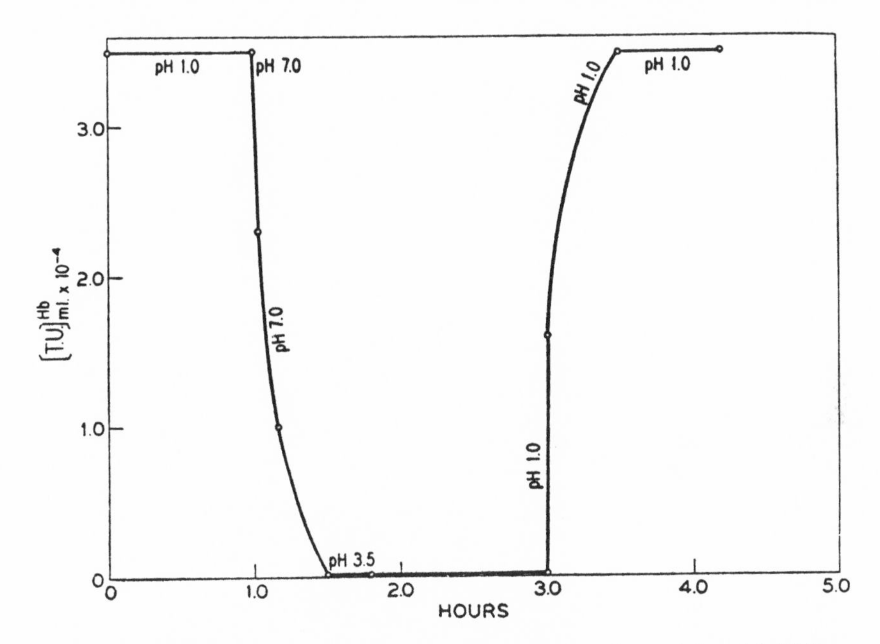


FIG. 5. Effect of standing at 6°C. and various pH on the activity of inhibitor-trypsin compound.

Solution of inhibitor-trypsin compound pH 5.0. 1.3 mg. P. N./ml. diluted 1/500 with 0.0025 N hydrochloric acid. pH changed as indicated. 1 ml. samples taken at various time intervals, added immediately to standard hemoglobin solution, and the amount of digestion determined after 10 minutes.

solution. Both dissociation and combination require measurable time intervals so that the reaction does not appear to be ionic. An experiment illustrating this inactivation and reactivation is shown in Fig. 5. This inactivating effect is also apparent when the activity is measured by the digestion of casein, clotting of blood, digestion of sturin, or by the activation of chymo-trypsinogen or trypsinogen in the presence of salt. The substance also inhibits chymo-trypsin but to a less marked extent.

### *Inhibitor-Trypsin Compound*

This substance is obtained in the form of hexagonal, many-faced crystals (Fig. 6). It consists of one molecule of inhibitor combined with one molecule of trypsin (Table III). These may be separated

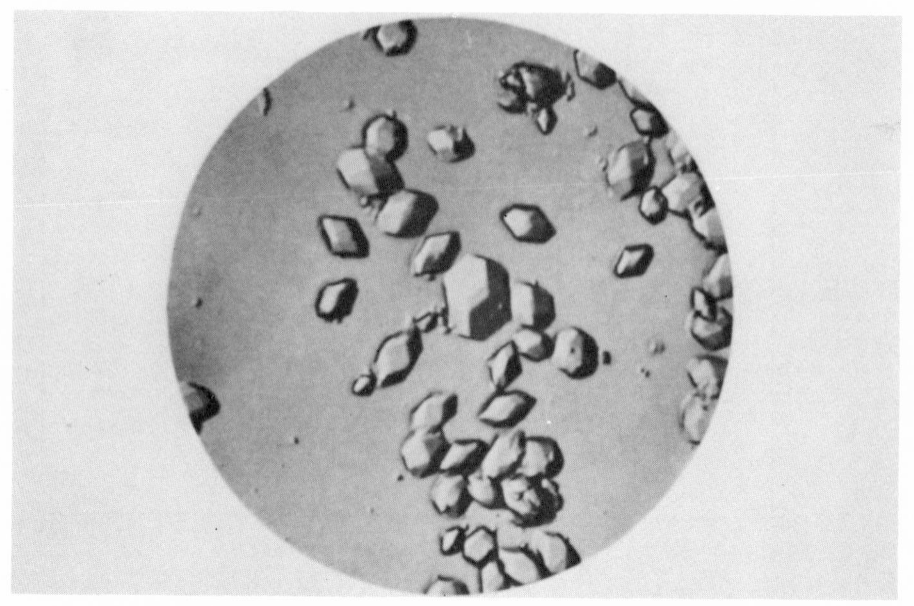

FIG. 6. Crystals of inhibitor-trypsin compound.

by precipitation with trichloracetic acid which precipitates the trypsin and leaves the inhibitor in solution. As described under "inhibitor," the compound when dissolved in acid solution and added to protein solutions possesses the full tryptic activity but if allowed to stand for a short time at pH 7.0 and then added to the protein solution it is completely inactive. It differs in one marked respect from trypsin in that it is not adsorbed by egg albumin under the conditions described by Waldschmidt-Leitz (11), whereas trypsin itself is adsorbed under these conditions. The molecular weight by osmotic pressure is 40,000.

TABLE III

*Composition of Inhibitor-Trypsin Compound (Corrected for Ash)*

| | Trypsin | Inhibitor | Inhibitor-trypsin compound | |
|---|---|---|---|---|
| | Observed | Observed | Observed | Calculated, assuming 1 molecule trypsin (36,500) + 1 molecule inhibitor (6,000) |
| C | 50.2 | 38.6 | 48.0 | 48.6 |
| H | 6.60 | 7.62 | 6.90 | 6.70 |
| N | 16.13 | 11.25 | 15.40 | 15.40 |
| O | 27.1 | 42.50 | 29.70 | 29.30 |

*Molecular weights*

| | Observed | Observed | Observed |
|---|---|---|---|
| | 36,500 ± 2,000 | 6,500 ± 1,000<br>Calculated* from per cent composition of trypsin, inhibitor, and inhibitor-trypsin compound, assuming molecular weight of trypsin 36,500<br>X<br>C 8,500<br>H (15,000)<br>N 6,400<br>O 7,400 | 40,000<br>Calculated from molecular weight trypsin + inhibitor<br>42,500 |

* $X$ = gm. inhibitor in 1 gm. mol inhibitor-trypsin compound

$C_c$ = per cent carbon in compound

$C_t$ = " " " " trypsin

$C_i$ = " " " " inhibitor

$$\text{Then } X = \frac{36{,}500}{100}\,\frac{C_t - C_c}{C_c - C_i}$$

$X$ is calculated from the percentages of the other elements in the same way as shown above for carbon.

## EXPERIMENTAL METHODS

The methods of preparing these compounds are given in the following sections. The yields reported represent average figures and may vary considerably in individual preparations but the proportion of precipitate to solvent specified in the text must be accurately adhered to. Weights of precipitates refer to the weight of the filter cake that is removed from the Büchner funnel. It is essential for the success of the preparation that these filtrations be as complete as possible. Large Büchner funnels must be used, the filter cake pressed with a spatula so as to fill all the cracks, and the filtration continued until little or no foam is drawn through the funnel. As a rule, the dry precipitate should form a layer not more than 2–3 mm. thick. The preparations are all quite stable in the form of a moist filter cake if kept in the ice box. Permanent dry preparations may be obtained by allowing the filter cake from saturated magnesium sulfate to stand near the coils of a mechanical refrigerator. Under these conditions the water evaporates and a mixture of the dry crystals with the anhydrous salt remains. Such dry preparations keep indefinitely.

pH determinations were made on a test plate by mixing 1 drop of the Clark indicator with 1 drop of the solution. Standards were prepared by mixing 1 drop of the indicator with 1 drop of the standard buffer. This method, of course, gives only apparent pH values which may be considerably removed from the true pH of the solution. The method, however, is perfectly adequate for reproducing the necessary conditions.

The saturated magnesium sulfate and ammonium sulfate solutions were prepared at 20°C.

The method of preparation varies somewhat depending upon the desired product. The first method describes a complete fractionation whereby all of the substances may be obtained starting from fresh inactive cattle pancreas. This is the best method for obtaining chymo-trypsinogen and fair yields of trypsinogen but does not always give good yields of inhibitor-trypsin compound. The second method is the most convenient and reproducible method for obtaining trypsin and inhibitor-trypsin compound but yields no trypsinogen. The third method is the most efficient for the preparation of inactive trypsinogen. The fourth method describes the conversion of trypsinogen into active trypsin and the fifth method describes the isolation and crystallization of trypsin from active pancreas.

### *I. Isolation of Crystalline Chymo-Trypsinogen, Trypsinogen, and Trypsin Inhibitor from Fresh Cattle Pancreas*

#### *1. Preliminary Purification and Concentration*[2]

Remove pancreas from cattle immediately after slaughter and immerse at once in enough ice cold 0.25 N sulfuric acid to cover the glands. Remove fat and connective tissue and mince in a meat chopper within a few hours. Suspend 3 liters

[2] The method is the same as that already described for the preparation of chymo-trypsinogen (2) up to the crystallization of this compound.

of minced pancreas in 6 liters of 0.25 N sulfuric acid at 5°C. and allow suspension to stand at about 5°C. for 18–24 hours. Strain the suspension through gauze, resuspend the residue in an equal volume of cold 0.25 N sulfuric acid, and strain through gauze immediately. Reject residue. Dissolve 242 gm. of solid ammonium sulfate in each liter of combined filtrate and washings. Filter through fluted paper (S. and S. No. 1450 ½) in cold room. Reject precipitate. Dissolve 205 gm. of solid ammonium sulfate in each liter of filtrate. Heavy precipitate forms. Remove foam and allow to settle for 2 days at 5°C. Decant supernatant solution and filter residue with suction through hardened paper (S. and S. No. 575) on a large funnel. Yield about 100 gm. precipitate Reject filtrate. Dissolve each 100 gm. precipitate in 300 ml. water, add 200 ml. saturated ammonium sulfate, stir in 5 gm. Standard Super Cel (Celite Corporation)[3] and filter *with suction* through soft paper (S. and S. No. 1450 ½). Reject precipitate. Add slowly 205 gm. solid ammonium sulfate to each liter of filtrate and filter with suction through hardened paper (S. and S. No. 575). *Precipitate A* about 90 gm. (mixture of crude chymo-trypsinogen, trypsinogen, and inhibitor). Reject filtrate.

### *2. Crystallization of Chymo-Trypsinogen*

Dissolve each 90 gm. of precipitate A in 135 ml. water, add 45 ml. saturated ammonium sulfate, then adjust to pH 5.0 (brick red color with 0.01 per cent methyl red solution on test plate) by addition drop by drop of about 2 ml. 5 N sodium hydroxide. Allow to stand for 2 days at 20–25°C. A heavy crop of chymo-trypsinogen crystals gradually forms. Filter with suction through hardened paper (*Filtrate Tg*). Wash crystalline filter cake with 0.25 saturated ammonium sulfate and finally with saturated ammonium sulfate, and store at 5°C. Yield about 25 gm. For recrystallization and activation see (2).

### *3. Crystallization of Trypsinogen*

Adjust filtrate and washings from chymo-trypsinogen crystallization (Filtrate Tg) to pH 3.0 (pink with 0.01 per cent methyl orange on test plate) with about 1 ml. 5 N sulfuric acid per 100 ml. filtrate. Dissolve 30.4 gm. of solid ammonium sulfate in each 100 ml. of filtrate and filter with suction through hardened paper. Reject filtrate. Dissolve precipitate (40 gm.) in 120 ml. water, add 80 ml. saturated ammonium sulfate and 2 gm. filter cell, and filter with suction through soft paper. Wash paper with 0.4 saturated ammonium sulfate. Reject precipitate. Add slowly 100 ml. saturated ammonium sulfate to each 100 ml. of combined filtrate and washings. Remove foam and filter with suction through hardened paper, size 18.5 cm. or larger. Reject filtrate. Wash precipitate on funnel with saturated magnesium sulfate in 0.02 N sulfuric acid to remove excess of ammonium sulfate. The washing with saturated magnesium sulfate must be done rapidly, otherwise the precipitate is partly dissolved. Saturated magnesium sulfate is

---

[3] This useful material was suggested by Dr. M. L. Anson.

poured on the precipitate to a height of about 5 mm., allowed to filter for a few minutes, then the excess of saturated magnesium sulfate is decanted, and filtration is continued until complete. Dissolve precipitate (30 gm.) in 30 ml. 0.4 M borate buffer pH 9.0 at 2–5°C. (in an ice water bath), add more borate buffer drop by drop to pH 8.0, measure solution, and add equal volume of saturated magnesium sulfate, mix and allow solution to stand in ice box at about 5°C. (*Solution B*). Short triangular prisms of trypsinogen appear in the course of 2–3 days. If the solution is inoculated with crystals of trypsinogen crystallization is much more rapid, but the crystals are not so well formed. (If crystallization is delayed more than 4–5 days, or if the material has become partly active during the preparation crystals of trypsin may appear.)

Filter the crystals with suction at 5°C. (*Filtrate C*). The precipitate (about 10 gm.) is washed on the funnel several times with cold 0.5 saturated magnesium sulfate made up in 0.1 M borate buffer pH 8.0 and finally with saturated magnesium sulfate made up in 0.1 N sulfuric acid at room temperature. The crystals are then dried in an electric refrigerator at 5°C. and stored in the ice box. The dried material generally contains about 40 per cent of trypsinogen protein and 60 per cent magnesium sulfate. For further purification of trypsinogen by means of trichloracetic acid see legend to Fig. 2. For activation into trypsin see Section IV.

### *4. Crystallization of Inhibitor-Trypsin Compound*

Combine Filtrates C and washings from several trypsinogen crystallizations (1000 ml.). Adjust to pH 3.0 with 5 N sulfuric acid and saturate with magnesium sulfate by stirring for 15 minutes at 25°C. with an excess of crystals of magnesium sulfate. Filter with suction through hardened paper. Reject filtrate. Dissolve precipitate (150 gm.) in 750 ml. water and add 750 ml. 5 per cent trichloracetic acid. Heat the mixture at 80°C. for 5 minutes, cool to 25°C., and filter with suction through hardened paper. Reject precipitate. Adjust filtrate to pH 3.0 with 5 M sodium hydroxide (about 3 ml. per 100 ml. of solution). Saturate with crystals of magnesium sulfate. Filter with suction through hardened paper. Reject filtrate. Dissolve precipitate (15 gm.) in 45 ml. 0.02 M hydrochloric acid, add 1.5 gm. of crystalline trypsin, allow to stand until the trypsin is dissolved, adjust to pH 8.0 by addition of 0.4 M borate pH 9.0. Allow to stand for 1 hour at 5°C. Adjust to pH 5.5 by addition of 5 N sulfuric acid, saturate with crystals of magnesium sulfate at 25°C., allow to stand for 2 days at 20–25°C. Hexagonal crystals of inhibitor-trypsin compound gradually appear mixed with amorphous precipitate. Filter with suction. Wash precipitate on paper with 0.5 saturated magnesium sulfate. This dissolves the amorphous precipitate. Residue on paper—crystals of inhibitor-trypsin compound. Yield about 5 gm. Filtrate and washings, when saturated with crystals of magnesium sulfate, on standing may yield more crystals of inhibitor-trypsin compound.

*Recrystallization of the Inhibitor-Trypsin Compound.*—Dissolve the filter cake of crystals (5 gm.) in 50 ml. M/10 acetate buffer pH 5.5, filter through Whatman No. 42 fluted paper. Saturate with crystals of magnesium sulfate and allow to stand 1 day at 20–25°C. Hexagonal crystals of the inhibitor-trypsin compound rapidly appear. Yield about 3 gm. filter cake.

### 5. *Isolation of Crystalline Trypsin Inhibitor and of Crystalline Trypsin from a Solution of Crystalline Inhibitor-Trypsin Compound*

Dissolve 1 gm. crystalline filter cake (of 3 times recrystallized inhibitor-trypsin compound) in 10 ml. water and add 10 ml. 5 per cent trichloracetic acid; allow to stand at 20°C. for 30 minutes until precipitation is about complete. Filter with suction (*Filtrate In*). The *precipitate* (*Ts*) is worked up for trypsin, as described below.

(*a*) *Crystallization of Trypsin Inhibitor.*—Heat the trichloracetic acid filtrate (In) for 5 minutes at 80°C., cool, and filter through fluted Whatman No. 42 paper. Reject precipitate. Adjust filtrate to pH 3.0 with 5 M sodium hydroxide. Dissolve 5.6 gm. of solid ammonium sulfate in every 10 ml. of filtrate. Filter with suction. Reject filtrate. Dissolve precipitate (0.25 gm.) in 2.5 ml. water, adjust to pH 5.5 with 0.4 M borate pH 9.0. Add equal volume of saturated ammonium sulfate. Filter through No. 42 filter paper. Wash paper with 0.5 saturated ammonium sulfate. Add more saturated ammonium sulfate to filtrate and washings combined until slight precipitate forms. The amorphous precipitate gradually changes into long hexagonal prisms. Allow to stand for 2 days at 20°C. Filter with suction. Filter cake 0.15 gm. of inhibitor crystals. Wash filter cake with saturated magnesium sulfate if it is desired to have the crystals free from ammonium salt, and recrystallize with magnesium sulfate.

*Recrystallization.*—Dissolve crystals (0.15 gm.) in 1.5 ml. of M/10 acetate buffer pH 5.5. Add 7.5 ml. saturated ammonium sulfate (or 7.5 ml. saturated magnesium sulfate plus a few crystals of solid magnesium sulfate). Allow to stand at 20°C. for 1 day. Crystals of inhibitor gradually appear. Yield about 0.1 gm. filter cake.

(*b*) *Crystallization of Trypsin.*—Wash the trichloracetic acid precipitate (Ts) on the filter paper with water to remove the free acid. Dissolve precipitate (0.7 gm.) in 20 ml. 0.02 M hydrochloric acid. Allow to stand 30 minutes at 20°C. Add 5 gm. solid ammonium sulfate. Filter through fluted Whatman No. 42 paper until clear. Dissolve 5 gm. of solid ammonium sulfate in the filtrate and filter with suction through 5 ½ cm. hardened paper. Reject filtrate. Wash precipitate on paper with saturated magnesium sulfate in 0.02 N sulfuric acid. Dissolve precipitate (0.5 gm.) in 0.25 ml. water. Cool to 5°C., add about 0.5 ml. borate buffer pH 9.0 until the solution reaches pH 8.0 (pink to 0.01 per cent phenol red but not to 0.01 per cent cresol red on test plate). Add 0.5 ml. saturated magnesium sulfate. Allow to stand at 5°C. Square prismatic crystals of trypsin rapidly appear. Filter with suction. Yield about 0.25 gm.

## *II. Isolation of Crystalline Trypsin and Inhibitor Trypsin Compound from Chymo-Trypsinogen Free Activated Pancreatic Extract*

Preliminary purification and concentration is the same as described under I for preparation of Solution B (Section I, 3).

### *1. Crystallization of Trypsin*

Inoculate Solution B (about 100 ml.) with trypsin crystals, allow to stand at 5°C. for several days. Precipitate of very small crystals of trypsin gradually forms. Occasionally a few triangular trypsinogen crystals may also appear. Filter with suction at 5°C. *Filtrate E.* Filter cake is washed on paper several times with 0.5 saturated magnesium sulfate at 5°C. and finally with saturated magnesium sulfate in 0.1 N sulfuric acid at room temperature. Yield 8 gm. of filter cake.

### *2. Recrystallization of Trypsin*

Dissolve filter cake (8 gm.) in 6 ml. 0.02 N sulfuric acid. Add a few drops of 5 N sulfuric acid if solution is incomplete. Cool to 5°C., add 12 ml. saturated magnesium sulfate and 6 ml. 0.4 M borate pH 9.0. Adjust to pH 8.0 with saturated potassium bicarbonate or 5 N sulfuric acid if necessary. Inoculate. Allow to stand for 1 day at 5°C. Yield 3 gm. filter cake.

### *3. Purification of Trypsin by Trichloracetic Acid*

When first crystallized trypsin sometimes has a slightly low specific activity due partly to the presence of some inhibitor or trypsinogen. The activity may be raised to the maximum value by repeated recrystallization or, more easily, by precipitation with trichloracetic acid followed by crystallization. The procedure for the latter method is the same as that described under Section I, 5 for the preparation of trypsin from inhibitor-trypsin compound except that the starting material is trypsin instead of inhibitor-trypsin compound.

### *4. Crystallization of Inhibitor-Trypsin Compound*

Adjust Filtrate E (Section II, 1) to pH 3.0 with 5 N sulfuric acid and saturate with crystals of magnesium sulfate at 25°C. Filter with suction through hardened paper. Reject filtrate. Dissolve precipitate (10 gm.) in 50 ml. M/16 hydrochloric acid and pour solution with stirring into a large beaker containing 250 ml. M/16 hydrochloric acid at 90°C. Cool after 1 minute in running cold water to 25°C. (use coils described elsewhere (1) if large quantities of solutions are used). Dissolve 24.2 gm. solid ammonium sulfate in each 100 ml. of solution, filter through fluted paper; dissolve 20.5 gm. of solid ammonium sulfate in each 100 ml. of filtrate, refilter with suction. Dissolve filter cake (3 gm.) in 12 ml. water, cool in ice water, add about 3 ml. 0.4 M borate pH 9.0 in order to bring the solution to pH

8.0, and pour with stirring into a large beaker containing 75 ml. boiling distilled water. Heavy precipitate forms. Cool after 1 minute in running cold water to 25°C. Dissolve 24.2 gm. of solid ammonium sulfate in each 100 ml. of suspension and filter with suction through hardened paper. Reject precipitate. Adjust filtrate to pH 3.0 by addition of several drops of 5 N sulfuric acid and then dissolve 20.5 gm. of solid ammonium sulfate in each 100 ml. of solution. Filter with suction on a large funnel. Reject filtrate. Wash precipitate with saturated magnesium sulfate.

Dissolve precipitate (1 gm.) in 5 ml. M/10 acetate buffer pH 5.5. Adjust to pH 5.5 with about 1 ml. 0.4 M borate pH 9.0. Filter through Whatman's No. 42 paper into a flask containing enough crystals of magnesium sulfate to saturate the solution. Wash filter paper with 4 ml. M/10 acetate buffer pH 5.5. Stir the solution after completion of filtration. Hexagonal crystals of inhibitor-trypsin compound rapidly appear. Allow to stand for 1 day at 20°C. to complete crystallization. Yield about 0.25 gm. filter cake. For recrystallization see Section I, 4.

## *III. Isolation of Trypsinogen from Fresh Beef Pancreas*

Preliminary purification and concentration the same as described for preparation of Precipitate A (Section I, 1).

Wash Precipitate A with saturated magnesium sulfate on the filter paper. Dissolve precipitate (80 gm.) in 80 ml. 0.4 M borate buffer pH 9.0 at 5°C., add 136 ml. saturated magnesium sulfate. Adjust to pH 8.0 with a few drops of saturated potassium bicarbonate or 5 N sulfuric acid, if necessary. Inoculate with trypsinogen crystals. Allow to stand 2–3 days at 5°C. Filter with suction and wash crystals of trypsinogen with magnesium sulfate as described in Section I, 3. Yield about 10 gm. filter cake. Adjust filtrate to pH 3.0, saturate with magnesium sulfate at 25°C., filter with suction, wash with saturated ammonium sulfate, and proceed for isolation of chymo-trypsinogen (partly active), trypsin, and inhibitor-trypsin compound as described in Section I, 2, 3, up to preparation of Solution B and then in Section II.

## *IV. Conversion of Trypsinogen into Active Trypsin and Crystallization of Trypsin*

If the trypsinogen is in the form of semi-dry filter cake then it is treated exactly as in the procedure described for recrystallization of trypsin (Section II, 2). Dried trypsinogen containing dry magnesium sulfate is treated as follows: Dissolve 10 gm. in about 50 ml. of water. Add a few drops of 5 N sulfuric acid if solution is incomplete. Saturate with excess of crystals of magnesium sulfate by stirring for 15 minutes at 25°C., filter with suction. Dissolve precipitate (20 gm.) in 20 ml. 0.4 M borate buffer pH 9.0 at 5°C. Add 34 ml. saturated magnesium sulfate. Adjust to pH 8.0 with a few drops of saturated potassium bicarbonate or 5 N sulfuric acid, if necessary. Inoculate with trypsin crystals. Allow to stand 2–3 days at 5°C., then filter, and recrystallize for trypsin. Yield about 5 gm. filter cake.

### *V. Isolation of Crystalline Trypsin from Active Pancreatic Extracts*

The starting point for this method may be either frozen pancreas which has been stored for some time until it is active or fresh pancreas which has been allowed to stand in the cold for several days until the extracts are active. The active pancreas is minced and extracted with two volumes of 0.25 N hydrochloric or sulfuric acid for 18–24 hours at 5°C. The acid extract is filtered and the filtrate is treated as described previously in Paragraph 2 of Table IV (1) for the preparation of Precipitate 3 and the procedure followed until Precipitate 5 is obtained. Precipitate 5 is then washed with saturated magnesium sulfate in 0.02 M sulfuric acid Each 30 gm. of precipitate is dissolved in 30 ml. 0.4 M borate buffer pH 9.0 at 2–5°C., the solution titrated to pH 8.0 by the addition of borate buffer, and an equal volume of saturated magnesium sulfate added. This solution corresponds to Solution B (II, 1) and is then treated as described under II, 1 for the crystallization of trypsin from Solution B.

Most of the activity determinations and some of the preparative work described in this paper were carried out by Miss Margaret R. McDonald.

## SUMMARY

Methods are described for the isolation and crystallization of trypsinogen, trypsin, a substance which inhibits trypsin, and an inhibitor-trypsin compound. Analyses and some of the properties of these compounds are given.

## REFERENCES

1. Northrop, J. H., and Kunitz, M., *J. Gen. Physiol.*, 1932, **16**, 267.
2. Kunitz, M., and Northrop, J. H., *J. Gen. Physiol.*, 1935, **18**, 433. For a general review see Northrop, J. H., *Harvey Lectures*, 1935, **30**, 229.
3. Kunitz, M., and Northrop, J. H., *Science*, 1934, **80**, 505 (preliminary note).
4. Reported at The National Academy of Sciences, May, 1935.
5. Kunitz, M., and Northrop, J. H., *Science*, 1934, **80**, 190.
6. Waldschmidt-Leitz, E., and Akabori, S., *Z. physiol. Chem.*, 1934, **228**, 224.
7. Holter, H., Northrop, J. H., and Kunitz, M., *Z. physiol. Chem.*, 1935, **235**, 1.
8. Willstätter, R., and Rhodewald, M., *Z. physiol. Chem.*, 1930, **188**, 107.
9. Dyckerhoff, H., Michler, H., and Tadsen, V., *Biochem. Z.*, Berlin, 1934, **268**, 17.
10. Anson, M. L., and Mirsky, A. E., *J. Gen. Physiol.*, 1933, **17**, 151.
11. Waldschmidt-Leitz, E., *Z. physiol. Chem.*, 1933, **222**, 148.
12. Northrop, J. H., and Kunitz, M., *J. Gen. Physiol.*, 1932, **16**, 313.

Part VI

# ENZYME SUBSTRATE INTERACTIONS: ACTIVE CENTER, TERNARY COMPLEX, SUBSTRATE ACTIVATION, ENZYME-SUBSTRATE COMPLEX, THREE-POINT ATTACHMENT, AND STEREOSPECIFIC HYDROGEN TRANSFER

# Editor's Comments on Papers 41 Through 47



Paper 41 by Juda Hirsch Quastel (1899– ) and Walter Reginald Wooldridge, using *Escherichia coli* suspensions, was published a year after Sumner's announcement of crystalline urease. It clearly shows the influence of Willstätter's ideas of enzymes as active groups on colloidal carriers. This paper develops the idea of the *active center*, or site of activation, a term coined earlier by Quastel (1926) with reference to surface structure in an early version of the strain theory. The present work, developing an approach that we encountered in Emil Fischer's research some thirty years earlier (Papers 24 and 25), is based on the demonstrated pertinence of *in vitro* studies of enzymatic processes to an understanding of *in vivo* events. It is yet another illustration of the fact that enzyme studies leading to valid interpretations were and are possible in the absence of knowledge—in this case any knowledge—of the chemical nature of the enzyme molecule. In this light it is remarkable that we find here the statement, well ahead of its time, that various reagents may affect enzymatic activity by reacting not with groups in the active center but rather by altering the surface structure as a whole. A further point to note is that as a mechanistic study, this paper is unusual for its time since it concerns dehydrogenases rather than hydrolases. In a later development of this work, compounds similar to substrates were tested for their effects on enzyme reactions, and this investigation led to the discovery that malonic acid is a powerful inhibitor of succinic dehydrogenase (Quastel and Wooldridge 1928), an obvservation that was central to the later demonstration in H. A. Krebs's laboratory of the cyclic nature of the reaction sequence that oxidizes pyruvate (Krebs and Johnson 1937). For an autobiographical sketch of Quastel, with an entertaining description of life in Cambridge at the time of this paper, see Quastel (1980). In this article Quastel mentions that he was the one who introduced the use of washed bacterial suspensions for systematic biochemical investigations. A biographical sketch of Quastel was written by Sung (1979).

On the basis of careful rate studies, again with resting *E. coli* cells, on the enzymatic formation of succinate, malate and aspartate from fumarate, Barnet Woolf deduced the first evidence for the occurrence of distinct complexes involving a specific enzyme and two substrates, which later came to be called *ternary complexes* (Paper 42). In the second slightly later paper included in this volume (Paper 43), Woolf expands this deduction into a theory of enzymatic action in which for the first time it is clearly stated that the binding of substrate or substrates and their subsequent transformation to product(s) proceed through various chemical transformations at the specific combining site of the enzyme. This view hence abolished the earlier notion that substrate binding and substrate activation and transformation were chemically and perhaps spatially separate processes. It is of great interest that, as part of this approach to a theory of enzyme action, the view is advanced, probably for the first time, again without reference to the chemical nature of enzymes, that the substrate is distorted or polarized as a consequence of its binding to the enzyme surface.

The note by Kurt Guenter Stern (1904–1956) entitled "Spectroscopy of an Enzyme Reaction" (Paper 44) constitutes the first direct observation of an enzyme substrate complex. This work rapidly followed a report published in the same journal that showed the prosthetic group of catalase to be identical or closely related to protohematin (Stern 1935). The present note, reproduced here along with the more detailed paper published about a year later (Paper 45) after the author had moved from England to the United States, is an early example of the now much more common practice of the publication of a preliminary note followed by that of the detailed report. The demonstration of an enzyme-substrate complex by optical methods was developed in extensive classical studies by Britton Chance (see, for example, Chance 1947).

The short note by Alexander George Ogston (1911– ) (Paper 46) on what has been called the *Ogston effect*—namely, the enzymatic formation of asymmetric products from apparently symmetric substrates—stimulated an enormous amount of thinking and experimental work on the nature of chemical asymmetry, on enzymatic reaction mechanisms, and on precursor-product relationships in metabolic pathways. Some of the points made in this classic paper have not, however, stood the test of further work. To quote Alworth (1972): "Unfortunately [the] emphasis on the concept of three-point combination between enzyme and substrate is misleading and promotes erroneous conclusions. Even the most wondrous of asymmetric enzymes cannot differentiate between paired groups that are chemically and geometrically equivalent—no matter how many combination points are involved."

Paper 47, by Fisher, Conn, Vennesland, and Westheimer, is one of the first in a series of classical enzyme mechanism studies on oxidation-reduction studies involving pyridine nucleotides. (Note that DPN, diphosphopyridine nucleotide, a term introduced by Otto Warburg, is an earlier designation for what is now called NAD.) A slightly earlier short report on yeast alcohol dehydrogenase by the same group (Westheimer et al. 1951) had demonstrated a direct enzyme-dependent transfer of hydrogen between coenzyme and substrate, and the present investigation reports these experiments in detail, with the additional result that this transfer is stereospecific. A paper from the same research group immediately following this one (not reproduced here) arrives at similar conclusions with lactic dehydrogenase from heart muscle (Loewus et al. 1953). This type of study was extended to a large number of NAD- and NADP-dependent enzymes. It was soon found that some enzymes transfer the substrate hydrogen to one side, some to the other side of the nicotinamide ring. This work has been amply reviewed (see, for example, Vennesland and Westheimer 1954; Vennesland 1955, 1956, 1958; Levy et al. 1962, Popják 1970). In these pioneering researches the stereospecificities of hydride transfer reactions were determined by the use of deuterium- and, later, of tritium-labelled substrates. In more recent years it has been possible to determine these stereospecificities by means of proton magnetic resonance measurements (Arnold et al. 1976). Detailed correlations between the stereospecificity of some dehydrogenases and the geometry of their active sites have been accomplished by means of X-ray crystallographic observations (Garavito et al. 1977).

41



Reprinted from pages 1224–1233, 1238–1239, and 1246–1251 of *Biochem. J.* **21**:1224–1251 (1927)

# EXPERIMENTS ON BACTERIA IN RELATION TO THE MECHANISM OF ENZYME ACTION.

By JUDA HIRSCH QUASTEL AND WALTER REGINALD WOOLDRIDGE

(*Beit Memorial Research Fellow*).

*From the Biochemical Laboratory, Cambridge.*

(*Received August 3rd, 1927.*)

[*Editor's Note:* The table of contents has been omitted.]

## Introduction.

It may be urged that resting or non-proliferating bacteria do not form suitable material for experiments on the mechanism of enzyme action; for the reason that the reactions to which such bacteria give rise may be due, not so much to the enzymes as to the circumstance that they form organised systems capable of effects not brought about by constituent members of the systems.

In fact, it may be urged, a living, though not proliferating, cell may possess some property which is expressed by the reactions we are investigating and which would not be possessed by the cell were it disorganised or dead. That this criticism is groundless is shown by the following evidence.

1. If a suspension of *B. coli* be shaken with toluene or ether, the organism is no longer capable of reproduction on the usual nutrient media and therefore, so far as we can ascertain, is dead. But it can still give rise to a large number of activations, *e.g.* with formic, lactic, and succinic acids, at rates not markedly different from those induced by the normal or untreated organism.

2. The effect of exposing *B. coli* to some abnormal treatment (resulting usually in the death of the organism) is to produce, as exposure is prolonged, a step by step elimination of the various activating mechanisms, as though each mechanism were differently susceptible to the treatment. Now, were the activations of the cell due to the latter acting as an organised whole, we should expect that at some particular point the cell would become disorganised and its activations would cease. Each mechanism would be eliminated at the same time and there would be no step by step degradation. Since this is contrary to what actually occurs we cannot suppose that the activations of the cell depend upon the survival of the latter as an intact organised system.

We have indeed *no* evidence to support a contention that the reactions of the organism are dependent upon its survival as a living unit. It is true that when an organism is killed certain of its activations (usually those of the sugars in the case of *B. coli*) disappear, but the simple interpretation is that the agent responsible for the death of the organism is also responsible for the disappearance of the activating mechanisms—not that the disappearance occurs because the organism has been killed.

There can be no doubt, however, that the power of the cell to proliferate is dependent upon the intactness of (among other things) many of the activating mechanisms with which we have to deal. This is entirely to be anticipated, for we cannot expect a cell to survive, whose cellular or intracellular structures have been so changed, by exposure to abnormal conditions, that they can no longer perform activations (which, on our views, are simply an expression of the nature of these structures).

In a previous communication by one of us [Quastel, 1926] the theory was advanced that the dehydrogenations[1] effected by bacteria are primarily due to polarisations of substrate molecules induced by electric fields which characterise particular centres—the "active centres"—of cellular and intracellular surfaces. The effect of polarisation may be to activate the molecule, *e.g.* it may become a hydrogen donator or acceptor, but since activation necessitates the uptake of a critical amount of energy, it follows that not all molecules which are polarised will become activated. This will depend on the nature

[1] Such dehydrogenations do not refer to reductions brought about by glutathione or substances which are active *in vitro*; they refer primarily to substances which must be activated by the cell before their oxidation can occur.

and intensity of the polarising field and upon the electrical and chemical nature of the substrate molecules. The mechanism by which a substrate, when activated, becomes a hydrogen acceptor or donator is discussed in earlier papers [Quastel, 1926; Quastel and Wooldridge, 1927] and will not be discussed further here.

The view there put forward makes it possible to interpret in a consistent manner such phenomena of biological oxidations as $\beta$-oxidation, asymmetric oxidation of the double bond, the $\alpha$-oxidation of propionic acid, the mode of oxidation of branched chains and so forth. It also indicates how it is possible to regard specificity of behaviour as belonging to the molecules themselves; it being unnecessary to postulate the existence of numerous specific enzymes related to oxidations. Thus one active centre may be able to effect the activation of a number of substrates, the chance of one molecule rather than another being activated depending upon its structure. But specificity of behaviour must depend not only on the chance of the molecule being activated at a particular centre but also on its power of obtaining access to that centre. If the structure of the molecule is such that, although it may become activated by the centre, it cannot reach the centre, or remain there for a sufficient period of time for activation to occur, it is clear that the molecule will not react. Specificity, then, is also a function of the accessibility of a molecule; adsorption and appropriate orientation are factors which help the access of a molecule to a centre.

The "active centre" is a part of, or a property of, cellular and intracellular surface structures; it is inseparable from surface structure, as it is a function of it. Some conception of its properties and mode of formation will be given in Section V. It is not necessary to believe that only surfaces of considerable magnitude are capable of possessing such centres. Any surface structure, however small, may possess such centres; this will depend upon the nature of the surface. This view is discussed more fully later.

If a surface structure be attacked by chemical or by physical means a change in the number and type of the active centres distributed over the surface will take place. The change will be dependent upon the extent to which the surface structure is changed. The centres which are least resistant to "shocks" (though the resistance must vary, we expect, to some extent with the type of "shock") will be the first to disappear as the surface structure is attacked. But it does not follow that an active centre must *entirely* disappear as a result of change. It may form a new centre, according to the nature of the change, with a smaller strength of field, or perhaps a field of quite different nature. As a surface structure is attacked, the general effect, we expect, will be that the least stable active centres are first eliminated; then come those relatively more stable and finally the most stable. Thus there will result a definite sequence of eliminations which will proceed step by step. The sequence will be the same, as a whole, for a large number of methods of treatment of the organism—but that it is unlikely *always* to be so is clear, for individual variations, due to particular effects of an attacking reagent, will on certain occasions make themselves manifest. Thus a certain reagent may react vigorously with a certain constituent group of the active centres, rendering inert the substrates which are made accessible to the centres by means of this group. Yet the elimination of this group may not influence very markedly the activating power of these centres, and so substrates whose accessibility is unaltered are still activated. Such a reagent would be liable to give a different sequence of eliminations from that produced by another reagent.

That the general effect of reagents is to give a sequence of eliminations which remains as a whole the same is illustrated by experiments described in our last paper [Quastel and Wooldridge, 1927][1] where it was noted that a variety of different treatments of *B. coli* gave the same general order of eliminations. Experiments described in Section III will further illustrate this.

The importance of accessibility—*e.g.* adsorption or orientation—of a substrate as a factor in the determination of specificity will be discussed later, but it is as well to point out now that although a molecule may become activated at a centre it may not, in this activated state, undergo the reaction in question until it has passed through a particular phase of internal (electronic) structure or through a particular phase of stability. Various factors, such as orientation or adsorption, may influence the time necessary for the activated molecule to reach this phase. This interval of time is of importance, for it is possible that the molecule may become deactivated by collision before it has time to react. If the substrate molecule, through some property of its own structure or of that of the active centre, is unable to remain in contact with the centre for a sufficiently long period of time, its chance of reacting becomes small, in spite of the fact that the centre may have sufficient power to produce an activation. On the other hand, if there be present in the centre a grouping (or set of groupings) which will hold the molecule for a sufficient period of time reaction will ensue. This grouping is the "active group" of an enzyme. On our views it is simply one of the factors influencing accessibility of a substrate and hence influencing the specific behaviour of the active centre. The centre, *as a whole*, activates; one or more of its groupings is responsible for the access of the substrate. It may be possible to modify or destroy the centre, without necessarily altering the groupings responsible for access, *i.e.* the substrate may still be adsorbed without activation occurring. Or again, it may be possible to eliminate a group responsible for access without reducing the field strength to such an extent that it can no longer activate the molecule. The importance of this point lies in the fact that we should be able to differentiate between two molecules, activated at the same centre but each requiring a different group to make it accessible. By eliminating such a group one

[1] The references [Quastel, 1926] and [Quastel and Wooldridge, 1927] will, from this point, be represented by the signs [1] and [2] respectively. These papers should be consulted for full details of the experimental technique adopted in this communication.

substrate will be rendered inert whilst the other molecule, whose access is unaltered, may still be capable of activation by the centre. Thus the centre would have the appearance of being composed of two specific enzymes.

Specificity, then, is a function of the following factors.

1. Nature of the polarising field at the active centre.
2. Structure of the substrate molecule.
3. Orientation and adsorption of the substrate molecule at the centre; these depend also on the chemical structure of the centre itself.

It will be seen that a centre may easily differ from a neighbouring centre. Each has an individuality of its own—each may, indeed, be regarded as a specific enzyme but with a certain limited range of specificity. The two fundamental ways in which an active centre differs from a truly specific enzyme are:

(*a*) a range of specificity which is limited—characteristic of each centre:

(*b*) the identity of an active centre with a surface structure.

Willstätter [1927] states: "It seems that we must consider an enzyme to be composed of a specifically active group and a colloidal carrier.... The colloidal carrier seems to vary somewhat in its nature, but to be necessary for the stability of the active group." This statement can be regarded as an approximation to the view of active centres, it being emphasised once more, however, that on this view activation is due to the centre as a whole, probably to a composite of groups, and not necessarily to a specifically active group. The latter, whilst forming an essential feature of the centre, may be important primarily in securing the access of the substrate to the activating field.

Let us now consider the evidence which leads us to suppose that activations (dehydrogenations) due to bacteria are associated with surface action.

## I. Evidence for surface action in the dehydrogenations produced by *B. coli*.

Whilst the evidence from work on enzymes points very clearly to the participation of surface action it is difficult to prove rigidly, by one particular experiment, that in the activations due to bacteria surface activity is fundamentally involved. It is only when the evidence as a whole is taken into account that there seems to be no escape from this conclusion, except by postulating a number of assumptions which are usually highly questionable and for which we have no evidence. By surface activity is meant activity by surfaces, either cellular or intracellular.

Quantitative measurements [1] have already indicated that the main site of reduction of methylene blue by a hydrogen donator is at the cell surface and for reasons given in that paper it was shown that *a site* of activation of hydrogen donators is also at the cell surface.

If a suspension of *B. coli* be shaken with toluene [2] a number of activations (notably those of the sugars and glutamic acid) are eliminated, whilst others (succinic acid, lactic acid, formic acid) are retained. We may try to explain this purely on permeability considerations. The argument would be, presumably, that the effect of the toluene is to act selectively on the cell membrane, rendering it impermeable to the sugars but not to succinic acid, etc. Now this selectivity of action cannot be attributed merely to mechanical blocking for whereas glycerol is made inactive, $\alpha$-glycerophosphoric acid is as active after as before treatment; again, methylene blue must still be freely permeable for the rates of action due to formic acid, etc., have not changed. We may then suppose that the selectivity of action is due to the formation of a film, at the surface, possessing selective permeabilities, but this cannot be true, for the formation of such a film would not allow methylene blue to penetrate it. Methylene blue does not diffuse, under the conditions of our experiments, from aqueous solution into toluene and we should expect a film of toluene at the cell surface to act like a lipoid layer which, as we know from the work of Loewe [1912], would adsorb but not dissolve the dyestuff, *i.e.* the latter would not penetrate into the cell.

We may suppose finally that there is a specific entrance into the cell for the sugars, glycerol and glutamic acid and another one for succinic acid, lactic acid, methylene blue, etc., and that toluene blocks the specific entrance for the sugars and not the other entrances. Now from the results of other experiments we can demonstrate differences between succinic acid and lactic acid, between glycerol and glucose, etc., so that if the interpretation of our results is simply to be that of specific attacks on pores specific for substrates we would have to assume a specific pore for almost every substrate—a conclusion which seems to be entirely out of the question.

Since the problem does not find a satisfactory solution simply in the consideration of changes in permeability, we may take the view that the toluene penetrates the cell and there affects certain specific enzymes, *i.e.* those related to the sugars, etc. But ether, benzene, chloroform, propyl alcohol, all have extremely similar effects to toluene, and since it is scarcely conceivable that such substances as these form definite chemical compounds with the enzymes, it is reasonable to suppose that they would act in the manner we would anticipate from such capillary active substances. We expect that they would form films upon the specific enzymes, rendering the latter inaccessible to methylene blue or to the substrates or to both. This clearly indicates, as we would naturally expect from other work on enzymes, that the specific enzymes for the dehydrogenations are associated with surfaces of some description. This does not apply only to the sugars, for experiments with propyl alcohol indicate that the other specific enzymes—those for succinic acid, lactic acid, etc.—are equally associated with surface action.

If the toluene does *not* enter the cell, thus affecting certain enzymes therein, we must imagine that it forms a film upon a particular patch or series of patches on the cell surface, these patches being particularly associated with the activations of the sugars, etc. If these patches were not so associated there seems to be no reason why the sugars and methylene blue should not enter the cell at parts of the surface not affected by toluene and so react with enzymes in the cell.

Were we to ascribe the effects of treatment of *B. coli* by nitrites [2] or by high concentrations of salts [2] or by the halogens or by allyl alcohol or by sodium acrylate (Tables I and II) to changes in permeability of the cell membrane alone, we should be placed in the difficult situation of having to account for the fact that whilst a variety of donators are prevented from entering the cell, methylene blue appears to be freely permeable always. For if it were not so, formic acid could not react at its normal rate when many of the other donators have been rendered inert. Again, it seems inconceivable that such a variety of attacks should change the permeabilities of the substrates in almost exactly the same order. The simple interpretation of these facts is that the main site of reduction of methylene blue is at the cell surface, supporting the conclusion drawn from quantitative evidence.

It seems, therefore, from various lines of evidence that surface activity is intimately associated with the activations of many hydrogen donators. The evidence is based upon experiments carried out with methylene blue, and *with this dye-stuff* it appears that the outer cell surface of *B. coli* is the main site of reduction; hence this surface must constitute *one* important site of the activation of hydrogen donators. Methylene blue has been valuable in demonstrating this fact and hence in indicating clearly the importance of surface action. But it would be absurd to consider that because, with methylene blue, the outer cell surface is the main site of reduction, activations cannot be performed within the cell. The facts that mechanical disintegration of bacteria appears to have a destructive action on certain activating mechanisms [1] and that lysis of the organism by bacteriophage ultimately removes the power of the cell to reduce methylene blue [Shwartzman, 1927; Gozony and Surányi, 1925] support our conclusion that the cell surface is a site of activations, but we cannot expect that the breakdown of the cell surface will necessarily involve the destruction of all intracellular surfaces and the elimination therefore of all activating power. Our thesis has been to show that surface activity, whether it be cellular or intracellular, is fundamentally connected with the activations we are investigating and it should therefore be borne in mind that when reference is made to surface activity, the term surface includes both cellular and intracellular surfaces.

Let us now consider briefly the phenomenon of activation and the conception of active centres.

## II. Activation.

It has been necessary for several reasons to consider that in homogeneous reactions a molecule cannot undergo transformation and dissociation unless it possess a certain minimum internal energy. "By internal energy is meant all the energy it possesses over and above what it would possess at very low temperatures except energy of translation, which, by the principle of relativity, can have nothing to do with the possibility of an internal change. Any molecule containing an amount of energy equal to or greater than a critical amount may be called an activated molecule. It may not be in a position to undergo the particular transformation we have in mind for this may depend upon the orientation of the several parts of the molecule, upon the location of the energy within the molecule and upon other factors of a similar nature" [Lewis and Smith, 1925].

Two types of chemical union are recognised: the electrostatic attraction between oppositely charged ions, this is a polar linkage; and the attraction which is produced through the sharing of an electron by two atomic nuclei, this is a non-polar linkage. We suppose that for the dissolution or formation of a non-polar linkage there must be primarily a shift of the electrons from their original orbits. This necessitates the uptake of energy, the process being termed activation. A critical amount of energy of activation must be taken up by a molecule for reaction to be possible.

Ions do not require activation in ionic reactions with which we are familiar, for such reactions consist simply of rearrangements of polar linkages. There is no necessity, for instance, to postulate activation of hydrogen or hydroxyl ions.

At present two fundamental methods of activation are recognised, collision and radiation. It is unnecessary, for the purposes of this paper, to go into the processes in detail, but it is important to note the following points.

1. An activated molecule may be deactivated by collision.

2. An activated molecule may not react until a particular phase of the internal structure of the molecule has been reached [Lindemann, 1922; Thomson, 1927]. We may expect a definite time interval between time of activation and time of transformation, the interval depending on the way in which activation has been effected.

### *Active centres.*

Much of the work on surface catalysis carried out within recent years[1] has gone to show that catalytic activity occurs at particular areas of the surface and that it varies in nature according to the nature of these areas. The areas, or active centres, constitute as a whole the active surface of the catalyst.

Adsorbed molecules at an active surface are regarded as in a distorted condition, or in a state of strain, the amount of distortion depending upon the molecular structure.

It has been shown that if a reaction catalysed by a surface occurs between two reactants both must be adsorbed and activated [see Hinshelwood, 1926].

[1] The subject is reviewed in the *Annual Reports of the Chemical Society*, **22**, 17. Owing to the existence of this report we refrain from quoting the literature concerned with surface catalysis and active centres. But fully to understand this section, the report should be read by those unacquainted with the literature.

The value of adsorption, of some description, at an active centre is obvious when it is considered that we cannot expect activation at the centre, in general, to be instantaneous. When it is considered, too, that deactivation by collision may occur it is sufficiently patent that appropriate orientation of a substrate molecule (which may lessen the time for activation to occur) and adsorption (which may give time for the activation) are factors of utmost importance in the consideration of the specificity of behaviour of active centres.

It seems to be fully established that the active areas occupy only a small fraction of the surface and that they vary in capacity both to adsorb reactants and to promote reaction.

Taylor [1925] conceives that the activity of metal surfaces is due to incomplete surface crystallisations, so that there are produced occasional groups of atoms associated with high energy and chemical unsaturation relative to the atoms in the regular crystal lattices. [In this connection see also Kistiakowsky, Flosdorf and Taylor, 1927.]

In this paper we shall adopt the view put forward previously by one of us [1] that the activations at biological surfaces occur at the active centres of these surfaces, the mechanism of activation being a polarisation of the substrate due to electric fields at the centres. The specificity of action at the centre depends upon the accessibility of the substrate to the centres, as well as upon the electrical nature of the substrate and the nature of the polarising field.

We are aware that a purely "chemical" view of activations is possible but we find such a view inadequate to interpret our results and unsatisfactory in its inability to give us a rational picture of the mechanism whereby activations are effected. Our own view can be regarded as quite definitely chemical, yet it has the considerable advantage that one is enabled to perceive more clearly the extent to which various factors can operate upon specificity of action, than by a view which calls for a new enzyme related to, and for the formation of a new unstable intermediate compound with, every new substrate which is found to be activated.

We will now enquire into some of the properties of the active centres concerned with the dehydrogenations effected by bacteria.

## III. Biological active centres.

Treatment with the most diverse reagents has the effect of eliminating the activating mechanisms of *B. coli*. This leads to the conclusion that different groupings may be involved in the activation of substrates and since any one enzyme, or active centre, can be eliminated by any of the treatments, if the latter be carried out for a sufficient period of time, we must conclude that an active centre is made up of a number of groupings.

Or, if we take the attitude that the effect of various reagents is not to attack specifically certain groups in the active centres but to alter surface structure as a whole, so that the active centres are changed or eliminated, we must still conclude that such centres are of a somewhat complex nature and not composed simply of one grouping, as, for instance, the amino- or hydroxyl-group.

[*Editor's Note:* Material has been omitted at this point.]

*The nature of the active centres.*

It seems, from the general results on treatment of *B. coli*, that we must regard an active centre as made up of a number of groupings each of which plays its part in determining the access of a substrate to the centre. If the surface structure as a whole is altered by a number of reagents we expect each reagent to give the same sequence of eliminations. Reagents of this class, which give the same order of eliminations, modify the fields (at the active centres) or affect the stability of the centres by a general change in the entire surface structure; reagents of the second class, which give pronounced departures from the typical order of eliminations, act by attacking particular groups at the centres. It appears to us likely that the average reagent has properties belonging to both classes.

Each centre acts as if it were a separate entity, possessing individual characteristics and, in a word, acting as if it were a separate enzyme. It differs from our usual conception of an enzyme in that it has a limited range of specificity and is inseparable from a surface structure. On such a view it is clearly as impossible for us to state the constitution of a biological active centre as it is to give the constitution of an active centre of a catalytically active alloy from only a knowledge of the constitution of this alloy. Yet it is to be remembered that a relatively small structure or colloidal aggregate may possess an active centre and it would be legitimate to state that the constitution of the structure or aggregate as a whole would be the constitution of the enzyme.

[*Editor's Note:* Material has been omitted at this point.]

## V. General considerations.

### *Formation of active centres at biological surfaces.*

There is, naturally, great vagueness at present as to the mechanism of formation of enzymes and their relationship to the remaining constituents of the cells, but it seems to be fairly widely held that the cell is able to elaborate at least two distinct classes of molecules, the very highly specialised molecules which exhibit enzymic behaviour, and the enzymically inert substances which together make up the protoplasmic and histological structures of the cell. This view, which calls for a sharp line of demarcation between the architectural units of the cell, proteins, nucleotides, etc., and the specific enzymes, which are not only being synthesised themselves in the cell but which are regulating the course of metabolism and the growth of the cell itself, seems to be greatly strengthened by the fact that a large number of enzymes can be secreted by the cell. This gives the impression that enzymes are simply products of the cell in much the same way as, let us say, adrenaline.

When it is considered (*a*) that the number of highly specialised enzyme molecules, which the cell is presumed to contain, must be very large indeed and that they must vary considerably in their type and constitution, (*b*) that the evidence, presented in this paper, concerning the dehydrogenations effected by bacteria is contrary to the supposition that activity is due to the presence of many highly specific enzyme molecules, (*c*) that it is extremely difficult to understand how a cell is able to cope with material to which hitherto it has not been accustomed, if its content of specific enzymes, though large, is yet limited, it will be granted that the view stated above stands in need at any rate of some emendation.

The hypothesis we put forward, that enzymic activity may be regarded as the property of the active centres of cellular and intracellular structures (and this includes the smaller structures capable of extraction from or secretion by the cell) leads to a considerable simplification of the above view. Precisely what enzymic behaviour a particular structure or colloidal aggregate in the cell may possess depends on the nature of the active centres which form a part of the structures or of the colloidal aggregates. Enzymes, on this view, are themselves part of and cannot be dissociated from the architectural units of the cell. This does not imply, of course, that only the histological structures are involved; the smaller colloidal aggregates are just as much part of the architecture of the organism. The conditions existing in the cell are such as to bring about just that arrangement or juxtaposition of molecules which makes for the formation of active centres on the normal material of the cell. Thus we may imagine that the protein, nucleotides, etc., are not only so arranged as to form the various substances of the cell but that the arrangement is such that active centres are formed on these particular substances. Enzymes, therefore, and cellular structures are inseparably connected. It is necessary to consider, now, how the active centres are formed.

Let us consider the formation of molecular structures at an interface. This may be accomplished in either of two ways.

1. As new molecules are produced they will orientate themselves in a definite way determined by the forces at an interface and will tend to arrange themselves into a crystal lattice pattern. Now if a number of dissimilar molecules are engaged in forming a surface structure we expect that there will be produced points of weakness or of loose fitting. It is at these points that

stray fields occur, the fields being balanced by surface forces and constituting the active centres of the surface structure.

2. We may conceive of the production of a colloidal aggregate which is represented in a highly diagrammatic manner in Fig. 1. Here various molecules, represented by rectangles, are joined to each other by arrows which represent the connecting links between molecule and molecule. The link may be brought about by electron sharing, or the positions of the molecules relative to each other may be determined by orientating forces in the phases surrounding the molecules. The molecules will tend to take up positions so that the potential energy of the system is at a minimum.

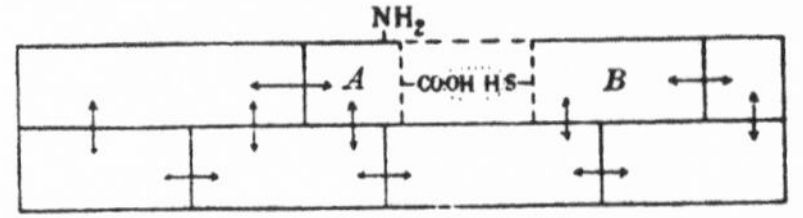


Fig. 1. Illustrating the colloidal aggregate (sectional).

Now let us consider two particular constituents, $A$ and $B$, of the aggregate. It may happen that were the molecule $A$ free from the restraint of the neighbouring molecules of the aggregate, it would form a structure represented by the skeleton scheme (Fig. 2). Here the external field of the molecule would be at its minimum; "affinities" between atom and atom in it are saturated as far as possible. But when forming part of a structure the action of the other molecules might tend to strain the configuration. This is illustrated in Fig. 3 where we suppose that two constituent groups $M$ and $M'$ of the molecule $A$ are held or attracted by neighbouring molecules in the aggregate, so that a strain occurs, pulling two other constituents $X$ and $Y$ considerably apart. Such a strain must cause the production of an external field between $X$ and $Y$, much greater than that which normally occurs. To take a concrete example,

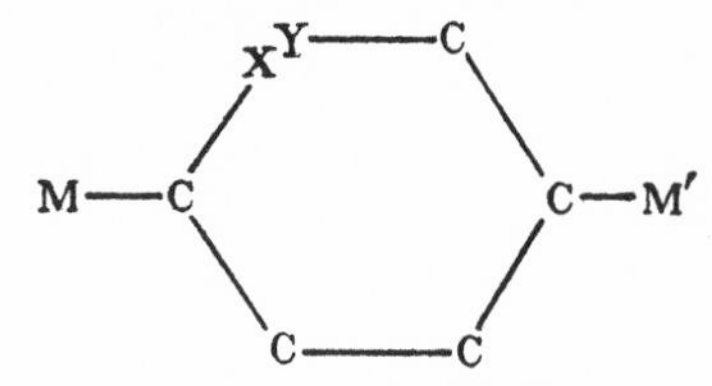


Fig. 2. Illustrating the normal character of molecule $A$ with external field between $X$ and $Y$ at a minimum.

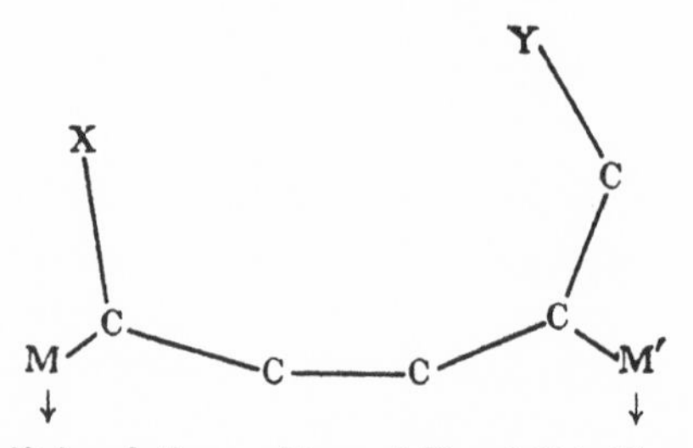


Fig. 3. Illustrating shift in relative positions of $X$ and $Y$ (with resultant production of increased external field) due to action of surface forces at $M$ and $M'$.

in Fig. 1, $X$ and $Y$ are represented by the amino- and carboxyl groups. These would normally have considerable affinity for each other and tend to approach to a critical distance. The strain imposed by the neighbouring molecules of the surface structure and surrounding phase causes a widening of the distance between the two groups and it is here that we may expect the local production of an external electric field. The carboxyl group at $A$ may link with, say, an —SH group at $B$, to form continuity of the surface, but this link, being perhaps much weaker than that normally between carboxyl and amino-group, will not necessarily reduce the field at $A$ to any considerable extent.

The active centre we would imagine in the molecule —$A$—CO—S—$B$— (with $NH_2$ attached to $A$)

to be made up at least of the groups —$NH_2$, —CO— and —S— and each of these groups will play its part in rendering a substrate accessible to the centre. The field due to these groupings may not be homogeneous and hence the orientation of a substrate at the centres, so that activation may occur, becomes a highly important consideration.

We have given this illustration of a possible formation of an active centre in order to make clear the difference between the centre and the usual conception of an enzyme. The actual composition of an active centre will in all probability be much more complex than in the illustration given. We may regard the entire aggregate as the enzyme, or the particular centre (at the link) as the enzyme. Each view is equally legitimate. But the residue of the aggregate, distinct from the region occupied by the centre, may be the seat of other active centres, so that the aggregate as a whole may have a much wider range of specificity than were the residue inert in this respect. Such an aggregate would be a relatively large colloidal particle and it would certainly be difficult to regard it as a *specific* enzyme. On the other hand, the residue may be of comparatively small dimensions and contain no other active centres. The specificity of action would be determined by the single centre and the range of specificity may be so small as to make the particle a highly specific enzyme. In this manner we regard the "soluble" enzymes[1] as related to the "insoluble." The distinction between them is simply one of degree.

### *Stability of active centres.*

In the first place there can be little doubt that the centre, represented, let us say, by the illustration (Fig. 1), will have a limited range of stability. Its stability simply rests upon the operation of forces at the interface and we expect any factor which disturbs these forces to result in a disturbance of the links between $A$ and $B$ and other molecules of the aggregate with the consequent elimination of the centre. Any agent which has the effect of altering surface structure will have a "poisoning" action on the centres, though the

[1] It is to be remembered that only heterogeneous catalytic systems are being discussed.

poisoning may be selective according to the nature of the centres. We expect the most unstable centres to disappear first, the more stable next and so on. The second type of poisoning action will be on the groupings constituting the centre itself, *i.e.* one of these, say —CHO, may link up with the poison eliminating the access of some particular substrate to the centre. But the field, though changed, may still be capable of activating other substrates accessible to the centre. We expect that such a poisoning action may be reversible.

The actual size of the aggregate can give us little information on the stability of the centres. It may be small enough to be classed as a "soluble" enzyme and yet have relatively considerable stability; it may be large enough to be classed as "insoluble" and yet be extremely unstable. We can be sure, however, that any means which will strengthen the forces at the interface in preventing fission of *A*—*B* will increase the stability of the centre. There appear to be at least the following ways of accomplishing this.

(*a*) By adsorption of the aggregate at an *appropriate* surface, which may tend to keep the molecules in position and prevent rupture by collision.

(*b*) By the adsorption or combination at the centre, itself, of some substrate, which is reversible in action and which may protect the centre. For instance, adsorption of sucrose by invertase has a protective action on the enzyme.

(*c*) By reducing the chance of rupture of *A*—*B* by impact, *e.g.* by increasing the viscosity of the medium in which the aggregate is present.

### *Co-enzymes.*

It is clear that any substance may be classed as a co-enzyme which (*a*) will tend to make a substrate accessible to the centre by combination with, or by facilitating a combination with, the substrate, the resulting combination possessing a lower energy of activation or increased adsorption or a more appropriate orientation at the centre, or which, (*b*) by some action at the centre itself, increases the activating power of the latter or the accessibility of the substrate. Such co-enzymes may not only make activity demonstrable where otherwise it could not be perceived but may increase the range of specificity of the active centre or enzyme.

### *Anti-enzymes.*

We cannot expect that a molecule of the type *A*—*B* (Fig. 1), whose enzymic activity is dependent upon surface forces and the nature of the union between *A* and *B*, *must* give rise to an antibody which will necessarily eliminate the activity of the enzyme. We may expect the production of antibodies which may act as precipitins, neutralising the charge of the aggregate as a whole, but such combination does not in the least indicate that the active centre will be destroyed.

SUMMARY.

1. An attempt based upon the conception of active centres is made to interpret the mechanism of enzyme action. This conception, which is a development of views put forward earlier [Quastel, 1926] on the mechanism of biological dehydrogenations, is discussed in relation to the specificity of enzyme action and to the behaviour of enzymes when exposed to various changes in chemical and physical environment.

2. Results are given which show that a hypothesis which claims that the activity of surfaces is due to the adsorption thereon of specifically active molecules is inadequate.

3. A suggestion is made as to the mode of formation of active centres (or enzymes) in the cell. It is shown upon what factors their stability depends and in what ways their activity is influenced.



We are greatly indebted to Sir F. G. Hopkins and to Sir W. B. Hardy for their interest in this work and for their ready advice. One of us (W. R. W.) is indebted to the Medical Research Council for a part-time grant.


REFERENCES.


Gozony and Surányi (1925). *Zentr. Bakt. Par. Abt. I Orig.* **95**, 353.
Hinshelwood (1926). Kinetics of chemical change (Oxf. Univ. Press), 161.
Kistiakowski, Flosdorf and Taylor (1927). *J. Amer. Chem. Soc.* **49**, 2200.
Lewis and Smith (1925). *J. Amer. Chem. Soc.* **47**, 1508.
Lindemann (1922). *Trans. Faraday Soc.* **17**, 598.
Loewe (1912). *Biochem. Z.* **42**, 190.
Myrbäck (1926). *Z. physiol. Chem.* **158**, 160.
Quastel (1926). *Biochem. J.* **20**, 166.
Quastel and Wooldridge (1927). *Biochem. J.* **21**, 148.
Shwartzman (1927). *Zentr. Bakt. Par. Abt. I Orig.* **101**, 62.
Taylor (1925). *Proc. Roy. Soc. Lond.* A, **108**, 105.
Thomson (1927). *Phil. Mag.* (VII), **3**, 241.
Willstätter (1927). *J. Chem. Soc.* 1374.

# 42


Reprinted from *Biochem. J.* **23**:472–482 (1929)

# SOME ENZYMES IN *B. COLI COMMUNIS* WHICH ACT ON FUMARIC ACID.

By BARNET WOOLF (*Beit Memorial Research Fellow*).

*From the Biochemical Laboratory, Cambridge.*

(*Received May 6th, 1929.*)

It was shown by Quastel and Woolf [1926] that when fumaric acid and ammonia are incubated with a suspension of *B. coli communis* at $p_H$ 7·4 in presence of an "inhibitor," such as 2 % propyl alcohol, there is a disappearance of free ammonia from the solution. Similarly, when *l*-aspartic acid is incubated under the same conditions, it is partially deaminated, and the time-course and final values of the ammonia uptake or output correspond with those required for the reversible reaction

$$COOH.CH = CH.COOH + NH_3 \rightleftharpoons COOH.CH_2.CH(NH_2).COOH$$

where the molecular equilibrium constant $K = \frac{[\text{Fumaric acid}]\,[\text{Ammonia}]}{[\text{Aspartic acid}]}$ has a value of about 0·04. In the absence of "inhibitors," there is also, under anaerobic conditions, an irreversible complete deamination of the aspartic acid, with production of succinic acid. It was shown by Cook and Woolf [1928] that the mechanism responsible for this reductive deamination is independent of the enzyme governing the aspartic-fumaric equilibrium, since it is present in strict aerobes and strict anaerobes, which do not bring about the latter reaction. When equilibrium has been attained, both aspartic acid and fumaric acid can be isolated, but, while the yields of aspartic acid agree satisfactorily with the theoretical, the yields of fumaric acid are invariably very low.

The presence in animal tissues of an enzyme capable of forming malic acid from fumaric acid was shown by Batelli and Stern [1911], who named it "fumarase." Einbeck [1919] reported that the reaction stopped when about three-quarters of the fumaric acid had been converted, and Dakin [1922] confirmed this and proved that the malic acid produced was exclusively the *laevo*-form. Clutterbuck [1927] followed polarimetrically the production of *l*-malic acid from fumaric acid in presence of muscle, and found that the reaction followed a linear course until the equilibrium was nearly reached, when it slowed down. Alwall [1929] completed the proof that a true chemical equilibrium was involved by showing that the same final state was reached when muscle tissue was allowed to act on *l*-malic acid.

The presence of fumarase in *B. coli* was reported by Quastel and Whetham [1924]. It is obvious that if this enzyme were present in the bacillus in any considerable quantity it would interfere with the aspartic-fumaric equilibrium, giving a complex malic-fumaric-aspartic equilibrium as the final state. Quastel and Whetham, however, found that when fumaric acid was incubated anaerobically with relatively large concentrations of organisms, only minute quantities of malic acid could be detected. They therefore concluded that there was only a small quantity of fumarase in *B. coli*, and in the previous work on the aspartic-fumaric equilibrium this conclusion was accepted, and it was taken for granted that the production of malic acid was so slow as not to interfere appreciably with the main reaction. It will be shown in this paper that the interpretation given by Quastel and Whetham of their results was erroneous, though justified by the facts known at the time they did their work; that, in fact, *B. coli* shows a very high fumarase activity; and that the equilibrium constant previously reported for the aspartic-fumaric equilibrium is really that of the complex malic-fumaric-aspartic equilibrium. It will also be shown that malic acid itself does not take up ammonia, that the addition of water and of ammonia to fumaric acid is due to two distinct enzymes, and that it is possible to eliminate the fumaric-malic reaction and so obtain the true constant of the equilibrium between aspartic acid, fumaric acid and ammonia. The bearing of these results on the "active centre" hypothesis of Quastel and Wooldridge will be discussed.

## Experimental.

All the work described was done with suspensions of "resting" *B. coli communis*. The organism was grown, either in Roux bottles containing 150 cc. of Cole and Onslow's tryptic broth, or on the surface of tryptic broth agar in Petri dishes. There was no apparent difference in the behaviour of the organisms obtained by the two methods. In each case, the nutrient medium was inoculated from an 18 hours old broth culture, and incubated at 37° for 2 days. When agar plates were used, the growth was washed off with normal saline and centrifuged; when broth was used, this was centrifuged from the organisms. The deposit of *B. coli* was then washed three times by centrifuging in normal saline, and finally suspended in saline and aerated for a few hours. It was stored at 0°, and, although generally used fresh, did not lose its activity after several months. The growth from one Roux bottle or one Petri dish generally corresponded to about 10 cc. of the suspension.

The following stock solutions were used: $M/2$ sodium fumarate, $M/2$ sodium *l*-aspartate, $M/2$ sodium *l*-malate, and $M$ ammonium chloride. The organic acids were weighed out and neutralised with sodium hydroxide, and all solutions were brought to $p_H$ 7·4. The buffer used was Clark and Lubs's phosphate buffer, $p_H$ 7·4, containing $M/20$ phosphate. Ammonia was estimated on 0·5 cc. samples by the method of Woolf [1928]. Malic acid was estimated polarimetrically as the molybdate compound, by the method of Auerbach and Kruger [1923]. A 5 cc. sample was added to 10 cc. of 14·2 % ammonium molybdate, then 1 cc. of glacial acetic acid was added, and the mixture was allowed to stand a few hours in the dark. It was then filtered through kieselguhr, exactly 5 cc. of water being used for wetting the filter-paper and washing, so that the total volume of fluid used was 21 cc. The presence of the molybdate helped to precipitate the bacteria, so that a crystal-clear filtrate was obtained. Another 5 cc. sample was added to 11 cc. of 6 % trichloroacetic acid, which was filtered in the same way with the addition of 5 cc. of water. The trichloroacetic acid acted as a protein precipitant and produced about the same degree of acidity as that due to the acetic acid in the molybdate mixture. The two solutions were examined polarimetrically in a 2 dm. tube, and the difference in rotation was proportional to the *l*-malic acid present. It was found that 10 mg. of malic acid, under these conditions, gave a rotation difference, with light from the mercury green line, of + 0·84°. This agrees with the value given by Needham [1927]. All the reaction mixtures were incubated at 37°, the mixtures without inhibitor being contained in filter-flasks evacuated at the water-pump, while those with inhibitor were placed in stoppered flasks, it having been previously ascertained that there was no difference in the course of the reaction anaerobically and aerobically.

## RESULTS.

The result of a typical experiment demonstrating the fumarase activity of the organism is shown in Fig. 1. The following reaction mixtures were made up, each containing in addition 50 cc. buffer solution and 2 cc. of *B. coli* suspension.

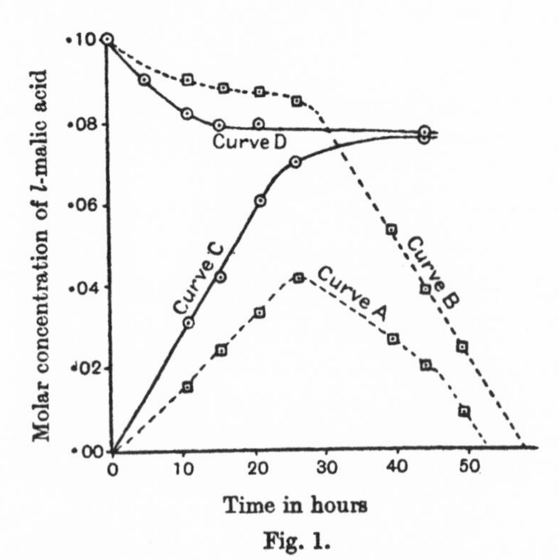

Fig. 1.

| Mixture | M/2 fumaric cc. | M/2 l-malic cc. | Propyl alcohol cc. | Water cc. |
|---|---|---|---|---|
| *A* | 20 | — | — | 28 |
| *B* | — | 20 | — | 28 |
| *C* | 20 | — | 2 | 26 |
| *D* | — | 20 | 2 | 26 |

Initial malic acid estimations were made, the mixtures were incubated, *A* and *B* being anaerobic, and at suitable intervals further samples were taken for estimation.

Curves *C* and *D* show that in presence of 2 % propyl alcohol an equilibrium is reached when about 76 % of the fumaric acid is converted into malic acid, or 24 % of the malic to fumaric. This equilibrium value agrees with that found by many workers for muscle. The initial portion of curve *C* is linear, as found by Clutterbuck [1927, 1928] for the fumarase of muscle and liver, which suggests that the enzyme is saturated with its substrate for nearly the whole course of the reaction. Curves *A* and *B* show that in the absence of an inhibitor the fumaric-malic equilibrium is masked by some other irreversible reaction which results in the destruction of the malic acid, and follows a linear course. The final portion of curve *A* is invariably parallel to the linear part of curve *B*, but sometimes the curves cross before this part is reached. The same phenomenon is observed in the absence of inhibitors for the aspartic-fumaric

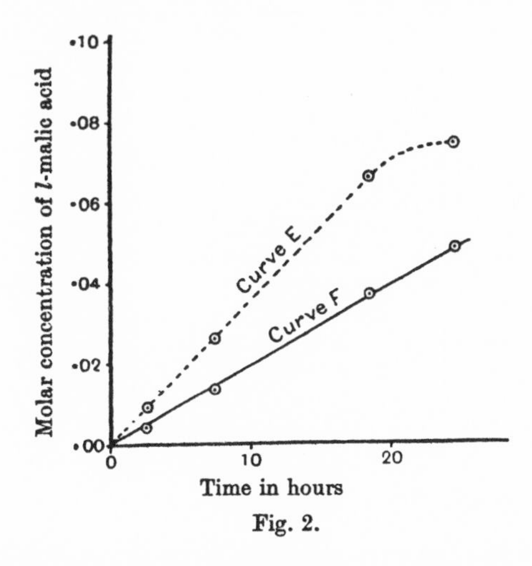


Fig. 2.

equilibrium [cf. Quastel and Woolf, 1926, figs. 1 and 2; Cook and Woolf, 1928, fig. 1], and, as in these cases, the chemical change involved is no doubt a reduction to succinic acid of at any rate part of the malic acid. The matter is under investigation. The figure also shows that the presence of propyl alcohol increases the velocity of the fumarase action. This activation is not always observed.

It was shown by Clutterbuck [1928] that phosphates had an activating action on animal fumarase. This has been confirmed for the enzyme of *B. coli*, as shown by Fig. 2. Curve *E* was obtained with a mixture similar in composition to mixture *C* of Fig. 1, and for curve *F* the buffer was replaced by water. The figure shows that the presence of $M/40$ phosphate at $p_H$ 7·4 almost doubles the reaction velocity. All the other reaction mixtures used in this work contained half their volume of buffer solution.

It is now possible to explain the results of Quastel and Whetham [1924]. Their reaction mixture was similar to that used for curve *A*, save that they used a much higher concentration of *B. coli*. When they examined their solution, at the end of 24 hours' incubation *in vacuo*, it had probably reached the final part of curve *A*, and most of the malic acid previously formed had disappeared, thus leading them to believe that there was very little fumarase in the organism.

Having established the presence in *B. coli* of considerable quantities of fumarase, it became necessary to re-examine the supposed aspartic-fumaric equilibrium. A typical result is shown in Figs. 3 and 4. The reaction mixtures were:

| Mixture | *M*/2 *l*-aspartic cc. | *M*/2 fumaric cc. | *M* $NH_4Cl$ cc. | Propyl alcohol cc. | Water cc. |
|---|---|---|---|---|---|
| *G* | 20 | — | — | 2 | 26 |
| *H* | — | 20 | 10 | 2 | 16 |
| *K* | — | 20 | — | 2 | 26 |

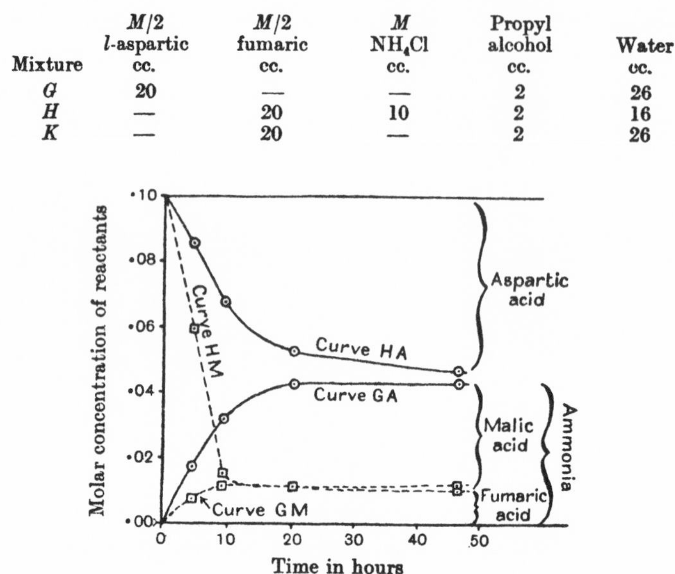


Fig. 3.

Ammonia and malic acid estimations were made on *G* and *H* and malic acid estimations on *K*; the solutions were then incubated and the estimations were repeated at suitable intervals. The results with *G* and *H* are shown in Fig. 3. Curves *GA* and *GM* represent the changes that take place when *M*/10 aspartic acid is incubated with the organism. The distance between *GA* and the base-line represents the molar concentration of free ammonia, found by estimation. This will of course be equal to the sum of the concentrations of fumaric and malic acids. The distance between curves *GA* and *GM* represents the concentration of malic acid, as found by estimation. Hence, by difference, the distance between *GM* and the base-line gives the amount of fumaric acid present, while the aspartic acid concentration is equal to the distance between *GA* and the horizontal line at the top of the figure. Similarly, for the other pair of curves, *HA* denoting free ammonia in *H* and the distance between *HA* and *HM* the malic acid. It will be seen that the ammonia curves *GA* and *HA* are similar to those given by Quastel and Woolf [1926], but that, at equilibrium, the fumaric acid concentration is only about a quarter of the ammonia concentration, the remaining three-quarters having changed to malic acid, in accordance with the requirements of the fumaric-malic equilibrium.

Fig. 4 is strong evidence that the addition of water and of ammonia to fumaric acid is the work of separate enzymes. It is conceivable that the two reactions could be effected by a single enzyme, which activates fumaric acid, possibly on the lines suggested by Quastel [1926], the activated fumaric acid then "accepting" either water or ammonia, in the form of their ions, which do not need "activation" (Quastel). If this were the case, and one started with a system containing the enzyme and fumaric acid only, then the rate at which malic acid was formed would be governed by the speed with which the

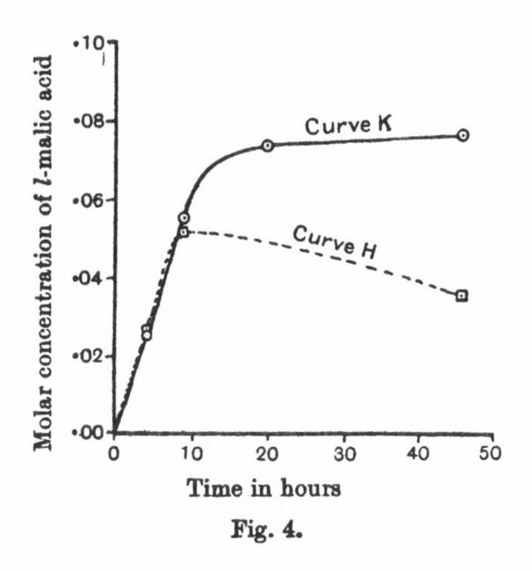


Fig. 4.

enzyme could activate fumaric acid. Now, if ammonia were added to the system, some of the fumaric acid which would have yielded malic acid would be diverted to form aspartic acid; there would be competition between water and ammonia for the activated fumaric acid. On the other hand, if two distinct enzymes were involved, then, so long as the fumaric acid was present in sufficient concentration to saturate the fumarase, the addition of ammonia would not affect the rate of formation of malic acid. Fig. 4 shows that this is in fact the case. Curve *H* shows the rate of malic acid formation from *M*/10 fumaric acid in presence of *M*/10 ammonia, *i.e.* the ordinate at any time equals the distance between curves *HA* and *HM* in Fig. 3. Curve *K* shows the production of malic acid in mixture *K*, which contains no ammonia, and it is clear that the initial rates of malic acid production in the two mixtures are identical.

When malic acid and ammonia are incubated with *B. coli* the final state is the same as that shown in Fig. 3, but the ammonia uptake is slower than with fumaric acid. It is, of course, possible that malic acid goes directly to aspartic acid without passing through the stage of fumaric acid. That this is unlikely is shown by Fig. 5. Mixture *L* contains *M*/10 fumaric acid and *M*/10 ammonia, together with a very small quantity of organism, while in mixture *M*, malic acid is substituted for the fumaric acid. The curves show the details of the beginnings of the ammonia uptakes. Fumaric acid immediately begins to react, but with malic acid there is a lag, indicating that fumaric acid must be formed before the ammonia uptake can begin.

The equilibrium constant $K_1$ for the reaction

$$COOH.CH = CH.COOH + H_2O \rightleftharpoons COOH.CH_2CH.OH.COOH$$

is given by the equation

$$K_1 = \frac{[\text{Fumaric acid}]\,[\text{Water}]}{[\text{Malic acid}]}.$$

The concentration of water can be taken as constant. Hence the ratio of fumaric acid to malic acid at equilibrium will be constant, whatever their initial concentrations may be. Now in the aspartic-fumaric equilibrium the constant $K_2$ is given by

$$K_2 = \frac{[\text{Fumaric acid}]\,[\text{Ammonia}]}{[\text{Aspartic acid}]}.$$

The value found for this constant by Quastel and Woolf [1926] was 0·04, but what they supposed was the fumaric acid concentration was really the sum of the fumaric and malic acid concentrations. The true concentration of fumaric acid is only a quarter of the value they used. It is this constancy of the

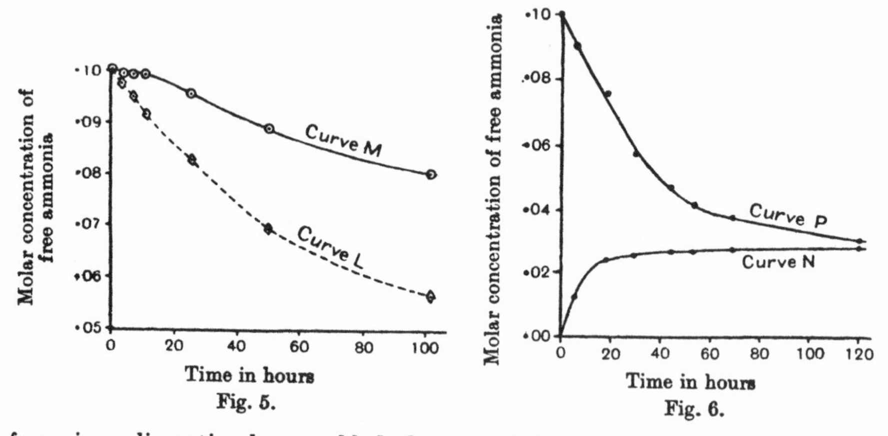


Fig. 5.

Fig. 6.

fumaric-malic ratio that enabled them to find the same value of K for the complex equilibrium when they varied the concentrations of their reactants. Substituting the true value of the fumaric acid concentration in the equation for $K_2$, one finds that $K_2$ is about 0·01. From this value it can easily be calculated that, if the fumaric-malic reaction could be entirely eliminated, the equilibrium mixture for the aspartic-fumaric equilibrium, when the reactants were present in the usual initial concentrations of $M/10$, would contain only about 28 % of free ammonia instead of the 47 % found for the mixed equilibrium.

Fig. 6 shows the result of an experiment in which these conditions are realised. The effect of cyclohexanol on the fumarase of *B. coli* is peculiar. If the organism is incubated with 2 % cyclohexanol for an hour or two, the fumarase is markedly activated, but if the incubation is continued for 15 hours, the fumarase activity is sometimes found to have entirely disappeared. This is not always the case, and it is probable that the time required for the destruction of the enzyme varies for different strains and preparations of the organism. When the fumarase is destroyed in this way it is found that malic acid will show no ammonia uptake, but ammonia uptake by fumaric acid still occurs. In the experiment to be described, 20 cc. of stock *B. coli* suspension were added to 78 cc. of saline and 2 cc. of cyclohexanol, and the mixture was incubated for 15 hours, when the fumarase was found to be completely destroyed. The following reaction mixtures were then made up:

| Mixture | *M*/2 *l*-aspartic cc. | *M*/2 fumaric cc. | *M* $NH_4Cl$ cc. | Buffer cc. | Cyclo-hexanol cc. | Treated *B. coli* cc. | Water cc. |
|---|---|---|---|---|---|---|---|
| *N* | 20 | — | — | 50 | 1·8 | 10 | 18·2 |
| *P* | — | 20 | 10 | 50 | 1·8 | 10 | 8·2 |

The mixtures were incubated, and ammonia and malic acid estimations made at intervals. The malic acid estimations were negative, and the ammonia results are shown in Fig. 6. It will be seen that the final value obtained is the true one for the aspartic-fumaric equilibrium.

This differential destruction of the mechanisms responsible for the two reactions is a further proof that two distinct enzymes are involved, and the absence of ammonia uptake by malic acid after the destruction of fumarase confirms the conclusion that malic acid itself is not converted into aspartic acid. It is convenient to have a name for the enzyme responsible for the aspartic acid reaction and it is proposed that it be provisionally termed *aspartase*.

It was found by Alwall [1928] that the fumarase in a succinoxidase preparation from muscle is destroyed by incubation at 50° for half an hour. Boiled *B. coli* shows no fumarase or aspartase activity, and preliminary experiments indicate that the inactivation temperature for each enzyme is in the neighbourhood of 50°

## Discussion.

*B. coli communis* is able to bring about several chemical reactions involving fumaric acid. One is the irreversible reaction shown in Fig. 1, which is inhibited by propyl alcohol. Possibly another is the oxidation of fumaric acid, though it seems probable that this may only take place *via* malic acid. Finally there are three reversible reactions: the fumaric-succinic-methylene blue equilibrium [Quastel and Whetham, 1924], and the equilibria governed by fumarase and aspartase. Each of these reactions seems to be catalysed by a distinct enzyme, which can effect one reaction only. Working with mammalian tissues, Alwall [1928] destroyed fumarase and left succinoxidase, and Clutterbuck [1928] destroyed succinoxidase and left fumarase. It has been shown in this paper that the fumarase of bacteria can be destroyed without elimination of aspartase. Quastel and Wooldridge [1927, 1] showed that the succinoxidase of

*B. coli* is destroyed by incubation with cyclohexanol for 5 minutes, whereas fumarase has been shown above to be far more resistant to this substance. They also showed that succinoxidase is little affected by one hour at 57°, a treatment that would destroy both fumarase and aspartase. Quastel and Wooldridge [1928] report that, after treatment with toluene, the affinity of the succinoxidase of *B. coli* for malonic acid is vastly increased, so that small concentrations have a large effect in retarding the reaction velocity. Experiments made on fumarase and aspartase show that they do not possess this property. Finally, repeated attempts have been made to demonstrate the presence of aspartase in muscle, with uniform lack of success, although the various preparations used showed marked fumarase activity.

It seems clear, therefore, that three reactions as similar as the addition across the double bond of fumaric acid of 2H, H and OH, and H and $NH_2$, require separate enzymes for their catalysis. Now according to the "activation" theory of Quastel [1926], as modified and extended by Quastel and Wooldridge [1927, 2] into the "active centre" theory, one would expect the essential happening in all these reactions to be the same, the activation of fumaric acid. The fumaric acid would be adsorbed or combined at an active centre, at which it would come under the influence of an intense electric field, becoming activated, so that its state could be diagrammatically represented as $COOH.CH_2.\check{C}.COOH$. The activated molecule would then react with ions from the solution, which do not need activation, and which would combine at the temporarily unsaturated or active carbon atom represented by the sign $\check{C}$. If this were the true mechanism, one would expect all three reactions to be effected by a single enzyme; or, if it were objected that the three reactions required fumaric acid activated to different extents, then it would be anticipated that the enzyme that could effect the most difficult of the three reactions —the one requiring the highest energy of activation or the most intense electric field—should also be able to bring about the other two reactions, while the enzyme catalysing the more difficult of these two reactions should also bring about the easier, and only the enzyme with the weakest field should be truly specific for one reaction only. But experiment shows that the contrary is the case; each enzyme is specific for one type of reaction. The differences in the enzymes seem not to be merely quantitative, as one would expect if the views of Quastel and Wooldridge were accepted as adequate, but there appear to be qualitative differences between them.

At the present stage it is only possible to give the merest indications of what the nature of these qualitative differences may prove to be. It will be noticed in curve *HA*, Fig. 3, that the initial portion is linear for a large portion of the reaction. This suggests that the enzyme is working at full saturation with its substrate. Now the reaction taking place involves two substances, fumaric acid and ammonia. If it were supposed that the ammonia were acting in solution, combining with activated molecules of fumaric acid at the enzyme when it came into contact with them by collision, then, by the law of mass action, it would be expected that, as the ammonia concentration fell, the reaction velocity would fall proportionately, even although the enzyme was saturated the whole time with fumaric acid. One would only expect to get a linear reaction if the enzyme were saturated with both the reactants—fumaric acid and ammonia. It follows that one must suppose that ammonia is combined at the enzyme as well as fumaric acid. Similar considerations apply to the reduction of methylene blue by succinic acid in presence of bacteria. Quastel [1926] gives a figure (Fig. 1) showing that the rate of decoloration is linear during the greater part of the reaction. This implies that the enzyme is saturated with methylene blue, and therefore that there is combination between enzyme and dye. Further evidence leading to the same conclusion is furnished by the work of Dixon [1926] on xanthine oxidase and Quastel and Wooldridge [1927, 1] on several of the dehydrogenating enzymes of *B. coli*. These workers used various dyes of different reduction potentials, and found that the rate of reaction was not dependent merely upon the ease of reduction of the indicator. Dixon points out that the presence of sulphonic groups in the dye tends to slow its reduction rate. These facts suggest that the chemical nature of the dye affects its affinity for the enzyme, and hence its rate of reaction. Quastel and Wooldridge themselves state that "the velocity of reduction is governed by the concentration of indicator as well as by the concentration of activated donator at the surface," but they do not seem to mean that the combination between enzyme and indicator is of the Michaelis type, nor do they seem to the writer to bear this condition sufficiently in mind in the development of their theory.

These considerations, and similar ones applied to the facts known about some other enzymes, point to the rather attractive hypothesis that one of the conditions for a reaction to occur at these enzymes is that all the substrates shall be combined there together—succinic acid and methylene blue for succinoxidase action, fumaric acid and ammonia with aspartase, and fumaric acid and water with fumarase. Whether there is also activation by electric fields as Quastel postulates, or whether the mechanism by which reaction is effected is of a different nature, seems to the writer still a very open question. This hypothesis has at any rate the merit that it is capable of being tested on strictly quantitative lines, and it is hoped to carry out further studies on fumarase and aspartase with this object in view.

## Summary.

1. The presence is demonstrated in *B. coli communis* of considerable quantities of fumarase, the enzyme governing the equilibrium between fumaric acid and *l*-malic acid. In the absence of inhibitors, such as 2 % propyl alcohol, the action of this enzyme is masked by an irreversible process resulting in the anaerobic destruction of malic acid. In presence of propyl alcohol the same equilibrium is attained as that given by animal fumarases.

2. The equilibrium previously reported between aspartic acid, fumaric acid and ammonia is really a complex malic-fumaric-aspartic equilibrium. The fumarase of *B. coli* can be destroyed by treatment of the organism with cyclohexanol, and then the true aspartic-fumaric equilibrium is obtained.

3. The three reversible changes involving fumaric acid—those to succinic, malic and aspartic acids—are catalysed by distinct enzymes. It is proposed to call the enzyme responsible for the third of these reactions *aspartase*.

4. The bearing of these results on enzyme theory is discussed, and a limited hypothesis on enzyme action is put forward.

It is a pleasure to express my gratitude to Sir F. G. Hopkins and Mr J. B. S. Haldane for their continued encouragement during the course of this work.

REFERENCES.

Alwall (1928). *Skand. Arch. Physiol.* **54**, 1.
—— (1929). *Skand. Arch. Physiol.* **55**, 91.
Auerbach and Kruger (1923). *Z. Nahr. Genussm.* **46**, 97.
Batelli and Stern (1911). *Biochem. Z.* **30**, 172.
Clutterbuck (1927). *Biochem. J.* **21**, 512.
—— (1928). *Biochem. J.* **22**, 1193.
Cook and Woolf (1928). *Biochem. J.* **22**, 474.
Dakin (1922). *J. Biol. Chem.* **52**, 183.
Dixon (1926). *Biochem. J.* **20**, 703.
Einbeck (1919). *Biochem. Z.* **95**, 296.
Needham (1927). *Biochem. J.* **21**, 739.
Quastel (1926). *Biochem. J.* **20**, 166.
—— and Whetham (1924). *Biochem. J.* **18**, 519.
—— and Wooldridge (1927, 1). *Biochem. J.* **21**, 148.
—— —— (1927, 2). *Biochem. J.* **21**, 1224.
—— —— (1928). *Biochem. J.* **22**, 689.
—— and Woolf (1926). *Biochem. J.* **20**, 545.
Woolf (1928). *Chem. Ind.* **47**, 1060.

# 43


Reprinted from *Biochem. J.* **25**:342–348 (1931)

# THE ADDITION COMPOUND THEORY OF ENZYME ACTION.

By BARNET WOOLF (*1851 Exhibition Senior Student*).

*From the Biochemical Laboratory, Cambridge.*

(*Received December 24th, 1930.*)

## 1. Introduction.

In 1929 the writer, as a result of a study of three enzymes in *B. coli communis* which act on fumaric acid, propounded "the rather attractive hypothesis that one of the conditions for a reaction to occur at these enzymes is that all the substrates shall be combined there together—succinic acid and methylene blue for succinoxidase action, fumaric acid and ammonia with aspartase, and fumaric acid and water with fumarase" [Woolf, 1929]. Since that time, further experimental findings, coupled with a thorough study of the work of others, have led the writer to develop this hypothesis into a definite theory of enzyme action, which it is the object of this paper to describe.

According to the view about to be put forward, the meaning of the term substrate is widened to include all the molecules participating in the catalysed reaction, including, for example, water in a hydrolysis, and the methylene blue or other hydrogen acceptor in a dehydrogenase action. The addition compound theory may then be stated as follows. An enzyme is a definite chemical compound which is able to form an unstable addition compound with all its substrates, each at its own specific combining group in the enzyme molecule. The process of catalysis then consists of a series of tautomeric changes in the enzyme-substrate complex, as a result of which, in a certain proportion of cases, the complex is able to dissociate into free enzyme plus the products of the catalysed reaction.

The line of reasoning on which this theory is based falls naturally into two sections, first an examination of all the known data of enzyme kinetics, and the induction from them of general laws, followed by the deduction from these laws of the probable mechanism of the catalytic process. A complete presentation of the first section is obviously impossible in this place. The full evidence for the conclusions given below will be set out, however, in a review by the writer on "Enzyme Kinetics," which is now in course of preparation for *Biological Reviews.*

## 2. Enzyme kinetics.

### *The enzyme-substrate compound.*

The hypothesis that the enzyme and substrate unite stoichiometrically was first put forward explicitly by Michaelis and Menten [1913], who deduced from it the well-known equation

$$V = \frac{U.[E].[S]}{K+[S]},$$

where $[E]$ and $[S]$ represent the concentrations of enzyme and substrate, $V$ the velocity of the catalysed reaction, $U$ the maximum velocity for unit enzyme concentration, and $K$, the "Michaelis constant," the concentration of the substrate at which half the maximum velocity is attained. The reciprocal of $K$ is the apparent affinity of the enzyme for the substrate.

This equation was found by Michaelis and Menten to hold with great accuracy for the action of saccharase on sucrose, and has since been found to fit the results obtained with many other enzymes. The full accounts of the many applications of this equation which are given by Kuhn [1925] and Haldane [1930] leave no doubt that the hypothesis of a stoichiometric enzyme-substrate compound is in good accord with the facts of enzyme kinetics.

### *The* $p_H$*-activity curve.*

Many enzymes give $p_H$-activity curves showing an optimum $p_H$ about which the curve is more or less symmetrical. Several authors have interpreted these curves as being due to the ionisation of the enzyme as an ampholyte, the isoelectric or un-ionised form of the enzyme alone being catalytically active, activity being determined not by the charge on the colloidal enzyme complex as a whole, but solely by the state of ionisation of one definite acidic and one basic group in the enzyme molecule. This view is supported by a considerable body of evidence, which is of two types: that based upon the shape of the curve, and that drawn from the effects of salts on the reaction rate.

Quantitative deductions from the shape of the curve can only be made when it covers eight or more $p_H$ units. On the ampholyte theory, the curve would fall away from the optimum in a Henderson-Hasselbalch curve if the affinity did not change with $p_H$. If the affinity falls as the $p_H$ recedes from the optimum, the curve will be steeper than a true dissociation curve, to an extent depending on the change in affinity with the $p_H$. These theoretical predictions have been realised, for instance by Josephson [1925] for $\beta$-glucosidase, and for saccharase, which is discussed by Kuhn [1925] and Haldane [1930]. Evidence from the effects of salts is given in such studies as those of Myrbäck [1926, 1] on saccharase and [1926, 2] amylase, and by Mann and Woolf [1930] on fumarase. These authors found that salts affected the velocity of the catalysed reaction in just the way that would be expected if the salt, or one of its ions, combined with the enzyme and altered the $p_K$ value of one or both of the critical ionisable groups, and hence the proportion of enzyme molecules in the isoelectric or active form at any given $p_H$.

Now there are many enzymes which do not give a $p_H$-activity curve of the symmetrical type, as for example the dehydrogenases [Dixon and Thurlow, 1924; Quastel and Whetham, 1924; and unpublished observations in this laboratory] and catalase [Michaelis and Pechstein, 1913]. Though it does not seem to have been pointed out before, the ampholyte type of curve is not characteristic of all enzymes, but only of those of a particular class—the enzymes which catalyse reactions in which water is one of the substrates. The non-hydrating enzymes mentioned above give a curve which rises in the acid range until a maximum velocity is reached, which persists unchanged until the enzyme begins to be destroyed by excessive alkalinity. Such a curve suggests that the enzyme acts as an ionised acid or an un-ionised base, and this suggestion is borne out by the fact that Michaelis and Pechstein [1913] found that the $p_H$-activity curve of liver catalase is a very accurate Henderson-Hasselbalch curve, which is shifted by salts in a manner analogous to the $p_K$-shifts found for the hydrating enzymes.

*The enzyme a chemical compound.*

All the kinetic results indicated above suggest strongly that an enzyme is a definite chemical structure. Further reasons for this conclusion are given by Willstätter [1927] in a summary of his work on enzyme purification, and by Haldane [1930] on several other grounds.

## 3. The addition compound theory.

*The activation of the substrate.*

It is obvious that the substrate must be distorted or activated in some way at the enzyme in order that reaction may occur, and various mechanisms, such as the electric field of Quastel [1926], have been proposed to account for this activation. All such theories are compelled to assume a double mechanism at the enzyme, the activating mechanism and some chemical grouping which will bind the substrate while it is being activated and undergoing reaction [*cf.* Quastel and Wooldridge, 1927, 1, 2]. It seems to the writer that this double mechanism is an unnecessary assumption. When the substrate is bound at the enzyme, becoming part of a temporary enzyme-substrate complex, the very chemical forces which bind it must surely produce sufficient distortion in its molecule to make it abnormally reactive. Until the chemical nature of enzymes is discovered, the nature of these forces can only be the subject of more or less plausible speculation. It seems likely, however, that since substrates are often stable, saturated molecules—*e.g.* succinic acid—the union with the enzyme is of the nature of a multiple co-ordination compound or chelate ring, which, as shown by Sidgwick [1927], involves the polarisation of the molecules which are so linked. But whether this or some other mechanism is the true one, it seems to the writer an economy of hypotheses to assume that the substrate-binding and substrate-activating processes are one and the same until evidence to the contrary is produced. This view is also put forward in a slightly different form by Haldane [1930]. It may be assumed, then, that when the enzyme and substrate are combined, the chemical structures of the molecules are so altered that the complex will readily undergo further tautomeric changes, which constitute the actual catalytic process. The nature of these changes is discussed below.

*The mode of action of a hydrating enzyme.*

A convenient illustrative enzyme is fumarase [*cf.* Woolf, 1929], which catalyses the reversible reaction

$$HOOC.CH:CH.COOH + H_2O \rightleftharpoons l\text{-}HOOC.CH_2.CH(OH).COOH$$

since the complication is absent of the hydrated product splitting into two molecules, and the forward and reverse reactions can be readily induced, so that the mechanism of both can be discussed. The kinetic results already considered show that the conversion of fumaric to malic acid can only occur at an enzyme molecule which is combined with fumaric acid and in the iso-electric form. It may therefore be assumed that for reaction to be effected there must be present at the enzyme a fumaric acid molecule, a hydrogen ion, and a hydroxyl ion. Now the reaction product, malic acid, is formed by the addition of these three bodies. It is therefore a natural step to assume that the enzyme can only act in the fully hydrated form because the hydrogen and hydroxyl ions are substrates for the reaction, as well as the fumaric acid.

An enzyme molecule can be pictured uniting with a molecule of fumaric acid. In the majority of cases, the substrate molecule will dissociate off unchanged. But while the substrate is at the enzyme, its structure is distorted, it is polarised by the chemical forces holding it there, in such a way that it is able to "accept" a hydrogen and a hydroxyl ion at the two carbon atoms of the structure —CH:CH—. If these two ions are also present at the enzyme at the same time, in a given proportion of cases the enzyme-fumaric-water addition compound will tautomerise so that the water ions leave the enzyme and add themselves across the double bond of the substrate, which will thus become free to dissociate off as malic acid. When this occurs, it is important to notice that the enzyme is left with neither a hydrogen nor a hydroxyl ion, *i.e.* in the zwitterion form.

There are thus four forms of the enzyme to consider—the dissociated acid, the dissociated base, the fully hydrated, and the zwitterion form. The last two together constitute the isoelectric form of the enzyme, which will be partitioned into hydrated and zwitterion forms in a constant ratio, which is independent of the $p_H$.

Consider now the reverse reaction. When a molecule of malic acid combines at the enzyme, it too will undergo polarisation so that the elements of water become dissociable as ions. If the enzyme already has a hydrogen or a hydroxyl ion, it will be unable to take up water from the substrate, and reaction cannot occur. If however it is in the zwitterion form, in a proportion of cases water will leave the substrate and combine with the enzyme, the substrate then dissociating off as fumaric acid and leaving the enzyme fully hydrated. The equilibrium can therefore be represented

$$[E^{\mathrm{H}}_{\mathrm{OH}}] + \text{Fumaric} \rightleftharpoons [E^{\mathrm{H}}_{\mathrm{OH}}].\text{Fumaric} \rightleftharpoons [E^{-}_{+}].\text{Malic} \rightleftharpoons [E^{-}_{+}] + \text{Malic}.$$

The mechanism here suggested thus conforms to the thermodynamic requirement that the reverse reaction in a reversible process shall pass through exactly the same stages, in the opposite order, as the forward process. The essential features of the process may be expressed as follows. The fumaroid skeleton, when combined with the enzyme, is so polarised that it acquires temporary acidic properties at one carbon atom and basic properties at the other. The complex thus becomes a dibasic acid and diacidic base, and the hydrogen and hydroxyl ions will partition themselves, statistically, between the acidic and basic groups of the enzyme and the substrate in a proportion determined by the relative affinity of the enzyme and the substrate for the water. It is obvious that from such a process a true thermodynamic equilibrium will be obtained.

In a system containing fumarase, fumaric acid and water there are the following forms of the enzyme to be considered:

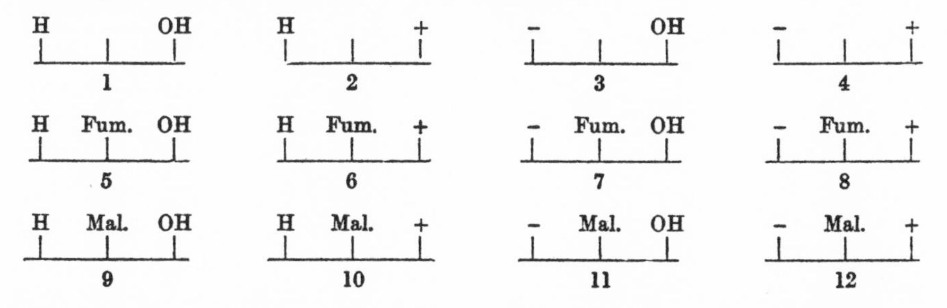


Form 5 is catalytically active in the forward reaction and form 12 in the reverse direction. All these forms are in equilibrium with one another and with fumaric acid, malic acid and the ions of water in the medium, and the affinity of each ionic species of the enzyme for fumaric acid and malic acid will, in general, not be the same. The state of affairs is therefore rather complex. It is however amenable to treatment by the law of mass action, and is then found to lead quantitatively to the kinetic results which are actually found by experiment, and which are set out in Section 2 of this paper.

The above line of argument applies to any hydrating enzyme, with the complication that when the catalysed reaction is a hydrolysis, the reverse reaction involves two substrates, each of which must combine with the enzyme, with its own characteristic affinity.

*The mode of action of a dehydrogenase.*

Succinoxidase will be taken as an example, since it is the only dehydrogenase whose thermodynamic reversibility has been proved experimentally, and methylene blue will be considered to be the hydrogen acceptor. The catalysed reaction then consists of the transfer of two hydrogen atoms from the donator to the acceptor. Many considerations render it probable, however, that the actual act of oxidation consists in the transfer of two electrons to the dye from the donator, which has previously lost two hydrogen ions, the process being completed by the acquisition of two hydrogen ions by the charged acceptor. One must therefore visualise the donator being polarised by its combination with the enzyme so as to acquire acidic properties at the structure —$CH_2.CH_2$—, the enzyme, in a given proportion of cases, taking up the two temporarily ionisable hydrogens from the donator in the same way as fumarase takes up the temporarily ionisable hydrogen and hydroxyl from malic acid. In order to be able to do this, the enzyme must itself be an ionised acid, which is precisely the conclusion that is indicated by the shape of the $p_H$-activity curve.

There is good kinetic evidence that the donator must combine with the enzyme. Reasons were given by Woolf [1929] for supposing that the acceptor must also combine. These reasons are reinforced by the work of Bertho [1929], who found that quinone, acting as hydrogen acceptor with the dehydrogenases of certain lactic acid bacteria, gave a good Michaelis curve. Unpublished observations by the writer show that the same is true for methylene blue and other dyes acting as hydrogen acceptors with a variety of dehydrogenases. In order to be catalytically active, therefore, a dehydrogenase molecule must have combined at it a molecule of donator and a molecule of acceptor, and be itself dissociated as an acid. If $DH_2$ and D represent the reduced and oxidised forms of the donator, $AH_2$ and A the corresponding forms of the acceptor, $E$ the enzyme and * an electron, the sequence of changes may then be formulated as follows:

$$[E^{**}] + DH_2 + A \rightleftharpoons [E^{**}]_{A}^{DH_2} \rightleftharpoons [EH_2]_{A}^{D^{**}} \rightleftharpoons [EH_2]_{A^{**}}^{D} \rightleftharpoons [E^{**}]_{AH_2}^{D} \rightleftharpoons [E^{**}] + D + AH_2.$$

In this scheme the enzyme is represented as a dibasic acid, but the same general result would be obtained if it were taken as monobasic, though the formulation would be a little more lengthy. The doubly charged enzyme combines with both substrates. The donator is activated, and parts with two hydrogen ions to the enzyme. The two excess electrons are then transferred from the donator to the acceptor, this being the actual act of oxidation. The acceptor then acquires two hydrogen ions, and both substrates dissociate off in the altered condition, leaving the enzyme in its initial state. Since all the intermediate forms are in dynamic equilibrium, it is probable that only a small proportion of the enzyme-substrate complexes formed will be completely transformed, and most will dissociate into enzyme and unchanged substrates. When the sequence of changes is read in the opposite order, it represents the reverse reaction—*e.g.* the reduction of fumaric to succinic acid by leuco-methylene blue. This mechanism, like that suggested for the hydrating enzymes, is therefore thermodynamically sound.

Similar considerations could be applied to other types of enzyme action, but these two examples suffice to indicate the scope and method of the theory. While it is claimed that the central idea is new, the writer wishes to acknowledge that he has freely drawn on the hypotheses of many previous workers, and adopted from them whatever elements seemed to him of value. It is claimed for this theory that it contradicts no known facts; that it gives a unitary explanation of the phenomena of enzyme kinetics, including $p_H$-activity curves, which no previous theory has professed to explain; that the mechanism of enzyme action proposed is a plausible one; and that it is a useful working hypothesis, in that it enables many quantitative predictions to be made, which can be tested by clear-cut experiments. It has certainly proved of extraordinary value as a working guide to the writer, and it is hoped to publish in the near future several enzyme studies suggested by the theory. The writer hopes that it may also prove of value to other workers in the field of enzyme action.

## Summary.

The theory is proposed that an enzyme is a definite chemical compound, which is able to form an addition compound with all its substrates, the actual process of catalysis consisting of a series of tautomeric changes in this enzyme-substrate complex, as a result of which, in a certain proportion of cases, it is able to dissociate into free enzyme and reaction products. This theory is shown to account in a quantitative manner for the observed effects of substrate concentration, $p_H$, and salts on the velocity of the catalysed reaction.

It is a pleasure to express my gratitude to Prof. J. B. S. Haldane for his constant and illuminating criticism, and to Sir F. G. Hopkins for his interest and encouragement.

## REFERENCES.

Bertho (1929). *Liebig's Ann.* **474**, 1.
Dixon and Thurlow (1924). *Biochem. J.* **18**, 976.
Haldane (1930). Enzymes. (Longmans, Green and Co., London.)
Josephson (1925). *Z. physiol. Chem.* **147**, 1.
Kuhn (1925). Oppenheimer's Die Fermente. (Thieme, Leipzig.)
Mann and Woolf (1930). *Biochem. J.* **24**, 427.
Michaelis and Menten (1913). *Biochem. Z.* **49**, 333.
—— and Pechstein (1913). *Biochem. Z.* **53**, 320.
Myrbäck (1926, 1). *Z. physiol. Chem.* **158**, 160.
—— (1926, 2). *Z. physiol. Chem.* **159**, 1.
Quastel (1926). *Biochem. J.* **20**, 166.
—— and Whetham (1924). *Biochem. J.* **18**, 519.
—— and Wooldridge (1927, 1). *Biochem. J.* **21**, 148.
—— —— (1927, 2). *Biochem. J.* **21**, 1224.
Sidgwick (1927). The electronic theory of valency. (Oxford.)
Willstätter (1927). *J. Chem. Soc.* 1359.
Woolf (1929). *Biochem. J.* **23**, 472.

# 44



Reprinted from *Nature* **136**:335 (1935)

## SPECTROSCOPY OF AN ENZYME REACTION

**Kurt G. Stern**

*Courtauld Institute of Biochemistry, Middlesex Hospital Medical School, London*

IF a suitable amount of monoethyl-hydrogen peroxide is added to an active preparation of liver catalase at *p*H 7, the enzyme spectrum disappears instantly, and two new absorption bands in the green range of the spectrum appear. In the course of a few minutes, at room temperature, the new absorption bands decrease in intensity, and simultaneously the original enzyme spectrum reappears. When the change is complete, the cycle may be repeated by the addition of fresh substrate. The time required for the reaction cycle at 4° is four times that required at 24°. The reappearance of the enzyme spectrum is accompanied by the disappearance of the titratable peroxide oxygen from the system. No gas is evolved in this reaction. Acetaldehyde and ethyl alcohol are among the most probable reaction products.

The spectrum of the intermediate compound is to be attributed to a combination of the enzyme with the substrate. The alternative explanation that one is dealing here with the reduced (ferrous-) form of catalase can be discarded because neither molecular oxygen nor ferricyanide reoxidise it to the original ferric form, and because it does not combine with carbon monoxide.

Enzyme spectrum (after K. Zeile[1]):

I. 650 . . . 646–620 (629) . . . 610; II. 550–530 (540) . . . 520 . . . 510–490 (500) mμ.

Spectrum of the enzyme-substrate compound:

I. 576–564 (570); II. 540–529 (534·5) mμ.

$5 \times 10^5$ molecules of the substituted peroxide per enzyme molecule are required to complete the transformation into the intermediate compound. It i interesting to note that according to J. B. S. Haldane' calculations[2], based on Zeile's and Euler's data $1{\cdot}9 \times 10^5$ molecules of hydrogen peroxide ar destroyed by one molecule of liver catalase pe second under optimal conditions. This coincidenc is still more striking since the enzyme is at leas 5,000 times more active towards hydrogen peroxid than towards ethyl-hydrogen peroxide.

With hydrogen peroxide as a substrate, no appare change of the enzyme spectrum is observed. Th inference is, that since the rate of formation and th stoichiometry of the two enzyme-substrate complex seem to be similar, the different catalytic efficienc towards the two substrates is due to the differer velocity constants of the breakdown of the inte mediate compounds.

A non-enzymatic model of the present observatio is found in the reaction of methæmoglobin wit hydrogen peroxide (Kobert[3], Haurowitz[4]) and wit monoethyl-hydrogen peroxide. Here the intermediat compounds are similar to that observed above, s far as the absorption spectrum is concerned. But on molecule of methæmoglobin will only break u $3{\cdot}6 \times 10^{-2}$ molecules of hydrogen peroxide per secon (cf. K. G. Stern[5]), and an excess of only 20 molecule of hydrogen peroxide (Haurowitz, l.c.) and of 8 mole cules of monoethyl-hydrogen peroxide will suffice t transform completely the spectrum of this catalys into that of the intermediate compounds.

[1] K. Zeile and H. Hellstroem, *Z. physiol. Chem.*, **192**, 171; 1930.
[2] J. B. S. Haldane, *Proc. Roy. Soc.*, B, **108**, 559; 1931.
[3] Kobert, *Pflüger's Archiv*, **82**, 603; 1900.
[4] F. Haurowitz, *Z. physiol. Chem.*, **232**, 159; 1935.
[5] K. G. Stern, *ibid.*, **215**, 35; 1933.

# 45



Reprinted from *J. Biol. Chem.* **114**:473–494 (1936)

# ON THE MECHANISM OF ENZYME ACTION

## A STUDY OF THE DECOMPOSITION OF MONOETHYL HYDROGEN PEROXIDE BY CATALASE AND OF AN INTERMEDIATE ENZYME-SUBSTRATE COMPOUND

By KURT G. STERN

(*From the Department of Physiological Chemistry, Yale University, New Haven*)

(Received for publication, March 23, 1936)

The classical methods of enzyme chemistry were of necessity restricted to the study of the catalysis as a whole. Either the rate of disappearance of the substrate or the rate of formation of the end-products was measured. The kinetics of these processes could in some instances be explained by the assumption that the catalyst combines with the substrate to form an unstable intermediate which may either reversibly dissociate into the two components or break down with the formation of free enzyme and the split-products (1, 2).

The spectroscopic observation of the formation and the breakdown of an intermediary enzyme-substrate compound during the action of catalase on monoethyl hydrogen peroxide (3) permits an experimental analysis of the two main phases of the enzyme reaction:

$$\text{Enzyme} + \text{substrate} \rightleftharpoons \text{enzyme-substrate compound} \quad (1)$$

$$\text{Enzyme-substrate compound} \rightarrow \text{enzyme} + \text{product molecules} \quad (2)$$

Reaction 1 lends itself to optical study. The over-all process (Reactions 1 + 2) is measured by volumetric determination of the substituted peroxide.

### EXPERIMENTAL

*Preparation of Enzyme Solutions*—Purified catalase solutions were prepared from horse liver[1] essentially according to Zeile and

[1] The author is greatly indebted to Chappel Brothers, Inc., Rockford, Illinois, for generously supplying the frozen horse liver used for these preparations.

Hellström (4). It was found expedient to carry out all operations of separation in a centrifuge which can accommodate four bottles of 500 cc. capacity each. Alumina gel (aluminum hydroxide, pure, moist (Eimer and Amend)) was used as adsorbent. Some data concerning the four enzyme solutions used in the present experiments are given in Table I.

For most of the experiments the undiluted enzyme preparations were used. These were dark brown in color and showed the typical catalase absorption spectrum in layers ranging from 2 to 5 cm., depending on their enzymatic activity. The secondary sodium phosphate used for the elution of the enzyme from the alumina gel adsorbate was neutralized by adding solid primary

TABLE I

*Data for Catalase Preparations*

| Catalase No. | Liver used | Eluate obtained | Activity* |
|---|---|---|---|
| | *gm.* | *cc.* | *k* |
| XXIX | 4,500 | 500 | 7125 |
| XXXIV | 3,700 | 200 | 2020 |
| XXXV | 3,700 | 250 | 5300 |
| XXXVI | 10,000 | 400 | 4525 |

* The activity was determined by measuring the rate of hydrogen peroxide decomposition by the highly diluted enzyme solution at 0°, pH 6.6, and at a total substrate concentration of 0.01 M. $k$ is obtained as the product of the monomolecular velocity constant, calculated from the amount of substrate destroyed within the first 5 minutes of the experiment, and of the dilution factor of the enzyme.

potassium phosphate. The enzyme solutions were stored over chloroform in the refrigerator (approximately at +2°) and filtered before use.

*Preparation of Substrate Solutions*—Solutions of monoethyl hydrogen peroxide were prepared by following essentially the directions given by Baeyer and Villiger (5) and by Rieche (6). The alkylation of hydrogen peroxide was found to give better yields if carried out with efficient cooling, the reaction temperature being kept near 10° by external cooling with an ice and water mixture. Throughout the alkylation process (requiring about 6 hours) and the subsequent neutralization with sulfuric acid, the reaction mixture was vigorously stirred with an electric stirrer.

The distillation of the substituted peroxide together with water was carried out under atmospheric pressure. The temperature inside the distillation flask was near 100°, whereas the oil bath was heated up to 150°. The distillation was discontinued after approximately one-half of the reaction mixture had been distilled over.

Since a sample of pure monoethyl hydrogen peroxide prepared according to the ether extraction method of Rieche and Hitz (7) gave essentially the same results as the dilute solutions obtained by single distillation, the distillates obtained as described above were used directly for the experiments after neutralization with solid secondary sodium phosphate. The amount of peroxide was determined by the iodometric method outlined below and was found to be between 1 and 2 M. These solutions were found to keep well for several months if stored in the refrigerator.

### *Volumetric Study of Decomposition of Monoethyl Hydrogen Peroxide by Catalase*

#### *Method*

For the estimation of monoethyl hydrogen peroxide Baeyer made use of an iodometric method. According to Rieche (8), however, the active oxygen of this peroxide may not be quantitatively determined with either hydriodic acid or titanous trichloride. The latter author therefore carried out his analysis by oxidation with chromic acid; the acetic acid formed was distilled off in the presence of an excess of phosphoric acid and determined by acidimetric titration. This procedure is not only somewhat tedious but is also not applicable to solutions containing foreign organic matter. Inasmuch as the object of most of the experiments reported in this paper was to follow the breakdown of the substituted peroxide as catalyzed by the enzyme and to study the effect of temperature, pH, etc., on the rate of this reaction, it was not necessary to assay accurately the absolute concentration of the peroxide but only to record the relative changes in concentration with time. It was found that the iodometric method as used by the author for studying the breakdown of hydrogen peroxide by catalase (9) was adequate for this purpose. The oxidation of hydriodic acid by the substituted peroxide is slower than by hydrogen peroxide even in the presence of molybdic acid. Except in the lowest concentrations of the peroxide used, it was found sufficient to wait for 1 hour between the addition of the peroxide to the acid potassium iodide solution plus a few drops of molybdic acid solution and the titration with thiosulfate with starch as internal indicator.

The relationship between the amount of iodine liberated and the concentration of monoethyl hydrogen peroxide was established in the following manner. In ten flasks a 1.1 M peroxide solution was introduced in amounts varying from 0.1 to 1.0 cc. and made up to 5 cc. by addition of water. These solutions were allowed to react

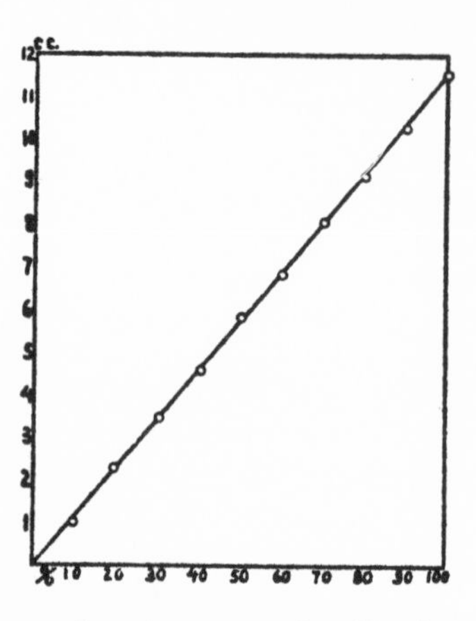

Fig. 1. Curve illustrating the proportionality between the amounts of ethyl hydrogen peroxide present and the amounts of 0.1 N thiosulfate required. The peroxide concentration (abscissa) is given in per cent of the maximum concentration used (0.2 M).

for 1 hour with 5 cc. of a 10 per cent potassium iodide solution plus 2 cc. of 33 per cent sulfuric acid plus 3 drops of a saturated aqueous solution of molybdic acid. The titration was carried out with 0.1 N thiosulfate solution. The determination was performed in duplicate. The curve in Fig. 1 was drawn through points representing the averages of the two individual values obtained. From the diagram it follows that a strictly linear relationship exists under the conditions of the experiment. Furthermore, it is to be concluded that if the reaction between the peroxide and hydriodic acid is not going to completion, the percentage deviation from the theoretical value remains constant in the concentration range tested. The method could therefore be used with confidence for the present purpose.

*Order of Reaction*—3 cc. of the undiluted Catalase XXXIV ($k = 2020$) and 1 cc. of M/15 phosphate buffer (pH 6.79) were placed in a 50 cc. Pyrex beaker and mechanically stirred at a moderate rate. At the experimental time = 0, 1.0 cc. of 1.1 M monoethyl hydrogen peroxide was quickly added. The reaction was stopped

TABLE II

*Kinetics of Decomposition of Monoethyl Hydrogen Peroxide by Catalase at Different Temperatures*

Each titration figure given was obtained by a separate experiment.

| Reaction stopped at time $t$ | 0.1 N thiosulfate required | 0.1 N peroxide decomposed | $k$ (monomolecular) | $k$ (zero order) |
|---|---|---|---|---|
| At 23°; 2 cc. Catalase XXXIV ($k$ = 2020) + 2 cc. M/15 phosphate buffer (pH 6.79) + 1 cc. 1.1 N $C_2H_5OOH$ | | | | |
| *min.* | *cc.* | *cc.* | $\frac{1}{t}\log\frac{a}{a-x}$ | $\frac{x}{t}$ |
| 0 | 10.27 | | | |
| 3 | 7.98 | 2.29 | 0.037 | 0.76 |
| 6 | 5.2 | 5.07 | 0.049 | 0.84 |
| 9 | 2.86 | 7.41 | 0.062 | 0.82 |
| 12 | 1.19 | 9.08 | 0.078 | 0.76 |
| 15 | 0.32 | 9.95 | 0.100 | 0.66 |
| At 0°; 3 cc. Catalase XXXIV + 1 cc. M/15 phosphate buffer (pH 6.79) + 1 cc. 1.1 N $C_2H_5OOH$ | | | | |
| 0 | 11.20 | | | |
| 3 | 9.25 | 1.95 | 0.028 | 1.95 |
| 6 | 8.38 | 2.82 | 0.021 | 1.41 |
| 12 | 6.53 | 4.67 | 0.020 | 1.17 |
| 24 | 3.44 | 7.76 | 0.021 | 0.97 |
| 48 | 1.04 | 10.16 | 0.022 | 0.64 |

at various intervals by addition of 3 cc. of 33 per cent sulfuric acid. The glass stirrer was lifted above the surface of the mixture and rinsed with a few cc. of distilled water. 4 cc. of 10 per cent potassium iodide solution and 3 drops of molybdic acid solution were added. After standing for 1 hour, the liberated iodine was titrated with 0.1 N thiosulfate. Experiments were conducted at room temperature (22°) and also at 0–1° (beaker immersed in a water-ice mixture). The solutions in the case of the latter series were all previously cooled. The results are given in Table II. From them were calculated both the monomolecular and the zero order reaction constants, as shown in the last two columns of the table.

The calculations show that whereas the course of the reaction at the low temperature may satisfactorily be described by the equation of the first order, the data obtained at 22° yield greatly increasing monomolecular reaction constants in time and appear to fit approximately the equation of the linear relation. No explanation at present can be offered for this observation.

*Effect of Temperature on Rate of Reaction*—For these experiments a water thermostat was used, the temperature of which

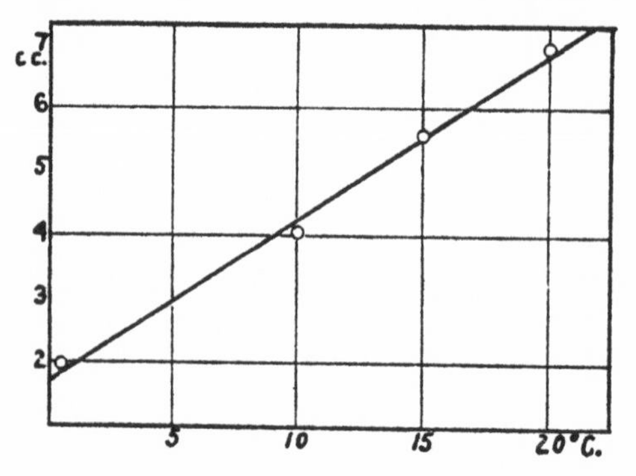


FIG. 2. Variation of the reaction velocity with the temperature. The ordinate represents the amounts of substrate decomposed in 5 minutes, expressed in cc. of 0.1 N thiosulfate.

could be adjusted to any desired value between +0.1° and +99° with a constancy of ±0.003°. The reaction time was 5 minutes throughout. The value for time = 0 was obtained as the average of four determinations in which the sulfuric acid had been added before the substrate. The actual runs at the four temperatures selected were performed in quadruplicate, and the average obtained of the four values was used for the plotting of Fig. 2. From Fig. 2 it appears that there exists a linear relationship between temperature and rate of reaction as expressed by turnover of substrate. $Q_{10}$ in the interval 0–10° equals 2.3 and between 10–20°, 2.19.

*Effect of Hydrogen Ion Concentration*—In these experiments

1.0 cc. of Catalase XXXV ($k$ = 5300) was added to 1.0 cc. of 1.1 M monoethyl hydrogen peroxide and 3.0 cc. of buffer mixtures ranging from pH 3.85 to 10.43. All the pH values given in this paper were obtained by measurements with the glass electrode in solutions of identical composition with that of the mixtures used for titration; these had been allowed to stand for at least 1 hour to allow the enzymatic reaction to go to completion. The author wishes to thank Mr. D. DuBois for performing the pH determinations with the modified glass electrode and circuit as devised by him.

The temperature in these experiments was kept near 0° by cooling with ice and water. Borate, phthalate, acetate, phosphate, and glycocoll buffer mixtures of a molar strength of 0.06

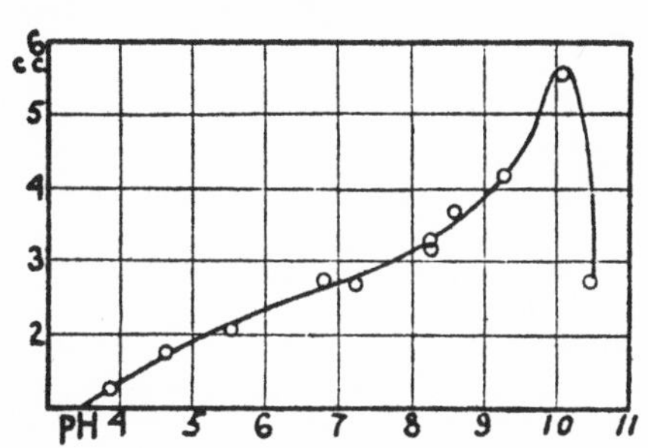


FIG. 3. Activity-pH curve. The ordinate represents the amounts of peroxide decomposed in 10 minutes, expressed in cc. of 0.1 N thiosulfate.

to 1.0 were used. No attempt was made to carry out all experiments at the same ionic strength, as a variation of this value did not appear to affect the reaction rate appreciably. All experiments were done in duplicate. The resulting pH-activity curve is given in Fig. 3. Whereas for hydrogen peroxide as substrate catalase will show an activity optimum between pH 6.5 and 9, slightly varying according to the source of enzyme material used and to the different investigators (10), this curve shows that the activity towards the substituted peroxide rises steadily with pH, attaining a sharp maximum near 10 and falling sharply again beyond this value, probably due to enzyme destruction at this hydrogen ion concentration. Control experiments showed that the rather alkaline optimum is not an artifact due to an instability of the peroxide at this pH. In spite of this finding, most of the present experiments were conducted near pH 7 in order to exclude the possibility of any secondary reactions caused by the high hydroxyl ion concentration and in order to facilitate a comparison with the enzymatic decomposition of hydrogen peroxide which is commonly studied at a pH slightly below 7.

*Affinity of Catalase for Monoethyl Hydrogen Peroxide*—The so called Michaelis constant ($K_m$), *i.e.* the substrate concentration at which the enzymatic reaction proceeds at half the maximal speed, was determined in a manner analogous to former experiments in which hydrogen peroxide was the substrate (11). The fact that monoethyl hydrogen peroxide has some affinity for

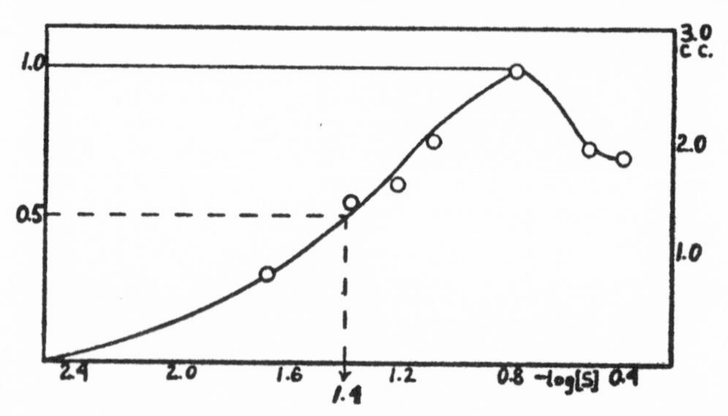


FIG. 4. Activity-$P_{[S]}$ curve. The abscissa represents the negative logarithm of the substrate concentration; the right ordinate, amounts of peroxide decomposed in 5 minutes, expressed in cc. of 0.1 N thiosulfate; the left ordinate, the rational measure, the maximal reaction rate being taken as 1.0.

catalase was borne out in the earlier work, when it could be shown that the inhibition of the catalase-hydrogen peroxide reaction by the substituted peroxide is of the competitive type.

Of the five series of experiments carried out with constant enzyme concentration and varying substrate concentration, the most satisfactory one was used for plotting the curve in Fig. 4. The reaction mixtures consisted of 3.0 cc. of M/15 phosphate buffer of pH 6.79, amounts of a 2.07 N monoethyl hydrogen peroxide solution (neutralized) varying from 0.05 to 1.0 cc., 1.0 cc. of Catalase XXXV ($k$ = 5300), and distilled water to make up a total volume of 5.0 cc. The reaction time in this series was 5 minutes and the temperature near 0°. The curve shows the variation of substrate turnover under these conditions with a variation of the total substrate concentration between 0.0207 and 0.414 N. As usual, not the absolute substrate concentration but the negative logarithm of this value is taken as the abscissa. This manner of plotting has the advantage that the experimental points corresponding to low substrate concentration are more evenly spaced than they would be by direct use of the concentration values. The Michaelis constant, which is the parameter of the left upward branch of the activity-$P_{[S]}$ curve, is found to be nearly $-\log [S] = 1.4$. The corresponding value for $[S]$ is 0.04 N. The $K_m$ value for hydrogen peroxide as a substrate was formerly found to be 0.066 N = 0.033 M. Since in the case of a mono-substituted peroxide the normality equals molarity, it follows that the affinity of catalase for both substrates is very similar.

The descending branch of the activity-$P_{[S]}$ curve at higher substrate concentrations is similar to that of the curves with hydrogen peroxide as substrate (11). It is a demonstration of the fact that at high substrate concentrations the enzyme is inhibited. For ethyl hydrogen peroxide sufficient data are not yet available to determine with accuracy the value of the parameter of the descending branch of the activity-$P_{[S]}$ curve. A graphical extrapolation, however, would yield a value for this "second Michaelis constant" of the order of magnitude found for hydrogen peroxide (0.4 M).

*Thermolability of Catalyst*—In order to ascertain that no thermostable non-enzymatic factor present in the enzyme preparation is catalyzing the breakdown of the substituted peroxide, 5 cc. of Catalase XXXV ($k$ = 5300) were pipetted in a test-tube and heated for 10 minutes in a boiling water bath. A dark brown clot formed during this procedure and was filtered off. 1 cc. of this solution when added to 3 cc. of phosphate buffer, pH 6.79, and 1 cc. of 1.1 N monoethyl hydrogen peroxide solution, did not cause a decrease of the peroxide titer within 10 minutes at 0°. In a control experiment with non-treated enzyme solution an amount of peroxide equivalent to 2 cc. of 0.1 N thiosulfate was decomposed. The catalysis is therefore due to a thermolabile agent.

*Inhibition by Cyanide*—In a mixture consisting of 1 cc. of 1.1 N monoethyl hydrogen peroxide, 3.9 cc. of phosphate buffer, pH 6.79, 1 cc. of Catalase XXXV ($k$ = 5300), and 0.1 cc. of 0.1 M sodium cyanide (neutral), no cleavage of the peroxide was observed within 10 minutes at 0°. The complete inhibition of the catalysis by 2 × $10^{-3}$ M HCN suggests that a catalyst containing trivalent iron is concerned in the reaction. Catalase has been shown to contain stabilized ferric iron combined with protoporphyrin as the prosthetic group of the enzyme (12). The reaction between catalase and hydrogen peroxide is already completely inhibited by 2 × $10^{-4}$ M HCN (11).

*Reaction Products*—Whereas hydrogen peroxide, when acted upon by catalase, yields molecular oxygen and water as the final reaction products, in the breakdown of monoethyl hydrogen peroxide no appreciable amounts of gas are released. Instead, a strong odor of acetaldehyde is noticed. The formation of aldehyde was qualitatively demonstrated by the well known condensation test in the presence of strong alkali and by the blue color produced by addition of sodium nitroprusside and piperidine. It should be mentioned, however, that even the untreated peroxide solution gave a somewhat positive aldehyde test.

In order to test for the possible formation of free acid, *e.g.* acetic acid, in the course of the process, 3 cc. of Catalase XXXIV (*k* = 2020) and 1 cc. of M/15 phosphate buffer, pH 6.79, were placed in a small crucible. An electric stirrer, a glass electrode, and a calomel electrode were inserted in the solution. The potential value was recorded, and 1.0 cc. of 1.1 N monoethyl hydrogen peroxide solution was added. The introduction of the buffered peroxide solution caused some shift in the pH. While the reaction proceeded, the potentials were recorded in short intervals. There was no detectable pH change within 5 minutes; therefore no acid is formed in the reaction.

According to Rieche (8), monoethyl hydrogen peroxide may break down in various ways, depending on the agent used; *e.g.*, on attack by alkali at higher temperatures only very little gas is formed, whereas under the action of formaldehyde and alkali much hydrogen and, moreover, formic acid, acetic acid, ethyl alcohol, and acetaldehyde are formed in a vigorous reaction.

It is planned to subject the enzymatic breakdown of the peroxide to further and quantitative analysis.

*Optical Study of Intermediate Enzyme-Substrate Compound. Technique*—The exact position of the absorption bands of the intermediate was determined with a Hilger wave-length spectrometer. The accuracy of the setting of the wave-length drum was checked by the emission lines of an electrical sodium burner (Zeiss). For the other observations calibrated pocket spectroscopes (Brown, Zeiss) were used. Such straight vision instruments of small dispersion permit a better definition of absorption bands and an easier spotting of weak bands than big spectroscopes of the Kirchhoff-Bunsen type. Most of the observations were made visually, but the whole spectral cycle was also photographically recorded by means of a miniature roll film camera of wide

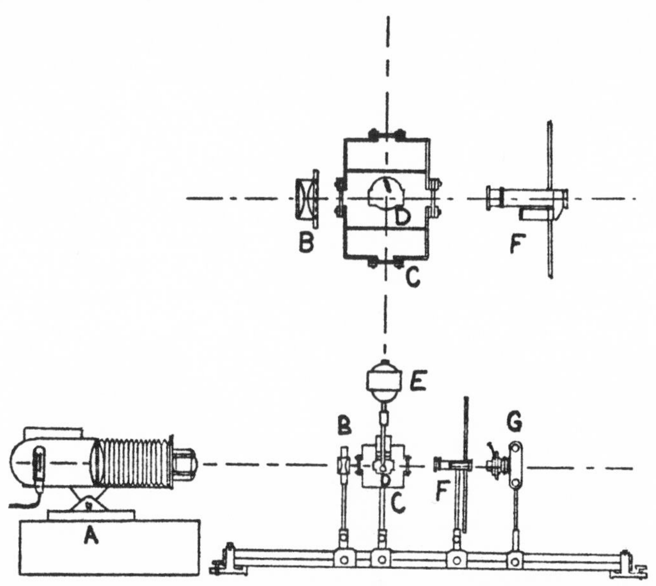


FIG. 5. Schematic representation of the arrangement used for optical study of the enzyme-substrate compound: *A* represents the projection lantern; *B*, condenser; *C*, trough for cooling; *D*, experimental vessel; *E*, motor and stirrer; *F*, pocket spectroscope; *G*, roll film camera.

aperture ($f = 2.9$) and supersensitive panchromatic film (Eastman).

In those experiments where rapid mixing of enzyme and substrate was desired, a Pyrex tubularly-shaped cell with fused on windows and an inserted glass stirrer was used. The enzyme, buffer, and substrate solutions were stratified above each other by means of different concentrations of sucrose and the motor stirrer started at 0 time as suggested by Stadie (13). It was ascertained that sucrose in the concentrations used is no inhibitor for the enzyme reaction. The details of the arrangement are shown in Fig. 5 which is largely self-explanatory.

*Qualitative Observations*—On direct visual observation in transmitted light and in the thickness of layer used for the spectroscopic experiments the enzyme solutions appear brown in color. Upon the addition of monoethyl hydrogen peroxide, there is a rapid color change to a greenish hue. Within the following

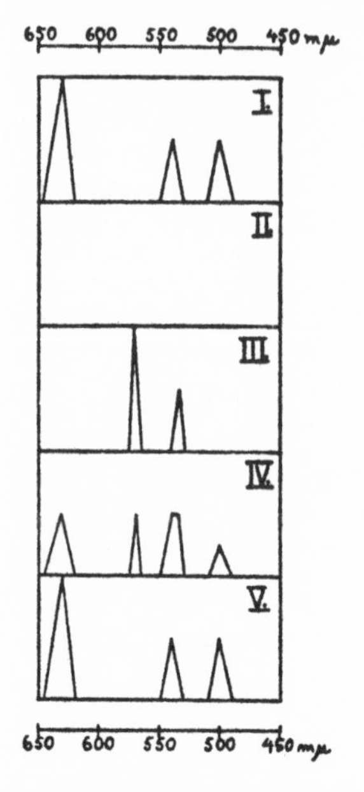


Fig. 6. Schematic representation of the spectroscopic cycle. *I* represents the spectrum of free enzyme; *II*, lag period during which a greenish color but no discrete absorption bands are noticed; *III*, spectrum of enzyme-substrate compound; *IV*, coexisting intermediate and free enzyme; *V*, restored enzyme spectrum. (After direct observation with the spectroscope. The heights of the bands indicate their visual intensity.)

seconds the red color of the intermediate catalase-peroxide compound appears. In the course of the breakdown of the compound which requires time of the order of minutes, the red tint fades and with the reformation of the free enzyme the original brown color is restored. The corresponding changes in light absorption, as observed with a small spectroscope, are represented schematically in Fig. 6. No specific absorption bands are visible during the greenish transition period (*II*) It is therefore possible that this color shade is due neither to the enzyme nor to an intermediate but to concomitant pigments in the enzyme solutions (biliverdin, hepatoflavin), the absorption bands of which are located in the far red and violet regions respectively.

The whole cyclic phenomenon may again be released by addition of fresh substrate. The restoration of the original enzyme spectrum is accompanied by the disappearance of titratable peroxide from the system.

*Position of Absorption Bands of Intermediate*—In order to prolong the visibility of the spectrum of the enzyme-substrate complex, the measurement was carried out at 0°. 5 cc. of Catalase XXIX ($k = 7125$) and 5 cc. of a 0.41 N monoethyl hydrogen peroxide were cooled separately in ice. A small absorption tube of the Baly type with fused on windows and ground joints (Schott and Genossen, Jena) was also cooled. The solutions were mixed and transferred at once to the tube. The tube was submersed in an ice-filled trough. In the Hilger spectrometer, the long wave absorption band of the intermediate was better defined and more intense than the short wave band. Below are given the data obtained together with the values reported by Zeile and Hellström (4) for the uncombined enzyme, with which values our own observations were in satisfactory agreement.

*Spectrum of Free Enzyme*—

I. 650...$\underbrace{646\text{–}620}_{629}$...610; II. $\underbrace{550\text{–}530}_{540}$...520...$\underbrace{510\text{–}490}_{500}$ mμ

*Spectrum of Enzyme-Substrate Compound*—

I. $\underbrace{576\text{–}564}_{570}$; II. $\underbrace{540\text{–}529}_{534.5}$ mμ

*Effect of Temperature on Rate of Formation of Intermediate*—The arrangement shown in Fig. 5 was used, with the exception of the film camera. The vessel with the electric stirrer in place was dipped into the trough filled with water of room temperature (22°). The bottom layer in the experimental vessel consisted of 15 cc. of a mixture of 10 parts of Catalase XXXVI ($k = 4525$) and 5 parts of a sucrose solution prepared by dissolving 100 gm.

of pure sucrose in 100 cc. of water. A layer of 2.5 cc. of a mixture of 20 parts of M/15 phosphate buffer (pH 6.79) and 5 parts of the sucrose solution was carefully stratified above the bottom layer. The top or third layer was formed by 2.5 cc. of 2.07 N monoethyl hydrogen peroxide solution containing no sucrose. The electric stirrer was started at 0 time, and the interval lapsing before the stronger absorption band of the enzyme-substrate compound at 570 $m\mu$ could be first detected by visual observation through the pocket spectroscope was recorded. For three runs the time intervals 4.5, 3.6, and 3.4 seconds were obtained, averaging 3.8 seconds. In the following series the side compartments of the external trough were filled with ice. They were separated from the middle compartment which contained water, by means of metal gauze. Before beginning the experiment the vessel containing the three layers of solution was cooled for at least 15 minutes. The temperature of the ice water surrounding the vessel was 2°. Under these conditions the time intervals recorded in three trials were 6.3, 6.4, and 6.3 seconds, averaging 6.3 seconds. It follows that from 2–22° the rate of formation of the intermediate increases 1.6 times. Assuming a linear relationship as in the case of the over-all reaction, $Q_{10}$ in this interval would be $\sqrt{1.6} = 1.26$. This value is smaller than $Q_{10}$ for the over-all reaction (Table III). It is preferable not to base calculations of the apparent critical increment of the combination of the enzyme with the substrate on this result which was obtained with a primitive technique. It is probable, however, that the apparent activation energy of this reaction is somewhat smaller than that of the over-all reaction, and more specifically of the breakdown of the enzyme-substrate compound.

*Effect of Hydrogen Ion Concentration*—With the external trough filled with water of room temperature (22°), similar experiments were conducted with buffers of different pH. It was found that variation of pH between 4.11 and 8.82 is without a significant effect on the rate of the formation of the enzyme-substrate compound. Nine readings at four different hydrogen ion concentrations yielded an average time value of 4.0 seconds, the highest figure obtained being 5.3 and the lowest 3.4 seconds.

The pH values given were obtained by measurements with the glass electrode after the experiment.

*Experiments on Reality of Observed Lag Period*—The time necessary for complete mixing under the conditions of the preceding experiments was determined by measuring the length of time required for uniform distribution throughout the system of a top layer containing methylene blue dissolved in water. The bottom layer consisted of 10 cc. of water and 5 cc. of sucrose solution; the middle layer contained 2.5 cc. of a mixture of 20 parts of water with 5 parts of sucrose solution. Direct visual observation yielded a time value of 1.6 and 1.4 seconds, respectively. If the spectroscope was used, values of 2 seconds were obtained in two experiments. It follows that although the time necessary for complete mixing was comparatively long, the above rates observed for the appearance of the intermediate were distinctly slower than would be warranted by the mixing time. However, it appeared desirable to decrease the mixing time considerably so as to make it negligible compared with the observed time lag in the formation of the intermediate. Dr. W. C. Stadie, who had used a similar arrangement for the study of carbonic anhydrase (13), was good enough to advise the author on this aspect. Consequently, the rate of the stirring device was increased. Furthermore, the specific gravity of the different layers was lowered by the use of more dilute sucrose solutions. In order to avoid the foaming which occurred at the great speed of stirring, it was necessary but also sufficient to connect the motor with a Morse telegraph switch and to press the switch down for the length of time necessary to start the stop-watch. The sucrose content of the bottom layer in the final experiments was 5 per cent, in the middle layer 2.5 per cent, and in the top layer 0. By means of methylene blue control experiments, it was found that a stirring time of 0.2 to 0.3 second was sufficient to obtain uniform mixing. In three runs with the enzyme and substrate mixture at pH 6.6 and 22° the time lapsing before the first appearance of the stronger absorption band of the intermediate was found to be 1.5, 2.3, and 2.0 seconds, respectively. Since in these last experiments the mixing time was small compared with the observed lag period, some confidence may be placed in the reality of its existence.

*Amount of Substrate Required for Complete Transformation into Intermediate*—It had been observed in preliminary experiments that if a small amount of monoethyl hydrogen peroxide was added to a highly active catalase solution, the absorption band of the enzyme with the center at 629 $m\mu$ was not completely extinguished but that a certain residual absorption in this region remained while the absorption bands of the enzyme-substrate compound appeared. The amount of substrate required for complete transformation into the intermediate was determined by adding the substrate solution from a burette to the enzyme solution at low temperature under spectroscopic observation. The titration has to be carried out rapidly, as the breakdown of the intermediate also yields amounts of free enzyme increasing with time.

2.0 cc. of Catalase XXIX ($k$ = 7125) required 2.95 and 3.0 cc. of 0.94 N monoethyl hydrogen peroxide (neutralized with phosphate buffer) for the complete temporary abolition of the enzyme absorption band in the red. In a later experiment, 15 cc. of Catalase XXXVI ($k$ = 4525) required 6.55 cc. of a 2.07 N peroxide solution. Both titrations were performed at 3–4°.

The number of enzyme molecules present may be computed by using the result of Haldane's (14) calculations, based on Zeile's (4) experimental data, according to which 1 molecule of liver catalase will decompose at 0° and at a substrate concentration of 0.01 M, $5.4 \times 10^4$ molecules of hydrogen peroxide per second. On Haldane's assumption that 1 enzyme molecule contains 1 iron atom and therefore one active center and with the molecular weight of catalase taken to be 68,000, as found by measuring the rate of diffusion of the enzyme (Stern (15)), it can readily be shown that under the conditions of the assay 1.26 mg. of enzyme will decompose 1 mM of hydrogen peroxide per second. On this basis Catalase XXIX ($k$ = 7125) contained 0.214 mg. of enzyme per cc. or $3 \times 10^{-6}$ mM of catalase. In the titration experiment cited above, 2 cc. of this enzyme solution or $6 \times 10^{-6}$ mM of catalase required 3 cc. of 1 N monoethyl hydrogen peroxide or 3 mM of the substrate for the optical end-point. Therefore 1 enzyme molecule requires $5 \times 10^5$ substrate molecules for the complete transformation into the intermediate. From the titration experiment with Catalase XXXVI (see above) a very similar figure may be obtained.

Obviously this ratio does not imply necessarily that 1 molecule of the enzyme-substrate compound consists of 1 catalase molecule and $5 \times 10^5$ peroxide molecules. It simply means that an excess of substrate of this order is required to shift the equilibrium in the reaction

$$\text{Enzyme} + \text{substrate} \rightleftharpoons \text{enzyme-substrate compound}$$

entirely to the right. In other words, the probability that each enzyme molecule, whether originally uncombined or whether released on the breakdown of the intermediate, will recombine immediately with fresh substrate molecules, approaches unity. The figure is therefore rather of statistical than of a stoichiometrical character. In particular, it does not indicate how many substrate molecules will combine simultaneously with 1 enzyme molecule.

The figure given may require a correction for two reasons. The lag period mentioned above has not been taken into consideration in the titration experiment. This would tend to make the ratio appear too large. On the other hand, if 1 enzyme molecule does not contain 1 but 4 iron atoms, as does hemoglobin, then the ratio given would have to be multiplied by 4.

*Formation of Intermediate While Enzyme Is Adsorbed*—As has been reported in a recent note (16), the spectrum of the enzyme remains unchanged while it is adsorbed on aluminum hydroxide gel or on silica gel. On addition of monoethyl hydrogen peroxide to the adsorbate in suspension, the peroxide is decomposed and the spectrum of the intermediate is observed as with freely dissolved enzyme. It follows that the combination of the enzyme with the substrate takes place at a grouping of the catalase molecule which is different from that which is attached to the adsorbent. Since the spectrum of the enzyme is not altered by the adsorption but is changed by the addition of substrate, the conclusion appears to be warranted that combination of enzyme and adsorbent takes place by means of the protein carrier of the enzyme, while the substrate combines with the hemin group of the enzyme, causing the specific light absorption in the visible range of the spectrum.

## DISCUSSION

*Specificity of Catalase*—Catalase has been considered to be the prototype of an enzyme exhibiting an absolute specificity. Heretofore the only substrate known to be attacked by this enzyme was hydrogen peroxide. It has been maintained that neither alkylated peroxides (17) nor substances of the type of perbenzoic or peracetic acid (18) are affected by catalase. This seemed to provide a significant contrast to peroxidase which in the presence of suitable oxygen acceptors may not only utilize hydrogen peroxide but also ethyl hydrogen peroxide and, to a lesser degree, peracetic acid (19, 20). That monoethyl hydrogen peroxide shows some affinity for catalase was demonstrated several years ago (11). It was then found that the addition of this compound to a system containing catalase and hydrogen peroxide will cause an inhibition of the decomposition of the latter. By varying the hydrogen peroxide concentration alone, it was shown that the inhibition is decreased by increasing the hydrogen peroxide concentration. It was therefore concluded that the inhibitory effect of ethyl hydrogen peroxide is of the competitive type. In those experiments, as in the work of the other authors, highly diluted catalase preparations were used. It was only when the substituted peroxide was added to very concentrated enzyme solutions that the decomposition of this substrate and the formation of a spectroscopically visible intermediate were discovered (3). This fact is proof that the specificity of catalase is of a relative and not of an absolute character. Hydrogen peroxide appears to represent the most readily attacked substrate, while the monoethyl hydrogen peroxide is affected at a much smaller rate. However, as the present study shows, the decomposition of this compound by catalase may be studied in a manner similar to the cleavage of hydrogen peroxide provided sufficiently active enzyme preparations are used.

The physiological function of catalase has hitherto been discussed in a purely speculative manner, inasmuch as all attempts to demonstrate the occurrence of its supposedly unique substrate, hydrogen peroxide, in the cells of higher animals have failed. The demonstration that a substituted organic peroxide may serve as a substrate of this enzyme together with the fact that sufficiently high catalase concentrations to effect such a catalysis have been found to exist in the liver of mammals suggests a new approach to the problem.

*Comparison between Hydrogen Peroxide and Ethyl Hydrogen Peroxide As Substrates of Catalase*—Two points of major interest emerge from the comparative study of the action of liver catalase on these two substrates. Even at low temperatures it has not yet been possible to observe the appearance of the spectrum of an intermediate when hydrogen peroxide is decomposed. The only explanation suggested at present is a greater rate of breakdown of the enzyme-substrate compound in this case (see Table III).[2] On decomposition of hydrogen peroxide molecular oxygen and water are the reaction products. Monoethyl hydrogen peroxide, on the other hand, yields no gaseous products on enzymatic cleavage but acetaldehyde and other unidentified compounds. The aldehyde formation from a simple organic peroxide is obviously of physiological interest.

TABLE III

*Catalysis of Hydrogen Peroxide and Ethyl Hydrogen Peroxide by Liver Catalase (Over-All Reaction)*

The data for hydrogen peroxide as substrate were taken from the literature (*cf.* Haldane and Stern (10); Stern (11)).

| | Hydrogen peroxide H—O—O—H | Ethyl hydrogen peroxide $C_2H_5$—O—O—H |
|---|---|---|
| Kinetics of over-all reaction at 0° | Monomolecular | Monomolecular |
| " " " " " 20° | " * | 0 order |
| pH, activity optimum | 6.5–9 | 10 |
| 1st Michaelis constant ($K_m$), M | 0.033 | 0.04 |
| 2nd " " " " | 0.4 | Same order |
| Temperature coefficient, $Q_{10}$ | 1.4 (0–20°) | 2.2 (0–20°) |
| No. of substrate molecules destroyed by 1 enzyme molecule at 0°, pH 6.6, total substrate concentration 0.02 N | $5.4 \times 10^4$ | $1.2 \times 10^2$† |

* The velocity constants are decreasing with time, owing to enzyme inactivation by the substrate.

† Calculated from the determination incorporated in the curve of Fig. 4.

In Table III are compiled some features of the catalysis of both substrates by liver catalase.

*Comparison between Catalase and Methemoglobin As Catalysts*—

[2] In order to observe the spectrum of the intermediate compound the enzyme must be kept saturated with substrate. Since 1 molecule of catalase, if saturated with substrate, splits $2 \times 10^5$ molecules of $H_2O_2$ per second at 0° (Haldane (14)), for a catalase preparation of $k = 7000$ a $H_2O_2$ concentration of 6 M would be required for an observation time of 10 seconds. This is experimentally not feasible because the reaction will proceed at an explosive rate and because the enzyme is quickly destroyed by high $H_2O_2$ concentrations.

The methemoglobin catalysis of hydrogen peroxide and ethyl hydrogen peroxide is more than a model reaction. Catalase and methemoglobin have an identical prosthetic group, namely parahematin (12). The much smaller catalytic efficiency of methemoglobin is due to the difference in the nature of the protein carrier. Methemoglobin will only decompose of the order of $10^{-2}$ hydrogen peroxide molecules per catalyst molecule per second (21). Figures for the ethyl hydrogen peroxide catalysis are not yet available. In contrast to catalase, methemoglobin will form intermediate compounds of a distinct absorption spectrum with both substrates. The pattern of the spectra of the two unstable complexes resembles closely that of the catalase-ethyl hydrogen peroxide compound. The complex of methemoglobin with hydrogen peroxide was discovered 36 years ago by Kobert (22). Its light absorption and composition were recently studied by Haurowitz (23). Haurowitz concludes that the molecular ratio in this case is unity and that the complex contains ferric iron to which the peroxide is linked by coordinative valencies. The complex of methemoglobin with ethyl hydrogen peroxide was recently described by the author (3) and independently by Keilin and Hartree (24). An excess of only 8 peroxide molecules per methemoglobin molecule is sufficient to suppress the absorption band of methemoglobin in the red, compared with an excess of the order of $10^5$ molecules in the case of catalase. The ratio of these figures resembles that of the catalytic activity of the enzyme and of the blood pigment.

*On Mechanism of Enzyme Reaction*—The present study sheds some light on the mechanism of an enzyme action. The catalyst operates by providing a new path of reaction which leads over an intermediate composed of enzyme and substrate molecules. This compound is unstable but has a mean span of life sufficient to allow for direct observation. It should be emphasized that the interpretation of the observations reported in the present paper is consistent with but not dependent on the validity of the evidence offered for the constitution of the enzyme (4, 12). It is felt that the interpretation is justified by the agreement of the data obtained by the optical and volumetric methods employed. The time required by the spectral cycle to go to completion equals the time required for complete decomposition of the substrate.

The close analogy with the non-enzymatic methemoglobin catalysis suggests a similar constitution of the intermediate compounds. In agreement with the interpretation by Haurowitz (23) of the methemoglobin-hydrogen peroxide complex, the catalase-ethyl hydrogen peroxide complex may be depicted as a covalency compound, where the peroxide molecule is linked to the coordinately tetravalent ferric iron of the hematin group of the enzyme. It appears probable that a similar intermediate occurs during the catalase-hydrogen peroxide catalysis, though hitherto attempts to demonstrate its formation have failed.

It is shown that the rate of formation of the enzyme-substrate compound is rapid compared with the rate of the over-all reaction. The kinetics of the latter are therefore governed by other steps in the series of reactions. The effect of hydrogen ion concentration and of temperature as observed in the study of the total process obviously concerns the later reaction phases. It is quite possible that the actual course of events is more complex than the spectroscopic findings would indicate. The product molecules formed by the breakdown of the intermediate may be radicals, initiating a chain reaction in which the original catalyst no longer participates (11, 25). The dependence of the rate of the over-all reaction on temperature indicates that at any given instant only a fraction of the molecules of the intermediate is in an activated state.

It remains to be seen to which extent the findings of this study apply to enzyme action in general.

SUMMARY

1. A study was made of the decomposition of monoethyl hydrogen peroxide by liver catalase. A volumetric procedure was used for the assay of the peroxide. In this manner, the kinetics and the effect of temperature, of pH, of varying the substrate concentration, and of cyanide on the enzyme reaction were studied. The results are compared with those obtained with hydrogen peroxide as substrate.

2. In the course of the enzymatic process there is formed an intermediate compound with a characteristic absorption spectrum. The intermediate is unstable; it breaks down to form free enzyme and reaction products. It exhibits the properties postulated by

Michaelis and Menten for an enzyme-substrate compound. A preliminary optical study of this compound has revealed that it is not a mere adsorption complex; that the rate of formation of the enzyme-substrate compound is great compared with that of the total reaction; that it has a smaller temperature coefficient than the over-all reaction; that it is independent of pH between 4 and 9.

BIBLIOGRAPHY

1. Henri, V., Lois générales de l'action des diastases, Paris (1903).
2. Michaelis, L., and Menten, M. L., *Biochem. Z.*, **49**, 333 (1913).
3. Stern, K. G., *Nature*, **136**, 335 (1935).
4. Zeile, K., and Hellström, H., *Z. physiol. Chem.*, **192**, 171 (1930).
5. Baeyer, A., and Villiger, V., *Ber. chem. Ges.*, **34**, 738 (1901).
6. Rieche, A., *Ber. chem. Ges.*, **62**, 218 (1929).
7. Rieche, A., and Hitz, F., *Ber. chem. Ges.*, **62**, 2473 (1929).
8. Rieche, A., Alkylperoxyde und Ozonide, Dresden, 23 (1931).
9. Stern, K. G., *Z. physiol. Chem.*, **204**, 259 (1932).
10. Haldane, J. B. S., and Stern, K. G., Allgemeine Chemie der Enzyme, Dresden (1932).
11. Stern, K. G., *Z. physiol. Chem.*, **209**, 176 (1932).
12. Stern, K. G., *J. Biol. Chem.*, **112**, 661 (1936).
13. Stadie, W. C., and O'Brien, H., *J. Biol. Chem.*, **103**, 521 (1933).
14. Haldane, J. B. S., *Proc. Roy. Soc. London, Series B*, **108**, 559 (1931).
15. Stern, K. G., *Z. physiol. Chem.*, **217**, 237 (1933).
16. Stern, K. G., *Science*, **83**, 190 (1936).
17. Bach, A., and Chodat, R., *Ber. chem. Ges.*, **36**, 1756 (1903).
18. Freer, P. C., and Novy, F. G., *J. Am. Chem. Soc.*, **27**, 161 (1902).
19. Grimmer, W., *Milchwirtschaft. Zentr.*, **44**, 246 (1915).
20. Wieland, H., and Sutter, H., *Ber. chem. Ges.*, **63**, 73 (1930).
21. Stern, K. G., *Z. physiol. Chem.*, **215**, 35 (1933).
22. Kobert, R., *Arch. ges. Physiol.*, **82**, 603 (1900).
23. Haurowitz, F., *Z. physiol. Chem.*, **232**, 159 (1935).
24. Keilin, D., and Hartree, E. F., *Proc. Roy. Soc. London, Series B*, **117**, 1 (1935).
25. Haber, F., and Willstätter, R., *Ber. chem. Ges.*, **64**, 2844 (1931).

# 46



Reprinted from *Nature* **162**:963 (1948)

# INTERPRETATION OF EXPERIMENTS ON METABOLIC PROCESSES, USING ISOTOPIC TRACER ELEMENTS

**A. G. Ogston**

*Department of Biochemistry, Oxford*

In two instances, the distribution of isotopic carbon in the product of a metabolic process has been used to infer that a symmetrical intermediate compound is not involved. Wood *et al.*[1] showed that isotopic carbon, introduced as carbon dioxide together with pyruvate, led to the formation of ketoglutarate which contained isotopic carbon only in the carboxy group next to the keto group; on these grounds, they excluded citrate as an intermediate. Shemin[2] found that when glycine is formed from serine, containing isotopic nitrogen and isotopic carbon in its carboxy group, the relative abundance of nitrogen-15 and carbon-13 in the glycine was the same as in the serine; he argued that amino-malonic acid is therefore not an intermediate in this process.

These conclusions seem to arise from the fallacy that, because symmetrical products arising from the *d*- or *l*-form of an optically active precursor cannot be distinguished, therefore the two identical groups of a symmetrical product formed from one optical antipode cannot be distinguished. On the contrary, it is possible that an asymmetric enzyme which attacks a symmetrical compound can distinguish between its identical groups. This power of distinction is illustrated for the case of amino-malonic acid in the accompanying formulæ. *a′*, *b′*, *c′* represent points in the enzyme which specifically combine with the groups *a*, *b*, *c* of the substrate. Evidently, decarboxylation could occur at *a′* but not at *b′*, or *vice versa*. The same conclusion follows if any three of the groups of the substrate specifically combine with the enzyme. If decarboxylation occurs only at *b′*, then all the isotopic carbon in the carboxy group of serine will appear in the glycine. The case of the formation of ketoglutarate from citrate is exactly parallel.

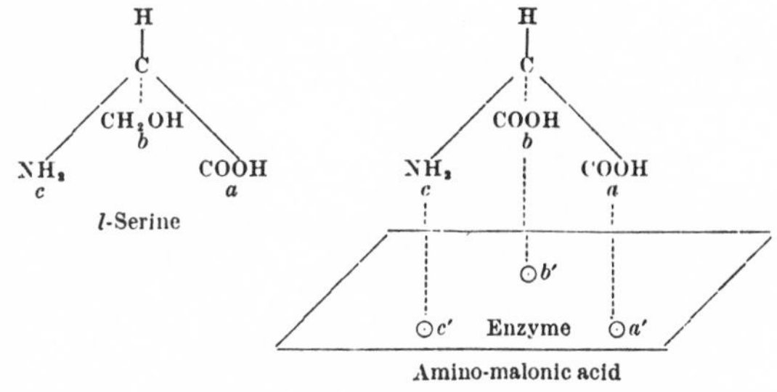


This argument depends on two conditions: (*a*) that the sites *a′* and *b′* are catalytically different; (*b*) that three-point combination occurs between the symmetrical substrate and the enzyme; such combination is, of course, necessary wherever a single optical antipode is formed enzymatically from an inactive precursor. Since neither is unlikely, it follows that the asymmetrical occurrence of isotope in a product cannot be taken as conclusive evidence against its arising from a symmetrical precursor.

[1] Wood, Werkman, Hemingway and Nier, *J. Biol. Chem.*, **139**, 483 (1941).

[2] Shemin, *J. Biol. Chem.*, **162**, 297 (1946).

# 47



Reprinted from *J. Biol. Chem.* **202**:687–697 (1953)

# THE ENZYMATIC TRANSFER OF HYDROGEN*

## I. THE REACTION CATALYZED BY ALCOHOL DEHYDROGENASE

BY HARVEY F. FISHER,† ERIC E. CONN,‡ BIRGIT VENNESLAND, AND F. H. WESTHEIMER

*(From the Departments of Biochemistry and Chemistry, University of Chicago, Chicago, Illinois)*

(Received for publication, November 17, 1952)

In preliminary communications (1, 2) the authors have reported that the reaction catalyzed by yeast alcohol dehydrogenase involves a direct transfer of hydrogen from ethanol to diphosphopyridine nucleotide (DPN). The essential experiments were performed with 1,1-dideuteroethanol, with which it was shown that the reaction proceeds according to Equation 1.

$$CH_3CD_2OH + \text{(pyridinium ring with } CONH_2\text{, } N^+\text{–R)} \underset{\text{dehydrogenase}}{\overset{\text{Alcohol}}{\rightleftharpoons}} CH_3CDO + H^+ + \text{(dihydropyridine ring with D, H, } CONH_2\text{, N–R)} \quad (1)$$

The present paper discusses the details of the tracer experiments, and the stereochemistry of the monodeutero reduced DPN.

### *Materials and Methods*

*1,1-Dideuteroethanol*—7 gm. (0.17 mole) of isotopically pure lithium-aluminum deuteride, 93 per cent active (purchased from Metal Hydrides, Incorporated), were suspended in 500 ml. of diethyl ether in a three-necked

* This investigation was supported in part by a research grant G-3222 from the National Institutes of Health, United States Public Health Service, and by the Dr. Wallace C. and Clara A. Abbott Memorial Fund of the University of Chicago. Part of this material is taken from a thesis submitted by H. F. Fisher in partial fulfilment of the requirements for the degree of Doctor of Philosophy.

† Predoctoral Fellow, Atomic Energy Commission, 1951–52. Present address, Department of Biochemistry, College of Physicians and Surgeons, Columbia University, New York.

‡ Present address, Division of Plant Nutrition, University of California, Berkeley.

flask equipped with a reflux condenser, dropping funnel, and stirrer, and stirred under nitrogen for 2 hours. Then 0.3 M of phenyl acetate in 40 ml. of ether was added to the slurry at such a rate as to maintain gentle refluxing (3) (samples of lithium-aluminum deuteride, purchased more recently from Metal Hydrides, Incorporated, have proved almost completely ether-soluble). Refluxing was continued for 2 hours, after which the reaction mixture was decomposed by cautious addition, first of wet ether and then of 50 ml. of water. The water-ether system was made alkaline to litmus, and an additional 0.1 M of sodium hydroxide was added in order to saponify any ethyl acetate which might have been formed during the reaction by ester exchange. The water-alcohol azeotrope was obtained by fractionating the heterogeneous mixture through an 18 inch Podbielniak column. The fraction boiling from 77.8–78.3° was dried over calcium oxide, the dry alcohol containing 2.06 atoms of deuterium per molecule. The yield, based on $LiAlD_4$, was 55 per cent.

*Yeast Alcohol Dehydrogenase*—Alcohol dehydrogenase from yeast was prepared by the method of Racker (4). The lyophilized protein usually had about 10 per cent of the activity of the preparation he described.

*Buffer*—Tris (tris(hydroxymethyl)aminomethane, or $\beta,\beta,\beta$-trihydroxy-*t*-butylamine) was purchased from the Commercial Solvents Corporation, recrystallized from acetone, and dried under a vacuum.

*DPN*—DPN of varying purity was purchased from the Sigma Chemical Company. When necessary, the purity was brought to 80 to 90 per cent by the method of Horecker and Kornberg.[1]

Both DPN and reduced DPN were measured by enzymatic assay with alcohol dehydrogenase. As pointed out by Racker (4), DPN may be completely reduced by an excess of ethanol at pH 9.0. The increase in optical density at 340 m$\mu$ was measured with a Beckman model DU quartz spectrophotometer. A molar extinction coefficient of $6.22 \times 10^6$ sq. cm. per mole (5) was used to make the calculations. The decrease in optical density which occurred when reduced DPN was oxidized with acetaldehyde at pH 7.4 was used similarly to determine reduced DPN. Per cent purity on a weight basis was calculated by assuming a molecular weight of 663 for oxidized DPN (*i.e.*, the material is assumed to exist as the free acid). The chemically reduced DPN was assumed to be present as a hydrated disodium salt and Ohlmeyer's correction for water and sodium was applied to give a molecular weight of 790. The enzymatically reduced DPN was assumed to be present as the di-Tris salt with a calculated molecular weight of 907.

*Chemical Reduction of DPN in $D_2O$*—The procedure of Ohlmeyer (6) was used for the preparation of chemically reduced DPN. In a typical experi-

[1] Horecker, B. L., and Kornberg, A., private communication.

ment, the reaction mixture contained 100 mg. of DPN, 104 mg. of $NaHCO_3$, and 50 mg. of sodium hydrosulfite in 8 ml. of $D_2O$ (99.98 per cent). The exchangeable hydrogen in the DPN and in the bicarbonate could not have diluted the $D_2O$ by more than 0.4 per cent. The reduced DPN isolated from the methanol-, ethanol-ether mixture of Ohlmeyer was dissolved in a few ml. of water which were removed under a vacuum. This procedure eliminates essentially all exchangeable deuterium from the reduced DPN. Its deuterium content was not changed by a repetition of the washing procedure.

*Enzymatic Reduction of DPN*—Similar procedures were used for the reduction of DPN by dideuteroethanol in a medium of $H_2O$ and for the reduction of DPN by normal ethanol in a medium of $D_2O$. In both cases the medium was buffered with a Tris-HCl mixture prepared by adjusting the pH of a solution of the amine with concentrated HCl. Since there was no appreciable quantity of metallic cations in the reaction mixture, the DPN was isolated as the amine salt. This salt was not precipitated by ethanol until most of the Tris had been converted to the acid form. Several fractions of reduced DPN were usually obtained by successive addition of acid and ethanol. The yield and purity of each of these fractions were not exactly reproducible under the conditions employed but the results were uniformly satisfactory, and the method is recommended for the preparation of solid reduced DPN free from inorganic salts.

A typical experiment was carried out as follows. A solution at pH 8.8 was prepared from 8 ml. of $H_2O$, 604 mg. of Tris, 0.052 ml. of concentrated HCl, 90 mg. of DPN (of 93 per cent purity), and 2.6 mg. of alcohol dehydrogenase. To this solution, 0.4 ml. of 1,1-dideuteroethanol and 1.6 ml. of $H_2O$ were added. From time to time, small aliquots were removed, diluted with buffer at pH 8.8, and assayed spectrophotometrically. Maximal light absorption at 340 m$\mu$ was reached within less than 30 minutes. At this time the flask was placed in a boiling water bath for 75 seconds to denature the enzyme. The flask was cooled, 20 ml. of ethanol were added, and the precipitate, which consisted mainly of denatured protein, was removed by centrifugation and discarded. The supernatant fluid was diluted with 84 ml. of ethanol, and the pH reduced to 7.4 by the careful addition with stirring of 0.4 ml. of 6 N HCl (the pH must not go below 7.0 because of the instability of reduced DPN in acid medium). The solution was kept at $-17°$ for 4 hours and then centrifuged. The precipitate was washed with cold alcohol, then with ether, and finally dried in a vacuum. The supernatant fluid was diluted with an additional 50 ml. of alcohol containing 0.12 ml. of 6 N HCl, and the isolation procedure repeated. Finally, a third crop of reduced DPN was similarly obtained. In order to wash out all exchangeable deuterium, each precipitate was dissolved in 2 ml. of a Tris buffer of pH 7.4 and reprecipitated with 30 ml. of cold alcohol. Yields of reduced DPN: Precipitate 1, 57 mg. (73 per cent purity); Precipitate 2, 12 mg. (82 per cent purity); Precipitate 3, 13 mg. (49 per cent purity).

*Enzymatic Oxidation of Reduced DPN*—A solution of pH 7.4 was prepared from 2.8 ml. of $H_2O$, 0.12 ml. of 6 N HCl, 181 mg. of Tris, 30 mg. of reduced DPN, and 3 mg. of alcohol dehydrogenase. To this solution, 0.3 ml. of a 2 per cent solution of acetaldehyde was added. After 30 minutes, spectrophotometric examination showed that the reaction was complete. The reaction flask was placed in a boiling water bath for 75 seconds, cooled, and 3 ml. of ethanol were added. The precipitate was removed by centrifugation and discarded. Then 0.04 ml. of 6 N HCl and 15 ml. of ethanol were added to the supernatant fluid, which was allowed to stand at −17° for 5 hours. The mixture (pH around 3.5) was centrifuged, and the precipitated DPN washed with alcohol and ether before it was dried in a vacuum.

*Acetaldehyde from Enzymatic Oxidation of Alcohol*—200 mg. of DPN were added to 9.6 ml. of a cold 0.17 M pyrophosphate buffer of pH 8.8. (Tris buffer was not used because of the ability of the amine to bind acetaldehyde.) Then 0.4 ml. of dideuteroethanol and 10 mg. of alcohol dehydrogenase were added, and the flask was stoppered and warmed to 18°. After 2 hours, the flask was cooled to 0°, an additional 5 mg. of enzyme were introduced, and the mixture was again warmed to 18° for 2 hours. Then the flask was connected through a Claisen distilling head to two absorption tubes, each of which contained 10 ml. of a saturated solution (about 0.4 gm. per 100 ml.) of "methone" (dimethyldihydroresorcinol). A stream of nitrogen was passed through the flask and the solution was heated to inactivate the enzyme. The stream of nitrogen was continued for an hour. Then the precipitate from the absorption tubes was collected by centrifugation and recrystallized from methanol. Yield, 2 mg., m.p. 142° (melting points for ethylidene bisdimethyldihydroresorcinol have been recorded from 140–143° (7, 8)). A weighed sample of the derivative was diluted, before analysis for deuterium, with some methone derivative of ordinary acetaldehyde.

*Deuterium Analyses*—The compounds containing deuterium were burned and the water was collected and converted to hydrogen over zinc at 400°, according to the method of Graff and Rittenberg (9). The deuterium content of the hydrogen was then determined with a Consolidated-Nier model No. 21-201 isotope ratio mass spectrometer (10).

*Calculations*—For pure compounds, the calculation of the atoms of deuterium per molecule is quite straightforward. DPN and reduced DPN, however, create rather special problems because the isolated solids always contain some impurities. In this work, the purity of the samples ranged from 40 to 90 per cent.

Two different methods were used to determine the deuterium content of this DPN. Both methods depend on an assumption that all the deuterium present in any sample is actually present in the DPN or reduced DPN, and not in the enzymatically inactive impurities. The validity of this assumption is supported by the fact that, when samples of reduced DPN of varying purity are isolated from the same experiment, the deuterium content of the samples is almost proportional to their assayed purity. Further proof of the assumption is provided by the experimental results themselves. Thus, all the deuterium present in a sample of enzymatically reduced DPN can be removed from the DPN by enzymatic oxidation with acetaldehyde. Finally, as described in Paper II, all the deuterium introduced into DPN by enzymatic reduction can be transferred completely to pyruvate to give lactate, which can be analyzed in the form of a pure derivative.

In the first method of determining the deuterium content of DPN, the sample was analyzed both for enzymatically active coenzyme and for total hydrogen (by combustion). The material was then further analyzed by the mass spectrometric method for the hydrogen-deuterium ratio, and the atoms of deuterium per molecule calculated. For example, in Experiment C′ (see Table I) a particular sample of reduced DPN assayed 73.6 per cent pure as the Tris salt (mol. wt. 907), and contained 6.36 per cent total hydrogen. Mass spectrometric analysis showed that this hydrogen contained 1.285 atom per cent excess deuterium. 1 gm. of the DPN salt contained $63.6 \times 0.01285/1.008 = 0.81$ mM of excess deuterium, and $736/907 = 0.81$ mM of DPN. Therefore there was 1.00 atom of deuterium per molecule of coenzyme.

In the second method (which was more useful when only limited amounts of DPN were available) a weighed sample of the DPN or reduced DPN salt was diluted with a known quantity of glycine (usually about 5 to 10 times the weight of the DPN), and the resulting mixture was burned and analyzed for deuterium. This method avoids the necessity of analyzing each sample for total hydrogen content, because in the diluted sample most of the total hydrogen comes from the glycine. Under these circumstances, no reasonable error in estimating the total hydrogen content of the coenzyme sample can significantly affect the result. However, it is still necessary to know accurately the purity of the nucleotide and to assume that whatever deuterium is present is incorporated in the coenzyme. For example, in one particular analysis (Experiment F), 1.99 mg. of a sample of oxidized DPN with a purity of 73 per cent were diluted with 8.90 mg. of glycine. The hydrogen of this sample contained 0.144 atom per cent excess deuterium. The glycine contains 0.592 mM of hydrogen. The DPN

(4.1 per cent H, mol. wt. 663) contains 1.99 × 0.73 × 0.041 = 0.060 mM of hydrogen. In addition, the impurities present in the DPN contain some hydrogen. On the basis of analyses of other DPN samples, it was crudely estimated that the hydrogen content of these impurities is 7 per cent. There are, therefore, 0.037 mM of hydrogen in the 0.537 mg of impurities. The sample, then, contains 0.689 mM of hydrogen and 0.689 × 0.00144 = 0.000992 mM of excess deuterium and the atoms of deuterium per molecule of DPN = (0.000992 × 663)/(1.99 × 0.73) =

TABLE I

*Deuterium Transfer with Alcohol Dehydrogenase*

| Experiment | Reductant | Oxidant | Medium | Substance isolated | Purity, per cent* | Atoms D per molecule* | |
|---|---|---|---|---|---|---|---|
| | | | | | | Direct | Diluted |
| A | $Na_2S_2O_4$ | DPN | $D_2O$ | Reduced DPN | 89 | 1.07 | 1.02 |
| | | | | " " | 88 | | 0.92 |
| | | | | " " | 60 | | 1.05 |
| B | $CH_3CH_2OH$ | DPN | $D_2O$ | Reduced DPN | 78 | 0.00 | 0.00 |
| | | | | " " | 40 | 0.06 | |
| C | $CH_3CD_2OH$ | DPN | $H_2O$ | Reduced DPN | 88 | 1.04 | |
| | | | | " " | 82 | 0.97 | |
| C′ | $CH_3CD_2OH$ | DPN | $H_2O$ | " " | 74 | 0.98 | |
| | | | | " " | 82 | 1.00 | |
| D | $CH_3CD_2OH$ | DPN | $H_2O$ | Acetaldehyde dimethone | 100 | | 1.00 |
| E | Reduced DPN from C′ | $CH_3CHO$ | $H_2O$ | DPN | 65 | 0.00 | |
| F | Reduced DPN from A | $CH_3CHO$ | $H_2O$ | DPN | 73 | | 0.44 |

* The figures given are the averages of several determinations.

0.45. An error of 20 per cent in the percentage of hydrogen assumed for the impurities in the DPN will cause an error of only 1 per cent in the calculated atoms of deuterium per DPN molecule.

*Accuracy*—The difficulties encountered in the measurement of the absolute value of the atom per cent excess deuterium in an organic compound have been discussed by Kirshenbaum (11). Although there is some difficulty arising from the fact that an empirical calibration curve for the mass spectrometer must be employed, the major source of error arises from the "memory" of the converter in which the hydrogen samples are prepared. If successive samples relatively high in deuterium are burned just after a sample with lower concentration of the heavy isotope, the percentage of deuterium found generally increases asymptotically with the number of samples analyzed. Until and unless two successive samples of a given mixture give the same analysis, the result must be regarded with suspicion. The data quoted in this paper are those obtained after successive samples gave concordant results. The accuracy was in all cases probably better than ±5 per cent, and was therefore more than sufficient to warrant the conclusions drawn.

### *Results*

The experimental results are summarized in Table I.

*Exchange*—Before the isotopic tracer method could be utilized to elucidate the mechanism of the enzymatic oxidation-reduction reaction, it was necessary to ascertain whether deuterium atoms attached to carbon in the nicotinamide and reduced nicotinamide rings would exchange with the hydrogen atoms of the solvent under the experimental conditions here employed. Reduced DPN prepared by sodium hydrosulfite reduction in $D_2O$ was found to contain 1.0 deuterium atom per molecule (Experiment A). The compound was dissolved in water and isolated again; its deuterium content was unchanged. Hydrogen exchange on the reduced carbon therefore does not occur under the mild conditions of these experiments.

*Enzymatic Reduction of DPN by Ethanol*—Experiments B and C are a complementary pair; in Experiment B, DPN was reduced with ordinary alcohol in $D_2O$; the resulting reduced DPN contained no deuterium, and therefore must have taken light hydrogen directly from the alcohol. In Experiment C, DPN was reduced with 1,1-dideuteroethanol in ordinary (light) water, and the reduced DPN contained 1 deuterium atom per molecule; the conclusion must be reached that the reaction involves direct transfer of deuterium from the alcohol to DPN. Experiment C′ was a duplicate of Experiment C, except that double the amount of enzyme and twice the reaction time were employed.

*Acetaldehyde*—The acetaldehyde produced from 1,1-dideuteroethanol contained 1 deuterium atom per molecule (Experiment D). This completes the evidence that the reaction occurs according to Equation 1. There is no certainty that the reduction occurs at carbon 6 of the nicotinamide ring, as pictured in this equation; the reduction may occur instead at another carbon atom of the ring. The formula for which Karrer *et al.* (12) have expressed a slight preference was here chosen. The conclusions reached, however, are valid, regardless of the particular site of the reduction.

*Stereochemistry of Reaction*—When DPN is reduced by 1,1-dideuteroethanol, a new asymmetric center is introduced (*e.g.* at carbon 6) in the dihydropyridine ring (Fig. 1). If the pyridine ring is regarded as lying in a horizontal plane, then it is proper to speak of the deuterium atom in reduced DPN as lying either above (Fig. 1, *A*) or below (Fig. 1, *B*) this plane; the other position is of course occupied by hydrogen. Since DPN is itself optically active, the two possible stereoisomeric 6-monodeutero reduced DPN molecules are diameric and not enantiomorphic. The possibility arises that the enzymatic reaction may be sterically specific in the sense that only one of the two possible diamers is formed. Evidence for the steric specificity of the reduction was obtained in the following experiments.

In Experiment E, monodeutero reduced DPN was prepared enzymatically, and was then reoxidized with an excess of acetaldehyde. The resulting DPN contained no excess deuterium (Experiment E). In this reaction, therefore, the enzyme selected the deuterium in preference to the hydrogen atom attached to the same carbon atom of the ring. Although there can be, and often is, a difference in the rates of cleavage of carbon-hydrogen as compared with carbon-deuterium bonds (13–15), the carbon-

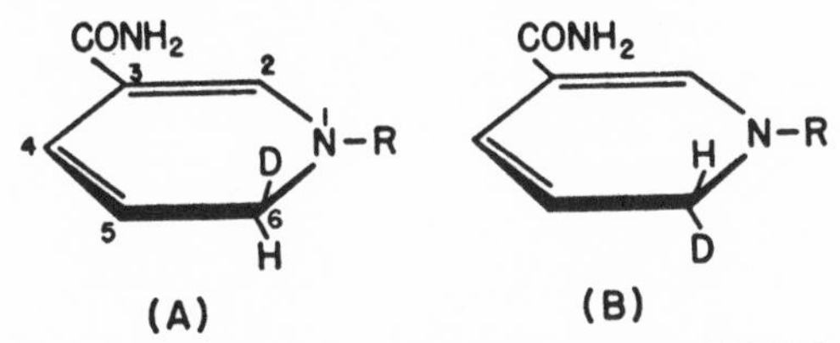


FIG. 1. Diamers of monodeutero reduced DPN

hydrogen bond is cleaved more rapidly (usually 6 times more rapidly) than the carbon-deuterium bond. But, in the present instance, a deuterium atom is removed from carbon in preference to a hydrogen atom. Clearly the preference for deuterium rather than hydrogen demonstrates the stereospecificity of the reaction. The enzyme can promote the addition (or withdrawal) of deuterium only on one side of the pyridine (or dihydropyridine) ring.

The stereospecificity is further demonstrated by the results obtained in the enzymatic oxidation of monodeuterated chemically reduced DPN. The latter would be expected to be a mixture of the diamers, A and B, since the relatively remote asymmetric centers (in the ribose residues) probably can account, at most, for a partial asymmetric synthesis. After enzymatic oxidation of monodeuterated chemically reduced DPN, the DPN isolated contained 0.44 atom of deuterium per molecule (Experiment F). Since chemically reduced DPN is enzymatically active, it is very unlikely that it differs structurally (*i.e.*, in the position of the carbon atom to which the deuterium is attached) from enzymatically reduced material. Since, however, reoxidation of chemically reduced deuterated DPN gives 0.44 deuterium atom per molecule, whereas reoxidation of enzymatically reduced material gives none, it is clear that the two samples are different and it follows that they must differ sterically. In fact, the chemically reduced monodeuterated DPN must contain about 56 per cent of the same diamer (*e.g.*, diamer A) as that obtained enzymatically, and about 44 per cent of the other diamer (*e.g.*, diamer B).

It is possible that the monodeuteroethanol prepared by the enzymatic reduction of acetaldehyde with deutero DPN will consist of only one of the two possible enantiomorphs; this matter is now under investigation.

## DISCUSSION

In their early kinetic studies of yeast alcohol dehydrogenase, Negelein and Wulff (16) assumed a combination between the enzyme and its substrates and measured the apparent dissociation constants of these compounds. More recently, Theorell and Bonnichsen (17) and Theorell and Chance (18) have shown that alcohol dehydrogenase prepared from horse liver combines with reduced DPN to form a compound with an absorption spectrum different from that of the unbound reduced DPN. Our experimental results are in accord with, and supplement, the conclusions reached by the above authors. On the other hand, our results completely exclude all mechanisms (19) which require electronic oxidation-reduction with hydrogen supplied to DPN from the solvent, since the hydrogen atom transferred to DPN by enzymatic reduction with ethanol is the same hydrogen atom that was removed from the $\alpha$ position of the ethanol molecule.

These results may be described as a "direct transfer of hydrogen," a phrase which implies that the hydrogen is donated from 1 molecule to the other without intermediary transfer to any other group. The possibility may be considered that deuterium is transferred from the ethanol to the protein itself, and then, in a subsequent step, from the protein to the coenzyme. It is improbable that a sulfhydryl group could serve as such a hydrogen donor and acceptor, since deuterium attached as a sulfhydryl (S-D) group would almost certainly exchange with the solvent. The function of sulfhydryl groups in some other capacity is not excluded, however. For example, Theorell and Bonnichsen (17) suggested that sulfhydryl groups in liver alcohol dehydrogenase act to bind the coenzyme to the protein, and Racker and Krimsky (20) have provided evidence that sulfhydryl groups in glyceraldehyde-3-phosphate dehydrogenase participate in the enzyme reaction. Direct hydrogen transfer has not yet been demonstrated with either of these two enzymes, but such direct transfer would not necessarily be incompatible with the sulfhydryl functions suggested above.

The possibility that the protein should be reduced in such a way as to contain non-exchangeable deuterium seems, *a priori*, highly unlikely. Furthermore, such a possibility appears to be ruled out by the experiments of Kaplan, Colowick, and Neufeld (21), who found that, with yeast alcohol dehydrogenase, ethanol will reduce either DPN or desamino DPN. Yet when desamino DPN and reduced DPN were incubated with the enzyme in the absence of alcohol, no reduction of desamino DPN by reduced DPN took place.

It seems most probable, therefore, that, in the action of yeast alcohol dehydrogenase upon alcohol and DPN, both substrate and coenzyme are simultaneously absorbed in adjacent and stereochemically defined positions, and that hydrogen is transferred directly from the substrate to the coenzyme.

SUMMARY

1. When DPN is equilibrated with $CH_3CD_2OH$ in the presence of yeast alcohol dehydrogenase, the reduced DPN formed contains one non-exchangeable deuterium atom per molecule.

2. The acetaldehyde formed in the same reaction contains one non-exchangeable deuterium atom per molecule.

3. When DPN is equilibrated with ordinary alcohol in the presence of alcohol dehydrogenase and in a medium of pure deuterium oxide, the reduced DPN formed contains no excess non-exchangeable deuterium.

4. When monodeutero reduced DPN, prepared enzymatically, is reoxidized enzymatically by acetaldehyde, the oxidized DPN formed contains no excess deuterium. When monodeutero reduced DPN, prepared chemically, is reoxidized enzymatically by acetaldehyde, the oxidized DPN formed contains about 0.44 deuterium atom per molecule.

5. From (1), (2), and (3) it is concluded that, in the enzymatic oxidation-reduction, hydrogen is transferred directly from the alcohol to the coenzyme.

6. From (4), it is concluded that the enzymatic reduction is stereospecific with respect to the reduced position of the dihydropyridine ring.

The deuterium oxide and the lithium-aluminum deuteride were purchased on allocation from the Atomic Energy Commission. The funds for the purchase of the mass spectrometer used in most of this research were supplied by the Atomic Energy Commission under contract No. At(11-1)-92.

The authors are indebted to Dr. Peter Ofner, who helped in the purification of DPN, to Dr. H. S. Anker, who assisted with some of the mass spectrometric analyses, and to Dr. William Saschek, who made the microanalyses for carbon and hydrogen.

*Addendum*—Pullman, San Pietro, and Colowick have obtained evidence that DPN is reduced at position 4 of the pyridine ring (private communication from Dr. Colowick). This is compatible with the results recently reported by M. E. Pullman (22). Although this may necessitate a change in our diagrams, it does not change the nature of our argument.

## BIBLIOGRAPHY


1. Westheimer, F. H., Fisher, H. F., Conn, E. E., and Vennesland, B., *J. Am. Chem. Soc.*, **73**, 2403 (1951).
2. Fisher, H. F., Ofner, P., Conn, E. E., Vennesland, B., and Westheimer, F. H., *Federation Proc.*, **11**, 211 (1952).
3. Brown, W. G., in Adams, R., Organic reactions, New York, **6**, 469 (1951).
4. Racker, E., *J. Biol. Chem.*, **184**, 313 (1950).
5. Horecker, B. L., and Kornberg, A., *J. Biol. Chem.*, **175**, 385 (1948).
6. Ohlmeyer, P., *Biochem. Z.*, **297**, 66 (1938).
7. Gee, A. H., and Chaikoff, I. L., *J. Biol. Chem.*, **70**, 151 (1926).
8. Radulescu, D., and Georgescu, V., *Chem. Zentr.*, **1**, 1455 (1927).
9. Graff, J., and Rittenberg, D., *Anal. Chem.*, **24**, 878 (1952).
10. Alfin-Slater, R. B., Rock, S. M., and Swislocki, M., *Anal. Chem.*, **22**, 421 (1950).
11. Kirshenbaum, I., Physical properties and analysis of heavy water, New York (1951).
12. Karrer, P., Kahnt, F. W., Epstein, R., Jaffe, W., and Ishii, T., *Helv. chim. acta*, **21**, 223 (1938).
13. Urey, H. C., and Teal, G. K., *Rev. Modern Phys.*, **7**, 34 (1935).
14. Reitz, O., *Z. physik. Chem.*, *Abt. A*, **179**, 119 (1937); *Abt. A*, **184**, 429 (1939).
15. Westheimer, F. H., and Nicolaides, N., *J. Am. Chem. Soc.*, **71**, 25 (1949).
16. Negelein, E., and Wulff, H. J., *Biochem. Z.*, **284**, 289 (1936).
17. Theorell, H., and Bonnichsen, R., *Acta chem. Scand.*, **5**, 1105 (1951).
18. Theorell, H., and Chance, B., *Acta chem. Scand.*, **5**, 1127 (1951).
19. Geissman, T. A., *Quart. Rev. Biol.*, **24**, 309 (1949).
20. Racker, E., and Krimsky, I., *J. Biol. Chem.*, **198**, 731 (1952).
21. Kaplan, N. O., Colowick, S. P., and Neufeld, E. F., *J. Biol. Chem.*, **195**, 107 (1952).
22. Pullman, M. E., *Federation Proc.*, **12**, 255 (1953).

Part VII

# TWO-SUBSTRATE MECHANISMS: TWO CASES OF DISACCHARIDE FORMATION, AND UTILIZATION OF DOUBLE-RECIPROCAL PLOTS

# Editor's Comments on Papers 48, 49, and 50

**48 DOUDOROFF, BARKER, and HASSID**
*Studies with Bacterial Sucrose Phosphorylase. I. The Mechanism of Action of Sucrose Phosophorylase as a Glucose-transferring Enzyme (Transglucosidase)*

**49 FITTING and DOUDOROFF**
*Phosphorolysis of Maltose by Enzyme Preparations from* Neisseria Meningitidis

**50 CLELAND**
*The Kinetics of Enzyme-catalyzed Reactions with Two or More Substrates or Products. III. Prediction of Initial Velocity and Inhibition Patterns by Inspection*

Paper 48, by Doudoroff, Barker, and Hassid, on the mechanism of bacterial sucrose phosphorylase, is an early and brilliant example of the use of radioisotopes for the study of enzyme mechanisms. The discovery that the enzyme catalyzed an exchange reaction between phosphate and glucose 1-phosphate (actually $\alpha$-D-glucose 1-phosphate) in the absence of fructose demonstrated that the overall reaction resulting in the formation of sucrose (an $\alpha$-glucosyl fructoside, see endnote 1, Part IV) could be divided into two distinct steps, with glucosyl-enzyme as an essential covalent intermediate. It is very interesting to note, as a historical footnote to important scientific discoveries, that H. A. Barker states in the autobiographical preface to the *1978 Annual Review of Biochemistry: "Almost as an afterthought we included a control with only* glucose-1-phosphate and $^{32}$Pi, and were surprised to find that more $^{32}$P was incorporated into glucose-1-phosphate in the absence of the sugars than in their presence. In fact we did not believe the first result, and concluded that there had been a mix up of the samples. However, repetition confirmed the initial observation" (Barker 1978). Five years later a comparable mechanistic study by Fitting and Doudoroff (Paper 49) on a very similar reaction, bacterial maltose phosphorylase, showed a quite different mechanism: The overall reaction leading to the formation of maltose (an α-glucosyl glucoside, see endnote 1, Part IV) could not be divided into two steps; no exchange reaction between glucose 1-phosphate (actually β-D-glucose 1-phosphate) and phosphate occurred in the absence of glucose. In marked contrast to the case of sucrose phosphorylase, this and other evidence indicated the formation of a ternary complex between reacting substrates and enzyme. Furthermore, it was concluded that the latter reaction, leading to maltose, proceeded with an inversion of the configuration of the glycosidic bond of β-D-glucose 1-phosphate, while the sucrose formed by the earlier reaction retained the configuration of the α-D-glucose 1-phosphate utilized. These two studies gave early and clear-cut evidence on completely different reaction mechanisms for very similar two-substrate reactions. It is of interest that a glucosyl-sucrose phosphorylase intermediate was isolated almost twenty-five years after the initial studies that indicated its existence (Voet and Abeles 1970).

About a year after the Fitting and Doudoroff paper, Daniel E. Koshland, Jr. (1953), published a review entitled "Stereochemistry and the Mechanism of Enzymatic Reactions." In this review the terms *single displacement* and *double displacement,* based on a survey of a number of substitution reactions, are discussed. These ideas were based on studies in the field of organic chemistry of reactions such as the Walden inversion, which had already been discussed in the Fitting and Doudoroff paper. A postulated front-side displacement mechanism has not withstood the test of time. This review stresses that the terms *single* and *double displacement* refer to the number of displacements on the reactant and not to the number of steps in the overall enzymatic reaction.

Exactly ten years after Koshland's introduction of the terms *single* and *double displacement,* William W. Cleland published three papers which, elaborating on studies by workers such as Alberty (1953, 1958, 1959), Dalziel (1957), Frieden (1959), and Bloomfield, Peller, and Alberty (1962), advanced kinetic criteria for distinguishing or excluding certain types of reaction mechanisms. The double-reciprocal plots of initial velocities against varying concentrations of one substrate for different fixed concentrations of the other can be shown to give a parallel family of curves for enzyme reactions that proceed via distinct partial reactions (as in the case of sucrose phosphorylase), but a family of curves intersecting in one point for enzyme reactions that proceed via a ternary complex (as in the case of maltose phosphorylase). The kinetic approach has been refined and extended to a large number of multisubstrate reactions. In addition, this approach draws heavily on the kinetics of inhibitory effects of substrates and products in relation to postulated reaction intermediates to enable one to study the sequence of addition or removal of substrates and products. An idea of the sophistication of this field is provided by the third of Cleland's original papers, included in this volume (Paper 50). It is rather entertaining to read Cleland's own comments on the publication history of these papers (Cleland 1977). Related ideas had been advanced independently by Wong and Hanes (1962). In the intervening years this field has burgeoned. The most recent specialized book (Purich 1979) contains a number of chapters which utilize kinetic analysis of enzyme reactions, including two by Cleland (1979a, 1979b).

# 48


Reprinted from *J. Biol. Chem.* **168**:725–732 (1947)

# STUDIES WITH BACTERIAL SUCROSE PHOSPHORYLASE

## I. THE MECHANISM OF ACTION OF SUCROSE PHOSPHORYLASE AS A GLUCOSE-TRANSFERRING ENZYME (TRANSGLUCOSIDASE)*

BY MICHAEL DOUDOROFF, H. A. BARKER, AND W. Z. HASSID

*(From the Department of Bacteriology and the Division of Plant Nutrition, College of Agriculture, University of California, Berkeley)*

(Received for publication, February 19, 1947)

It has been shown that sucrose phosphorylase preparations from *Pseudomonas saccharophila* catalyze the reversible reaction between glucose-1-phosphate and certain ketose sugars. This reaction results in the formation of sucrose and analogous disaccharides together with the liberation of inorganic phosphate (1–5). In the experiments reported here, it was observed that when glucose-1-phosphate and radioactive inorganic phosphate are added to enzyme preparations in the absence of ketose sugars, a rapid interchange of phosphate occurs between the organic and inorganic fractions.

This indicated that the enzyme can liberate inorganic phosphate from glucose-1-phosphate without the production of an equivalent amount of glucose. The following reaction was postulated,

$$\text{Glucose-1-phosphate} + \text{enzyme} \leftrightarrows \text{glucose-enzyme} + \text{phosphate} \quad (1)$$

The glucose-enzyme bond presumably retains the energy of the phosphoric ester linkage. This observation suggested that an analogous reaction would occur between the enzyme and sucrose, with fructose being liberated instead of phosphate,

$$\text{Glucose-1-fructoside} + \text{enzyme} \leftrightarrows \text{glucose-enzyme} + \text{fructose} \quad (2)$$

Reactions (1) and (2) can account for the ability of the enzyme to substitute a glycosidic linkage for a phosphoric ester bond.

The postulated mechanism is strongly supported by experiments in which virtually phosphate-free enzyme preparations were found capable of synthesizing the sucrose analogue, glucosidosorboside, directly from sucrose and sorbose,

$$\text{Glucose-1-fructoside} + \text{sorbose} \leftrightarrows \text{glucose-1-sorboside} + \text{fructose} \quad (3)$$

In like manner, sucrose was produced from its synthetic analogue, glucosidoketoxyloside and fructose,

$$\text{Glucose-1-ketoxyloside} + \text{fructose} \leftrightarrows \text{glucose-1-fructoside} + \text{ketoxylose} \quad (4)$$

---

*Supported in part by a grant from the Corn Industries Research Foundation.

These results show that sucrose phosphorylase not only catalyzes the substitution of glycosidic linkages for a phosphate linkage but also causes an exchange between equivalent glycosidic linkages.

## EXPERIMENTAL

### *Reaction between Glucose-1-phosphate and Enzyme*

The sucrose phosphorylase preparation was obtained by methods previously described, except that it was reprecipitated six times with 0.63 saturated ammonium sulfate (6, 2). A dilute solution of the enzyme was incubated at 30° with 0.1 M potassium glucose-1-phosphate and 0.033 M potassium phosphate containing $P^{32}$. Parallel experiments were conducted in which fructose and glucose were added separately to the above mixture.

TABLE I

*Exchange of $P^{32}$ between Inorganic Phosphate and Glucose-1-phosphate*

| Experiment No. | Reaction mixture | Radioactivity found in fractions, counts per min. per micromole after 60 min. at 30° | |
|---|---|---|---|
| | | Inorganic phosphate | Glucose-1-phosphate |
| 1 | 0.1 M glucose-1-phosphate + 0.033 M inorganic phosphate | 1098 (±40) | 0 (±1) |
| 2 | Same as (1) but with enzyme preparation | 859 (±40) | 119 (±3) |
| 3 | Same as (2) but with 0.06 M fructose | 886 (±40) | 99 (±3) |
| 4 | Same as (2) but with 0.12 M glucose | 1096 (±40) | 7 (±1) |

Fructose was used since it is known to participate in a reversible reaction involving glucose-1-phosphate and inorganic phosphate, while glucose has been previously shown to have a strong inhibitory action on sucrose phosphorylase (6). After incubation, the enzyme was inactivated by boiling, and the inorganic phosphate was precipitated as the barium salt, redissolved with acid, and reprecipitated for analysis. The remaining traces of radioactive inorganic phosphate were removed from the solution of glucose-1-phosphate by repeated additions of inorganic phosphate and its removal with barium. Chemical analyses were then made on the inorganic and organic fractions and radioactivity was determined with a Geiger counter. A control experiment, in which glucose-1-phosphate and radioactive inorganic phosphate were incubated together in the absence of enzyme, showed that no non-enzymatic exchange occurs between these compounds.

The results of the experiment with sucrose phosphorylase are presented n Table I. No appreciable formation of free glucose from glucose-

1-phosphate could be detected during the course of the experiment. Sucrose was formed in the presence of fructose, but the rate of its formation was not determined.

The experiment shows that $P^{32}$ appeared rapidly in the glucose-1-phosphate fraction, even though a dilute enzyme preparation was used. In the presence of fructose, less $P^{32}$ was found in glucose-1-phosphate. This is due, at least in part, to a dilution of the radioactive phosphate with inactive phosphate liberated as a result of the synthesis of sucrose. The principal reason, however, must be the competition of fructose with phosphate for combination with glucose.

The presence of glucose almost completely stops the exchange of $P^{32}$. This is consistent with the observation that glucose inhibits phosphorylase activity and supports the view that glucose competes with glucose-1-phosphate for combination with the enzyme.

*Interconversion of Disaccharides in Absence of Phosphate*

In order to show that the enzyme can liberate fructose from sucrose without the formation of either free glucose or glucose-1-phosphate, an experiment was devised in which its ability to produce glucosidosorboside from sucrose in the absence of glucose-1-phosphate was tested. To prevent the accumulation of glucose-1-phosphate in the reaction mixtures, the enzyme preparations were rendered virtually free of inorganic phosphate. The initial steps of obtaining the enzyme from dried cells of *Pseudomonas saccharophila* were identical with the previously described method. After three reprecipitations of the enzyme with ammonium sulfate from phosphate buffer, the protein precipitate was redissolved in phosphate-free 0.05 M citrate buffer at pH 6.7 containing 0.01 M KCl. The enzyme was precipitated with recrystallized ammonium sulfate at 0.63 saturation. After three further reprecipitations from citrate buffer, carried out over a period of 2 days, the enzyme was redissolved in the same buffer mixture and used in the experiments. Such preparations were found to retain strong phosphorylase activity. The rate of glucosidosorboside formation is low, regardless of whether glucose-1-phosphate or sucrose is used as substrate for the enzyme. For this reason, the experiments were of fairly long duration and a relatively high concentration of enzyme had to be used. Enzyme preparations from $\pm 1$ gm. of dry cells were used per 20 ml. of final experimental solutions. As a result, an appreciable amount of reducing sugar was formed from sucrose in the course of the experiments. This phenomenon has been noted previously and is ascribable to traces of invertase in the preparation. Less than $10^{-5}$ M phosphate was found to be present in the final reaction mixtures.

Sucrose was estimated from the reducing value obtained on hydrolysis with invertase. Neither glucosidosorboside nor glucosidoketoxyloside is attacked to an appreciable extent by invertase. These compounds, therefore, could be estimated in phosphate-free mixtures from the difference between reducing values obtained upon hydrolysis with invertase and hydrolysis with 0.2 N HCl for 5 minutes at 100°. In experiments in which phosphate was added to the reaction mixture, the glucose-1-phosphate which was produced was measured as inorganic phosphate released on 7 minutes hydrolysis with 0.1 N HCl. The glucosidosorboside produced in experiments conducted in the presence of phosphate was estimated from the difference between initial and final reducing sugar values obtained

TABLE II

*Production of Glucosidosorboside from Sucrose*

| Experiment No. | Additions to enzyme preparation (different preparations used for Experiments 1 and 2) | Time of incubation at 30° | Total sucrose disappearing | Sucrose phosphorolyzed | Glucosidosorboside produced |
|---|---|---|---|---|---|
| | | *hrs.* | *mg. per ml.* | *mg. per ml.* | *mg. per ml.* |
| 1, a | 0.05 M sucrose | 2 | 0.3 (±0.3) | 0 | 0 |
| b | 0.05 " " 0.14 M sorbose | 2 | 2.9 (±0.3) | 0 | 2.7 (±0.3) |
| c | 0.05 " " 0.14 " " | 4 | 4.9 (±0.3) | 0 | 4.4 (±0.3) |
| d | 0.05 " " 0.14 " " 7 × $10^{-4}$ M $Na_2HPO_4$ | 2 | 3.0 (±0.3) | 0.2 | 2.2 (±0.3) |
| e | 0.05 M sucrose, 0.14 M sorbose, 7 × $10^{-3}$ M $Na_2HPO_4$ | 2 | 3.7 (±0.3) | 2.4 | 0.6 (±0.3) |
| f | 0.05 M sucrose, 0.14 M sorbose, 7 × $10^{-3}$ M $Na_2HPO_4$ | 4 | 5.5 (±0.3) | 2.3 | 2.5 (±0.3) |
| 2, a | 0.06 M sucrose, 0.11 M sorbose | 3 | 4.9 (±0.3) | 0 | 4.6 (±0.3) |
| b | 0.06 " " 0.11 " " 0.12 M glucose | 3 | 0.6 (±0.3) | 0 | 0.3 (±0.3) |

on hydrolysis with invertase and corrected for the decrease in reducing value due to the formation of glucose-1-phosphate.

When sucrose and sorbose were added to the enzyme preparations in the absence of phosphate, sucrose disappeared and glucosidosorboside was formed (see Experiment 1, Table II). Since at equilibrium considerably less of the latter than of the former would be expected (3), the rate of transformation must decrease rapidly in the course of the experiment. The initial rate of conversion must, therefore, be greater than that which was observed for the first 2 hours.

The initial rate of glucosidosorboside formation by the same enzyme preparation could be measured with a fair degree of accuracy when glucose-1-phosphate was used as substrate in place of sucrose, by determining the rate of the evolution of inorganic phosphate. In the presence of 0.14 M sorbose, and 0.025 M glucose-1-phosphate, the initial rate of disaccharide synthesis was found to be 3.24 mg. per 2 hours per ml. at 30°. At this concentration, glucose-1-phosphate is not limiting to the rate of reaction. When the concentration of glucose-1-phosphate was reduced to 0.002 M, the initial rate was found to be 2.72 mg. per 2 hours per ml., while at 0.001 M concentration, the rate was 2.14 mg. per 2 hours. It can be roughly computed from these data that with $10^{-5}$ M glucose-1-phosphate the rate would be in the neighborhood of 0.06 mg. of glusosidosorboside formed per 2 hours per ml. Since the maximum limit of inorganic phosphate present as impurity in the "phosphate-free" reaction mixtures was found to be $10^{-5}$ M, this would be the maximum concentration of glucose-1-phosphate which could be attained. Since the observed rate of glucosidosorboside formation from sucrose approaches the maximum rate of synthesis from glucose-1-phosphate, it is clear that the interconversion of disaccharides cannot depend on a preliminary phosphorolysis and a subsequent utilization of the accumulated glucose-1-phosphate.

This view is further supported by the fact that the addition of phosphate to reaction mixtures does not increase the rate of interconversion of disaccharides but decreases it. This inhibitory effect of phosphate is entirely in accord with the postulated hypothesis for the mechanism of enzyme action. Not only the phosphate, itself, but also the fructose liberated in the phosphorolysis must compete with sorbose for the enzyme-glucose complex.

The hypothesis also requires that glucose will inhibit the conversion of sucrose to glucosidosorboside by competing with the glucose portion of the sucrose molecule for a position on the enzyme. The competition between glucose and sucrose has already been shown in studies of the phosphorolytic breakdown of sucrose (6). That the rate of interconversion of the disaccharides is indeed decreased by the addition of glucose is demonstrated by Experiment 2 (Table II).

The quantitative discrepancy between the sucrose disappearance and glucosidosorboside formation in both Experiments 1 and 2 (Table II) is due to the formation of reducing sugar, as explained earlier.

Since the direct interconversion of related disaccharides is obviously a reversible process, an experiment was devised in which sucrose would be produced rather than consumed. This was particularly important, since the evidence for the identity of the non-reducing disaccharide produced from sucrose and sorbose was entirely circumstantial and not based on the isolation and identification of the sugar. Sucrose, on the other hand, can be identified with a fair degree of certainty with invertase.

A phosphate-free enzyme preparation was allowed to act on a mixture containing 2 per cent fructose and 2 per cent α-D-glucosido-β-D-ketoxyloside which had been previously synthesized from glucose-1-phosphate and ketoxylose (5). Sucrose was estimated with invertase and glucosidoketoxyloside by acid hydrolysis in the same manner as was glucosidosorboside in the previous experiments.

After 1 hour of incubation at 30°, 2.5 (±0.3) mg. of sucrose were formed per ml. and a total of 4.8 (±0.3) mg. per ml. was found after 3 hours. Approximately equivalent amounts of glucosidoketoxyloside disappeared.

The initially observed rate of sucrose synthesis was found to be less than half of the maximum rate observed when glucose-1-phosphate was used as substrate in place of glucosidoketoxyloside in the presence of fructose. Since the reversible reaction between glucose-1-phosphate and ketoxylose is known to be slow (3), it may be inferred that the rate of decomposition of glucosidoketoxyloside rather than the rate of sucrose synthesis limited the total rate of interconversion of the disaccharides.

## DISCUSSION

There can be no doubt that both the phenomena of isotope exchange between inorganic and organic phosphate and the interconversion of disaccharides in the absence of phosphate are due to one and the same enzyme, which has been called "sucrose phosphorylase," if the following considerations are taken into account: (1) Both processes are carried out vigorously by sucrose phosphorylase preparations which are virtually devoid of most other enzymes. (2) Both reactions are inhibited by glucose. (3) Most significantly, the presence of phosphate depresses the rate of interconversion of disaccharides, indicating that phosphate competes with the ketoses for the enzyme which catalyzes this process.

It appears, then, that the rôle of the enzyme is to combine reversibly with the glucose residue of glucose-1-phosphate or of those disaccharides which can act as substrates, and to release the esterified phosphate or glycosidically bound sugar. The enzyme-glucose complex must retain the energy of the phosphate or glycosidic bond. The sucrose phosphorylase may, therefore, be considered as a glucose-transferring system which can react with a rather remarkable variety of substrates. In addition to reacting with phosphate and carbonyl groups of certain ketoses, the enzyme has been found to catalyze the addition of glucose to the secondary alcoholic group of at least one aldose (7).

The reversible phosphorolysis of sucrose can now be interpreted as consisting of the following set of reactions.

$$\text{Glucose-1-fructoside (sucrose)} + \text{enzyme} \underset{}{\overset{\pm\ \text{fructose}}{\rightleftharpoons}} \text{glucose-enzyme}$$

$$\text{glucose-enzyme} \overset{\pm\ \text{phosphate}}{\rightleftharpoons} \text{glucose-1-phosphate} + \text{enzyme}$$

The direct interconversion of disaccharides may be illustrated by the production of sucrose from glucosidoketoxyloside:

$$\text{Glucose-1-ketoxyloside} + \text{enzyme} \underset{}{\overset{\pm\text{ ketoxylose}}{\rightleftharpoons}} \text{glucose-enzyme}$$

$$\text{glucose-enzyme} \overset{\pm\text{ fructose}}{\rightleftharpoons} \underset{\text{(sucrose)}}{\text{glucose-1-fructoside}} + \text{enzyme}$$

The demonstration that the accumulation of glucose-1-phosphate is not necessary for the conversion of disaccharides does not preclude the possibility that phosphate does enter into the enzymatic reaction, possibly as a firmly bound coenzyme. It is impossible to tell whether the protein itself transfers the "energy-rich glucose" or whether a coenzyme which is closely associated with the protein is involved. It is possible, for instance, to visualize a carbohydrate residue which would be directly concerned with the glucose transfer and would act as a glucose-carrying coenzyme.

The direct exchange of glycosidic linkages has been observed with other bacterial enzymes and discussed in previous papers (8, 9). Thus, neither the production of dextran nor of levan from sucrose requires the accumulation of phosphoric esters. The enzymes which catalyze these reactions appear simply to exchange a glucosidofructoside linkage for a 1:6-glucosidoglucose bond in one case or for a 2:6-fructosidofructose bond in the other.

Sucrose phosphorylase seems to belong to the same class of enzymes, although it has the additional power to attack the phosphate bond. It is quite possible that the better known phosphorylases which are involved in the synthesis of starch and glycogen are essentially similar in their mode of action. Since the transfer of glucose may be compared to the well known transmethylation and transamination reactions, as well as to the transfer of hydrogen atoms in biological oxidations, it would seem appropriate to consider sucrose phosporylase as a "transglucosidase." The general type of enzyme involved in the exchange of glycosidic bonds might then be called "transglycosidase."

It is quite possible that the production of many disaccharides and polysaccharides in plant and animal tissues may depend on transfers of "energy-rich" sugar residues without the intermediate accumulation of phosphoric esters (9). The remarkable versatility of the sucrose phosphorylase suggests that one and the same enzyme might, in some cases, account for the formation of a number of different compounds.

The direct synthesis of sucrose from glucosidoketoxyloside adds a third mechanism of biological synthesis of sucrose to those already studied in our laboratory (1, 8). It is, in reality, but a variant of the mechanism postulated from the indirect evidence of the reversible nature of levan synthesis.

The strong competition of glucose with both glucose-1-phosphate and sucrose would indicate that the carbonyl atom of glucose is not greatly involved in the affinity of sucrose phosphorylase for the glucose portion of these compounds. Unpublished experiments have shown that xylose, xylose-1-phosphate, arabinose, galactose, and galactose-1-phosphate have a very much weaker inhibitory action on sucrose phosphorylase than does glucose. It seems probable, therefore, that the alcoholic groups away from the carbonyl group, possibly together with the pyranose ring structure, are important in determining the specificity of enzyme-substrate combination.

## SUMMARY

1. A rapid exchange of $P^{32}$ was found to occur between inorganic phosphate and glucose-1-phosphate in the presence of sucrose phosphorylase of *Pseudomonas saccharophila* and in the absence of ketose sugars.

2. Virtually phosphate-free preparations of sucrose phosphorylase were found capable of interconverting sucrose and its analogue, glucosidosorboside.

3. By applying the same principle, sucrose was synthesized directly from glucosidoketoxyloside and fructose with the aid of the enzyme.

4. Sucrose phosphorylase may be considered as a versatile "transglucosidase," representing a class of enzymes which may be referred to as "transglycosidases."

## BIBLIOGRAPHY

1. Doudoroff, M., Kaplan, N., and Hassid, W. Z., *J. Biol. Chem.*, **148**, 67 (1943).
2. Hassid, W. Z., Doudoroff, M., and Barker, H. A., *J. Am. Chem. Soc.*, **66**, 1416 (1944).
3. Doudoroff, M., Hassid, W. Z., and Barker, H. A., *Science*, **100**, 315 (1944).
4. Hassid, W. Z., Doudoroff, M., Barker, H. A., and Dore, W. H., *J. Am. Chem. Soc.*, **67**, 1394 (1945).
5. Hassid, W. Z., Doudoroff, M., Barker, H. A., and Dore, W. H., *J. Am. Chem. Soc.*, **68**, 1465 (1946).
6. Doudoroff, M., *J. Biol. Chem.*, **151**, 351 (1943).
7. Doudoroff, M., Hassid, W. Z., and Barker, H. A., *J. Biol. Chem.*, **168**, 733 (1947).
8. Doudoroff, M., and O'Neal, R., *J. Biol. Chem.*, **159**, 585 (1945).
9. Doudoroff, M., *Federation Proc.*, **4**, 242 (1945).

Reprinted from *J. Biol. Chem.* **199**:153–163 (1952)

# PHOSPHOROLYSIS OF MALTOSE BY ENZYME PREPARATIONS FROM NEISSERIA MENINGITIDIS*

By CHARLOTTE FITTING† AND MICHAEL DOUDOROFF

*(From the Department of Bacteriology, University of California, Berkeley, California)*

(Received for publication, May 17, 1952)

In previous studies on the utilization of maltose and glucose by *Neisseria meningitidis* (meningococcus) it has been found that the disaccharide is utilized more rapidly than glucose (1, 2). Phosphate esterification was observed with growing cultures, washed cells, and cell-free extracts of this organism in the presence of maltose but not of glucose (3). The properties of the phosphate ester appeared to be similar to those of glucose-1-phosphate (4). α-D-Glucose-1-phosphate had been found previously to be the product of the phosphorolysis of sucrose by *Pseudomonas saccharophila* (5) and *Pseudomonas putrefaciens* (6). It is also the product of the combined action of the enzymes amylomaltase and phosphorylase of *Escherichia coli* on maltose (7). It seemed likely, therefore, that a mechanism similar to that of *P. saccharophila* or *E. coli* was involved in the decomposition of maltose by the meningococcus.

Further investigation of this problem has led to the conclusion that the meningococcal enzyme catalyzes the direct phosphorolysis of maltose but that the phosphoric ester produced is β-D-glucose-1-phosphate (8, 9). The mechanism of action of the enzyme appears to be very different from that of sucrose phosphorylase. The detailed studies on the reversible phosphorolytic cleavage of maltose by the meningococcal enzyme are presented in this paper.

## *Materials and Methods*

*The organism* used for this study was *N. meningitidis*. This old non-virulent laboratory strain (No. 69) of a type I meningococcus was preserved in the desiccated state (10). This material was plated on Trypticase Soy agar (Baltimore Biological Laboratory) and the cells collected after an incubation period of not more than 24 hours (4). The cells were washed several times and then were either ground up with alumina (Blue Label R. R. Alundum, electrically fused crystalline alumina 60 mesh, Norton

* This work was supported in part by the United States Public Health Service and a contract between the Office of Naval Research and the University of California.

† Public Health Service Research Fellow of the National Microbiological Institute. Present address, University Laboratory of Physical Chemistry Related to Medicine and Public Health, Harvard University, Boston 15, Massachusetts.

Company, Worcester, Massachusetts) or were lyzed by repeated freezing and thawing. Then the extracts were fractionated with ammonium sulfate. The material which precipitated between 0.4 and 0.7 saturation was redissolved in either phosphate or tris(hydroxymethyl)aminomethane (Commercial Solvents Corporation) buffer at pH 7.0. Repeated reprecipitations with ammonium sulfate apparently freed these extracts of interfering hydrolases.

In addition to the methods mentioned previously (4), the following procedures were used.

*Phosphoric esters of sugars* were identified and estimated quantitatively by the methods described by Umbreit *et al.* (11).

*Glucose* was analyzed directly by using it as the substrate for oxidation with notatin (12). Glucose was also estimated from the decrease in reducing value after fermentation with *Torula monosa.*

*Maltose* was determined from the difference between reducing values obtained after fermenting first with suspensions of *T. monosa* and then with *Saccharomyces cerevisiae* (7).

*Radioactivity* of substrates was determined by means of a Victoreen Geiger-Müller tube with a window thickness of 1.5 mg. per sq. cm.

Further details of experimental procedures are given in the experimental section.

## EXPERIMENTAL

### *Phosphorolysis and Synthesis of Maltose*

When maltose and phosphate were incubated together with the enzyme from the meningococcus, 1 mole of phosphate was esterified and 1 mole of glucose was produced for each mole of maltose decomposed. The data substantiating this finding are given in Table I (Experiment 1).

The identification of the ester was accomplished in the following way: The mixture was deproteinized with trichloroacetic acid. The resulting supernatant solution was subjected to barium fractionation. The water-soluble, ethanol-insoluble barium salt was reprecipitated until it was apparently free of inorganic phosphate. This crude barium salt was converted to the potassium salt with sulfuric acid followed by potassium hydroxide. The ester was non-reducing to Fehling's or ferricyanide solution. It was hydrolyzed in 7 minutes in N HCl at 100°, yielding 1 mole of glucose per mole of inorganic phosphate. Since these properties agree with those of α-D-glucose-1-phosphate, which previously had been found as a product of sucrose phosphorolysis (5) and maltose breakdown (7) in other enzyme systems, an attempt was made to establish the identity of the two esters.

The crude ester solution was used as a substrate for sucrose synthesis with the sucrose phosphorylase from *P. saccharophila*. No evidence of a reaction between the compound and fructose was obtained while α-D-glucose-1-phosphate was rapidly deesterified in the presence of fructose. That the crude ester contained no inhibitory substances for the enzyme was shown by the fact that the addition of the unknown ester to reaction mixtures with α-D-glucose-1-phosphate and fructose had no effect on sucrose synthesis. It seemed likely that the phosphorolysis of maltose would be a reversible process. The ester, as well as α-D-glucose-1-phosphate, was tested, therefore, as substrate for the meningococcal enzyme both in the

TABLE I
*Analysis of Maltose Phosphorolysis and Maltose Synthesis*

| Experiment No. | | Analyses of reaction components in μM per ml. | | | |
|---|---|---|---|---|---|
| | | Maltose | $P_o$ | Glucose | $P_7$ |
| 1 | Initial | 28.0 | 35.9 | 0 | 0 |
| | After 7 hrs. at 37° | 19.2 | 28.7 | 9.2 | 7.2 |
| | Substrate disappeared | 8.8 | 7.2 | | |
| | Products formed | | | 9.2 | 7.2 |
| 2 | Initial | 0 | 0.3 | 15.0 | 8.7 |
| | After 10 hrs. at 37° | 7.2 | 7.1 | 7.8 | 1.9 |
| | Substrate disappeared | | | 7.2 | 6.8 |
| | Products formed | 7.2 | 6.8 | | |

*Experiment 1*—Maltose was analyzed as the reducing sugar remaining after fermentation with *T. monosa*. The glucose value was obtained from the difference in total reducing values before and after fermentation with *T. monosa*.

*Experiment 2*—Maltose analyses as in Experiment 1. Glucose value obtained from the amount of oxygen consumed with notatin.

$P_o$ = orthophosphate; $P_7$ = phosphate hydrolyzed in N acid at 100° for 7 minutes.

presence and absence of glucose. The results were clear cut. The product of maltose phosphorolysis was deesterified in the presence of glucose, but not in its absence. In either test system, α-D-glucose-1-phosphate was unattacked. The above considerations led to the conclusion that the unknown ester must be β-D-glucose-1-phosphate. This compound had been chemically synthesized and characterized by Wolfrom *et al.* (13). It was kindly supplied to us by Dr. W. Z. Hassid and Dr. E. W. Putman. When tested with the enzyme, the synthetic compound behaved like the unknown ester. The experiments with sucrose phosphorylase and the maltose phosphorylase from the meningococcus are summarized in Table II.

To identify further the ester with β-D-glucose-1-phosphate, its ammonium salt was chromatographed on a cellulose column with 50 per cent aqueous ethanol. The effluent was converted to the barium salt. Its properties were compared with those of the synthetic compound (14). The barium salt had an optical rotation of $[\alpha]_D^{20} = +10°$ in water, $c = 1$ (synthetic compound, $[\alpha]_D^{20} = +12°$). The per cent phosphorus was found to be 6.93 and 6.88 for the two salts, respectively. Neither compound gave a reducing value before hydrolysis. After hydrolysis of the natural ester, the ratio of glucose to phosphorus was determined as 0.92.

The product of the reverse reaction was identified as maltose by chromatography and by fermentation. It had the same $R_F$ value on paper when developed either with a mixture of butanol, ethanol, and water or with

TABLE II

*Enzymatic Identification of Sugar Ester from Phosphorolysis of Maltose*

| Enzyme system | Reaction mixture | After 60 min. at 37° inorganic phosphate liberated |
|---|---|---|
| | | *μM per ml.* |
| Sucrose phosphorylase | 4.4 μM X* | 0 |
| | 4.4 " " + 12 μM fructose | 0 |
| | 10 " α-D-glucose-1-phosphate | 0 |
| | 10 " " + 12 μM fructose | 3.3 |
| Maltose " | 10 " " | 0 |
| | 10 " " + 30 μM glucose | 0 |
| | 1.0 " X | 0 |
| | 1.0 " " + 2 μM glucose | 0.36 |
| | 20 " β-D-glucose-1-phosphate | 0 |
| | 20 " " + 250 μM glucose | 16.4 |

* X is the unknown ester isolated as the water-soluble, ethanol-insoluble barium salt from the phosphorolysis of maltose by the meningococcal enzyme system.

phenol. It was not fermented by *T. monosa*. It was, however, fermented completely by *S. cerevisiae*. The balance for the reverse reaction is given in Experiment 2 of Table I.

Next, the equilibrium constant for the synthesis of maltose was determined at pH 7.0 and at 37°, as $K$ = (maltose) (phosphate)/(β-D-glucose-1-phosphate) (glucose). Some difficulties were encountered in the early experiments in which the reaction was allowed to proceed from either direction starting with only two substrates, *i.e.* maltose and phosphate, or β-D-glucose-1-phosphate and glucose. Under these conditions the mixtures must be incubated for very long periods of time during which hydrolytic reactions involving maltose and β-D-glucose-1-phosphate occur. It was found that the methods used for the determination of glucose were not sufficiently sensitive in the presence of the other components of the mixtures for the accurate determination of this compound. It will be seen that small errors in these analyses would greatly affect the $K$ value, since the determination of maltose was also dependent upon the glucose analyses. In early experiments values between 3 and 9 were obtained. The value of 7 appeared to be the most plausible (9). However, a careful reinvestigation indicates that the true value lies between 4.25 and 4.5. These values were obtained by incubating mixtures of all four components for a short period of time during which hydrolytic side reactions were negligible. The concentrations of maltose, glucose, β-D-glucose-1-phosphate, and enzyme were kept constant, while different amounts of inorganic phosphate were added to the mixtures. After 80 minutes of incubation at 37°, the reactions were stopped by the addition of trichloroacetic acid, the inorganic phosphate in the reaction mixtures was removed with magnesia mixture, and the

TABLE III

*Equilibrium Constant for Synthesis of Maltose*

| Incubation time at 37° | Concentration of reaction components in μM per ml. | | | | K |
|---|---|---|---|---|---|
| | Maltose | Phosphate | β-Glucose-1-phosphate | Glucose | |
| *min.* | | | | | |
| 0 | 12.50 | 15.90 | 3.00 | 12.50 | |
| 80 | 12.20 | 15.60 | 3.30 | 12.80 | 4.5 |
| 0 | 12.50 | 11.60 | 3.00 | 12.50 | |
| 80 | 12.66 | 11.76 | 2.84 | 12.34 | 4.25 |

changes from the initial value in the concentration of β-D-glucose-1-phosphate were carefully determined. In this way the direction of the reaction from any arbitrarily chosen initial ratios of the components was established. The final concentration of the other components was then computed from the change in the β-D-glucose-1-phosphate value. The results are presented in Table III. Since maltose hydrolysis was not considered in these computations, the $K$ values may be slightly high.

### *Specificity of Enzyme*

It had been shown previously (4) that the following compounds cannot replace maltose as the substrate for the enzyme: α-methyl glucoside, trehalose, isomaltose, dextran from *Leuconostoc*, cellobiose, gentiobiose, and type I meningococcus polysaccharide. Soluble starch was also found to be inactive. Arsenate could be substituted for phosphate, causing a decomposition of maltose to 2 molecules of glucose. This arsenolysis is analogous to that found for sucrose (15) and starch (16). Sucrose phosphorylase also causes the arsenolytic decomposition of α-D-glucose-1-phosphate. For comparison, the ability of the meningococcal enzyme to produce glucose from β-D-glucose-1-phosphate in the presence of arsenate was tested. The reaction was found to proceed very slowly as compared with that of mal-

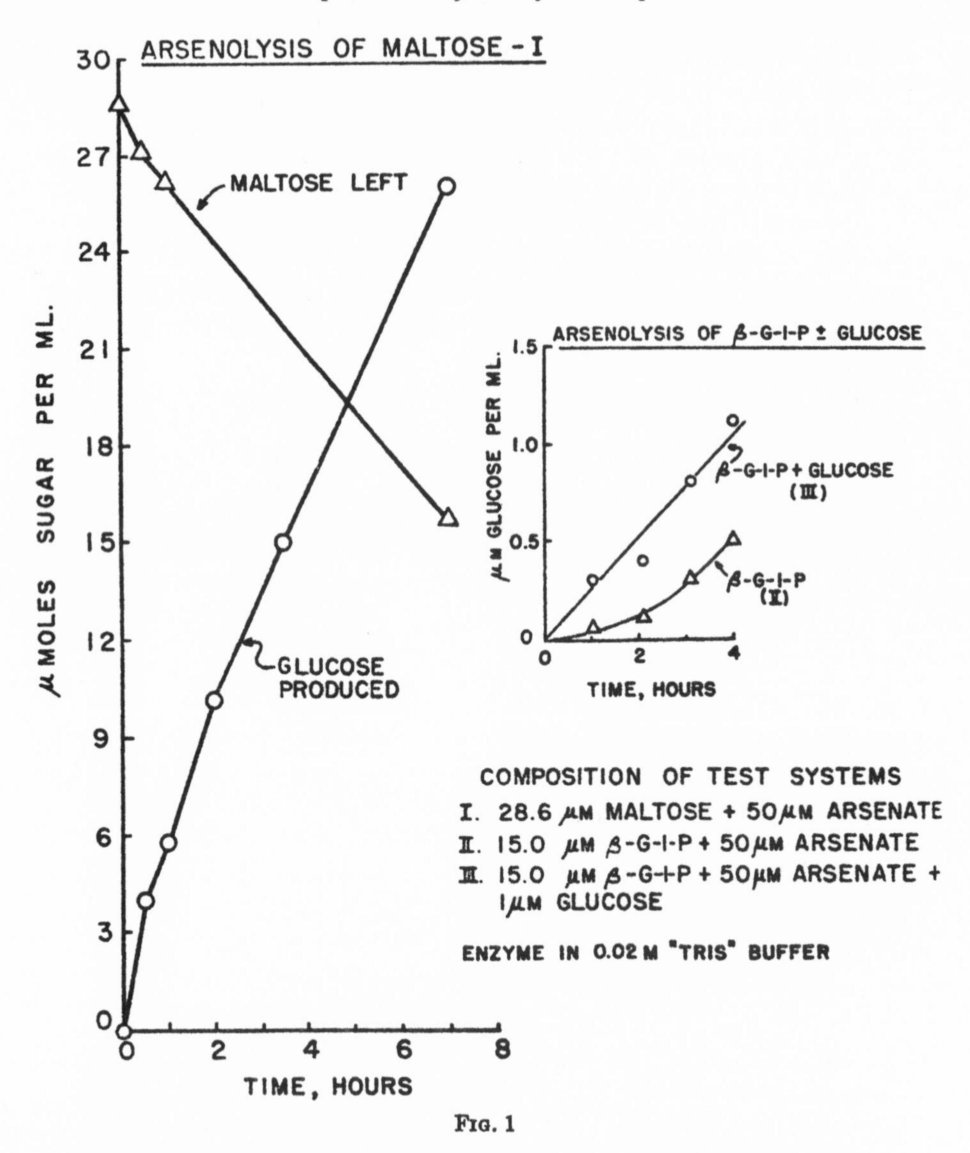


Fig. 1

tose. It appeared to be autocatalytic. This observation was taken as evidence that β-D-glucose-1-phosphate was not directly arsenolyzed, but that it was first converted to maltose with the traces of glucose produced from the hydrolysis of the ester. The maltose was then arsenolyzed, liberating more and more glucose as the reaction proceeded. Indeed, the addition of a catalytic amount of glucose decreased the initial lag in the glucose production. The above observations are graphically presented in Fig. 1. The striking difference in the rates of arsenolysis is emphasized by the difference in scales used for plotting glucose production from maltose and β-D-glucose-1-phosphate, respectively.

A number of sugars were tested as possible substrates in the reverse reaction with β-D-glucose-1-phosphate. The addition of the following compounds did not cause an appreciable dephosphorylation of the ester: D-fructose, D-galactose, L-glucose, D-mannose, D-ribose, D-arabinose, L-arabinose, α-methyl glucoside, α-D-glucose-1-phosphate, maltose, trehalose, cellobiose, and D-gluconic acid. In the presence of D-xylose, however, inorganic phosphate was liberated from β-D-glucose-1-phosphate. Since the expected product of this reaction would be a maltose analogue, α-D-glucosidoxylose, the test mixture was chromatographed on paper and developed with solvents used for maltose. A disaccharide having approximately the same $R_F$ value as maltose was detected. It was found to be a reducing sugar yielding approximately equal quantities of glucose and xylose on hydrolysis. Its structure and properties are now being studied in collaboration with Dr. W. Z. Hassid and Dr. E. W. Putman and will be reported elsewhere.

### *Studies on Mechanism of Maltose Phosphorolysis*

The ability of the maltose phosphorylase to catalyze exchanges similar to those observed with sucrose phosphorylase (17) was tested. Radioactive phosphate ($P^{32}$) was incubated with the enzyme, together with the inactive β-D-glucose-1-phosphate, both in the presence and absence of glucose. The reaction was stopped by the addition of trichloroacetic acid; the inorganic phosphate was removed with magnesia mixture, and the supernatant fluid was treated three times with inorganic phosphate followed by magnesia mixture. The final solution was analyzed for β-D-glucose-1-phosphate and for $P^{32}$ content. In another experiment, radioactive glucose labeled equally in all carbon atoms with $C^{14}$ (kindly supplied to us by Dr. W. Z. Hassid) was incubated with inactive maltose both in the presence and absence of phosphate. The reaction was stopped by boiling. Glucose was then fermented with *T. monosa*. The carbon dioxide evolved was measured and collected in KOH. It was then converted to barium carbonate and its specific activity was determined. The remaining maltose was fermented with *S. cerevisiae* and the activity of the carbon dioxide analyzed in the same manner. The specific activities of the sugars were computed from the activities of the carbon dioxide and from the quantities of carbon dioxide evolved from known amounts of the sugars. Unlike sucrose phosphorylase, maltose phosphorylase did not catalyze an appreciable exchange between labeled glucose and maltose in the absence of phosphate. Nor did any exchange occur between labeled inorganic phosphate and β-D-glucose-1-phosphate in the absence of glucose. The addition of small amounts of phosphate to the first mixture and of glucose to the second increased the exchanges (Table IV). This finding is in contrast to the effect of the addition of phosphate and of fructose to the analogous reactions with sucrose phosphorylase. The absence of exchange between inorganic phosphate and β-D-glucose-1-phosphate in the maltose phosphorylase system is in complete agreement with the observation that the latter compound is not directly arsenolyzed.

TABLE IV

*Exchange Experiments with Maltose Phosphorylase*

| Experiment No. | Labeled substrate | Reaction mixture | C.p.m. per μM | | | |
|---|---|---|---|---|---|---|
| | | | Initial | | After 2 hrs. at 37° | |
| | | | $P_o$ | Glucose | $P_7$ | Maltose |
| 1 | $P^{32}$ | 10 μM β-glucose-1-phosphate + 8 μM $P^{32}$ | 4100 | | 12 | |
| | | 10 " " + 8 " " + 1 μM glucose | 4100 | | 145 | |
| 2 | Glucose $C^{14}$ | 8 μM maltose + 18 μM glucose-$C^{14}$ | | 7800 | | 54 |
| | | 8 " " + 18 " " + 2 μM $P_o$ | | 7800 | | 841 |

## DISCUSSION

The finding of β-D-glucose-1-phosphate as a major product of a biological reaction is of great interest, not only because it suggests that this compound may be of importance in cellular metabolism but also because the reaction in which it is produced involves a Walden inversion on the 1st carbon atom of glucose. The occurrence of such an inversion makes it possible to visualize direct syntheses of β-linked polysaccharides from α-D-glucose-1-phosphate or α-linked sugars in other biological systems. While this is the first case of a direct phosphorolysis of maltose, a mechanism for converting maltose to α-D-glucose-1-phosphate and glucose with the formation of a polysaccharide as an intermediate is also known (7). Thus artificial combinations of the enzymes involved could be used for the interconversion of the two optical isomers of the ester in the presence of glucose.

It is interesting to contrast the equilibrium constant for the synthesis of maltose which is in the neighborhood of 4.4 with that for the synthesis of sucrose which is about 0.05. Assuming that the energy contents of the α and β forms of glucose-1-phosphate are not too different, the difference of the energy levels of the sucrose and maltose linkages is emphasized (approximately 2700 calories).

The absence of exchange reactions between phosphate or arsenate and β-D-glucose-1-phosphate on the one hand, and between maltose and glucose

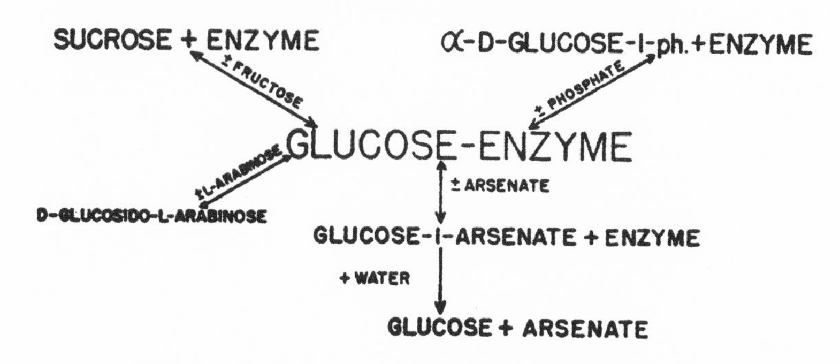


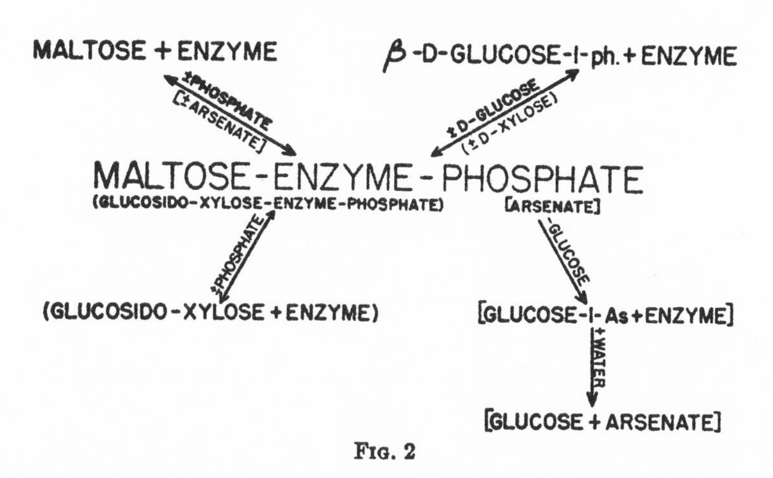


Fig. 2

on the other, distinguishes the mechanisms of catalyses of the phosphorolyses of maltose and of sucrose. The enzyme from *P. saccharophila* has been shown to be a transglucosidase capable of accepting the glucosidic moiety of sucrose or α-D-glucose-1-phosphate and transferring it to any of a number of suitable acceptors. The α linkage is preserved in this transfer. In the case of maltose phosphorylase, the apoenzyme itself does not appear to be a carrier of the glucosidic moiety. Both donor and acceptor seem to be necessary components for the catalysis of the glucose transfer to take place. Such a situation might be analogous to the catalysis of hydrogen transfers between substrates and the pyridine nucleotide coenzymes by dehydrogenases. A scheme which accounts for the necessity of the three components would involve the formation of a maltose-enzyme-phosphate complex interconvertible with a β-D-glucose-1-phosphate-enzyme-glucose complex as an intermediate (Fig. 2). The apoenzyme would then have at least three affinities, one for the glucosidic portion of its substrates, one for the reducing glucose residue, and one for the phosphate group. In such a scheme, D-xylose can take the place only of glucose, while arsenate can take the place only of phosphate. It has been suggested that, in the formation of the sucrose phosphorylase-glucose complex, a Walden inversion occurs to form β-glucose-1-enzyme (18). The phosphorolysis of sucrose presents then a double Walden inversion. The observed single inversion in the phosphorolysis of maltose may be explained on the basis of a three component system in which the phosphate transfer occurs between the substrates rather than between the apoenzyme and each substrate separately. The occurrence of the inversion would indicate that the carbon-oxygen bond of the phosphate ester is broken, as has been shown to be the case with sucrose phosphorylase in the work of Cohn (19).

## SUMMARY

In studies with enzyme preparations from *Neisseria meningitidis* the following observations have been made.

1. Maltose is reversibly phosphorolyzed with a Walden inversion to yield β-D-glucose-1-phosphate and glucose.
2. The equilibrium constant for the synthesis of maltose is in the neighborhood of 4.4.
3. D-Xylose can be substituted for glucose in the reverse reaction to form a reducing disaccharide, presumably α-D-glucosidoxylose.
4. The maltose phosphorylase does not catalyze the exchange reactions between phosphate or arsenate and β-D-glucose-1-phosphate on the one hand, or between glucose and maltose on the other, in the absence of glucose and phosphate, respectively. These findings are in direct contrast to those made with the sucrose phosphorylase system.

The implications of the above observations are discussed.

## BIBLIOGRAPHY

1. Fitting, C., and Scherp, H. W., *J. Bact.*, **61**, 203 (1951).
2. Fitting, C., and Scherp, H. W., *J. Bact.*, **63**, 545 (1952).
3. Fitting, C., and Scherp, H. W., *Bact. Proc.*, 120 (1950).
4. Fitting, C., and Scherp, H. W., *J. Bact.*, **64**, 287 (1952).
5. Doudoroff, M., *J. Biol. Chem.*, **151**, 351 (1943).

6. Doudoroff, M., Wiame, J. M., and Wolochow, H., *J. Bact.*, **57**, 423 (1949).
7. Doudoroff, M., Hassid, W. Z., Putman, E. W., Potter, A. L., and Lederberg, J., *J. Biol. Chem.*, **179**, 921 (1949).
8. Fitting, C., and Doudoroff, M., *Federation Proc.*, **11**, 212 (1952).
9. Fitting, C., and Doudoroff, M., *Bact. Proc.*, 144 (1952).
10. Scherp, H. W., and Fitting, C., *J. Bact.*, **58**, 1 (1949).
11. Umbreit, W. W., Burris, R. H., and Stauffer, J. F., Manometric techniques and related methods for the study of tissue metabolism, Minneapolis (1945).
12. Monod, M. J., and Torriani, A. M., *Ann. Inst. Pasteur*, **78**, 65 (1950).
13. Wolfrom, M. L., Smith, C. S., Pletcher, D. E., and Brown, A. E., *J. Am. Chem. Soc.*, **64**, 23 (1942).
14. Reithel, F. J., *J. Am. Chem. Soc.*, **67**, 1056 (1945).
15. Doudoroff, M., Barker, H. A., and Hassid, W. Z., *J. Biol. Chem.*, **170**, 147 (1947).
16. Katz, J., Hassid, W. Z., and Doudoroff, M., *Nature*, **161**, 96 (1948).
17. Doudoroff, M., Barker, H. A., and Hassid, W. Z., *J. Biol. Chem.*, **168**, 725 (1947).
18. Koshland, D. E., in McElroy, W. D., and Glass, B., Phosphorus metabolism, Baltimore, **1**, 536–546 (1951).
19. Cohn, M., *J. Biol. Chem.*, **180**, 771 (1949).

# 50



Reprinted from *Biochim. Biophys. Acta* **67**:188–196 (1963)

# THE KINETICS OF ENZYME-CATALYZED REACTIONS WITH TWO OR MORE SUBSTRATES OR PRODUCTS

## III. PREDICTION OF INITIAL VELOCITY AND INHIBITION PATTERNS BY INSPECTION*

W. W. CLELAND

*Department of Biochemistry, College of Agriculture, University of Wisconsin, Madison, Wisc. (U.S.A.)*

(Received May 22nd, 1962)

SUMMARY

A general method for predicting initial velocity and dead end and product inhibition patterns by inspection of the mechanism is presented. This method is applicable to any non-random mechanism without alternate reaction sequences. Several examples of the application of the method are given.

### INTRODUCTION

The previous two papers[1,2] have described a nomenclature and general theory for handling the kinetics of enzymic reactions with more than one substrate or product. With the approach outlined there, one can use kinetic studies as a tool for determining enzymic mechanisms and for studying the properties of active sites on enzymes. Many of these studies involve determining initial velocity, or product or dead end inhibition patterns. While these are easily deduced for a given mechanism from the overall rate equation by simplification and rearrangement, this takes considerable time, particularly if the rate equation is not available and must be derived. Furthermore, kinetic studies can only show consistency or inconsistency of the experimental data with possible mechanisms. In order to be sure that one has correctly determined the mechanism it is necessary to examine every possible mechanism that is at all plausible, and reject those that do not fit. If one is thorough about this, one ends up analyzing a very large number of mechanisms and deriving a very large number of rate equations. All of this labor would be tremendously reduced if one could look at a mechanism and tell by inspection what the initial velocity or the various inhibition patterns were. In this paper, rules will be given for this type of analysis, together with examples and several applications. The notation and nomenclature used will be those described in the previous papers[1,2].

---

* Published with the approval of the Director of the Wisconsin Agricultural Experiment Station.

## THEORY

It will be assumed that experimental data are to be plotted as reciprocal plots according to the general equation:*

$$\frac{1}{v} = (\text{slope})\left(\frac{1}{A}\right) + (\text{intercept})$$

where $v$ is initial velocity, $A$ is substrate concentration, and the slope and intercept may be functions of the concentrations of other substrates, activators, products, dead end inhibitors, etc., but are constant for any particular reciprocal plot. When the concentration of any of these other variables is changed, the slope of the line, the intercept, or both may be changed, and the change may be linear, parabolic, hyperbolic, or of greater complexity. The type of variation of slope and intercept with the concentration of the other variable constitutes the desired initial velocity or inhibition pattern.

The intercept of a reciprocal plot represents the reciprocal velocity at infinite substrate concentration. Thus, if changing the concentration of the other variable affects the rate even at saturation, the intercepts will vary as a function of this variable; if the variable no longer can influence the rate when the substrate is saturating, the intercepts do not vary.

The slope of a reciprocal plot, on the other hand, is a measure of how fast the reaction slows down as the substrate concentration decreases from infinity to zero. The net rate of the step where the substrate adds to the enzyme, which in a steady-state mechanism is equal to the overall rate, can be expressed:

$$v = k_1 A\,(E) - k_2\,(EA)$$

where $E$ represents the enzyme form that reacts with $A$. When the concentration of $A$ is infinite the steady-state concentration of $E$ is zero, and the actual rate of reaction is determined only by how fast $EA$ is converted back to $E$ by the rest of the reaction sequence. As the concentration of $A$ is decreased, the steady-state concentration of $E$ rises and that of $EA$ drops, finally becoming zero when $A$ becomes zero; so that when $A$ is very small, the rate of reaction is represented only by $k_1A(E)$. If changing the concentration of a variable other than the substrate specifically tends to increase the steady-state concentration of $E$, or decrease that of $EA$, then the rate of reaction decreases less rapidly as $A$ is lowered, and the slope of the reciprocal plot is decreased. If changing the concentration of the other variable specifically decreases the concentration of $E$ or increases that of $EA$, then the slope of the reciprocal plot is increased.

With these considerations in mind, we can now state two fundamental rules:

1. A compound affects the intercept of a reciprocal plot when it combines reversibly with an enzyme form other than the one the variable substrate combines

---

* If $A/v$ is plotted *versus* $A$, the slopes become intercepts and *vice versa*, and the rules given in this paper yield the proper patterns if this is kept in mind. Because of the fundamental nature of slope and intercept variation, data should be plotted only by one of these two equations. Since in practice one uses statistical methods to fit the experimental points to the actual hyperbola $v = VA/(K + A)$ (see ref. 3) and uses the graphical plot only to test for linearity, to determine if any point should be discarded, and to visualize the results after the fitted curve is plotted, it makes little difference which plot is used, and the more familiar double reciprocal plot will be used here.

with, and thereby changes the reaction velocity in a manner which can not be eliminated by saturation with the variable substrate.

2. A compound affects the slope of a reciprocal plot when it and the variable substrate either combine with the same form of the enzyme, or are separated in the reaction sequence by a series of reversible steps along which they can interact in such a manner that a change in the concentration of the compound specifically alters the net rate of the step involving the addition of the variable substrate in a manner which can be eliminated by a change in the concentration of the variable substrate. Release of a product at zero concentration (but not at finite concentration), or addition of a substrate at infinite concentration (saturation) are considered irreversible steps for purposes of this analysis.

To use these rules, one simply analyzes for slope and intercept effects separately, and then combines the results to determine the pattern. If the compound whose effects are being analyzed adds to only one enzyme form, the effect is always linear; that is, a replot of slopes or intercepts *versus* concentration for inhibitors, or *versus* reciprocal concentration for substrates and activators, is a straight line. If the compound in question reacts with two (or more) enzyme forms, the effects on slopes or intercepts are determined separately for each point of combination, and the combined effects are given by Rule 3:

3. When combination by a compound at two or more points in a reaction sequence gives by Rules 1 or 2 a multiple effect on slopes or intercepts, the resultant effects will be parabolic (or of higher degree for more than two points of combination) if: (a) the effects produced by all combinations are the same (inhibition or activation), (b) the points of combination are separated in the reaction sequence by reversible steps along which interaction may take place so that an increase in the concentration of the compound specifically causes, as the result of combination at one point, an increase in the steady-state concentration of the enzyme form reacting with this compound at the other (or next) point of combination (and so forth, if there are more than two points of combination). If no reversible sequences connect the points of combination, or if combination at one point does not affect the steady-state concentrations of the enzyme forms combined with at other points, then the resulting effects are linear. When analyzing intercept effects, it should be remembered that the variable substrate is saturating, and its combination with the enzyme is an irreversible step.

In a mechanism without alternate reaction sequences, when a compound reacts with two different enzyme forms to cause inhibition at one point and activation at another, and both combinations should affect either slope or intercept, then the resulting effects are neither linear nor parabolic. Replots of the slopes or intercepts would be a hyperbola having the vertical axis as one asymptote, and having as the other the expected plot in the absence of the activation if the replot is *versus* the concentration, or the inhibition if the replot is *versus* the reciprocal concentration of the compound under consideration. Such situations result when the compound is a substrate other than the variable one and can act also as a dead end inhibitor.

### *Initial velocity patterns*

Initial velocity patterns are usually obtained by making reciprocal plots for one substrate (variable substrate) at different fixed concentrations of one of the others (changing fixed substrate), while keeping all other substrates, if there are any, at constant concentration. If the mechanism does not include alternate reaction sequences, and no substrate reacts with more than one enzyme form, then according to Rule 1 the intercepts must always be a linear function of the reciprocal concentration of the changing fixed substrate. Analysis of the slope effects, however, may or may not show a reversible connection between the points of combination of the variable and changing fixed substrates. As a result there are two possible initial velocity patterns: parallel lines when no reversible connection exists, or lines intersecting to the left of the vertical axis when such a connection does exist. Whether the lines cross above, below, or on the horizontal axis depends on the ratio of certain kinetic constants (such as the ratio $K_a/K_{ia}$ in Ordered Bi Bi) rather than on the mechanism. A replot of slopes *versus* the reciprocal concentration of the changing fixed substrate is linear.

As an example of this analysis, consider the Bi Uni Uni Bi Ping Pong mechanism[1]:

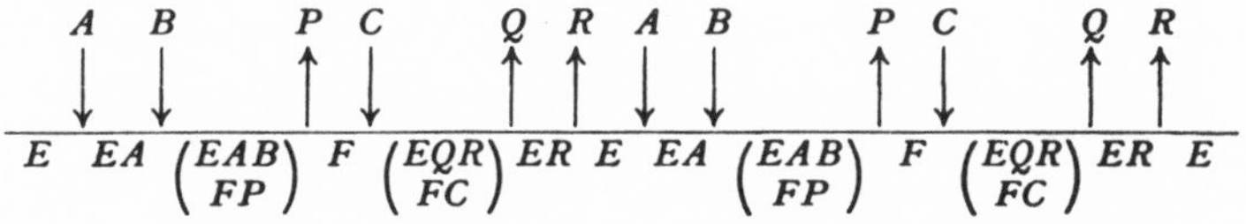

$$\begin{array}{ccccccccccccccccccccccccc}
 & A\downarrow & & B\downarrow & & P\uparrow & & C\downarrow & & Q\uparrow & & R\uparrow & & A\downarrow & & B\downarrow & & P\uparrow & & C\downarrow & & Q\uparrow & & R\uparrow & \\
\hline
E & & EA & & \begin{pmatrix} EAB \\ FP \end{pmatrix} & & F & & \begin{pmatrix} EQR \\ FC \end{pmatrix} & & ER & & E & & EA & & \begin{pmatrix} EAB \\ FP \end{pmatrix} & & F & & \begin{pmatrix} EQR \\ FC \end{pmatrix} & & ER & & E
\end{array} \qquad (1)$$

When $A$ is the variable and $B$ the changing fixed substrate, $B$ specifically alters the rate of conversion of $E$ to $EA$ by reacting with $EA$ and converting it into the central complex $(EAB)$. Thus the slope is affected, and the intersecting pattern is observed. When $B$ is the variable and $A$ the changing fixed substrate, a high concentration of $A$ converts $E$ to $EA$ more rapidly and specifically increases the concentration of $EA$, and thereby the rate of reaction between $EA$ and $B$. The slope is affected as before. When $A$ is the variable and $C$ the changing fixed substrate, however, the situation is different. Changing the concentration of $C$ does not specifically change the concentration of either $E$ or $EA$ (although these will both change by the same ratio as the result of the general redistribution of the enzyme among its forms as the concentration of $C$ changes), because there is no reversible sequence between the points of combination of $A$ and $C$. The release of $P$ blocks the sequence on one side, and the release of $Q$ and $R$ blocks the sequence on the other side. The parallel pattern is thus observed.

If the concentration of $P$ is not zero, but has a finite value, then a reversible connection is established between $C$ and $A$. An increase in the concentration of $C$ lowers the steady-state concentration of $F$; this lowers the rate of reaction between $F$ and $P$ and the steady-state concentration of $(EAB\text{–}FP)$. As the concentration of this central complex falls, so does the rate of its decomposition to form $EA$, and the concentration of $EA$. A rise in the concentration of $C$ thus specifically lowers the steady-state concentration of $EA$, and raises the net rate of combination of $E$ with $A$. A decrease in slope results.

If $B$ were saturating in the system, however, the reversible connection between $A$ and $C$ would again be broken. The steady-state level of $EA$ would be zero already, and changing the concentration of $C$ would have no effect. The parallel pattern would be observed, even though $P$ were present. If $B$ were the variable and $C$ the changing fixed substrate (or *vice versa*), the presence of $P$ would always establish a reversible link and the intersecting pattern, regardless of the concentration of $A$.

This type of analysis also tells whether any substrates can be changed together with their concentrations at constant ratio. Any substrates not reversibly connected can be varied together as variable substrate without giving a non-linear reciprocal plot or as changing fixed substrate without giving non-linear replots of slopes and intercepts. Thus in Bi Uni Uni Bi, $A$ and $C$ or $B$ and $C$ can be varied together, but not $A$ and $B$. If $A$ and $B$ are varied together as variable substrate, the reciprocal plots ($1/v$ *versus* $1/A$ or $1/B$) are parabolas. If $A$ and $B$ are varied together as changing fixed substrate, the replots of slopes and intercepts (*versus* $1/A$ or $1/B$) are likewise parabolas.

This theory will also predict what happens when a substrate is added two or more times during the reaction sequence. When such a substrate is the variable substrate it gives linear reciprocal plots if the points of addition of the substrate are not reversibly connected, but parabolas if the two points are reversibly connected (or higher degree functions if there are more than two points of addition reversibly connected). If such a substrate is the changing fixed substrate, replots of slopes and intercepts will be linear if the two points of addition are not reversibly connected, and parabolic if they are. It must be remembered, however, that the intercepts represent the points at which the variable substrate is saturating, and thus a parabolic intercept replot is observed only when there is a reversible sequence between the points of addition of the changing fixed substrate which does not include the addition of the variable substrate. For instance in the following partial reaction sequences, with $B$ as variable substrate, replots of slopes *versus* $1/A$ are always parabolic, but replots of intercepts will be parabolic only for the first two mechanisms:

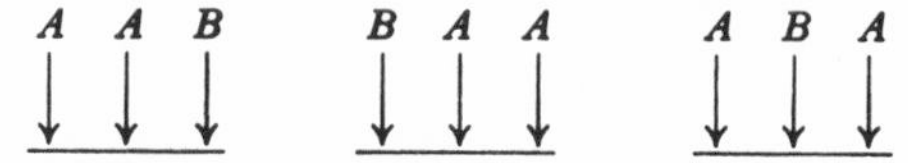


If $A$ is the variable substrate, reciprocal plots ($1/v$ *versus* $1/A$) will always be parabolas except in the third case, where the lines would be straight if $B$ were saturating, and parabolic if it were not.

### *Product inhibition patterns*

Products that react with only one enzyme form always give linear inhibitions, which may be competitive (slopes only vary), uncompetitive (intercepts only vary), or non-competitive (both vary). In non-competitive cases, the lines cross to the left of the vertical axis, and as with the similar initial velocity pattern, the intersection point may be above, below, or on the horizontal axis, depending on the ratio of certain kinetic constants. If two or more molecules of a product are generated by a reaction so that the product normally combines with more than one enzyme form, the total effect is the sum of the various slope and intercept effects predicted separately. In accordance with Rule 3, if slope or intercept effects result from two points of combination, the effect is parabolic if the two points are reversibly connected, and linear if they are not.

A product inhibitor is capable of interacting along a reversible series of steps with variable substrates both upstream and downstream of it in the reaction sequence, so the application of Rules 1, 2 and 3 is straightforward. For example, consider the following analysis of product inhibition in the Bi Uni Uni Bi Ping Pong mechanism (Mechanism 1):

*P* as inhibitor, *A* varies: *P* reacts with *F*, and *A* with *E* in the sequence, so the intercept will vary with *P*. *P* and *A* can interact along the reversible sequence *E–EA–(EAB)–F*, so the slope varies also, and the inhibition is non-competitive. Saturation with *B* has no effect on the intercept term, but interupts the sequences between *A* and *P* so that the slope term no longer varies with *P* and the inhibition uncompetitive. Saturation with *C* eliminates inhibition by *P*, since *P* reacts with *F*, and when *C* is very large the steady-state concentration of *F* remaining in the system is nearly zero.

*P* as inhibitor, *B* varies: *P* and *B* react with different enzyme forms, so the intercept varies. The sequence *EA–(EAB)–F* is reversible, so the slope varies also, giving non-competitive inhibition. Saturation with *A* has no effect, but saturation with *C* eliminates inhibition by *P*, as before.

*P* as inhibitor, *C* varies: *P* and *C* react with the same enzyme form only, so the slope but not the intercept varies, giving competitive inhibition. Saturation with *A* or *B* has no effect.

*Q* as inhibitor, *A* varies: *Q* and *A* react with different enzyme forms, so the intercept varies. The sequence *ER–E* is interrupted by liberation of *R* and the sequence *E–EA–(EAB)–F–(EQR)–ER* by liberation of *P*, so the slope does not vary and the inhibition is uncompetitive. Saturation with *B* or *C* has no effect since neither reacts with *ER*.

*Q* as inhibitor, *B* varies: *Q* and *B* react with different enzyme forms, so the intercept varies. The sequences between them are blocked by release of *P* and *R* as before, so the inhibition is uncompetitive. Saturation with *A* or *C* has no effect.

*Q* as inhibitor, *C* varies: *Q* and *C* react with different enzyme forms, so the intercept varies. The sequence *F–(ERQ)–ER* is reversible so the slope varies also, giving non-competitive inhibition. Saturation with *A* or *B* has no effect.

*R* as inhibitor, *A* varies: *A* and *R* both react only with *E*, so the slope and not the intercept varies, giving competitive inhibition.

*R* as inhibitor, *B* varies: *R* and *B* react with different enzyme forms, so the intercept varies. The sequence *E–EA* is reversible, so the slope varies also and the inhibition is non-competitive. Saturation with *A* eliminates inhibition by *R*, since the concentration of *E* is reduced to zero. Saturation with *C* has no effect.

*R* as inhibitor, *C* varies: *R* and *C* react with different enzyme forms, so the intercept varies. The sequences between *R* and *C* are rendered irreversible by release of *P* and *Q*, so the slope does not vary and the inhibition is uncompetitive. Saturation with *A* eliminates inhibition by *R*. Saturation with *B* has no effect.

This type of analysis will also successfully predict inhibition patterns in mechanisms such as the Theorell–Chance mechanism (Mechanism 2) where product inhibition by *P* against *B* as variable substrate is competitive.

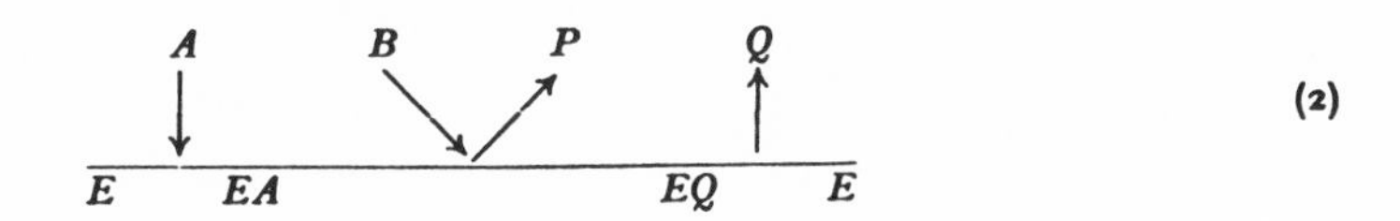

Saturation with $B$ overcomes this inhibition completely by converting all of the $EA$ into $EQ$ and lowering the steady-state concentration of $EA$ essentially to zero. There is no inhibition by $P$ under these conditions, since the inhibitory effect of $P$ is to raise the steady-state level of $EA$. The intercepts of reciprocal plots are thus not affected and the inhibition is competitive. These rules also predict the competitive product inhibition by either product in the Rapid Equilibrium Random Bi Bi mechanism[1], where saturation with the variable substrate lowers the equilibrium concentrations of the enzyme forms reacting with the inhibitor to zero and thus eliminates the inhibition.

*Dead end inhibition patterns*

Compounds which are not substrates or products but which react with one or more enzyme forms to give complexes which can not participate in the reaction give dead end inhibition. Because dead end inhibitors can not back up reaction sequences, as can products, they influence the slopes of reciprocal plots only when the point of combination of the variable substrate follows the point of addition of inhibitor in the reaction sequence and is reversibly connected with it. Dead end inhibitors often add to more than one enzyme form. According to Rule 3, however, the inhibitions are always linear because adding the inhibitor at one point never increases the concentration of any of the other normal enzyme forms. Only if the inhibitor combines with an $EI$ complex to form $EI_2$ is a parabolic inhibition observed. The following analysis of dead end inhibition for the Bi Uni Uni Bi Ping Pong mechanism (Mechanism 1) will serve to illustrate the method.

$I$ reacts with $E$, $A$ varies: $I$ and $A$ react only with the same enzyme form, so the inhibition is competitive.

$I$ reacts with $E$, $B$ varies: $I$ and $B$ react with different enzyme forms, so the intercept varies. The sequence $EI$–$E$–$EA$ is reversible, so the slope varies, giving non-competitive inhibition. Inhibition by $I$ is eliminated by saturation with $A$, but saturation with $C$ has no effect.

$I$ reacts with $E$, $C$ varies: No reversible sequence connects $E$ and $F$, so the inhibition is uncompetitive. Saturation with $A$ eliminates inhibition by $I$, as before.

$I$ reacts with $EA$, $A$ varies: $EAI$ and $E$ are not part of a reversible sequence, so the inhibition is uncompetitive. Saturation with $B$ eliminates inhibition by $I$.

$I$ reacts with $EA$, $B$ varies: The inhibition is competitive, since both $B$ and $I$ react only with $EA$.

$I$ reacts with $EA$, $C$ varies: The release of $P$ interrupts the reversible sequence between $EAI$ and $F$, so the inhibition is uncompetitive. Saturation with $B$ eliminates the inhibition.

$I$ reacts with $(EAB)$, $(EQR)$, or $ER$: No substrates react with these enzyme forms and reversible sequences are blocked by release of products, so the inhibition is always uncompetitive.

$I$ reacts with $F$, $A$ varies: There is no reversible sequence connecting $FI$ and $E$ so the inhibition is uncompetitive. Saturation with $C$ lowers the concentration of $F$ to zero and eliminates the inhibition.

$I$ reacts with $F$, $B$ varies: The situation is the same, and the inhibition is uncompetitive, and is eliminated by saturation with $C$.

*I* reacts with *F*, *C* varies: The inhibition is competitive, since *C* and *I* both react only with *F*.

Note the large number of uncompetitive inhibitions predicted by the above analysis, and the fact that these inhibitions result from combination with enzyme forms other than central complexes. This author feels that dead end combination with central complexes is extremely unlikely, and that all uncompetitive inhibitions observed experimentally can be explained by combination with non-central transitory complexes or stable enzyme forms. The belief that uncompetitive inhibition results from combination with a central complex is one of the very unfortunate results of trying to extend a theory designed for unireactant mechanisms to multireactant ones. This author is aware of no uncompetitive inhibitions observed for true Uni Uni reactions.

### *Mixed dead end and product inhibition*

If a product also reacts in dead end fashion with an enzyme form other than the one it normally reacts with, it gives mixed product and dead end inhibition. The inhibitory effects are predicted separately by Rules 1 and 2, and then the net effects, if multiple ones are predicted, are determined by Rule 3. In order for parabolic effects to be observed, however, the dead end combination must precede the normal combination point with reversible steps in between. Combination at the normal point in the sequence then backs up the reaction and results in an increase in the concentration of the enzyme form that combines to form the dead end complex.

As an example of this type of analysis, consider the following multiple inhibitions by *P* in the Bi Uni Uni Bi mechanism (Mechanism 1) when *B* is the variable substrate. If *P* combines only with *F*, we have linear non-competitive inhibition, as found above. If *P* also reacts with *E*, there will be additional effects on both slope and intercept, since *P* and *B* still combine with different enzyme forms, and there is a reversible sequence between *E* and *EA*. The two points of addition of *P* are reversibly connected, but with the addition of *B* occurring in between. The slope is thus a parabolic, but the intercept a linear function of inhibitor, and we have S-parabolic I-linear non-competitive inhibition. Saturation with *A* eliminates the dead end inhibition by *P* and leaves only the linear non-competitive product inhibition. Saturation with *C* eliminates the product inhibition, but leaves the dead end inhibition, which is also linear non-competitive.

If *P* reacts with *EA* as well as with *F*, an additional effect results only on the slope, since *B* and *P* are both combining with *EA*. The two points of addition of *P* are reversibly connected, so the slope is a parabolic function of inhibitor concentration, and we have again S-parabolic I-linear non-competitive inhibition. Saturation with *A* now has no effect, while saturation with *C* eliminates the product inhibition and leaves only linear competitive inhibition due to the dead end combination with *EA*.

If *P* reacts with the central *(EAB–FP)* complex as well as with *F*, only an additional effect on the intercept is predicted. Since the two points of addition of *P* are reversibly connected, the variation is parabolic and we have S-linear I-parabolic non-competitive inhibition. Saturation with *A* has no effect, but saturation with *C* eliminates the product inhibition, leaving linear uncompetitive inhibition.

When $P$ reacts in dead end fashion with $F$, $(EQR)$, or $ER$ in addition to the expected combination with $F$, additional effects on only the intercept are predicted. Since the two points of addition of $P$ are not reversibly connected, or are the same, the intercept will still be a linear function of inhibitor concentration. Saturation with $A$ has no effect, while saturation with $C$ eliminates all inhibition when both combinations are with $F$, and leaves linear uncompetitive inhibition when the dead end combination of $P$ is with $(EQR)$ or $ER$.

If we look at multiple interactions by $P$ in the same mechanism with $C$ as variable substrate, one other interesting point is brought out. Combination of a dead end inhibitor with $E$, $EA$, or $(EAB)$ would normally not give an effect on the slope, since the release of $P$ is an irreversible step. Dead end inhibition by $P$ does give an effect on the slope here, however, since the presence of $P$ in the system keeps the reaction sequence reversible. Thus combination of $P$ with any of these forms gives parabolic slope variation, although saturation with $B$ reduces this to linear if the combination is with $E$ or $EA$, and saturation with $A$ does likewise if combination is with $E$.

## DISCUSSION

The value of the methods outlined in this paper is too obvious to need further elaboration here. They have been used routinely in this laboratory for exploring possible mechanisms, and they have proven of special value to students trying to learn something useful about kinetics without getting lost in equations and algebra. While only non-random mechanisms without alternate reaction sequences (excepting Rapid Equilibrium Random mechanisms) have been treated in this paper, preliminary analysis indicates that the general approach can probably be extended to cover alternate product and similar useful types of experiments. This will require a more sophisticated development of the theory of slope and intercept variation than the simple analysis given here, however, and will be left for the future.

## ACKNOWLEDGEMENT

Supported in part by a grant from the National Science Foundation (G 14388).

## REFERENCES

1 W. W. Cleland, *Biochim. Biophys. Acta*, 67 (1963) 104.
2 W. W. Cleland, *Biochim. Biophys. Acta*, 67 (1963) 173.
3 G. N. Wilkinson, *Biochem. J.*, 80 (1961) 324.

Part VIII

# ACTIVE SITE STUDIES: INDUCED FIT, AND REPORTER GROUPS

# Editor's Comments on Papers 51 and 52



Countess: Will your answer serve to fit all questions?
Clown: As fit as ten groats is for the hand of an attorney,...as the nun's lip to the friar's mouth, nay as the pudding to his skin.

[*All's Well That Ends Well*, William Shakespeare, c. 1602]

One would not today refer to Koshland's paper (Paper 51) on protein synthesis if it had confined itself to this topic. The part of the paper dealing with protein synthesis is quite outdated and has hence been omitted here. However, the discussion contains the celebrated induced fit variation or elaboration of Emil Fischer's famous key-lock theory of enzyme action (Paper 24). It is for this almost incidental section, included here, that this Koshland paper is extensively quoted. Emil Fischer, as emphasized by Fruton (1970, p. 402), never explicitly stated his theory to require the enzyme molecule that acts on the substrate to be a rigid structure. Koshland explicitly stated the idea that flexibility was a prerequisite for the activity of many, though not necessarily of all enzymes.

Koshland (1973) subsequently stressed that the notion of induced fit is just one manifestation of protein flexibility. This flexibility is associated with phenomena concerning proteins apart from enzyme-substrate interactions in a strictly limited sense. Examples of such phenomena are allosteric behavior, and changes in enzyme activities due to covalent modifications of the enzyme molecule. Papers on allostery are included in this volume (Part IX). Covalent modifications have been discussed by Segal (1973) and by Stadtman (1970). Similar changes occur in numerous non-enzymic proteins such as those concerned with hormonal action (receptor proteins), with the induction of genetic transcription (derepression), with immunological processes, with some types of cellular transport, with muscular contraction, and so on. There is by now extensive evidence for conformational changes due to substrates and other ligand molecules, reviewed in detail by Citri (1973). The biological importance of substrate-induced fit in providing enzyme specificity and catalytic control has been discussed by Jencks (1975). It is important to recognize that the notion of induced fit began to deemphasize exclusive preoccupation with the active site as an explanation of various aspects of enzyme function, particularly those of regulation (see Part X). Atkinson (1970) in this context has stressed the importance of the whole protein molecule, and the incorrectness of assuming "...the active site...recognized as being protein in nature...to be carried on a nondescript and passive mass of protein...This attitude [is] somewhat like that of investigators who might seek an explanation of Michelangelo's genius by undertaking a detailed anatomical study of his hands."

Once it was realized that all enzymes consist of proteins which frequently act in association with various more or less tightly bound smaller substances it was possible to become more specific than hitherto about the chemistry and function of the active center or active site. This region of the enzyme molecule had now to be studied in terms of the chemistry of amino acid side chains and of their interaction with each other, with the substrate or substrates, and with cofactors. A review by Koshland, Strumeyer, and Ray, presented at a symposium in 1962 and rather too long to be included in this volume, surveys the evidence that only a few among the twenty-odd protein amino acids actually participate in the great variety of enzymatic reactions. It is of interest that in the discussion to this presentation, J. Kallos first introduces the idea of reporter groups as an approach to the study of active sites. This topic is treated in more detail in the article by Burr and Koshland (Paper 52).

One can have an intuitive feeling for the notion of a conformational change induced in the enzyme molecule by substrate without having more than a vague picture of some kind of complementarity between active site and substrate or substrates. It must hence be mentioned that Linus Pauling already in 1946 had postulated that the active site is complementary not towards the substrate in its ground state, but rather towards the substrate in its activated or transition state. This postulate facilitated an understanding of enzymatic rate accelerations. It has since been utilized more elaborately in studies of enzymatic catalysis (see, for example, Lienhard 1973) and in studies of the inhibitory activity of so-called transition state analogs of substrates (see, for example, Wolfenden 1972). Studies on the chemistry and behavior of the active site of an enzyme hence are related to studies on the efficiency of enzymatic catalysis. This topic will be discussed in Part IX. We will come to discuss some X-ray crystallographic studies of active sites as three-dimensional structures in Part XII.

# 51

Reprinted from pages 99–101 and 104 of *Natl. Acad. Sci. (USA) Proc.* **44**:98–104 (1958)

## APPLICATION OF A THEORY OF ENZYME SPECIFICITY TO PROTEIN SYNTHESIS*

**D. E. Koshland, Jr.†**

*Biology Department, Brookhaven National Laboratory, Upton, New York*

[*Editor's Note:* In the original, material precedes these excerpts.]

Let us, therefore, first examine the manner in which an enzyme exerts its specificity. It is apparent that this specificity does not arise from inductive effects in the substrate molecules. A change from a hydroxy to a methoxy group at the C-3 position of glucose will have negligible electrostatic effects on the basic or acidic properties of the C-6 hydroxyl, and yet this change is enough to block completely the action of hexokinase.[9] To explain this specificity, a "key-lock theory" was proposed.[10, 11] In essence this theory said that the enzyme was a rather rigid negative of the substrate and that the substrate had to fit into this negative to react. Hence a modification which had minor electrostatic effects but which would so increase the bulk or change the shape of the molecule that it could not fit on the enzyme surface would prevent reaction. This explanation is supported by the demonstration of steric repulsions in organic reactions and by the fact that the kinetics of enzyme action are compatible with an enzyme-substrate complex. It is capable of explaining almost all the observed specificity patterns.

On close examination,[12] however, it is clear that this hypothesis does not explain all cases, and, as usual, it is the anomalies which lead to revisions of our theories. The anomalies fall, in general, in the class that smaller analogous compounds react extremely slowly or not at all. For example, ribose-5-phosphate is hydrolyzed by 5′-nucleotidase much less rapidly than adenylic acid.[13] Since the ribose-5-phosphate

*Research carried out at Brookhaven National Laboratory under the auspices of the U.S. Atomic Energy Commission.

†Visiting professor of chemistry, Cornell University, 1957–58.

is simply adenylic acid without the purine, it could be argued that an attractive group is eliminated and hence the affinity to the enzyme is less (which it probably is). However, when this lower affinity is compensated for by comparing the velocities at enzyme saturation, the ribose-5-phosphate still reacts at $^1/_{100}$ the rate of adenylic acid. The key-lock theory would certainly predict equal rates at saturation levels. Similarly, the failure of alpha methyl glucoside to act as a substrate for amylomaltase[14] is inconsistent with this hypothesis. Alpha methyl glucoside has the same structure as the natural substrate maltose, except that it is smaller, i.e., two hydrogens on the aglycon carbon take the place of the second pyranose ring. As would be expected from its smaller, but otherwise similar, structure, alpha methyl glucoside should be absorbed on the enzyme surface. That it does is shown by its action as a competitive inhibitor. However, it is not hydrolyzed by the enzyme. In this case certainly, lack of access to the enzyme surface cannot explain its failure to be a substrate. Numerous similar anomalies can be found.[12]

The explanation that we suggest to explain these phenomena is as follows: (*a*) a precise orientation of catalytic groups is required for enzyme action; (*b*) the substrate may cause an appreciable change in the three-dimensional relationship of the amino acids at the active site; and (*c*) the changes in protein structure caused by a substrate will bring the catalytic groups into the proper orientation for reaction, whereas a non-substrate will not. This set of postulates has been called "the induced fit" theory for brevity and to emphasize that, while the idea of a fit is retained from the key-lock theory, the fit in this case occurs *only after* the changes induced by the substrate itself.

A simplified schematic representation is shown in Fig-

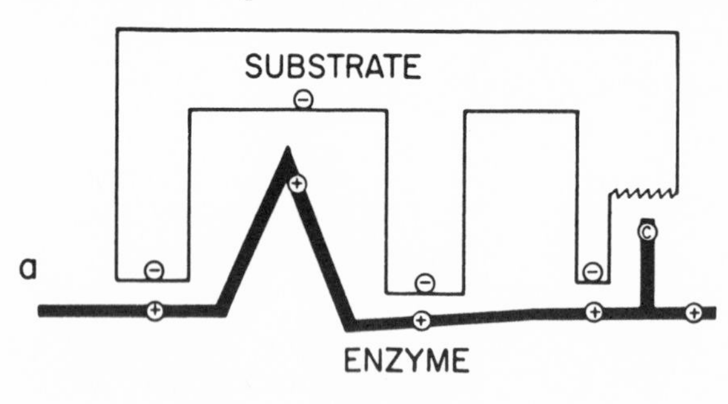


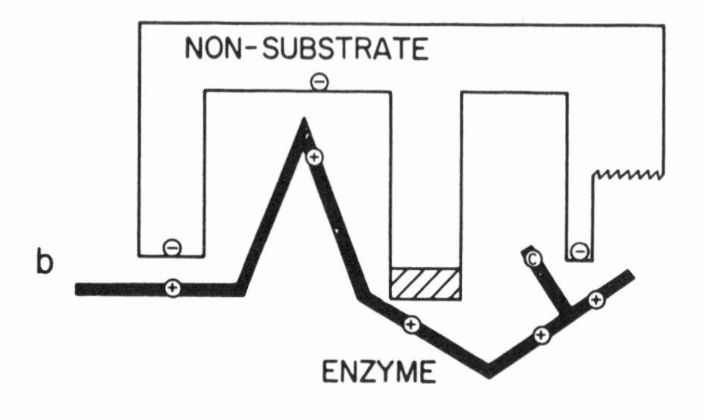


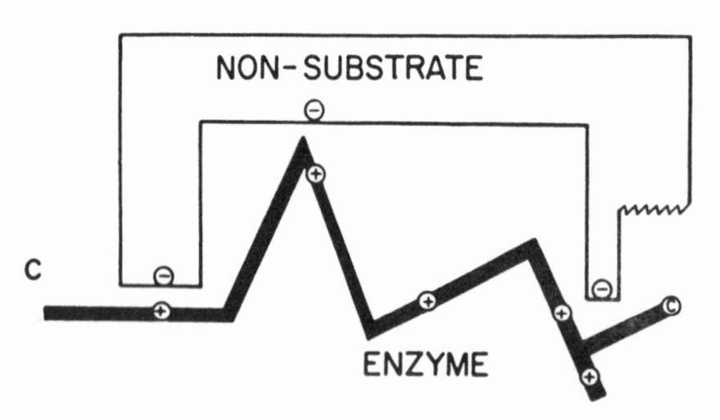


FIG. 1.—Interaction of enzyme with (*a*) substrate, (*b*) compound too large to be substrate, and (*c*) compound too small to be substrate. Circled pluses and minuses indicate any mutually attractive groups on enzyme and substrate. Circled *C* stands for catalytic group, and jagged line for bond to be broken.

ure 1. In Figure 1, *a*, a catalytic group, *C*, is aligned with the bond to be broken (shown by the jagged line). In Figure 1, *b*, increasing the size of a group well removed from the bond to be broken destroys the alignment; and in Figure 1, *c*, the complete removal of the same group likewise destroys the alignment. Thus this theory explains why a substrate can be converted to a non-substrate by a decrease in, as well as by an increase in, its size.

Furthermore, there is independent support for the various postulates. First, the flexible nature of portions, if not all, of the protein chain is adduced from many sources, and these changes can be caused by small molecules, charged or uncharged. A typical example is the reversible denaturation of a number of enzymes by urea.[15] This denaturation is accompanied by changes in viscosity, optical rotation, and sedimentation constant, which certainly are indicative of an alteration in the geometry of the amino acids of the protein relative to each other. The urea undoubtedly causes these changes by competing for the internal hydrogen bonds of the protein.[16] The reversible nature of this denaturation shows that, upon removal of urea, the protein returns to its natural conformation. The urea, because of its particular structure, is attracted non-specifically to many peptide links, whereas a hydrogen-bonded substrate is attracted to only one or a small number of sites. It is certainly reasonable, however, to expect that the specifically attracted substrate will cause similar conformation changes in its immediate environment. Charged or hydrocarbon-type compounds are observed to cause analogous alterations in protein structure.

Second, there is support for the postulate that a precise orientation of catalytic groups is important in enzyme catalysis. The evidence comes from analogies to both heterogeneous and homogeneous catalysis in simple chemical systems and also from the observed behavior of the enzymes themselves.[17] Finally, the idea that the substrate is necessary to promote the proper orientation of catalytic groups can explain phenomena other than their specificity, for example, substrate-promoted isotopic exchanges.[18]

[*Editor's Note:* Material has been omitted at this point.]

[9] A. Sols and R. K. Crane, *J. Biol. Chem.*, **210**, 581, 1954.

[10] E. Fisher, *Ber.*, **27**, 2985, 1894.

[11] The "key-lock" theory of specificity is also frequently referred to as a "template" theory. In this article it will be mentioned as a "key-lock theory" to distinguish it from the template theory of protein synthesis, which, of course, deals with a quite different problem.

[12] D. E. Koshland, Jr., in *The Enzymes*, ed. P. Boyer, H. Lardy, and K. Myrback (rev. ed., New York: Academic Press), Vol. 1 (in press).

[13] L. A. Heppel and R. J. Hillmoe, *J. Biol. Chem.*, **188**, 665, 1951.

[14] H. Wiesmeyer and M. Cohn, *Federation Proc.*, **16**, 270, 1957.

[15] J. I. Harris, *Nature*, **177**, 471, 1956; H. Neurath, J. A. Rupley, and W. J. Dreyer, *Arch. Biochem. and Biophys.*, **65**, 243, 1956; E. P. Kennedy and D. E. Koshland, Jr., *J. Biol. Chem.*, **228**, 419, 1957.

[16] W. Kauzmann, *Mechanism of Enzyme Action*, ed. W. D. McElroy and B. Glass (Baltimore: Johns Hopkins Press, 1954, p. 70.

[17] D. E Koshland, Jr., *J. Cell. Comp. Physiol.*, **47**, Suppl. 1, p. 217, 1956.

[18] D. E. Koshland, Jr., *Discussions Far. Soc.*, **20**, 142, 1956.

# 52

Reprinted from *Natl. Acad. Sci. (USA) Proc.* **52**:1017–1024 (1964)

## *USE OF "REPORTER GROUPS" IN STRUCTURE-FUNCTION STUDIES OF PROTEINS**

BY MERRILL BURR AND D. E. KOSHLAND, JR.

THE ROCKEFELLER INSTITUTE, AND THE BIOLOGY DEPARTMENT OF BROOKHAVEN NATIONAL LABORATORY

*Communicated by D. D. Van Slyke, August 12, 1964*

Correlation of structure and function in protein molecules has taken enormous strides in recent years. X-ray crystallography has made it possible to delineate the three-dimensional structure of hemoglobin and myoglobin,[1, 2] and protein modification methods have identified amino acids at the active sites of a number of enzymes.[3–5] It would appear that these techniques can be applied in other cases and thus both the three-dimensional structures of many proteins and the positions in these structures of essential amino acids will be known. It is, therefore, not premature to ask what further information will be needed to complete the correlation of function and structure and whether tools can be devised to provide this information.

One problem which remains, even after the active site has been located, is the description of the steric relations between the various residues at the active site and the various parts of the substrate molecule. This problem is illustrated in Figure 1. If $E$ represents an essential residue, e.g., the serine in chymotrypsin, it seems clear that part of the substrate must be in contact with this residue but the alignment of other parts of the substrate is not determined. Thus, a requirement that the amino

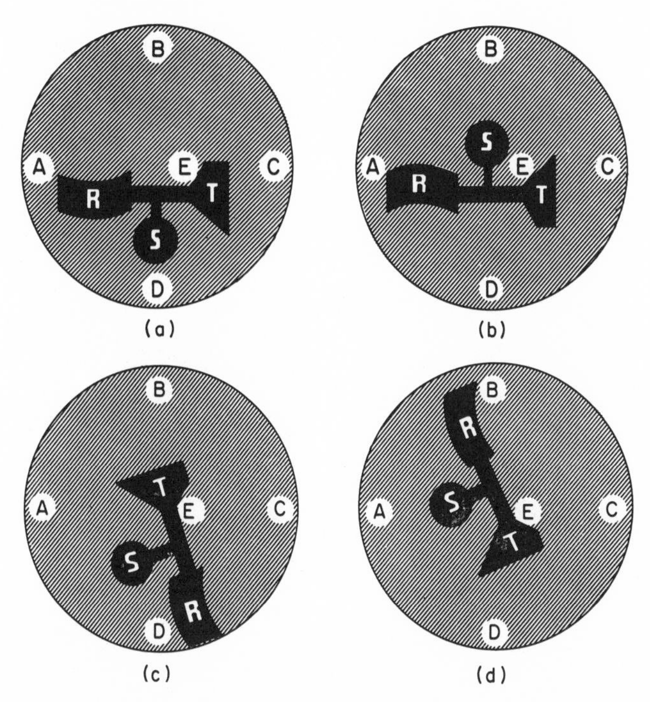


FIG. 1.—Orientations of substrate on a hypothetical enzyme surface. *A, B, C, D, E* represent positions of amino acid residues. All orientations place an essential residue *E* adjacent to the *T* moiety of the substrate. The position of the substrate in (*a*) and (*b*) are achieved by a 180° out-of-plane rotation, whereas in (*d*) the position is achieved by movement of the substrate only within the plane of the enzyme surface.

acid residue *E* of Figure 1 must be near the *T* moiety of the substrate still leaves a number of alternatives for the location of the *R* and *S* portions of the substrate.

A second problem which is related to the first, but which poses separate experimental and theoretical difficulties arises from the need to determine the contributions to binding and catalysis made by the interaction of specific parts of the substrate with specific parts of the protein. The successful solution of the first problem, e.g., the finding that the *R* moiety of the given substrate is adjacent to the amino acid residue, *B*, in the *ES* complex will be a giant step, but it will not establish the contribution of this interaction unless *R* or *B*, or both, can be modified and the binding and velocity constants for the new modifications correlated with the original values. To make such a comparison the orientations of the new substrates and inhibitors on the protein surface must be known, and it cannot be presumed *a priori* that their orientation will be analogous to the original system. Niemann and co-workers[6] have used the concept of variable orientation of substrates in an attempt to correlate the kinetic constants of a wide variety of chymotrypsin substrates. Also, the substrate for β-amylase has been shown to act as its own competitive inhibitor by binding incorrectly to the active site.[7]

The logical consequence of such considerations can be illustrated by reference to Figure 1. A given substrate, for example, might be bound to an enzyme in the orientation of Figure 1*a*. Modification of either the *S* group of this substrate or the *D* group of the protein might produce a new alignment similar to that of

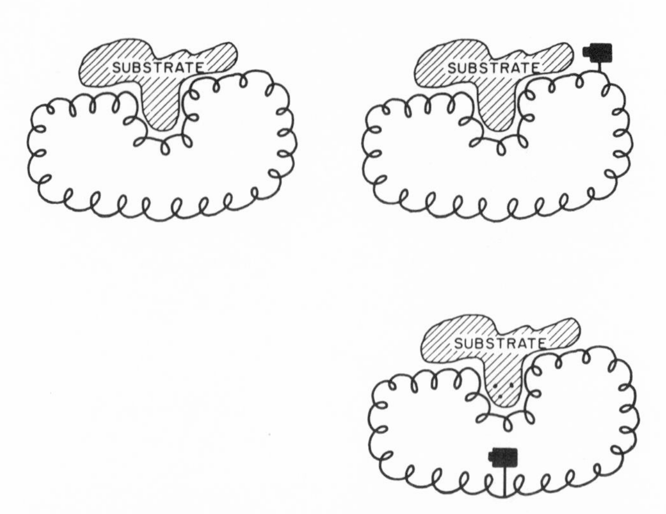


FIG. 2.—Schematic representation of enzyme-substrate complex in native protein, protein containing reporter group (solid black area) adjacent to substrate binding area and reporter group distant from substrate binding area.

Figure 1*b*. Interpretation of kinetic constants for the modified system would, therefore, be incorrect if the orientation of Figure 1*a* was assumed throughout. Because of this alternative a ready and objective criterion will be needed to establish orientations in modified systems relative to a standard system.

A third problem relates to the role of flexibility in enzyme action. It seems increasingly apparent that conformation changes play a vital role in the specificity process[8] and these changes make the structure-function correlation more complex. For example, it will be important to know which parts of the protein undergo conformational changes on interaction with the substrate. Moreover, the flexibility hypothesis[9] assumes that the alignment of catalytic groups is involved and, thus, conformation changes may be identified not only with the initial binding process, but also with the subsequent catalytic steps. It will be desirable to identify the conformational changes with the appropriate steps in the enzyme action.

To aid the solution of these and related problems, a new technique has been developed based on the rationale of Figure 2. Briefly stated, it involves the introduction of an environmentally sensitive residue into specific positions of the protein molecule. The environmentally sensitive group will be referred to as a "reporter" group since it is designed to "report" changes in its environment to an appropriate detector. Two situations are shown in which a reporter group might register environmental changes. In one, the reporter moiety is placed next to the active site so that a direct interaction with substrate is possible. In the other, the reporter group is placed at a point distant from the active site. In the former case the absorption of the substrate or inhibitor[10] to the active site changes the environment of the reporter group by direct contact. In the latter case, direct contact with substrate is excluded but a substrate-induced conformation change may trigger changes in conformation in the neighborhood of the reporter group.

To be of maximum utility, the position occupied by the reporter group must be known, and the simplest method of achieving this is to attach it by a covalent bond

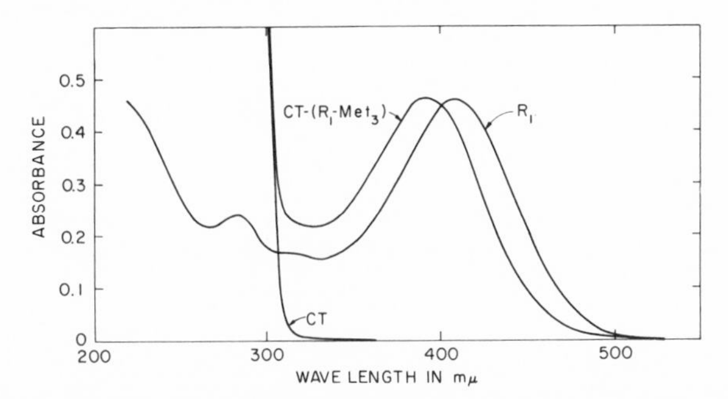


FIG. 3.—Spectra of chymotrypsin (CT), 2-acetamido-4-nitrophenol ($R_1$), and chymotrypsin which has been reacted with 2-bromoacetamido-4-nitrophenol. The nomenclature CT-($R_1$-$Met_3$) reflects the fact that the reporter group ($R_1$) is attached to the methionine three residues from the active serine of the chymotrypsin.

to a specific residue in the protein. The original reagent must, therefore, contain both a "reporter moiety," which is sensitive to changes in environment, and a "positioning" group which will guide and bond the reporter moiety to the appropriate residue in the protein. One such reagent, 2-hydroxy-5-nitrobenzyl bromide, has already been described.[11] In this paper a second reagent is described, together with some preliminary experiments which indicate that this approach will be of general utility in the problems outlined above.

The reagent, 2-bromoacetamido-4-nitrophenol (I), was designed based on the knowledge (*a*) that the nitrophenol chromophoric group is highly sensitive to its environment and absorbs in a region of the spectrum in which proteins are transparent, (*b*) that chymotrypsin contains a methionine residue which is three residues from the active serine in the primary sequence,[12] (*c*) that this methionine residue can be modified by iodoacetamide without modification of any other residue in the protein,[13] and (*d*) that this methionine can be carboxymethylated with only minor changes in the activity of the enzyme.[13]

Treatment of 2-amino-4-nitrophenol with bromoacetyl bromide in acetone by classical procedures[14] produced the compound which was crystallized from ethanol and found to melt with decomposition at 212–213°. NMR and infrared spectra were consistent with this structural formula as was the carbon-hydrogen analysis. On treatment of $8 \times 10^{-5}$ *M* chymotrypsin for 18 days in 20 per cent methanol at pH 3, 20° with $9 \times 10^{-4}$ *M* compound I, 0.6 moles of reporter group were introduced per mole of chymotrypsin. The spectra of the free reagent, of the native chymotrypsin, and of the labeled protein are shown in Figure 3. The enzyme has 40 per cent of its native activity as tested by the method of Schwert and Takenaka.[15]

Difference spectra were obtained by comparing solutions of the labeled enzyme in the presence of substrate and buffer with solutions of labeled enzyme in the presence of buffer only. Although this reporter group is sensitive to pH (a fact which can be used to advantage in some studies), this property was not of interest in the present studies. The solutions were carefully buffered and pH measurements were made before and after mixing of substrate with enzyme to ensure that pH changes were less than 0.01 pH units. In Figure 4*e*, the difference spectrum obtained with a change in pH of 0.01 pH units is shown to indicate the spectrum to be expected from such a source.

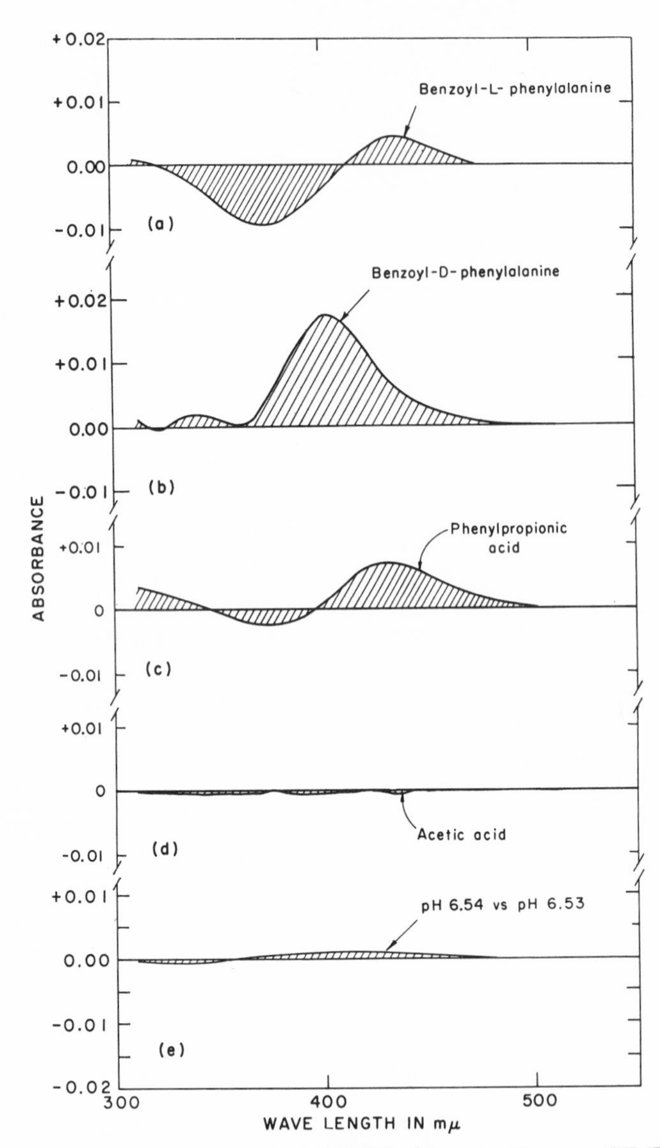


Fig. 4.—Difference spectra of chymotrypsin labeled with a reporter group CT-($R_1$-$Met_3$). Conditions: $7.4 \times 10^{-5}$ *M* enzyme, 0.05 *M* phosphate buffer; (*a*)–(*d*) $1.5 \times 10^{-3}$ *M* substrate, pH 6.49; (*e*) spectrum calculated for pH change of 0.01 pH units based on measured spectra for a change of 0.10 and 0.02 pH units.

Some representative spectra shown in Figure 4 establish that significant difference spectra are obtained in the presence of some substrates of chymotrypsin. In each case the free acid was used (substrate for the back reaction) to obviate pH errors resulting from hydrolysis. Controls with free dyes and substrates showed no difference spectra. A control with acetic acid (Fig. 4*d*) showed essentially no difference spectrum indicating that ionic strength effects or a nonspecific carboxyl interaction were not responsible for the difference spectra of Fig. 4*a*–*c*. Benzoyl phenylalanine, the ethyl ester of which is an excellent substrate for the enzyme, gives pronounced difference spectra, those of the D and L isomers being significantly different from each other. Acetyl-L-phenylalanine and benzoyl-L-alanine give similar but less pronounced differences. Considering even these few examples, it is clear that the spectrum is a function of a specific interaction of substrate and enzyme. Thus, the first desired goal has been achieved—an environmentally sensitive group has been introduced in a sensitive region of the protein and has been shown to report changes which are related to specific structural features of the substrate.

Even from these early studies some of the ramifications and limitations of this method are indicated. They are as follows:

(1) A relative correlation of substrate structure and orientation of the protein may be obtained without knowledge of the three-dimensional structure of the protein. Moreover, the approximate relative positions of a residue of the protein and a portion of the substrate may be obtained. Using Figure 1 as an illustrative example, all substrates with a bulky $R$ group might cause a signal change in a reporter group attached to residue $A$ on the protein, whereas substrates with a bulky $T$ group might interact with a reporter group attached to residue $C$. Thus, new specificity information will become available even if X-ray data are not completed, and the specificity information will become more precise and meaningful, rather than become obsolete, when the three-dimensional structure of the protein is known. It should be emphasized that the ease with which orientations can be assigned to substrates is not yet known.

(2) Either a direct interaction with substrate or an indirect conformation change can cause a signal from the reporter group. Hence the observation of such a signal does not *a priori* determine which type of change is occurring. Knowledge of the structure of the protein as revealed by X-ray data may resolve this dilemma, e.g., the reporter group may be positioned so far from the active site that direct interaction is impossible. In the absence of such information, techniques now in existence for detecting conformational changes in solution may be used.[8]

(3) A reporter group should be particularly useful in stopped flow and temperature-jump experiments. By focusing on the reporter group, direct interactions and conformational changes may be correlated with individual steps in the enzymatic reaction.

(4) A reporter moiety can be attached either to an essential or to a nonessential residue. Perhaps the maximum information will be obtained by attaching the reporter group to a residue which is sensitive to changes at the active site but which has only a minor effect on the enzyme action, such as the methionine in chymotrypsin. The active enzyme can, in this case, be examined during essentially normal functioning. However, it is possible that modification of an essential catalytic group may occur without, for example, change in the binding process. In this case the reporter moiety could be used to study the binding phenomenon without the complication of the simultaneous catalytic process.

(5) The positioning group of the reagent does not have to have absolute specificity, but its effectiveness will depend in part on fairly high specificity and in part on the characteristics of the individual protein under study. For example, previous work had indicated that an iodoacetamide derivative would attack only the surface methionine residue in chymotrypsin under acid conditions and this appears to be

true also for 2-bromoacetamido-4-nitrophenol. This reagent should also attack sulfhydryl residues in proteins containing cysteine, and preliminary experiments with glyceraldehyde phosphate dehydrogenase indicate that such reaction occurs. Since the SH reactivity is greater than that of a thioether, the reagent may even be specific for SH in the presence of methionine under proper experimental condition.

If reaction with several residues occurs, chromatography may be able to separate the monosubstitution products as has been done in the cases of derivatives of ribonuclease by Crestfield, Stein, and Moore,[5] and Hirs.[4]

Moreover, "positioning groups" of different specificity can guide the "reporter" group to different protein sites. The previously reported reagent, for example, was specific for tryptophan but contained the same reporter group. By judicious choice of reagents the regions of the protein which are affected or unchanged by substrate may be "mapped out."

(6) Permutations on the same principle will allow application to many problems, e.g., antibody-antigen interactions, association of subunits, etc. Moreover, different types of reporter groups are possible, e.g., fluorescing groups or groups generating electron spin resonance signals.

*Summary.*—A new technique has been developed for the correlation of function with structure in protein molecules. A group which is sensitive to changes in environment and which can transmit a signal to an appropriate detector is introduced into a specific position in the protein. This group may report changes in its environment caused either by direct interaction with substrate or by a conformational alteration in the protein structure. An application of this method is described in which the compound 2-bromoacetamido-4-nitrophenol is reacted with the methionine residue near the active serine of chymotrypsin. On mixing substrates and inhibitors with the modified protein, characteristic difference spectra are obtained. Some of the applications and limitations of the method are discussed.

* Support of the National Science Foundation is gratefully acknowledged. A portion of the research was carried out at Brookhaven National Laboratory, Upton, New York, under the auspices of the U.S. Atomic Energy Commission.

[1] Perutz, M. F., M. G. Rossman, A. F. Cullis, H. Muirhead, G. Will, and A. C. T. North, *Nature,* **185,** 416 (1960).

[2] Kendrew, J. C., H. C. Watson, B. E. Stranderg, R. E. Dickerson, D. C. Phillips, and V. C. Shore, *Nature,* **190,** 666 (1961).

[3] Balls, A. K., and E. F. Jansen, *Advan. Enzymol.,* **13,** 321 (1952); Koshland, D. E., Jr., *Advan. Enzymol.,* **22,** 45 (1960).

[4] Hirs, C. H. W., in *Enzyme Models and Enzyme Structures,* Brookhaven Symposia in Biology, No. 15 (1962), p. 154.

[5] Crestfield, A. M., W. H. Stein, and S. Moore, *J. Biol. Chem.,* **238,** 2413, 2421 (1964).

[6] Hein, G. E., and C. Niemann, *J. Am. Chem. Soc.,* **84,** 4495 (1963).

[7] Thoma, J. A., and D. E. Koshland, Jr., *J Am. Chem. Soc.,* **82,** 3329 (1960).

[8] Koshland, D. E., Jr., in *Synthesis and Structure of Macromolecules,* Cold Spring Harbor Symposia on Quantitative Biology, vol. 28 (1963), p. 473; Gerhart, J. C., and A. B. Pardee, in *Synthesis and Structure of Macromolecules,* Cold Spring Harbor Symposia on Quantitative Biology, Vol. 28 (1963), p. 491; Monod, J., J. Changeux, and F. Jacob, *J. Mol. Biol.,* **6,** 306 (1963); Inagami, T., and T. Murachi, *J. Biol. Chem.,* **238,** PC1905 (1963); Theorell, H., in *Célébration du Cinquantenaire* (Paris: Soc. de Chim. Biol., 1964), p. 13.

[9] Koshland, D. E., Jr., these Proceedings, **44,** 98 (1958).

[10] Clearly, most of the statements in which substrates are discussed are equally true for inhibitors, haptens, etc., but for clarity the phrase "and inhibitor" will not be included each time.

[11] Koshland, D. E., Jr., Y. D. Karkhanis, and H. G. Latham, *J. Am. Chem. Soc.*, **86**, 1448 (1964).
[12] Hartley, B. S., in *Enzyme Models and Enzyme Structures*, Brookhaven Symposia in Biology, No. 15 (1962), p. 85.
[13] Koshland, D. E., Jr., D. H. Strumeyer, and W. J. Ray, Jr., in *Enzyme Models and Enzyme Structures*, Brookhaven Symposia in Biology, No. 15 (1962), p. 101.
[14] Newbery, G., and M. A. Phillips, *J. Chem. Soc.*, 3049 (1928).
[15] Schwert, G. W., and Y. Takenaka, *Biochim. Biophys. Acta*, **16**, 570 (1955).

Part IX

# THE EFFICIENCY OF ENZYMATIC CATALYSIS: STERIC COMPRESSION, COMPARISONS OF ENZYMATIC AND NONENZYMATIC REACTION RATES, ADDITIVITIES OF VARIOUS FACTORS, AND TERMINOLOGIES

# Editor's Comments on Papers 53 Through 56



The next set of papers constitutes a small selection from the large area that tries to obtain a chemical explanation for the specific rate acceleration of enzymatic processes. In the words of Jencks (1969): "It is not generally appreciated how little is understood about the mechanism by which enzymes bring about their extraordinary and specific rate accelerations. . . . there are few instances in which we have even an elementary notion of what provides the driving force for the reduction in the free energy of activation, of the meaning of the remarkable specificity, or of the relationship between catalysis and specificity in enzyme catalyzed reactions." In spite of this admission of our limited knowledge, every step in the explanation of the mechanism of enzymatic catalysis has substituted a sense of awe, based on reason, for a sense of mystery, based on ignorance.

It is agreed that the catalytic efficiency of any given enzyme is the composite of a variety of factors. The following papers, arranged chronologically, address themselves to one or more of these factors. Bruice and Pandit (Paper 53) elegantly demonstrate in their model of an enzyme-substrate interaction that nucleophilic catalytic hydrolysis of an ester group is helped enormously when the participating carboxylate anion is present on the same molecule as the ester group. Moreover, the particular type of intramolecular proximity of these two groups will affect the efficiency of the catalytic process (see also Bruice 1970). Koshland (Paper 54) establishes by a quantitative approach that if one wishes to explain enzymatic rate acceleration, it does not suffice to postulate that substrate and enzyme be brought into proximity to each other and in a proper orientation. The participation of additional factors is needed. Paper 55 by Bender, Kézdy, and Gunter for the first time achieves a complete quantitative dissection of catalysis of an enzyme, α-chymotrypsin, by a calculation of the quantitative contributions of five different factors that, in addition to intramolecular and conformational effects, include general basic catalysis by imidazole, general acidic catalysis by imidazole, and the particular type of amide bond cleavage that applies to this enzyme. The paper ends with the potent statement: "This combination produces an enzyme."

The article by Jencks and Page (Paper 56) in a masterly fashion summarizes the field of enzymatic rate accelerations by concentrating on the exponential scholastic proliferation of the burdensome and completely equivalent terms prevalent in this field.

Reprinted from *Natl. Acad. Sci. (USA) Proc.* **46**:402–404 (1960)

# *INTRAMOLECULAR MODELS DEPICTING THE KINETIC IMPORTANCE OF "FIT" IN ENZYMATIC CATALYSIS**

By Thomas C. Bruice and Upendra K. Pandit

DEPARTMENT OF PHYSIOLOGICAL CHEMISTRY, JOHNS HOPKINS UNIVERSITY SCHOOL OF MEDICINE

*Communicated by Albert L. Lehninger, February 25, 1960*

This communication describes a few of the intramolecular models which have been studied in this laboratory in order to ascertain the kinetic effect of steric compression on an intramolecular catalysis of the hydrolysis of an ester bond. In enzyme catalysis the bond making and breaking processes occur within a complex of enzyme and substrate (ES) presumably *via* the participation of particular functional groups at the "active site." For the case of certain esteratic enzymes an intracomplex nucleophilic displacement of —OR′ from R − CO − OR′ to give an acyl enzyme as an intermediate is involved. Thus, for the enzyme ficin, a carboxyl anion has been suggested to be the intracomplex participant.[1] The reaction may be pictured schematically as (I):

$$\underset{\text{ES}}{\begin{array}{cc}\text{Enzyme}\ldots\ldots & \text{R}\\ | & |\\ \text{COO}^- & \text{C=O}\\ & |\\ & \text{OR}'\end{array}} \xrightarrow[k_1]{-\text{OR}'} \underset{\text{Acyl Enzyme}}{\begin{array}{cc}\text{Enzyme}\ldots\ldots & \text{R}\\ | & |\\ \text{CO—O—} & \text{CO}\end{array}} \xrightarrow[k_2]{H_2O} \begin{array}{c}\text{Enzyme}\\ |\\ \text{COO}^-\end{array} + \begin{array}{c}\text{R}\\ |\\ \text{C=O}\\ |\\ \text{OH}\end{array} \quad \text{(I)}$$

The similarity between (I) and an intramolecular nucleophilic catalysis of ester hydrolysis (as II) has been pointed out:[2–6]

$$\left\langle\begin{array}{l}\text{—COOR}\\ \\ \text{—COO}^-\end{array}\right. \xrightarrow[k_1]{-\text{OR}} \left\langle\begin{array}{l}\text{—CO}\\ \quad\quad \text{O}\\ \text{—CO}\end{array}\right. \xrightarrow[k_2]{H_2O} \left\langle\begin{array}{l}\text{—COOH}\\ \\ \text{—COOH}\end{array}\right. \quad \text{(II)}$$

When the intramolecular model (II) is compared with its bimolecular counterpart (i.e., the catalysis of ester hydrolysis by acetate anion), it is found that II is much more efficient, due to a decrease in translational entropy in the formation of the transition state.[2,6] By analogy, the high efficiency of enzymic reactions must be due, at least in part, to a similar phenomenon, in which the ES complex brings the reacting groups together. It is commonly accepted that the enzyme must also align the reacting groups in a particularly favorable steric conformation to allow the reaction to proceed at an optimum rate. This can be best accomplished by placing the ground state of the reacting species in approximately the same steric configuration as the transition state (steric compression). In the model II this can be accomplished by at least three different types of steric compression. The esters A, B, C, and D (Table 1) have been chosen to illustrate the kinetic effect of each.

TABLE 1

THE RELATIVE RATES OF ANHYDRIDE FORMATION ($k_1$) AND SOLVOLYSIS ($k_2$) IN THE INTRAMOLECULAR NUCLEOPHILIC CATALYSIS OF THE HYDROLYSIS OF FOUR MONO ESTERS OF DIBASIC ACIDS*, †

| Ester | $\frac{k_1}{k_1 \text{ glut.}}$ | $\frac{k_2}{k_2 \text{ glut.}}$ |
|---|---|---|
| 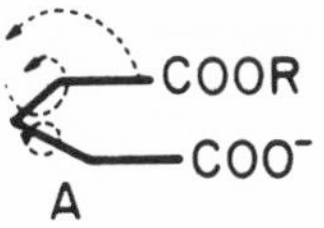 A (COOR, COO⁻) | 1.0 | 1.0 |
| 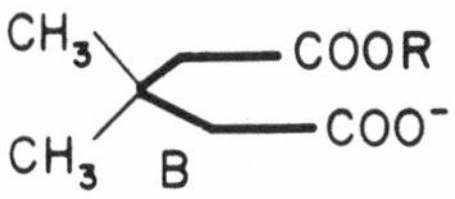 B ($CH_3$, $CH_3$; COOR, COO⁻) | 20 | 0.07 |
| 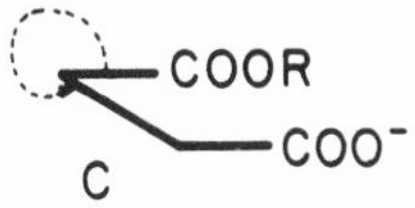 C (COOR, COO⁻) | 230 | 1.46 |
| 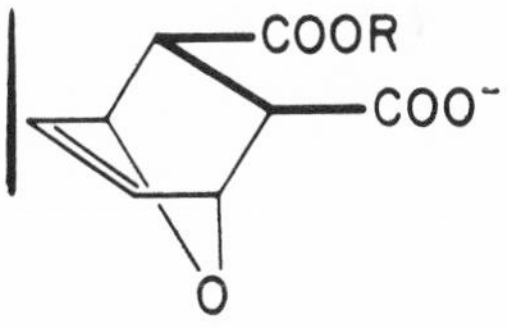 D (COOR, COO⁻; O) | 53,000 | 5.2 |

* The esters employed were the mono-*p*-bromophenyl and *p*-methoxyphenyl. Their rates of solvolysis were followed spectrophotometrically at constant pH between pH 2.6 and 7.0. Values of $k_1$ when plotted versus pH gave the typical sigmoid curve for a mono basic acid (carboxyl anion participation).[3-5] The values of $k_1$ represent the rate of hydrolysis of the mono anion. For the mono-*p*-bromo ester of glutaric acid (standard of reference for relative rates) $k_1 = 4.44 \times 10^{-3}$ min$^{-1}$ (30°; $\mu = 0.65$ M; solvent dioxane-water 50-50, v/v).

† The values of $k_2$ were obtained titrimetrically at pH 6.5 and 35° in 28.5% ethanol water v/v, $\mu = 0.65$ M. For glutaric anhydride $k_2 = 0.23$ min$^{-1}$.

In B the $\beta,\beta$-dimethyl substitution hinders rotation of the $COO^-$ and COOR groups away from each other (gem-dimethyl effect[7]) and thus sterically compresses the ground state somewhat toward the transition state. The nucleophilic displacement in C is favored over that in A because fewer bonds need be frozen into position upon entering the transition state in C as compared to A (cf. Brown, Brewster, and Schecter rule[8] dealing with relative stabilities of five versus six membered rings with exocyclic double bonds). In D, the attacking carboxyl group and the ester bond are locked together in an eclipsed conformation by the endoxo bridge. The ester

D possesses a stereochemistry very similar to that of the transition state in contrast to A which permits free rotation of the reacting groups around the three carbon-carbon bonds.

It is also noteworthy that the relative rate ($k_2$) for anhydride solvolysis (akin to the deacylation step in I) is not decreased with increasing compression. This may be easily explained by assuming that the attack of $OH^-$ ion on the anhydride molecule is the rate limiting step

$$\left\langle\begin{matrix}\text{—CO}\diagdown \\ \quad\quad \text{O} \\ \text{—CO}\diagup\end{matrix}\right. + \text{OH}^- \xrightarrow[k_3]{\text{slow}} \left\langle\begin{matrix}\text{OH}\ \ \text{O}^- \\ \text{—C}\diagdown \\ \quad\quad \text{O} \\ \text{—C}\diagup \\ \| \\ \text{O}\end{matrix}\right. \xrightarrow[k_4]{\text{fast}} \left\langle\begin{matrix}\text{—COOH} \\ \\ \text{—COO}^-\end{matrix}\right.$$

Thus, the compression which increases $k_1$ (the rate of anhydride formation) need not have an influence on anhydride opening, since the effect does not enter into the rate determining step ($k_4$). A completely analogous type of steric control could be operative in enzymic reactions involving double displacements.

These experimental results point to the tremendous enhancement of rate that an enzyme could achieve by fixing the reacting species in a steric conformation closely resembling that of the transition state for the reaction. A detailed account of "The Effect of Geminal Substitution, Ring Size and Rotamer Distribution on the Intramolecular Catalysis of Ester Hydrolysis" will be submitted to the *Journal of the American Chemical Society.*

* This work was supported by grants from The National Institutes of Health, The National Science Foundation, and the Upjohn Company.

[1] Bernhard, S. A., and H. Gutfreund, *Biochem. J.*, **63**, 1 (1956).

[2] Zimmering, P. E., E. W. Westhead, and H. Morawetz, *Biochem. Biophys. Acta.*, **25**, 376 (1957).

[3] Bruice, T. C., and J. M. Sturtevant, *ibid.*, **30**, 208 (1958); *J. Am. Chem. Soc.*, **81**, 2860 (1959).

[4] Bruice, T. C., *J. Am. Chem. Soc.*, **81**, 5444 (1959).

[5] Schmir, G. L., and T. C. Bruice, *ibid.*, **80**, 1173 (1958).

[6] Bender, M. L., and M. C. Neveu, *ibid.*, **80**, 5388 (1958).

[7] Brown, R. F., and N. M. van Gulick, *J. Org. Chem.*, **21**, 1046 (1956).

[8] Brown, H. C., *J. Org. Chem.*, **22**, 439 (1957).

# 54



Reprinted from *J. Theor. Biol.* **2**:75–86 (1962)

## The Comparison of Non-Enzymic and Enzymic Reaction Velocities†

D. E. KOSHLAND, JR.

*Department of Biology, Brookhaven National Laboratory, Upton, N.Y., U.S.A.*

(*Received* 22 *May* 1961)

A method of comparing enzymic and non-enzymic velocities is obtained by assuming that the only function of a hypothetical enzyme is to bring the substrates and catalysts into a close and properly oriented proximity. Equations are derived which can be applied to measured values in specific cases. Using rather common reaction conditions, it is seen that a small acceleration and perhaps even a deceleration would occur if the only function of an enzyme were an oriented juxtaposition of two substrate molecules. On the other hand, factors as large as $10^{17}$ can be calculated for the ratio of enzymic to non-enzymic velocity if the enzyme causes the oriented juxtaposition of two substrates plus three catalysts.

### 1. Introduction

Studies on enzyme mechanisms have already made clear the fact that non-enzymic analogs can be obtained for essentially every enzymic process. Whether these analogs are true models for the enzymic process is, however, not nearly so clear. Imidazole can catalyze ester hydrolysis, but so can hydroxide ion. Iron salts can catalyze hydrogen peroxide decomposition, but so can dust. An understanding of enzyme action, therefore, has two facets. First, one must exclude those analogs which have no counterpart in the enzymic mechanism. Second, one must assess the relative importance of those factors which do contribute to the large enzymic velocities. Quantitative calculations of model systems will assist in the elucidation of enzyme mechanism in two ways: (*a*) they will allow a determination of whether a particular analog is capable by itself of accounting for the enzymic velocity and (*b*) they will allow a calculation of the maximum contribution of a particular model system. In this and subsequent papers, some attempts at the calculation of these factors will be presented.

The early demonstration that enzymes adsorb substrates can lead to the postulate that the enforced juxtaposition of these substrates in "the ternary complex" might contribute importantly to the high enzymic velocities of reaction. The elegant work on "neighboring group" accelera-

† Research carried out at Brookhaven National Laboratory under the auspices of the U.S. Atomic Energy Commission.

tions by Winstein and co-workers (e.g. Winstein & Lucas, 1939) and on intramolecular catalysis by carboxyl and imidazole groups by Bender, Bruice and others (cf. Bender & Neveu, 1959; Bruice, 1959; Gaetjens & Morawetz, 1961) have reinforced the possible importance of this effect. Yet there is no theory to estimate what should be expected in such a ternary complex and, therefore, no theory to establish upper limits on the contribution of this process. This paper discusses a method to assess the relative contribution of this juxtaposition and to establish a basis for comparison of the enzymic and non-enzymic velocities.

## 2. Rationale of the Calculation

It is relatively easy to measure the velocity constant for a non-enzymic reaction in aqueous solution. The techniques for the determination of the rate constants are well known from the conventional procedures of physical organic chemistry (Hammett, 1940; Hine, 1956). It is also relatively simple to measure the "turnover number" of an enzyme, i.e. the velocity when the active sites are saturated (Neilands & Stumpf, 1958). The comparison of these two observed quantities is, therefore, a major goal of the theoretical treatment. The comparison must be based on an analysis of the factors which, from experience with physical organic systems, might be important in the "ternary complex".

One of these factors is the relative concentration of the two substrates or of the substrates and the catalytic agents. In solution it is usually not practical to have the reagents in concentrations above 1 M for solubility reasons, and in the enzymic reaction the concentrations of reagents are considerably less than this. The probability of the collision between the two substrates is correspondingly small and the probability of termolecular or higher order collisions is very small indeed. On the other hand, the enzyme complex, by its affinity for each substrate, can maintain the two substrates in close proximity and hence increase the probability of collision (at least in that small area known as the active site). This contribution, which will be called the "proximity" effect, refers only to the property of the enzyme which increases the concentration of the reagents relative to each other.

In addition to this proximity effect, the binding to the enzyme may serve the function of orienting the substrates relative to one another. In the hexokinase reaction, for example, it seems probable that the terminal phosphate of ATP will be within 1 or 2 angstroms of the 6-hydroxyl oxygen of the glucose molecule. The enzyme, therefore, may not only maintain a high concentration of glucose in the vicinity of the ATP, but may orient the molecules to give a maximum probability for reaction. A method which may allow the calculation of the contribution of this "orientation" effect is the second goal of this treatment.

The derivation which follows is designed to deduce the enzymic velocity if only the proximity effect and the orientation effect were operating on the enzyme surface. This calculation should not be taken to imply that these effects are the only ones operating in enzyme action, or to imply that others such as the "compression" effect suggested by Bruice & Pandit (1960) are excluded. Rather it is a method of evaluating the potential contribution of these effects and a basis for comparing enzymic and non-enzymic velocities.

## 3. Mathematical Derivation

### Proximity Effect

Let us suppose that a chemical reaction of the type $A + B \rightarrow C + D$ is proceeding in aqueous solution as shown schematically in Fig. 1 and the

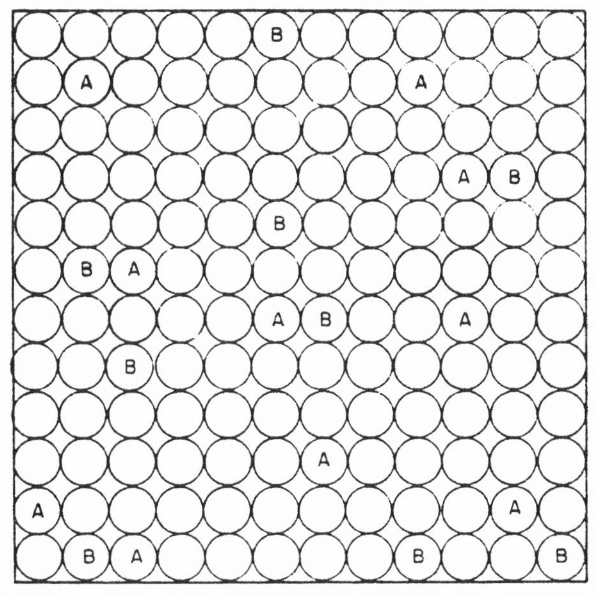


FIG. 1. Schematic representation of molecules reacting in the absence of enzyme in a liter of water molecules. A and B are assumed to be the same size as the water molecules.

velocity of the reaction is given by the kinetic expression of equation 1.

$$v_0 = k_0 \, (A) \, (B) \qquad (1)$$

In this expression, $k_0$ is the usual bimolecular constant which can be evaluated by conventional procedures and (A) and (B) represent concentrations of the reactants in moles per liter. If the substrate molecules, A and B are the size of water molecules, and if there are no special attractions of A and B to each other and to water molecules, the concentration of A–B pairs, i.e. configurations of A and B in juxtaposition as shown in the middle of Fig. 1, will be given by equation (2).

$$\text{Concentration of A–B pairs} = \frac{(A)\,(B)\,12}{55{\cdot}5} \tag{2}$$

In this expression 12 is included to account for the presence of 12 "nearest neighbors" in a close-packed assemblage. The water lattice undoubtedly has fewer nearest neighbors, but this is a small correction which will automatically cancel out in most cases. Hence the most unfavorable case as far as the enzyme is concerned is chosen. If it is assumed that such a juxtaposition of molecules A and B is a prerequisite for reaction (the usual assumption in chemical kinetics) and that the velocity of reaction after this configuration has been achieved is $k_0'$, it is clear that $k_0'$ and $k_0$ are related by equation (3).

$$k_0 = \frac{k_0'\,12}{55{\cdot}5} \tag{3}$$

(This equation assumes that the value for $k_0$ is measured at concentrations of A and B sufficiently dilute so that the probability of having more than one of the 12 nearest neighbors occupied by a molecule of the second substrate, B, is very small. This is an easily attainable condition in the experimental determination of velocity constants.)

We shall now calculate the concentration of A B pairs at the active site of an enzyme in aqueous solute. In an enzyme catalyzed reaction in which the concentrations of A and B are sufficiently great to saturate the active sites of the enzyme, a situation as shown schematically in Fig. 2 would obtain. The observed velocity of the enzyme catalyzed reaction would be given by the expression of equation (4)

$$v_e = k_E\,(E_t) \tag{4}$$

in which $k_E$ is the "turnover number" of the enzyme and $(E_t)$ is the enzyme concentration in moles per liter.

If the only catalytic contribution of the enzyme were the proximity effect, the velocity of the enzymic reaction shown in Fig. 2 would be expressed by the relation $v_e = k_0'\,(E_t)$ and the ratio of the enzymic to the non-enzymic rates would be given by equation (5).

$$\frac{v_e}{v_0} = \frac{(E_t)\,55{\cdot}5}{(A)\,(B)\,12} \tag{5}$$

It is desirable to extend the derivation of equation (5) to a case involving the additional molecules R, S, T. These might represent catalysts or a combination of catalysts and additional substrate molecules. Applying the type of reasoning used above, it can be shown that the concentration of

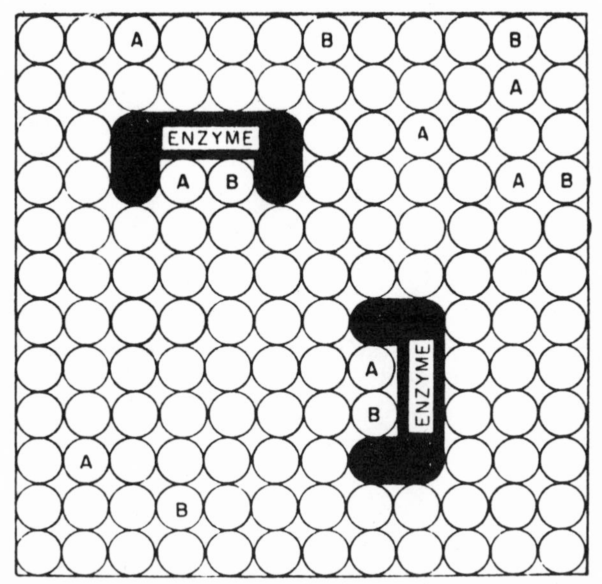


FIG. 2. Schematic representation of molecules reacting on an enzyme surface in a liter of water molecules.

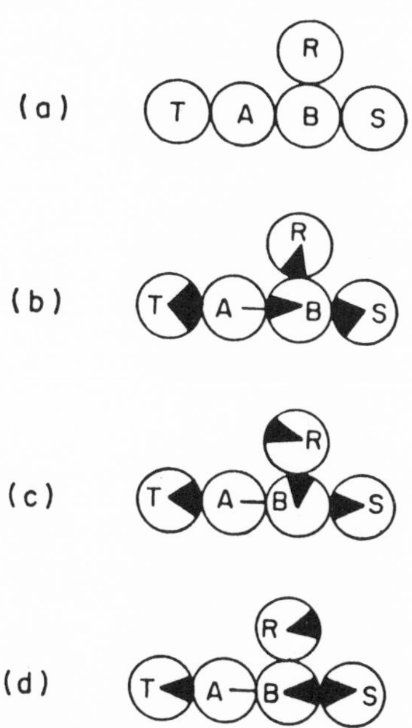


FIG. 3. Some possible arrangements of molecules in solution (all molecules the size of water molecules).

(a) Arrangement of molecules A, B, R, S, T, in water lattice without regard to any preferred orientation within the molecules.

(b) Arrangement of molecules A, B, R, S, T. Proper arrangement requires that contact point of A be tangential to black wedge portion ($1/\theta_B$) of B, that appropriate contact point of B be tangential to black wedge portion of S, etc.

(c) and (d) Non-reactive arrangements of molecules A, B, R, S, T. T and S properly oriented, but B and R are not.

arrangements of the type shown in Fig. 3 (a) per liter in the absence of enzyme would be given by equation (6) and the calculated $v_e/v_o$ ratio would be given by equation (7).

Concentration of A, B, R, S, T arrangements, Fig. 3 (a), per liter

$$= \frac{(A)(B) \times 12 \times (R) \times 11 \times (S) \times 10 \times (T) \times 11}{55{\cdot}5 \times 55{\cdot}5 \times 55{\cdot}5 \times 55{\cdot}5} \quad (6)$$

$$\frac{v_e}{v_0} = \frac{(E_t)(55{\cdot}5)^4}{(A)(B)(R)(S)(T) \times 12 \times 11 \times 11 \times 10} \quad (7)$$

In these expressions the numbers 11 and 10 are less than 12 to allow for the fact that some of the nearest neighbor positions are already occupied.

As examples, the groups R, S, and T might be the side chains of amino acids, such as the imidazole group of histidine, prosthetic groups, such as heme, or bound metals such as $Mg^{++}$. In this case the equivalent groups in Fig. 3 (a) would be the effective catalytic group without its covalent linkage to the protein surface. It should be emphasized that the Fig. 3 (a) represents a particular constellation of five molecules in which, for example, the position of B with its neighbors S, A, and R is fixed and not a random association of five molecules.

## Orientation Effect

The previous equations were based only on the proximity effect, i.e. the effect of placing A, B, etc., in juxtaposition without any consideration of orientation. If it is now supposed that there is a preferred orientation for reaction, a situation similar to that shown schematically in Fig. 3 (b) would exist. In this diagram the wedge-shaped areas are taken to indicate the fraction of the total solid angle within which reaction can occur. If in each case this fraction is represented by a term $1/\theta'_B$, $1/\theta'_R$, $1/\theta'_S$, etc., the expression which includes both this orientation and proximity, will be given in equation (8).

$$\frac{v_e}{v_0} = \frac{(E_t)(55{\cdot}5)^4\, \theta'_B\, \theta'_R\, \theta'_S\, \theta'_T}{(A)(B)(R)(S)(T) \times 12 \times 11 \times 11 \times 10 \times f_{AB}\, f_{BR}\, f_{BS}\, f_{AT}} \quad (8)$$

The f symbols in equation (8) represent the factors to correct for the number of molecules in nearest neighbor positions that can react. Thus if A has 12 nearest neighbor positions, there are 12 possible constellations of A–B when a single one of these positions is occupied by B. If A has no preferred orientation and can act with any of these B molecules, the rate will obviously be 12 times that of a reaction in which A has an orientation such that it can react with only one of the 12 positions.

If molecule A can react with B equally well in any direction, i.e. $f_{AB} = 1$, and B can react only in the cross-hatched area $\theta'_B$, the $v_e/v_0$ ratio for these two molecules only will be $(E_t)$ $(55 \cdot 5)$ $\theta'_B/(A)$ $(B) \times 12$. However, if there is a preferred orientation in the A molecule as well, so that only one of the 12 nearest neighbors of A can react at a given time, $f_{AB} = 1/12$ and the $v_e/v_0$ ratio will equal $(E_t)$ $(55 \cdot 5)$ $\theta'_B/(A)$ $(B)$. For simplicity in the subsequent discussion, equation (8) might be written as equation (9) in which $\theta_B = \theta'_B/12\ f_{AB}$, $\theta_R = \theta'_R/11\ f_{BR}$, $\theta_S = \theta'_S/10\ f_{BS}$, and $\theta_T = \theta'_T/11\ f_{AT}$.

$$\frac{v_e}{v_0} = \frac{(E_t)}{(A)} \times \frac{55 \cdot 5 \theta_B}{(B)} \times \frac{55 \cdot 5\ \theta_R}{(R)} \times \frac{55 \cdot 5\ \theta_S}{(S)} \times \frac{55 \cdot 5\ \theta_T}{(T)} \tag{9}$$

Each of the terms in equation (9), therefore, can be in a sense considered to represent a proximity component, e.g. $(55 \cdot 5)/(A)$ and an orientation component, e.g. $\theta_A$. In this expression, the nearest neighbors correction has been shifted from the proximity component to the orientation component for simplicity and because this correction is in a sense really a part of the orientation problem. Generalization to larger or smaller number of reacting species is obvious.

## 4. Discussion

One of the most important features of the above equations is that they allow the calculation of a $v_e/v_0$ ratio from observable experimental quantities. A comparison of observed and calculated $v_e/v_0$ ratios and hence of the contribution of orientation and proximity can be made quantitatively. Specific comparisons will be made in subsequent papers. However, some general conclusions can be drawn by some very simple calculations.

### Role of the Proximity and Orientation Effects for Two Substrates

Let us consider a reaction between two substrates catalyzed by an enzyme. A calculation using equation (8) gives the increase in velocity to be expected if proximity and orientation were the only contributors to the enzyme action. We shall refer to this hypothetical enzyme as the $E_{op}$ enzyme to indicate that only orientation and proximity factors are presumed to operate in its catalysis. Assuming $f_B = 1/12$ (only one of A's 12 nearest neighbors can react), $\theta'_B = 10$ (reaction can occur for 10% of the B's surface), $(E_t) = 10^{-6}$ M, $(A) = (B) = 1 \cdot 0$, $v_e/v_0$ becomes $5 \cdot 5 \times 10^{-4}$. In other words, the calculated enzyme velocity under these conditions would occur at only a fraction of the non-enzymic reaction. Since experience shows that just the opposite is true, this calculation clearly indicates that proximity and orientation of *two substrates* cannot be the main source of enzyme catalysis in general. It might be argued that the $\theta$ factor might be fantastically high, a possibility in some cases, but certainly in some well-known reactions, e.g. those involving water, $\theta'_B$ factors of 10 or less are reasonable.

Examination of the mathematics reveals the reason for the low ratio under the above conditions. It is the low concentration of enzyme ($10^{-6}$ M) which limits the number of juxtaposed A, B configurations per liter of solution to this value. In other words, the relative concentrations of A and B at the active site are very high, but there are very few active sites per liter. If the substrates were both $10^{-3}$ M, a not unusual experimental situation, the ratio would rise to 550 indicating that the proximity effect makes an important contribution to the overall rate under these conditions. However, since factors of $10^{11}$ to $10^{13}$ are not uncommon for the observed enzymic–non-enzymic ratios at $10^{-3}$ M substrate concentrations (Koshland, 1956) the proximity effect for two substrates is clearly not adequate to explain the high enzymic velocities.

How insignificant the proximity effect can be in certain cases might best be illustrated by considering the hydrolysis of a phosphate ester, a reaction which has been measured both enzymically and non-enzymically. For 0·1 M ester in a liter of aqueous solution, there are 0·1 moles of phosphate ester and let us presume that three of the nearest neighbor water molecules are in position to react. If the enzyme surface contains a water site for one water molecule, even if the enzyme were 0·1 M, the rate would be reduced by a factor of three due to the nearest neighbor factor. A 0·1 M enzyme concentration would be impossible, however, and a more usual $10^{-6}$ M enzyme concentration would reduce the rate by a further factor of $10^5$. Since phosphatases bring about rapid reaction at pH 7 where phosphate esters are *per se* very unreactive, it is clear that these considerations support the previously mentioned conclusions.

## Role of Proximity and Orientation Effects for Two Substrates and Three Catalysts

A rather different picture is obtained, however, if the situation of two substrates *and* three catalysts is calculated. Assuming substrates and catalysts are all $10^{-3}$ M, $\theta$ factors are 10, and $(E_t) = 10^{-5}$ M, equation (9) gives a ${}^{v}E_{op}/v_o$ ratio of $9 \cdot 5 \times 10^{17}$. Even if a hydrolytic reaction is involved, the ratio would be $1 \cdot 7 \times 10^{13}$. These are clearly in the range of enzymic accelerations. In other words, if the reaction involves a combination of two substrates and three catalytic groups, proximity and orientation alone could account for a large part of the great accelerations caused by enzymes. In qualitative terms, the difference between the latter value and the similar value for two substrates lies in the probabilities of a chance orientation of five molecules versus the chance orientation of two molecules. The chance orientation of five molecules is so improbable that the binding of substrates to catalysts on enzymes can increase the number of such configurations by many powers of 10 even at low enzyme concentrations. On the other hand, the chance collision of two molecules is much more likely and hence the concentration of juxtaposed and oriented substrates may be considerably greater than the *ca.* $10^{-5}$ M concentrations of active sites at usual enzyme concentrations.

## Limitations of the General Derivation

The derivation of equation (9) involved certain assumptions and the usefulness of the final equation will depend on the limitations of these assumptions. Hence, it may be worth considering these limitations.

(i) *Volume of the substrate and catalytic molecules*

It was assumed that A, B, R, S, T, etc., had the same volume as water molecules. It is clear that this will only be true in a small number of cases.

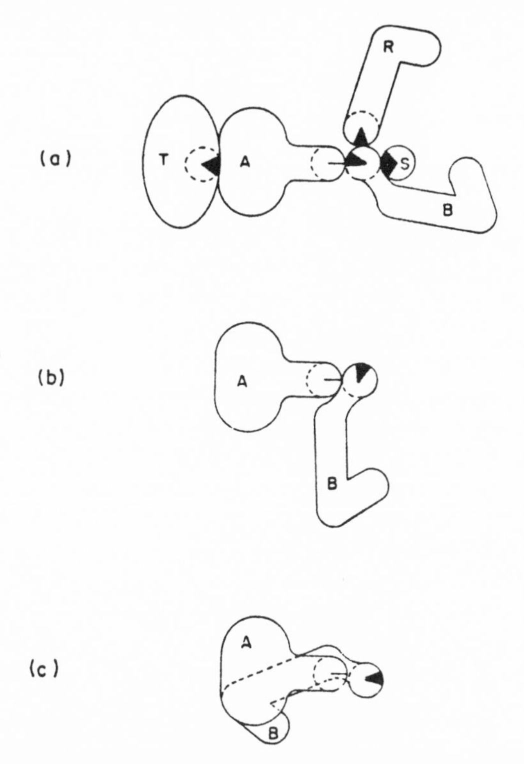


FIG. 4. Some possible arrangements of molecules in solution. (Molecules not necessarily the size of water molecules.)
(a) Reactive arrangement of A, B, R, S, T.
(b) Non-reactive, but possible, arrangement of A and B with reactive atoms oriented in analogy to Fig. 3 (d).
(c) Sterically impossible arrangement with reactive atoms oriented in analogy to Fig. 3 (d).

However, this difference is not serious as can be seen from the following considerations.

Let us suppose that molecules considerably larger than $H_2O$, e.g. ATP and glucose in the hexokinase reaction, are under consideration. Although the total chemical structure of ATP occupies far more volume than the water molecule, the reactive group, in this case the terminal phosphorus atom, is quite comparable in size to a water molecule. The same argument holds for the reactive C-6 hydroxyl of the glucose. These two areas must be juxtaposed and oriented in ways exemplified in Fig. 4 and the probability for such orientations will only be slightly affected by the existence of the large appendage to the active atom or atoms. Thus, configurations of the active atoms such as in Fig. 4 (a) and (b) will exist in direct analogy to the case in which the reactants are only the size of water molecules. However, configurations of the type in Fig. 4 (c), which would be possible with the small molecules, would be impossible with bulkier substrates. Moreover, the total lattice arrangements of A, B, R, S, and T will be affected if all of these molecules are bulky. This is very similar to the excluded volume problem in polymer configurations. It can be minimized by measuring the non-enzymic velocity at fairly low substrate concentrations. Even so, it will require refinements of the calculations when ratios are to be determined to high accuracy. Since the present calculations are only to outline order of magnitude comparisons, this excluded volume can be safely considered small. It can probably be estimated with fair accuracy with molecular models when more detailed calculations are desired.

(ii) *Preferred orientations in solution*

In the derivation of equation (9) it is also assumed that all orientations and positions in the water lattice are equally probable. This may not necessarily be the case, however. For example, a positive metal ion might have a far greater probability of being a nearest neighbor of a negative phosphate ester than a water molecule. In this case, the calculated proximity effect contributed by the enzyme would be too high. On the other hand, an ATP–metal ion complex might contain a preponderant configuration of metal ion at an unreactive position, e.g. the non-terminal phosphate groups In this case the calculated proximity effect would be too low. These effects are certainly important and, unlike the volume effect, may actually be very large. Some correction for this factor may be made by the study of complexes, such as the magnesium–ATP complex studies of Cohn and Hughes (1960). In some cases preferred configurations can be deduced to be of minor consequence, and in others the magnitude of the deviation may be roughly calculable from model studies. The value so obtained would then be used as a correction on the present calculation for the random case which serves as a base for comparison. Since, in general, deviations which lead to low results are as likely as those leading to high results, the general conclusions are valid even though care must be taken to assess the role of preferred orientations in specific cases.

### Time of Formation of the Complex

In the calculations above it has been assumed that a value for the turnover number of the enzyme saturated with respect to all of its substrates and catalytic groups can be obtained. It might be said that this is an improbable situation and that the time required to absorb all of these molecules on the enzyme surface simultaneously.will always be appreciable. If the three catalytic groups are part of the enzyme structure itself, e.g. the side chains of amino acid residues, however, it is easy to understand how the five-group constellation can be formed readily on the enzyme surface. Even if one or more of the catalyst molecules must be absorbed, however, the absorption process does not have to be rate limiting. Metals are, for example, tightly absorbed by proteins and usually associate at diffusion velocities. Even if saturation with all substrates and catalysts is not obtained, a calculation to saturation velocity is possible if data for a reasonable Lineweaver-Burk plot can be obtained. This is not to say that the formation of the complex or the dissociation of the products may not be rate-limiting in certain cases, but only that in most cases either it is not limiting or correction to saturation by the usual enzyme kinetics is possible.

### Size of the $\theta$ Factors

The calculations emphasize the importance of the $\theta$ values and the question arises as to the size of these factors. It is, of course, well known that some reactions, such as the Walden inversion mechanism, impose stereochemical restraints. Phenomena observed with neighboring group effects, elimination reactions, hydrolysis of phosphates and phosphonates have all been explained by a required steric pathway for reaction. In some cases, e.g. the Walden inversion, one could estimate that the attack would have to occur within a region comprising 10% of the solid angle of the carbon atom. In cases of elimination reactions, the fraction of solid angle leading to ready attack might be even smaller. On the other hand, in hydrolysis reactions, the orientation of the water molecule would seem to be of minor importance and $\theta$ factors of 2 or 1 would seem reasonable. The paucity of data in this area is alarming and these calculations emphasize the importance of acquiring such physical chemical data in order to make refined comparisons of enzymic and non-enzymic velocities. Nevertheless, the assumptions of $\theta$ in the above cases seem reasonable, and sufficient to indicate that large effects may be observed even when modest orientation requirements per group are used. If a very conservative estimate, i.e. all $\theta$ factors are 1 and all f factors are about 1/10, is made, the calculation for the case of the three catalysts and two substrates would give $9{\cdot}5 \times 10^{13}$. This is less than the previous calculation by $10^4$, but still indicates a pronounced contribution by the proximity effect. It also appears quite possible that the $\theta$ factors may be larger than 10 in some cases.

## 5. Conclusions

The previous calculations show that orientation and proximity in the ternary complex are not sufficient to explain enzyme velocities for *two substrates* and *an enzyme* in a ternary complex unless the $\theta$ factors are fantastically large, e.g. greater than $10^5$. In the case of hydrolytic reactions even this factor would not be sufficient to explain the velocities of enzymic reactions of reasonable turnover number. Since $\theta$ factors of this sort seem unlikely in view of existing physical organic data, it seems safe to conclude that factors other than the oriented juxtaposition of two substrates are involved in enzyme action.

The "other factors" might be either some as yet undisclosed form of catalysis or the required orientation of substrates plus catalytic groups. If the total of substrates plus catalysts is 5 or more, it is seen that even moderate orientation factors can give very large catalysis and it is at least conceivable that this combination of orientation and proximity may be the only factor in some enzymic processes. At the least, it would seem that such a combination of proximity and orientation effects would play an important role in enzymic processes. Indeed there are indications from other types of studies (Koshland, 1958) that a combination of several catalytic groups may be a particularly suitable mechanism for explaining many features of enzyme catalysis. In that case, equation (9) provides a basis for the quantitative comparison of the enzymic velocity and the velocity expected on the basis of the non-enzymic measurements. Comparison of these quantities in specific cases can then aid in the evaluation of the role of orientation and proximity in individual enzyme catalyzed reactions.

### REFERENCES

Bender, M. L. & Neveu, M. C. (1959). *J. Amer. chem. Soc.* **81**, 1590.
Bruice, T. C. (1959). *J. Amer. chem. Soc.* **81**, 5444.
Bruice, T. C. & Pandit, V. K. (1960). *Proc. nat. Acad. Sci., Wash.* **46**, 402.
Cohn, M. & Hughes, T. R., Jr. (1960). *J. Biol. Chem.* **235**, 3250.
Gaetjens, E. & Morawetz, H. (1961). *J. Amer. chem. Soc.* **83**, 1738.
Hammett, L. F. (1940). "Physical Organic Chemistry". New York: McGraw-Hill.
Hine, J. (1956). "Physical Organic Chemistry". New York: McGraw-Hill.
Koshland, D. E., Jr. (1956). *J. cell comp. Physiol.* **47**, Suppl. 1, 217.
Koshland, D. E., Jr. (1958). *In* "The Enzymes", Vol. 1, revised edition, p. 305. (Boyer, P. D., Lardy, H. & Myrback, K., eds.) New York: Academic Press.
Neilands, J. B. & Stumpf, P. K. (1958). "Enzyme Chemistry". New York: John Wiley.
Winstein, S. & Lucas, H. J. (1939). *J. Amer. chem. Soc.* **61**, 1576, 1581, 2845.

# 55



Reprinted from *Am. Chem. Soc. J.* **86**:3714–3721 (1964)

## The Anatomy of an Enzymatic Catalysis. $\alpha$-Chymotrypsin[1]

By Myron L. Bender, Ferenc J. Kézdy, and Claude R. Gunter

*Department of Chemistry, Northwestern University, Evanston, Ill.*

Received February 12, 1964

The advantages of defining specificity in terms of kinetic specificity are presented. The kinetic specificity in the deacylation of a series of acyl–$\alpha$-chymotrypsins shows an additive specificity of two parts of the acyl group. The temperature dependence of four deacylation reactions involving acyl–$\alpha$-chymotrypsins of varying specificity was determined. The enthalpies of activation of this series are essentially constant whereas a wide variation in the entropies of activation is found, varying from −13.4 e.u. in the deacylation of a specific acyl–enzyme to −36 e.u. in the deacylation of a nonspecific acyl–enzyme. The kinetic specificity of deacylation (and acylation) is interpreted in terms of a variation in rotational entropy of activation, a specific substrate being rigidly fixed in a conformation resembling the transition state and a nonspecific substrate being free to rotate in many conformations. The standard free energy *vs.* reaction coordinate diagrams for the $\alpha$-chymotrypsin-catalyzed hydrolysis of N-acetyl-L-tryptophan ethyl ester and of N-acetyl-L-tryptophan amide show that the standard free energy of the acyl-enzyme intermediate, N-acetyl-L-tryptophanyl–$\alpha$-chymotrypsin, is intermediate in energy between the reactant and the product. Thus, the acyl–enzyme intermediate is both kinetically and thermodynamically favored in the over-all pathway. Five factors can quantitatively account for the difference between the enzymatic and nonenzymatic (hydroxide ion catalyzed) rate constants of hydrolysis of N-acetyl-L-tryptophan amide: (1) the intramolecular character of the enzymatic process and the concomitant increase in effective concentration of the catalytic group(s); (2) general basic catalysis by imidazole; (3) the change in rate-determining step of the amide hydrolysis to an alcoholysis; (4) the freezing of the substrate in a conformation resembling the transition state; and (5) the general acidic catalysis by imidazole (Table V). This combination of rate factors can account for the enzymatic reactivity of $\alpha$-chymotrypsin.

### Introduction

The previous discussion has provided an over-all pathway for catalysis by $\alpha$-chymotrypsin, and a catalytic mechanism for each step of the catalytic process.[2] The pathway involves, besides the usual enzyme–substrate complexes, the formation of a covalent intermediate, an acyl–enzyme. The catalytic mechanism of each step involves intracomplex or intramolecular bifunctional catalysis involving imidazole.

In the present paper an attempt is made to dissect the kinetic factors contributing to the difference between the enzymatic hydrolysis of an amide substrate (the usual physiological substrate of $\alpha$-chymotrypsin) and a corresponding nonenzymatic hydrolysis of the same amide, taken as a saponification reaction at the same pH. In order to carry out a complete discussion, specific amide substrates must be used, and therefore our treatment will begin with a discussion of specificity, that is, the ability of an enzyme to catalyze selectively a particular reaction.

The specificity of enzyme reactions is usually discussed in terms of "fit" in the binding of a substrate to an enzyme, "fit" being a euphemism for a stereospecific maximization of the forces existing between two molecules. The classical theory of Emil Fischer[3] discussed

(1) This research was supported by grants from the National Institutes of Health. Paper XXXIII in the series: The Mechanism of Action of Proteolytic Enzymes.

(2) M. L. Bender and F. J. Kézdy, *J. Am. Chem. Soc.*, **86**, 3704 (1964).

specificity in these terms by means of a template-type pattern. One may consider specificity in terms of binding and thus in terms of relative Michaelis constants, but this procedure is open to many criticisms. One must first assume that the Michaelis constants are true equilibrium constants of binding, and do not contain any rate constants of subsequent steps. Such an assurance is difficult to obtain because of the difficulty in determining whether the Michaelis constant is an equilibrium constant ($k_{-1}/k_1$) or a steady-state constant ($(k_{-1} + k_2)/k_1$) and secondly whether it is a true equilibrium constant ($k_{-1}/k_1$) or an apparent equilibrium constant ($k_{-1}k_3/k_1(k_2 + k_3)$). Even if the above ambiguities could be overcome and measurement of the gross binding could be accomplished straightforwardly, determination of the stereospecificity of binding, including the problem of nonproductive binding,[4,5] is difficult to carry out. Finally it must be realized that binding of a substrate to a protein has no inherent relation to catalytic activity, for D-substrates bind very well to α-chymotrypsin, but for all practical purposes do not react; furthermore, substrates bind to chymotrypsinogen, and again no reaction is observed. The problem of interpretation of Michaelis constants is therefore a complicated one, and the use of Michaelis constants to define specificity is difficult at best.

On the other hand, one can infer stereospecific bindings favorable to reaction from the rate constants of the enzymatic process, and define in this fashion a "kinetic specificity." Catalysis implies kinetics; therefore specificity of catalysis may be defined in terms of a "kinetic specificity," the exceptional reactivity of the enzymatic process relative to that of the nonenzymatic reaction. The use of rates of α-chymotrypsin-catalyzed hydrolyses for this purpose does not eliminate complications of interpretation, unfortunately, for the two-step catalytic process discussed previously results in three possible rate constants: $k_2$, the rate constant of acylation; $k_3$, the rate constant of deacylation; and $k_{cat}$, the over-all catalytic rate constant defined as $1/k_{cat} = 1/k_2 + 1/k_3$. However, this complication has been resolved in enough cases of interest to make feasible the following discussion.[6]

## Experimental

**Materials.**—The enzyme, the determination of the normality of its solution, the buffers, N-acetyl-L-tryptophan ethyl ester, N-acetyl-L-tyrosine ethyl ester, N-acetyl-DL-tryptophan *p*-nitrophenyl ester *trans*-cinnamoyl-α-chymotrypsin,[7] and have been described.[8,9] Acetyl-α-chymotrypsin was prepared in the following way. Five ml. of a stock α-chymotrypsin solution ($3 \times 10^{-3}$ *M*) was pipetted into pH 3.5 acetate buffer; 250 μl. of *p*-nitrophenyl acetate, $1.5 \times 10^{-6}$ *M*, was added. The reaction was allowed to proceed for 24 hr. at 25° and 24 hr. at 5°, and was centrifuged at 13,000 r.p.m. for 1 hr. The liberated *p*-nitrophenol was removed by chromatography on a Sephadex column, made by equilibrating 1 g. of Sephadex G-25 in pH 3.5 aceta buffer for 60 hr. The amount of free α-chymotrypsin in t acetyl-α-chymotrypsin was shown to be less than 3% by tit tion with N-*trans*-cinnamoylimidazole. The acetyl-α-chym trypsin was stored at 5° and used within 48 hr. of preparati The product was shown to give the same deacylation rate co stant as that prepared by the method of Balls and Wood[10] a that prepared at high pH with *p*-nitrophenyl acetate.[11]

**Kinetic Measurements.**—The determination of the *k* of the α-chymotrypsin-catalyzed hydrolysis of N-acetyl- tryptophan ethyl ester utilized the experimental methods a calculations described previously.[9] The kinetics of the chymotrypsin-catalyzed hydrolysis of N-acetyl-L tyrosine eth ester was determined by measurement of initial rates of reacti at 300 mμ. The difference spectrum of N-acetyl-L-tyrosi ethyl ester *vs*. N-acetyl-L-tyrosine at pH 8.6 and 3.1% acet nitrile–water, $\Delta\epsilon_{300}$, = −113. The $k_{cat}$'s were calculated usi Lineweaver and Burk[12] plots.

The kinetics of the deacylation of *trans*-cinnamoyl-α-chym trypsin have been described previously.[7] One method of t determination of the deacylation of acetyl-α-chymotrypsin us the internal indicator of the hydrolysis of N-acetyl-L-tyrosine eth ester to monitor the reappearance of enzyme. The measur ments were carried out in a Cary 14PM recording spectroph tometer according to Dixon and Neurath.[13] The first-order a pearance of the free enzyme from the acyl–enzyme is seen in t increasing rate of hydrolysis of N-acetyl-L-tyrosine ethyl est which approaches a limiting value as the deacylation of the acy enzyme approaches completion.[13]

A second method of following the deacylation of acetyl- chymotrypsin does not involve an internal indicator of the k netics but rather an external one. Acetyl-α-chymotrypsin w allowed to deacylate for a specified length of time at pH 8. The reaction was then stopped by adding sufficient hydrochlor acid to lower the pH to 1.5 where the acetyl group does not h drolyze. After quenching at pH 1.5 (for periods of less than hour, which lead to no significant denaturation), a rate assay the amount of free enzyme was carried out at pH 5.0 using exce N-acetyl-DL-tryptophan *p*-nitrophenyl ester. The slope of t (zero-order) straight line obtained in the rate assay is propo tional to the free enzyme concentration. From these zero-ord rates, the rate constant of deacylation of acetyl-α-chymotryps may be obtained.

## Results

The temperature dependence of the deacylatio rate constant, $k_3$, of two acyl–enzymes, acetyl-α chymotrypsin and *trans*-cinnamoyl-α-chymotrypsi were measured, as shown in Table I and in Fig. These deacylations were carried out at pH 8.6 to 8. in the pH-independent region of the pH–$k_3$ profile these reactions, and thus any small temperature de

TABLE I

TEMPERATURE DEPENDENCE OF THE DEACYLATION OF TWO ACYL-α-CHYMOTRYPSINS[a,e]

| Temp., °C. | $k_3 \times 10^3$, sec.$^{-1}$ | Temp., °C. | $k_3 \times 10^3$, sec.$^{-1}$ | Temp., °C. | $k_3 \times 10$ sec.$^{-1}$ |
|---|---|---|---|---|---|
| *trans*-Cinnamoyl-α-chymotrypsin[d] | | Acetyl-α-chymotrypsin[b,d] | | Acetyl-α-chymotrypsin[c,d] | |
| 11.5 | 4.52 | 6.9 | 1.95 | 11.3 | 13.9 |
| 20 | 8.18 | 9.2 | 2.73 | 16.0 | 15.4 |
| 25 | 12.5 | 14.5 | 3.80 | 20.0 | 22 4 |
| 30 | 22.4 | 18.0 | 4.58 | 25.0 | 33.6 |
| 35 | 34.3 | 25.0 | 6.80 | 31.5 | 46.5 |
| | | 27.9 | 9.85 | | |

[a] pH of buffers at 25° was 8.6–8.8. [b] Deacylation in the a sence of any added substrate. [c] Deacylation in the presence $2.33 \times 10^{-3}$ *M* N-acetyl-L-tyrosine ethyl ester. [d] 3.2% (v./v acetonitrile–water and 0.1 *M* Tris–HCl buffer. [e] The errors the rate constants are ±5%.

(3) E. Fischer, *Ber.*, **27**, 2985 (1894); *Z. physiol. Chem.*, **26**, 60 (1898).

(4) H. T. Huang and C. Niemann, *J. Am. Chem. Soc.*, **74**, 4634, 5963 (1952); G. E. Hein and C. Niemann, *ibid.*, **84**, 4497 (1962).

(5) S. A. Bernhard and H. Gutfreund, "Proc. Intern. Symp. Enzyme Chemistry," Maruzen, Tokyo, 1957, p. 124.

(6) Nonproductive binding[4,5] may also complicate enzymatic rate constants. However, in the following discussion, only $k_3$ involving a covalently-bound substrate is considered, where nonproductive binding should be minimal. This problem will be considered in a future publication.

(7) M. L. Bender, G. R. Schonbaum, and B. Zerner, *J. Am. Chem. Soc.*, **4**, 2562 (1962).

(8) B. Zerner, R. P. M. Bond, and M. L. Bender, *ibid.*, **86**, 3674 (1964).

(9) M. L. Bender, G. E. Clement, F. J. Kézdy, and H. d'A. Heck, *ibid.*, **86**, 3680 (1964).

(10) A. K. Balls and H. N. Wood, *J. Biol. Chem.*, **219**, 245 (1956).

(11) F. J. Kézdy and M. L. Bender, *Biochemistry*, **1**, 1097 (1962).

(12) H. Lineweaver and D. Burk, *J. Am. Chem. Soc.*, **56**, 658 (1934).

(13) G. H. Dixon and H. Neurath, *J. Biol. Chem.*, **225**, 1049 (1957).

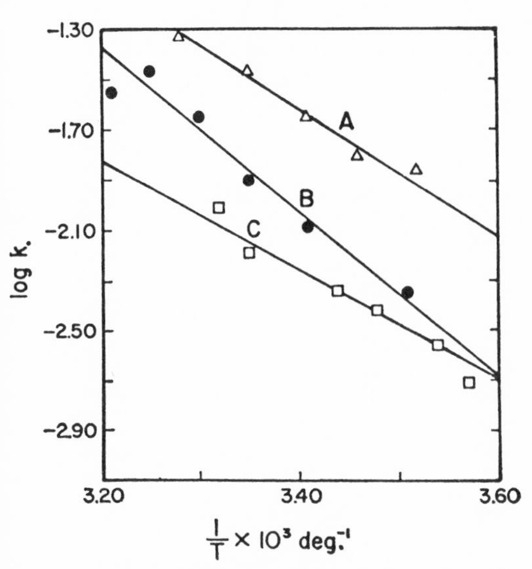


Fig. 1.—The kinetics of deacylation of some acyl-chymotrypsins at pH 8.6 to 8.8 in 3.2% (v./v.) acetonitrile-water: A, acetyl-$\alpha$-chymotrypsin in the presence of $2.33 \times 10^{-3}$ *M* N-acetyl-L-tyrosine ethyl ester; B, *trans*-cinnamoyl-$\alpha$-chymotrypsin; C, acetyl-$\alpha$-chymotrypsin.

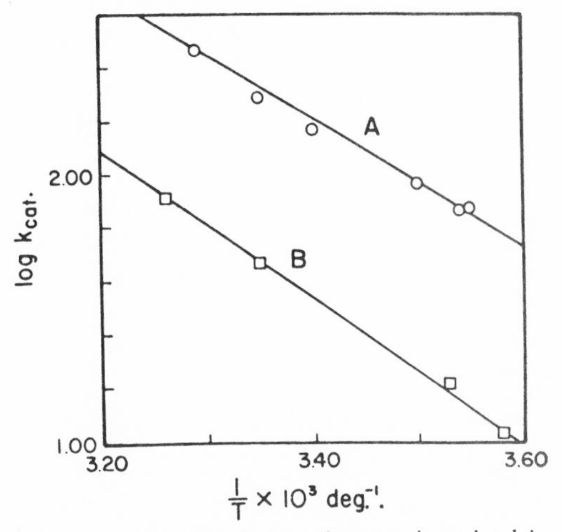


Fig. 2.—The kinetics of the $\alpha$-chymotrypsin-catalyzed hydrolysis of N-acetyl-L-tyrosine ethyl ester (A) and N-acetyl-L-tryptophan ethyl ester (B) at pH 8.55 to 8.70 in 1.8% (v./v.) acetonitrile-water.

pendence of the p$K_a$ of the deacylation reaction will have no effect on these experiments.

The temperature dependence of the deacylation of acetyl-$\alpha$-chymotrypsin was carried out using two different experimental procedures. The first involves the deacylation in the absence of any added substrate. The analysis of this reaction is performed on aliquots of the reaction mixture which are quenched by lowering the pH; the free enzyme of the aliquot is then measured by a rate assay at a low pH. The second method involves the deacylation in the presence of N-acetyl-L-tyrosine ethyl ester, a specific substrate of $\alpha$-chymotrypsin, whose hydrolysis serves as a measure of the regeneration of free enzyme and thus as a measure of the deacylation reaction. The absolute magnitude of the deacylation rate constant in the latter procedure is about five times that in the former procedure at a given temperature. A similar increase in the rate constant of the deacylation of acetyl-$\alpha$-chymotrypsin in the presence of indole has been noted.[14] In both these accelerations, the presence of a (portion of a) specific substrate appears to promote the deacylation reaction significantly.

The temperature dependence of the catalytic rate constants of the $\alpha$-chymotrypsin-catalyzed hydrolysis of two specific substrates, N-acetyl-L-tyrosine ethyl ester and N-acetyl-L-tryptophan ethyl ester, as also measured, as shown in Table II and Fig. 2. These hydrolyses were also carried out in a pH independent region of pH 8.55 to 8.70 so that any small temperature dependence of the p$K_a$ of the hydrolysis will have no effect on the results. The activation parameters, calculated from Fig. 1 and 2, are listed in Table IV.

## Discussion

Table III presents kinetic constants of a representative group of $\alpha$-chymotrypsin-catalyzed reactions of interest to the problem of specificity and binding. The catalytic rate constants are directly determined from experiment, using the Michaelis-Menten kinetic analysis. The individual rate constants, $k_2$ and $k_3$ were determined according to methods[15] discussed in detail in various papers of this series. The composite constant, $k_{cat}$, is not the constant of choice for a discussion of specificity. The acylation rate constant, $k_2$, describes the reaction of an adsorbed substrate molecule with the enzyme to give an acyl-enzyme and the first product (*e.g.*, alcohol or ammonia). Specificity may be discussed in terms of relative $k_2$ values, but such comparisons must be restricted to substrates containing similar leaving groups, such as methyl and ethyl esters, *p*-nitrophenyl esters, or the N-acylimidazoles. A consideration of the most extensive groups, the methyl and ethyl esters, shows a variation of $10^5$-fold between the most reactive compound, N-acetyl-L-tyrosine ethyl ester, and the least reactive compound, methyl cinnamate.[16] There is a considerable kinetic specificity in these reactions, above that predicted

Table II

Temperature Dependence of the Catalytic Rate Constants of the $\alpha$-Chymotrypsin-Catalyzed Hydrolysis of Two Specific Substrates[a,b]

| Temp., °C. | $k_{cat}$, sec.$^{-1}$ | Temp., °C. | $k_{cat}$, sec.$^{-1}$ |
|---|---|---|---|
| N-Acetyl-L-tyrosine ethyl ester | | N-Acetyl-L-tryptophan ethyl ester | |
| 8.6 | 75.5 | 6.3 | 10.7 |
| 9.6 | 74.9 | 10.9 | 16.7 |
| 12.9 | 94.2 | 25.0 | 46.2 |
| 20.8 | 151 | 34.0 | 83.2 |
| 25.0 | 193 | | |
| 31.0 | 295 | | |

[a] pH of Tris-HCl buffer at 25° was 8.55-8.70; 1.8% (v./v.) acetonitrile-water. [b] The errors are ±5% in all cases.

(14) R. J. Foster, *J. Biol. Chem.*, **236**, 2461 (1961).

(15) B. Zerner and M. L. Bender, *J. Am. Chem. Soc.*, **86**, 3669 (1964).

(16) The aryloxy group of *p*-nitrophenyl acetate must bind in approximately the same position as the *acyl* group of N-acetyl-L-tyrosine ethyl ester, since the binding of aromatic groups is of great importance. These two enzyme-substrate complexes are thus stereochemically different, and any comparison of the $k_2$ kinetic specificities of these two compounds are meaningless. Fortunately, this complication does not occur in the $k_3$ step.

TABLE III

KINETIC CONSTANTS OF SOME α-CHYMOTRYPSIN-CATALYZED REACTIONS

| | Substrate | $k_{cat}$, sec.$^{-1}$ | $k_2$, sec.$^{-1}$ | $k_3$, sec.$^{-1}$ | $k_{OH^-}$ of ethyl ester $\times 10^2$,[f] $M^{-1}$ sec.$^{-1}$ | Rel. $k_3$ (corrected)[g] | Ref. |
|---|---|---|---|---|---|---|---|
| 1 | N-Acetyl-L-tyrosine ethyl ester | 193 | 4000 | 193[b] | 76 | 3540 | 17 |
| 2 | N-Acetyl-L-phenylalanine ethyl ester | 173 | 2000 | 173[a] | 76 | 3180 | 18 |
| 3 | N-Acetyl-L-tryptophan ethyl ester | 46.5 | 374 | 46.5 | 76 | 942 | 8 |
| 4 | N-Acetylglycine methyl ester | 0.013 | 0.013 | 2.29[d] | 85 | 38.2 | 19 |
| 5 | Methyl hippurate | .143 | .192 | 0.567[c] | 93 | 4.9 | 19 |
| 6 | Methyl hydrocinnamate | .018 | .020 | .18[e] | 7.7 | 32.4 | 24–26 |
| 7 | Methyl cinnamate | .0073 | .0282 | .0125 | 1.19 | 14.7 | 20 |
| 8 | *p*-Nitrophenyl cinnamate | .0125 | | .0125 | 1.19 | 14.7 | 21 |
| 9 | N-*trans*-Cinnamoylimidazole | .0125 | | .0125 | 1.19 | 14.7 | 21 |
| 10 | *p*-Nitrophenyl acetate | .0068 | 4.8 | .0068 | 9.5 | 1.0 | 22, 28 |
| 11 | *p*-Nitrophenyl trimethylacetate | .00013 | 0.37 | .00013 | 0.0223 | 8.1 | 23 |
| 12 | N-Benzoylimidazole | | | .0002 | 0.51 | 0.54 | 27 |
| 13 | Methyl acetate | 0 (?) | 0 (?) | .0068 | 9.5 | 1.0 | 22 |

[a] The *p*-nitrophenyl and ethyl esters give essentially identical $k_{cat}$ indicating that $k_{cat} = k_3$.[25] [b] The magnitude of $K_m$(app) indicates that $k_2/k_3 = 18$.[15] [c] Value determined from the turnover of the corresponding *p*-nitrophenyl ester. [d] Value determined from the turnover and presteady state of the corresponding *p*-nitrophenyl ester. [e] Values determined from the methanolysis action. [f] Estimated from data in the National Bureau of Standards Circular 510, "Tables of Chemical Kinetics," corrected to aqueous solution and 25°. [g] $(k_{3(i)})(k_{OH}(\text{acetate}))/(k_3(\text{acetate}))(k_{OH}(i))$.

from electronic and steric factors operating in the alkaline hydrolysis of the corresponding esters.

The deacylation rate constant, $k_3$, describes the reaction of the acyl–enzyme with water to produce the carboxylic acid and regenerate the enzyme. *All* $k_3$ values in Table III may be compared with one another, leading to an extensive list of data covering a range of over $10^6$-fold in rate constant. In the acylation step, an adsorbed substrate molecule reacts while in the deacylation step a substrate molecule bound covalently to the enzyme reacts. There is still considerable kinetic specificity in the deacylation reaction; in fact, the range of rate constants is somewhat greater in the deacylation reaction involving a chemisorbed substrate than in the acylation reaction involving a physically adsorbed substrate. This similarity in kinetic specificities is expected if the acylation and deacylation reactions are mechanistically similar to one another.[2]

Since the data available for deacylation rate constants are more extensive than for acylation rate constants, let us consider the kinetic specificities exhibited in deacylation, $k_3$, and assume that such specificities are representative of both steps. Of the total spread in the relative rates of acylation and deacylation, a small part may be attributed to the electronic and steric factors manifest in the corresponding nucleophilic reactions. A correction has been made in Table III for these effects in the relative rates of deacylation, by using the relative saponification rate constants for the corresponding ethyl esters, on the basis that the Hammett ρ-constants (susceptibility to electronic effects) of the deacylation reaction and saponification reactions are almost identical.[27] The corrected rates have been normalized, setting the acetate derivative as 1. Of the million-fold variation seen in the deacylation series, about $10^2$-fold must be attributed to normal electronic effects, (steric effects are not fully accounted for), leaving approximately a $10^4$-fold rate enhancement which must be attributed to specificity by the enzyme (column relative $k_3$ (corrected) of Table III). Thus in deacylation, a complete spectrum of specificity is still seen.

The contributions to this specificity are interesting. Two rate comparisons involve the addition of an N-acetylamino group to the molecule, the relative $k_3$'s of 4 *vs.* 10, and 2 *vs.* 6. These comparisons result in rate ratios of 38.2 and 98, respectively. Two other rate comparisons involve the addition of a benzyl group to the molecule, the relative $k_3$'s of 6 *vs.* 10, and 2 *vs.* 4. These comparisons result in rate ratios of 32.4 and 83. The contributions of an N-acetylamino group and of a benzyl group to the specificity are thus independent of one another, within a factor of approximately two. These contributions indicate that specificity is a relative and not an absolute quantity. Further, the spectrum of specificity may be analyzed in terms of the independent contributions of two important components to the total specificity.

Let us assume that the kinetic specificity observed in deacylation is the full specificity exhibited by α-chymotrypsin and inquire how one may explain this phenomenon. The rate enhancement of $10^4$-fold exhibited in the specificity by α-chymotrypsin is certainly explainable in chemical terms, since series of related reactions differing by as much as $10^{14}$-fold are seen in simple organic chemical reactions.[29]

A simple, chemical explanation for the $10^4$-fold spread in the deacylation rate constants is depicted in Fig. 3 which illustrates the idea that although the acyl groups of the acyl–enzyme are covalently attached to the oxygen atom of serine, the placement of the car-

(17) L. W. Cunningham and C. S. Brown, *J. Biol. Chem.*, **221**, 287 (1956).
(18) B. R. Hammond and H. Gutfreund, *Biochem. J.*, **61**, 187 (1955).
(19) J. P. Wolf, III, and C. Niemann, *Biochemistry*, **2**, 493 (1963).
(20) M. L. Bender and B. Zerner, *J. Am. Chem. Soc.*, **84**, 2550 (1962).
(21) M. L. Bender, G. R. Schonbaum, and B. Zerner, *ibid.*, **84**, 2540 (1962).
(22) F. J. Kézdy and M. L. Bender, *Biochemistry*, **1**, 1097 (1962).
(23) M. L. Bender and G. A. Hamilton, *J. Am. Chem. Soc.*, **84**, 2570 (1962).
(24) K. J. Laidler and M. L. Barnard, *Trans. Faraday Soc.*, **52**, 497 (1956).
(25) B. Zerner and M. L. Bender, *J. Am. Chem. Soc.*, **86**, 356 (1963).
(26) M. L. Bender, G. E. Clement, C. R. Gunter, and F. J. Kézdy, *ibid.*, **86**, 3697 (1964).
(27) M. Caplow and W. P. Jencks, *Biochemistry*, **1**, 883 (1962).
(28) G. H. Dixon, W. J. Dreyer, and H. Neurath, *J. Am. Chem. Soc.*, **78**, 4810 (1956).

(29) M. L. Bender in "Technique of Organic Chemistry," A. Weissberger, Ed., 2nd Ed., Vol. VIII, Part II, John Wiley and Sons, Inc., New York, N. Y., 1963, p. 1471.

SPECIFIC SUBSTRATE

Ground State Transition State

NON-SPECIFIC SUBSTRATE

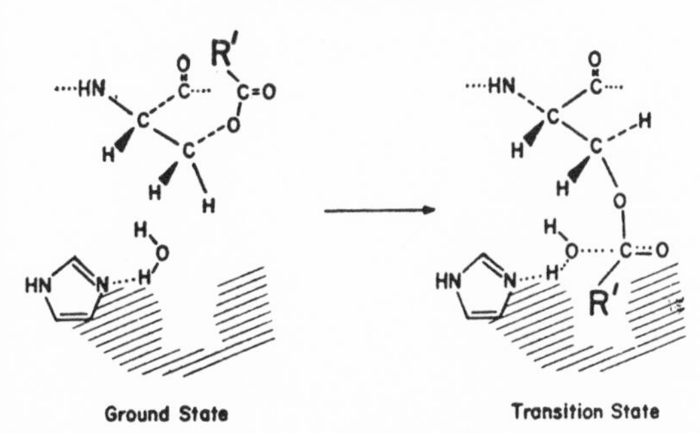


Fig. 3.—Specificity in the deacylation of acyl–α-chymotrypsins.

bonyl group of the ester in the ground state with respect to the water molecule and the catalytic entities of the protein is highly variable. Starting from the rigid polypeptide backbone of the protein (dotted lines from the NH and C=O groups) the side chain bearing the acyl group of the acyl enzyme may rotate around four single bonds in the ground state. In the transition state the rotation around these four bonds must be frozen so that the molecule may react with the water molecule and the catalytic entities on the enzyme surface. Specificity is then interpreted in terms of an interaction of the group *R* of the acyl group with the enzyme surface, *rigidifying* the whole set of bonds so that the carbonyl portion of the acyl group occupies the *correct* position for reaction, even in the ground state. Thus for a specific substrate, the reaction will be facile because the ground state of the acyl–enzyme resembles the transition state of the reaction, while for a nonspecific substrate the reaction will be slow because the ground state of the acyl–enzyme does not resemble the transition state.[30]

The requirement for the positioning of the carbonyl portion of the acyl group with respect to the catalytic components of the active site predicts that the specificity should be manifest in differences in the rotational entropy of activation measuring the correct positioning of the carbonyl group. The activation parameters of four deacylation reactions spanning the entire specificity spectrum have therefore been determined, as shown in Table IV.

(30) *Cf.* I. B. Wilson in "The Mechanism of Enzyme Action," W. D. McElroy and B. Glass, Ed., Johns Hopkins Press, Baltimore, Md., 1954, p. 656.

TABLE IV

THE ACTIVATION PARAMETERS OF THE DEACYLATION OF SOME ACYL-α-CHYMOTRYPSINS[a,f]

| Acyl–Enzyme | Rel. $k_3$ (corrected)[b] | $\Delta F^*$, kcal./mole | $\Delta H^*$, kcal./mole | $-T\Delta S^*$,[c] kcal./mole | $\Delta S^*$, e.u. |
|---|---|---|---|---|---|
| N-Acetyl-L-tyrosyl– | 3540 | 14.3 | 10.3 | 4.0[e] | −13.4 |
| N-Acetyl-L-tryptophanyl– | 942 | 17.9 | 12.0 | 5.9 | −19.8 |
| *trans*-Cinnamoyl– | 14.7 | 20.1 | 11.2 | 8.9 | −29.6 |
| Acetyl– | 1 | 20.4 | 9.7 | 10.7[d] | −35.9 |

[a] At pH 8.6 so that the temperature dependence of the ionization of p$K_a$ 7 does not affect the results. [b] These values reflect the specificity imposed by the enzyme (see Table III). [c] $T\Delta S^*$ calculated at 298°K. [d] A $\Delta H^*$ of 15.1 and $T\Delta S^*$ of −4.7 kcal./mole have previously been reported for the deacylation of acetyl-α-chymotrypsin,[28] but these measurements were carried out in the presence of N-acetyl-L-tyrosine ethyl ester. [e] A $\Delta H^*$ of 10.9 and $T\Delta S^*$ of −3.2 kcal./mole were previously reported.[31] [f] Error analysis: $\Delta H^*$ ± 1 kcal.; $\Delta S^*$ ± 5 e.u.

The activation parameters for the deacylation of the specific acyl–enzymes, N-acetyl-L-tyrosyl– and N-acetyl-L-tryptophanyl–α-chymotrypsin, are based on the catalytic rate constants for the hydrolysis of the corresponding ethyl esters since deacylation is the rate-determining step of these hydrolyses.[8] The activation parameters of the deacylation of the relatively nonspecific acyl–enzymes, *trans*-cinnamoyl- and acetyl-α-chymotrypsin, were determined directly.

The acyl–enzymes in Table IV are arranged in order of descending specificity according to Table III. The enthalpies of activation are essentially constant for both specific and nonspecific substrates, whereas a more or less regular increase in $-T\Delta S^*$ occurs in going from specific to nonspecific substrates. Thus the kinetic specificity of deacylation is not caused by a difference in enthalpies of activation, but almost solely results from a large difference in entropies of activation. The difference in entropies of activation between the most specific and the least specific substrate is 23 e.u. The assignment of this difference in entropies of activation to differences in rotational entropies owing to torsional oscillations around single bonds is reasonable. Approximate calculations give values of 2.5 to 7.5 e.u. per single bond for the entropy associated with torsional oscillations.[32] Experimental determinations of the activation entropy associated with the freezing of rotation of one bond is 4 to 6 e.u.[33] Using the value of 6 e.u. per bond, it is thus calculated that zero to four bonds must be frozen in going from the ground state to the transition state in the various reactions of Table IV, a reasonable number. In terms of Fig. 3, none of the four single bonds which determine the position of the carbonyl group needs to be frozen in the activation process for a specific substrate since they are already correctly frozen in the ground state. On the other hand, all four of these single bonds need to be frozen in the activation process for a completely nonspecific substrate. Thus the entropy of activation of the latter process should be approximately 24 e.u. more negative than for the former process, as observed experimentally.

Both the kinetic parameters and the activation parameters of the deacylation of acetyl-α-chymotryp-

(31) J. E. Snoke and H. Neurath, *J. Biol. Chem.*, **182**, 577 (1950).

(32) H. A. Scheraga, "Protein Structure," Academic Press, Inc., New York, N. Y., 1961, p. 40.

(33) E. G. Foster, A. C. Cope, and F. Daniels, *J. Am. Chem. Soc.*, **69**, 1893 (1947); P. D. Bartlett and R. R. Hiatt, *ibid.*, **80**, 1402 (1958).

sin differ depending on the presence of N-acetyl-L-tyrosine ethyl ester.[14,34] This observation may be explained on the basis of the binding of the phenolic group of the N-acetyl-L-tyrosine ethyl ester to the active site, leading to a partial fixation of the acetyl moiety, a lowering of the rotational entropy of activation, and thus an increase in the rate constant of the reaction. This picture is born out in the fact that the $\Delta S^*$ of deacylation of acetyl-α-chymotrypsin in the presence of N-acetyl-L-tyrosine ethyl ester is −27.2 e.u. while that in the absence of this compound is −35.9 e.u.

This discussion of the specificity of deacylation is unique in that it is based on reactions in which the substrate is covalently bound to the enzyme. This fact reduces the number of variables pertinent to the specificity, and provides a starting point for a detailed analysis of the meaning of specificity. The interpretation given here is essentially a modern physicochemical description of what Fischer called the lock and key theory, more recently called the template theory of specificity. Of course, there is no act of inserting a key in a lock in the deacylation step, but the requirement of a rigid correct fit between the substrate and enzyme, leading to the maximal rate of reaction, is implicit in both Fischer's early description and the present refinement.

The findings of an additive specificity pattern, that the specificity is related to the entropy of activation, and that the specificity as defined both in rate and in activation entropy terms may be enhanced by an added, extraneous molecule may possibly be interpreted in ways other than that presented here. One such interpretation is embodied in the "induced-fit" theory[35] which would explain the specificity of the deacylation step by varying degrees of induction of the correct conformation of the catalytic entities of the active site by various acyl groups. Likewise, the "induced-fit" theory would explain the increase in rate of deacylation by the addition of an extraneous reagent by its induction of a better conformation of the catalytic entities of the active site.

The specificity of deacylation may also be interpreted in terms of the introduction of strain into that part of the substrate which is to undergo reaction.[36,37] Presumably the binding forces available to the enzyme produce such a strain and an enhanced reaction, the magnitude of which would then be proportional to the magnitude of the binding forces and thus result in a variable specificity.

Distortions of the substrate to produce a reactive species have been discussed from two points of view, involving static and dynamic enzymes. Within the context of the transition state theory, distortion of the substrate must be viewed as a change in conformation stabilized by the enzyme which approximates the conformation of the transition state. Bruice has made an ingenious suggestion in the fremework of the distortion hypothesis to explain the efficiency of α-chymotrypsin,[38] based on the adsorption of an ester substrate on the enzyme in a lactone conformation. However, whereas

(34) *Cf.* T. Inagami and T. Murachi, *J. Biol. Chem.*, **238**, PC 1905 (1963).

(35) D. E. Koshland, Jr., *Proc. Natl. Acad. Sci. U. S.*, **44**, 98 (1958).

(36) R. Lumry in "The Enzymes," Vol. I, P. D. Boyer, H. Lardy, and K. Myrback, Ed., Academic Press, Inc., New York, N. Y., 1959, Chapter 4.

(37) W. P. Jencks, *Ann. Rev. Biochem.*, **32**, 658 (1963).

(38) T. C. Bruice, *J. Polymer Sci.*, **49**, 100 (1961).

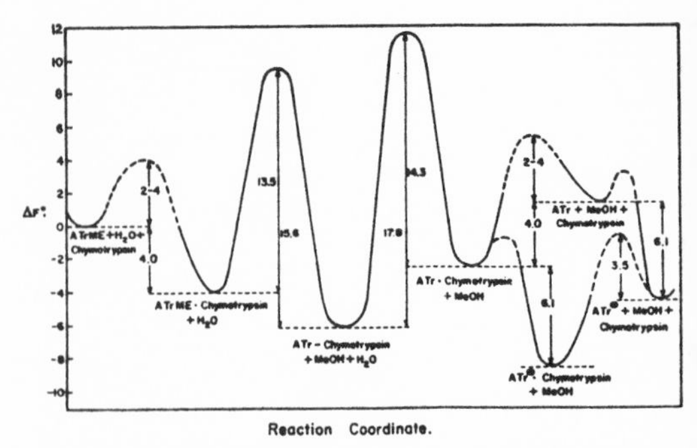


Fig. 4.—Standard free energy *vs.* reaction coordinate diagram for the α-chymotrypsin-catalyzed hydrolysis of N-acetyl-L-tryptophan methyl ester at 25°.

the *cis* form of an ester (a lactone) is exceptionally reactive, the *cis* form of an amide (a lactam) is not exceptionally reactive. Therefore, it is difficult to accept this specific suggestion as an important contributor to enzymatic efficiency.

It is clearly not possible to differentiate between the three suggestions given above to explain the specificity of deacylation. The detailed discussion of the rotational explanation of the specificity is given for two reasons. One is that to the present authors it is the simplest explanation requiring the least extrapolation from current chemical theory. The second is that it has not been described before in this form. The explanation of specificity given in this paper is satisfying in that it explains the magnitude of the activation entropy effect in a precise manner, whereas entropy changes in an induced fit might be much larger. However, definitive experiments to distinguish between these possibilities must still be sought.

## Conclusions

At this time it is possible to discuss an over-all pathway, a mechanism of each step, and specificity of each step for catalysis by α-chymotrypsin. Let us now consider how these factors account for the kinetics of the hydrolysis of the specific amide substrate, N-acetyl-L-tryptophan amide.

First consider the α-chymotrypsin-catalyzed hydrolysis of the analogous specific ester substrate, N-acetyl-L-tryptophan methyl ester, all of whose individual rate constants have been determined.[2] The standard free energy *vs.* reaction coordinate diagram for this enzymatic process, Fig. 4, is well defined, there being only two minor areas of uncertainty—those representing the activation energies of the presumably diffusion-controlled reactions involving enzyme–substrate formation from enzyme and substrate, and enzyme–product formation from enzyme and product. In going from the methyl ester reactant to the carboxylate ion product, there are two intermediate energy minima, occurring at successively lower standard free energies, so that from reactant to product, each step in the process leads to a lower standard free energy.

The standard free energy *vs.* reaction coordinate diagram for the α-chymotrypsin-catalyzed hydrolysis of the corresponding amide, N-acetyl-L-tryptophan amide, Fig. 5, is normalized so that the product acid is identical in standard free energy in both Fig. 4 and 5.

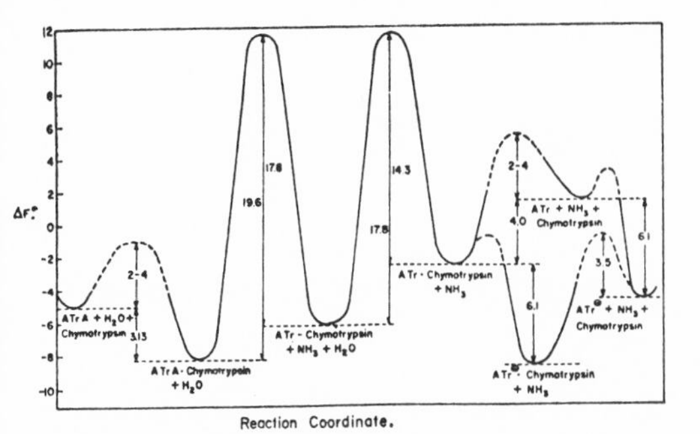


Fig. 5.—Standard free energy *vs.* reaction coordinate diagram for the α-chymotrypsin-catalyzed hydrolysis of N-acetyl-L-tryptophan amide at 25°.

From the N-acetyl-L-tryptophanyl-α-chymotrypsin intermediate to the product, the diagrams for the hydrolysis of the methyl ester and amide are identical with one another. Representative amide and methyl ester pairs indicate that the amide is about 5 kcal./mole more stable than the corresponding ester.[39] The standard free energy of the amide with respect to the rest of the diagram is confirmed by the equilibrium measurements of amide formation, leading to a difference in free energy between the amide and the carboxylic acid of 6.4 kcal./mole,[40] which is also found when setting the amide 5 kcal./mole lower than that of the ester. Knowing this relationship, the standard free energies of activation of $k_2$ from the corresponding rate constant and of the $K_s$ of the amide[19] completely fixes the standard free energy *vs.* reaction coordinate diagram for the amide hydrolysis. Again, the free energy of the important minima of the amide reaction, the acyl–enzyme intermediate, and the carboxylate ion product successively decreases. Thus, the acyl–enzyme intermediate is not a high energy intermediate with respect to the amide reactant, and one may say, not only kinetically[2] but also thermodynamically, that the acyl–enzyme intermediate is favored in the over-all pathway of the reaction.

Let us now compare in some detail the nonenzymatic (hydroxide ion-catalyzed) hydrolysis and the α-chymotrypsin-catalyzed hydrolysis of N-acetyl-L-tryptophan amide. The second-order rate constant for the reaction of hydroxide ion with acetamide is $3.78 \times 10^{-5}$ $M^{-1}$ sec.$^{-1}$ in water at 25°. From the hydrolytic rate constants of the esters N-acetyl-L-tryptophan ethyl ester and ethyl acetate, it is concluded that the N-acetyl-L-tryptophanyl substituent increases the saponification by eightfold, and therefore the second-order rate constant for the hydrolysis of the (unsubstituted) amide group of N-acetyl-L-tryptophan amide is $3 \times 10^{-4}$ $M^{-1}$ sec.$^{-1}$.[41] The rate constant for the α-chymotrypsin-catalyzed hydrolysis of N-acetyl-L-tryptophan amide at pH 8, the pH of maximal rate in water at 25°, is $4.4 \times 10^{-2}$ sec.$^{-1}$ [19] Thus, the nonenzymatic and enzymatic reactions differ not only quantitatively, but also in kinetic form, the former being a second-order process and the latter being a first-order process. However, let us inquire as to the factors which account for the difference between the nonenzymatic and enzymatic processes.

The difference between the nonenzymatic and enzymatic processes must be attributed to more than one cause. Differences in reactivity in organic chemical reactions may frequently be expressed in terms of three (additive) contributions to the difference in standard free energies of activation: polar, steric, and resonance effects.[42] In seeking the explanation to the difference between two forms of catalysis, other factors than these will be involved, but certainly more than one factor must be considered.

The comparison of an intramolecular general basic catalysis by imidazole (the enzymatic process) with an intermolecular hydroxide ion catalysis (the nonenzymatic process) is a knotty problem. It may be resolved by (1) comparing the efficacy of intermolecular general basic catalysis by imidazole with intermolecular hydroxide ion catalysis, and by (2) using the hypothesis that an intramolecular general basic catalysis by imidazole is equivalent to intramolecular general basic catalysis by 10 $M$ imidazole. The first step of this process is given by the rate constants of the hydrolysis of haloacetate esters catalyzed by hydroxide ion and by imidazole as an intermolecular general base. This comparison indicates that 1 $M$ imidazole = $1.6 \times 10^{-6}$ $M$ hydroxide ion.[43] The second step of this process depends on the observation, made many times before, that the *effective* concentration of an intramolecular catalytic group is the equivalent of *ca.* 10 $M$ of the corresponding intermolecular catalyst.[44] On this basis, the second-order hydroxide ion-catalyzed hydrolysis of rate constant, $3 \times 10^{-4}$ $M^{-1}$ sec.$^{-1}$, may be transformed into an intramolecular first-order imidazole-catalyzed hydrolysis of rate constant, $4.8 \times 10^{-9}$ sec.$^{-1}$. That this transformation has meaning is seen in a calculation of the rate constant of the intramolecular general base-catalyzed hydrolysis of *p*-nitrophenyl 3-nitrosalicylate[45] from data on intermolecular reactions. In this reaction a phenoxide group of p$K_a$ 6.02 is the intramolecular general basic catalyst. The rate constant of the hydroxide ion-catalyzed hydrolysis of *p*-nitrophenyl 3-nitrobenzoate is 21.2 $M^{-1}$ sec.$^{-1}$. General basic catalysis by a base of p$K_a$ 6.02 (mechanistically B + $H_2O$) is $0.58 \times 10^{-6}$ less efficient than is catalysis by hydroxide ion.[43] Using this factor and the hypothesis that the effective concentration of intramolecular phenoxide ion is the equivalent of *ca.* 10 $M$ of intermolecular phenoxide ion, the intramolecular rate constant for the hydrolysis of *p*-nitrophenyl 3-nitrosalicylate is calculated to be $1.26 \times 10^{-4}$ sec.$^{-1}$. The experimental rate constant is $1 \times 10^{-4}$ sec.$^{-1}$, in reasonable agreement. Thus, the calculated conversion of intermolecular hydroxide ion catalysis of N-acetyl-L-tryptophan amide hydrolysis to intramolecular general basic catalysis by imidazole has a solid analogy.

In comparing the nonenzymatic and enzymatic reactions, one compares a nonenzymatic reaction involving one intermediate, a tetrahedral addition com-

(39) F. H. Carpenter, *J. Am. Chem. Soc.*, **82**, 1111 (1960).

(40) H. Morawetz and P. S. Otaki, *ibid.*, **85**, 463 (1963).

(41) This constant cannot be measured directly since hydroxide ion cannot distinguish between the two amide linkages in the molecule.

(42) R. W. Taft, Jr., in "Steric Effects in Organic Chemistry," M. S. Newman, Ed., John Wiley and Sons, Inc., New York, N. Y., 1956, Chapter 13.

(43) W. P. Jencks and J. Carriuolo, *J. Am. Chem. Soc.*, **83**, 1746 (1961).

(44) M. L. Bender, *Chem. Rev.*, **60**, 53 (1960); *cf.* D. E. Koshland, Jr., *J. Theoret. Biol.*, **2**, 75 (1962).

(45) M. L. Bender, F. J. Kézdy, and B. Zerner, *J. Am. Chem. Soc.*, **85**, 3017 (1963).

pound,[46] to an enzymatic reaction involving at least three intermediates, two enzyme–substrate complexes and one enzyme–substrate compound and, if one counts possible tetrahedral intermediates, two more.[47] Thus one large activation process is distributed among several smaller activation processes. This description of catalysis has been often considered in the past; it is, however, interesting to see it come to light.

The chemistry which makes possible the stepwise pathway described above consists of both complex and compound formation between the enzyme and substrate. Complex formation between reactant and catalyst does not, in itself, facilitate reaction. If, however, complex formation not only lowers the ground-state energy but also lowers the transition-state energy (to a greater extent than the ground state), then complex formation may contribute to the over-all catalytic process.

Compound formation between the enzyme and the acyl group of the substrate is a key contributor to the stepwise process leading to the lowered multiple activation process. The acyl–enzyme intermediate is slightly lower than the amide substrate in standard free energy and is thus a reasonable intermediate in the catalytic process, from a thermodynamic point of view. However, a kinetic factor is also involved in the conversion of the amide substrate to the acyl–enzyme intermediate. In amide hydrolysis catalyzed by α-chymotrypsin, this step, an alcoholysis reaction since the nucleophile is a serine hydroxyl group, is the rate-determining step. On the other hand, in amide hydrolysis catalyzed by hydroxide ion, the rate-determining step is a hydrolysis. The *alcoholysis* of a carboxylic acid derivative, in general, occurs at a rate which is approximately 100-fold faster than the *hydrolysis* of the corresponding compound.[26] Thus, the conversion of the rate-determining step of the saponification of an amide from hydrolysis to alcoholysis should increase the rate constant by $10^2$-fold. Implicit in this calculation is the assumption that the concentration of the serine hydroxyl group in acylation equals the concentration (55 $M$) of water in hydrolysis.

The enzyme imposes a kinetic specificity in its reactions, accounted for previously, in terms of a steric facilitation manifest in the lowered (rotational) entropy of activation. Whether one accepts this explanation or others, this rate enhancement amounts to 942 in the case of N-acetyl-L-tryptophanyl derivatives using the acyl group, acetate, as a standard. Although this calculation was carried out with deacylation rate constants, data indicate that this kinetic specificity is also operative in acylation. Thus in both acylation and deacylation steps, kinetic specificity will increase the rate constant of hydrolysis another factor of $10^3$. This specificity factor is valid only on the assumption that the calculation made previously in converting the reaction from an intermolecular process to an intramolecular process produced an intramolecular process with complete freedom of rotation (or the equivalent in entropy).

Finally, consideration must be given to the fact that the enzymatic catalysis has the characteristics of both general basic and general acidic catalysis. Catalysis by the imidazole(s) on the enzyme is postulated to involve two proton transfers, one to an imidazole group and one from an imidazole group, as well as a nucleophilic attack, all occurring in a concerted cyclic process.[2] In the calculation, proton abstraction by imidazole (general basic catalysis) has been accounted for, but proton donation by imidazole (general acidic catalysis) has not. The effect of an intramolecular proton transfer on the hydrolysis of an ester may be seen in the hydrolyses of the compounds $CH_3CH_2C(O)SCH_2CH_2NH^+(CH_3)_2$ and $CH_3CH_2C(O)SCH_2CH_2N^+(CH_3)_3$. The former hydrolyzes 240 times faster than the latter,[48] which may be most reasonably interpreted in terms of the internal proton transfer which is available to the former but not the latter compound. Therefore a factor of $10^2$-fold will be attributed to the general acidic catalysis function of imidazole in the enzymatic process, assuming additivity of the general basic and acidic components.

We have thus far considered five aspects of the enzymatic process: (1) its intramolecular character and the concomitant increase in effective concentration of the catalytic group(s); (2) the general basic catalysis by imidazole; (3) the change in rate-determining step of the hydrolysis of an amide to an alcoholysis; (4) the freezing of the substrate in a conformation resembling the transition state; and (5) the general acidic catalysis by imidazole. A summation of the effects of each factor on our comparison between the nonenzymatic and enzymatic processes is shown in Table V. It is seen that these five factors successfully account for the total enzymatic rate constant. The workings of the enzyme α-chymotrypsin can thus be discussed on a straightforward chemical basis, where no single factor is all-important and where each factor is based qualitatively on solid chemical analogy and each factor except the last is based quantitatively on solid chemical analogy. This combination produces an enzyme.

TABLE V

KINETIC FACTORS RESPONSIBLE FOR THE DIFFERENCE BETWEEN THE HYDROXIDE ION AND α-CHYMOTRYPSIN-CATALYZED HYDROLYSES OF N-ACETYL-L-TRYPTOPHAN AMIDE

| | |
|---|---|
| Rate constant of hydroxide ion catalysis | $3 \times 10^{-4}$ $M^{-1}$ sec.$^{-1}$ |
| (1) Conversion to an intermolecular general base-catalyzed reaction involving imidazole ($1.6 \times 10^{-6}$) | $4.8 \times 10^{-10}$ $M^{-1}$ sec.$^{-1}$ |
| (2) Conversion to an intramolecular general base-catalyzed reaction involving imidazole (10 $M$) | $4.8 \times 10^{-9}$ sec.$^{-1}$ |
| (3) Change in rate-determining step ($10^2$) | $4.8 \times 10^{-7}$ sec.$^{-1}$ |
| (4) Freezing of the substrate specificity ($10^3$) | $4.8 \times 10^{-4}$ sec.$^{-1}$ |
| (5) General acidic catalysis by imidazole ($10^2$) | $4.8 \times 10^{-2}$ sec.$^{-1}$ |
| Total calcd. enzymatic rate on the basis of the above five factors | $4.8 \times 10^{-2}$ sec.$^{-1}$ |
| Experimental rate constant | $4.4 \times 10^{-2}$ sec.$^{-1}$ |

(46) M. L. Bender, *J. Am. Chem. Soc.*, **73**, 1626 (1951).

(47) Tetrahedral addition intermediates have not been depicted in Fig. 4 and 5, but their omission does not affect the over-all conclusions.

(48) B. Hansen, *Acta Chem. Scand.*, **12**, 324 (1958).

Reprinted from *Biochem. Biophys. Res. Commun.* **57**:887–892 (1974)

# "ORBITAL STEERING", ENTROPY, AND RATE ACCELERATIONS

W. P. Jencks[1] and M. I. Page[2]

[1] Graduate Department of Biochemistry,
Brandeis University, Waltham, Massachusetts 02154

and

[2] Department of Chemistry, The Polytechnic,
Huddersfield HD1 3DH

Received February 25,1974

Summary

Some important conceptual and quantitative differences are described between the "orbital steering" and entropy descriptions of the rate accelerations in intramolecular and enzymic reactions that may be brought about by geometric constraints other than distortion. The treatments differ by a factor of $10^3$ - $10^4$ in the maximum rate acceleration that may be obtained from these constraints. The estimation of a "proximity factor" without taking adequate account of the translational and overall rotational entropy terms gives a misleading value for this factor. The conclusion is reaffirmed that increasing the probability of reaction by restricting the free translational and rotational movement of reacting groups can play a large role in the catalytic power of enzymes.

Dafforn and Koshland (D and K) have recently compared "orbital steering" and entropy as ways of describing the rate accelerations caused by certain geometric restrictions in intramolecular and enzymic reactions and have concluded that the statistical mechanical treatments [i.e. entropy] are an adjunct but not a satisfactory substitute for the concepts of proximity and orbital steering (1). They further pointed out that the two approaches should give similar results if properly applied and concluded, from an analysis of cyclopentadiene dimerization, that this is indeed the case (1). We reluctantly add still another paper to the literature on this subject in order to present an extension of this comparison (Table I) that leads us to different conclusions.

The comparisons of Table I show that there are important

TABLE I

| "Orbital Steering" | Entropy |
|---|---|
| 1. Most of the rate increase in unstrained intramolecular (and some enzymic) reactions compared to corresponding bimolecular reactions in solution is attributed to an orientation factor, with an angular preference far greater than previously estimated. This orientational preference is exhibited after reacting molecules are brought into proximity with their reactive atoms in contact (I) and is measured by the product of the orientation factors $\theta_A$ and $\theta_B$ (1,2). 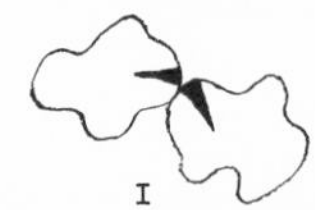 I | This rate increase is attributed to the improbability of two molecules in dilute solution losing their freedom of translation and rotation in three dimensions upon forming a transition state or product. This probability is greatly increased in an intramolecular or enzymic reaction when the reacting groups are brought together so that translational and rotational motions are constrained. The increase in probability is measured by a difference in entropy (3). Although an exact separation of translational and rotational entropy terms is not possible in solution, the contributions of both terms are comparable and a large part of the entropy loss has already taken place when the reacting atoms are in contact (I). |
| 2. The maximum rate acceleration from orientation has been estimated, from the decrease in the rotational partition functions upon conversion of two molecules to a single transition state, to be about $10^4$ for typical molecules (4). However, this calculation is inconsistent with the above definition of "orbital steering," because the two molecules in I have already lost their freedom to rotate independently. | The maximum rate acceleration from the loss of 20-25 units of (overall) rotational entropy,* which is equivalent to the decrease in rotational partition functions, is about $10^4$ - $10^5$ for typical molecules (3). |
| 3. The spatial (as opposed to orientational) requirements for a reaction are incorporated into a "proximity factor" for the formation of I that is estimated to be about $55/n_n \cong$ 10 M, equivalent to 4.5 e.u. (1). The loss of translational freedom of two reacting molecules is accommodated in the proximity factor (4), as is also the loss of free (overall) rotation of polyatomic, nonlinear molecules (I). | The spatial requirements are accommodated in the loss of translational entropy of the reacting molecules. This loss is typically 30 e.u. in the gas phase and 5 ± 5 e.u. less in solution, which corresponds to a maximum factor of $10^4$ - $10^7$ M (3). A spatial requirement closer to $10^7$ M than to 10 M has also been estimated from distance distribution functions (6). |
| 4. The entropy of reaction of two bromine atoms to form $Br_2$ (-13 e.u. (5)) is taken as a model for the proximity effect because Br atoms have no orientational requirement for reaction (1,4). | The spatial requirements and the loss of translational entropy in the bromine reaction are the same as for other reactions (-31 e.u., corresponding to a factor of $10^7$ M), but are offset by an atypical gain of rotational entropy upon formation of the $Br_2$ molecule (we might call this "orbital anti-steering") (5). |
| 5. The maximum total advantage in an intramolecular or enzymic reaction from "orbital steering" and "proximity" is approximately $10^4 \times 10$ M = $10^5$ M (4). | A conservative estimate of the maximum total advantage that may arise from the restriction of translational and rotational motions in an intramolecular or enzymic reaction in solution is about $10^8$ M (3). |
| 6. It is emphasized that the "orientation" of orbital steering is not the gross orientation of substrates and catalytic groups on the enzyme surface or the juxtapositioning of reacting atoms (2). | It is precisely these factors that are responsible for the major part of the rate acceleration from entropic factors in intramolecular and enzymic reactions. Most of the entropy of a rotation is lost with a relatively mild orientational restriction (e.g. an 80% loss upon conversion of a free internal rotation, corresponding to 7 e.u., to a vibrational frequency of 300 $cm^{-1}$ which, for carbon atoms, corresponds to an angle of approximately 30°). The juxtapositioning of reacting atoms requires the loss of 3 degrees of translational freedom, corresponding to ~30 e.u. We believe that one of the most important functions of an enzyme is to increase the probability for the formation of the transition state simply by bringing about the snug juxtapositioning of reacting atoms, with a resulting restriction of low frequency stretching and other motions that replace translation (3,7). |

TABLE I Continued

| | |
|---|---|
| 7. Experimental support for "orbital steering" factors of $10^4$ was adduced from comparisons of the rate constants for a series of acid-catalyzed esterification and lactonization reactions (8). However, orbital steering is defined as the rate acceleration that may be obtained from orientation factors after the reacting atoms are in contact (I). Consequently, these results provide no quantitative support for "orbital steering," because the reacting atoms in the compounds examined are not initially in contact. | The intramolecular reaction of succinate half esters II is faster by a factor of $10^5$ M than 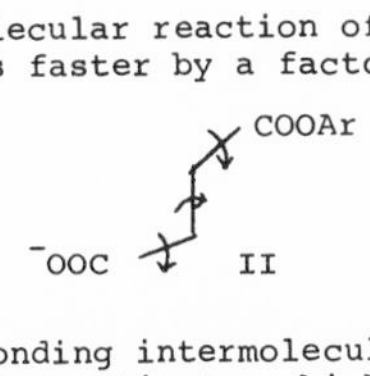  the corresponding intermolecular reaction (3,9). A conservative estimate, which ignores the differences in strain energy (3,10), increases this factor to $10^8$ M if the three internal rotations of II are frozen into favorable positions and a bicyclic compound with these rotations frozen exhibits a rate increase of $10^8$ M (3,9). Most of the rate increase of $10^5$ M results from increasing the probability of reaction by preventing independent translational motions and from a mild restriction on the rotation of the reacting groups. It clearly does not result from an orientational restriction of juxtaposed reacting atoms and, hence, cannot be explained by the "proximity" - "orbital steering" approach. |
| 8. The basis for the conclusion that "orbital steering" and entropy calculations give similar results is the assignment of an orientation factor of $10^6 - 10^7$ to the cyclopentadiene reaction. This factor is stated to be "not unreasonable" compared to a factor of $10^4$ for esterification, but no calculations are given (1, Figure 2, footnote f). | (i) No more entropy can be lost upon freezing a rotation than was present in the rotation initially. The complete loss of the rotational entropy of the reacting cyclopentadiene molecules in the transition state corresponds to 24 e.u. or a factor of $10^5$ (3). The published values of the rotational partition functions and entropies for esterification (4) are closely similar to those for cyclopentadiene dimerization (3).<br><br>(ii) Once the reactive atoms are in contact (as required by the definition of "orbital steering") all rotations are already lost. The complete freezing of remaining bending motions (e.g. 4 rocking modes each of 300 $cm^{-1}$) would give a loss of 5.6 e.u. and a rate factor of 20.<br>(iii) If only two of the four reacting atoms are initially in contact (as shown in Figure 2 of D and K, (1)), the complete freezing of one free internal rotation (moment of inertia $11 \times 10^{-39}$ g $cm^2$) and 4 bending motions as above could give a loss of 14 e.u. and a rate factor of $10^3$, not $10^6 - 10^7$. |
| 9. Support for the hypothesis that orbital steering can introduce large rate effects in intramolecular and enzymic reactions is adduced from the observation that certain structural changes in cyclic reaction systems cause rate decreases (11). | It is important to keep in mind that the magnitude of rate decreases in constrained systems cannot be equated with the rate increase that may be obtained by optimal constraint of an unconstrained system. For example, the rate of a reaction that occurs on only one side of a group can be decreased by $>10^6$ by constraining a previously unconstrained reactant to the unreactive side (III), but can be increased by only a factor of 2 by constraining it to the reactive side (IV).  III 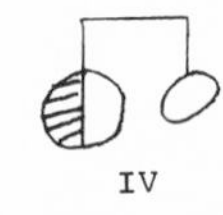 IV |
| 10. It is suggested that if each θ factor is $10^3 - 10^5$, a combination of two substrates and two catalytic groups could produce a factor of $10^9 - 10^{15}$, just what is needed to bridge the gap in enzymic and nonenzymic rates (2). | There is no doubt that additional catalytic groups in the active site of an enzyme can cause rate accelerations from entropic effects (7). However, the rate factor must be related to comparable nonenzymic reactions, which are 4th order rather than 2nd order in this case (12). Since the rate constants for these hypothetical 4th order reactions are not reported, the comparison is incomplete. It should be noted that the addition of a molecule of catalyst to a solution reaction ordinarily requires the loss of 3 degrees of translational and of rotational freedom; consequently, 4th and 5th order reactions are rare. |

*Entropy is expressed in units of calories $mol^{-1}$ °$K^{-1}$, abbreviated as e.u. Standard states are taken as 1 M at 25° throughout.

differences between the "orbital steering" and entropy approaches, with respect to both their definitions and their quantitative description of reaction rates. The most important differences are in (i) the maximum total rate acceleration that can be expected from these effects in an unstrained intramolecular or enzymic reaction (ca. $10^5$ vs $10^8$ M) and (ii) the attribution of these effects principally to a high degree of orientational restriction that occurs <u>after</u> the reacting atoms are juxtaposed in "orbital steering" and to a snug juxtapositioning of the reacting atoms accompanied by a relatively mild orientational restriction according to the entropy calculations. The entropy calculations are not novel and are in agreement with experiment (3,13). If orbital steering is identified with rotational entropy the differences appear in the assignment of a factor of up to ~10 M (4.5 e.u.) to "proximity" (4) and a factor of up to $10^7$ M (30 e.u.) to translational entropy (3). Other identifications do not account for the missing 25 e.u. We note further that the application and "instinctive understanding" (1) of entropy and probability requirements for a reaction are no more difficult than the correct application and understanding of "proximity" and "orbital steering."

We conclude from the comparisons summarized in Table I that the "proximity" and "orbital steering" treatments, when carefully defined and rigorously applied, may be an adjunct, but are not a satisfactory substitute for the concept of entropy or the methods of statistical mechanics. We would be willing to accept a definition of "orbital steering" based on rotational and perhaps even vibrational partition functions of the reactants and transition state (4) (these terms are equivalent to $S_{rot}$ and $S_{vib}$), but believe that it can only lead to confusion to suggest "orbital steering" effects that are larger than the total loss of rotational entropy in a reaction and to include the loss of overall rotational freedom in the definition of "proximity."

The effects of geometric restrictions on the rates of intramolecular and enzyme catalyzed reactions have been described over the years by the terms entropy loss (3,10,14), approximation, orientation, anchimeric assistance (15), propinquity, rotamer distribution (9), proximity, orbital steering (1,4), stereopopulation control (16), distance distribution function (6), togetherness (7), other terms that are not appropriate for the open literature, and FARCE (Freezing At Reactive Centers of Enzymes (17)). This exponential scholastic proliferation is becoming burdensome to students and could soon fill the presently available journals. We believe that it would be desirable to reverse this growth by dropping some of the more colorful and imprecisely defined of these terms and accordingly offer to initiate this process by withdrawing the term "togetherness."

Acknowledgment. We are grateful to the National Science Foundation (GB 4648) and the National Institute of Child Health and Human Development of the National Institutes of Health (HD 01247) for financial support. Contribution No. 933 from the Graduate Department of Biochemistry, Brandeis University.

REFERENCES

1. Dafforn, A. and Koshland, Jr., D.E. (1973) Biochem. Biophys. Res. Commun. 52, 779.
2. Storm, D.R. and Koshland, Jr., D.E. (1970) Proc. Nat. Acad. Sci. U.S. 66, 445.
3. Page, M.I. and Jencks, W.P. (1971) Proc. Nat. Acad. Sci. U.S. 68, 1678.
4. Dafforn, A. and Koshland, Jr., D.E. (1971) Proc. Nat. Acad. Sci. U.S. 68, 2463.
5. Page, M.I. (1972) Biochem. Biophys. Res. Commun. 49, 940.
6. DeLisi, C. and Crothers, D.M. (1973) Biopolymers 12, 1689.

7. Jencks, W.P. and Page, M.I. (1972) Proc. 8th FEBS Meeting, Amsterdam 29, 45.
8. Storm, D.R. and Koshland, Jr., D.E. (1972) J. Amer. Chem. Soc. 94, 5805.
9. Bruice, T.C. and Pandit, U.K. (1960) J. Amer. Chem. Soc. 82, 5858; (1960) Proc. Nat. Acad. Sci. U.S. 46, 402; Bender, M.L. and Neveau, M.C. (1958) J. Amer. Chem. Soc. 80, 5388; Gaetjens, E. and Morawetz, H. (1960) J. Amer. Chem. Soc. 82, 5328; Bruice, T.C. and Turner, A. (1970) J. Amer. Chem. Soc. 92, 3422; Bruice, T.C., in The Enzymes, Vol. II, ed. P.D. Boyer, H. Lardy and K. Myrbäck (Academic Press, New York, N.Y., 1970), 3rd ed., p. 217.
10. Page, M.I. (1973) Chem. Soc. Revs. 2, 295.
11. Storm, D.R. and Koshland, Jr., D.E. (1972) J. Amer. Chem. Soc. 94, 5815.
12. Jencks, W.P. (1969) Catalysis in Chemistry and Enzymology, McGraw-Hill Book Co., New York, N.Y., p. 19.
13. Kistiakowsky, G.B. and Lacher, J.R. (1936) J. Amer. Chem. Soc. 58, 123; Wassermann, A. (1941) Proc. Roy. Soc. A178, 370; O'Neal, H.E. and Benson, S.W. (1970) Internat. J. Chem. Kinetics 2, 423.
14. Westheimer, F.H. (1962) Advan. Enzymol. 24, 455.
15. Winstein, S., Lindegren, C.R., Marshall, H., and Ingraham, L.L. (1953) J. Amer. Chem. Soc. 75, 147.
16. Milstien, S. and Cohen, L.A. (1970) Proc. Nat. Acad. Sci. U.S. 67, 1143.
17. Nowak, T. and Mildvan, A.S. (1972) Biochemistry 11, 2813.

Part X

# REGULATORY ENZYMES: ALLOSTERY

# Editor's Comments on Papers 57 and 58

**57 MONOD, WYMAN, and CHANGEUX**
*On the Nature of Allosteric Transitions: A Plausible Model*

**58 PANAGOU et al.**
*A Monomeric, Allosteric Enzyme with a Single Polypeptide Chain. Ribonucleotide Reductase of* Lactobacillus leichmannii

The realization that many enzyme molecules exhibit the type of regulatory behavior known as *allostery* has given a new kind of biological emphasis to the *in vitro* study of enzymes and enzyme systems. At the very same time the discovery of this phenomenon has stimulated chemical or structural explanations. Thus the investigation of allosteric enzymes has made enzymology more biological and more chemical at one and the same time. Paper 57 by Monod, Wyman, and Changeux, by introducing the so-called symmetry model and a new and by now widely used nomenclature (*oligomer, protomer,* etc.), has exerted a profoundly stimulating effect on studies of enzyme function, enzyme mechanisms and protein behavior. In addition, it has stimulated the formulation of other models to explain allosteric properties, notably the so-called sequential model due to Koshland, Nemethy, and Filmer (1966).

Monod, Wyman, and Changeux wrote their paper shortly after it was demonstrated that the classical allosteric enzyme, aspartate transcarbamylase (now called *aspartate carbamoyltransferase*) can be separated into what were later called *catalytic* and *regulatory subunits.* From this example they speculated, toward the end of their paper, that perhaps in general "a protein should contain as many different subunits (peptide chains) as it bears stereospecifically different receptor sites." This view cannot be held any more. Paper 58 by Panagou, Orr, Dunstone, and Blakley on ribonucleotide reductase from *Lactobacillus leichmannii* was the first to give clear-cut evidence for a single-chain protein that has both types of binding sites.

# 57


Reprinted from *J. Mol. Biol.* **12**:88–118 (1965)

# On the Nature of Allosteric Transitions: A Plausible Model

JACQUES MONOD, JEFFRIES WYMAN AND JEAN-PIERRE CHANGEUX

*Service de Biochimie Cellulaire, Institut Pasteur, Paris, France and Istituto Regina Elena per lo Studio e la Cura dei Tumori, Rome, Italy*

(Received 30 December 1964)

*"It is certain that all bodies whatsoever, though they have no sense, yet they have perception; for when one body is applied to another, there is a kind of election to embrace that which is agreeable, and to exclude or expel that which is ingrate; and whether the body be alterant or altered, evermore a perception precedeth operation; for else all bodies would be like one to another."*

*Francis Bacon*
*(about 1620)*

## 1. Introduction

Ever since the haem-haem interactions of haemoglobin were first observed (Bohr, 1903), this remarkable phenomenon has excited much interest, both because of its physiological significance and because of the challenge which its physical interpretation offered (cf. Wyman, 1948,1963). The elucidation of the structure of haemoglobin (Perutz *et al.*, 1960) has, if anything, made this problem more challenging, since it has revealed that the haems lie far apart from one another in the molecule.

Until fairly recently, haemoglobin appeared as an almost unique example of a protein endowed with the property of mediating such indirect interactions between distinct, specific, binding-sites. Following the pioneer work of Cori and his school on muscle phosphorylase (see Helmreich & Cori, 1964), it has become clear, especially during the past few years, that, in bacteria as well as in higher organisms, many enzymes are electively endowed with specific functions of metabolic regulation. A systematic, comparative, analysis of the properties of these proteins has led to the conclusion that in most, if not all, of them, *indirect* interactions between *distinct* specific binding-sites (allosteric effects) are responsible for the performance of their regulatory function (Monod, Changeux & Jacob, 1963).

By their very nature, allosteric effects cannot be interpreted in terms of the classical theories of enzyme action. It must be assumed that these interactions are mediated by some kind of molecular transition (allosteric transition) which is induced or stabilized in the protein when it binds an "allosteric ligand". In the present paper, we wish to submit and discuss a general interpretation of allosteric effects in terms of certain features of protein structure. Such an attempt is justified, we believe, by the fact that, even though they perform widely different functions, the dozen or so allosteric systems which have been studied in some detail do appear to possess in common certain remarkable properties.

Before summarizing these properties, it will be useful to define two classes of allosteric effects (cf. Wyman, 1963):

(a) "*homotropic*" effects, i.e. interactions between *identical* ligands;

(b) "*heterotropic*" effects, i.e. interactions between *different* ligands.

The general properties of allosteric systems may then be stated as follows:

(1) Most allosteric proteins are polymers, or rather oligomers, involving several identical units.

(2) Allosteric interactions frequently appear to be correlated with alterations of the *quaternary* structure of the proteins (i.e. alterations of the bonding between subunits).

(3) While heterotropic effects may be either positive or negative (i.e. co-operative or antagonistic), homotropic effects appear to be always co-operative.

(4) Few, if any, allosteric systems exhibiting *only* heterotropic effects are known. In other words, co-operative homotropic effects are almost invariably observed with at least one of the two (or more) ligands of the system.

(5) Conditions, or treatments, or mutations, which alter the heterotropic interactions also simultaneously alter the homotropic interactions.

By far the most striking and, physically if not physiologically, the most interesting property of allosteric proteins is their capacity to mediate homotropic co-operative interactions between stereospecific ligands. Although there may be some exceptions to this rule, we shall consider that this property characterizes allosteric proteins. Furthermore, given the close correlations between homotropic and heterotropic effects, we shall assume that the same, or closely similar, molecular transitions are involved in both classes of interactions. The model which we will discuss is based upon considerations of molecular symmetry and offers primarily an interpretation of co-operative homotropic effects. To the extent that the assumptions made above are adequate, the model should also account for heterotropic interactions and for the observed correlations between the two classes of effects.

We shall first describe the model and derive its properties, which will then be compared with the properties of real systems. In conclusion, we shall discuss at some length the plausibility and implications of the model with respect to the quaternary structures of proteins.

## 2. The Model

Before describing the model, since we shall have to discuss the relationships between subunits in polymeric proteins, we first define the terminology to be used as follows:

(a) A polymeric protein containing a *finite*, relatively small, number of *identical* subunits, is said to be an *oligomer*.

(b) The *identical* subunits associated within an oligomeric protein are designated as *protomers*.

(c) The term *monomer* describes the fully dissociated protomer, or of course any protein which is not made up of *identical* subunits.

(d) The term "subunit" is purposely undefined, and may be used to refer to any chemically or physically identifiable sub-molecular entity within a protein, whether identical to, or different from, other components.

Attention must be directed to the fact that these definitions are based exclusively upon considerations of identity of subunits and do not refer to the number of different peptide chains which may be present in the protein. For example, a protein made up of two different peptide chains, each represented only once in the molecule, is a monomer according to the definition. If such a protein were to associate into a molecule which would then contain two chains of each type, the resulting protein would be a dimer (i.e., the lowest class of oligomer) containing two protomers, each protomer in turn being composed of two different peptide chains. Only in the case where an oligomeric protein contains a single type of peptide chain would the definition of a protomer coincide with the chemically definable subunit. An oligomer the protomers of which all occupy exactly equivalent positions in the molecule may be considered as a "closed crystal" involving a fixed number of asymmetric units each containing one protomer.

The model is described by the following statements:

(1) Allosteric proteins are oligomers the protomers of which are associated in such a way that they all occupy equivalent positions. This implies that the molecule possesses at least one axis of symmetry.

(2) To each ligand able to form a *stereospecific* complex with the protein there corresponds one, and only one, site on each protomer. In other words, the symmetry of each set of stereospecific receptors is the same as the symmetry of the molecule.

(3) The conformation of each protomer is constrained by its association with the other protomers.

(4) Two (at least two) states are reversibly accessible to allosteric oligomers. These states differ by the distribution and/or energy of inter-protomer bonds, and therefore also by the conformational constraints imposed upon the protomers.

(5) As a result, the affinity of one (or several) of the stereospecific sites towards the corresponding ligand is altered when a transition occurs from one to the other state.

(6) When the protein goes from one state to another state, its molecular symmetry (including the symmetry of the conformational constraints imposed upon each protomer) is conserved.

Let us first analyse the interactions of such a model protein with a single ligand (F) endowed with differential affinity towards the two accessible states. In the absence of ligand, the two states, symbolized as $R_0$ and $T_0$, are assumed to be in equilibrium. Let $L$ be the equilibrium constant for the $R_0 \rightleftharpoons T_0$ transition. In order to distinguish this constant from the dissociation constants of the ligand, we shall call it the "allosteric constant". Let $K_R$ and $K_T$ be the microscopic dissociation constants of a ligand F bound to a stereospecific site, in the R and T states, respectively. *Note that by reason of symmetry and because the binding of any one ligand molecule is assumed to be intrinsically independent of the binding of any other, these microscopic dissociation constants are the same for all homologous sites in each of the two states.* Assuming $n$ protomers (and therefore $n$ homologous sites) and using the notation $R_0, R_1, R_2, \ldots R_n$; $T_0, T_1, T_2,$

$\ldots T_n$, to designate the complexes involving 0, 1, 2,...$n$ molecules of ligand, we may write the successive equilibria as follows:

$$R_0 \longleftrightarrow T_0$$

$$\begin{array}{llll} R_0 + F \longleftrightarrow R_1 & & T_0 + F \longleftrightarrow T_1 \\ R_1 + F \longleftrightarrow R_2 & & T_1 + F \longleftrightarrow T_2 \\ \cdots\cdots & & \cdots\cdots \\ R_{n-1} + F \longleftrightarrow R_n & & T_{n-1} + F \longleftrightarrow T_n \end{array}$$

Taking into account the probability factors for the dissociations of the $R_1, R_2 \ldots R_n$ and $T_1, T_2 \ldots T_n$ complexes, we may write the following equilibrium equations:

$$T_0 = LR_0$$

$$R_1 = R_0\, n \frac{F}{K_R} \qquad T_1 = T_0\, n \frac{F}{K_T}$$

$$R_2 = R_1 \frac{n-1}{2} \frac{F}{K_R} \qquad T_2 = T_1 \frac{n-1}{2} \frac{F}{K_T}$$

$$\cdots\cdots \qquad \cdots\cdots$$

$$R_n = R_{n-1} \frac{1}{n} \frac{F}{K_R} \qquad T_n = T_{n-1} \frac{1}{n} \frac{F}{K_T}$$

Let us now define two functions corresponding respectively to:

(a) the fraction of protein in the R state:

$$\bar{R} = \frac{R_0 + R_1 + R_2 + \ldots + R_n}{(R_0 + R_1 + R_2 + \ldots + R_n) + (T_0 + T_1 + T_2 + \ldots + T_n)}.$$

(b) the fraction of sites actually bound by the ligand:

$$\bar{Y}_F = \frac{(R_1 + 2R_2 + \ldots + nR_n) \quad + \quad (T_1 + 2T_2 + \ldots + nT_n)}{n[(R_0 + R_1 + R_2 + \ldots + R_n) + (T_0 + T_1 + T_2 + \ldots + T_n)]}.$$

Using the equilibrium equations, and setting

$$\frac{F}{K_R} = \alpha \quad \text{and} \quad \frac{K_R}{K_T} = c$$

we have, for the "function of state" $\bar{R}$:

$$\bar{R} = \frac{(1+\alpha)^n}{L(1+c\alpha)^n + (1+\alpha)^n} \tag{1}$$

and for the "saturation function" $\bar{Y}_F$:

$$\bar{Y}_F = \frac{Lc\alpha(1+c\alpha)^{n-1} + \alpha(1+\alpha)^{n-1}}{L(1+c\alpha)^n \quad + \quad (1+\alpha)^n}. \tag{2}$$

In Fig. 1(a) and (b), theoretical curves of the $Y_F$ function have been drawn, corresponding to various values of the constants $L$ and $c$. In such graphs the co-operative homotropic effect of the ligand, predicted by the symmetry properties of the model, is expressed by the curvature of the lower part of the curves. The graphs illustrate the fact that the "co-operativity" of the ligand depends upon the values of $L$ and $c$. The co-operativity is more marked when the allosteric constant $L$ is large (i.e. when the

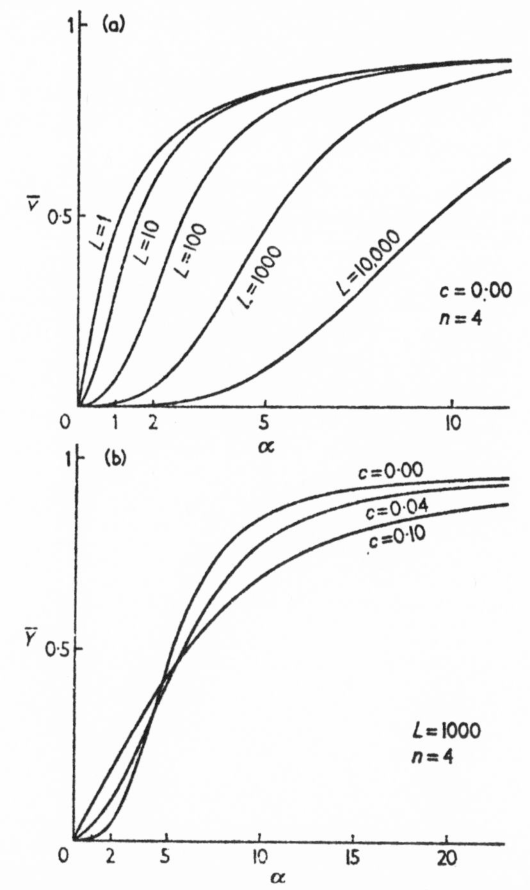


FIG. 1(a) and (b). Theoretical curves of the saturation function $\bar{Y}$ (equation (2)) drawn to various values of the constants $L$ and $c$, with $n = 4$ (i.e. for a tetramer).

$R_0 \rightleftharpoons T_0$ equilibrium is strongly in favour of $T_0$) and when the ratio of the microscopic dissociation constants ($c = K_R/K_T$) is small.†

It should be noted that for $c = 1$ (i.e. when the affinity of both states towards the ligand is the same) and also when $L$ is negligibly small, the $\bar{Y}_F$ function simplifies to:

$$\bar{Y}_F = \frac{\alpha}{1+\alpha} = \frac{F}{K_R + F}$$

that is, to the Michaelis–Henri equation.

The model therefore accounts for the homotropic co-operative effects which, as we pointed out, are almost invariably found with allosteric proteins. Let us now analyse the properties of the model with respect to heterotropic interactions between different allosteric ligands. For this purpose, consider a system involving three stereospecific ligands, each binding at a different site. Assume that one of these ligands is the substrate (S) and, for simplicity, that it has significant affinity only for the sites in one of the two states (for example R). Assume similarly that, of the two other ligands,

† When $c$ is very small, equation (2) simplifies to:

$$\bar{Y}_F = \frac{\alpha(1+\alpha)^{n-1}}{L + (1+\alpha)^n}.$$

one (the inhibitor I) has affinity exclusively for the T state, and the other (the activator A) for the R state. Let $\bar{Y}_S$ be the fractional saturation of the enzyme with S.

According to the model, heterotropic effects would be due exclusively to displacements of the spontaneous equilibrium between the R and T states of the protein. The saturation function for substrate in the presence of activator and inhibitor may then be written as:

$$\bar{Y}_S = \frac{\alpha(1+\alpha)^{n-1}}{L' + (1+\alpha)^n} \tag{3}$$

where $\alpha$ is defined as above and $L'$ is an "apparent allosteric constant", defined as:

$$L' = \frac{\sum_0^n T_I}{\sum_0^n R_A}$$

where $\sum_0^n T_I$ and $\sum_0^n R_A$ stand respectively for the sum of the different complexes of the T state with I and of the R state with A. Following the same derivation as above, it will be seen that:

$$L' = L\frac{(1+\beta)^n}{(1+\gamma)^n}$$

with $\beta = \dfrac{I}{K_I}$ and $\gamma = \dfrac{A}{K_A}$, where $K_I$ and $K_A$ stand for the microscopic dissociation constants of activator and inhibitor with the R and T states respectively. Substituting this value of $L'$ in equation (3) we have:

$$\bar{Y}_S = \frac{\alpha(1+\alpha)^{n-1}}{L\dfrac{(1+\beta)^n}{(1+\gamma)^n} + (1+\alpha)^n}. \tag{4}$$

This equation† expresses the second fundamental property of the model, namely, that the (heterotropic) effect of an allosteric ligand upon the saturation function for another allosteric ligand should be to modify the homotropic interactions of the latter. When the substrate itself is an allosteric ligand (as assumed in the derivation of equation (4)), the presence of the effectors should therefore result in a change of the *shape* of the substrate saturation curve. As is illustrated in Fig. 2, the inhibitor increases the co-operativity of the substrate saturation curve (and also, of course, displaces the half-saturation point), while the activator tends to abolish the co-operativity of substrate (also displacing the half-saturation point). Both the activator and the inhibitor, as well as the substrate, exhibit co-operative homotropic effects.

The model therefore accounts for both homotropic and heterotropic interactions and for their interdependence. Its main interest is to predict these interactions solely on the basis of symmetry considerations. No particular assumption has been, or need be, made about the structure of the specific sites or about the structure of the protein, except that it is a symmetrically bonded oligomer, the symmetry of which is *conserved* when it undergoes a transition from one to another state. It is therefore a fairly stringent, even if abstract model, since co-operative interactions are not only allowed but even required for any ligand endowed with differential affinity towards the two states of the protein, and heterotropic interactions are predicted to occur between any ligands showing homotropic interactions.

† A much more complicated, albeit more realistic, equation would apply if the ligands were assumed to have significant affinity for *both* of the two states.

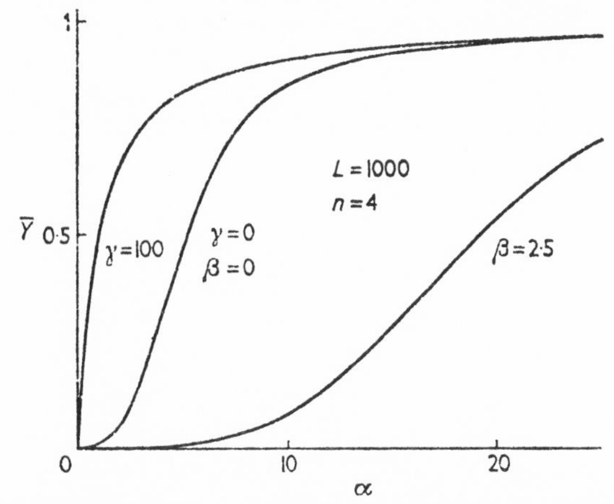


FIG. 2. Theoretical curves showing the heterotropic effects of an allosteric activator ($\gamma$) or inhibitor ($\beta$) upon the shape of the saturation function for substrate ($\alpha$) according to equation (3).

## 3. Application to the Description of Real Systems

### (a) *The kinetics of allosteric systems*

In Fig. 3, results for the fractional saturation of haemoglobin by oxygen at different partial pressures (Lyster, unpublished work) have been fitted to equation (2). While the fit is satisfactory, we feel that strict quantitative agreement is neither sufficient

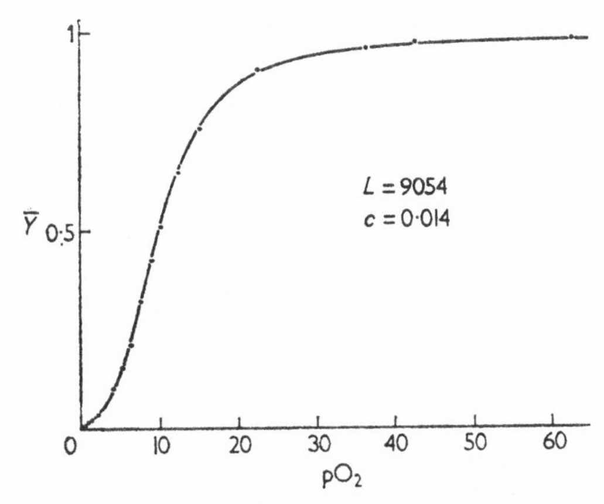


FIG. 3. Saturation of haemoglobin with oxygen. Results (points) obtained by R. W. J. Lyster (unpublished work) with horse haemoglobin (4·6%) in 0·6 M-phosphate buffer (pH 7) at 19°C. Solid line drawn to equation (2) using the values of the constants $L$ and $c$ given on the graph.

nor necessary as a test of the basic assumptions of the model. It must be borne in mind that in almost all enzyme systems, the saturation functions with respect to substrate or effectors cannot be determined directly, and are only inferred from kinetic measurements. (This of course does not apply to the case of haemoglobin just cited.) Very often it is difficult to judge to what extent the inference is correct, and the interpretation of kinetic results in terms of saturation functions sometimes depends upon assumptions about the mechanism of the reaction itself. It is to be expected, then, that most real systems will exhibit appreciable deviations from the theoretical functions, as indeed is very often the case for the much simpler Michaelis–Henri saturation law.

We shall therefore discuss only the most characteristic qualitative predictions of the model in its application to real systems.

In any enzyme system, activating or inhibitory effects are measured in terms of variations of the two classical kinetic constants ($K_M$ and $V_M$), as a function of the concentrations of substrate (S) and effector(s) (F). Two classes of effects may then be expected in allosteric systems.

(a) "K *systems*." Both F and S have differential affinities towards the T and R states (i.e. both F and S are *allosteric* ligands). Then evidently the presence of F will modify the apparent affinity of the protein for S, and conversely.

(b) "V *systems*." S has the same affinity for the two states. Then there is no effect of F on the binding of S, nor of S on the binding of F. F can exert an effect on the reaction only if the two states of the protein differ in their *catalytic* activity. Depending on whether F has maximum affinity for the active or for the inactive state, it will behave as an activator (positive $V$ system) or as an inhibitor (negative $V$ system).

It should be noted that this classification of allosteric systems is compatible with other mechanisms and does not depend upon the specific properties of the model.

The following predictions, however, are based on the distinctive properties of the model.

(a) In an allosteric enzyme system, an *allosteric* effector (i.e. a specific ligand endowed with different affinities towards the two states) should exhibit co-operative homotropic interactions.

(b) In those systems in which an allosteric effector modifies the apparent affinity of the substrate, the substrate also should exhibit co-operative homotropic interactions.

(c) In those systems in which the effector does not modify the affinity of the substrate, the latter should not exhibit homotropic co-operative interactions.

As may be seen from inspection of Table 1, where the properties of a number of systems have been summarized, all four classes of effects (positive and negative $K$ and $V$ systems) have been found among the dozen or so allosteric enzymes adequately studied. In inspecting Table 1, it should be borne in mind that the published data concerning allosteric enzymes are very heterogeneous and often do not provide the kind of information which we are now seeking. Reasonably adequate kinetic data are available, however, for the systems numbered 1 to 8, 10, 13, 14, 16, 18, 19 and 20. In all but two of these 15 systems, homotropic co-operative interactions of at least one of the ligands have been observed. Three of these systems (18, 19 and 20) show no $K$ effect of the inhibitor and no co-operative interactions of substrate, while the $K$ for systems 2 to 8 and 16 show evidence of homotropic interactions for *both* substrate and effector(s), as predicted by the model.†

It is somewhat difficult to judge whether systems 13 and 14 represent true exceptions or not. One of these (glycogen synthetase, no. 14) is a "positive $K$ system", where the occurrence of homotropic interactions might easily be missed. The other (glutamine—F6P transaminase, no. 13) is a negative $K$ system which has not yet been studied extensively. The possible significance of these exceptions will be considered in the general discussion.

† Attention must be directed to the fact that the homotropic effect of a ligand may not be expressed in the absence of an antagonistic ligand. For example, the co-operative interactions of G–1–P, in the case of phosphorylase *b*, are visible only in the presence of ATP (Madsen, 1964).

Table 1

*Summary of properties of various allosteric systems*†

| Enzyme | Substrate | Inhibitor | Activator | *V* System | *K* System | Subunits | References |
|---|---|---|---|---|---|---|---|
| 1. Haemoglobin (vertebrates) (invertebrates) | Oxygen + | | | | | + | Bohr, 1903; Wyman, 1963; Manwell, 1964 |
| 2. Biosynthetic L-threonine deaminase (*E. coli* K12) and (yeast) | L-Threonine + | L-Isoleucine + | L-Valine + | | + | (+) | Umbarger & Brown, 1958*a*; Changeux, 1961,1962,1963,1964*a*,*b*; Freundlich & Umbarger, 1963; Cennamo *et al.*, 1964 |
| 3. Aspartate transcarbamylase (*E. coli*) | Aspartate + Carbamyl phosphate | CTP + | ATP | | + | + | Gerhart & Pardee, 1962,1963,1964 |
| 4. Deoxycytidylate aminohydrolase (ass spleen) | dCMP + | dTTP + | dCTP + | | + | | Scarano *et al.*, 1963,1964; Scarano, 1964; Maley & Maley, 1963,1964 |
| 5. Phosphofructokinase (guinea pig heart) | Fructose-6-phosphate ATP | ATP (+) | 3′-5′ AMP | | + | | Passoneau & Lowry, 1962; Mansour, 1963; Vinuela *et al.*, 1963 |
| 6. Deoxythymidine kinase (*E. coli*) | Deoxythymidine ATP + or GTP − | (dTTP) | dCDP | (+) | + | | Okazaki & Kornberg, 1964 |
| 7. DPN-isocitric dehydrogenase (*N. crassa*) | D-Isocitrate + DPN | (α-Ketoglutarate) | Citrate | | + | | Sanwal *et al.*, 1963,1964 |
| 8. DPN-isocitric dehydrogenase (yeast) | D-Isocitrate + DPN | | 5′ AMP | | + | | Hataway & Atkinson, 1963 |

| | | | | | | | |
|---|---|---|---|---|---|---|---|
| 9. Homoserine dehydrogenase (*R. rubrum*) | Homoserine (−)<br>Aspartate semialdehyde<br>TPN–TPNH | L-Threonine | L-Isoleucine (+)<br>L-Methionine | | + | + | Sturani *et al.*, 1963; Datta *et al.*, 1964 |
| 10. L-Threonine deaminase (*C. tetanomorphum*) | L-Threonine | | ADP + | | + | | Hayaishi *et al.*, 1963 |
| 11. Acetolactate synthetase (*E. coli*) | Pyruvate (−) | L-Valine | | | + | | Umbarger & Brown, 1958*b* |
| 12. "Threonine" aspartokinase (*E. coli*) | Aspartate (−) | L-Threonine | | | + | | Stadtman *et al.*, 1961 |
| 13. L-Glutamine-D-fructose-6-P transaminase (rat liver) | L-Glutamine −<br>D-Fructose-6-P | UDP-N acetyl-glucosamine − | | | + | | Kornfeld *et al.*, 1964 |
| 14. Glycogen synthetase (yeast) (lamb muscle) | UDP-glucose − | | Glucose-6-P − | | + | | Algranati & Cabib, 1962; Traut & Lipmann, 1963 |
| 15. Glutamate dehydrogenase (beef liver) | Glutamate | ATP<br>GTP<br>DPNH<br>Oestrogens +<br>Thyroxine | ADP<br>Leucine +<br>Methionine | + | (+) | + | (Ref. in Tomkins *et al.*, 1963) |
| 16. Phosphorylase *b* (rabbit muscle) | Glucose-1-P +<br>Glycogen<br>$P_i$ (+) | ATP | 5′ AMP + | + | + | + | Helmreich & Cori, 1964; Madsen, 1964; Schwartz (personal communication); Ullmann *et al.*, 1964 |
| 17. UDP-N acetyl-glucosamine-2-epimerase (rat liver) | UDP-N acetyl-glucosamine | CMP-N acetyl-neuraminic acid + | | + | | | Kornfeld *et al.*, 1964 |
| 18. Homoserine dehydrogenase (*E. coli*) | Homoserine −<br>Aspartate semialdehyde<br>TPN-TPNH | L-Threonine + | | + | | + | Patte *et al.*, 1963; Cohen *et al.*, 1963; Patte & Cohen, 1964 |

| Enzyme | Substrate | Inhibitor | Activator | V System | K System | Subunits | References |
|---|---|---|---|---|---|---|---|
| 19. "Lysine"-aspartokinase (*E. coli*) | Aspartate − ATP | L-Lysine + | | + | | | Stadtman *et al.*, 1961; Patte & Cohen, 1964 |
| 20. Fructose-1-6-diphosphatase (frog muscle) (rat liver) | Fructose-1-6-diphosphate (−) | 5′ AMP + | | + | | | Krebs, 1964; Salas *et al.*, 1964; Taketa & Pogell, 1965 |
| 21. ATP-PRPP-pyrophosphorylase (*S. typhimurium*) | ATP − PRPP | Histidine | | + | | + | Martin, 1962 |
| 22. "Tyrosine" 3-deoxy-D-arabinoheptulosonic-acid-7-phosphate synthetase (*E. coli*) | Phosphoenol-pyruvate − D-Erythrose-4-P − | L-Tyrosine | | + | | | Smith *et al.*, 1962 |
| 23. "Phenylalanine" 3-deoxy-D-arabino-heptulosonic-acid-7-phosphate synthetase (*E. coli*) | Phosphoenol-pyruvate − D-Erythrose-4-P − | L-Phenylalanine | | + | | | Smith *et al.*, 1962 |
| 24. Acetyl-CoA carboxylase (rat adipose tissue) | Acetyl CoA ATP, $CO_2$ | | Citrate + | | | + | Martin & Vagelos, 1962 |

† The + and − signs against the name of the substrate(s) and effector(s) of each system indicate whether or not co-operative homotropic effects occur with the corresponding compound. A blank implies no relevant data, while (+) or (−) implies uncertainty. The + signs in the "*K*" and "*V*" columns indicate whether *K* or *V* effects have been observed. In the "subunit" column we have noted with a + those systems for which some evidence (direct or indirect) of the existence of subunits (not necessarily proved to be identical) has been obtained.

Note that (a) this summary is not claimed to be complete; (b) many of the systems listed have been described only recently and as yet incompletely; (c) the properties assigned to many systems represent our (rather than the original authors') interpretation of the data. We therefore assume responsibility for interpretative mistakes.

Let us now examine some of the more specific predictions of the model. According to the theory developed above, the *V* systems are described by the "function of state" ($\bar{R}$ or $1-\bar{R}$), assuming that the two states differ in their catalytic activity towards the substrate. We shall mostly discuss the properties of the *K* systems, of which there are more examples and for which the predictions of the model are particularly interesting and characteristic.

According to the model, the complex kinetics of such systems simply result from displacements of the R $\rightleftharpoons$ T equilibrium, and their properties are described by equation (3). We shall examine only a few typical experimental situations and compare them with predictions based on the model.

Consider first a *K* system involving a substrate and an allosteric inhibitor. Assume that the R state binds the substrate, and the T state binds the inhibitor. We may expect that in any such system, the allosteric constant will be very different from 1. In other words, one of the two states (R or T) will be greatly favoured. Threonine deaminase of *E. coli* is a *K* system, threonine being the substrate, and isoleucine the inhibitor (Umbarger, 1956). In the presence of inhibitor and substrate, the rate–concentration curve for *both* is S-shaped. In the *absence* of inhibitor, the substrate saturation curve is *still* S-shaped. According to the model, this indicates that the favoured state (in the absence of both ligands) is the one that has minimum affinity for threonine and maximum affinity for isoleucine. It is therefore expected that the saturation curve for inhibitor *in the absence of substrate* should be Michaelian, exhibiting no co-operative effect. This prediction has been verified experimentally (Changeux, 1964*a*). Moreover, as shown in Fig. 4, the co-operativity of the inhibitor increases with the concentration of substrate.

More generally, in any *K* system, we expect heterotropic effects to be expressed essentially as alterations of the homotropic "co-operativity" of any one allosteric ligand when in the presence of another. As a measure of homotropic effects, it is convenient to use the Hill approximation:

$$\bar{Y} = \frac{\alpha^{\underline{n}}}{Q + \alpha^{\underline{n}}}$$

where $Q$ is a constant and $\underline{n}$ (the Hill coefficient) is *not* the number of interacting sites (which we write $n$), but an interaction coefficient. It has been shown by one of us (Wyman, 1963) that under certain conditions the Hill coefficient can be interpreted as measuring the free energy of interaction between sites. As it may be seen from Table 2, the Hill coefficients for the substrate of the allosteric system deoxycytidine deaminase are modified, in the expected direction, when the concentration of the other ligands (activator or inhibitor) varies. Another specific prediction of the model has been verified in the case of threonine deaminase, namely, the fact that a *true* competitive inhibitor (allothreonine), i.e. a substrate analogue (able to inhibit the enzyme by binding at the same site as the substrate), should exert the same effect as the substrate itself (L-threonine) as an antagonist of the allosteric inhibitor (isoleucine) (Changeux, 1964*a*). Another prediction, concerning the effect of analogues, is that at very low concentrations of substrate low concentrations of analogue should activate, rather than inhibit, the enzyme. This is observed with aspartic transcarbamylase (Gerhart & Pardee, 1963) (Fig. 5) and also with threonine deaminase (Changeux, 1964*a*).

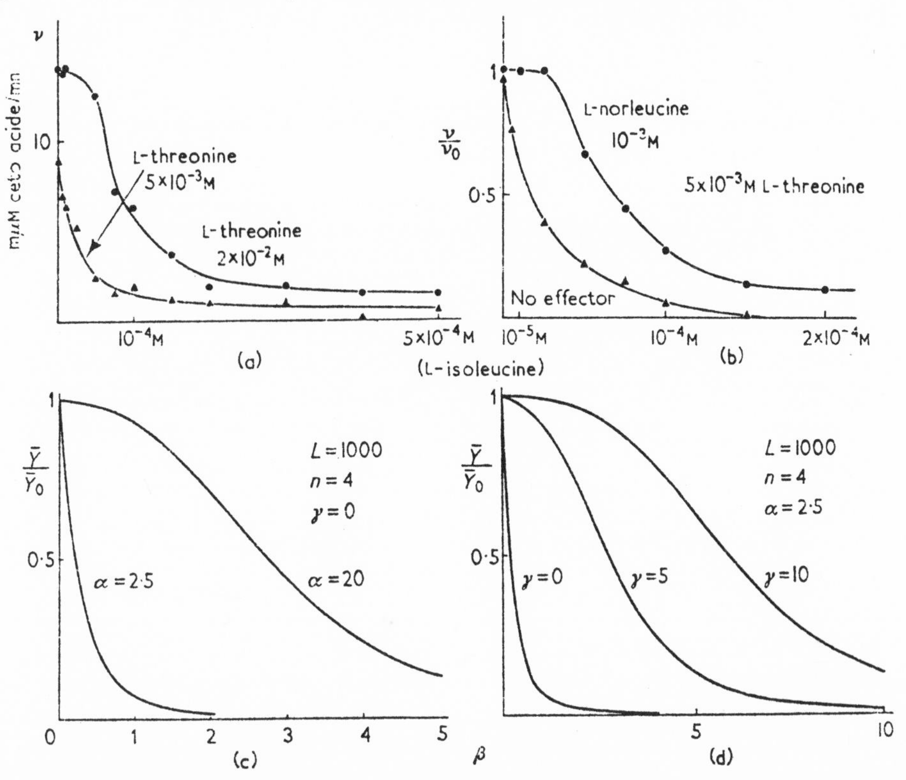


FIG. 4. Effects of the allosteric inhibitor L-isoleucine upon the activity of L-threonine deaminase.
(a) In the presence of two different concentrations of the substrate (L-threonine).
(b) At low concentration of substrate in the presence or absence of the allosteric activator L-norleucine.
Compare with theoretical curves (c and d) describing similar situations according to equation (3). Note that at low concentrations of substrate the co-operative effect of the inhibitor is scarcely detectable either in the theoretical or in the experimental curves. An increase of the concentration of substrate, or the addition of an activator, both reveal the co-operative effects of the inhibitor.

TABLE 2

*Hill coefficients of homotropic interactions with respect to substrate* (n), *inhibitor* (n′) *and activator* (n″) *observed with dCMP deaminase*

(*From Scarano* et al., *1963; Scarano, 1964*)

| | | **n** |
|---|---|---|
| Substrate (dCMP) | No effector | 2·0 |
| | + dTTP 1·25 μM | 3·0 |
| | ,, 2·25 μM | 4·1 |
| | ,, 10·00 μM | 3·9 |
| | + dCTP 100·00 μM | 1·0 |
| | | **n′** |
| Inhibitor (dTTP) | Substrate concentration 4 mM | 3·4 |
| | | **n″** |
| Activator (dCTP) | Substrate concentration 67 μM | 2·0 |

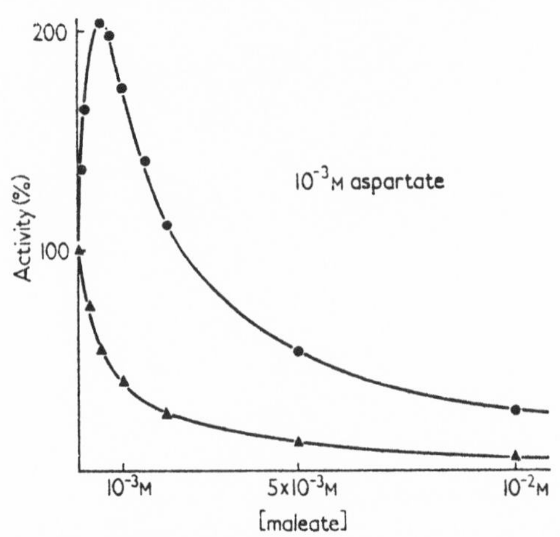


**Fig. 5. Effect of a substrate analogue (maleate) upon the activity of aspartic transcarbamylase at relatively low concentration of substrate (aspartate).**

**Upper curve: native enzyme. Lower curve: desensitized enzyme.**

**Note the large increase of activity at low maleate concentration which occurs with the native enzyme, but not with the desensitized enzyme (data from Gerhart & Pardee, 1963).**

The effect of an allosteric *activator* in a $K$ system should be, according to the model, to decrease or abolish the substrate–substrate interactions. This has been observed in several different systems. As illustrated in Fig. 6, the effect is particularly striking because, as expected, at saturating concentration of activator it results in converting the S-shaped rate–concentration curve for substrate into a Michaelian hyperbola. Moreover, of course, the presence of an activator should increase the co-operativity of an inhibitor, and conversely. Both effects are observed (see Figs 4 and 7).

It is clear from the model and the equations that the homotropic interactions of an allosteric ligand are independent of the absolute values of the microscopic dissociation constants. One may therefore expect that two sterically closely analogous ligands could bind to the same sites with the same interaction coefficient, even though their affinities might be widely different. For example, with haemoglobin, the functionally significant steric features of the prosthetic groups must be virtually the same, whether the haems are bound to oxygen or to carbon monoxide. Therefore, although the affinity of carbon monoxide for the haem is known to be nearly 250 times that of oxygen, we should expect the interaction coefficients to be the same for both, as indeed they are (Wyman, 1948). When, however, the binding of two analogous ligands depends very much on steric factors, it may be expected that the *ratios* of the affinities of each ligand towards the two states of the protein (i.e. the constant $c$ in equation (1)) will be different. If so, the two ligands might bind to the same sites with widely different interaction coefficients. This appears to be the case, according to the observations of Okazaki & Kornberg (1964) for various triphosphonucleosides which act as phosphoryl donors in the deoxythymidine–kinase reaction. With ATP, for example, the rate–concentration curve is strongly co-operative, whereas with dATP the curves exhibit scarcely any evidence of homotropic effects. Furthermore, this enzyme, as shown by the same authors, is allosterically activated by CDP (Fig. 6(c)). It is easily seen that if this effect conforms to the model, activation should be observed only with those substrates that show evidence of homotropic effects (ATP), and not with those that do not (dATP). This, actually, is the observed result.

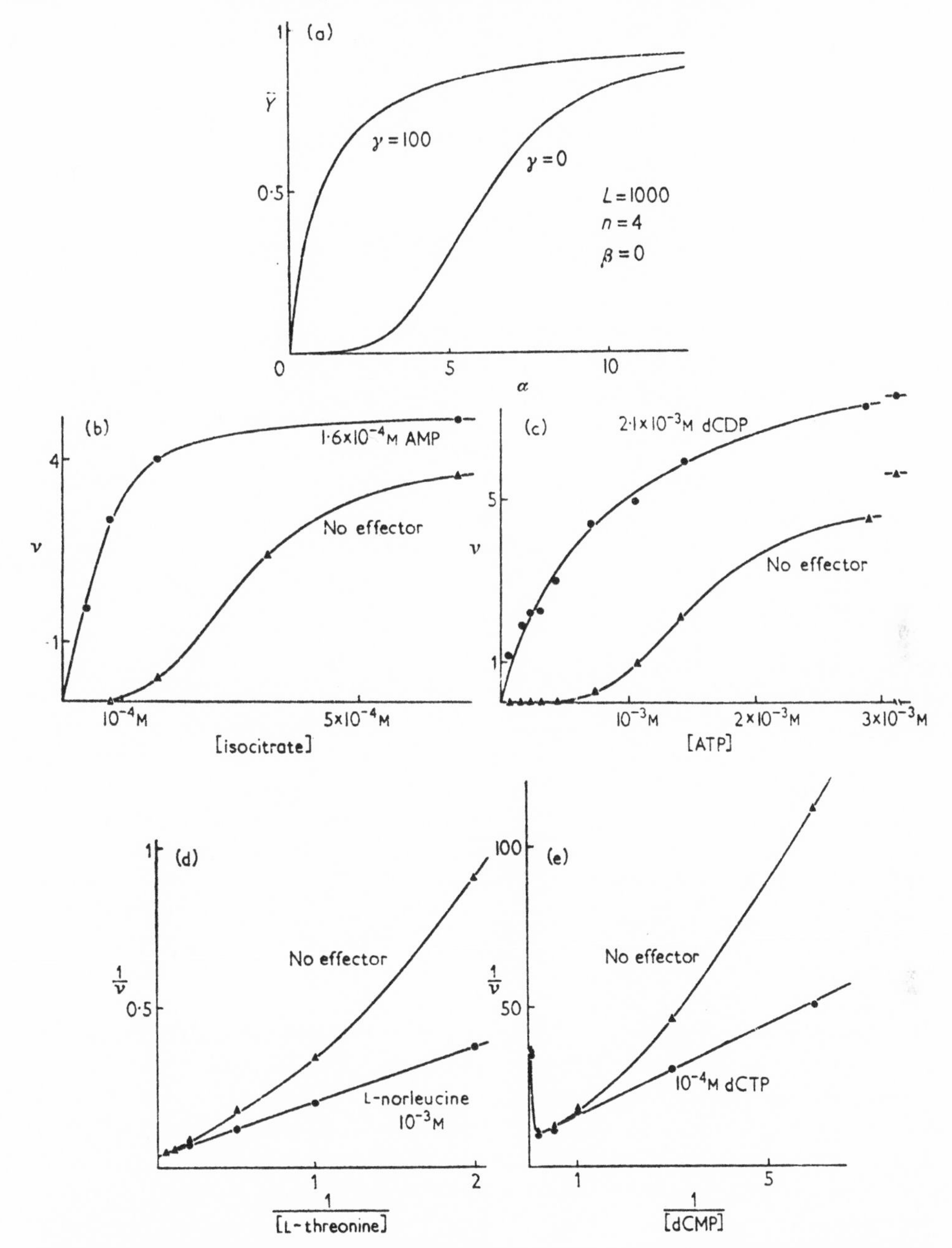


**Fig. 6.** Activity of various allosteric enzymes as a function of substrate concentration in the presence or absence of their respective activators.

(a) Theoretical curve according to equation (3).

(b) DPN-isocitrate dehydrogenase from *Neurospora crassa* (results from Hataway & Atkinson, 1963).

(c) Deoxythymidine kinase from *Escherichia coli* (results from Okazaki & Kornberg, 1964).

(d) Biosynthetic L-threonine deaminase from *E. coli* (Lineweaver–Burk plot) (results from Changeux, 1962,1963).

(e) dCMP deaminase from ass spleen (Lineweaver–Burk plot) (results from Scarano *et al.*, 1963).

Note that in all these instances, the presence of the allosteric activator abolishes the co-operative interactions of the substrate.

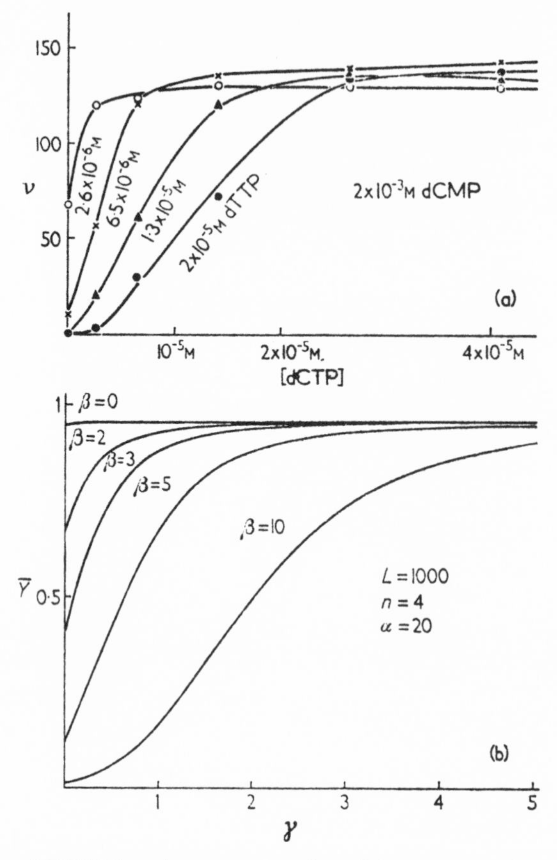


**Fig. 7.** (a) Activity of dCMP deaminase as a function of the concentration of its all ostericactivator dCTP in the presence of substrate (dCMP) at near saturating concentration and at various concentrations of the allosteric inhibitor dTTP (results from Scarano, personal communication).

(b) Theoretical curves of equation (3) corresponding to a similar situation.

Note that the co-operative effects of the activator are revealed only at relatively high concentration of the inhibitor.

Since, again, the homotropic interactions are independent of absolute affinities, certain conditions or agents may modify the affinity of an allosteric ligand without altering its interaction coefficient. This is apparently the case for the Bohr effect shown by haemoglobin: as is well known, the oxygen saturation curves obtained at different values of pH can all be superimposed by a simple, adequately chosen, change of the abscissa scale. In terms of the model, this would mean that the binding of the "Bohr protons" does not alter the equilibrium between the two hypothetical states of the protein. Hence also the Bohr protons themselves would *not* be allosteric ligands, and their own binding is not expected to be co-operative. This, again, appears to be the case, at least for human and horse haemoglobin (Wyman, *loc. cit.*).

In the preceding paragraph, we have discussed only the more straightforward predictions of the model. It should be pointed out that the model could also account for more complicated situations, and for certain effects which were not considered here. For example, it seems possible that, in some instances, the phenomenon of inhibition by excess substrate might be due to an allosteric mechanism (rather than to the classically invoked direct interaction between two substrate molecules at the active site).

This effect could be described on the basis of our model by assuming two states with different affinities for the substrate, the one with higher affinity being catalytically inactive. The equation for such a situation would be of the form:

$$\frac{V}{V_m} = \frac{L\,S/K_a\,(1+S/K_a)^{n-1}}{L(1+S/K_a)^n + (1+S/K_I)^n}$$

with $K_I$ (dissociation constant of S with the inactive state) smaller than $K_a$ (dissociation constant with the active state).

### (b) *Desensitization and dissociation*

One of the most striking facts about allosteric enzymes is that their regulatory properties may be lost as a result of various treatments, without loss (indeed often with increase) of activity (Changeux, 1961; Gerhart & Pardee, 1961). That it should be so is understandable on the basis of the model, since conservation of the interactions should depend upon the integrity of the whole native structure, including in particular the inter-protomer binding, whereas conservation of activity should depend only on the integrity of the active site. Also, according to the model, the homotropic and heterotropic interactions should in general be simultaneously affected, if at all, by alterations of protein structure. This was first observed with threonine deaminase (Changeux, 1961) and ATCase (Gerhart & Pardee, 1962), and similar observations have since then been made with several other systems. These observations constituted the main initial basis for the assumption that regulatory interactions in general may be indirect (Changeux, 1961; Monod & Jacob, 1961; Monod *et al.*, 1963).

According to the model, loss of the interactions would follow from any structural alteration that would make one of the two states (R or T) virtually inaccessible. Now, one of the events most likely to result from various treatments of the protein is that quaternary (inter-protomer) bonds may be broken, completely or partially. One may therefore expect that:

(a) Under any condition, or following any alteration, such that the protein is (and remains) dissociated, *both* types of interactions should disappear.

(b) Conversely treatments, or mutations, which abolish the interactions should frequently be found to result in stabilization of a monomeric state.

These expectations are verified by observations made with at least two different systems (Gerhart & Pardee, 1963,1964; Patte, Le Bras, Loviny & Cohen, 1963; Cohen & Patte, 1963).

Furthermore, since it is assumed from the model that in one of the two alternative states (R) the protomers are less constrained and therefore closer to the conformation of the monomer than in the other state (T), we expect that, under conditions where the protein is monomeric, it may exhibit high affinity for the ligand which stabilizes the R state, and little or no affinity for the ligand which stabilizes the T state. Hence, if the experiment can be performed, one may deduce *which* of the two states (R or T) is stabilized by a given ligand.

If conditions can be set up such that reversible dissociation of the protein actually occurs, one may expect that an allosteric ligand (i.e. any ligand exhibiting homotropic interactions) should now prove to act as a specific associative or dissociative agent. Actually, there is now clear evidence that under conditions where human haemoglobin shows a detectable amount of dissociation (low pH, high ionic strength), dissociation is favoured by oxygenation (Antonini, Wyman, Belleli & Caputo, unpublished experiments, 1961; Benesch, Benesch & Williamson, 1962; Gilbert & Chionione, recent unpublished experiments). Lamprey haemoglobin, in the oxygenated form, exists primarily as a monomer under all conditions, but when deoxygenated shows a strong tendency to polymerize (see Table 3) (Briehl, 1963; Rumen, 1963). Myoglobin, which may be thought of as an isolated (and therefore relaxed) protomer of haemoglobin, has a much higher oxygen affinity, as would be expected on the basis of these two facts regarding human and lamprey haemoglobin.

**Table 3**

*Sedimentation coefficients of oxygenated and reduced lamprey haemoglobin (from Briehl, 1963)*

| $t°$ C | pH | Haemoglobin concentration (E. 275) | Sedimentation coefficient ($S_{20,w}$) | |
|---|---|---|---|---|
| | | | Oxygenated | Reduced |
| 5·5 | 6·8 | 15·7 | 2·02 | 3·68 |
| 5·0 | 7·3 | 21·0 | 1·90 | 2·98 |

Similarly, Changeux (1963) has found that in the presence of urea (1·5 M) threonine deaminase is reversibly dissociable. As expected, under these conditions, all three types of allosteric ligands active in the system, namely the substrate (threonine or analogue of threonine), the activator (valine) and the inhibitor (isoleucine) powerfully affect the dissociation, the inhibitor favouring the associated state, whereas both the substrate and the activator appear to stabilize the dissociated state. Hence, under normal conditions, the substrate and the activator presumably stabilize an R state, while the inhibitor favours a T state.

The observations of Datta, Gest & Segal (1964) on homoserine dehydrogenase from *Rhodospirillum rubrum* provide a further striking example of the effects of allosteric ligands upon dissociation of the protein. This enzyme is activated by both methionine and isoleucine, and inhibited by threonine. Both activators, as well as the substrate, promote dissociation of the protein, whereas the inhibitor favours an aggregated state.

We may conclude from the preceding discussions that the characteristic, unusual, apparently complex functional properties of allosteric systems can be adequately systematized and predicted on the basis of simple assumptions regarding the molecular symmetry of oligomeric proteins. In the next section, we shall examine the structural implications and the plausibility of these assumptions from a more general point of view.

## 4. Quaternary Structure and Molecular Symmetry of Oligomeric Proteins

### (a) *Geometry of inter-protomer bonding*

The first major assumption of the model is that the association between protomers in an oligomer may be such as to confer an element of symmetry on the molecule. The plausibility of this assumption has already been pointed out by Caspar (1963) and by Crick & Orgel (1964, and unpublished manuscript). We will analyse the implications of this assumption in terms of the possible or probable modes of bonding between protomers. Although next to nothing is known, from direct evidence, regarding this problem, the following statements would seem to be generally valid.

(a) A large number, probably a majority, of enzyme proteins are oligomers involving several *identical* subunits, i.e. protomers (see: Schachman, 1963; Reithel, 1963; Brookhaven Symposium, 1964, *Subunit Structure of Proteins*, no. 17).

(b) In most cases the association between protomers in such proteins does not appear to involve covalent bonds.

(c) Yet most oligomeric proteins are stable as such (i.e. do not dissociate into true monomers, or associate into superaggregates), over a wide range of concentrations and conditions.

(d) The specificity of association is extreme: monomers of a normally oligomeric protein will recognize their identical partners and re-associate, even at high dilution, in the presence of other proteins (e.g. in crude cell extracts).

These properties indicate that within oligomeric proteins the protomers are in general linked by a *multiplicity* of non-covalent bonds, conferring both specificity and stability on the association. Clearly also the steric features of the bonded areas must play a major part.

Let us now distinguish between two *a priori* possible modes of association between two protomers. For this purpose we define as a "binding set" the spatially organized collection of all the groups or residues of *one* protomer which are involved in its binding to *one* other protomer. Considered together, the two linked binding sets through which two protomers are associated will be called the *domain of bonding* of the pair.

The two modes of association which we wish to distinguish may then be defined as follows.

(a) *Heterologous associations:* the domain of bonding is made up of two *different* binding sets.

(b) *Isologous associations:* the domain of bonding involves two *identical* binding sets.

These definitions imply the following consequences.†

(1) In an isologous association (Figs 8 and 9), the domain of bonding has a two-fold axis of rotational symmetry. Along this axis, homologous residues (i.e. identical residues occupying the same position in the primary structure) face each other (and may form unpaired "axial" bonds). Anywhere else, within the domain of bonding, any bonded group-pair is represented *twice*, and the two pairs are symmetrical with respect to each other. Put more generally: in an isologous association, any group which contributes to the binding in one protomer furnishes precisely the same contribution in the other protomer. Isologous associations will therefore tend to give rise to "closed" i.e. *finite* polymers since, for example, an isologous dimer can further polymerize only by using "new" binding sets (i.e. areas and groups not already satisfied in the dimer). Note that this mode of association can give rise only to *even numbered* oligomers.

† The validity of the statements that follow can be visualized and demonstrated best with the use of models. The interested reader may find it helpful to use a set of dice for this purpose.

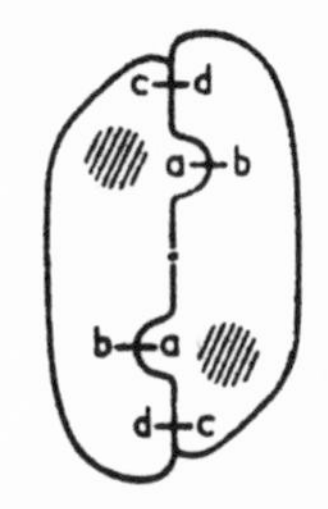

I- Isologous association

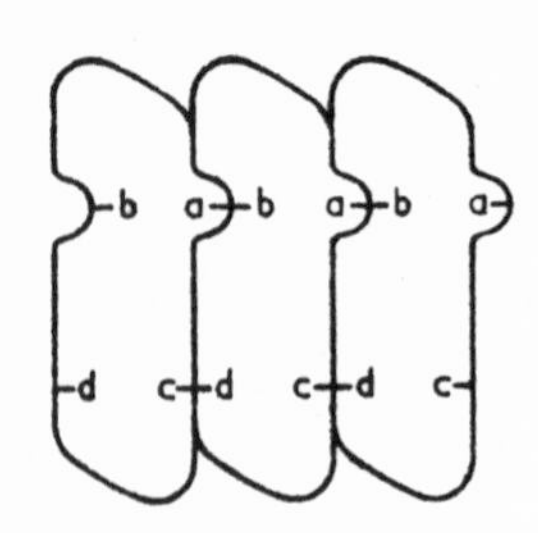

II-Heterologous association

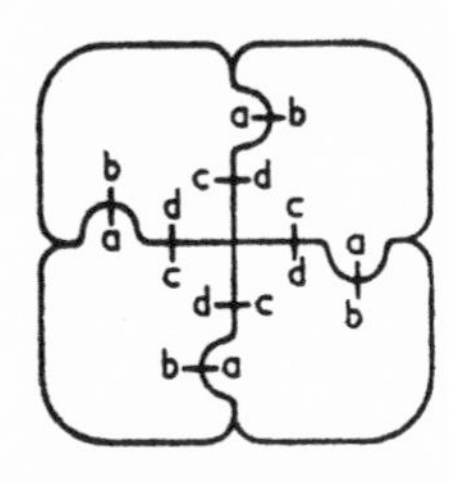

III-Heterologous tetramer

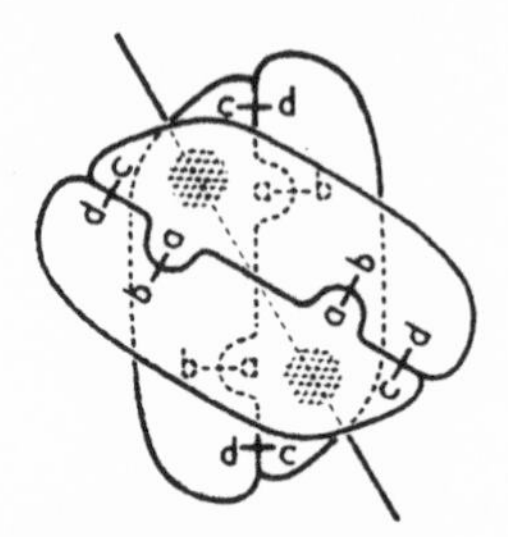

IV-Isologous tetramer
(pseudotetrahedral)

**Fig. 8. Isologous and heterologous associations between protomers.**

**Upper left:** an isologous dimer. The axis of symmetry is perpendicular to the plane of the Figure.

**Upper right:** "infinite" heterologous association.

**Lower left:** "finite" heterologous association, leading to a tetramer with an axis of symmetry perpendicular to the plane of the Figure.

**Lower right:** a tetramer constructed by using isologous associations only. Note that two different domains of bonding are involved.

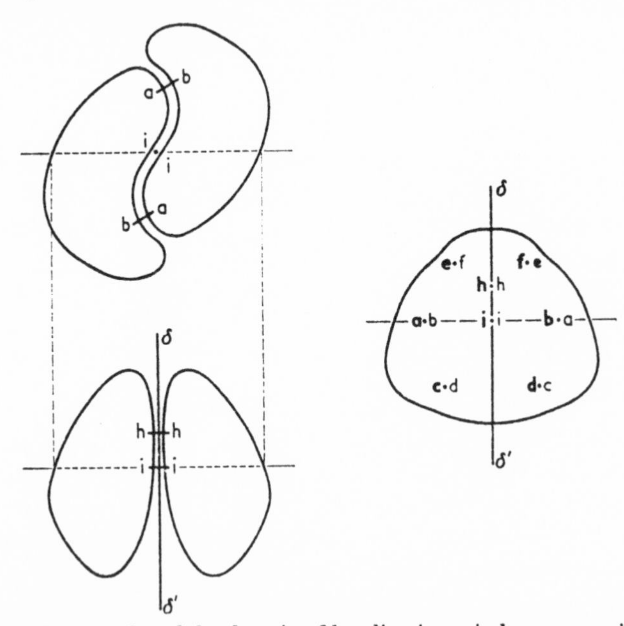


Fig. 9. Topography of the domain of bonding in an isologous association.

**Upper left:** represented in a plane perpendicular to the axis of symmetry.

**Lower left:** the same viewed in a plane of the axis of symmetry.

**Right:** projection of the domain of bonding in a plane of the axis of symmetry.

**hh** and ii, axial bonds; ab, ba, cd, dc, ef, fe, antiparallel bonds. It should be understood that in this Figure the bonding residues a, b, c, etc. are supposed to project from under and from above the plane of the Figure.

(2) In a heterologous association (Fig. 8), the domain of bonding has no element of symmetry; each bonded group-pair is unique. Heterologous associations would, in general, be expected to give rise to polydisperse, eventually large, helical polymers except, however, in two cases.

(a) If polymerization is stopped at some point by steric hindrance, giving rise to a "hinged helix". Such aesthetically unpleasant structures should have less stability than "closed" structures.

(b) If a "closed" structure can be achieved such that any binding set which is used by one protomer is also satisfied in all the others. This is impossible of course in a dimer, but it is possible for trimers, tetramers, pentamers, etc., provided that the angles defined by the domains of bonding are right or nearly so. Such an oligomer would necessarily possess an axis of symmetry.

On the basis of these considerations, it is reasonable to assume that, if an oligomeric protein possesses a wide range of stability, it consists of a closed structure where all the protomers use the same binding sets; which implies, as we have just seen, that the molecule should possess at least one axis of symmetry.

Direct experimental evidence on this important problem is available for haemoglobin. As is well known, although made up of four subunits, haemoglobin is not, strictly speaking, a tetramer, since the $\alpha$ and $\beta$ chains are not identical. For our present purposes, however, we may consider the four subunits as equivalent protomers. The work of Perutz *et al.* (1960) has shown that these are associated into a pseudotetrahedral structure which possesses a twofold axis of symmetry.

Three further examples of oligomers possessing an element of molecular symmetry have recently been provided. Green & Aschaffenburg (1959) find that $\beta$-lactoglobulin (a dimer) has a dyad axis. Lacticodehydrogenase M4 (Pickles, Jeffery & Rossmann, 1964) and glyceraldehyde-phosphate-dehydrogenase (Watson & Banaszak, 1964), both tetrameric, appear to possess one (at least) axis of symmetry.

From the preceding discussion, and on the strength of these examples, it appears that oligomeric proteins are not only capable of assuming molecular symmetry, but also that this may be a fairly general rule.

Assuming this conclusion to be correct, it is of interest to enquire which mode of association (isologous or heterologous) may be most frequently used in Nature. For the reasons pointed out above, stable dimers, of which many examples are known, must represent isologous associations. Moreover, it may be pointed out that a symmetrical (isologous) dimer can further polymerize into a *closed* structure in two ways only.

(a) By again using isologous associations, thereby forming an isologous tetramer. Isologous polymerization, however, must stop at this point, since no further closed structure could be built by polymerization of such a tetramer.

(b) By using heterologous associations when the next closed structure would necessarily consist of three isologous dimers, and hence be an hexamer.

It follows from these remarks that (1) the exclusive use of isologous associations can lead to dimers and tetramers only; (2) the use of *both* isologous and heterologous domains of bonding should lead to even-numbered oligomers containing a *minimum* of six protomers; (3) the exclusive use of heterologous domains of bonding could lead to oligomers containing any number of protomers (except two). On this basis, the apparently rather wide prevalence of dimers and tetramers among oligomeric enzymes suggests rather strongly that the quaternary structures of these proteins are mostly built up by isologous polymerization.

### (b) *Protomer conformation*: "*quaternary constraints*"

The formation of stable, specific associations involving multiple bonds and strict complementarity between protein protomers is likely to imply in most cases a certain amount of re-arrangement of the tertiary structures of the monomers. Certain observations seem to confirm this assumption.

(1) The artificially prepared monomers of enzymes that are normally oligomeric generally exhibit functional alterations, suggesting that the structure of the active site in each protomer depends upon a conformation which exists only in the native oligomeric associated state (see Brookhaven Symposium, 1964, *Subunit Structure of Proteins, 17*).

(2) The rate of reactivation of oligomeric enzymes inactivated by dissociation into monomers is markedly dependent on temperature (alkaline phosphatase, Levinthal, Signer & Fetherolf, 1962; β-galactosidase, Perrin, manuscript in preparation; phosphorylase b, Ullmann, unpublished work). Since the association reaction does not involve the formation of covalent bonds, the temperature dependence of the rate is to be attributed, presumably, to a "conformational" transition state.

(3) The phenomenon of intra-cistronic complementation between different mutants of the same protein appears, as pointed out by Crick & Orgel (1964), to be due to a repair of altered structures which results from association between differently altered monomers of a (normally oligomeric) protein. Note that this interpretation necessarily implies, as pointed out by the authors, that the domain of bonding has an axis of symmetry.

It is reasonable therefore to consider that the conformation of each protomer in an oligomer is somewhat "constrained" by, and dependent upon, its association with other protomers. (An excellent discussion of this concept, as applied to haem proteins, is given by Lumry (1965).) In a symmetrical oligomer, all the protomers are engaged by the same binding sets and submitted to the same "quaternary constraints"; they should therefore adopt the same conformations. By contrast, in any non-symmetrical association, identical monomers would, as protomers, assume somewhat *different* conformations and cease to be truly equivalent. Thus, symmetry of bonding is to be regarded as a condition, as well as a result, of the structural equivalence of subunits in an oligomer.

These remarks justify the assumption that the specific biological properties of an oligomer depend in part upon its quaternary structure, and that the protomers will be functionally as well as structurally equivalent if, and only if, they are symmetrically associated within the molecule.

The last assumption of the model, namely, that in an allosteric transition the symmetry of quaternary bonding, and therefore the equivalence of the protomers, should tend to be conserved, may now be considered. Let us analyse the meaning and evaluate the possible range of validity of this postulate.

Consider a symmetrical oligomer (for simplicity, a dimer) wherein the conformation of each protomer is constrained and stabilized by the quaternary bonds (T). If these constraints were relaxed (i.e. the bonds broken) each protomer would tend towards an alternative conformation (R), involving certain tertiary bonds which were absent in the other configuration. The transition may be written:

$$\mathrm{TT} \xrightleftharpoons{\Delta F_1} \mathrm{X} \xrightleftharpoons{\Delta F_2} \mathrm{RR}$$

where TT and RR stand for two symmetrical configurations and X for one (or several) non-symmetrical intermediate states. To say that symmetry should "tend to be conserved" is to imply that the occurrence of the R $\rightleftharpoons$ T transition in one of the protomers should facilitate the occurrence of the same transition in the other. This would be the case of course if the intermediate state(s) X were less stable than either one of the symmetrical states; but it would also be the case, even if the X state were more stable than one of the symmetrical states, provided only that the $\Delta F$ of the first transition (from one of the symmetrical states to the intermediate state) were more positive than the second.

It is easy to see that the dissociation of a symmetrical oligomer should in general satisfy this condition. This may conveniently be symbolized as in Fig. 10, where each subunit is represented as an arrow and only a minimum number of bonds is shown—

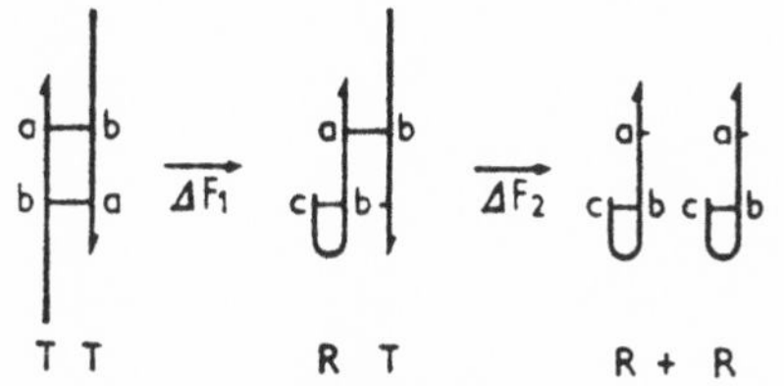

FIG. 10

actually two symmetrical (antiparallel) inter-protomer bonds (ab and ba) and one intra-chain bond (cb) the presence or absence of which is taken to characterize two distinct conformations (R and T) available to each subunit.

Although the symmetry of the protomers would not be conserved after dissociation into monomers, their equivalence would be, and the transition itself is symmetrical since it involves the breaking (or formation) of symmetrical bonds and symmetrical suppression (or creation) of identical quaternary constraints. The free energy of each of the two transitions may then be considered to involve two contributions: one ($\Delta F_b$), assignable to the breaking and formation of individual bonds, the other ($\Delta F_x$) associated with the freedom gained or lost by the protomers in respect to one another. By reason of symmetry, $\Delta F_b$ would be the same for both transitions, while $\Delta F_x$ would not. Since, in the example chosen, the second transition involves dissociation, the entropy gained in this step would be larger than in the first, and the sum of the two contributions would give $\Delta F_1 > \Delta F_2$, satisfying the condition of co-operativity. A ligand able to stabilize either the R or the T state would in turn exert homotropic co-operative effects upon the equilibrium.

There are examples in the literature of co-operative effects of this kind. The best illustration may be the muscle phosphorylase conversions which involve, as is well known, the formation of a tetrameric molecule (phosphorylase *a*) from the dimeric phosphorylase *b*. The conversion occurs when phosphorylase *b* is phosphorylated (in the presence of ATP and phosphorylase kinase). As expected since it is a tetramer, phosphorylase *a* contains four phosphoryl groups. Krebs & Fischer (personal communication) have observed that, when the amount of ATP used in the reaction is sufficient to phosphorylate only a fraction of the (serine) acceptor residues, a *stoicheiometric amount* of fully phosphorylated tetrameric phosphorylase *a* is formed, while the excess protein remains dimeric and unphosphorylated. Another striking illustration of co-operative effects upon dissociation is provided by the work of Madsen & Cori (1956), who observed that phosphorylase *a* would dissociate into monomers in the presence of parachloromercuribenzoate, and showed that when the amounts of mercurial used were insufficient to dissociate all the protein, the remaining non-dissociated fraction did not contain any mercuribenzoate.

Reversible allosteric transitions however do not, in the majority of known cases, involve actual dissociation of the protomers. A transition between two undissociated symmetrical states of an oligomer would, nevertheless, be co-operative if it were adequately symbolizable, for example as in Fig. 11, which expresses the assumption

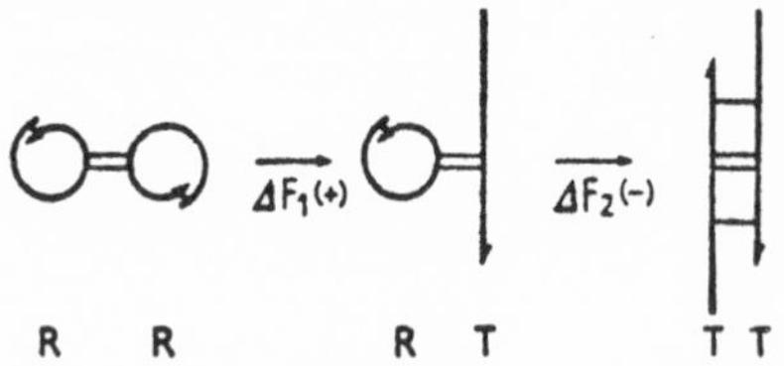

$\Delta F_1(+)$ $\Delta F_2(-)$

R R R T T T

FIG. 11

that one of the alternative conformations (T in this case) is stable only when held by quaternary bonds which could be formed only at the price of breaking *symmetrically* certain tertiary bonds present in the other configuration (R).

In such a system, the $\Delta F$ of the first transition would be positive, the second negative, and the intermediate state (RT) therefore less stable than either one of the symmetrical states. Such a system could be very highly co-operative, and the strong homotropic interactions observed with many real systems† suggests that they may conform to such a pattern.

However, Fig. 12 symbolizes a much more general pattern of symmetrical transitions which is interesting to consider.

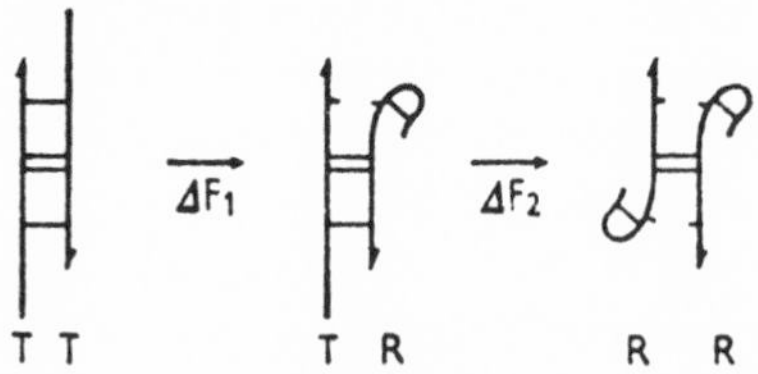

$\Delta F_1$ $\Delta F_2$

T T T R R R

FIG. 12

Here again the free energy assignable to the formation and breaking of individual bonds is the same in both transitions. Whether the RR ⇌ TT transition will be co-operative, non-co-operative, or anti-co-operative, should then depend entirely on the entropy term associated with the degrees of mutual freedom gained or lost by the protomers in each transition. If these entropy terms were equal in absolute value and of the same sign for both transitions (or if they were negligible) the system would be non-co-operative. In general, however, one would expect these two terms to be

† That is, when the Hill coefficient ($n$) approaches the value corresponding to the actual number of protomers.

unequal and of significant magnitude in at least one of the transitions. The system would then be co-operative whenever the second entropy term was more positive than the first, and anti-co-operative in the reverse case.

The first possibility appears more likely on general grounds, since it seems reasonable to believe that certain degrees of mutual freedom, in a symmetrical dimer, may be held by either one of two (or more) symmetrical quaternary bonds, and liberated only when *both* are broken. Such a system would be closely comparable to a dissociating system, and it is interesting to note in this respect that in certain allosteric systems actual dissociation is observed under certain "extreme" conditions, whereas it is not seen under more normal conditions (see p. 104).

The possibility should also be considered that certain allosteric transitions might not involve a non-symmetrical intermediate. Such transitions would have to involve the initial breaking of axial bonds, eventually perhaps leading to, or allowing, the symmetrical breaking of symmetrical bonds as pictured in Fig. 13. Such a mechanism would necessarily be co-operative.

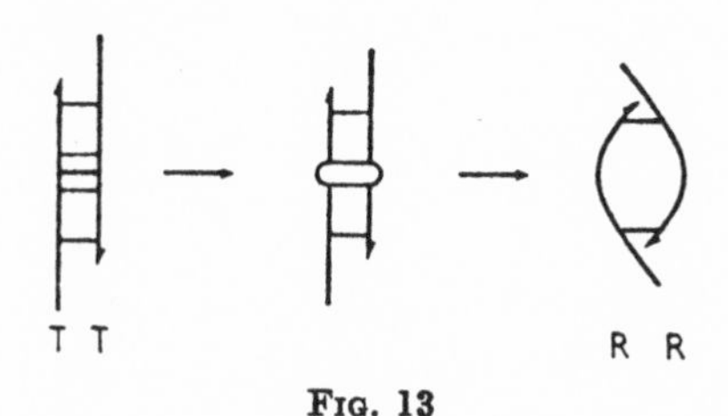


FIG. 13

It is impossible to say, at the present time, whether the co-operative homotropic effects observed with real systems are better described by one or other pattern of symmetrical transitions. One might hope, however, to identify or eliminate some of these mechanisms by adequate thermodynamic and kinetic studies (using fast-mixing techniques) of the transition itself. It is clear in any event that none of these descriptions could apply to a non-symmetrically bonded oligomer, the protomers of which would have to assume different conformations and could not therefore undergo co-operatively the same transitions. On this basis, the fact that allosteric ligands appear invariably to exert co-operative homotropic effects may be taken as experimental evidence that the transitions which they stabilize occur in a symmetrical structure; indeed it was pointed out several years ago by one of us (Allen, Guthe & Wyman, 1950) that the symmetry properties of the oxygen saturation function for haemoglobin appeared to reveal the existence of elements of structural symmetry in the molecule itself. This inference was proved correct when the structure of haemoglobin was elucidated. Moreover, the recent work of Muirhead & Perutz (1963) and Perutz, Bolton, Diamond, Muirhead & Watson (1964) has shown that while the *quaternary* structure of haemoglobin is very significantly different in the oxygenated *versus* the reduced state, the molecular symmetry of the tetramer is conserved in both states. These observations would give a virtually complete illustration of the model if the X-ray pictures also showed some evidence of concomitant alterations of the tertiary structure of the protomers. This has not been observed; but it is reasonable to assume that a functionally significant allosteric transition need not involve more than a very small structural alteration of the protomers. In other words, given the very close and numerous intra-chain interactions, it would not be surprising that the quaternary constraints, even if strong, should not be expressed at the present level of resolution (5·5 Å) of the X-ray pictures. It is also possible that the quaternary constraints might not force any significant *sensu stricto* "conformational" alteration of the protomers, but only, for example, a (symmetrical) redistribution of charge within the molecule. We wish to point out that the assumptions of the model would remain valid also in such a case, and that the adjective "conformational" which we have used extensively (for lack of a better one) to qualify allosteric transitions, should be understood in its widest connotation.

## 5. General Discussion

In the preceding discussion we have tried to show, first that the functional properties of regulatory enzymes could be accounted for on the assumption that the quaternary structures of oligomeric proteins involve an element of symmetry in many, if not most, proteins made up of identical subunits (that is, presumably, in the majority of enzymes). We may now consider the problem in reverse and ask why molecular evolution should have so frequently favoured the appearance and maintenance of oligomeric globular proteins.

That it should be so must mean that there are functional advantages of some kind, inherent in the oligomeric state, and absent or difficult to achieve in the monomeric state. If most or all oligomeric proteins were endowed with the property of mediating allosteric interactions, especially homotropic interactions, we might believe that we had an answer to the question. Actually most of the enzymes known to be oligomeric are not, or at least are not known to be, allosteric. One should note, however, that the capacity to mediate physiologically significant interactions might be more frequent and widespread among proteins than has been realized so far. As we have seen, these properties are frequently very labile and may easily be lost during extraction and purification of an enzyme. Furthermore, it is conceivable that the effector for certain proteins may be an unknown or simply an improbable metabolite, if not, in some cases, another cellular protein (cf. Lehninger, 1964).

It probably remains true, however, that most oligomeric proteins are not endowed with specific regulatory functions. One must therefore presume that there are some other, more general, advantageous properties associated with the oligomeric state.

This problem may be related to the even more general question: Why should enzymes be so large, as compared with the size of their stereospecific sites? It seems reasonable to believe that two factors in particular contribute to determining a minimum size for enzymes. One is the requirement of fixing a very precise position in space for the several residues which together constitute the stereospecific site. Not only does this involve the necessity of a peptide chain with enough degrees of freedom (i.e. long enough) to allow the precise relative arrangement of these residues, but also the use of a further length of peptide to freeze these degrees of freedom, thereby conferring enough rigidity (i.e. specificity) upon the site. Another factor probably is the requirement that a given protein should *not* tend to associate more or less indiscriminately with other cellular proteins. As Pauling has pointed out, proteins are inherently "sticky", and the structure of enzymes must have been selected against the tendency to form random aggregates. Such a "purpose" may be, in part, fulfilled by decreasing the surface–volume ratio, and also by putting the polar groups on the surface, thereby increasing the solubility.

Now, association between monomers may evidently also contribute both to the fixation of an adequate structure and to a decrease in the surface–volume ratio, as well as to the covering-up of the hydrophobic areas of the monomers. Moreover, it is evidently more economical to achieve such results, whenever possible, by associating monomers rather than by increasing the unit molecular weight (i.e. the molecular weight per active centre).

These selective factors should therefore have favoured in general the appearance of closed (i.e. symmetrical) oligomers, since "open" structures (potentially infinite and polydisperse) would be disadvantageous in the case of most enzymes. Isologous (rather than heterologous) polymerization may have been frequently preferred for the same reasons, since this type of association leads to closed structures exclusively and, in the process of evolution from a monomeric to a polymeric state, it is evidently easier to start at the dimer stage (at which a heterologous association is still necessarily open), rather than right away at a higher stage.

However, the most decisive factor in the emergence and selective maintenance of symmetrical oligomeric proteins may have been the inherent co-operativity of their structure. To illustrate this point, consider schematically the events which may lead to the formation of a primitive dimer from a monomer.

On the surface of a protein monomer, any particular area contains a variety of randomly distributed groups, many of which may possess inherent chemical affinity for another one in the area. Since the *distance* between any two such groups is necessarily the same in two individual monomers, antiparallel association of the two pairs whenever possible would satisfy simultaneously two such valencies, creating a dimer involving two bonds and possessing a dyad axis. Furthermore, since this applies to *any* pair of groups capable of forming a bond, the monomers have a choice of *any one* of the mutually attractive pairs to achieve such a structure. Even so, the primitive dimer may not be formed, or might remain very unstable, because of the presence, within the area of contact of the protomers, of mutually repulsive groups. These pairs of groups would be distributed symmetrically about the dyad axis defined by the first two, mutually attractive, pairs. Therefore any mutation of *one* residue, conferring upon it the capacity to form a bond with its partner, would result in *two* new bonds being achieved in the dimer. Because of the interactions through "quaternary constraints" between the conformation of each protomer and the structure of their common domain of bonding, any such mutational event would affect symmetrically and co-operatively the functional properties of each of the two protomers. It is clear that, because of these reciprocal interactions, the same general reasoning applies to any mutation which might, even very discretely, affect the conformation of the protomers, including in particular the steric features of the domain of bonding which must of course play an important part in the stability of the association. Thus the structural and functional effects of single mutations occurring in a symmetrical oligomer, or allowing its formation, should be greatly amplified as compared with the effects of similar mutations in a monomer or in a non-symmetrical oligomer. In other words, because of the inherent co-operativity of their structure, symmetrical oligomers should constitute particularly sensitive targets for molecular evolution, allowing much stronger selective pressures to operate in the random pursuit of functionally adequate structures.

We feel that these considerations may account, in part at least, for the fact that most enzyme proteins actually are oligomeric; and if this conclusion is correct, the homotropic co-operative effects which seem at first to "characterize" allosteric systems should perhaps be considered only as one particular expression of the advantageous amplifying properties associated with molecular symmetry.

The same general argument may account for the fact that (apart from one or two possible exceptions) allosteric proteins have invariably been found to mediate *both* heterotropic and homotropic interactions, which implies of course that they are oligomeric. It should be clear from the discussion of the model that heterotropic interactions could *a priori* be mediated by a monomeric protein possessing two (necessarily different) binding sites, associated with two different "tautomeric" states of the molecule. If, for example, one of the states were stabilized by the substrate and the other by some other specific ligand, the latter would act as a competitive inhibitor. The saturation function ($\bar{Y}_s$) would then simplify into:

$$\bar{Y}_s = \frac{\alpha}{L(1+\beta)+1+\alpha}$$

which we write only to indicate that, for $n = 1$ (i.e. for a monomer) the model *formally* allows heterotropic effects to occur, but not of course, homotropic effects.

Just as the effect of a single amino-acid substitution will be greater in a symmetrical oligomer than in a monomer, the stabilization by a specific ligand of an alternative conformation, implying a significant increase of potential, may be possible in an oligomer when it would not be, for lack of co-operativity, in a monomeric protein. The fact that *both* heterotropic and homotropic interactions disappear when an allosteric protein is "desensitized" as a result of various treatments may be considered to illustrate this point, and actually constitutes one of the main experimental justifications of the model. It might be said in other words, that the molecular symmetry of allosteric proteins is used to amplify and effectively translate a very low-energy signal.†

In addition, it is clear that the sigmoidal shape of the saturation curve characteristic of homotropic interactions may in itself offer a significant physiological advantage, since it provides the possibility of threshold effects in regulation. This property is of course essential in the case of haemoglobin, and it seems very likely that it has an important role in most, if not all, regulatory enzymes. Selection, in fact, must have operated on these molecules, not only to favour the structures which allow homotropic interactions, but actually to determine very precisely the energy of these interactions according to metabolic requirements.

The selective "choice" of oligomers as mediators of chemical signals therefore seems to be justified (*a posteriori*) by the fact that certain desirable physical and physiological properties are associated with symmetry, and therefore inaccessible to a monomeric protein.

We should perhaps point out here again that in the present discussion, as in the model, we accept the postulate that a monomeric protein or a protomer does not possess more than *one stereospecific site* able to bind a given ligand. That this postulate

† Consider for example an allosteric system with an intrinsic equilibrium constant ($L = T_0/R_0$) of 1000. Assume, that the R state has affinity $1/K_R$ for a ligand F, and set $F/K_R = \alpha$. In the presence of the ligand, the ratio of the two states will be: $\frac{\sum T}{\sum R} = \frac{1000}{(1+\alpha)^n}$. Taking $\alpha = 9$, for example, we would have, for a tetramer, $\frac{\sum T}{\sum R} = 0{\cdot}1$. In order to reach the same value for the $T/R$ ratio with a monomeric system, the concentration of F would have to be more than one thousand times larger.

does apply to stereospecific sites is amply documented (cf. Schachman, 1963) and need not be discussed at length here. It is obvious of course that, lacking symmetry, a monomer or an individual protomer cannot present two or more *identical* elements of tertiary structure of any kind.

The postulate, however, does not apply to *group-specific*, as opposed to *stereospecific*, ligands. Homotropic interactions of various kinds (not necessarily co-operative) may therefore occur in the binding of group-specific ligands (such as SH reagents, detergents, ions, etc.) whether the protein is monomeric or not. As is well known, the vast literature on the denaturation of proteins is replete with descriptions of multimolecular effects exerted by various group-specific reagents. It may be worth noting in this respect that in the last analysis, the co-operative effects of such reagents are accounted for by the simultaneous attack of numerous bonds occupying functionally similar (although not geometrically symmetrical) positions in the molecule.

The significance of this generalization may be made clear by considering the melting of double-stranded DNA. This is a typically co-operative phenomenon the co-operativity of which is evidently dependent upon and expresses the (helical) symmetry of the "domain of bonding" between the two strands in the Watson–Crick model. In the last analysis therefore, the axial symmetry requirement for homotropic co-operative effects to occur with a globular protein, when *stereospecific* ligands are concerned, reflects essentially the fact that, in general, only one stereospecific site able to bind such a ligand exists on a protein monomer or protomer.

Gerhart (1964) and Schachman (1964) have recently reported the successful separation, from crystalline aspartic transcarbamylase, of two different subunits, one of which bears the specific receptor for aspartate, and the other the receptor for CTP. It is very tempting to speculate on the possibility that this remarkable and so far unique observation may in fact correspond to a general rule, namely, that a protein should contain as many different subunits (peptide chains) as it bears stereospecifically different receptor sites. The emergence and evolution of such structures, by association of primitively distinct entities, would be much easier to understand than the acquisition of a new stereospecific site by an already existing and functional enzyme made up of a single type of subunit.

We have so far not discussed one of the major assumptions of the model, namely, that allosteric effects are due to the displacement of an equilibrium between discrete states assumed to exist, at least potentially, apart from the binding of a ligand. The main value of this treatment is to allow one to define, in terms of the allosteric constant, the contribution of the protein itself to the interaction, as distinct from the dissociation constants of the ligands. This distinction is a useful and meaningful one, as we have seen, and its validity is directly justified by the fact that the affinity of a ligand may vary widely without any alteration of its homotropic interaction coefficient (cf. page 103). But it should be understood that the "state" of the protein may not in fact be exactly the same whether it is actually bound, or unbound, to the ligand which stabilizes it. In this sense particularly, the model offers only an over-simplified first approximation of real systems, and it may prove possible in some cases to introduce corrections and refinements by taking into consideration more than two accessible states.

We feel, however, that the main interest of the model which we have discussed here does not reside so much in the possibility of describing quantitatively and in detail the complex kinetics of allosteric systems. It rests rather on the concept, which we have tried to develop and justify, that a general and initially simple relationship between symmetry and function may explain the emergence, evolution and properties of oligomeric proteins as "molecular amplifiers", of both random structural accidents and of highly specific, organized, metabolic interactions.

This work has benefited greatly from helpful discussions and suggestions made by our friends and colleagues Drs R. Baldwin, S. Brenner, F. H. C. Crick, F. Jacob, M. Kamen, J. C. Kendrew, A. Kepes, L. Orgel, M. F. Perutz, A. Ullmann. We wish to thank Mr F. Bernède for his kindness in performing many calculations with the computer.

The work was supported by grants from the National Institutes of Health, National Science Foundation, Jane Coffin Childs Memorial Fund, Délégation Générale à la Recherche Scientifique et Technique and Commissariat à l'Energie Atomique.

## REFERENCES

Allen, D. W., Guthe, K. F. & Wyman, J. (1950). *J. Biol. Chem.* **187**, 393.

Algranati, I. & Cabib, E. (1962). *J. Biol. Chem.* **237**, 1007.

Benesch, R. E., Benesch, R. & Williamson, M. E. (1962). *Proc. Nat. Acad. Sci., Wash.* **48**, 2071.

Bohr, C. (1903). *Zentr. Physiol.* **17**, 682.

Briehl, R. W. (1963). *J. Biol. Chem.* **238**, 2361.

Caspar, D. L. D. (1963). *Advanc. Protein Chem.* **18**, 37.

Cennamo, C., Boll, M. & Holzer, H. (1964). *Biochem. Z.* **340**, 125.

Changeux, J. P. (1961). *Cold Spr. Harb. Symp. Quant. Biol.* **26**, 313.

Changeux, J. P. (1962). *J. Mol. Biol.* **4**, 220.

Changeux, J. P. (1963). *Cold Spr. Harb. Symp. Quant. Biol.* **28**, 497.

Changeux, J. P. (1964*a*). Thèse Doctorat ès Sciences, Paris. To be published in *Bull. Soc. Chim. Biol.*

Changeux, J. P. (1964*b*). *Brookhaven Symp. Biol.* **17**, 232.

Cohen, G. N. & Patte, J. C. (1963). *Cold Spr. Harb. Symp. Quant. Biol.* **28**, 513.

Cohen, G. N., Patte, J. C., Truffa-Bachi, P., Sawas, C. & Doudoroff, M. (1963). Colloque International C.N.R.S. *Mécanismes de régulation des activités cellulaires chez les microorganismes.* Marseille: in the press.

Crick, F. & Orgel, L. L. (1964). *J. Mol. Biol.* **8**, 161.

Datta, P., Gest, H. & Segal, H. (1964). *Proc. Nat. Acad. Sci., Wash.* **51**, 125.

Ferry, R. M. & Green, A. A. (1929). *J. Biol. Chem.* **81**, 175.

Freundlich, M. & Umbarger, H. E. (1963). *Cold Spr. Harb. Symp. Quant. Biol.* **28**, 505.

Gerhart, J. C. (1964). *Brookhaven Symp. Biol.* **17**, 232.

Gerhart, J. C. & Pardee, A. B. (1961). *Fed. Proc.* **20**, 224.

Gerhart, J. C. & Pardee, A. B. (1962). *J. Biol. Chem.* **237**, 891.

Gerhart, J. C. & Pardee, A. B. (1963). *Cold Spr. Harb. Symp. Quant. Biol.* **28**, 491.

Gerhart, J. C. & Pardee, A. B. (1964). *Fed. Proc.* **23**, 727.

Green, D. W. & Aschaffenburg, R. (1959). *J. Mol. Biol.* **1**, 54.

Hataway, J. A. & Atkinson, D. E. (1963). *J. Biol. Chem.* **238**, 2875.

Hayaishi, O., Gefter, M. & Weissbach, H. (1963). *J. Biol. Chem.* **236**, 2040.

Helmreich, E. & Cori, C. F. (1964). *Proc. Nat. Acad. Sci., Wash.* **51**, 131.

Kornfeld, S., Kornfeld, R., Neufeld, E. F. & O'Brien, P. J. (1964). *Proc. Nat. Acad. Sci., Wash.* **52**, 371.

Krebs, H. (1964). *Proc. Roy. Soc.* B, **159**, 545.

Lehninger, A. (1964). Centenaire Société de Chimie Biologique, *Bull. Soc. Chim. Biol.*, in the press.

Levinthal, C., Signer, E. R. & Fetherolf, K. (1962). *Proc. Nat. Acad. Sci., Wash.* **48**, 1230.

Lumry, R. (1965). *Nature*, in the press.

Madsen, N. B. (1964). *Biochem. Biophys. Res. Comm.* **15**, 390.

Madsen, N. B. & Cori, C. F. (1956). *J. Biol. Chem.* **223**, 1055.

Maley, F. & Maley, G. F. (1963). *Science*, **141**, 1278.

Maley, G. F. & Maley, F. (1964). *J. Biol. Chem.* **239**, 1168.

Mansour, T. E. (1963). *J. Biol. Chem.* **238**, 2285.
Manwell, C. (1964). *Oxygen in the Animal Organism*. London: Pergamon Press.
Martin, D. B. & Vagelos, P. R. (1962). *J. Biol. Chem.* **237**, 1787.
Martin, R. G. (1962). *J. Biol. Chem.* **237**, 257.
Monod, J., Changeux, J. P. & Jacob, F. (1963). *J. Mol. Biol.* **6**, 306.
Monod, J. & Jacob, F. (1961). *Cold Spr. Harb. Symp. Quant. Biol.* **26**, 389.
Muirhead, H. & Perutz, M. F. (1963). *Nature*, **199**, 633.
Okazaki, R. & Kornberg, A. (1964). *J. Biol. Chem.* **239**, 269.
Passoneau, J. & Lowry, O. (1962). *Biochem. Biophys. Res. Comm.* **7**, 10.
Patte, J. C. & Cohen, G. N. (1964). *C. R. Acad. Sci. Paris*, **259**, 1255.
Patte, J. C., Le Bras, G., Loviny, T. & Cohen, G. N. (1963). *Biochim. biophys. Acta*, **67**, 16.
Perutz, M. F., Bolton, W., Diamond, R., Muirhead, H. & Watson, H. C. (1964). *Nature*, **203**, 687.
Perutz, M. F., Rossmann, M. G., Cullis, A. F., Muirhead, H., Will, G. & North, A. C. T. (1960). *Nature*, **185**, 416.
Pickles, B., Jeffery, B. A. & Rossmann, M. G. (1964). *J. Mol. Biol.* **9**, 598.
Reithel, F. J. (1963). *Advanc. Protein Chem.* **18**, 124.
Rumen, N. M. (1963). *Fed. Proc.* **22**, 681.
Salas, M., Vinuela, E., Salas, J. & Sols, A. (1964). *Biochem. Biophys. Res. Comm.* **17**, 150.
Sanwal, B., Zink, M. & Stachow, C. (1963). *Biochem. Biophys. Res. Comm.* **12**, 510.
Sanwal, B., Zink, M. & Stachow, C. (1964). *J. Biol. Chem.* **239**, 1597.
Scarano, E. (1964). Sixth Intern. Congress Biochem., New York: Pergamon Press, in the press.
Scarano, E., Geraci, G., Polzella, A. & Campanile, E. (1963). *J. Biol. Chem.* **238**, 1556.
Scarano, E., Geraci, G. & Rossi, M. (1964). *Biochem. Biophys. Res. Comm.* **16**, 239.
Schachman, H. (1963). *Cold Spr. Harb. Symp. Quant. Biol.* **28**, 409.
Schachman, H. (1964). Sixth Intern. Congress Biochem., New York: Pergamon Press, in the press.
Smith, L. C., Ravel, J. M., Lax, S. & Shive, W. (1962). *J. Biol. Chem.* **237**, 3566.
Stadtman, E. R., Cohen, G. N., Le Bras, G. & de Robichon-Szulmajster, H. (1961). *J. Biol. Chem.* **236**, 2033.
Sturani, E., Datta, F., Hugues, M. & Gest, H. (1963). *Science*, **141**, 1053.
Taketa, K. & Pogell, B. M. (1965). *J. Biol. Chem.* **240**, 651.
Tomkins, G. M., Yielding, K. L., Talal, N. & Curran, J. F. (1963). *Cold Spr. Harb. Symp. Quant. Biol.* **28**, 461.
Traut, R. & Lipmann, F. (1963). *J. Biol. Chem.* **238**, 1213.
Ullmann, A., Vagelos, P. R. & Monod, J. (1964). *Biochem. Biophys. Res. Comm.* **17**, 86.
Umbarger, H. E. (1956). *Science*, **123**, 848.
Umbarger, H. E. & Brown, B. (1958*a*). *J. Biol. Chem.* **233**, 415.
Umbarger, H. E. & Brown, B. (1958*b*). *J. Biol. Chem.* **233**, 1156.
Vinuela, E., Salas, M. L. & Sols, A. (1963). *Biochem. Biophys. Res. Comm.* **12**, 140.
Watson, H. C. & Banaszak, L. J. (1964). *Nature*, **204**, 918.
Wyman, J. (1948). *Advanc. Protein Chem.* **4**, 407.
Wyman, J. (1963). *Cold Spr. Harb. Symp. Quant. Biol.* **28**, 483.

Reprinted from *Biochemistry* **11**:2378–2388 (1972)

# A Monomeric, Allosteric Enzyme with a Single Polypeptide Chain. Ribonucleotide Reductase of *Lactobacillus leichmannii*†

Democleia Panagou, M. D. Orr,‡ J. R. Dunstone, and R. L. Blakley*

ABSTRACT: Ribonucleoside triphosphate reductase has been purified from *Lactobacillus leichmannii* on a large scale as well as on the conventional scale. The final preparation has been shown homogeneous by behavior in plateau gel filtration, velocity sedimentation as examined by several criteria, equilibrium sedimentation, and polyacrylamide gel electrophoresis. The mean value of the molecular weight determined by all methods is 76,000. Equilibrium sedimentation under various denaturing conditions and polyacrylamide gel electrophoresis in the presence of 0.1% sodium dodecyl sulfate indicated that the protein cannot be disaggregated into subunits. The amino acid analysis is consistent with a molecular weight of 76,000 and indicates the presence of eight methionine residues. Cleavage with cyanogen bromide gave the expected nine peptides as shown by polyacrylamide gel electrophoresis and by gel filtration. The only detectable N-terminal residue is serine and the only C-terminus lysine. No aggregation of the enzyme occurred in the absence or presence of allosteric modifiers under the experimental conditions employed, except under denaturing conditions.

The ribonucleotide reductase of *Lactobacillus leichmannii* catalyzes reduction of GTP, ATP, CTP, ITP, and to a much smaller extent, UTP by dithiols such as dihydrolipoate or by reduced thioredoxin, a low molecular weight protein with two sulfhydryls (Blakley and Vitols, 1968). This reduction is absolutely dependent on the presence of deoxyadenosylcobalamin (coenzyme $B_{12}$). Several of the deoxynucleoside triphosphate products specifically activate reduction of particular ribonucleotides. For example, dGTP specifically activates ATP reduction and dATP specifically activates CTP reduction. On the basis of kinetic studies it has been suggested (Goulian and Beck, 1966; Beck, 1967; Vitols *et al.*, 1967a) that the deoxynucleotides produce such activation by binding at a site other than the catalytic site, that is at an allosteric or regulatory site. Kinetic evidence has also been obtained (Vitols *et al.*, 1967a) which suggests that ribonucleoside triphosphates also bind at such a site.

Binding of deoxynucleotide modifiers at the regulatory site has a profound effect on the interaction of cobamides with the enzyme. Thus, binding of cob(II)alamin ($B_{12r}$) to the active site is greatly enhanced by such modifiers, each of which probably also determines a specific conformation of the enzyme–cobamide complex (Hamilton *et al.*, 1971; Yamada *et al.*, 1971). Exchange of hydrogen between water and the cobalt-bound methylene group of 5′-deoxyadenosylcobalamin is dependent on a modifier nucleotide (Hogenkamp *et al.*, 1968), and degradation of deoxyadenosylcobalamin in presence of enzyme and dithiol to 5′-deoxyadenosine and cob(II)alamin, presumably *via* an active intermediate, occurs only in presence of a modifier (Hamilton *et al.*, 1971; Yamada *et al.*, 1971).

Many enzymes subject to regulatory control, including most of those studied intensively, have proven to be oligomeric proteins. However, estimates of the molecular weight of ribonucleotide reductase of *L. leichmannii* (Goulian and Beck, 1966; Vitols *et al.*, 1967b) indicate that this enzyme is unlikely to contain many subunits, and indeed suggest that it might be monomeric. It was therefore of interest to redetermine the molecular weight of the lactobacillus reductase, to determine whether any aggregation or gross conformational changes occur in the presence of modifiers and to establish the number of polypeptide chains constituting the catalytically active enzyme.

## Materials

Materials were obtained commercially as follows: sodium dodecyl sulfate and urea, Fisher (both chemicals were recrystallized from 95% ethanol); guanidine hydrochloride (enzyme grade), ammonium sulfate, imidazole, and diisopropyl fluorophosphate (DFP) treated bovine pancreatic carboxypeptidase A, Mann Research Laboratories; 3,3-dimethylglutaric acid, Tris, serononin creatinine sulfate, didimethylaminonaphthalenesulfonyl chloride (dansyl chlo-

† From the Department of Biochemistry, College of Medicine, University of Iowa, Iowa City, Iowa, and from the Departments of Biochemistry and of Physical Biochemistry, John Curtin School of Medical Research, Australian National University, Canberra, Australia. *Received February 28, 1972.* This is contribution IX in a series on cobamides and ribonucleotide reduction. This work was supported in part by U. S. Public Health Service Grant CA-11165, and by a grant from the National Institutes of Health General Research Support to the University of Iowa College of Medicine.

‡ Australian National University Scholar.

* To whom correspondence should be addressed at the Department of Biochemistry, College of Medicine, University of Iowa.

ride), danyslamino acids, ribonuclease A, cytochrome *c*, 1,4-dithiothreitol, and DFP-treated hog pancreatic carboxypeptidase B, Sigma Chemical Co.; *N*-ethylmorpholine, Eastman Organic Chemicals (this chemical was redistilled before use); cyanogen bromide, Aldrich Chemical Co.; tritiated water, New England Nuclear Corp.; nucleotides, P-L Biochemicals; human serum albumin, Pentex Inc.; human γ-globulin, Commonwealth Serum Laboratories, Melbourne, Australia; protamine sulfate, Sigma and Krishell Laboratories; crystalline rabbit muscle creatine kinase, Worthington; ovalbumin, Sigma and Worthington.

## Methods

*Purification of Ribonucleotide Reductase.* The enzyme was prepared essentially according to the method previously described (Vitols *et al.*, 1967a) but with several modifications which increased the purification while permitting adaption to a large scale procedure. The method has been used extensively on the small scale (120 g of cell paste) and several times on the large scale (4–5 kg). In the following, quantities applicable to both scales are given. The early stages of the first two large-scale preparations were carried out at the New England Enzyme Center, Boston.

The organism and its growth and harvesting were as previously described (Blakley, 1965; Vitols *et al.*, 1967a), except as follows. The organism was maintained by monthly transfers in stab cultures prepared by addition of agar (1.5 g/100 ml) to the liquid medium used for growth. The original medium (Blakley, 1965) was modified by decreasing the casein hydrolysate to 1.5 g/l. and adding L-tyrosine (50 mg/l.), the amino acid which analyses suggested most likely to be present in limiting amounts. Casamino Acids low in cyanocobalamin must be used in order to avoid repression of reductase synthesis (Ghambeer and Blakley, 1966). The minimum concentration of added cyanocobalamin necessary for optimum growth was determined with each batch of casamino acids and varied from 50 to 0 ng per l. When the absorbance at 660 nm reached 1.1, the culture was chilled rapidly (below 20° within 15 min) and was kept below 20° during harvesting which was completed within 1.5 hr. Since experiments showed that the reductase content of cells in suspension slowly declined even at 5°, the cell paste was not washed but immediately frozen in liquid nitrogen or Dry Ice. Cells were grown in the laboratory in volumes of medium up to 60 l. Large batches (two 860-l. portions) were grown by the Grain Processing Co., Muscatine, Iowa. In both cases the yield was about 2 g of cell paste/l.

Frozen cells were thawed in the cold room (5°) overnight or in a few hours by standing the containers in water at room temperature. The thawed cells were suspended in five to eight volumes of 0.1 M potassium phosphate buffer (pH 7.3) and disrupted either in a Ribi cell fractionator at 20,000 psi or, on the large scale, by two passages through a Manton-Gaulin mill. The insoluble material was removed by centrifugation (2 hr at 25,000*g*) at 5° and all further steps were carried out at 5°.

The soluble extract was dialyzed overnight against 0.01 M potassium phosphate buffer (pH 7.3; at least four volumes), and diluted with the same buffer to a protein concentration of 10 mg/ml. A protamine sulfate solution was prepared by stirring 2 g of the soild/100 ml of water and adjusting to pH 5.6 at room temperature with 5 N KOH. It is not necessary to remove undissolved material. The dialyzed enzyme solution in batches that could be centrifuged at one time (1–4 l.) was adjusted to pH 5.6 with 5 N acetic acid. The freshly prepared protamine sulfate solution (at 25°) was then run into the vigorously stirred enzyme solution (at 0–5°) during a 15- to 20-min period. Stirring was continued for 15 min after all the protamine sulfate was added and the mixture then centrifuged for 10 min (25,000*g*). The optimum volume of protamine sulfate solution to be added must be determined in a preliminary trial with a series of 10-ml portions of enzyme solution, and varies from 0.05 to 0.16 volume depending on the batch of cells and the batch of protamine sulfate. The supernatant from the centrifugation was readjusted to pH 7.3 with 5 N NaOH as soon as possible and treated with 0.1 volume of 0.1 M Na EDTA (pH 7.3).

Enzyme was next precipitated from the combined solutions obtained in the previous step by addition of solid ammonium sulfate at 1°. A total of 360 g/l. was added slowly with constant stirring, and stirring was continued at 0° for 1 hr after all the crystals had dissolved. After recovering the precipitate by centrifugation (15 min at 25,000*g*), a minimum volume of 0.01 M Tris-acetate buffer (pH 7.3), containing 1 mM EDTA, was used to partially dissolve and suspend the precipitate. The mixture was dialyzed overnight against the same buffer, with the use of dialysis tubing previously soaked for at least 2 hr in 0.1 M EDTA (pH 7).

The deep yellow dialyzed solution was next subject to gel filtration on Sephadex G-100 equilibrated with 0.01 M Tris-acetate buffer (pH 7.3). In small-scale preparations a 50- to 100-ml sample was passed through a 4 × 140 cm column with collection of 15-ml fractions. On the large scale a 500-ml sample was applied to a 16 × 130 cm column with collection of 250-ml fractions. The latter column was constructed from Plexiglass tubing and the gel was packed over a disk of porous polypropylene sheet ($^3/_{16}$ in. thick, Labpor sheet, Bel Art Products). Fines must be thoroughly removed from the Sephadex so that a flow rate of about 500 ml/hr is obtained with the large column or 30 ml/hr with the small column. At this flow rate the enzyme is eluted from the column within 2 days and is associated with the second of two large, yellow protein peaks. A third fainter yellow band and sometimes an orange band elute after the enzyme.

The active fractions from the column were combined and treated with 0.1 volume of 0.1 M EDTA (pH 7.0) and the enzyme precipitated from the solution by solid ammonium sulfate as previously, except that this time 430 g of solid was added per l. After suspension in, and dialysis against, Tris-acetate buffer as before the solution was combined with other batches brought to the same stage of treatment.

The enzyme was next passed through another G-100 Sephadex column of the same dimensions as in the earlier step, but this time equilibrated with 0.01 M imidazole acetate buffer (pH 5.0), containing 1 mM EDTA. Before the enzyme was passed through this column it was dialyzed against 0.01 M imidazole acetate buffer (pH 5.0), containing 1 mM EDTA for about 16 hr. A considerable pale yellow precipitate that formed during dialysis was removed by centrifugation and discarded, before the dialyzed solution was placed without delay on the column. Other conditions were as for the first Sephadex step. Some inactive protein precedes the enzyme from the column and is incompletely separated from the enzyme.

As soon as possible, active fractions were combined and treated with 0.1 volume of 0.1 M EDTA, and the pH adjusted to 7.0 with 1 N NaOH. Care must be taken not to exceed this pH since at higher pH the enzyme denatures in the presence of imidazole (Orr *et al.*, 1972). The proteins were then frac-

TABLE I: Purification of Ribonucleotide Reductase from *L. leichmannii*.[a]

| Purification Step | Vol (ml) | Total Protein (g) | Total Act. (Units) | Sp Act. (Units/mg) |
|---|---|---|---|---|
| Dialyzed extract | 34,500 | 480 | 1,900,000 | 4.0 |
| Protamine sulfate treated extract[b] | 44,900 | 259 | 1,670,000 | 6.6 |
| First ammonium sulfate precipitation[c] | 2,540 | 85.7 | 1,430,000 | 17.5 |
| First gel filtration[c] | 17,270 | 34.5 | 1,602,000[e] | 46 |
| Second ammonium sulfate precipitation[c] | 806 | 19.0 | 1,050,000 | 55 |
| Second gel filtration[d] | 7,300 | 6.6 | 572,000 | 87 |
| Third ammonium sulfate precipitation[d] A | 173 | 3.12 | 378,000 | 121 |
| B | 175 | 1.78 | 154,000 | 86 |

[a] From two 860 liter cultures. Approximately 4 kg of cell paste. [b] Carried out in 14 batches. Results refer to combined product of all batches. [c] Carried out in 5 batches. Results refer to combined product of all batches. [d] Carried out in 2 batches. A refers to the combined second fractions from both batches. B refers to the combined third fractions from both batches. [e] The activity at this step frequently seemed high compared with the preceding step. The explanation of this is presently unknown.

tionally precipitated with ammonium sulfate, the first fraction being precipitated by addition of 290 × *V* g of solid ammonium sulfate in the manner described previously, where *V* l. is the volume of the enzyme solution after pH adjustment. After recovery of the precipitate by recentrifugation the second fraction was precipitated from the supernatant by addition of 50 × *V* g of ammonium sulfate, and from the supernatant from the second fraction the third was precipitated by 90 × *V* g of ammonium sulfate. The precipitates were each dissolved without delay in a minimum volume of 0.1 M sodium 3,3-dimethylglutarate buffer (pH 7.3) and were dialyzed overnight against either the same buffer (for keeping at 0°) or against 0.01 M dimethylglutarate buffer (pH 7.3) preparatory to freeze-drying.

The enzyme solution lost activity slowly at 0°, the activity declining to 50% over a period of 2–3 weeks. About 20% loss in activity occurred on freeze-drying, but there was no further loss in activity from the frozen-dried powder over several months at 0°.

Results for a large-scale preparation are shown in Table I. The fraction containing the majority of the activity in the final step (usually the second fraction) had a specific activity of 100–190 units/mg. These fractions were judged to be 70–80% pure on the basis of analytical polyacrylamide gel electrophoresis (Orr *et al.*, 1972). Of many attempts to remove the residual contaminating proteins none was successful except preparative electrophoresis, which was used to purify 40- to 50-mg samples (Orr *et al.*, 1972). The resulting preparation contained no impurities detectable by polyacrylamide gel electrophoresis, although it may have contained some denatured reductase (Orr *et al.*, 1972) and was used in all the experiments reported in this paper. Preparative electrophoresis of the material referred to in A of the final step of Table I gave a preparation with specific activity 160 units/mg.

*Assay of Ribonucleotide Reductase.* The colorimetric method was used to estimate dATP production under conditions previously defined (Orr *et al.*, 1972). A unit of activity corresponds to formation of 1 μmole of dATP/hr.

*Radioactivity Determination.* This was carried out in a Tri-Carb scintillation counter in vials containing 10 ml of scintillation fluid prepared as described by Bray (1960).

*Analytical Gel Filtration.* The zonal methods of Andrews (1964) and Determann (1966) were used at 4° for molecular weight estimations on Sephadex G-100 equilibrated with 0.05 M potassium phosphate (pH 7.3), containing 1 mM EDTA ($\Gamma/2 = 0.13$) or 0.1 M sodium dimethylglutarate buffer, pH 7.3 ($\Gamma/2 = 0.27$). Samples of 10 mg of each protein in a volume of 1 ml of the appropriate buffer were applied to a column (1 × 100 cm). Collection of the first fraction was commenced immediately after the sample was applied. Protein was eluted into weighed tubes at flow rates between 10 and 21 ml per cm² per hr and the volume of each fraction (approximately 0.4 ml) determined from its mass. Elution profiles were determined by measurement of absorbance at 280 nm and in some cases by activity determinations. The molecular weight values for the reference compounds were obtained from the literature as follows: horse cytochrome *c*, 12,400 (Margoliash, 1962); ribonuclease A, 13,700 (Hirs *et al.*, 1956); trypsin inhibitor, 21,500 (Wu and Sheraga, 1962); ovalbumin, 45,000 (Warner, 1954); human serum albumin monomer, 69,000 (Oncley *et al.*, 1947); human γ-globulin, 160,000 (Phelps and Putnam, 1960).

Homogeneity of the enzyme was examined by gel filtration experiments that produced a plateau region in the elution profile (Winzor and Scheraga, 1963). A column (0.6 × 30 cm) of Sephadex G-100 was used. Collection of fractions and other details were as in the zonal experiments but the sample volume was 10 ml.

*Sedimentation Velocity Measurements.* Except where stated otherwise, sedimentation velocity measurements were made in a Spinco Model E analytical untracentrifuge with schlieren optics at 50,740 rpm. Solutions were buffered with 0.1 M sodium dimethylglutarate (pH 7.3) and temperature was controlled with the RTIC unit. The weight-average sedimentation coefficients, $\bar{s}$, were calculated from the rates of movement of the square roots of the second moments of the schlieren patterns (Goldberg, 1953). When comparisons between experiments at different temperatures were required, the $\bar{s}$ values were corrected to 20° in water giving the values $\bar{s}_{20,w}$. Concentrations were determined refractometrically using peak areas (obtained by trapezoidal integration) and assuming a specific refractive increment of 0.0018 dl g$^{-1}$. In the determination of the concentration dependence of the sedimentation coefficient mean plateau region concentrations were used.

As a test of homogeneity an analysis of the schlieren patterns was carried out (Baldwin, 1954). A sedimentation coefficient distribution (a plot of $g^*(S)$ *vs.* $S$) was obtained at different times during an experiment. The terms $g^*(S)$, an apparent differential distribution function, and $S$, a reduced coordinate with the same dimensions as the sedimentation coefficient, have been clearly defined (Baldwin, 1954; Signer and Gross, 1934). The curves of $g^*(S)$ *vs.* $S$ were very nearly symmetrical. The range of $S$ decreased progressively with time indicating the significant contribution of diffusion to the spreading of the boundary. These effects were eliminated by extrapolating to infinite time (Baldwin, 1959). The quantity $(S - \bar{s})^2$, where $\bar{s}$ is the weighted mean of the distribution of $S$, was plotted against $1/te^{\bar{s}\omega^2 t}$ at fixed values of $g^*(S)_{\max}$ (where $\omega$ is the angular velocity). Since the weighted means of the distributions of $S$ did not vary significantly from the corresponding weight-average sedimentation coefficients the single symbol $\bar{s}$ has been used throughout. While the method accounts for the diffusional spreading of the boundary it does not account for the concentration dependence of the sedimentation coefficient. This should not markedly affect the analysis because of the relative closeness of the measured sedimentation coefficient to that at infinite dilution (Baldwin, 1959).

A boundary analysis procedure (Creeth and Pain, 1967) based on a simplified form (Van Holde, 1960) of an expression for the height and area of a gradient curve as a function of time (Fujita, 1959) accounts for small linear concentration-dependence effects and provides a further test of homogeneity and a method for determining apparent diffusion coefficients. This method has been applied to the schlieren patterns from the sedimentation velocity experiments. The apparent diffusion coefficients, $D^*$, so obtained have been corrected to 20° in water giving the values $D^*_{20,w}$.

*Equilibrium Sedimentation.* Experiments in the absence of denaturants and in urea were performed at 14,210 rpm; Kel-F polymer oil was used as an inert base fluid; volumes of polymer oil, equilibrium diffusate and solution were taken so that the diffusate column just overlapped the solution column (1.5 mm) at both ends; most experiments were carried out in the 0.1 M sodium dimethylglutarate buffer (pH 7.3); the An-J equilibrium rotor and schlieren optics were used. Separate experiments with solutions of different initial concentrations were performed. Equilibrium was usually reached in 15 hr (33 hr in 7 M urea). Measurements (20–30) were made of the refractive index gradient at equal intervals throughout the solution columns using a Gaertner toolmakers' microscope, type M2001, As-P.

The molecular weight at any point in the cell can be calculated from the slope, at that point, of a graph of $\ln c$ *vs.* $r^2/2$ or of $\ln [(1/r)(dc/dr)]$ *vs.* $r^2/2$ by multiplying by the constant factor $RT/(1 - \bar{v}\rho)\omega^2$ where $c$ is the protein concentration at distance $r$ from the center of rotation, $R$ is the gas constant, $T$ is the absolute temperature, $\bar{v}$ is the partial specific volume of the protein, $\rho$ is the density of the solvent, and $\omega$ the angular velocity of the rotor. The former method gives point weight-average molecular weights and the latter point z-average molecular weights. In practice these graphs were almost linear, so apparent weight-average ($\bar{M}_w$) and z-average ($\bar{M}_z$) molecular weights were obtained from the slopes of the best straight lines through the data, the slopes being calculated by the method of least squares. Point concentrations were computed (Schachman, 1957) using initial concentrations obtained by measurements with a differential refractometer (Cecil and Ogston, 1951).

Sedimentation equilibrium experiments in the presence of guanidine hydrochloride were performed by the method of Yphantis (1964) at 30,000 rpm in an AN-H rotor and the use of Rayleigh optics. The sample was 0.1 ml of a solution containing 0.2–0.5 mg of protein/ml and equilibrium diffusate was placed in the second sector of the cell. Fluorocarbon was not used in these experiments, and equilibrium was reached after 20 hr. Photographs were analyzed with a Nikon microcomparator by measuring vertical fringe displacement, $D$, in microns, as a function of $r$ at small intervals throughout the solution. The positions of five fringes, three black and two white, were measured at each value of $r$, and the average displacement of these five fringes was used in a plot of $\ln D$ *vs.* $r^2$ after correction for the average displacement of five fringes in a base-line photograph. The slope of this plot, *i.e.*, $d(\ln D)/dr^2$, was obtained by computing the least-squares fit to the data on a Linc 8 computer and inserted into the usual equation to obtain the weight-average molecular weight.

*Partial Specific Volume of the Reductase.* The partial specific volume of the reductase, required for calculation of molecular weight from the sedimentation data, was deduced from the amino acid composition by the method outlined by Schachman (1957). A value of 0.73 was obtained. Since there is considerable evidence that for many proteins the value of $\bar{v}$ in denaturants is the same or very slightly lower than the value in dilute salt solutions (Castellino and Barker, 1968; Kielley and Harrington, 1960; Woods *et al.*, 1963; Marler *et al.*, 1964), the same value for $\bar{v}$ was used for calculation of molecular weight in 7 M urea and in 6 M guanidine hydrochloride in the presence of 0.01 M 2-mercaptoethanol.

*Amino Acid Analysis.* Since ribonucleotide reductase is insoluble in water, samples were dialyzed against 0.01 M sodium phosphate buffer (pH 7.3). Samples of approximately 1 mg in 1 ml of 6 N HCl were hydrolyzed in evacuated Pyrex tubes for periods of 24, 48, 96, and 140 hr at 110°. Two to six samples were hydrolyzed at each time interval. Analyses were performed by the method of Spackman *et al.* (1958) on a Beckman Model 120B amino acid analyzer. An 8-cm column of Aminex-A5 and a 50-cm column of Aminex-A4 were used. Results for serine, threonine, methionine and tyrosine were plotted *vs.* time of hydrolysis and extrapolated to zero time (Lindberg, 1967). Values at the longest period of hydrolysis were used for leucine, valine, and isoleucine. Results for other amino acids were the mean for all determinations. Alkaline hydrolysis for determination of tryptophan was carried out in 4.2 N NaOH in the presence of 25 mg of acid-washed soluble starch according to the method of Hugli and Moore (1971). Analysis was carried out on a 12.5-cm Aminex-A5 column with 0.25 M sodium citrate (pH 5.28) as eluting buffer. For more accurate determination of cystine, these residues in the protein were oxidized by performic acid and the cysteic acid determined on the analyzer after hydrolysis. The procedure of Hirs (1967a) was followed except that after treatment of the protein with reagent the solution was diluted with 200 ml of cold water and freeze-dried. The dried protein was hydrolyzed in 6 N HCl for analysis on the analyzer.

*Spectroscopic Determination of Tyrosine and Tryptophan.* The procedure of Edelhoch (1967) was used. The tryptophan content was determined from the absorbance of the protein (1.0 mg/ml) in 6.0 M guanidine hydrochloride at 288 and 280 nm after correction for absorbance due to tyrosine, cystine, and reagents. The tyrosine content was calculated from the change in absorbance at 295 or 300 nm after adjustment to pH 11.0 with 5 N NaOH, and extrapolation to zero time.

*Determination of Carboxyl-Terminal Amino Acids.* Two methods were used for identification of the amino acid residue at the carboxyl terminus of the polypeptide chain. (1) The first was by means of selective tritiation carried out by a slight modification of the procedure of Holcomb *et al.* (1968). An electrophoretically pure sample of ribonucleotide reductase (8.0 mg) was dialyzed against 0.01 M ammonium acetate, the dialyzed solution freeze-dried and the residue dissolved in 0.5 ml of tritiated water (0.5 Ci). Pyridine (1 ml) and acetic anhydride (0.5 ml) were added and the mixture incubated at 37° for 4 hr. The reagents were removed by dialysis with stirring against 2 l. of water in a stoppered vessel overnight, with several changes of water. The suspension of precipitated protein was transferred to a hydrolysis tube in which it was freeze-dried. Hydrolysis of the protein was performed with 1.0 ml of 6 N HCl at 110° for 24 hr. After evaporation of solvent under reduced pressure the residue was dissolved in 3.0 ml of 0.2 M citrate buffer (pH 2.2). After removal of samples of hydrolysate for determination of total radioactivity, 1.5 ml of the solution was transferred to a 135-cm column (type B Chromobeads) of a Technicon NC1 amino acid analyzer. The effluent line from the column was first passed through a 1-ml anthracene-packed flow cell in a Nuclear-Chicago Unilux-IIA scintillation counter before entering the mixer assembly of the analyzer. The apparatus had been calibrated so that the delay time between recording of counts (over 4-min periods) in the scintillation counter and recording of ninhydrin color on the analyzer chart for the same material was known to be 18–20 min, thus permitting determination of specific radioactivity for each amino acid. (2) The carboxy-terminal amino acid was also determined by identifying amino acids released by carboxypeptidase according to the procedures outlined by Ambler (1967). Ribonucleotide reductase was dialyzed against 0.2 M *N*-ethylmorpholine buffer (pH 8.5) and treated with a solution of carboxypeptidase A (carboxypeptidase to reductase ratio, 1:30), or carboxypeptidase B (carboxypeptidase to reductase, 1:30 or 1:90), or a mixture of carboxypeptidases A and B (A–B–reductase, 1:2:40). In some experiments urea was also present as a denaturant in the reaction mixture. During incubation of the reaction mixture at 37° samples were removed at intervals and adjusted to pH 2.0–3.0 (indicator paper) by addition of Dowex 50 $H^+$ resin (100–200 mesh). The suspension was stirred mechanically for 15 min and then poured into a glass column (0.9 × 20 cm) where the resin was allowed to drain and was then twice washed with two bed volumes of water. The adsorbed amino acids were eluted with 5 M ammonia, the solution was evaporated to dryness and the residue dissolved in citrate buffer (pH 2.2) and applied to the column of the amino acid analyzer. The analysis for each amino acid was corrected for the amount found in a control experiment performed under identical conditions except that the reductase was omitted. Reductase concentration in the complete reaction mixture was estimated from amino acid analysis of a sample of the stock reductase solution after acid hydrolysis, and from the amino acid composition of the reductase given in the Results section.

*Determination of Amino-Terminal Amino Acid.* The amino-terminal amino acids were identified by two methods. (i) Carbamylation was carried out as described by Stark (1967) on approximately 40 mg of reductase in 4 M guanidine hydrochloride. In control experiments the entire procedure was carried out except for addition of potassium cyanate and protein, respectively; amino acid analyses obtained in this control were deducted from values found in the complete experiment. Protein concentration in the reaction mixture was determined as in the preceding section. (ii) Dansylation was carried out according to Gray (1967) and the dansylamino acids were identified by thin-layer chromatography (Hartley, 1970) on Cheng Chin polyamide plates (Gallard-Schlesinger Chemical Manufacturing Corp., New York).

*Cyanogen Bromide Cleavage and Gel Electrophoresis of Peptides.* The reductase (6.5 mg) in 6 M guanidine hydrochloride, about 0.2 M 2-mercaptoethanol and 4 mM EDTA (total volume 3 ml) was adjusted to pH 8.1 by addition of 1 N NaOH. Solid dithiothreitol was added to a concentration of 0.25 M and reduction of cystine residues was allowed to proceed for 4.5 hr under nitrogen. Aminoethylation of cysteine residues was accomplished by adding ethylenimine (cautiously adjusted to pH 11 with HCl) in 13-fold excess over all thiol groups present. After a reaction time of 1 hr at room temperature the reaction mixture was dialyzed against distilled water and freeze-dried (Schroeder *et al.*, 1967). The residue was dissolved in 70% formic acid and solid cyanogen bromide was added (40 moles/mole of protein). After the reaction mixture had stood at room temperature for 24 hr it was diluted ten times with water and the mixture freeze-dried. The number of cyanogen bromide peptides was determined by electrophoresis on polyacrylamide gel in the presence of sodium dodecyl sulfate.

Electrophoresis in the presence of sodium dodecyl sulfate was carried out according to Weber and Osborn (1969), except that the gel was prepared from 15% acrylamide and 0.5% bisacrylamide. After destaining, the gels were scanned at 600 nm in a Gilford microdensitometer, Model 2000, and the curves analyzed on a DuPont curve resolver. When the reductase was electrophoresed the reference proteins were creatine kinase which has subunits of mol wt 40,000 (Dawson *et al.*, 1967) and serum albumin, molecular weight of monomer 68,000 (Tanford *et al.*, 1967). Polyacrylamide gel electrophoresis in triethanolamine–*N*-tris(hydroxymethyl)methyl-2-aminoethanesulfonic acid (buffer system I) was carried out as previously described (Orr *et al.*, 1972).

*Cyanogen Bromide Cleavage and Gel Filtration of Peptides.* Approximately 70 mg of the reductase was reduced by approximately 40 mM dihydrolipoate in a solution containing 0.2 M Tris·HCl buffer (pH 8.6), 0.135 mM EDTA, and 6 M guanidine hydrochloride in a total volume of 12.0 ml. The mixture was flushed with nitrogen and the reaction was allowed to proceed for 60 min.

A 15-fold excess of ethylenimine over the total sulfhydryl groups was added to the mixture. The pH of the reagent had previously been adjusted cautiously to pH 11.0 with 6 N HCl. After a period of 1 hr, the solution was dialyzed against water, and freeze-dried. Hydrolysis and amino acid analysis of a portion of the product indicated the presence of 8.6 moles of *S*-aminoethylcysteine/mole of enzyme (*cf.* 9 moles of half-cystine/mole of unmodified enzyme, Table IV). The residue was dissolved in 70% trifluoroacetic acid and 500 mg of CNBr was added. After 24 hr the solution was freeze-dried, and amino acid analysis showed a very small amount of residual methionine and no methionine sulfone present, corresponding to 98% cleavage. The residue was dissolved in 6 ml of 5% formic acid and the solution applied to a column of Sephadex G-50 (fine, 200 × 2.5 cm) which had been equilibrated with 5% formic acid. The elution rate was 15 ml/hr and 3-ml fractions were collected. They were analyzed by measuring the absorbance at 280 nm and by reaction with ninhydrin after alkaline hydrolysis (Hirs, 1967b).

The number of peptides under each peak was determined by high-voltage paper electrophoresis and by thin-layer chromatography on cellulose sheets. For the latter, two solvents were used: (A) isopropyl alcohol–5% formic acid (6:4.5,

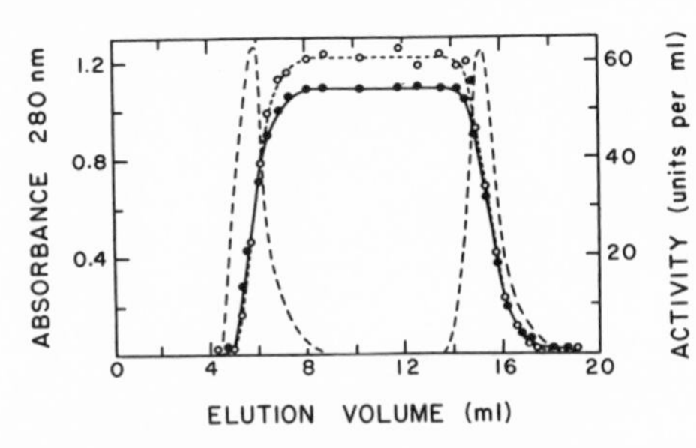


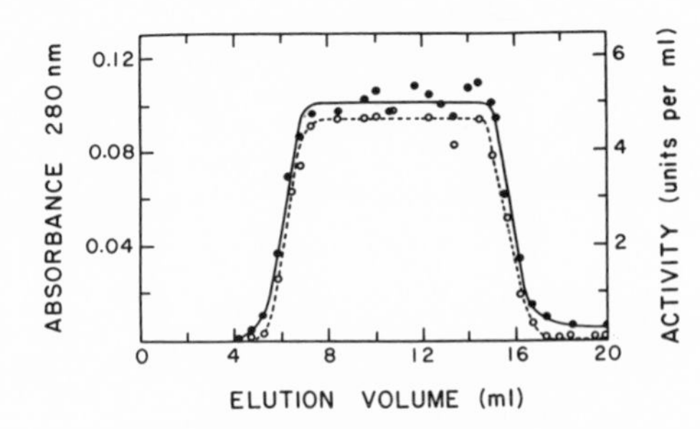


FIGURE 1: Homogeneity of ribonucleotide reductase by analytical gel filtration. The protein concentration was 1.2 mg/ml in A (left) and 0.09 mg/ml in B (right). (●) Absorbance at 280 nm; (○) activity (units/ml). The superimposed broken curves in A show the first differential of the absorbance curve.

v/v) and (B) *sec*-butyl alcohol–ammonia–water (6:3:1, v/v). Electrophoresis was carried out in a Gilson Model D-2 electrophorator in pyridine–acetic acid (pH 6.5) (Bennett, 1967) or formic acid–acetic acid (pH 2.0) (Smith, 1960). The chromatograms were stained for arginine with the phenanthrenequinone reagent (Yamada and Itano, 1966), for histidine with the Pauly reagent (Easley, 1965) or with *p*-bromoaniline (Elliott, 1959), and with the ninhydrin–cadmium reagent (Yamada and Itano, 1966).

## Results

*Homogeneity of the Ribonucleotide Reductase Preparation.* Reductase purified by preparative electrophoresis has previously been shown to give a single band in analytical electrophoresis on polyacrylamide gel in buffer systems operating over the pH range 6–8.5 (Orr *et al.*, 1972). However, the total purification achieved in the preparation was only 30- to 40-fold and since it was important to demonstrate that this was a single enzyme capable of catalyzing the reduction of several ribonucleotides rather than a collection of isoenzymes of differing specificity, the homogeneity was further examined by several other methods.

In studies on the reductase by zonal gel filtration, elution patterns of protein and enzyme activity were both symmetrical and had identical elution volumes. In plateau experiments, leading and trailing edges provided first differential curves that departed only slightly from the enantiographic pattern expected for a pure protein (Figure 1A). The elution volume and pattern did not change significantly at lower protein concentration (Figure 1B).

In sedimentation velocity experiments a single, apparently symmetrical schlieren peak was observed (Figure 2), the area under which remained constant (after correction for radial dilution) for the duration of the experiment (3–4 hr), a result indicating the absence of gross heterogeneity. An apparent content of 3–4% of slower sedimenting material was seen in some preparations, but as no corrections were made for the Johnston–Ogston effect (1946) the true proportion of these components is probably smaller.

The sedimentation coefficient of the reductase varied linearly with concentration over the range 0.03–0.3 g dl$^{-1}$, and from the least-squares treatment of the experimental data the following expression for the concentration dependence of $s_{20,w}$ was derived. $s_{20,w} = 5.2(1 - 0.09C) \times 10^{-13}$, where $C$ is the concentration of protein in g dl$^{-1}$. This relationship was used in the boundary analysis of Creeth and Pain (1967) where the apparent diffusion coefficient was calculated at various times. Since the apparent diffusion coefficient was constant with time, the analysis provided additional evidence for homogeneity.

The apparent diffusion coefficient varied linearly with concentration in the range 0.03–0.3 g dl$^{-1}$ and least-squares treatment of the experimental data gives the following expression for the concentration dependence of $D^*_{20,w}$. $D^*_{20,w} = 6.4(1 + 0.4C) \times 10^{-7}$, where $C$ equals concentration in g dl$^{-1}$.

The results of the boundary analysis of Baldwin (1954) are shown in Figure 3 where $(S - \bar{s})^2$ is plotted against $1/te^{\bar{s}\omega^2 t}$ for various values of $g^*(S)/g^*(S)_{max}$ and the extrapolations to infinite time are shown by the dashed lines. The values of $(S - \bar{s})^2$ at infinite time are close to zero, indicating the absence of measurable heterogeneity in terms of sedimentation coefficient. The plots of $g^*(S)_{max}$ *vs.* $S$ were symmetrical which gives a further indication of the homogeneity.

Examination of the reductase by equilibrium sedimentation permitted the calculation of weight-average and z-average molecular weights from the slopes of the plots log ($c$ *vs.* $r^2$)/2 and log $[(1/r)(dc/dr)]$ *vs.* $r^2/2$, respectively. The linearity of the analytical plots and the similarity of the slopes from which the $\bar{M}_w$ and $\bar{M}_z$ were calculated also indicated the homogeneity of the preparations.

*Molecular Weight of Ribonucleotide Reductase.* Comparison of the elution volume of the reductase (initial concentrations of 7, 0.7, and 0.07 mg per ml) from Sephadex G-100 with those experimentally determined with proteins of known

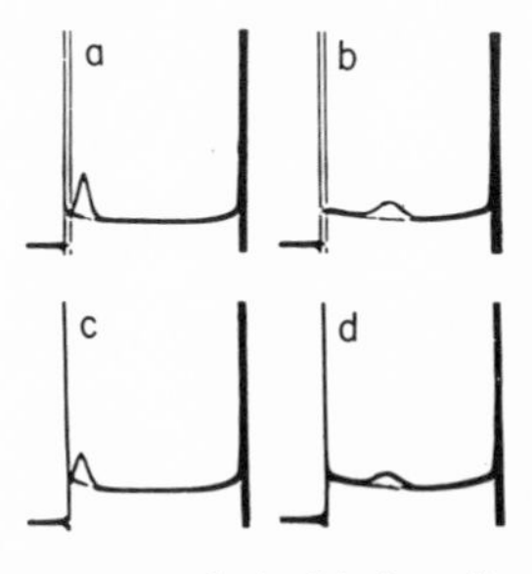


FIGURE 2: Schlieren patterns obtained during sedimentation velocity determinations on ribonucleotide reductase. For a and b: concentration, 0.2 g dl$^{-1}$; schlieren phase-plate angle 70°, temperature 4.1°. Time after reaching speed in a, 20 min; in b, 108 min. In c and d: concentration 0.13 g dl$^{-1}$; schlieren phase-plate angle 65°, temperature 5°. Time after reaching speed in c, 19 min; in d, 99 min.

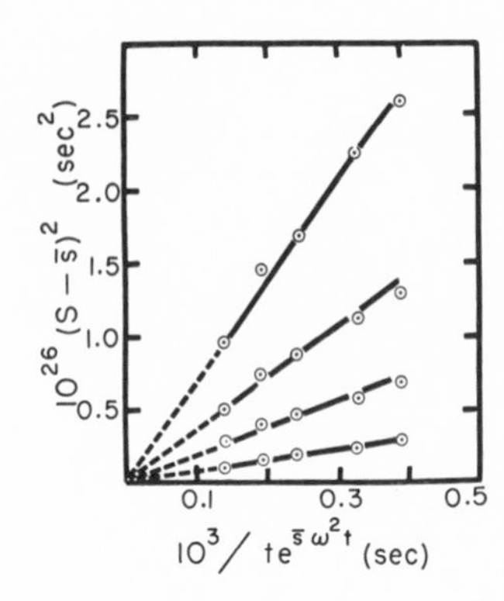


FIGURE 3: Boundary analysis of Baldwin (1954) applied to data for sedimentation of ribonucleotide reductase. For details, see the Methods section.

TABLE III: Molecular Weight of Ribonucleotide Reductase under Various Conditions as Determined by Sedimentation Equilibrium.

| Conditions | Mol Wt |
|---|---|
| 0.1 M Dimethyl glutarate (pH 7.3) | 76,000 |
| 6 M Guanidine·HCl–0.01 M 2-mercapto-ethanol–0.1 M dimethyl glutarate (pH 7.3) | 74,300 |
| 0.1 M Glycine·HCl–0.01 M 2-mercapto-ethanol (pH 2.0) | 79,950 |
| 0.1 M Sodium carbonate buffer–0.01 M 2-mercaptoethanol (pH 11.0) | 74,300 |
| 7 M Urea–0.01 M 2-mercaptoethanol–0.1 M dimethyl glutarate (pH 7.3) | 76,000 |
| 0.1% Sodium dodecyl sulfate–0.1% mercap-toethanol–0.01 M sodium phosphate buffer (pH 7.0) | 70,000 |

molecular weight (Andrews, 1964) indicated a molecular weight for the reductase of about 70,000.

The relationships given in the preceding section were used to calculate the $s_{20,w}$ and the $D^*_{20,w}$ at infinite dilution. The partial specific volume ($\bar{v}$) calculated as described in Methods was 0.73. Substitution of these values into the Svedberg equation gave an apparent molecular weight of 73,000. Molecular weights calculated from equilibrium sedimentation at four different initial concentrations (0.03, 0.08, 0.13, and 0.20 g dl$^{-1}$) are shown in Table II. The weight-average and the z-average molecular weights are identical within the expected experimental variation of the method. The average value of molecular weight of the reductase from all estimations by gel filtration, sedimentation diffusion, and equilibrium sedimentation is 76,000.

*Number of Polypeptide Chains in Ribonucleotide Reductase.* When the reductase was examined by sedimentation equilibrium in the presence of 6 M guanidine hydrochloride and 0.01 M 2-mercaptoethanol, the plot of ln $D$ *vs.* $r^2$ was linear indicating homogeneity under these conditions. From determinations of the slope of this plot in three experiments, values calculated for the weight-average molecular weight were 70,940, 78,880, and 73,140 giving a mean value of 74,300.

Equilibrium sedimentation was also carried out under the following conditions: in 0.1% sodium dodecyl sulfate, in glycine·HCl buffer (pH 2.0), in carbonate buffer (pH 11.0), and in 7 M urea. Mercaptoethanol (0.01 M) was present in all denaturing conditions since, except at pH 2.0, it prevented the formation of aggregated material which otherwise occurred. At pH 2.0 the plot indicated the presence of a minor component of molecular weight about 118,000 even with mercaptoethanol present. In all other cases the plots showed that a single major species was present with molecular weight as shown in Table III. Clearly this evidence indicates that the protein does not disaggregate into subunits when the polypeptide chain is unfolded.

When the reductase was treated with 0.1% sodium dodecyl sulfate and 0.1% mercaptoethanol and subsequently electrophoresed on polyacrylamide gel in the presence of dodecyl sulfate, only one band, with mobility less than that of serum albumin and of creatine kinase, was observed. Calculation of the molecular weight from the mobility gave a value of approximately 80,000. This result provides confirmation that the reductase contains no subunits.

TABLE II: Molecular Weight of Ribonucleotide Reductase by Equilibrium Sedimentation.

| Initial Concn (g dl$^{-1}$) | Concn Range (g dl$^{-1}$) | App Mol Wt $\bar{M}_w$ | App Mol Wt $\bar{M}_z$ |
|---|---|---|---|
| 0.2 | 0.06–0.5 | 73,000 | 72,000 |
| 0.13 | 0.05–0.3 | 80,000 | 73,000 |
| 0.08 | 0.02–0.17 | 82,000 | 77,000 |
| 0.03 | 0.01–0.09 | 79,000 | 79,000 |

*Amino Acid Analysis and Cyanogen Bromide Cleavage.* Table IV shows the number of residues of each amino acid found per molecule of enzyme assuming 9.0 half-cystine residues/mole. These values represent the best estimate obtained from a large number of determinations, as indicated in the Methods section. The integral numbers of residues shown in column 2 correspond to a molecular weight of 75,966 which must be increased on account of an unknown number of amide groups. It should be noted in particular that there are 8 methionine residues/76,000 so that if these are in a single polypeptide chain, 9 different peptide fragments should be produced by cyanogen bromide treatment.

*Peptides Formed by Cyanogen Bromide Treatment.* When the peptides produced by cyanogen bromide cleavage were subjected to electrophoresis on polyacrylamide gel in the presence of sodium dodecyl sulfate and mercaptoethanol ten bands were detected after staining with Coomassie Blue, that is, one in excess of the expected number for a single chain. This is seen clearly in the densitometer scan and the corresponding resolution into individual curves (Figure 4). Amino acid analysis of the material after cyanogen bromide treatment indicated that 0.3 mole of methionine residue remained per 76,000 g of protein, so that it is likely that peptide I and perhaps peptide II contained a methionine residue at which cleavage did not occur. If both peptides I and II, which together constituted about 10% of the total protein, contained unreacted methionine, another band (presumably X) may have contained two peptides.

The elution pattern from the gel filtration of the peptides

TABLE IV: Amino Acid Composition of Ribonucleotide Reductase.

| Amino Acid | Column 1[a] | Column 2[b] |
|---|---|---|
| Lysine | 43.4 | 43 |
| Histidine | 8.1 | 8 |
| Arginine | 33.4 | 33 |
| Aspartic acid | 78.1 | 78 |
| Threonine | 31.6[c] | 32 |
| Serine | 54.4[c] | 54 |
| Glutamic acid | 78.2 | 78 |
| Proline | 30.6 | 31 |
| Glycine | 55.6 | 56 |
| Alanine | 54.7 | 55 |
| Cystine (half) | 9.0 | 9 |
| Valine | 47.5[d] | 48 |
| Methionine | 7.7 | 8 |
| Isoleucine | 39.4[d] | 39 |
| Leucine | 59.3[d] | 59 |
| Tyrosine | 23.8 | 24 |
| Phenylalanine | 24.7 | 25 |
| Tryptophan | 9.0 | 9 |

[a] Mean values (except as indicated). Results calculated on the basis of 9 half-cystines. [b] Nearest integral values. [c] Obtained by extrapolation to zero hydrolysis time. [d] Obtained by extrapolation to maximum value.

produced by cyanogen bromide action contained nine peaks (Figure 5). Peak I was eluted in the void volume as determined by gel filtration of Blue Dextran and was probably uncleaved protein. Peak II was probably some incompletely cleaved material, possibly with intact disulfide bridges since aminoethylation was only 95.5% complete. This peak remained at the origin both in high-voltage paper electrophoresis at pH 2.0 or pH 6.5 and in thin-layer chromatography in a number of solvents.

On high-voltage electrophoresis the material from peaks

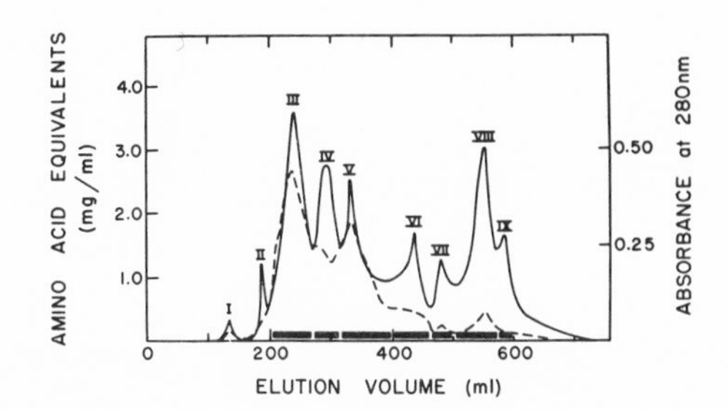


FIGURE 5: Gel filtration of peptides formed by cyanogen bromide treatment of ribonucleotide reductase. The continuous curve indicates amino acid equivalents calculated from ninhydrin color after alkaline hydrolysis: broken curve, absorbance at 280 nm. Horizontal bars indicate pooled fractions. Other details are given in Methods.

VIII and IX moved toward the cathode with identical mobilities, both at pH 2.0 and 6.5. The material from both was arginine and ninhydrin positive but histidine negative. Since the remaining peptides stayed at the origin during electrophoresis, the peptides in peaks VIII and IX were different from that in the other peaks. Peak VIII was shown by thin-layer chromatography on cellulose in solvent B to contain two peptides and peak IX one which had different mobility from either peptide of peak VIII (Figure 10). Since some material remained at the origin, especially from the peaks containing large peptides, two-dimensional thin-layer chromatography, first in solvent A and then in B, was employed to further resolve and distinguish the peptides in peaks III to VII. As seen in Figure 6 these peaks contained a total of six major peptides. Of these, all were arginine positive and all except peptide 4 histidine positive. These data indicate that peaks III to IX contained a total of nine different peptides in major amounts.

*End-Group Determinations.* Determinations of the carboxyl terminus by the selective tritiation method clearly indicated that lysine was the only C terminus present. The specific activities (counts per minute per micromole) of various amino

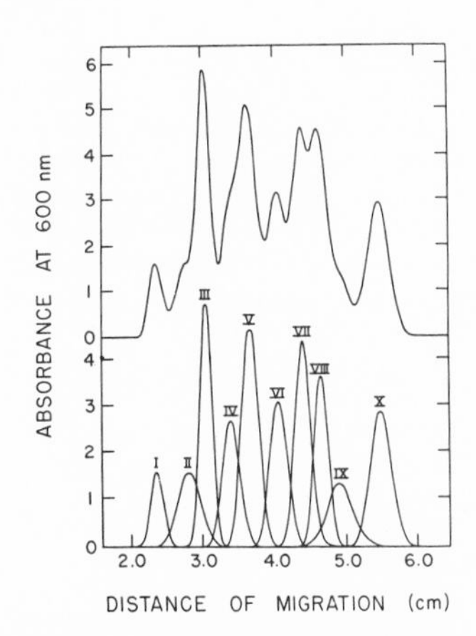


FIGURE 4: Polyacrylamide gel electrophoresis of peptides formed by cyanogen bromide treatment of ribonucleotide reductase. The upper curve is the densitometer scan of the gel and the lower curves are the component curves obtained by the curve resolver.

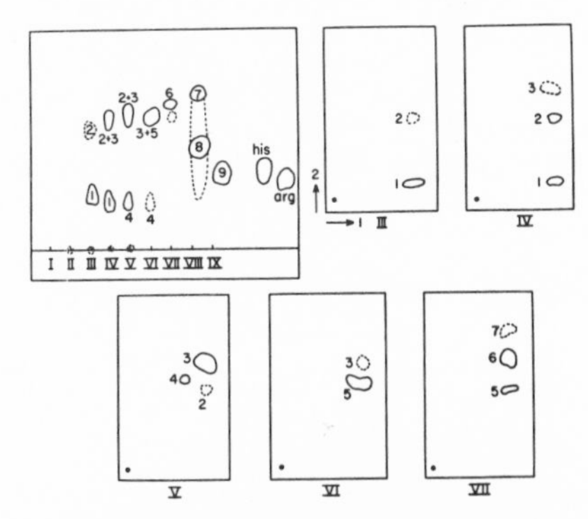


FIGURE 6: Chromatographic separation of peptides formed by cyanogen bromide treatment of ribonucleotide reductase. Samples in the one-dimension chromatogram at upper left were the fractions from the gel filtration described in Figure 5 and the solvent was solvent B. The other diagrams represent two-dimensional chromatograms of individual fractions developed with solvent A in the first direction and solvent B in the second. Peptides were detected with phenanthrene quinone.

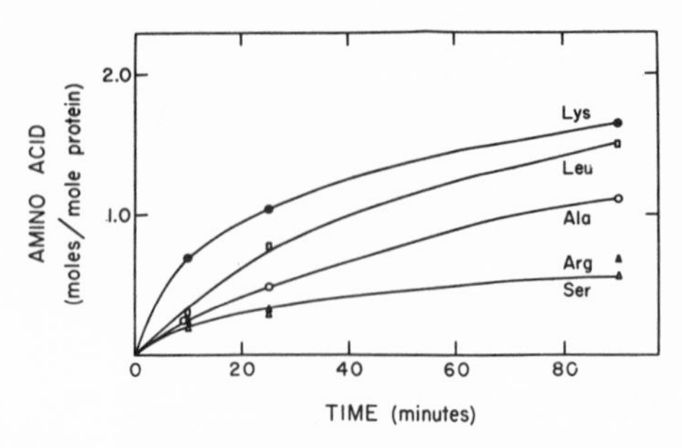


FIGURE 7: Release of amino acids from ribonucleotide reductase by carboxypeptidase B. The reductase (40) mg) was treated with 0.45 mg of carboxypeptidase under the conditions described in Methods.

TABLE V: Amino Acids Released from the N Terminus of Ribonucleotide Reductase by the Carbamylation Method.

| Amino Acid | Mole of Amino Acid/ Mole of Protein[a] |
|---|---|
| Arginine | 0.041 |
| Aspartic acid | 0.020 |
| Threonine | 0.015 |
| Serine | 0.740 |
| Glutamic acid | 0.079 |
| Glycine | 0.021 |
| Alanine | 0.015 |

[a] Values corrected according to the recovery factors of Stark (1967).

acids released after selective tritiation were as follows: Lys, 13,900; Asp, 3950; Glu, 1,900. All other values were less than 650. The much smaller amount of radioactivity introduced into glutamic and aspartic acids is probably due to mixed anhydride formation at nonterminal residues (Holcomb *et al.*, 1968) and has been observed in several other proteins which do not have glutamic or aspartic acid at the C terminus (Holcomb *et al.*, 1968; Hsieh *et al.*, 1971). In order to confirm terminal lysine and eliminate the possibility of terminal glutamic or aspartic acid the reductase was treated with carboxypeptidases A and B. Lysine was the amino acid released at the fastest rate in the presence of both carboxypeptidases, and glutamic and aspartic acids were not released at a significant rate with carboxypeptidase A alone or with A and B. Treatment with carboxypeptidase A alone resulted in release of rather small amounts of amino acids. After 15-min incubation the amounts of amino acids released (moles per mole of reductase) were: Lys, 0.111; Ser, 0.047; Leu, 0.061. After 60 min they were: Lys, 0.29; Ser, 0.178; Leu, 0.135. However, carboxypeptidase B rapidly released not only lysine but several other amino acids although in smaller amounts (Figure 7). Liberation of leucine and alanine in this experiment was favored by carboxypeptidase A present in commercial preparations of carboxypeptidase B, but the negligible rate of their release by carboxypeptidase A alone indicates that they could not have been present at the C terminus. To ensure that all C termini were accessible, treatment with carboxypeptidase B was also carried out in urea but this did not result in change in the pattern of amino acids released. The results shown in Figure 7 indicate that the C terminus is Ala-Leu-Lys.

Determination of the amino terminus by carbamylation of the protein, cyclization, isolation of the resulting hydantoins, and their subsequent hydrolysis to amino acids yielded serine as the only N-terminal residue (Table V). Recovery of lysine from the carbamylated protein after hydrolysis with 6 N HCl at 10° for 24 hr was 20%, which is within the range expected to be formed from homocitrulline during hydrolysis of a fully carbamylated protein (Stark and Smyth, 1963), so that carbamylation was probably complete.

Thin-layer chromatography of the hydrolysis products from the dansylated protein revealed the presence of five fluorescent products which were identified by comparison to authentic materials as dimethylaminonaphthalenesulfonamide (dansylamine), dimethylaminonaphthalenesulfonic acid (dansyl-OH), $N^{\epsilon}$-dansyllysine, *O*-dansyltyrosine, and $N^{\alpha}$-dansylserine. The first two products result from reaction of the reagent with ammonia and water, respectively, and the $N^{\epsilon}$-dansyllysine and *O*-dansyltyrosine result from reaction of reagent with nonterminal lysine and tyrosine residues. The only amino-terminal residue detected was therefore serine. When authentic dansylserine was added to the hydrolytic products prior to chromatography, the authentic material reinforced the fluorescence of the product in this part of the chromatogram, whereas when authentic dansylthreonine was added it was clearly resolved from this product. In the absence of the added authentic compounds only one product was found in this area of the chromatogram.

*Sedimentation and Electrophoresis of the Reductase in the Presence of Modifiers.* Although the evidence presented above indicates that the native reductase consists of a single polypeptide chain, formation of an oligomeric protein might occur in the presence of modifiers of the enzyme activity, that is deoxyribonucleoside triphosphates. It has been shown that polymerization of an enzyme in the presence of modifier can provide, at least in theory, a mechanism for allosteric behavior (Nichol *et al.*, 1967; Frieden, 1967). Moreover, the ribonucleoside diphosphate reductase complex of *Escherichia coli* undergoes aggregation in presence of the negative modifier dATP (Brown and Reichard, 1969a).

Gel filtration data reported above for the native enzyme in absence of modifiers indicated that no detectable polymer was present under the conditions employed. Thus the elution volume was unaltered over a 100-fold concentration range in the zonal experiments and over a 10-fold range in the plateau experiments and the data showed no evidence of more than one species of protein or enzyme activity. The weight-average sedimentation coefficient of the reductase determined by equilibrium sedimentation did not decrease on dilution and in fact the concentration dependence was negative down to a concentration of 0.03 g $dl^{-1}$ and there was a small negative (or no) concentration dependence of molecular weight down to 0.01 g $dl^{-1}$. These data also indicate the absence of aggregated forms.

In the presence of 1–5 mM dGTP, which produces maximum activation of ATP reduction, the sedimentation coefficient of the reductase did not differ significantly from the value obtained in absence of dGTP, and plots of $g^*(S)_{max}$ *vs.* $S$ for the treated reductase were identical with the distribution for the enzyme in absence of nucleotide. The z-average molecular weight from sedimentation equilibrium experiments was

78,000, a value within the range for the reductase in absence of nucleotides. Similarly, the substrate ATP (10 mM) did not change the sedimentation coefficient in velocity experiments. When reductase was electrophoresed on polyacrylamide gel in the presence of 1 mM dGTP or GTP and buffer system I, only a single band of protein was revealed by staining. The mobility of this band was increased slightly (18%) with respect to Bromophenol Blue, probably due to the the higher charge density of the enzyme–nucleotide complex.

## Discussion

The homogeneity of the reductase is important since the molecular weight and amino acid analysis are in conflict with previously reported values (Blakley *et al.*, 1965; Goulian and Beck, 1966). By every method used, including polyacrylamide gel electrophoresis (Orr *et al.*, 1972), the maximum amount of contaminants or aggregates was a few per cent. Also the specific activity of the enzyme used in this work was 2.5 to 4 times higher than that used in the work of Goulian and Beck (1966) even after allowance for the suboptimal assay conditions used by the latter, and about 10–20 times higher than that used in the work of Blakley *et al.* (1965). As previously noted (Orr *et al.*, 1972), the enzyme after preparative electrophoresis often did not have the maximum specific activity, but this was probably due to the presence of variable amounts of denaturated reductase rather than to contamination with other protein species. There seems little doubt, therefore, that the higher purity of the preparation used in the present work has permitted more reliable assessment of the properties of the enzyme.

Estimates of the molecular weight in the present study ranged from 70,000 (zonal gel filtration) through 72,000–82,000 (sedimentation) with a mean from all determination of 76,000. The lower value from gel filtration may have been due to enzyme–Sephadex interaction or to a lower frictional ratio for the reductase compared with reference proteins (Siegel and Monty, 1966). Clearly, both values reported previously are erroneous, and the discrepancy between them can not be explained by aggregation–disaggregation phenomena as had been suggested (Larsson and Reichard, 1967) since no evidence for aggregation was found.

Sedimentation equilibrium studies in 6 M guanidine hydrochloride, in 7 M urea, in buffer at pH 2 or 11 or in 0.1% sodium dodecyl sulfate indicated that these denaturants cause no change in molecular weight, and electrophoretic mobility in polyacrylamide gel in the presence of dodecyl sulfate confirmed this. These data therefore strongly suggest that the enzyme consists of a single polypeptide chain.

Since no amino acid is present in an amount less than 8 residues/76,000 molecular weight, determination of a minimum chemical molecular weight from the amino acid analysis was not very accurate, but the value obtained on the assumption of nine half-cystine residues is remarkably close to 76,000. Molecular weight determination from the number of tryptic peptides was unsuccessful due to incomplete hydrolysis of the polypeptide chain. The acid-insoluble core has a molecular weight of about 20,000 (gel filtration, unpublished results) and contained about nine basic residues. Why tryptic digestion did not proceed further is uncertain, but may be related to the many acidic residues present (about 60). However, cleavage of the reductase by cyanogen bromide appeared from polyacrylamide gel electrophoresis in the presence of dodecyl sulfate and from gel filtration to yield the nine peptides expected for a single polypeptide chain of mol wt 76,000. The finding of a single N terminus (Ser) and one type of C terminus (Lys) is also consistent with the absence of subunits.

Several large proteins with a single polypeptide chain are known: DNA polymerase (109,000) (Jovin *et al.*, 1969), leucyl tRNA synthetase (110,000) (Hayashi *et al.*, 1970), phosphorylase subunit (90,000) (Hedrick *et al.*, 1969), myosin subunit (212,000) (Gersham *et al.*, 1969), and $\beta$-galactosidase subunit (135,000) (Steers *et al.*, 1965). The molecular weight of 76,000 for ribonucleotide reductase is therefore not exceptionally high for a single polypeptide chain.

Most of the enzymes showing allosteric behavior (that is altered activity in the presence of ligands that bind at a site other than the catalytic center) are oligomeric, or have such high molecular weights that it is probable that they are oligomeric. Because of this association of oligomeric structure with allosteric behavior, allosteric mechanisms are usually discussed in terms of interaction between protomers. Such mechanisms are inappropriate for the ribonucleotide reductase of *L. leichmannii*, which, according to the evidence presented above, has no quaternary structure. Yet the enzyme shows specific activation of the reduction of its various ribonucleotide substrates by appropriate deoxyribonucleoside triphosphates, and nonlinear kinetics under appropriate conditions. Equilibrium dialysis studies indicated that the modifiers bind tightly to the enzyme (Morley, 1968), yet have negligible inhibitory effect (Vitols *et al.*, 1967a) so that binding probably occurs at a modifier site. Frieden (1964) has shown that a monomeric enzyme that catalyzes even a one substrate reaction and has a single modifier site can give nonlinear kinetics of several types and can therefore show all aspects of allosteric behavior except the multisite cooperative interactions characteristic of the oligomeric enzymes. It seems very probable that the reductase is such an enzyme except that it catalyzes a multireactant process. To our knowledge it is the first well-documented example of a monomeric enzyme subject to regulatory control.

Ribonucleotide reductase of *E. coli* is in a category between *L. leichmannii* reductase and the classical type of oligomeric protein that shows cooperative allosteric behavior. Thus the *E. coli* reductase consists of two different subunits but these are loosely associated, separate during purification, and reassociate only in the presence of a relatively high concentration of $Mg^{2+}$ (Brown and Reichard, 1969a). One subunit appears to carry the catalytic site (Brown *et al.*, 1969) and the other is regulatory in function (Brown and Reichard, 1969b) but only the associated subunits are catalytically active. The functions that are performed by a single polypeptide chain in the case of the *L. leichmannii* enzyme are therefore carried out by two larger chains in the case of the *E. coli* enzyme.

## Acknowledgments

We are indebted to Barbara Craker, Barbara Humphrey, Kathleen Kelly, Mary Schrock, and Frances Reifenstahl for assistance in the preparation of various batches of enzyme.

## References

Ambler, R. P. (1967), *Methods Enzymol. 11*, 155.
Andrews, P. (1964), *Biochem. J. 91*, 222.
Baldwin, R. L. (1954), *J. Amer. Chem. Soc. 76*, 402.
Baldwin, R. L. (1959), *J. Phys. Chem. 63*, 1570.
Beck, W. S. (1967), *J. Biol. Chem. 242*, 3148.

Bennett, J. C. (1967), *Methods Enzymol. 11*, 330.
Blakley, R. L. (1965), *J. Biol. Chem. 240*, 2173.
Blakley, R. L., Ghambeer, R. K., Nixon, P. F., and Vitols, E. (1965), *Biochem. Biophys. Res. Commun. 20*, 439.
Blakley, R. L., and Vitols, E. (1968), *Annu. Rev. Biochem. 37*, 201.
Bray, G. A. (1960), *Anal. Biochem. 1*, 279.
Brown, N. C., Eliasson, R., Reichard, P., and Thelander, L. (1969), *Eur. J. Biochem. 9*, 512.
Brown, N. C., and Reichard, P. (1969a), *J. Mol. Biol. 46*, 25.
Brown, N. C., and Reichard, P. (1969b), *J. Mol. Biol. 46*, 39.
Castellino, F. J., and Barker, R. (1968), *Biochemistry 7*, 2207.
Cecil, R., and Ogston, A. G. (1951), *J. Sci. Instrum. 28*, 253.
Creeth, J. M., and Pain, R. H. (1967), *Progr. Biophys. 17*, 219.
Dawson, D. M., Eppenberger, H. M., and Kaplan, N. D. (1967), *J. Biol. Chem. 242*, 210.
Determann, H. (1966), *Protides Biol. Fluids Proc. Colloq. 14*, 563.
Easley, C. W. (1965), *Biochim. Biophys. Acta 107*, 386.
Edelhoch, H. (1967), *Biochemistry 6*, 1948.
Elliott, D. C. (1959), *in* Data for Biochemical Research, Dawson, R. C., Elliott, D. C., Elliott, W. H., and Jones, K. M., Ed., London, Oxford University Press, p 230.
Frieden, C. (1964), *J. Biol. Chem. 239*, 3522.
Frieden, C. (1967), *J. Biol. Chem. 242*, 4045.
Fujita, H. (1959), *J. Phys. Chem. 63*, 1092.
Gersham, L. C., Stracher, A., and Dreizen, P. (1969), *J. Biol. Chem. 244*, 2726.
Ghambeer, R. K., and Blakley, R. L. (1966), *J. Biol. Chem. 241*, 4710.
Goldberg, R. J. (1953), *J. Phys. Chem. 57*, 194.
Goulian, M., and Beck, W. S. (1966), *J. Biol. Chem. 241*, 4233.
Gray, W. R. (1967), *Methods Enzymol. 11*, 139.
Hamilton, J. A., Yamada, R., Blakley, R. L., Hogenkamp, H. P. C., Looney, F. D., and Winfield, M. E. (1971), *Biochemistry 10*, 347.
Hartley, B. S. (1970), *Biochem. J. 119*, 805.
Hayashi, H., Knowles, J. R., Katze, J. R., Lapointe, J., and Soll, D. (1970), *J. Biol. Chem. 254*, 1401.
Hedrick, J. L., Smith, A. J., and Bruening, G. E. (1969), *Biochemistry 8*, 4012.
Hirs, C. H. W., Moore, S., and Stein, W. H. (1956), *J. Biol. Chem. 219*, 623.
Hirs, C. H. W. (1967a), *Methods Enzymol. 11*, 59.
Hirs, C. H. W. (1967b), *Methods Enzymol. 11*, 325.
Hogenkamp, H. P. C., Ghambeer, R. K., Brownson, C., Blakley, R. L., and Vitols, E. (1968), *J. Biol. Chem. 243*, 799.
Holcomb, G. N., James, S. A., and Ward, D. N. (1968), *Biochemistry 7*, 1291.
Hsieh, W. T., Gundersen, L. E., and Vestling, C. S. (1971), *Biochem. Biophys. Res. Commun. 43*, 69.
Hugli, T. E., and Moore, S. (1971), *Fed. Proc., Fed. Amer. Soc. Exp. Biol. 30*, 1181.
Johnston, J. P., and Ogston, A. G. (1946), *Trans. Faraday Soc. 42*, 789.
Jovin, T. M., Englund, P. T., and Bertsch, L. L. (1969), *J. Biol. Chem. 244*, 2996.
Kielley, W. W., and Harrington, W. F. (1960), *Biochim. Biophys. Acta 41*, 401.
Larsson, A., and Reichard, P. (1967), *Progr. Nucl. Acid Res. 7*, 303.
Lindberg, U. (1967), *Biochemistry 6*, 323.
Margoliash, E. (1962), *J. Biol. Chem. 237*, 2161.
Marler, E., Nelson, C. A., and Tanford, C. (1964), *Biochemistry 3*, 279.
Morley, C. G. D. (1968), Ph.D. Thesis, Australian National University, Canberra, A. C. T., Australia.
Nichol, L. W., Jackson, W. J. H., and Winzor, D. J. (1967), *Biochemistry 6*, 2449.
Oncley, J. L., Scatchard, G., and Brown, A. (1947), *J. Phys. Colloid Chem. 51*, 184.
Orr, M. D., Blakley, R. L., and Panagou, D. (1972), *Anal. Biochem. 45*, 68.
Phelps, R. A., and Putnam, F. W. (1960), *in* The Plasma Proteins, Vol. 1, Putnam, F. W., Ed., New York, N. Y., Academic, p 171.
Schachman, H. K. (1957), *Methods Enzymol. 4*, 32.
Schroeder, W. A., Shelton, J. R., and Robberson, B. (1967), *Biochim. Biophys. Acta 147*, 590.
Siegel, L. M., and Monty, K. J. (1966), *Biochim. Biophys. Acta 112*, 346.
Signer, R., and Gross, H. (1934), *Helv. Chim. Acta 17*, 726.
Smith, I. (1960), Chromatographic and Electrophoretic Techniques, Vol. 2, New York, N. Y., Interscience, p 170.
Spackman, D. H., Stein, W. H., and Moore, S. (1958), *Anal. Chem. 30*, 1190.
Stark, G. R. (1967), *Methods Enzymol. 11*, 125.
Stark, G. R., and Smyth, D. G. (1963), *J. Biol. Chem. 238*, 214.
Steers, E., Craven, G. R., Anfinsen, C. B. (1965), *J. Biol. Chem. 240*, 2478.
Tanford, C., Kawahara, K., and Lapanje, S. (1967), *J. Amer. Chem. Soc. 89*, 729.
Van Holde, K. E. (1960), *J. Phys. Chem. 64*, 1582.
Vitols, E., Brownson, C., Gardiner, W., and Blakley, R. L. (1967a), *J. Biol. Chem. 242*, 3035.
Vitols, E., Hogenkamp, H. P. C., Brownson, C., Blakley, R. L., and Connellan, J. (1967b), *Biochem. J. 104*, 58c.
Warner, R. C. (1954), *Proteins 2*, 435.
Weber, K., and Osborn, M. (1969), *J. Biol. Chem. 244*, 4406.
Winzor, D. J., and Scheraga, H. A. (1963), *Biochemistry 2*, 1263.
Woods, E. F., Himmelfarb, S., and Harrington, W. F. (1963), *J. Biol. Chem. 238*, 2374.
Wu, Y. U., and Scheraga, H. A. (1962), *Biochemistry 1*, 698.
Yamada, S., and Itano, H. A. (1966), *Biochim. Biophys. Acta 130*, 538.
Yamada, R., Tamao, Y., and Blakley, R. L. (1971), *Biochemistry 10*, 3959.
Yphantis, D. A. (1964), *Biochemistry 3*, 297.

Part XI

# METHODS: ION EXCHANGE CHROMATOGRAPHY, GEL FILTRATION, AFFINITY CHROMATOGRAPHY, AND STABILIZATION BY GLYCEROL

# Editor's Comments on Papers 59 Through 63



From the early work of Payen and Persoz and of Berthelot (alcohol precipitation), through that of Brücke and of Danilevsky (selective adsorption of enzymes), of Buchner (new method for enzyme extraction from yeast cells), up to the present, enzymology has depended for progress on protein purification methods. It is somewhat ironic that the first crystallizations of enzymes by Sumner and by Northrop and Northrop and Kunitz, which established once and for all that enzymes are proteins, were accomplished by methods such as acetone precipitation and salting out with ammonium sulfate—the latter method, as we saw (see comments on Kühne, Part III) used on proteins already in the 1880s—rather than by any of the numerous adsorption methods introduced and used much later by Willstätter, which had led him to the conclusion that enzymes are not proteins.[1]

It soon became apparent that parameters such as differential solubility and adsorption would not work or suffice for all purifications, particularly when trace amounts of a given enzyme had to be separated from various other proteins that had similar physicochemical properties. Successful enzyme purification depends on the availability of different methods, each of which, applied in succession, separates proteins into various groups on the basis of parameters such as charge differences, size differences, and so on.

Only a few papers dealing with methods can be included in this book. Almost all of these papers use column chromatography. This technique was invented as early as 1905 for the purification of plant pigments (hence the name, *color writing*) by the part Russian, part Italian botanist Mikhail Semenovich Tswett (1872–1919). The method was rejected as unsuitable for separating small molecules by organic chemists such as Willstätter (in this incarnation the great expert on plant pigments) (see Robinson 1960). Its use by enzymologists until the middle 1950s was not extensive because of various limitations in the available chromatographic media (see Zechmeister and Cholnoky 1938; Zechmeister 1950; Porter 1955). A great advance was the introduction of column chromatography on hydroxyapatite (Tiselius, Hjerten, and Levin 1956). However, it was the invention of ion exchange chromatography that revolutionized this field. The background to the development of this method is most interesting. During World War II Waldo Cohn and his coworkers developed this method as part of the Plutonium Project under the Manhattan Project for the separation of radioisotopes produced by fission for the production of plutonium (Tompkins, Khym, and Cohn 1947; Johnson, Quill, and Daniels 1947). After the war Cohn applied the method to the separation of mononucleotides (Cohn 1949, 1957). The available ion exchangers were various synthetic products, some of which were soon used in classical studies for the separation of amino acids (see Moore and Stein 1952, 1963), but their application to protein separations was rather limited (see Samuelson 1953; Zittle 1953; Hirs 1955).

The invention of cellulose-based ion exchangers, announced in 1954 in a short note (Paper 59) by Sober and Peterson revolutionized the field of preparative protein purification. Two longer papers that developed from this note (Peterson and Sober 1956; Sober et al. 1956) are models of thoroughness and versatility but cannot be reproduced here for lack of space. These two papers deal with a classical topic, the separation of blood serum proteins, but the application to enzyme purification, already treated in the preliminary 1954 note, soon followed. Today practically all enzyme-purification procedures include one or more steps that use column chromatography on DEAE-cellulose, CM-cellulose, or related ion exchange materials which differ in the exchange group or in the supporting matrix.

A further application of column chromatography, elaborated within a few years of the cellulose-based ion exchangers, utilizes the phenomenon of so-called molecular sieving in what has been called *gel filtration*. The first substances used extensively for this purpose were synthetic cross-linked dextran gels. The first two papers in this area, by Porath and Flodin, and by Porath, are reproduced here (Papers 60 and 61). A large number of further papers by the same groups at Uppsala appeared at about the same time (Björk and Porath 1959; Lindner, Elmqvist, and Porath 1959; Gelotte and Krantz 1959; Porath 1960; Gelotte 1960). Other supports such as polyacrylamide have since been used as well. The molecular sieving technique introduced quite a new parameter, namely size differentiation, into routine enzyme purification and has become a standard method in every enzyme laboratory. The technique is also widely used for desalting, for molecular weight estimations, for studies of binding between various molecules, and so on.

Although selective adsorption or elution of a given enzyme by the use of substrates or of substances related to substrates had been employed sporadically for quite a few years, a standardized technique and a new term, *affinity chromatography*, introduced by Cuatrecasas, Wilchek, and Anfinsen (Paper 62), very greatly stimulated this approach to enzyme purification. This procedure frequently gives extensive purification in one step, a result that was inconceivable by most other techniques. This approach is particularly useful when only small amounts of starting material or extracts with very low activities are available.

The purification of many enzymes has been thwarted, in spite of the existence of mild methods, by instability. Glycerol is the most common agent now in use for the stabilization of enzymes against inactivation by heat and at times by cold. As pointed out in Paper 63 by Jarabak, Seeds, and Talalay, the use of glycerol was prevalent in the days before refrigeration. This paper is an early and influential example on the virtue of the simple addition of about 20 percent of glycerol for the study of a labile enzyme, in this case a cold-inactivated enzyme. By now enzyme purifications and enzyme studies are often routinely carried out in solutions containing up to 50 percent of glycerol. On occasion glycerol and other polyhydric compounds may also stimulate enzyme activity (Buss and Stalter 1978).

## NOTE

1. Among our early papers the use of ammonium sulfate for enzyme isolation is considered, for example, by Buchner in 1897 (Paper 21).

# 59



Reprinted from *Am. Chem. Soc. J.* **76**:1711–1712 (1954)

## CHROMATOGRAPHY OF PROTEINS ON CELLULOSE ION-EXCHANGERS

**Herbert A. Sober and Elbert A. Peterson**

*Laboratory of Biochemistry, National Cancer Institute, National Institutes of Health, Bethesda, Maryland*

Received January 20, 1954

Previous studies in this Laboratory[1] with an egg-white protein mixture and a bovine plasma albumin–hemoglobin mixture on a strong cation-exchange resin, as well as the experiments of other workers[2] with relatively stable, low molecular weight crystalline proteins on a weak cation-exchanger, were performed with commercial resins.

(1) H. A. Sober, G. Kegeles and F. J. Gutter, *Science*, **110**, 564 (1949); H. A. Sober, G. Kegeles and F. J. Gutter, This Journal, **74**, 2734 (1952).

(2) S. Moore and W. H. Stein, *Ann. Rev. Biochem.*, **21**, 521 (1953); C. A. Zittle, *Advances in Enzymology*, **14**, 319 (1953).

We have now found that suitable adsorbents possessing anion- or cation-exchange properties permitting adsorption and elution of relatively large amounts[3] of protein under mild conditions can be prepared from α-cellulose powder. Strongly alkaline cellulose was treated with chloroacetic acid to form a cation-exchanger (CM-cellulose) or with 2-chloro-N,N-diethylethylamine to form an anion-exchanger (DEAE-Polycel).[4] These adsorbents are white or almost white, contain from 0.2–2.0 meq./g. of acidic or basic groups, and exhibit very desirable physical and mechanical characteristics.

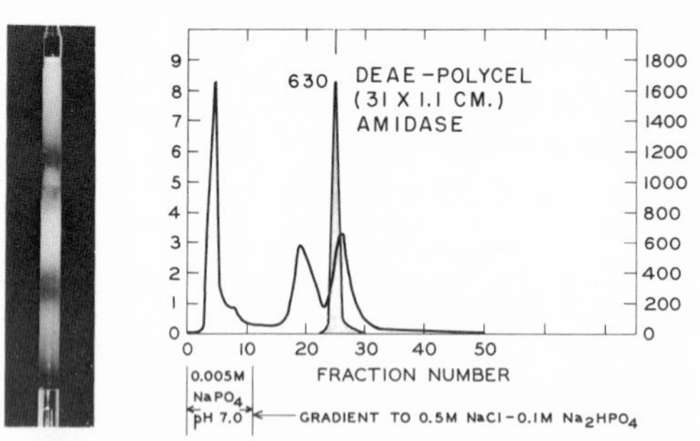


Fig. 1.—A, left: 3.5 g. CM-cellulose column, 21 × 1.1 cm. buffered initially at $pH$ 5.1 with 0.02 $M$ sodium phosphate; flow rate of 2 ml./hr.

B, right: 5.0 g. DEAE-cellulose buffered at $pH$ 7.0 with 0.005 $M$ sodium phosphate; load was 270 mg. of dialyzed, lyophilized kidney fraction in 2.7 ml. of same buffer. Fraction volume was 6 ml. Gradient was produced by continuous introduction of 0.1 $M$ $Na_2HPO_4$–0.5 $M$ NaCl into a constant volume reservoir initially containing 100 ml. of 0.005 $M$ sodium phosphate, $pH$ 7.0, and was begun at fraction 11. Left ordinate is optical density at 280 mμ and represents protein (solid line). Right ordinate is amidase activity and is represented by the shaded area. Specific activity of the amidase preparation was 80 μmole leucine amide split/hr./$D_{280}$. Fraction 25 had a specific activity of 630.

In Fig. 1A is shown the visible banding obtained by the partial development of a CM-cellulose column with $pH$ 6.5, 0.02 $M$ sodium phosphate after the addition of 380 mg. of a dialyzed water extract of pig heart acetone powder. Aspartic–glutamic transaminase activity was associated with a tan band, second from the bottom. The same enzyme activity has been purified from the acetone powder 11-fold to a purity index[5] of 0.025 by chromatography with $pH$ 6.75,0. 02 $M$ phosphate buffer on another adsorbent prepared by precipitation of $Ca_3(PO_4)_2$ within the cellulose fiber.[6]

Amidase activity purified 18-fold over the original kidney homogenate by $(NH_4)_2SO_4$ fractionation was not retained by CM-cellulose buffered at $pH$ 6.8. However, an additional 11-fold purification resulted because of the retardation of inactive protein. The elution diagram resulting from the application of the amidase preparation to the anion-exchanger, DEAE-Polycel, followed by development with a $pH$ gradient is shown in Fig. 1B. As can be seen the shaded area containing the amidase activity was part of a larger peak of non-specific protein. A subsequent chromatogram developed with a flatter gradient, while still not providing homogeneous material, resulted in further purification and the separation of two distinct amidase activities differing in their relative rates toward leucine and alanine amides.[7]

Experiments on the anion-exchanger with highly purified calf spleen preparations containing ribonuclease, deoxyribonuclease and cathepsin have resulted in 12-fold purification of ribonuclease as well as separation from the other activities.[7] Preliminary experiments with horse serum have indicated that the resolving power of these cellulose ion-exchangers is greater than that afforded by conventional electrophoresis.

In the chromatographic fractionation of these heart, kidney and spleen enzymes, although decreased stability to dialysis and lyophilization was found, recovery of activity varied from 50–100%. Essentially quantitative nitrogen recoveries were obtained in the serum studies.

(3) The capacity of the CM-cellulose for crystalline horse carbon monoxide hemoglobin in 0.01 $M$ sodium phosphate at $pH$ 6.0 is about 500 mg./g. The adsorbed protein can be eluted by raising the $pH$ to 7.5.

(4) We have recently become aware of the work of C. L. Hoffpauir and J. D. Guthrie (*Textile Research Journal*, **20**, 617 (1950)) who have modified cotton fabrics in a similar manner to produce anion- and cation-exchangers. F. C. McIntire and J. R. Schenk (This Journal, **70**, 1193 (1948)) and E. B. Astwood and co-workers (*ibid.*, **73**, 2969 (1951)) have reported the cation-exchange properties of polysaccharide acid esters and of oxidized cellulose, respectively.

(5) D. E. Green, L. F. Leloir and V. Nocito, *J. Biol. Chem.*, **161**, 599 (1945).

(6) Columns of this type have been used successfully by V. E. Price and R. E. Greenfield of this Laboratory in obtaining highly active crystalline catalase from rat liver in excellent yields.

(7) The experiments with the kidney and spleen enzymes will be reported in more detail with S. M. Birnbaum and M. E. Maver, respectively.

Reprinted from *Nature* **183**:1657–1659 (1959)

# GEL FILTRATION: A METHOD FOR DESALTING AND GROUP SEPARATION

By Dr. JERKER PORATH
**Institute of Biochemistry, University of Uppsala**

AND

Dr. PER FLODIN
**Research Laboratories, Pharmacia, Uppsala**

WE wish to report a simple and rapid method for the fractionation of water-soluble substances. The method is based on a column procedure similar to chromatography in which the stationary phase is comprised of a new type of gel. These gels consist of hydrophilic chains which are cross-linked. They are devoid of ionic groups, the polar character being almost entirely due to the high content of hydroxyl groups. While water-insoluble, the gels are nevertheless capable of considerable swelling. The fractionation depends primarily on differences in molecular size although phenomena have been observed which indicate the influence of other factors.

The idea that gels might be used for the zone separation of substances differing in molecular size is not a new one. Synge and Tiselius[1], for example, have suggested such a method, based on the effect of friction during electro-osmotic migration through gels. In a homologous series of uncharged substances the friction may be expected to rise with increasing molecular weight. Thus the higher homologues should be retarded more strongly than the lower ones. This is, in fact, the case as was experimentally confirmed by Mould and Synge[2].

On filtration through a bed of starch particles, high and low molecular weight substances also behave differently. In this case, however, it is the lower molecular weight substances which are retarded. We frequently encountered this type of phenomenon when using starch as a supporting medium in zone electrophoresis[3]. It appears likely that the cause of this behaviour is a restricted diffusion of the kind observed by Biget[4]. The restriction is related to the effect of friction observed when solutes are electro-osmotically transported through gels.

Lathe and Ruthven's meticulous and thorough investigation[5] gives a good conception of the possibilities which are inherent in starch filtration. However, starch itself has several undesirable properties. Among these are instability, poorly defined composition and the high resistance to flow of starch

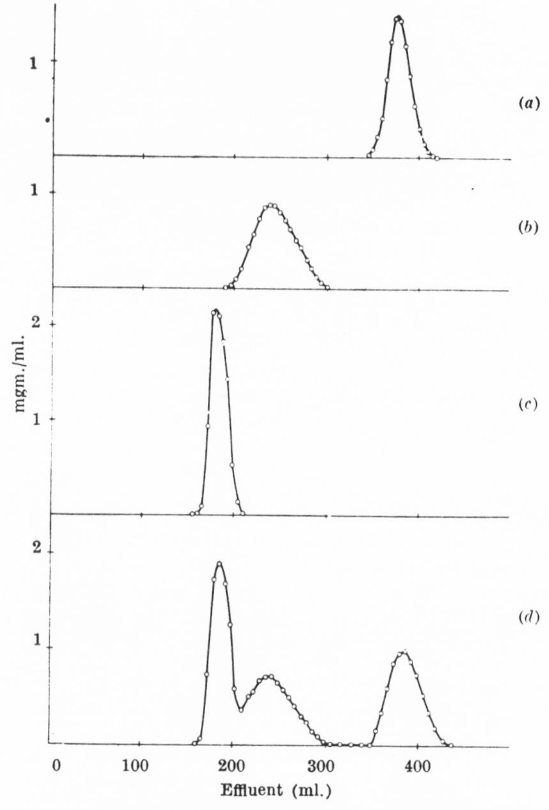


Fig. 1. The diagrams illustrate 'molecular sieving'. In experiments *a*, *b* and *c*, 5 ml. of 1 per cent aqueous solutions of glucose and dextran fractions of $\overline{M}_n$ 1,000 and 20,000, respectively, were filtered through a dextran gel column of the size 4·0 × 36·5 cm. Diagram 1*d* shows the result of a similar experiment with a mixture of the substances mentioned

columns. Together, these properties have caused the method to be of little practical importance so far.

The above disadvantages seem to be eliminated in the case of certain synthetic gels such as dextran gel. The latter is a material originally synthesized in an attempt to obtain an ideal convection depressor[6] for preparative zone electrophoresis. It is prepared by cross-linking dextran in such a way that the polysaccharide chains form a macromolecular network of great stability. A bed of particulate water-swollen dextran gel has very interesting properties. Aqueous solutions filter rapidly through it at ordinary pressures, and no significant quantity of the gel material is dissolved whether in alkaline, neutral or acid medium. Macromolecules are completely excluded from the gel phase and migrate without retention in the interstitial fluid. However, substances of low or intermediate molecular weight penetrate the gel particles to an extent which is in most cases determined by their molecular dimensions and the degree of cross-linkage of the gel. The separating ability of columns packed with dextran gels is based on these facts.

We have studied dextran gels ('Sephadex' obtained from Pharmacia, Uppsala, Sweden) with varying degrees of cross-linkage graded on the basis of water regain[8], and also other hydrophilic gels derived, for example, from starch and polyvinyl alcohol. Like dextran gels, the latter are chemically stable and as a rule have satisfactory mechanical properties. Moreover their separation properties may be varied within wide limits. More detailed reports will be published later. A few simple examples of the use of the method will be given here.

Some experiments with dextran fractions[7] and glucose are shown in Fig. 1. A glass tube with an inner diameter of 4·0 cm. was packed with 100 gm. of dextran gel suspended in distilled water. The final height of the bed was 36·5 cm. The particle size of the dried form of the gel was between 100 and 200 mesh and its water regain was 2·5 gm. per gm. of dry weight[8]. A filter paper was placed above the flat surface of the bed to prevent disturbances. The test solution was applied and when it had completely entered the bed the column was washed with distilled water at a constant rate of 110 ml./hr. The carbohydrate content was determined using the anthrone method. The recoveries in experiments 1*a*, 1*b*, 1*c* and 1*d* were 103, 95, 101 and 97 per cent, respectively. The elution volume of the substances and their distribution are independent of whether the substances are filtered through the columns singly or together. The oligosaccharide fraction has a wider distribution than the other substances due to polydispersity. On filtration it is fractionated, and molecular weight determinations of the material from different parts of the peak indicate that the higher homologues are located at the beginning of the peak.

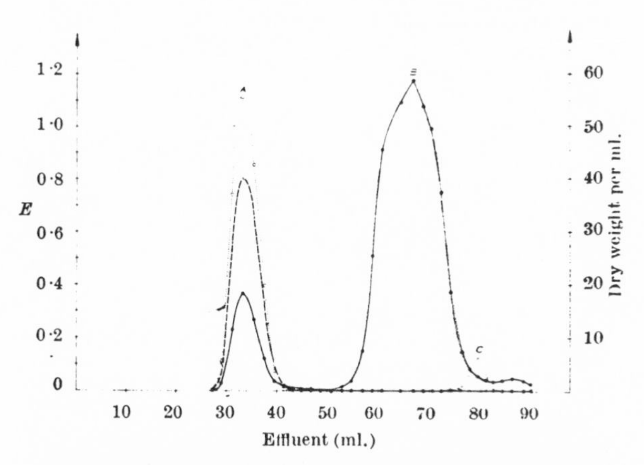


Fig. 2. Diagram showing the separation of ammonium sulphate from serum proteins as obtained after filtration of the solution of the mixture through a dextran gel column (2·5 × 17·5 cm.). Fractions (2 ml. each) were collected and analysed by the method of Folin–Lowry (0·01 ml. aliquots, — × —) and by ultra-violet light absorption at 280 mμ (— ○ —). The weight distribution of the substances in the effluent (— ● —) was determined by weighing 1-ml. aliquots after evaporation *in vacuo*

An experiment with serum proteins performed in order to demonstrate desalting is shown in Fig. 2. 5 ml. of a saturated ammonium sulphate solution was added to 5 ml. of fresh human serum and the precipitate was removed by centrifugation. 3 ml. of the supernatant was applied to the top of a bed (2·5 × 17·5 cm.) of the same type of dextran gel as in the previous experiment. The substances were eluted with distilled water at a constant rate of 25 ml./hr. A qualitative analysis for sulphate ions in peak *A* was negative, showing that the proteins had been freed effectively from salt. No proteins could be demonstrated in peak *B*, which contains the entire amount of ammonium sulphate. Peak *C* contains a low molecular weight substance with an absorption maximum at 292 mμ. The yield of material absorbing in ultra-violet at 280 mμ was 102 per cent, that is, it was quantitative within the limits of error.

The result of a gel filtration is often similar to that obtained by dialysis. The unavoidable dilution is naturally a disadvantage but it may be kept within reasonable limits by a suitable choice of experimental conditions. It should be mentioned that the rate of elution and the concentration of solutes can be varied greatly without causing a serious spreading of the zones or altering the position of the peaks. It is necessary in each problem of separation to compromise with respect to those factors which condition the separation efficiency and the rate of elution.

As a desalting method gel filtration presents many advantages relative to dialysis. For example, we have never observed complications corresponding to the clogging of dialysis membranes. The method is especially suited for changing a protein solution from one buffer system to another, a step which is often necessary prior to ion exchange chromatography and electrophoresis.

Gel filtration may be performed with great rapidity and nevertheless causes the complete separation of two or more substances or classes of substances if the differences in molecular size are sufficient. The scale of a procedure such as the desalting of a protein solution may be increased upwards without difficulty so that considerable quantities may be handled.

Since after each complete operation the gel is automatically regenerated, gel filtration is well suited for cyclic purification procedures.

[1] Synge, R. L. M., and Tiselius, A., *Biochem. J.*, **46**, xli (1950).
[2] Mould, D. L., and Synge, R. L. M., *Biochem. J.*, **58**, 571 (1954).
[3] Flodin, P., and Porath, J., *Biochim. Biophys. Acta*, **13**, 175 (1954).
[4] Biget, A.-M., *Ann. Chim.*, 12th Sér., **5**, 66 (1950).
[5] Lathe, G. H., and Ruthven, C. R. J., *Biochem. J.*, **62**, 665 (1956).
[6] Flodin, P., and Kupke, D. W., *Biochim. Biophys. Acta*, **21**, 368 (1956).
[7] Granath, K. A., *J. Coll. Sci.*, **13**, 308 (1958).
[8] Kressman, T. R. E., in Calmon, C., and Kressman, T. R. E., "Ion Exchangers in Organic and Biochemistry", 108 (Interscience Pub., New York, 1957).

# 61


Reprinted from *Clin. Chim. Acta* **4**:776–778 (1959)

# FRACTIONATION OF POLYPEPTIDES AND PROTEINS ON DEXTRAN GELS*

J. PORATH

*Institute of Biochemistry, University of Uppsala (Sweden)*

A new separation method similar to elution chromatography has been developed and studied over a period of about two years by the author in collaboration with FLODIN AND GELOTTE[1-3]. The procedure is briefly as follows. Cross-linked dextran gel in the form of small beads is allowed to swell in water or an aqueous salt solution, and is then packed into a column. The solution containing the substances to be separated is filtered into the bed. Development is performed with the same solution as was used when packing the column.

In general, substances of large molecular size do not enter the gel phase but move only in the interstitial fluid. These compounds thus emerge in the effluent immediately after the passage of a quantity of liquid equal to the void volume of the column. Compounds of low molecular weight can pass into the gel grains and are consequently retarded on the column. The retention is determined by the molecular dimensions of the solute and by the network of the gel but may also depend on other factors.

Perhaps the method should be regarded as belonging to the realm of chromatography but we use the term gel-filtration, since we believe that sorption is not the sole and sometimes not even the decisive factor in the fractionation. A "molecular sieve" effect may be caused by restricted diffusion. The method has wide application, but only a few examples can be mentioned here.

### *Fractionation of retroplacental serum*

Five ml of retroplacental serum was fractionated on a column of dextran gel of an intermediate degree of cross-linking (Sephadex G 50) (74 · 3 cm) in phosphate–borate buffer of pH 8.4 and ionic strength 0.1 at a rate of approximately 85 ml/h. The fractions were analyzed by the ninhydrin and FOLIN-LOWRY reactions and by measuring the absorption at 260 mμ. Four distinct peaks were obtained (Fig. 1). The fastest moving material, peak A, contained all the substances giving a positive reaction with the FOLIN-LOWRY method. The ninhydrin-positive peak B probably corresponds to hexosamine containing polysaccharides (positive Molish reaction) with free amino groups (no increase in ninhydrin colour after hydrolysis). The peak C contains amino acids as found by paper chromatography after desalting. Peak D has always been observed in all kinds of sera hitherto fractionated on dextran gels. Samples taken over the entire peak show a maximum around 292 mμ. The absorption curve indicates that the peak probably contains uric acid.

In one experiment the material in peak A was subjected to a second filtration under the same conditions. Only one single peak was then obtained, and this again occurred in the position expected for large proteins.

---

* Paper presented at the 7th Colloquium on "Protides of the Biological Fluids", Bruges, May 1959.

*Fractionation of pituitary hormones*

The method is suitable for determining free and protein-bound amino acids and peptides in biological extracts. LINDNER, ELMQVIST AND PORATH[4] utilized gel filtration for the enrichment of posterior pituitary hormones. By gentle extraction of acetone-dried posterior lobe powder of pig pituitaries with 0.1 $N$ acetic acid at room temperature, the pressor and the oxytocic hormones were dissolved in a high molecular form. By gel filtration (Sephadex G 25) at pH 5 all free peptides, amino acids and other contaminants of low molecular weight could be quantitatively removed. If the material

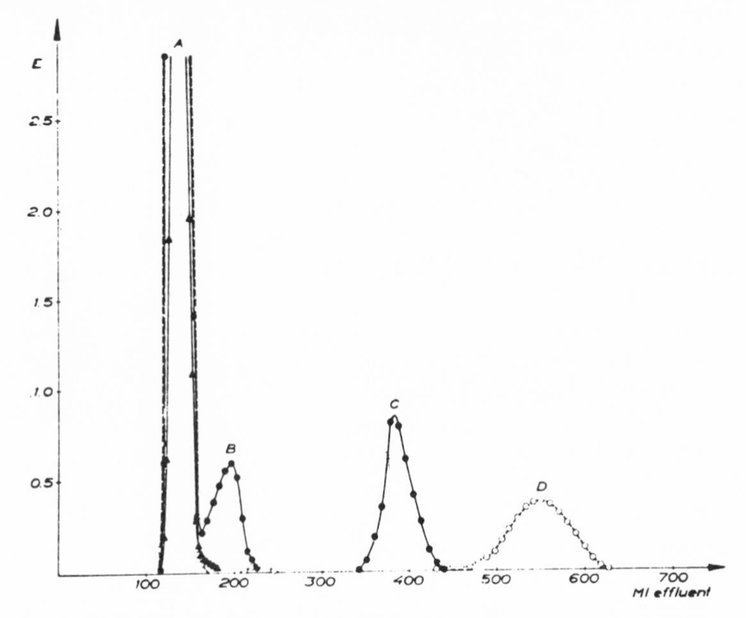


Fig. 1. Separation of 5 ml retroplacental serum on gel of a low degree of cross-linking in phosphate-borate buffer. ●—●—● Ninhydrin color; ▲—▲—▲ FOLIN-LOWRY color; ○—○—○ O.D. at 280 mμ.

of the first peak was dissolved in $N$ formic acid and subjected to a second filtration in acidic medium two main peaks again appeared in the same positions. The second peak was then quite small in size and contained all the vasopressin and oxytocin as free peptides. Thus by two consecutive operations a very efficient purification could be achieved with virtually no loss of activity. The fractionation could be done with less than 0.1 g of the acetone powder or with 100 g or more. It is noteworthy that a separation which can be achieved on a gram scale can be reproduced in minute detail on a milligram scale and very likely also with kilogram quantities.

*Fractionation of proteins*

In the two examples mentioned comparatively highly cross-linked dextran gels were used. Since the gels act as molecular sieves, gels with coarse-meshed network should be suitable for purification of large peptides and small proteins. The decrease in relative flow rate through the gel beds restricts the use of gels of low degrees of cross-linkage. However, some fractionation experiments have been made with gels of a sufficiently low degree of cross-linkage to permit small proteins to enter the gel phase. Proteolytic enzymes in pancreatic juice were thus purified by GELOTTE[5], and BJÖRK AND PORATH achieved a considerable purification of cholinesterase, phosphodiesterase and lecitinase in snake venoms. The latter fractionations were performed in dextrin gel as well as dextran gel. A protein phosphatase from cow spleen, which has been extensively studied by GLOMSET and others, can be effectively retarded on polyvinyl alcohol gels. This is particularly encouraging since the molecular weight of this enzyme is as high as about 23,000.

Some of the most obvious advantages of the gel filtration method can be summarized as follows: (a) The reproducibility is excellent. (b) With electrolytes present high recovery (usually quantitative) is achieved. (c) The electrolyte concentration and pH is usually the same or nearly the same in the free solution and the gel phase since no charged groups are fixed to the polymer matrices of the gels. (d) A gel bed retains its original properties even after daily use for periods of months. (e) Gel filtration can be used on a small or large scale. (f) Filtration can often be made rapidly without sacrificing resolution power. (g) The gel beds are stable in the pH-range 2–10 (and possibly at even higher pH). (h) By series operation involving the use of gels of different degrees of cross-linkage a very considerable purification may be achieved with molecular size as the selective factor. (i) Protein solutions can rapidly be freed from salts.

SUMMARY

A bed of particulate dextran gel (Sephadex) acts as a molecular sieve thus permitting rapid removal of salts, amino acids and other compounds of low molecular weight from protein solutions. By use of dextran gels of different degrees of cross-linkage in consecutive operations, very efficient purifications of peptides of intermediate size can also be achieved. Even proteins are sometimes retarded on dextran gel and may thus be considerably purified.

REFERENCES

1 J. PORATH AND P. FLODIN, *Nature*, 183 (1959) 1657.
2 P. FLODIN, in preparation.
3 J. PORATH, *Biochim. Biophys. Acta*, in the press.
4 E. B. LINDNER, A. ELMQVIST AND J. PORATH, in preparation.
5 B. GELOTTE, *J. Chromatog.*, in the press.
6 W. BJÖRK AND J. PORATH, *Acta Chem. Scand.*, in the press.

# 62

Reprinted from *Natl. Acad. Sci. (USA) Proc.* **61**:636–643 (1968)

*Laboratory of Chemical Biology, National Institute of Arthritis and Metabolic Diseases*

## *SELECTIVE ENZYME PURIFICATION BY AFFINITY CHROMATOGRAPHY*

By Pedro Cuatrecasas, Meir Wilchek,* and Christian B. Anfinsen
*Laboratory of Chemical Biology, National Institute of Arthritis and Metabolic Disease*

The purification of proteins by conventional procedures is frequently laborious and incomplete, and the yields are often low. Enzyme isolation based on a highly specific biological property—strong reversible association with specific substrates or inhibitors—has received only limited attention.[1–4]

In affinity chromatography, the enzyme to be purified is passed through a column containing a cross-linked polymer or gel to which a specific competitive inhibitor of the enzyme has been covalently attached. All proteins without substantial affinity for the bound inhibitor will pass directly through the column, whereas one that recognizes the inhibitor will be retarded in proportion to its affinity constant. Elution of the bound enzyme is readily achieved by changing such parameters as salt concentration or pH, or by addition of a competitive inhibitor in solution.

The successful application of the method requires that the adsorbent have a number of favorable characteristics. Thus, the unsubstituted matrix or gel should show minimal interaction with proteins in general, both before and after coupling to the specific binding group. It must form a loose, porous network that permits easy entry and exit of macromolecules and which retains favorable flow properties during use. The chemical structure of the supporting material must permit the convenient and extensive attachment of the specific ligand under relatively mild conditions, and through chemical bonds that are stable to the conditions of adsorption and elution. Finally, the inhibitor groups critical in the interaction must be sufficiently distant from the solid matrix to minimize steric interference with the binding processes.

In this report the general principles and potential application of affinity chromatography are illustrated by results of its application to the purification of staphylococcal nuclease, α-chymotrypsin, and carboxypeptidase A. The solid matrix used in these studies was Sepharose (a "beaded" form of the cross-linked dextran of highly porous structure, agarose[5]) which displays virtually all the desirable features listed above. Activation of the Sepharose by treatment with cyanogen bromide[6, 7] results in a derivative that can be readily coupled to unprotonated amino groups of an inhibitory analog. The resultant Sepharose-inhibitor gel is a highly stable structure which has nearly ideal properties for selective column chromatography.

*Experimental Procedure.—Materials:* Sepharose 4B was obtained from Pharmacia, cyanogen bromide from Eastman, pdTp and benzoyl-L-tyrosine ethyl ester from Calbiochem.

Staphylococcal nuclease (Foggi strain) was obtained by modification[8] of techniques described by Fuchs *et al.*[9] The following purified enzymes were purchased from Worthington: α-chymotrypsin (CDS 7LC); α-chymotrypsin, DFP-treated (CD-DIP 204); chymotrypsinogen A (CGC 8CC); subtilisin VIII (66B 3080); trypsin, DFP-treated

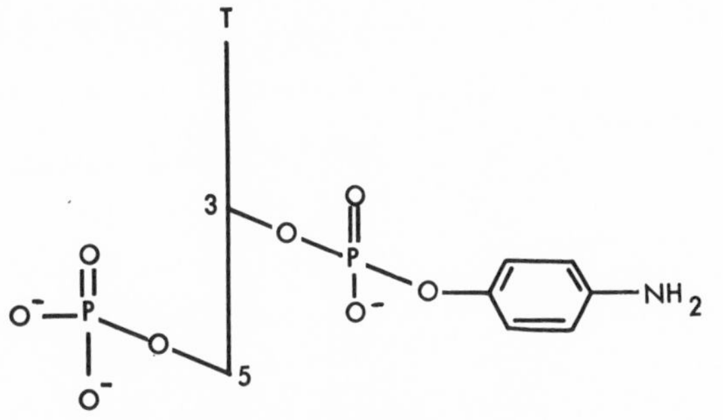

(TDIP 7HA); pancreatic ribonuclease A (RAF 8BA); carboxypeptidase A (COA DFP 8FB); carboxypeptidase B (COB DFP 7GA). Beef pancreas acetone powder (26B-8890) was obtained from Sigma.

3′-(4-Amino-phenylphosphoryl)-deoxythymidine - 5′ - phosphate (Fig. 1) was synthesized from pdTp.[10] The following compounds were prepared by classical methods of peptide synthesis: L-tyrosyl-D-tryptophan, D-tryptophan methyl ester, ε-amino caproyl-D-tryptophan methyl ester, and β-phenylpropionamide.



FIG. 1.—Structure of nuclease inhibitor used for attachment to Sepharose.

*Preparation of substituted Sepharoses:* Cyanogen bromide activation of Sepharose was based on procedures previously described.[6,7] Sepharose (decanted) is mixed with an equal volume of water, and cyanogen bromide (100 mg per ml of settled Sepharose) is added in an equal volume of water. The pH is immediately adjusted to, and maintained at, 11 by titration with 4 $N$ NaOH. When the reaction has ended (about 8 min), the Sepharose is washed with about 20 vol of cold 0.1 $M$ $NaHCO_3$ on a Buchner funnel under suction (about 2 min). The washed Sepharose is suspended in cold 0.1 $M$ $NaHCO_3$, pH 9.0, or another appropriate buffer, in a volume equal to that of the original Sepharose, and the inhibitor is quickly added in a solution representing 5–15% of the final volume. This mixture is stirred gently at 4° for 24 hr, after which it is washed extensively with water and buffer.

The quantity of inhibitor coupled to the Sepharose can be controlled by the amount of inhibitor added to the activated Sepharose, or by adding very large amounts of inhibitor to yield maximal coupling, followed by dilution of the final Sepharose-inhibitor with unsubstituted Sepharose. Furthermore, to increase the amount of inhibitor coupled to the Sepharose, it is possible to repeat the activation and coupling procedures on the already substituted material, provided the inhibitor is stable at pH 11 (for 10 min). In the cases reported here the amount of inhibitor that was coupled was easily estimated by calculating the amount of inhibitor (spectroscopically measured) which was not recovered in the final washings. Alternatively, acid hydrolysis of the Sepharose, followed by amino acid analysis, could be used for quantitation in those cases where amino acids or peptides are coupled to the Sepharose.

The operational capacity of the Sepharose columns for adsorption of protein was determined by two methods. In the first, an amount of enzyme in excess of the theoretical capacity was added to a small column (about 1 ml). After washing this column with buffer until negligible quantities of protein emerged in the effluent, the enzyme was rapidly removed (i.e., by acetic acid washing) and its amount determined. In the second method, small aliquots of pure protein were added successively to the column until significant protein or enzymatic activity emerged; the total amount added, or that which was subsequently eluted, was considered the operational capacity.

*Results.—Staphylococcal nuclease:* This extracellular enzyme of *Staphylococcus aureus*, which is capable of hydrolyzing DNA and RNA, is inhibited, competitively, by pdTp (see ref. 11 for recent review). Thymidine 3′,5′-di-*p*-nitrophenylphosphate is a substrate for this enzyme, which rapidly releases *p*-nitrophenylphosphate from the 5′-position and, slowly, *p*-nitrophenol from the 3′-position.[12] The presence of a free 5′-phosphate group, however, endows various synthetic derivatives with strong inhibitory properties.[12] 3′-(4-Amino-phenylphosphoryl)-deoxythymidine-5′-phosphate[10] (Fig. 1) was selected as an ideal derivative for coupling to Sepharose for affinity chromatography, since it has strong affinity for nuclease ($K_i$, $10^{-6}$ *M*), its 3′-phosphodiester bond is not cleaved by the enzyme, it is stable at pH values of 5–10, the pK of the amino group is low, and the amino group is relatively distant from the basic structural unit (pTp-X) recognized by the enzymatic binding site.[12]

This inhibitor could be coupled to Sepharose with high efficiency, and the resulting inhibitor-Sepharose shows a high capacity for nuclease (Table 1). Columns containing this substituted Sepharose completely and strongly adsorbed samples of pure or partially purified nuclease (Fig. 2). If the amount of nuclease applied to such columns does not exceed half of the operational capacity of the Sepharose, virtually no enzyme activity escapes in the effluent, even after washing with a quantity of buffer more than 50 times the bed volume of the column. Elution of nuclease could be effected by washing with buffers having a pH inadequate for binding (less than 6). The yields of protein and activity were invariably greater than 90 per cent. Acetic acid (pH 3) was a convenient eluant since, with this solvent, the protein emerged sharply (in a few tubes) and the material could be directly lyophilized. The purity of the nuclease obtained in these studies was confirmed by its specific activity, immunodiffusion,[9] and disc gel electrophoresis.[9]

Such columns can be used for rapid and effective large-scale purification of nuclease. For example, 110 mg of pure nuclease could be obtained from a crude nuclease concentrate (the same sample used in Fig. 2) with a 20-ml Sepharose column in an experiment completed in 1.5 hours. When the total concentration of protein in the sample exceeded 20–30 mg per ml, small amounts of nuclease appeared in the first peak of protein impurities, especially if very fast flow rates were used (400 ml/hr). However, with such flow rates, nuclease could still be completely extracted if more dilute samples were applied. The columns used in these experiments could be used repeatedly, and over protracted periods, without detectable loss of effectiveness.

Nuclease treated with cyanogen bromide, which is enzymatically inactive, was not adsorbed or retarded by the nuclease-specific Sepharose columns. Spleen phosphodiesterase, which, like staphylococcal nuclease, hydrolyzes DNA and RNA to yield 3′-phosphoryl derivatives but which is not inhibited by the 5′-phosphoryl nucleotides, passed unretarded through these columns.

A dramatic illustration of the value of these techniques in enzyme purification

TABLE 1. *Efficiency of coupling of 3′-(4-amino-phenylphosphoryl)-deoxythymidine-5′-phosphate (Fig. 1) to Sepharose, and capacity of the resulting adsorbent for staphylococcal nuclease.*

| | μMoles of Inhibitor/ml Sepharose | | Mg of Nuclease/ml Sepharose | |
|---|---|---|---|---|
| Expt. | Added | Coupled* | Theoretical† | Found |
| A | 4.1 | 2.3 | 44 | ... |
| B | 2.5 | 1.5 | 28 | 8 |
| C | 1.5 | 1.0 | 19 | ... |
| D | 0.5 | 0.3 | 6 | 1.2 |

* Determined by the procedures described in the text.
† Assuming equimolar binding.

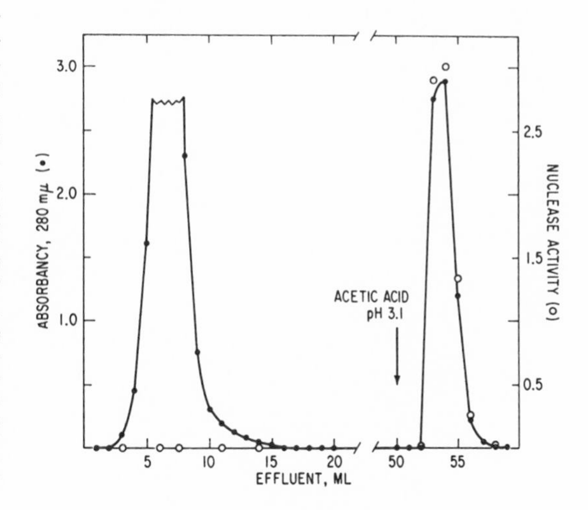


FIG. 2.—Purification of staphylococcal nuclease by affinity adsorption chromatography on a nuclease-specific Sepharose column (0.8 × 5 cm)(sample B, Table 1). The column was equilibrated with 0.05 *M* borate buffer, pH 8.0, containing 0.01 *M* $CaCl_2$. Approximately 40 mg of partially purified material (containing about 8 mg of nuclease) was applied in 3.2 ml of the same buffer. After 50 ml of buffer had passed through the column, 0.1 *M* acetic acid was added to elute nuclease. DNase activity is expressed as the change in absorbancy at 260 mμ caused by 1 μl of sample.[13] 8.2 mg of pure nuclease and all of the original activity was recovered. The flow rate was about 70 ml per hour.

was afforded by the one-step purification obtained by passing a crude culture of *Staphylococcus aureus* through such a nuclease-specific Sepharose column after removal of cells by centrifugation.[14] After adjustment of 500 ml to pH 8, addition of 50 ml 1 *M* $CaCl_2$ was then added to ensure a calcium ion concentration consistent with complex formation. The medium, containing about 6 μg of nuclease per ml, was passed through a 1-ml column with nearly complete adsorption of nuclease activity, which was subsequently eluted with acetic acid.

Affinity columns can also be of use in the separation of active and inactive nuclease derivatives, as from samples alkylated with iodoacetic acid[15] or subjected to proteolytic digestion.[16] The specific Sepharose adsorbent can also be used effectively as an insoluble inhibitor to stop enzymatic reactions (Fig. 3).

*α-Chymotrypsin:* A number of proteolytic enzymes, of which α-chymotrypsin and carboxypeptidase A are examples, are capable of binding, but not hydrolyzing, significantly, the enantiomeric substrate analog. Therefore, the techniques described here should be applicable to a large number of enzymes of this class.

D-tryptophan methyl ester was the inhibitor coupled to Sepharose in experiments with α-chymotrypsin (Fig. 4). When this inhibitor is coupled directly to Sepharose, incomplete and unsatisfactory resolution of the enzyme results (Fig. 4*B*). However, dramatically stronger adsorption of enzyme occurs if a 6-carbon chain (ε-amino caproic acid) is interposed between the Sepharose matrix and the inhibitor (Fig. 4*C*). This illustrates the marked steric interference that results when the inhibitor is attached too closely to the supporting gel. The importance of specific affinity for the enzyme binding site is illustrated by the absence of adsorption of DFP-treated α-chymotrypsin (Fig. 4*D*). Impurities in commercial α-chymotrypsin constituting 4–12 per cent, could be detected in different lots by these techniques. Greater than 90 per cent of the activity and protein added to these affinity columns could be recovered, and the columns could be used repeatedly without detectable loss of effectiveness.

It is notable that significant retention was obtained with an affinity adsorbent

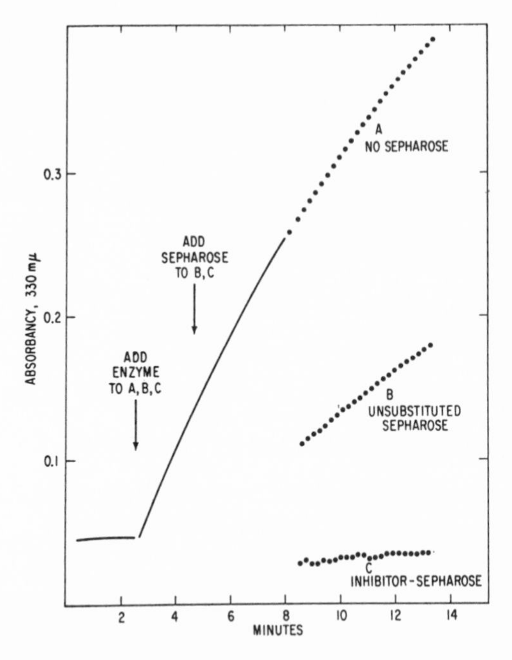


FIG. 3.—Stopping of nuclease-catalyzed reaction by addition of Sepharose inhibitor (sample *A*, Table 1). Ten μg of nuclease was added to each of three samples containing 1.5 ml of 0.05 *M* Tris buffer, pH 8.8, 0.01 *M* $CaCl_2$, and 0.1 mM synthetic substrate, thymidine 3′-5′-di-*p*-nitrophenyl-phosphate.[10, 12] The change in absorbancy at 330 mμ, representing release of *p*-nitrophenylphosphate, was recorded continuously in cuvette *A*. At 4.6 min, 0.5 ml of untreated Sepharose or of Sepharose coupled with inhibitor were added to *B* and *C*, respectively. After a 2-min centrifugation to remove the Sepharose, the changes in absorbancy were recorded (8 min). The difference in activity between *A* and *B* is due to dilution.

that contained a relatively weak inhibitor, N-ε-amino caproyl-D-tryptophan methyl ester. N-acyl-D-tryptophan esters have $K_i$ values of about 0.1 mM,[17] some 100 times greater than that of the inhibitor used for purification of staphylococcal nuclease. These results suggest that unusually strong affinity constants will not be an essential requirement for utilization of these techniques.

The relatively unfavorable affinity constant of α-chymotrypsin for the D-tryptophan methyl ester derivatives was compensated for by coupling a very large amount of inhibitor to the Sepharose. About 65 μmoles of the compound were present per milliliter of Sepharose during the coupling procedure, and there was an uptake of 10 μmoles per ml, with a resultant effective concentration of inhibitor, in the column, of about 10 mM. Such a high degree of substitution

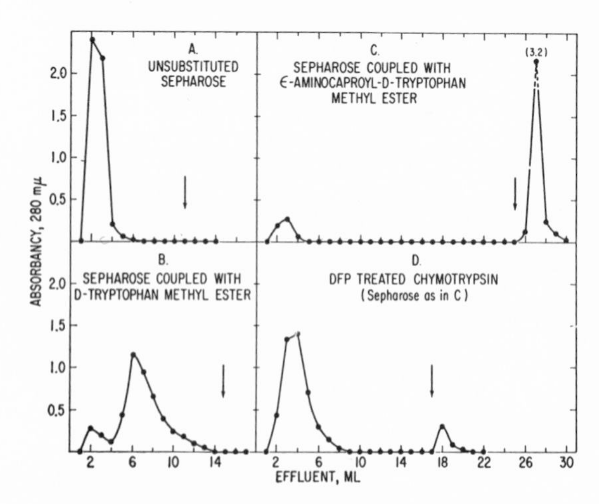


FIG. 4.—Affinity chromatography of α-chymotrypsin on inhibitor Sepharose columns. The columns (0.5 × 5 cm) were equilibrated and run with 0.05 *M* Tris-Cl buffer, pH 8.0, and each sample (2.5 mg) was applied in 0.5 ml of the same buffer. One-milliliter fractions were collected, the flow rate was about 40 ml per hour, and the experiments were performed at room temperature. α-Chymotrypsin was eluted with 0.1 *M* acetic acid, pH 3.0 (*arrows*). Peaks preceding the arrows in *B*, *C*, *D* were devoid of enzyme activity.[18]

occurred despite the relatively high pK of the amino group of the ε-amino caproyl derivative. The coupling was done in 0.1 *M* $NaHCO_3$ buffer, pH 9.0; higher pH values could not be safely used because of the probability of hydrolysis of the ester bond. The highly substituted Sepharose derivatives retained very good flow properties.

Figure 5 illustrates the effect of pH and ionic strength on the chromatographic patterns. Although stronger binding appears to occur with buffer of lower ionic strength (0.01 *M*), this should not be used since some proteins, such as pancreatic ribonuclease, will adsorb nonspecifically to unsubstituted or inhibitor-coupled Sepharose under those conditions. Figure 5 shows the patterns obtained with a number of other enzymes, emphasizing the specificity of the α-chymotrypsin-specific Sepharose columns. Very small chymotryptic impurities

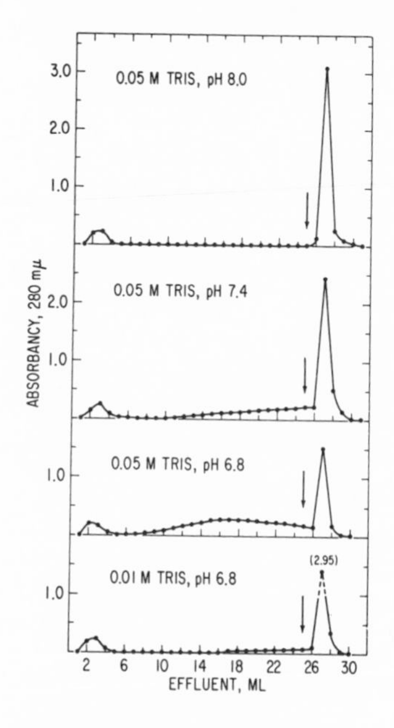


Fig. 5.—Effects of pH and ionic strength on affinity adsorption of α-chymotrypsin on a column (0.5 × 5 cm) of Sepharose coupled with ε-amino caproyl-D-tryptophan methyl ester. A sample containing 2.5 mg of α-chymotrypsin in 0.5 ml of buffer was applied to the column. Elution of α-chymotrypsin was performed with 0.1 *M* acetic acid, pH 3.0 (*arrow*). Other conditions were as in Fig. 2. The first peak (tubes 2–4) was devoid of chymotrypsin activity,[18] and the specific activity of the subsequently eluted protein was constant.

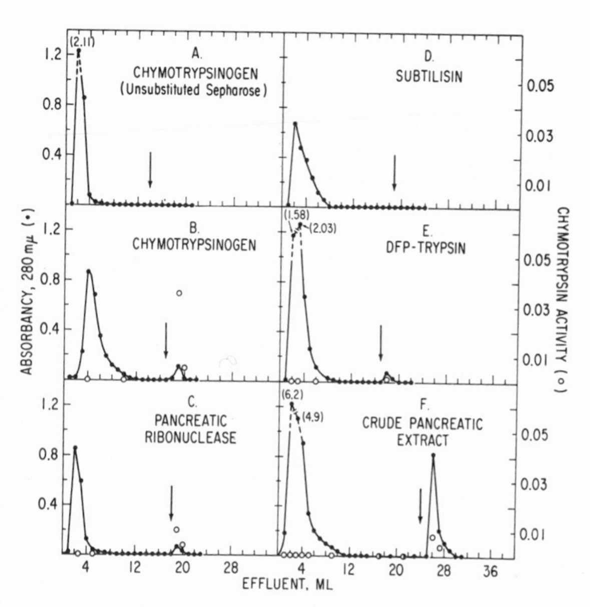


Fig. 6.—Chromatographic patterns obtained by passing several enzyme preparations through a column (0.5 × 5 cm) of Sepharose coupled with ε-amino caproyl-D-tryptophan methyl ester. The column was equilibrated and run with 0.05 *M* Tris-Cl buffer, pH 8.0. Approximately 3 mg of protein (*A* through *E*), dissolved in 0.5 ml of the same buffer, was applied to each column. Chymotrypsin was eluted with 0.1 *M* acetic acid (*arrow*). The sample used in *F* consisted of 1 ml of a supernatant (280 mμ absorbancy, about 15) obtained by dissolving about 100 mg of an acetone powder of bovine pancrease in 3 ml of Tris-Cl, pH 8.0, followed by centrifugation for 20 min at 4000 × *g*. Enzyme activity is expressed as the change in absorbancy (at 256 mμ) per minute per 5 μl of sample, with benzoyl-L-tyrosine ethyl ester as substrate.[18] Other conditions were as in Figs. 4 and 5.

can be readily removed from other enzymes by these techniques. This may be a useful way of removing chymotryptic impurities from other proteases used for structural studies, where even small traces can lead to unexpected cleavages.

It is of interest that chymotrypsinogen A is very slightly but significantly retarded by the chymotrypsin-specific column (Figs. 6*A* and *B*), suggesting that this precursor is capable of weakly recognizing this substrate analog. A proteolytic enzyme of broad substrate specificity, subtilisin, is not adsorbed (Fig. 6*D*). A small amount of chymotryptic-like protein could be readily separated from a crude pancreatic digest (Fig. 6*F*). This material contained all of the chymotryptic activity[18] of the crude digest, but the exact nature of this material or the reason for the low specific activity were not determined.

Unlike the results obtained with the nuclease-specific Sepharose system (Fig. 3), the chymotrypsin-specific Sepharose derivative was relatively ineffective in stopping the hydrolysis of benzoyl-L-tyrosine ethyl ester when added to a reaction mixture. Attempts to elute α-chymotrypsin from a column, like that shown in Figure 4, with a 0.01 *M* solution of the above substrate were unsuccessful. However, elution did occur with a 0.018 *M* solution of β-phenylpropionamide, an inhibitor with a $K_i$ of 7 mM.[19] If the Sepharose column was equilibrated and developed with 0.0058 *M* β-phenylpropionamide solution, α-chymotrypsin was only moderately retarded. These results indicate, again, that the processes involved in the separations are clearly related to the functional affinity of the enzymatic binding site for specific structural substances.

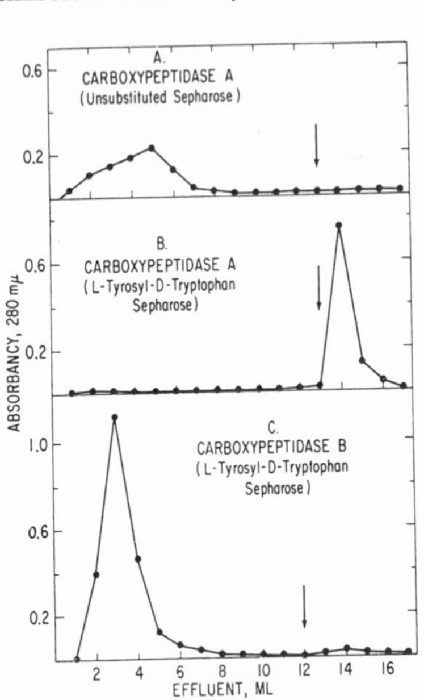


FIG. 7.—Affinity chromatography of carboxypeptidase A on a column (0.5 × 6 cm) of Sepharose coupled with L-tyrosine-D-tryptophan. The buffer used was 0.05 *M* Tris-Cl, pH 8.0, containing 0.3 *N* NaCl. About 1 mg of pure carboxypeptidase A (*A*, *B*) and 1.8 mg of carboxypeptidse B (*C*), in 1 ml of the same buffer, were applied to the columns. Elution was accomplished with 0.1 *M* acetic acid (*arrow*).

*Carboxypeptidase A:* A specific adsorbent for this enzyme was prepared by coupling the dipeptide, L-tyrosine-D-tryptophan, to Sepharose. In the coupling procedure, about 60 μmoles of the dipeptide inhibitor were added per milliliter of Sepharose, and approximately 8 μmoles per ml of Sepharose were coupled. Figure 7 illustrates that this enzyme was strongly adsorbed by a column containing such a substituted Sepharose. The yields obtained upon elution were again quantitative.

*Discussion.*—In recent years there has been considerable interest in the covalent attachment of biologically active compounds (i.e., enzymes, antibodies, and antigens) to insoluble polymers.[20] These materials, especially the derivatives of cellulose, have found use in the purification of antibodies,[21] nucleotides,[22] complementary strands of nucleic acids,[23] certain species of transfer RNA,[24] and enzymes.[1-3]

The principles and procedures, as illustrated and outlined in this communication, should be of value in the purification and isolation of a great many biologically active proteins or polypeptides which can reversibly and specifically bind to small molecules. If the latter are not chemically altered during the reversible adsorption process (e.g., an inhibitor), and if an amino group can be introduced in a region of its structure in such a way that binding to the macromolecule is unaffected, the procedures outlined here should be directly applicable.

*Summary.*—Principles and techniques for selective enzyme purification by affinity adsorption to inhibitor-Sepharose columns are presented and illustrated by experiments performed on staphylococcal nuclease, α-chymotrypsin, and carboxypeptidase A. Inhibitory substrate analogs linked to Sepharose provide adsorbents on which enzymes can be purified rapidly and completely in a single step.

The authors acknowledge the collaborative assistance of Dr. Sheldon Schlaff in many of these studies, and we are grateful for his efforts in the further extension of these methods.[15] We wish to express our gratitude to Mrs. Juanita Eldridge for technical assistance.

* Fellow in the Visiting Program of the U.S. Public Health Service (1967–1968). On leave of absence from the Weizmann Institute of Science, Rehovoth, Israel.

[1] Lerman, L. S., these PROCEEDINGS, **39**, 232 (1953).
[2] Arsenis, C., and D. B. McCormick, *J. Biol. Chem.*, **239**, 3093 (1964).
[3] *Ibid.*, **241**, 330 (1966).
[4] McCormick, D. B., *Anal. Biochem.*, **13**, 194 (1965).
[5] Hjertén, S., *Biochim. Biophys. Acta*, **79**, 393 (1964).
[6] Axén, R., J. Porath, and S. Ernback, *Nature*, **214**, 1302 (1967).
[7] Porath, J., R. Axén, and S. Ernback, *Nature*, **215**, 1491 (1967).
[8] Moravek, L., C. B. Anfinsen, J. Cone, H. Taniuchi, manuscript in preparation.
[9] Fuchs, S., P. Cuatrecasas, and C. B. Anfinsen, *J. Biol. Chem.*, **242**, 4768 (1967).
[10] Wilchek, M., P. Cuatrecasas, and C. B. Anfinsen, unpublished.
[11] Cuatrecasas, P., H. Taniuchi, and C. B. Anfinsen, in *Brookhaven Symposia in Biology*, in press.
[12] Cuatrecasas, P., M. Wilchek, and C. B. Anfinsen, manuscript in preparation.
[13] Cuatrecasas, P., S. Fuchs, and C. B. Anfinsen, *J. Biol. Chem.*, **242**,1541 (1967).
[14] In collaboration with Dr. G. Omenn.
[15] Schlaff, S., unpublished.
[16] Taniuchi, H., C. B. Anfinsen, and A. Sodja, these PROCEEDINGS, **58**, 1235 (1967), and unpublished data.
[17] Huang, H. T., and C. Niemann, *J. Am. Chem. Soc.*, **73**, 3228 (1951).
[18] Hummel, B. C. W., *Can. J. Biochem. Physiol.*, **37**, 1393 (1959).
[19] Foster, R. J., and C. Niemann, *J. Am. Chem. Soc.*, **77**, 3365 (1955).
[20] Silman, I., and E. Katchalski, *Ann. Rev. Biochem.*, **35**, 873 (1966).
[21] Moudgal, N. R., and R. R. Porter, *Biochim. Biophys. Acta*, **71**, 185 (1963).
[22] Sander, E. G., D. B. McCormick, and L. D. Wright, *J. Chromatog.*, **21**, 419 (1966).
[23] Bautz, E. K. F., and B. D. Holt, these PROCEEDINGS, **48**, 400 (1962).
[24] Erhan, S., L. G. Northrup, and F. R. Leach, these PROCEEDINGS, **53**, 646 (1965).

Reprinted from *Biochemistry* 5:1269–1279 (1966)

# Reversible Cold Inactivation of a 17β-Hydroxysteroid Dehydrogenase of Human Placenta: Protective Effect of Glycerol*

Joseph Jarabak, A. Elmore Seeds, Jr.,† and Paul Talalay

ABSTRACT: Purified preparations of the pyridine nucleotide linked 17β-hydroxysteroid dehydrogenase of human placenta lose activity rapidly in buffered aqueous solutions at temperatures below 10–15°. The rate of the cold inactivation is reduced by the addition of low concentrations of 17β-estradiol and of certain pyridine nucleotides, and by high concentrations of some polyvalent anions. The addition of 20% or more of glycerol totally protects the enzyme against cold inactivation. The loss of enzyme activity which occurs at low temperatures may be largely restored by warming the enzyme solutions at 30° under carefully controlled conditions but is no longer reversible after prolonged cooling. Cold treatment of the enzyme preparations is accompanied by the formation of enzymatically inactive high molecular weight components which may be separated from the active enzyme by gel filtration on Sephadex or by electrophoresis on polyacrylamide gels.

The process of cold inactivation appears to involve complex structural changes, including probable alterations in the configuration of the native protein as well as a series of aggregations resulting in the formation of multiple polymeric species. The properties of other cold-inactivated enzymes and the protective effect of glycerol are discussed.

A method for obtaining highly purified preparations of the pyridine nucleotide linked 17β-hydroxysteroid dehydrogenase of human placenta has been described (Jarabak *et al.*, 1962). This enzyme, which may function as either a dehydrogenase or a pyridine nucleotide transhydrogenase (Talalay and Williams-Ashman, 1958, 1960; Jarabak *et al.*, 1962), catalyzes the following freely reversible dehydrogenation reaction[1]

$$17\beta\text{-estradiol} + NAD^+ (NADP^+) \rightleftharpoons \text{estrone} + NADH (NADPH) + H^+$$

Difficulties were encountered during the development of satisfactory methods for the purification of this enzyme because of its profound instability under conditions commonly employed in enzyme fractionation. Addition of pyridine nucleotides, 17β-estradiol, or glycerol stabilized the enzyme (Langer and Engel, 1958; Talalay *et al.*, 1958; Jarabak *et al.*, 1962; Talalay, 1962). The effect of glycerol was especially striking in this respect. Whereas dilute solutions of partially purified 17β-hydroxysteroid dehydrogenase lost more than 90% of their activity upon storage for 24 hr near 0°, the same solutions in 50% glycerol could be stored for many months at this temperature and could be heated at 67° for several hours without the slightest loss in enzymatic activity. Success in obtaining highly purified preparations of this enzyme depended upon carrying out all the steps of the procedure in mixtures of glycerol and water of varying composition (Jarabak *et al.*, 1962). In these media, the enzyme was isolated in good yield by a method involving salt fractionation, heat treatment, and ion exchange chromatography on DEAE- and Ecteola-celluloses. The final product was substantially homogeneous by chromatographic and electrophoretic criteria. Other glycols and their ethers, sucrose, and appropriate concentrations of ethanol were also found to stabilize the enzyme (Jarabak, 1962; Jarabak *et al.*, 1962; Talalay, 1962). Closer examination of the behavior of the enzyme led to the realization that it had the interesting property of losing activity rapidly upon exposure to temperatures below about 10–15°, but that the addition of glycerol protected against this cold inactivation. Upon warming the enzyme under carefully controlled conditions, a substantial fraction of the lost activity could be restored. This paper describes some characteristics of the reversible cold inactivation of the placental 17β-hydroxysteroid dehydrogenase and the protective effect of glycerol and other agents.

## Experimental Section

The 17β-hydroxysteroid dehydrogenase was purified

* From the Department of Pharmacology and Experimental Therapeutics and the Department of Medicine, The Johns Hopkins University School of Medicine, Baltimore, Maryland. *Received December 6, 1965.* These studies were supported in part by Grant AM 07422 from the U. S. Public Health Service.

† Present address: Department of Gynecology and Obstetrics, The Johns Hopkins University School of Medicine, Baltimore, Md. Graduate Training Grant NIH 2 T1 HD-23-04.

[1] Abbreviations used: NAD⁺, nicotinamide–adenine dinucleotide; NADP⁺, nictotinamide-adenine dinucleotide phosphate; NADH, reduced nicotinamide-adenine dinucleotide; NADPH, reduced nicotinamide–adenine dinucleotide phosphate.

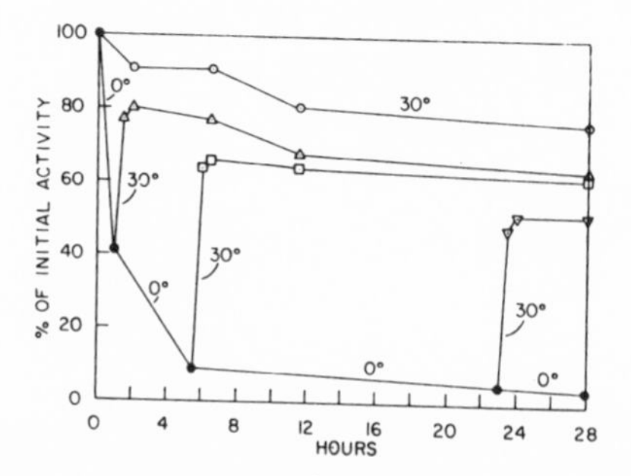


FIGURE 1: Inactivation of placental 17β-hydroxysteroid dehydrogenase at 0° and reactivation at 30°. The enzyme (120,000 units/ml; 1.28 mg of protein/ml) was diluted to a protein concentration of 25.6 μg/ml by the addition of 0.005 M potassium phosphate at pH 7.0. The final glycerol concentration was 1%. One portion of the diluted enzyme was maintained at 30°, while another portion was rapidly cooled and stored at 0°. At the indicated times (1, 5.5, and 23 hr) aliquots were removed from the cold solution, rapidly warmed to 30°, and maintained at this temperature. Periodic measurements of activity were carried out, and the results are expressed in terms of the original activity of the solution. No apparent reactivation occurs during the assay as judged by linearity with time.

according to the procedure of Jarabak *et al.* (1962). Details of chromatographic and electrophoretic purity were given at that time. With the exception of the experiment described in Figure 4, all other studies were carried out on a single purified preparation which was stored for many months at 4° in a medium containing 50% glycerol, 0.005 M potassium phosphate, 0.001 M EDTA, and 0.007 M β-mercaptoethanol, at pH 7.0. The protein concentration of this preparation was 1.28 mg/ml. The enzymatic activity varied somewhat during the course of this work, and is given with the appropriate protocols. Unless otherwise stated, activity measurements were carried out with 3-acetylpyridine–adenine dinucleotide as hydrogen acceptor according to Jarabak *et al.* (1962). The assay cuvets contained in a final volume of 3.0 ml: 440 μmoles of sodium pyrophosphate buffer at pH 10.2, 25 mg of crystalline bovine serum albumin (usually added as 0.5 ml of a 5% solution), 0.3 μmole of 17β-estradiol in 0.04 ml of 95% ethanol, 1.9 μmoles of 3-acetylpyridine–adenine dinucleotide, and sufficient enzyme to produce a change in absorbance at 363 mμ of between 0.001 and 0.040/min in a cuvet of 1.0-cm light path. This normally required about 0.5 μg of protein. The final pH was between 9.3 and 9.4. All activity measurements were carried out at 25°, irrespective of the earlier temperature history of the enzyme. The enzymatic reactions were initiated by the addition of enzyme (in a small volume) to the otherwise complete reaction system which had been equilibrated at 25 ± 0.5° The reaction rate was determined from changes in absorbance at 363 mμ against a blank cuvet from which nucleotide was omitted. The initial reaction rates were strictly constant over the period examined (the initial 3–5 min). When different aliquots of a single enzyme preparation were assayed in this system, the observed initial reaction rates were strictly proportional to enzyme concentration over a wide range. One unit of enzyme activity is defined as the quantity producing an absorbance change of 0.001/min at 363 mμ under the stated conditions.

Reagent grade urea was recrystallized from ethanol. Spectroscopic quality glycerol was used without further purification. The concentrations of glycerol are expressed by volume. Sephadex G-100 and G-200 beads (Pharmacia) were thoroughly equilibrated with appropriate buffer systems, and the fine particles were removed by decantation before packing of the columns. The void volumes of the Sephadex columns were determined with Blue Dextran 2000 (Pharmacia). Disk electrophoresis on polyacrylamide gels was carried out according to the general directions of Ornstein (1964) and Davis (1964). More detailed information on the electrophoresis is given in the legend of Figure 7.

## Results

*Cold Inactivation and Its Reversal.* The time course of inactivation of 17β-hydroxysteroid dehydrogenase upon storage at 0 and 30° in 1% glycerol is shown in Figure 1. The activities of all samples were assayed at 25°. The enzyme stored at 0° undergoes a rapid initial loss of activity which is followed by a prolonged and much more gradual inactivation. The enzyme preparation stored at 30° lost activity at the same rate as the slow phase of the enzyme stored in the cold. At 0°, more than 50% of the initial activity was lost within the first hour of storage, and only 3–4% of the activity remained after 28 hr. When aliquots of the cold-inactivated enzyme were warmed to 30°, they rapidly regained a substantial fraction of their initial activity. Upon continuing storage at 30°, the reactivated enzyme lost activity at the same slow rate as the control enzyme which had been stored at 30° from the beginning of the experiment. After prolonged storage in the cold the activity of the enzyme cannot be restored on warming. Although the addition of glycerol arrests the cold-inactivation process, the enzyme may be readily reactivated by warming at 30° in the presence of 20% glycerol. All the measurements shown in Figure 1 were made in the assay system containing 3-acetylpyridine–adenine dinucleotide, but entirely parallel changes in activity were observed with NAD or NADP as hydrogen acceptors.

*Effect of Temperature on Rate of Inactivation and Reactivation.* Examination of the effect of temperature on the inactivation of the enzyme reveals that little loss in activity is observed in 1 hr until the temperature is lowered to below about 11–12° (Figure 2A and 2B). The initial rapid phase of inactivation is highly dependent on temperature whereas the rate of the subsequent slower inactivation appears to be little affected by temperature changes within the range examined (Figure 2A). The biphasic time course of the inactivation is more clearly

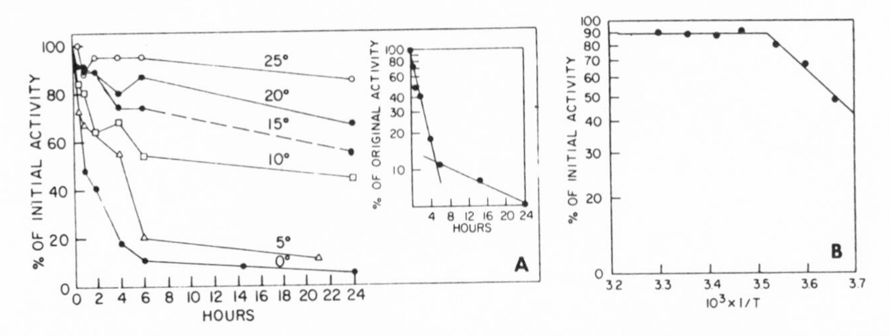

FIGURE 2: Effect of temperature on the rate of inactivation of the 17β-hydroxysteroid dehydrogenase. (A) The enzyme preparation described in Figure 1 was diluted with 0.005 M potassium phosphate buffer at pH 7.0 to a final protein concentration of 25.6 μg/ml. Aliquots were assayed immediately after dilution and after various times of storage at the indicated temperatures. The insert is a semilogarithmic plot of the activity as a function of time for the sample stored at 0°. (B) The fraction of the initial enzyme activity remaining after 1 hr of storage at various temperatures is plotted *vs.* the reciprocal of the absolute temperature (°K). The results obtained from Figure 2A at 1 hr have been used for this plot.

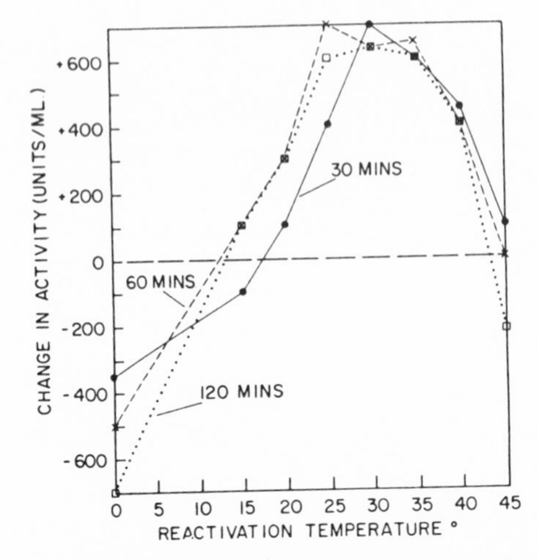

FIGURE 3: Reactivation of cold-inactivated 17β-hydroxysteroid dehydrogenase as a function of temperature and time. The enzyme solution was diluted to a protein concentration of 25.6 μg/ml as described in Figure 1. After storage at 0° for 1 hr, the activity of the diluted enzyme had fallen from 2200 units/ml to 1000 units/ml. The graph shows the change in activity from the latter level after incubation of the enzyme for 30, 60, and 120 min at the indicated temperatures.

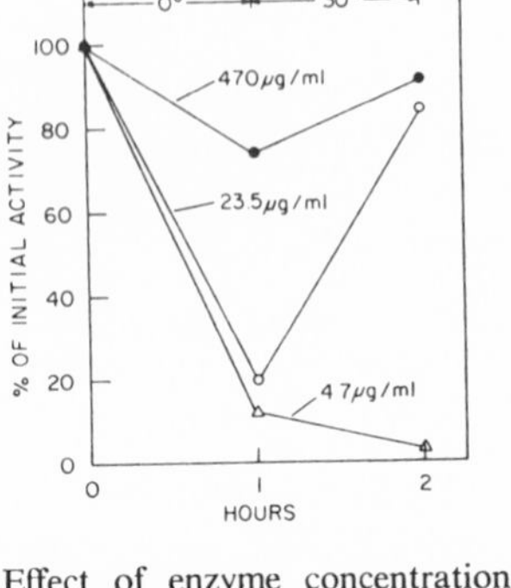

FIGURE 4: Effect of enzyme concentration on cold inactivation and subsequent reactivation on warming. The preparation of 17β-hydroxysteroid dehydrogenase used in these experiments had an initial protein concentration of 18.8 mg/ml (1.06 × 10[6] units/ml). Dilutions were made so that the enzyme contained 0.005 M potassium phosphate and 1% glycerol at pH 7.0. The initial assays were performed immediately upon dilution. The solutions were rapidly cooled at 0° and appropriate aliquots were assayed after 1 hr, warmed to 30°, and reassayed after standing for 1 hr at the latter temperature. Samples which were not cooled retained more than 90% of their original activity after 2 hr.

shown on a semilogarithmic plot (see insert in Figure 2A).

The optimum temperature of reactivation of the cold-inactivated enzyme is 25 to 35° (Figure 3). Above about 11–12° the cold-inactivated enzyme gains activity under the conditions described. Figure 3 also shows that the rate and ultimate extent of reactivation are temperature dependent.

*Effect of Protein Concentration.* The rate and extent of cold inactivation of the enzyme and its reversal are also profoundly affected by the protein concentration. Figure 4 shows that only 25% of the activity was lost in 1 hr at 0° when the protein concentration was 470 μg/ml, whereas nearly a 90% loss of activity occurred at a concentration of 4.7 μg/ml. The activity of the very dilute cold-inactivated enzyme could not be restored on

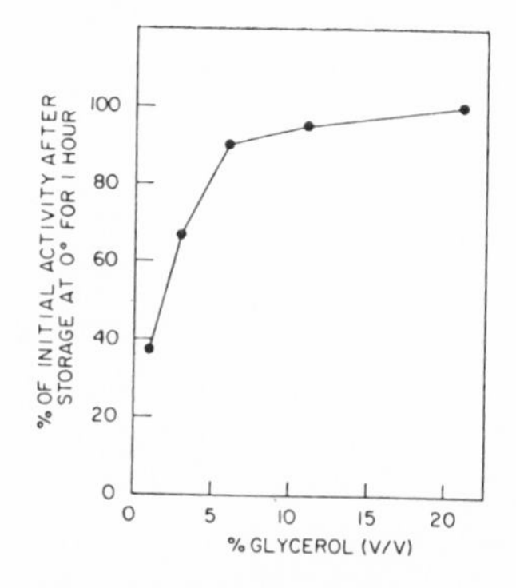


FIGURE 5: Protective effect of varying concentrations of glycerol on the cold inactivation of the placental 17β-hydroxysteroid dehydrogenase. The enzyme preparation described in Figure 1 was diluted to a protein concentration of 25.6 μg/ml with a medium containing 0.005 M potassium phosphate at pH 7.0 and various concentrations of glycerol. The enzyme was assayed immediately after dilution, cooled rapidly to 0°, and assayed for a second time after storage for 1 hr at this temperature. All assays were conducted at 25°.

TABLE I: Protection of 17β-Hydroxysteroid Dehydrogenase against Cold Inactivation by Various Substances.[a]

| Additions | Concentration | Residual Activity after Storage at 0° 1 hr (%) | 24 hr (%) |
|---|---|---|---|
| No additions | | 41 | 3 |
| Pyridine nucleotides | | | |
| NADP | 1 μM | 89 | 71 |
| NADP | 10 μM | 91 | 84 |
| NAD | 1 μM | 63 | 5 |
| NAD | 100 μM | 96 | 60 |
| 3-Acetylpyridine-AD | 1 μM | 50 | 3 |
| 3-Acetylpyridine-AD | 100 μM | 82 | 19 |
| 3-Pyridine aldehyde-AD | 100 μM | 82 | 21 |
| 3-Thionicotinamide-AD | 100 μM | 90 | 80 |
| Steroids | | | |
| 17β Estradiol | 1 μM | 88 | 42 |
| 17β-Estradiol | 10 μM | 93 | 83 |
| Potassium phosphate, pH 7.0 | 100 mM | 54 | 7 |
| Potassium phosphate, pH 7.0 | 500 mM | 87 | 41 |
| Potassium phosphate, pH 7.0 | 1 M | 87 | 87 |

[a] The enzyme preparation described in Figure 1 was diluted to 25.6 μg of protein/ml in the appropriate medium. The activity was assayed immediately after dilution and the enzyme cooled rapidly to 0°. The activity was measured again after 1 and 24 hr. All media contained a final concentration of 1% glycerol and 5 mM potassium phosphate (except when phosphate was added). The 17β-estradiol was added in ethanol to give a final concentration of 2%. All assays were carried out with 3-acetylpyridine–adenine dinucleotide (3-Acetylpyridine-AD) as acceptor. The final concentrations of protective nucleotides were too low to affect the assay system in the cuvets.

warming, whereas at the higher concentration the process was almost fully reversible.

*Effect of Composition of Medium on Cold Inactivation.* The composition of the medium influences not only the rate but also the extent of the cold inactivation of the enzyme. In contrast to the rapid loss of activity which occurs at 0° in 1% glycerol, the presence of 50% glycerol stabilized the enzyme for many months. The efficiency of various concentrations of glycerol in protecting the enzyme against cold inactivation is shown in Figure 5. The purified enzyme is very stable in 50% glycerol at room temperature and at higher temperatures (Talalay, 1962; Jarabak *et al.*, 1962). Although the enzyme appears to be quite soluble in almost pure glycerol, it loses activity in this medium. For example, after 24 hr at 23° in 99.5% glycerol, the enzyme had lost 27% of its activity while after 144 hr under these conditions, the loss was 85%.

Table I summarizes the effect of 17β-estradiol, various pyridine nucleotides, and potassium phosphate on the stability of the enzyme after storage for 1 and 24 hr at 0°. The protective effects of the pyridine nucleotides appear to be related to their Michaelis constants (Talalay *et al.*, 1958; Talalay and Williams-Ashman, 1960). High concentrations of potassium phosphate efficiently protected the enzyme against cold inactivation. In other experiments, it was found that high concentrations of potassium sulfate (1.0 M) also exercised a protective effect, but similar concentrations of potassium chloride or potassium nitrate greatly accelerated the loss of enzyme activity in the cold. 17β-Estradiol (Table I), and equivalent concentrations of testosterone, offered some degree of protection against cold inactivation. It is of interest that diethylstilbestrol (1–10 μM) also protected the enzyme against cold inactivation although it is not oxidized by the enzyme and actually competitively inhibits the oxidation of 17β-estradiol (Adams *et al.*, 1962).

The addition of urea to dilute solutions of the enzyme accelerates the loss of activity at 23 and 0°, but the cold inactivation is stimulated to a much greater extent by this reagent. Table II shows that the inactivating effect of urea increases with concentration and can be counteracted by the incorporation of glycerol into the solu-

TABLE II: Effect of Temperature on the Stability of the Placental 17β-Hydroxysteroid Dehydrogenase in Urea and in Glycerol.[a]

| | Residual Activity after Storage (%) | | | | | | | |
|---|---|---|---|---|---|---|---|---|
| | 1% Glycerol | | 21% Glycerol | | 51% Glycerol | | | |
| Additions | 1 hr at 0° | 1 hr at 23° | 1 hr at 0° | 1 hr at 23° | 1 hr at 0° | 24 hr at 0° | 1 hr at 23° | 24 hr at 23° |
| None | 40 | 95 | 100 | 100 | 100 | 100 | 100 | 100 |
| Urea, 1 M | 5 | 74 | 44 | 95 | | | | |
| Urea, 4 M | 0 | 0 | 0 | 28 | 87 | 66 | 101 | 99 |
| Urea, 8 M | | | | | | | 0 | |

[a] The enzyme preparation described in Figure 1 was diluted 50-fold with 0.005 M potassium phosphate buffer at pH 7.0 and urea in the concentrations indicated. The activity was assayed immediately, the solutions were rapidly cooled at 0°, and activity measurements were made after 1 and 24 hr of storage at this temperature.

tions. This is exemplified by the finding that in 1 hr at 23° the entire activity was lost in 4 M urea and 1% glycerol, but the entire activity was retained in a solution containing 4 M urea and 51% glycerol.

*Sephadex Gel Filtration.* The nature of the structural changes which accompany the cold inactivation of the 17β-hydroxysteroid dehydrogenase was examined by filtration of the enzyme on Sephadex gels. When a purified preparation of the enzyme which had been stored in 50% glycerol was applied to a Sephadex G-100 column (operated at 23°, previously equilibrated with 0.01 M potassium phosphate buffer at pH 7.0), the protein was eluted in two distinct peaks. The first was eluted in the void volume and was enzymatically inactive, whereas the second peak was eluted at 1.25–1.35 void volumes and contained the entire enzymatic activity (Figure 6A). It is not clear whether the material eluted in the void volume represents an impurity present in the enzyme, an inactive transformation product of the enzyme formed during filtration, or a combination of both such materials. If the same enzyme preparation was first dialyzed for 18 hr at 0° against 0.01 M potassium phosphate at pH 7.0, and then applied to the column again operated at 23°, the elution pattern shown in Figure 6B was obtained. The enzymatically inactive high molecular weight fraction appearing with the void volume was greatly increased in quantity whereas both the total activity and the amount of protein in the peak appearing at about 1.3 void volumes was considerably diminished. These findings suggest that the enzyme is capable of existing in at least two forms of which only the low molecular weight species is catalytically active.

When the lower molecular weight fractions obtained from the column operated at 23° (Figure 6A) were combined and reapplied to the same column, the protein once again became distributed between the two peaks, indicating that there may be a tendency for the enzyme to aggregate in dilute solution, or upon passage through the Sephadex column.

The experimental design was then modified to permit study of both cold inactivation and reactivation on warming. The purified enzyme was directly applied to a Sephadex G-100 column of similar dimensions, but maintained at 4° and developed with 0.2 M potassium phosphate buffer at pH 7.0 (Figure 6C). After determination of the protein concentration and enzymatic activity of the cold fractions, these were incubated at 30° for 1 hr and their activity again assayed. Figure 6C shows that reactivation of the enzyme on warming occurs primarily in the fractions containing the low molecular weight species, although some increases in enzyme activity resulting from reactivation of higher molecular weight forms cannot be excluded.

The interpretation of the experimental findings with the Sephadex columns is somewhat complicated because equilibrium conditions may not always exist during the gel filtration. If during the operation of the Sephadex columns the environmental conditions existing in the columns induce changes in the distribution of molecular species, the resolution of these species may be affected. It should also be pointed out that a part of the reactivation that occurs upon warming of the fractions shown in Figure 6C may reflect activity lost by the lower molecular weight material during storage in the cold after elution from the column.

The total recovery of enzyme activity in Figure 6B is considerably higher than that in the experiment illustrated in Figure 6C (before warming). This difference may be related to the fact that the column described in Figure 6B was operated at 23°, and some recovery of activity may have occurred during passage of the enzyme through the Sephadex bed at this temperature. It may be concluded from these experiments with Sephadex filtration that cold inactivation is associated with the formation of material of high molecular weight, but that other complicating factors preclude a quantitative analysis of this phenomenon.

*Gel Electrophoresis.* When gel filtration was performed on Sephadex G-200, the high molecular weight material produced on cooling was found to contain several components, which were not sharply resolved. However, electrophoresis of the cold-inactivated enzyme on polyacrylamide gel separated these components and permitted a more detailed analysis of the

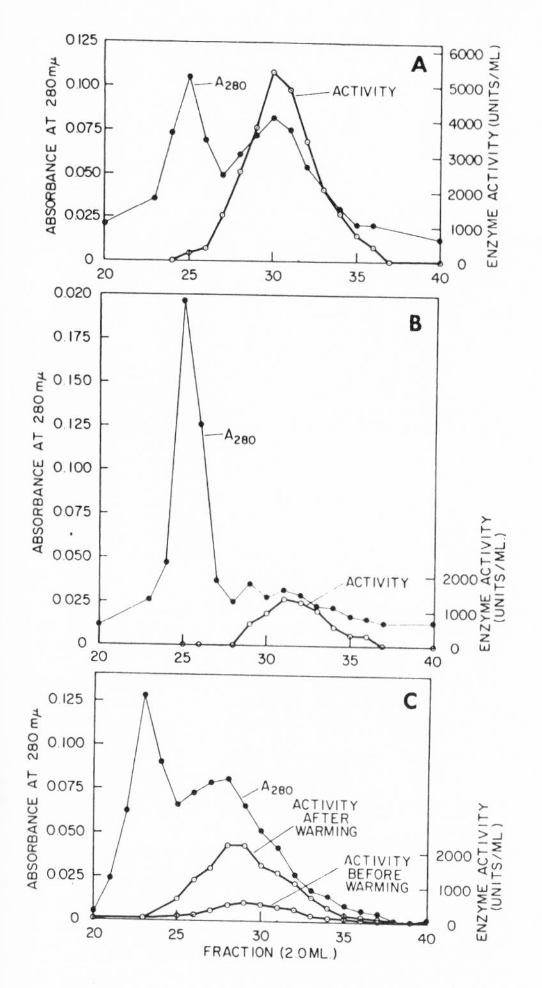


FIGURE 6: Filtration of 17β-hydroxysteroid dehydrogenase on a Sephadex G-100 column. The distribution of enzyme activity and absorbance at 280 mμ is shown. (A) A column of Sephadex G-100 (470 × 20 mm) was equilibrated with 0.01 M potassium phosphate buffer at pH 7.0. The column was operated at 23°. Enzyme (1 ml; 1.28 mg of protein and 80,000 units of activity) in 50% glycerol, 0.005 M potassium phosphate, 0.001 M EDTA, and 0.007 M β-mercaptoethanol at pH 7.0 was applied directly to the column. The column was developed by the addition of 0.01 M potassium phosphate buffer at pH 7.0. Fractions of 2.0-ml volume were collected at a flow rate of 0.8 ml/min. The exclusion volume of the gel was 50 ml. (B) Approximately the same quantity of protein as in A was applied to the same column after the enzyme had been dialyzed for 18 hr at 0° *vs.* 0.01 M potassium phosphate buffer at pH 7.0. The elution was carried out at 23° in the same manner as described in part A. (C) Another column of Sephadex G-100 (470 × 20 mm) was equilibrated with 0.2 M potassium phosphate buffer at pH 7.0. The column was operated at 4°. The enzyme (1 ml) described in part A was applied and the column was developed with 0.2 M potassium phosphate buffer at pH 7.0. Fractions of 2.0-ml volume were collected at a flow rate of 0.7 ml/min. The exclusion volume of the gel was 46 ml. All of the enzymatically active fractions had appeared from the column within 110 min after application of the enzyme. The fractions were stored at 4° until after their protein concentration and enzymatic activity had been determined. All of the initial assays had been completed within 270 min from the time of the application of the enzyme to the column. The fractions were then warmed for 1 hr at 30° and their activity again measured.

changes occurring during the cold-inactivation process and subsequent warming. An additional advantage of this method over Sephadex gel filtration was that small amounts of very dilute protein solutions could be analyzed since these solutions are initially concentrated during the electrophoresis upon passage through the stacking gel. Consequently, the entire process of cold inactivation and reactivation by warming could be examined at enzyme levels similar to those used in Figure 1.

The enzymatic activity is associated with a single band which migrates relatively rapidly upon electrophoresis on polyacrylamide gel (indicated by the arrow in Figure 7). As the enzyme is cooled for increasing periods, and becomes progressively inactivated, the band of active enzyme diminishes in intensity. After 14 days of storage at 0°, the enzyme is totally inactive and the band corresponding to active enzyme is no longer detectable. During the early stages of cold inactivation, a diffuse zone of less rapidly migrating protein appears just behind the band of active enzyme. As cooling is continued, this diffuse band also becomes less intense and is replaced by a series (at least 12) of sharply defined components which migrate even more slowly than the diffuse zone. On warming the enzyme stored in the cold for 4–24 hr, activity is restored and two changes in the electrophoretic pattern are noted. The band associated with enzymatic activity becomes more intense, and the multiple, sharply defined, more slowly moving components are also intensified.

When the distance of migration of these individual slow moving components is plotted against the logarithm of their order, beginning with the band representing the active enzyme, a strictly linear relation is obtained. Since the migration of components is probably a function of both size and charge, no unequivocal interpretation of these findings can be drawn, but it seems reasonable to suggest that the slow moving sharp bands represent simple multiples of molecular weight of the active enzyme. The formation of similar multiple protein bands was observed also after treatment of the enzyme with urea, sodium dodecyl sulfate, or *p*-(hydroxy)-mercuribenzoate.

When bovine serum albumin was polymerized by treatment with ethanol (Hartley *et al.*, 1962), multiple bands were observed on gel electrophoresis, and once again the distance of migration of these bands was a linear function of the logarithm of their order.

One simple interpretation of the electrophoretic patterns shown in Figure 7 is the following. On cooling the enzyme undergoes a series of conformational changes. This altered enzyme now migrates in a diffuse band just behind the active enzyme. On further and prolonged cooling the molecules of altered conformation aggregate and form a series of species of increasing molecular weight represented by the sharp slowly migrating bands. These high molecular forms are no longer converted to active enzyme on warming. If the enzyme is warmed during the early stages of cold inactivation, some of the molecules of altered conformation may revert to the native conformation and add to the intensity of the band of active enzyme, while others undergo a series of aggregations with the formation of the slowly moving bands. In the absence of quantitative information on the distribution of protein in the electrophoretic gels, this interpretation is necessarily tentative.

Sephadex filtrations of the cold-inactivated enzyme followed by polyacrylamide electrophoresis has confirmed the interpretation that the materials which migrate more slowly than the active enzyme on electrophoresis are those eluted in the exclusion volume on Sephadex G-100.

## Discussion

*Cold-Inactivated Enzymes.* Numerous reports dating from the very beginning of quantitative studies on enzymes have dealt with the influence of temperature on these catalysts and on the rates of the reactions which they promote (Dixon and Webb, 1964). It is commonly accepted that, with a few exceptions, elevated temperatures result in protein denaturation and loss of enzymatic activity, and that the temperature coefficient of this process is very high. This observation led to the widespread and often correct belief that strict maintenance of low temperatures (0–5°) is an essential prerequisite for success in enzyme isolation and fractionation. Uncritical adherence to this dogma can no longer be recommended since the past few years have brought to light some 12 examples of cold-sensitive enzymes (Table III) which display far greater stability near room temperature than in the icebox, and this property may be far more common than has been hitherto suspected. It is to be recommended that enzymes displaying unusual instability during the ordinary isolation procedures should be examined for cold sensitivity.

The temperature at which an enzyme is assayed and the temperature at which it is stored prior to assay are frequently different and may be varied at will. It is possible therefore to distinguish experimentally the effects of these two parameters on the activity of an enzyme. The experiments recorded in this paper deal exclusively with the influence of storage temperature on enzymatic activity, all assays being performed at a fixed temperature. We feel it has been possible to maintain the

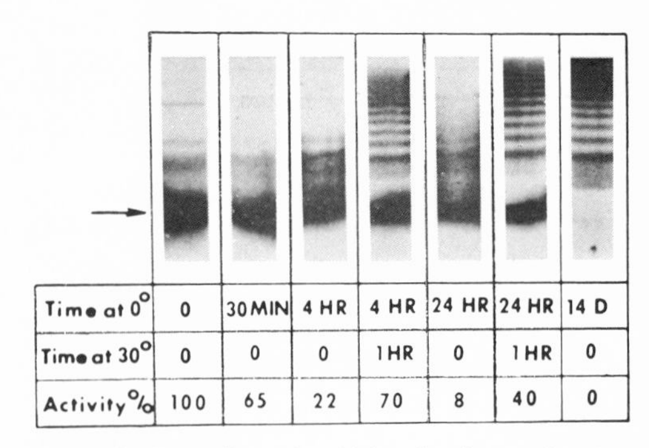

| Time at 0° | 0 | 30 MIN | 4 HR | 4 HR | 24 HR | 24 HR | 14 D |
|---|---|---|---|---|---|---|---|
| Time at 30° | 0 | 0 | 0 | 1 HR | 0 | 1 HR | 0 |
| Activity % | 100 | 65 | 22 | 70 | 8 | 40 | 0 |

FIGURE 7: The effect of cold inactivation and subsequent warming on the polyacrylamide gel electrophoresis patterns of the 17β-hydroxysteroid dehydrogenase. Disk electrophoresis was performed in a cold room at 4° by a modification of the procedure described by Davis (1964). Both the 3.5% (stacking) and the 7.5% cross-linked gels were polymerized in solutions containing 20% glycerol. The Tris–glycine buffer of pH 8.3 used for the electrophoresis also contained 20% glycerol. The enzyme described in Figure 1 was diluted to a protein concentration of 64 μg/ml by the addition of 0.005 M potassium phosphate buffer of pH 7.0. The enzyme was assayed immediately and stored at 0°. The assays were repeated after 30 min, 4 hr, 24 hr, and 14 days (designated 14 D) at 0°. Portions (25.6 μg of protein or 0.4 ml) were brought to a glycerol concentration of 30%, and were placed on top of the polyacrylamide stacking gel. The Tris–glycine buffer containing brom phenol blue was then added and the electrophoresis started. The initial current was 0.5 ma for each gel, and was increased to 1.0 ma as soon as the dye had entered the 7.5% cross-linked gel. The total time of electrophoresis was about 3.5 hr. The temperature of the buffers never rose above 10°. The gels were fixed and stained with Amido Schwartz dye in acetic acid. Aliquots of the cold-inactivated enzyme were removed after 4 and 24 hr at 0°; they were warmed at 30° for 1 hr, assayed, and then subjected to electrophoresis as described. The direction of migration is from top to bottom. The electrophoretic patterns are shown above together with the activities of the solutions just prior to electrophoresis. An aliquot of the sample cooled for 14 days remained completely inactive upon warming for 1 hr at 30°, and the electrophoretic pattern did not change after warming. The conditions of the electrophoresis were selected in order to minimize changes in enzyme activity resulting from cold inactivation or heat reactivation. This was achieved by using a glycerol concentration of 20% and maintaining temperatures below 10° during electrophoresis. In separate experiments, the gels were sectioned prior to staining, the enzyme was eluted, and the activity was assayed spectrophotometrically with 3-acetylpyridine–adenine dinucleotide. Only the band indicated with the arrow contained 17β-hydroxysteroid dehydrogenase activity. Histochemical tests (Goldberg, 1963) with intact gels also located the enzymatic activity in this single band, when NAD, NADP, or 3-acetylpyridine–adenine dinucleotide was used as hydrogen acceptor.

TABLE III: Cold-Inactivated Enzymes.

| Enzyme | Source | Reference |
|---|---|---|
| Urease | Jack bean | Hofstee (1949) |
| Glutamate dehydrogenase | *Neurospora crassa* (mutant) | Fincham (1957) |
| Adenosine triphosphatase | Beef heart mitochondria | Pullman *et al.* (1960) |
| Adenosine triphosphatase | Yeast | Racker *et al.* (1963) |
| Carbamyl phosphate synthetase | Frog liver | Raijman and Grisolia (1961) |
| D(−)β-Hydroxybutyrate dehydrogenase (DPN) | *Rhodospirillum rubrum* | Shuster and Doudoroff (1962) |
| Glutamate decarboxylase | *Escherichia coli* | Shukuya and Schwert (1960) |
| Nitrogen-fixing enzyme | *Clostridium pasteurianum* | Dua and Burris (1963) |
| Arginosuccinase | Steer liver | Havir *et al.* (1965) |
| Glycogen phosphorylase | Rabbit muscle | Graves *et al.* (1965) |
| Pyruvate carboxylase | Chicken liver mitochondria | Scrutton and Utter (1965) |
| Glucose 6-phosphate dehydrogenase | Human erythrocyte | Kirkman and Hendrickson (1962) |

separation of the effects of assay temperature and storage temperature because the enzyme undergoes no significant changes in activity during the assay procedure, irrespective of its earlier storage temperature. Enolase (Rosenberg and Lumry, 1964) and pyruvic kinase (Kayne and Suelter, 1965), which undergo inactivation at lowered assay temperatures, were not included in Table III because of insufficient information concerning the effect of storage temperature on their activities.

The concept of cold inactivation is naturally relative to a specific reference temperature. From an operational viewpoint we are here concerned with enzymes which maintain their activity much better on storage between about 20 and 30° than between 0 and 20°. The more complex changes which occur during freezing and thawing of enzyme solutions will not be considered.

The cold inactivation of the 17β-hydroxysteroid dehydrogenase has certain notable characteristics: it is partially reversed by warming, it may be reduced by certain anions as well as cofactors and substrates, it is accompanied by changes in both molecular weight and configuration, and it is prevented by glycerol. A comparison of the cold inactivation of the 17β-hydroxysteroid dehydrogenase with that of the enzymes which appear in Table III reveals many similarities, but particularly striking are the changes in molecular weight and configuration and the protection by glycerol or related compounds.

*Changes in Molecular Weight and Configuration.* In every case which has been examined, cold inactivation of susceptible enzymes was accompanied by changes in molecular weight. Dissociation into subunits on cooling has been observed in the case of pyruvate carboxylase (Scrutton and Utter, 1965), carbamyl phosphate synthetase (Grisolia and Raijman, 1964), arginosuccinase (Havir *et al.*, 1965), erythrocyte glucose 6-phosphate dehydrogenase (Kirkman and Hendrickson, 1962), and in the adenosine triphosphatase of beef heart mitochondria (Penefsky and Warner, 1965). In contrast to these enzymes, Graves *et al.* (1965) have clearly shown by ultracentrifugation that glycogen phosphorylase *b* undergoes cold inactivation accompanied by the appearance of a component or components which migrate with a sedimentation constant of about 20 S, whereas the active enzyme has a sedimentation constant of approximately 8 S. On rewarming the enzyme, a single component with a sedimentation constant of about 8 S is again obtained, and the high molecular weight material is almost totally obliterated. Hofstee (1949) suggested, on the basis of indirect evidence, that urease also aggregates in the cold. The experiments described in this paper indicate that the placental 17β-hydroxysteroid dehydrogenase undergoes a series of aggregation reactions on cooling. This conclusion is based on the behavior of the enzyme on Sephadex molecular sieves and on polyacrylamide gel electrophoresis, and is confirmed by preliminary experiments in the analytical ultracentrifuge. The placental enzyme displays the hitherto unique property of forming a large number of molecular aggregates on cooling.

Although cooling can induce either aggregation or dissociation into subunits, the extent of these physical changes does not necessarily correlate well with the enzyme activity. The kinetics of cold inactivation frequently do not obey simple rate laws, thus indicating that the physical changes may be quite complex. Havir *et al.* (1965) and Scrutton and Utter (1965) have intimated that other changes, probably of a conformational nature, precede physical changes in the size of the protein. Several findings also indicate that aggregation may not be the earliest cold-induced change in the placental enzyme. If aggregation were the sole cause for loss of enzyme activity, this might be expected to occur more rapidly at high protein concentrations, as is the case with phosphorylase *b* (Graves *et al.* (1965), whereas the opposite is observed for the placental enzyme. It has been shown that preparations of cold-inactivated 17β-hydroxysteroid dehydrogenase can be separated into high and low molecular weight components by Sephadex gel filtration. On warming these fractions, reactivation was observed largely, and perhaps exclusively, in the low molecular weight species (Figure 6C). This observation suggests that conformational changes occur upon cooling, and that these changes precede and are possibly a necessary prerequisite for the formation of molecular aggregates. Furthermore, the finding that inactivation of the 17β-hydroxysteroid dehydrogenase by such chemically dissimilar agents as sodium dodecyl sulfate, *p*-(hydroxy)mercuribenzoate, and urea is accompanied by the formation of molecular aggregates, suggests that there is no direct relation between aggregation and cold inactivation, but that cold and the chemical agents all bring about a disruption of the native configuration of the enzyme and that this favors the subsequent aggregation.

*Stabilization by Glycerol and Other Substances.* Varying concentrations of glycerol exert a profound stabilizing effect on the placental 17β-hydroxysteroid dehydrogenase. Glycerol protects the enzyme not only against the deleterious effects of cooling and urea, as shown in the present studies, but also against heat inactivation (Langer and Engel, 1958; Jarabak *et al.*, 1962). This protective influence is not confined to glycerol, but is shared by other polyhydroxylic compounds such as glycols, glycol ethers, and sucrose, as well as certain inorganic salts in high concentration (Table II; Jarabak, 1962; Talalay, 1962). The addition of glycerol has stabilized enzymes against cold inactivation in each instance where this question has been adequately examined. Racker *et al.* (1963) have shown that the dinitrophenol-stimulated adenosine triphosphatases of beef heart mitochondria and of yeast are both cold-sensitive enzymes which could be stabilized by the addition of 20–50% glycerol. Meyerhof and Ohlmeyer (1952) partially purified yeast adenosine triphosphatase in the presence of glycerol, long before its cold sensitivity was recognized. Similarly, the 17β-hydroxysteroid dehydrogenase of placenta was also purified in glycerol–water mixtures, before we became aware of its cold inactivation. Glycerol also stabilized the cold-sensitive carbamyl phosphate synthetase (Novoa and Grisolia, 1964). Complete protection against cold inactivation of pyruvate carboxylase is provided by 1.5 M sucrose (Utter *et al.*, 1964). The addition of 10% by volume of methanol, propylene glycol, or dimethyl sulfoxide completely protected phosphorylase *b* against cold inactivation (Graves *et al.*, 1965). Dua and Burris (1963) could find no stabilizing effect of glycerol and sucrose on the cold-inactivated nitrogen-fixing enzyme in extracts of *Clostridium pasteurianum*, but the concentrations used (0.1–0.001 M) were much lower than those required for significant stabilization of other enzymes.

Although the stabilizing effect of steroids and pyridine nucleotides can be attributed to the maintenance of the configuration at the active site, the mechanism of the stabilizing effect of glycerol and related compounds against cold, heat, and urea inactivation is at present obscure. The question also arises whether the stabilizing effect of high concentrations of certain anions is in any way related to the protective effect of glycerol. One suggestion which has been advanced to account for the protective effect of various organic solvents is that they may have the common property of stabilizing networks of "structured" water molecules which are essential to the maintenance of the proper spatial configuration of the protein in the native state (Talalay, 1962). The protective effect of anions might also be considered within this framework in view of their capability of inducing changes in the structure of the solvent medium. In a recent study, Fridovich (1963) has given a detailed analysis of the inhibition of acetoacetic decarboxylase by anions, and has analyzed the effect of replacing the aqueous medium by 50% glycerol or 50% ethylene glycol. His explanation for the specificity of anion inhibition is based on the effects on the structure of water produced by a combination of the ions with a specific binding site on the protein. He also states that the effects of replacing water partially by glycols is in accord with this proposal. Recent measurements of the anomalous dielectric constants of glycerol–water mixtures have been interpreted as indicating the existence of "packages" of glycerol and water of different composition in these solutions (McDuffie *et al.*, 1962). Thus the formation of water–glycerol structures around the protein molecules is not unreasonable, and may account for the stabilizing forces protecting the protein against inactivation. Evidence for the structured nature of water and the importance of this structure in the maintenance of protein configuration has been reviewed in a recent symposium (Frank, 1965; Klotz, 1965). The validity of such suggestions based on the stabilization of the hydrated structure of proteins must await a far better understanding of the structure of water surrounding biological macromolecules at different temperatures than is presently available (Jencks, 1965).

Of the bonds believed to play an important part in maintaining the three-dimensional structure of proteins, only hydrophobic bonds become significantly weaker as the temperature is lowered (Kauzmann, 1959; Scheraga *et al.*, 1962; Tanford, 1962). Therefore, it is perhaps only to be expected that some proteins become less stable when the temperature is lowered. Brandts (1964) has found that the temperature of maximum stability of chymotrypsinogen is 10°. Extensive studies have led him to conclude that the temperature dependence of hydrophobic bonding is an important factor in accounting for this. He has suggested that "...those proteins having a large number of hydrophobic interactions which are solvated during the unfolding process would be expected to have a maximum stability at a higher temperature than those which exhibit less hydrophobic character in their transition." In the light of this, it would be desirable to obtain information on the influence of glycerol on the strength of hydrophobic bonds of proteins in solution.

The behavior of proteins in nonaqueous solvents is attracting increasing attention. The thoughtful and interesting review of this subject by Singer (1962) created order out of many isolated observations. Numerous studies attest to the fact that the internal configuration of some protein molecules may undergo profound changes in nonaqueous systems, whereas no evidence for such changes can be obtained in other systems or with other proteins. For instance, Kay and Brahms (1963) showed that in 30–60% ethylene glycol, myosin A undergoes profound changes in internal configuration, a reduction in helicity, and an increase of about 2.5 times in adenosine triphosphatase activity. However, other fibrous muscle proteins do not change the proportion of their helical content in ethylene glycol. It would, therefore, be of great interest to study the conformational changes which cold-sensitive enzymes undergo during inactivation, and the influence of glycerol and related substances on these processes.

Although the use of glycerol in enzymology was widespread in earlier days, this agent was only rarely employed following the introduction of refrigeration. However, a number of recent reports have stressed the advantage of purifying enzymes in glycerol–water mixtures. Many of these enzymes are probably not cold inactivated, but they appear to enjoy considerably enhanced stability in this medium. In this connection, mention might be made of the purification of glucose 6-phosphate dehydrogenase from mammary gland (Levy, 1963), RNA polymerase from *M. lysodeikticus* (Nakamoto *et al.*, 1964), RNA polymerase from testis (Ballard and Williams-Ashman, 1966), 3α-hydroxysteroid dehydrogenase of *Pseudomonas testosteroni* (Boyer *et al.*, 1965), 20α-hydroxysteroid dehydrogenase of corpus luteum (Wiest and Wilcox, 1961), steroid $\Delta^4$-5β-dehydrogenase (Davidson and Talalay, 1966), and glycerokinase (Bublitz and Kennedy, 1954). In the last case, the protection by glycerol occurs at quite low concentrations, probably because glycerol is one of the substrates for this enzyme.

## References


Adams, J. A., Jarabak, J., and Talalay, P. (1962), *J. Biol. Chem. 237*, 3069.

Ballard, P., and Williams-Ashman, H. G. (1966), *J. Biol. Chem. 241* (in press).

Boyer, J., Baron, D. N., and Talalay, P. (1965), *Biochemistry 4*, 1825.

Brandts, J. F. (1964), *J. Am. Chem. Soc. 86*, 4291, 4302.

Bublitz, C., and Kennedy, E. P. (1954), *J. Biol. Chem. 211*, 951.

Davidson, S. J., and Talalay, P. (1966), *J. Biol. Chem. 241*, 906.

Davis, B. J. (1964), *Ann. N. Y. Acad. Sci. 121*, 404.

Dixon, M., and Webb, E. C. (1964), Enzymes, 2nd ed., New York, N. Y., Academic, p 158.

Dua, R. D., and Burris, R. H. (1963), *Proc. Natl. Acad. Sci. U. S. 50*, 169.

Fincham, J. R. S. (1957), *Biochem. J. 65*, 721.

Frank, H. S. (1965), *Federation Proc. 24, Suppl. 15*, S-1.

Fridovich, I. (1963), *J. Biol. Chem. 238*, 592.

Goldberg, E. (1963), *Science 139*, 602.

Graves, D. J., Sealock, R. W., and Wang, J. H. (1965), *Biochemistry 4*, 290.

Grisolia, S., and Raijman, L. (1964), *Advan. Chem. 44*, 128.

Hartley, R. W., Jr., Peterson, E. A., and Sober, H. A. (1962), *Biochemistry 1*, 60.

Havir, E. A., Tamir, H., Ratner, S., and Warner, R. C. (1965), *J. Biol. Chem. 240*, 3079.

Hofstee, B. H. J. (1949), *J. Gen. Physiol. 32*, 339.

Jarabak, J. (1962), Ph.D. Dissertation, University of Chicago.

Jarabak, J., Adams, J. A., Williams-Ashman, H. G., and Talalay, P. (1962), *J. Biol. Chem. 237*, 345.

Jencks, W. P. (1965), *Federation Proc. 24, Suppl. 15*, S-50.

Kauzmann, W. (1959), *Advan. Protein Chem. 14*, 1.

Kay, C. M., and Brahms, J. (1963), *J. Biol. Chem. 238*, 2945.

Kayne, F. J., and Suelter, C. H. (1965), *J. Am. Chem. Soc. 87*, 897.

Kirkman, H. N., and Hendrickson, E. M. (1962), *J. Biol. Chem. 237*, 2371.

Klotz, I. M. (1965), *Federation Proc. 24, Suppl. 15*, S-24.

Langer, L. J., and Engel, L. L. (1958), *J. Biol. Chem. 233*, 583.

Levy, H. R. (1963), *J. Biol. Chem. 238*, 775.

McDuffie, G. E., Jr., Quinn, R. G., and Litovitz, T. A. (1962), *J. Chem. Phys. 37*, 239.

Meyerhof, O., and Ohlmeyer, P. (1952), *J. Biol. Chem. 195*, 11.

Nakamoto, T., Fox, C. F., and Weiss, S. B. (1964), *J. Biol. Chem. 239*, 167.

Novoa, W. B., and Grisolia, S. (1964), *Biochim. Biophys. Acta 85*, 274.

Ornstein, L. (1964), *Ann. N. Y. Acad. Sci. 121*, 321.

Penefsky, H. S., and Warner, R. C. (1965), *J. Biol. Chem. 240*, 4694.

Pullman, M. E., Penefsky, H. S., Datta, A., and Racker, E. (1960), *J. Biol. Chem. 235*, 3322.

Racker, E., Pullman, M. E., Penefsky, H. S., and Silverman, M. (1963), Proceedings of the Fifth International Congress of Biochemistry, Moscow, 1961; Vol. V, New York, N. Y., Macmillan, p 303.

Raijman, L., and Grisolia, S. (1961), *Biochem. Biophys. Res. Commun. 4*, 262.

Rosenberg, A., and Lumry, R. (1964), *Biochemistry 3*, 1055.

Scheraga, H. A., Némethy, G., and Steinberg, I. Z. (1962), *J. Biol. Chem. 237*, 2506.

Scrutton, M. C., and Utter, M. F. (1965), *J. Biol. Chem. 240*, 1.

Shukuya, R., and Schwert, G. W. (1960), *J. Biol. Chem. 235*, 1658.

Shuster, C. W., and Doudoroff, M. (1962), *J. Biol. Chem. 237*, 603.

Singer, S. J. (1962), *Advan. Protein Chem. 17*, 1.

Talalay, P. (1962), *in* On Cancer and Hormones, Chicago, Ill., University of Chicago, p 271.

Talalay, P., Hurlock, B., and Williams-Ashman, H. G. (1958), *Proc. Natl. Acad. Sci. U. S. 44*, 862.

Talalay, P., and Williams-Ashman, H. G. (1958),

*Proc. Natl. Acad. Sci. U. S. 44*, 15.
Talalay, P., and Williams-Ashman, H. G. (1960), *Recent Progr. Hormone Res. 16*, 1.
Tanford, C. (1962), *J. Am. Chem. Soc. 84*, 4240.
Utter, M. F., Keech, D. B., and Scrutton, M. C. (1964), *Advan. Enzyme Regulation 2*, 49.
Wiest, W. G., and Wilcox, R. B. (1961), *J. Biol. Chem. 236*, 2425.

Part XII

# ENZYME STRUCTURE: AMINO ACID SEQUENCE AND THREE-DIMENSIONAL STRUCTURE

# Editor's Comments on Papers 64 Through 70

**64 DAVIDSON et al.**
*Amino-acid Sequence of Glyceraldehyde 3-Phosphate Dehydrogenase from Lobster Muscle*

**65 FLEMING**
*On a Remarkable Bacteriolytic Element found in Tissues and Secretions*

**66 BLAKE et al.**
*Structure of Hen Egg-white Lysozyme: Å Three-dimensional Fourier Synthesis at 2 A Resolution*

**67 JOHNSON and PHILLIPS**
*Structure of Some Crystalline Lysozyme-inhibitor Complexes Determined by X-ray Analysis at 6 Å Resolution*

**68 RICHARDS**
*On the Enzymic Activity of Subtilisin-Modified Ribonuclease*

**69 KARTHA, BELLO, and HARKER**
*Tertiary Structure of Ribonuclease*

**70 WYCKOFF et al.**
*The Structure of Ribonuclease-S at 3.5 Å Resolution*

Our selection of papers has by and large tried to balance studies on enzymatic activity with studies of enzymes as chemical individuals. The biological role of enzymes as catalysts in intermediary metabolism has received but little emphasis here and can form the topic of other Benchmark volumes. It is hardly necessary to emphasize that while enzymatic mechanisms can be studied from the viewpoint of the active site as a chemical structure (an approach exemplified in Parts IV, VI, and VIII), from the viewpoint of kinetics (Part IV, again, and Part VII), and from that of catalytic mechanisms (Part IX), the enzyme is a unit, and all these approaches tell valid but partial tales. A complete understanding of enzymatic activity requires a correlation of these and other approaches. Limited correlations, as between kinetics and catalytic mechanisms, are of course possible. However we are, for example, far from being able to predict the activity of a given protein molecule from its structure or to say, in fact, whether it will behave as an enzyme. Conversely, we cannot predict what type of enzyme structure will be necessary to catalyze a given chemical reaction. An excellent survey of the difficulties associated with attempts to make predictive correlations between various enzyme parameters was written by Luisi (1979). Although in these final sections a great number of papers could have been selected on various topics concerned with an understanding of enzyme structure and function, it was considered that these two areas could perhaps be best brought together in terms of studies on the structure of some representative enzyme molecules. The elucidation of enzyme structure, as an exercise in protein chemistry, is essential for the understanding of enzyme function and of evolutionary relationships.

There are two main approaches to such structural studies: the determination of the primary structure by chemical methods, and the determination of the secondary, tertiary, and often quaternary structure by X-ray crystallographic methods.

The first enzymes whose primary structures were determined are bovine pancreatic ribonuclease with 124 amino acid residues (Smyth, Stein, and Moore 1963 and references therein to their earlier work and to that in the laboratories of Anfinsen and of Gross and Witkop), and hen egg-white lysozyme with 129 amino acid residues (Canfield 1963). Paper 64 by Davidson, Sajgo, Noller, and Harris is the first report of the determination of the primary structure of an intracellular enzyme, lobster muscle glyceraldehyde 3-phosphate dehydrogenase, an enzyme consisting of 333 amino acid residues. Ever larger enzyme molecules have since been analyzed. The largest by far is $\beta$-galactosidase of *Escherichia coli*, a single-chain protein with 1021 amino acid residues (Fowler and Zabin 1977, 1978).

The papers on the results of X-ray crystallographic three-dimensional analysis of enzymes selected for this volume deal with two hydrolytic enzymes: hen egg-white lysozyme, the first enzyme for which such an analysis was carried out, and bovine pancreatic ribonuclease. Lysozyme, which has played such a fundamental role in initiating structure-function studies on enzymes, was discovered accidentally by Alexander Fleming, whose experience with the appearance of lysed bacterial colonies on Petri plates led him a few years later to the accidental discovery of penicillin. Parts of Fleming's elegant and thorough lysozyme paper, remarkable for already including egg-white and tears as sources of this enzyme, are reproduced in this book (Paper 65). The structure determination of lysozyme given here, from D. C. Phillips's laboratory, lists the workers in alphabetical order (Paper 66). Immediately following this paper the one by Johnson and Phillips was published, the first in which X-ray crystallographic studies have given direct evidence concerning the active site of an enzyme and the groups of a substrate—in this case a large substrate—bound to the enzyme (Paper 67). A profusely illustrated exposition of this work was published by Phillips about one year later (Phillips 1966).

Pancreatic ribonuclease, which, just as lysozyme, does not play a central role in metabolic pathways, has also been the subject of classical structure-function studies, and it is, furthermore, the only enzyme as we shall see in Part XIII of this book, that has been synthesized by the methods of organic chemistry. This enzyme has lent itself to a great variety of highly rewarding structure-function studies. One type of study, of profound impact on our thinking about protein folding, particularly as this is related to genetic information, was made possible by the spontaneous renaturation of the denatured enzyme molecule (see Anfinsen 1973). Quite a different approach began with the observation that careful treatment with the bacterial proteolytic enzyme subtilisin can cause a break in the ribonuclease chain at a distance of 20 amino acid residues from the N-terminal end of the molecule. The enzymic activity of the modified ribonuclease (known as ribonuclease-S) in which the so-called S-peptide remains attached by noncovalent bonds to the remaining 104-amino acid part of the molecule, is essentially unchanged. When the S-peptide is removed, activity is lost, but when the two fractions are mixed, full activity is restored. We have included here the first report, by F. M. Richards, of studies on the restoration of activity when the two fractions are mixed (Paper 68). A later, detailed study was published by Richards and Vithayathil (1959). Similar studies were subsequently performed by Anfinsen and collaborators on staphylococcal nuclease by selective cleavage with trypsin (see Anfinsen 1973). A paper on the three-dimensional structure of intact ribonuclease by Kartha, Bello, and Harker (Paper 69) and one on the three-dimensional structure of ribonuclease-S by Wyckoff et al. (Paper 70) are also reproduced here.

It is important to stress that two histidine residues—of which one, His-12, is on the S-peptide, and the other, His-119, is on the so-called S-protein—are essential for activity at the active site of the molecule; the diagrams in Papers 69 and 70 (perhaps slightly obscured by the detail in the latter) show that these two residues are in fact in close proximity in intact ribonuclease (also called ribonuclease-A) and in ribonuclease-S.

It is of some interest that pancreatic ribonuclease was first crystallized by the same Moses Kunitz (1939, 1940) who had, a very short time earlier, performed the monumental studies with Northrop on the crystallization of a large number of gastrointestinal proteases. It has been pointed out by Richards and Wyckoff (1971) that much of the initial interest in bovine pancreatic ribonuclease was due to the fact that in the early 1950s well over 1 kg of a very pure preparation of the crystalline enzyme had been prepared at Armour, Inc., and made available at a nominal fee to interested research workers.

Other early papers that could not, for reasons of space, be included in this volume, present the X-ray crystallographic analysis of carboxypeptidase A (Lipscomb et al. 1966; Ludwig et al. 1967) and of chymotrypsin (Mathews et al. 1967; Sigler et al. 1968). The field has since grown by leaps and bounds.

# 64



Reprinted from *Nature* **216**:1181–1185 (1967)

# Amino-acid Sequence of Glyceraldehyde 3-Phosphate Dehydrogenase from Lobster Muscle

by

B. E. DAVIDSON*
M. SAJGÒ†
H. F. NOLLER‡
J. IEUAN HARRIS

MRC Laboratory of Molecular Biology,
Hills Road, Cambridge

The protein sub-unit of glyceraldehyde 3-phosphate dehydrogenase consists of a single chain of 333 amino-acid residues, corresponding to a molecular weight of 36,000. Amino-acid sequence studies show that the active enzyme-NAD complex with a molecular weight of 146,000 is composed of four protein chains of identical sequence.

GLYCERALDEHYDE 3-phosphate dehydrogenase (GPDH) plays an important part in carbohydrate metabolism. In the presence of NAD and phosphate it catalyses the reversible oxidative phosphorylation of glyceraldehyde 3-phosphate to 1,3-diphosphoglyceric acid and as such it participates in the glycolytic conversion of glucose to pyruvic acid in most living organisms. Since its original isolation from yeast by Warburg and Christian[1] the crystalline enzyme has been prepared[2–5] from a wide range of species and has been the subject of much chemical and kinetic investigation[6,7].

* Aided by a grant from CSIRO (Australia), 1965–66; present address: Russell Grimwade School of Biochemistry, University of Melbourne, Parkville 3052, Victoria, Australia.

† EMBO fellow, 1966–67; present address: Institute of Biochemistry, Karolina út 29, Budapest, Hungary.

‡ Research fellow, US Public Health Service, 1965–66; present address: Laboratoire de Biochimie Génétique, 24 Quai de l'Ecole de Médicine, 1211 Genève 4, Switzerland.

A chemical study of the enzyme from pig muscle led Harris and Perham[8] to conclude that GPDH from any given species is composed of four similar, and probably identical, protein chains, each containing approximately 330 amino-acid residues (corresponding to a molecular weight of 36,000), including one catalytically active cysteine[9,10]. Among NAD-linked dehydrogenases, muscle GPDH is unique in having high affinity for NAD, and the active tetramer has the capacity to bind up to four moles of the coenzyme[11]. It has been shown that the binding sites are not equivalent and that a marked conformational change[12,13] involving some form of co-operative interaction between the protein chains[14] occurs when the first molecule of NAD is bound to the apo-enzyme. The enzyme–NAD complex would thus have a molecular weight of between 144,000 and 147,000 according to its content of bound NAD, and values recently obtained[15] by sedimentation-equilibrium methods are in agreement with those predicted from the chemical data[8].

Studies of haemoglobin[16] and of monomeric enzymes such as lysozyme[17], chymotrypsin[18] and ribonuclease[19] have shown how a knowledge of the three-dimensional structure of an enzyme can be used, in conjunction with chemical and kinetic information, to investigate mechanisms of enzyme catalysis. Similarly, in our view, a detailed knowledge of the three-dimensional structure of the protein chains in GPDH, and of the specific arrangement by which these chains associate and interact to form the quaternary structure of the active tetramer, will be necessary before the mechanism of enzyme catalysis can be fully understood. This detailed structure can be obtained by interpretation of X-ray diffraction analysis of crystals in conjunction with a knowledge of the amino-acid sequence of the enzyme protein. Of all the GPDHs which have been crystallized, the enzyme from lobster muscle[5] has proved to be the most suitable for analysis by X-ray diffraction methods. The crystal structure of lobster GPDH is being studied by Dr H. C. Watson and his colleagues in this laboratory[20]. We have therefore determined the amino-acid sequence of the protein sub-unit in the enzyme and a preliminary account of this work is given in this communication. A separate study[8] of the amino-acid sequence of GPDH from pig muscle (J. I. Harris and R. N. Perham, unpublished results) will be published elsewhere.

## Determination of Amino-acid Sequence

GPDH was prepared from lobster tail muscle as previously described[5]. The amino-acid composition, based on a molecular weight of 36,000 (ref. 8) and given in Table 1, indicates that the protein sub-unit consists of about 330 residues. This composition is in good agreement with the earlier data of Allison and Kaplan[5], except that it indicates the presence of an additional residue of cysteine and of arginine.

Studies of the amino-acid sequence were carried out chiefly with the N-trifluoroacetyl-S[$^{14}$C]-carboxymethyl-enzyme. This derivative was prepared by carboxymethylation of the five cysteine residues with [2-$^{14}$C] iodoacetic acid in 8 molar urea at *p*H 8·2 (ref. 8), followed by trifluoroacetylation[21] of the twenty-eight lysine residues in the continued presence of a large excess of S-ethyltrifluorothioacetate in 8 molar urea at *p*H 10·0 (unpublished results of Harris and Perham). Digestion with trypsin (1 per cent) was carried out in a *p*H-stat for 10 h at *p*H 8·5 and 25° C. The average number of peptide bonds which were hydrolysed by trypsin in these conditions was approximately equivalent to the total number of arginine residues (nine) per mole of protein.

The resulting mixture of peptide fragments gave rise to five separate components by gel-filtration on 'Sephadex *G*-50' in 0·1 molar ammonium bicarbonate (*p*H 8·0). The smaller components yielded pure peptide fractions by ionophoresis on paper, and the larger components were purified by ion-exchange chromatography on DEAE-cellulose in 8 molar urea containing *tris*-hydrochloric acid buffer at *p*H 7·5. In this manner, ten unique peptides (nine with C-terminal arginine, and one, the C-terminal peptide in the protein chain, with C-terminal alanine) were identified among the products of trypsin digestion. These peptides contained 3, 3, 4, 10, 14, 14, 32, 34, 43, and from 175–180 amino-acid residues, respectively, giving a total of about 335 residues.

Table 1. AMINO-ACID COMPOSITION OF THE PROTEIN MONOMER IN GLYCERALDEHYDE 3-PHOSPHATE DEHYDROGENASE (GPDH) FROM LOBSTER MUSCLE

| | No. of residues/36,000 g of protein | | |
|---|---|---|---|
| Amino-acid | 1 | 2 | 3 |
| Lysine | 28 | 27 | 28 |
| Histidine | 4–5 | 5 | 5 |
| Arginine | 8 | 9 | 9 |
| Cysteine | 4 | 5* | 5 |
| Aspartic acid | 35 | 31 | 32 |
| Threonine | 20 | 20† | 20 |
| Serine | 23 | 26† | 25 |
| Glutamic acid | 25 | 25 | 24 |
| Proline | 11 | 13 | 12 |
| Glycine | 32 | 30 | 30 |
| Alanine | 34 | 31 | 32 |
| Valine | 37 | 37‡ | 38 |
| Methionine | 10 | 10* | 10 |
| Isoleucine | 19 | 18‡ | 18 |
| Leucine | 18 | 19 | 18 |
| Tyrosine | 9 | 9 | 9 |
| Phenylalanine | 15 | 15 | 15 |
| Tryptophan | 4 | (3)§ | 3 |

1, Data of Allison and Kaplan[5] recalculated for a molecular weight of 36,000.
2, Analysis of performic acid oxidized protein.
3, Calculated from the amino-acid sequence (Fig. 1).

* Estimated as cysteic acid and methionine sulphone, respectively.
† Estimated by extrapolation to zero time of hydrolysis.
‡ Estimated by extrapolation to infinite time of hydrolysis.
§ Not determined.

Purified peptides which contained N-trifluoroacetyl-lysine were separately redigested with trypsin after removing the blocking group with *M* piperidine at 0° C (ref. 21). The resulting mixtures, which now contained lysine peptides in addition to the peptides containing C-terminal arginine, were purified by suitable combinations of gel-filtration, paper ionophoresis at various *p*Hs and paper chromatography. The amino-acid sequences of the purified peptides were determined by standard methods with major reliance on stepwise analysis by the dansyl–Edman procedure[22]. Overlaps of lysine residues were obtained from the sequences of peptides produced with chymotrypsin, pepsin, papain or cyanogen bromide; the lysine diagonal method[23] was occasionally used to facilitate the purification of these peptides. Tryptic peptides which contain arginine occupy C-terminal positions in the original lysine-blocked fragments, and the relative order of these fragments along the protein chain was established from the sequences of peptides containing arginine, which were obtained from pepsin or chymotrypsin digests of the S-[$^{14}$C] carboxymethyl-protein. The N-terminal trypsin peptide in the protein chain was found to be N-acetyl.Ser.Lys. The acetyl group was identified as acetylhydrazide after hydrazinolysis of the N-substituted peptide, and by mass spectrometry. Peptides which contained tryptophan were detected on paper by Ehrlich's reagent and the number of tryptophan residues in each of the purified peptides was established by the dansyl–Edman method (residues 192 and 309 (Fig. 1)) or by spectrophotometric analysis (residue 83). By these methods three residues of tryptophan were found in peptide fragments in contrast to the value of four obtained by spectrophotometric analysis[5] of the intact enzyme protein. The reason for this discrepancy has not been established, but it is possible that the value obtained by direct analysis of the protein is high because of the presence of bound NAD. Amide groups have been provisionally assigned from the electrophoretic mobilities[24] of peptides which contained aspartic and glutamic acids.

These results enable us to propose the following provisional amino-acid sequence for the protein sub-unit of lobster GPDH (Fig. 1). It consists of a single chain of 333 amino-acid residues (including N-terminal N,acetyl-serine), corresponding to a molecular weight of 36,003. The active tetramer would thus have a molecular weight of 144,012, or 146,650 if it contains four moles of bound NAD (compare pig GPDH, refs. 8 and 15).

Details of the experimental results on which the sequence is based cannot be given in this short communication. We therefore present the following evidence in support of its validity.

(1) The amino-acid composition derived from the proposed sequence is in excellent agreement with the composition calculated from the experimentally determined amino-acid analysis of the enzyme protein (Table 1). (2) All the tryptic peptides which occurred in major yield were purified and are unambiguously accommodated in the sequence. The sequence of 333 residues thus represents a complete recovery of the analysed parts. (3) All the lysine and arginine residues have been unequivocally overlapped, many with peptides derived from more than one type of digest. (4) Stepwise analysis by the Edman method, coupled with the identification of successively released amino-acid residues as their dansyl derivatives, was carried out through all the peptide bonds in the sequence (in many instances on more than one type of fragment, and on separate occasions), with the exception of the bonds between residues 1–2, 2–3, 83–84, 98–99, 135–136, 158–159, 175–176, 190–191 and 255–256. Sequences obtained by the dansyl–Edman method were in agreement in all cases with the amino-acid compositions of the analysed peptides and in most cases the sequences obtained in this way were confirmed by analysis of other fragments obtained by digestion with suitable proteolytic enzymes. (5) Finally, the validity of the sequence of the lobster enzyme receives additional support from a comparison with the sequence of the pig muscle enzyme (unpublished results of Harris and Perham). In many instances sequence homologies between the two proteins provide a valuable check on the accuracy of the two independently determined sequences. Further confirmatory evidence will continue to be sought in the course of work which is being undertaken to establish the identity of other reactive groups in the enzyme.

## Discussion of Methods

The use of methods based on the reversible blocking of lysine residues has greatly facilitated the sequence analysis of lobster GPDH. With proteins of smaller size (for example, ribonuclease)[25] the conventional first step has been to isolate and to work out the sequence of all the peptides from a total trypsin digest of the complete protein. The order in which the various trypsin peptides occur in the protein chain is then established from the sequences of a selection of different peptide fragments containing lysine and arginine; these are obtained by

Acetyl. Ser. Lys. Ile. Gly. Ile. Asp. Gly. Phe. Gly. **Arg**(10). Ile. Gly. **Arg**. Leu. Val. Leu.

**Arg**. Ala. Ala. Leu(20). Ser. Cys. Gly. Ala. Gln. Val. Val. Ala. Val. Asn(30). Asp. Pro. Phe. Ile.

Ala. Leu. Glu. Tyr. Met. Val(40). Tyr. Met. Phe. Lys. Tyr. Asp. Ser. Thr. His. Gly(50). Val. Phe.

Lys. Gly. Glu. Val. Lys. Met. Glu. Asp(60). Gly. Ala. Leu. Val. Val. Asp. Gly. Lys. Lys. Ile(70).

Thr. Val. Phe. Asn. Glu. Met. Lys. Pro. Glu. Asn(80). Ile. Pro. Trp. Ser. Lys. Ala. Gly. Ala.

Glu. Tyr(90). Ile. Val. Glu. Ser. Thr. Gly. Val. Phe. Thr. Thr(100). Ile. Glu. Lys. Ala. Ser. Ala.

His. Phe. Lys. Gly(110). Gly. Ala. Lys. Lys. Val. Val. Ile. Ser. Ala. Pro(120). Ser. Ala. Asp. Ala.

Pro. Met. Phe. Val. Cys. Gly(130). Val. Asn. Leu. Glu. Lys. Tyr. Ser. Lys. Asp. Met(140). Thr. Val.

Val. Ser. Asn. Ala. Ser. •**Cys**(148). Thr. Thr(150). Asn. Cys. Leu. Ala. Pro. Val. Ala. Lys. Val. Leu(160).

His. Glu. Asn. Phe. Glu. Ile. Val. Glu. Gly. Leu(170). Met. Thr. Thr. Val. His. Ala. Val. Thr.

Ala. Thr(180). Gln. •**Lys**(182). Thr. Val. Asp. Gly. Pro. Ser. Ala. Lys(190). Asp. Trp. **Arg**. Gly. Gly. **Arg**.

Gly. Ala. Ala. Gln(200). Asn. Ile. Ile. Pro. Ser. Ser. Thr. Gly. Ala. Ala(210). Lys. Ala. Val. Gly.

Lys. Val. Ile. Pro. Glu. Leu(220). Asp. Gly. Lys. Leu. Thr. Gly. Met. Ala. Phe. **Arg**(230). Val. Pro.

Thr. Pro. Asp. Val. Ser. Val. Val. Asp(240). Leu. Thr. Val. **Arg**. Leu. Gly. Lys. Glu. Cys. Ser(250).

Tyr. Asp. Asp. Ile. Lys. Ala. Ala. Met. Lys. Thr(260). Ala. Ser. Glu. Gly. Pro. Leu. Gln. Gly.

Phe. Leu(270). Gly. Tyr. Thr. Glu. Asp. Asp. Val. Val. Ser. Ser(280). Asp. Phe. Ile. Gly. Asp. Asn.

**Arg**. Ser. Ser. Ile(290). Phe. Asp. Ala. Lys. Ala. Gly. Ile. Gln. Leu. Ser(300). Lys. Thr. Phe. Val.

Lys. Val. Val. Ser. Trp. Tyr(310). Asp. Asn. Glu. Phe. Gly. Tyr. Ser. Gln. **Arg**. Val(320). Ile. Asp.

Leu. Leu. Lys. His. Met. Gln. Lys. Val(330). Asp. Ser. Ala(333). COOH

Fig. 1. Amino-acid sequence of glyceraldehyde 3-phosphate dehydrogenase from lobster muscle. The sequences of the tryptic peptide containing Cys-148 (residues 139 to 158); the N-terminal dipeptide (X.Ser).Lys, and the C-terminal tetrapeptide Val(Asp,Ser).Ala were determined previously (Allison and Harris[28]; W. S. Allison, unpublished results).

alternative methods of fragmentation such as pepsin, chymotrypsin or cyanogen bromide. Trypsin digests of larger proteins contain on average a correspondingly larger number of peptides. A digest of lobster GPDH, for example, contains a total of more than forty peptide fragments (including products of partial digestion of slowly hydrolysed bonds, and of cleavages at bonds other than those involving the twenty-eight lysine and nine arginine residues), and it has proved difficult to obtain all the necessary peptides in pure form from complex mixtures such as these[8,28].

In order to reduce the number of peptides to be fractionated from such a mixture, we attempted to limit the action of trypsin to the nine arginyl bonds in lobster GPDH by reacting the ε-amino groups of the lysine residues with S-ethyltrifluorothioacetate[21,23] (or with maleic anhydride[26], unpublished results). This procedure proved to be highly advantageous in that trypsin digests of the resulting lysine-blocked proteins, like their counterparts from the pig enzyme (unpublished results of Harris and Perham), were completely soluble, and readily amenable to fractionation by standard methods. The ten major peptide components which were obtained in this way (see earlier section) could then be studied separately.

Peptides which contained blocked lysines were unblocked to regenerate lysine residues, and the unblocked peptides were then separately redigested with trypsin. These digests were relatively easy to fractionate because in each case they contained only those tryptic peptides that occur between two consecutive arginine residues in the protein chain. The isolation of the tryptic peptides from lobster GPDH was thus accomplished more simply by a series of separate fractionation steps from lysine-blocked fragments than by direct fractionation of the complex mixture of more than forty peptides which is produced by digestion of the unblocked protein with trypsin. Similarly, peptides which overlap lysine residues are also obtained more easily (for example, by the lysine diagonal method[23,26]) from suitable digests of lysine-blocked fragments than from the considerably more complex mixtures of peptides that occur in similar digests of the whole protein. In this way the sequences of N-trifluoroacetyl- (and of maleyl[26]) peptides can be determined in a series of separate operations. The complete sequence of the protein chain is then established by overlapping arginine residues with peptides obtained from suitable digests of the whole protein as described earlier.

The overall strategy which has been developed for determining the amino-acid sequence of lobster GPDH is clearly applicable to other proteins of similar, or even larger, size. Many other important enzymes such as yeast and liver alcohol dehydrogenases, lactic dehydrogenase and aldolase, to mention but a few, are known to consist of sub-units comparable in size to those of GPDHs, and we suggest that the method of reversible lysine-blocking could be used to similar advantage in the sequence analysis of these enzymes. Ideally, it is desirable that the protein to be studied should contain a sufficient number of lysine and arginine residues, and the method will work best if there are from two to three times as many lysines as there are arginines; moreover, it is also advantageous if arginine residues are fairly evenly distributed along the protein chain. This is exemplified by the case of lobster GPDH where the uneven distribution of arginine residues gave rise to one large lysine-blocked fragment (residues 18 to 193, Fig. 1) containing 176 amino-acid residues, including sixteen lysines. This fragment was obtained in low yield partly because of its susceptibility to chymotrypsin-like cleavages during trypsin digestion, and also because of severe losses during chromatography on DEAE-cellulose. Consequently, in order to determine the sequences of some of its constituent lysine-containing peptides, it proved in some circumstances to be more convenient to obtain additional amounts of these peptides from a tryptic digest of the unblocked protein.

## Relationships between Structure and Activity

The amino-acid sequence of lobster GPDH presented here provides strong evidence for the chemical identity of the four protein chains comprising the active enzyme molecule. At first sight the active enzyme could therefore be envisaged as consisting of four structurally equivalent monomer chains, each chain (α) containing a reactive cysteine which in conjunction with NAD and phosphate would be capable of promoting the catalytic reaction within the quaternary structure of the tetramer.

X-ray diffraction studies[20] of crystals of the enzyme–NAD complex suggest, on the other hand, that the tetrameric molecule could consist of structurally identical pairs of sub-units. This concept is supported by recent spectrophotometric evidence which indicates that the molecule exhibits chemical as well as structural asymmetry (S. A. Bernhard, personal communication). In this respect, the GPDH–NAD complex is similar in structure to haemoglobin[16] in which two chemically different sub-units are related in pairs to form the tetrameric molecule.

One way of reconciling these results with the identity of chemical sequence would be to postulate that crystals of the apoenzyme possess a symmetrical tetrameric ($\alpha_4$) structure which changes to the $\alpha_2\alpha_2'$ type of structure as the result of conformational changes which occur when NAD binds to the protein sub-units. Conformational changes involving co-operative interactions between the monomer chains are known to occur when NAD interacts with the apoenzyme in solution[12–14]. Attempts to compare the crystal structure of the NAD-enzyme with that of the apoenzyme have not so far been possible because the apoenzyme, unlike that of lactic dehydrogenase[27], does not seem to crystallize in the absence of the coenzyme.

In a previous study of lobster GPDH, Allison and Harris[28] showed that Cys-148 reacts selectively with iodoacetic acid in the native enzyme. Moreover, the amino-acid sequence around this reactive cysteine (residues 143–155, including a second but unreactive cysteine in position 152) was shown to be identical with the sequences around the corresponding reactive cysteine in GPDHs from several other species[9,10,28]. It has now emerged that the other three cysteines in lobster GPDH (positions 22, 129, 249) are not homologous with cysteine residues in other GPDHs. For example, in pig GPDH the corresponding positions are occupied by serine, methionine and valine, respectively (ref. 8 and unpublished results of Harris and Perham); the positions of the other two cysteines in the pig enzyme, on the other hand, are occupied by Val-243 and Ser-280 in the lobster enzyme. These variable cysteines are therefore unlikely to be specifically involved either in the catalytic activity or in maintaining the tertiary structure of active muscle GPDH.

Cys-148 participates in the catalytic reaction by attaching the substrate in thioester linkage to the enzyme (for example, refs. 9 and 10). In the absence of substrate (but in the presence, or absence, of NAD) it can also form an intrachain disulphide bond with Cys-152 (ref. 8). This shows that the two sulphydryl groups can occur in close proximity in the three-dimensional structure, and that the apparently unreactive sulphydryl group of Cys-152 could also prove to be an essential feature of the active molecule.

Another glimpse of the three-dimensional structure of the apoenzyme is provided by the observation that the acetyl group which is bound initially to Cys-148 (in the pig and rabbit muscle enzymes) during the enzyme-catalysed hydrolysis of *p*-nitrophenylacetate[9,10] and acetylphosphate[29], respectively, is able to transfer to a specific lysine residue[29,30] (shown to be Lys-182 in the pig enzyme)[31] at a higher *p*H. The same reaction has now been shown to occur with lobster GPDH (unpublished results of B. E. Davidson) and a comparison of the sequences around the two reactive amino-acids in the pig[32] and lobster enzymes reveals that forty-two of the forty-eight residues in this segment of the chain (positions 143–190) are identical in the two species. Moreover, the changes that have occurred are all of a conservative nature and could all have arisen as the result of a single base change in the *Escherichia coli* code[33].

The remarkable conservation of the amino-acid sequence around the reactive cysteine and lysine residues in enzymes from such distantly related species as pig and lobster suggests that the three-dimensional conformation of this part of the protein chain has also been conserved, possibly because it contributes significantly to the structure of the "catalytic centre" of the enzyme. In this context the transfer of an acetyl group from the sulphydryl group of a cysteine in position 148 to the ε-amino group of a lysine in position 182 shows that these two groups can approach each other very closely within the three-dimensional structure of the tetrameric apoenzyme. Little is as yet known about the detailed arrangement of the protein chains within the tetramer and it therefore remains to be established whether the acetyl transfer reaction occurs within or between monomers and, more specifically, to what extent (if any) the interaction of amino-acid side chains from more than one monomer may be involved in the catalytic reaction itself.

The amino-acid sequence of the protein monomer in lobster GPDH provides a framework for the precise interpretation of the results of further chemical, kinetic and X-ray crystallographic studies on the active tetrameric molecule in solution, and in the crystalline state. GPDHs from all sources are likely to possess a similar three-dimensional structure and hence a common reaction mechanism. It is to be hoped that the combined results of chemical and X-ray diffraction studies now in progress will lead to the identification of the amino-acid side chains that bind the substrate, NAD and phosphate to the protein sub-unit, as a step towards elucidating the mechanism of enzyme catalysis.

We thank Dr W. S. Allison for the approximate amino-acid compositions of a number of tryptic peptides from lobster GPDH, which he isolated in this laboratory during 1964. We also thank Mrs Beryl Preston for technical assistance.

Received November 21, 1967.

1 Warburg, O., and Christian, W., *Biochem. Z.*, **303**, 40 (1939).
2 Caputto, R., and Dixon, M., *Nature*, **156**, 630 (1945).
3 Cori, G. T., Slein, M. W., and Cori, C. F., *J. Biol. Chem.*, **159**, 565 (1945).
4 Elodi, P., and Szorényi, E., *Acta Physiol. Acad. Sci. Hung.*, **9**, 339 (1956).
5 Allison, W. S., and Kaplan, N. O., *J. Biol. Chem.*, **239**, 2140 (1964).
6 Velick, S. F., and Furfine, C., in *The Enzymes* (edit. by Boyer, P. D., Lardy, H., and Myrbäck, K.), **7**, 243 (Academic Press, New York, 1963).
7 Colowick, S. P., Van Eys, J., and Park, J. H., in *Comprehensive Biochemistry* (edit. by Florkin, M., and Stotz, E. H.), **14**, 1 (Elsevier Publishing Co., Amsterdam, 1966).
8 Harris, J. I., and Perham, R. N., *J. Mol. Biol.*, **13**, 876 (1965).
9 Harris, J. I., Meriwether, B. P., and Park, J. H., *Nature*, **197**, 154 (1963).
10 Perham, R. N., and Harris, J. I., *J. Mol. Biol.*, **7**, 316 (1963).
11 Ferdinand, W., *Biochem. J.*, **92**, 578 (1964).
12 Listowsky, I., Furfine, C. S., Bethiel, J. J., and England, S., *J. Biol. Chem.*, **240**, 4253 (1965).
13 Havsteen, B. H., *Acta Chem. Scand.*, **19**, 1643 (1965).
14 Kirschner, K., Eigen, M., Bittman, R., and Voigt, B., *Proc. US Nat. Acad. Sci.*, **56**, 1661 (1966).
15 Harrington, W. F., and Karr, G. M., *J. Mol. Biol.*, **13**, 885 (1965).
16 Cullis, A. F., Muirhead, H., Perutz, M. F., Rossmann, M. G., and North, A. C. T., *Proc. Roy. Soc.*, **265A**, 15 (1961).
17 Phillips, D. C., *Proc. US Nat. Acad. Sci.*, **57**, 484 (1967).
18 Mathews, B. W., Sigler, P. B., Henderson, R., and Blow, D. M., *Nature*, **214**, 652 (1967).
19 Wyckoff, H. W., Hardman, K. D., Allewell, N. M., Inagami, T., Johnson, L. N., and Richards, F. M., *J. Biol. Chem.*, **242**, 3984 (1967).
20 Watson, H. C., and Banaszak, L. J., *Nature*, **204**, 918 (1964).
21 Goldberger, R. F., and Anfinsen, C. B., *Biochem.*, **1**, 401 (1962).
22 Gray, W. R., and Hartley, B. S., *Biochem. J.*, **89**, 59P (1963).
23 Perham, R. N., and Jones, G. M. T., *Eur. J. Biochem.*, **2**, 84 (1967).
24 Offord, R. E., *Nature*, **211**, 591 (1966).
25 Smyth, D. G., Stein, W. H., and Moore, S., *J. Biol. Chem.*, **238**, 227 (1963).
26 Butler, P. J. G., Harris, J. I., Hartley, B. S., and Leberman, R., *Biochem. J.*, **103**, 78P (1967).
27 Rossmann, M. G., Jeffery, B. A., Main, P., and Warren, S., *Proc. US Nat. Acad. Sci.*, **57**, 515 (1967).
28 Allison, W. S., and Harris, J. I., *Second F.E.B.S. Meeting*, Vienna 1965, *A* 205, 140.
29 Park, J. H., Agnello, C. F., and Mathew, E., *J. Biol. Chem.*, **240**, P.C. 3232 (1965); **241**, 769 (1966).
30 Polgàr, L., *Acta Physiol. Acad. Sci. Hung.*, **25**, 1 (1964); *Biochim. Biophys. Acta*, **118**, 276 (1966).
31 Harris, J. I., and Polgàr, L., *J. Mol. Biol.*, **14**, 630 (1965).
32 Perham, R. N., *Biochem. J.*, **99**, 14C (1966).
33 Crick, F. H. C., *Cold Spring Harb. Symp.*, **31**, 1 (1967).

# 65



Reprinted from pages 306–307 and 312–315 of *R. Soc. (London) Proc.*, ser. B, **93**:306–317 (1922)

## *On a Remarkable Bacteriolytic Element found in Tissues and Secretions.*

By ALEXANDER FLEMING, M.B., F.R.C.S.

(Communicated by Sir Almroth Wright, F.R.S. Received February 13, 1922.)

(From the Laboratory of the Inoculation Department, St. Mary's Hospital.)

In this communication I wish to draw attention to a substance present in the tissues and secretions of the body, which is capable of rapidly dissolving certain bacteria. As this substance has properties akin to those of ferments I have called it a "Lysozyme," and shall refer to it by this name throughout the communication.

The lysozyme was first noticed during some investigations made on a patient suffering from acute coryza. The nasal secretion of this patient was cultivated daily on blood agar plates, and for the first three days of the infection there was no growth, with the exception of an occasional staphylococcus colony. The culture made from the nasal mucus on the fourth day showed in 24 hours a large number of small colonies which, on examination, proved to be large gram-positive cocci arranged irregularly but with a tendency to diplococcal and tetrad formation. It is necessary to give here a very brief description of this microbe as with it most of the experiments described below were done, and it was with it that the phenomena to be described were best manifested. The microbe has not been exactly identified, but for purposes of this communication it may be alluded to as the *Micrococcus lysodeikticus.*

The fully developed colony of the coccus may be 2 or 3 mm. in diameter; it is round, opaque, raised, and has a bright lemon yellow colour; it grows luxuriantly on all the ordinary culture media, and growth takes place well at room temperature, or in the incubator at 37° C.; it is aerobic and facultatively anaerobic; it does not liquefy gelatin or coagulated albumin.

### Preliminary Experiments Showing the Action of the Lysozyme.

In the first experiment nasal mucus from the patient, with coryza, was shaken up with five times its volume of normal salt solution, and the mixture was centrifuged. A drop of the clear supernatant fluid was placed on an agar plate, which had previously been thickly planted with *M. lysodeikticus*, and the plate was incubated at 37° C. for 24 hours, when it showed a copious growth of the coccus, except in the region where the nasal mucus had been placed. Here there was complete inhibition of growth, and this inhibition extended for a distance of about 1 cm. beyond the limits of the mucus.

This striking result led to further investigations, and it was noticed that one drop of the diluted nasal mucus added to 1 c.c. of a thick suspension of the cocci caused their complete disappearance in a few minutes at 37° C.

These two preliminary experiments clearly demonstrate the very powerful inhibitory and lytic action which the nasal mucus has upon the *M. lysodeikticus*. It will be shown later that this power is shared by most of the tissues and secretions of the human body, by the tissues of other animals, by vegetable tissues, and, to a very marked degree, by egg white.

[*Editor's Note:* A section entitled "Further Observations on the Effect of the Lysozyme on Bacteria" has been omitted here. This section is followed by one entitled "Observations on the Properties of the Lysozyme and on the Conditions Governing its Action." The following paragraphs are taken from this section.]

### *Distribution of the Lysozyme in the Body.*

In the first experiments it was found that nasal mucus contained a large amount of lysozyme, and it was later found that tears and sputum were very potent in their lytic action. It was also found that this property was possessed by a very large number of the tissues and organs of the body. The lysozyme-content of the tissues was investigated by placing small portions of tissue not larger than a split pea in tubes containing 1 c.c. of a thick suspension of the *M. lysodeikticus* incubating the tubes at 45° C., and noting whether any lysis took place as evidenced by a clearing of the opacity of the suspension. Some of these tissues were obtained from the postmortem room, others from laboratory workers or from the operating theatre. The results obtained can be summed up by saying that all the tissues and organs possessed some lytic power, even a few hairs from the head causing solution of the cocci. While in these tests no attempt was made at an exact quantitative estimation, it was noticed that lysis proceeded very much more rapidly with some tissues than with others. Briefly, it may be said that epidermal structures, the lining membrane of the respiratory tract and especially the connective tissues (whether fibrous, fatty or cartilaginous) contained large amounts of lysozyme affecting *M. lysodeikticus.* The rapidity of the lysis with cartilage was so striking that an attempt was made to estimate more accurately the amount of lysozyme in this tissue. A small portion of cartilage from the patella (deep to the articular surface) was weighed and ground up in a mortar with a measured volume of salt solution. This was allowed to extract for 6 hours when it was centrifuged and the supernatant fluid was added in various dilutions to a suspension of the *M. lysodeikticus.* It was found that with an extract corresponding to one part of the original cartilage in 1,300 parts of normal salt solution, there was complete lysis of the cocci in 5 minutes at 45° C. which shows that cartilage has approximately one-tenth the lysozyme-content of tears.

The presence of lysozyme was sought for in certain physiological and pathological fluids, and the results are set forth in Table I.

Table I.

| Fluids containing lysozyme. | Fluids not containing lysozyme. |
|---|---|
| Tears. | Normal urine. |
| Sputum. | Cerebro-spinal fluid. |
| Nasal mucus. | Sweat (one sample only tested.) |
| Saliva. | |
| Blood serum. | |
| Blood plasma. | |
| Peritoneal fluid. | |
| Pleural effusion. | |
| Hydrocœle fluid. | |
| Ovarian cyst fluid. | |
| Sebum. | |
| Pus from acne pustule. | |
| Sero pus from a "cold" abscess in the popliteal space. | |
| Urine containing much albumin and pus. | |
| Semen (very weak). | |

In connection with the lysozyme-content of the blood, it is to be noted that, in addition to its being present in the leucocytes, in the plasma, and in the serum, it is also present in rather large amount in the fibrin of the blood clot. It is conceivable that this is a protective mechanism for open wounds, which rapidly become covered with a layer of fibrin and leucocytes, both of which are rich in lysozyme.

The lysozyme-content of tears, sputum, nasal mucus, saliva, and blood serum of the same individval were tested. The specimens were all collected at the same time and were tested about 4 hours afterwards. The titrations were carried out by making serial dilutions of the various fluids and adding to these dilutions a measured quantity of a thick suspension of *M. lysodeikticus*, after which the tubes were incubated at 45° C., and readings were made at intervals of 15, 30, and 60 minutes. The results are set out in Table II:—

Table II.—The Lysozyme-Content of various Fluids taken from the same Individual at the same Time.

| Material examined. | Time of incubation at 45° C. | Dilution of fluid, 1 in :— | | | | | | |
|---|---|---|---|---|---|---|---|---|
| | mins. | | 10 | 30 | 90 | 270 | 810 | 2430 |
| Blood serum | 15 | | + | + | ± | 0 | 0 | 0 |
| | 30 | | + | + | + | ± | trace | 0 |
| | 60 | | + | + | + | + | ± | 0 |
| | | | 100 | 300 | 900 | 2,700 | | |
| Saliva | 15 | | + | ± | 0 | 0 | | |
| | 30 | | + | + | ± | 0 | | |
| | 60 | | + | + | ± | 0 | | |
| | | 500 | 1,500 | 4,500 | 13,500 | 40,500 | 121,500 | |
| Nasal mucus | 15 | + | + | + | ± | ± | 0 | |
| | 30 | + | + | + | + | ± | 0 | |
| | 60 | + | + | + | + | ± | 0 | |
| Sputum | 15 | + | + | + | ± | ± | 0 | |
| | 30 | + | + | + | + | ± | 0 | |
| | 60 | + | + | + | + | ± | 0 | |
| Tears | 15 | + | + | + | + | 0 | 0 | |
| | 30 | + | + | + | + | ± | 0 | |
| | 60 | + | + | + | + | + | ± | |

+ signifies complete clearing of the fluid.
± " partial " "
0 " no " "

It will be seen from the above Table that tears, sputum, and nasal mucus are very rich in lysozyme to the *M. lysodeikticus*, while saliva and blood serum are relatively weak. Fluids from a number of different individuals have been tested and the relative amounts of lysozyme contained in these have been found to be comparatively constant, except in the case of saliva, which seems to vary considerably, although it never approaches in lysozyme-content tears, sputum, or nasal mucus.

*The Question as to whether Lysozyme exists in Tissues other than Human Tissues.*

Only a limited amount of work has been done in this direction, but it is sufficient to show that lysozyme is very widespread in nature. Rabbit and guinea-pig tissues were examined and it was found that nearly all of these contained some lysozyme for the *M. lysodeikticus*, but in general the lysis was not nearly so marked as it was with the corresponding human tissues. It may be noted that the lachrymal secretion of both these animals contained no lysozyme for the *M. lysodeikticus*, against which the human tears are so powerful. The tissues of a dog were much more lytic than those of the rabbit and guinea-pig, but even they were not so active as human tissues.

It was found that egg-white was very rich in lysozyme for the *M. lysodeikticus*, there being, after incubation for 24 hours, lysis visible to the naked eye when a dilution as great as 1 in 50,000,000 was employed. Egg-white also contains lytic substances for many other bacteria. It was found also that commercial dried egg albumin was very rich in lysozyme.

In the vegetable kingdom it was found that turnip had a very definite, though not very strong lytic action on *M. lysodeikticus*. Several of the other common table vegetables were tested, but they appeared to be devoid of lytic activity.

[*Editor's Note:* The paper ends with a consideration of the question: "Does the Lysozyme act on Bacteria other than the *M. lysodeikticus*?" This part has been omitted here.]

Reprinted from *Nature* **206**:757–761 (1965)

# STRUCTURE OF HEN EGG-WHITE LYSOZYME

## A Three-dimensional Fourier Synthesis at 2 Å Resolution

By DR. C. C. F. BLAKE, DR. D. F. KOENIG*, DR. G. A. MAIR, DR. A. C. T. NORTH†, DR. D. C. PHILLIPS†, and DR. V. R. SARMA

THE X-ray analysis of the structure of hen egg-white lysozyme[1] (*N*-acetylmuramide glycanohydrolase, *EC* 3.2.1.17), which was initiated in this Laboratory in 1960 by Dr. R. J. Poljak, has now produced a Fourier map of the electron density distribution at 2 Å resolution in which the structure of the molecule can be clearly seen.

Comparison of this map with the primary structure, determined by Jollès *et al.*[2,3], by Canfield[4,5] and in part by Brown[6], has made possible the location of each of the 129 amino-acid residues of which the molecule is composed, many of which, including the four disulphide bridges, can be identified unambiguously in the X-ray image. The arrangement of groups in the molecule is very complicated, and many interactions between them will be revealed only by detailed analysis of the map. This preliminary account of the work is intended, therefore, to indicate the quality of the results, and to show the general structure of the molecule. Further work which is described in the following article[7] has also revealed the location of a site at which competitive inhibitor molecules are bound in the crystal structure so that it is possible, even at this stage, to identify amino-acid residues which may be involved in the binding of substrates.

### X-ray Methods

Crystals of hen egg-white lysozyme chloride grown at *p*H 4·7 (ref. 8) are tetragonal with unit cell dimensions, $a = b = 79{\cdot}1$, $c = 37{\cdot}9$ Å, and space group $P4_12_12$ or $P4_32_12$ (ref. 9). Each unit cell contains eight lysozyme molecules (one per asymmetric unit), molecular weight about 14,600, together with 1 M sodium chloride solution which constitutes about 33·5 per cent of the weight of the crystal[10].

A preliminary investigation[11], in which the method of isomorphous replacement was used, showed that the space group is in fact $P4_32_12$ and made possible a calculation of the electron density distribution at 6 Å resolution.

In the extension of this analysis to higher resolution it has been necessary to find new and better isomorphous heavy-atom derivatives. Those originally used contained: (1) mercuri–iodide ions; (2) chloropalladite ions; and (3) *ortho*-mercuri hydroxytoluene *para*-sulphonic acid (MHTS) and of these only the MHTS derivative was found to be suitable for work at higher resolution. An extensive search, using methods already described[11], revealed, however, that derivatives containing (4) $UO_2F_5^{\equiv}$ (ref. 12); and (5) an ion derived from $UO_2(NO_3)_2$, probably $UO_2(OH)_n^{(n-2)-}$, were very suitable for this purpose. In addition derivatives including (6) *para*-chloromercuri benzene sulphonic acid (PCMBS) and (7) $PtCl_6^{=}$ (ref. 13) were found to be satisfactory for use at low resolution. All these derivatives except (1) and (2) were used in a re-examination of the structure at 6 Å. The necessary diffraction measurements were made with the Hilger and Watts, Ltd., linear diffractometer[14] adapted to measure three reflexions at a time[15,16] and the data were processed and the other calculations made on the Elliott 803 *B* computer at the Royal Institution. The phases were calculated by the phase probability method[17], adapted to make proper allowance for anomalous scattering of the copper *Kα* X-rays by the various heavy atoms[18]. The availability of uranium derivatives was particularly fortunate in that the anomalous scattering effect is large for uranium and indeed a particular search was made for them with that in mind. The mean figure of merit obtained in these calculations was 0·97 as compared with the 0·86 obtained before, and the greater part of this improvement has been shown to arise from the improved treatment of anomalous scattering rather than from the inclusion of additional heavy-atom derivatives.

Fig. 1 shows a model of the electron-density distribution at 6 Å resolution obtained in this new work. The outline of the molecule is certainly clearer in the new map than in the old, and within the molecule there is improved continuity suggesting the course of a folded polypeptide chain; but the two maps are, in fact, very similar (root mean square difference in electron density 0·12 eÅ$^{-3}$) and it remains impossible to determine the structure from them in any detail. Nevertheless a tentative general interpretation was possible. The model can be divided roughly into two parts; on the left-hand side of Fig. 1 the density has the appearance of a single chain without cross-connexions while to the right the arrangement appears to be much more complex. In general agreement with the picture, the primary structure, shown in Fig. 2, includes some relatively long lengths of polypeptide chain free from cross-connexions together with a region in which two disulphide bridges and two proline residues are close together.

### 2 Å Resolution

In preparation for the high-resolution investigation, the various heavy-atom derivatives were investigated in the centro-symmetric *hk*0 and *h*0*l* projections at 3 and 2 Å resolution. Comparison of the various sign predic-

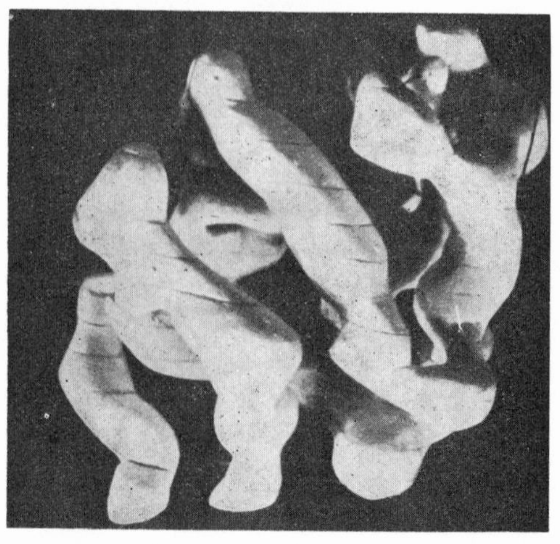

Fig. 1. Solid model of the lysozyme electron-density greater than about 0·5 electrons/Å$^3$ at 6 Å resolution

* Present address: Brookhaven National Laboratory, Upton, Long Island, New York.

† Medical Research Council External Staff.

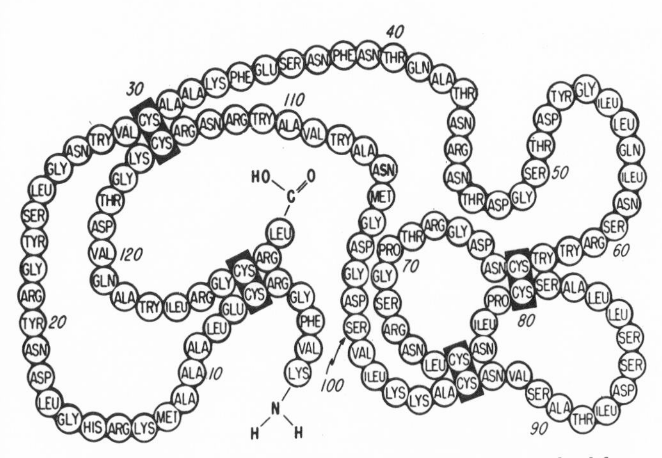


Fig. 2. Sequence of amino-acid residues in hen egg-white lysozyme reproduced from Canfield and Liu (ref. 5)

tions and examination of the radial distributions of the intensity differences[19] showed that only derivatives (3), (4) and (5) were suitable for high-resolution work. Accordingly 2 Å data were collected from the native crystals and for crystals containing these heavy atoms. The same experimental methods were used; but in order to minimize the effects of irradiation damage each crystal was exposed for only 20 h or less to the β-filtered X-ray beam from a copper tube operated at 40 kV, 20 m.amp. About 16 crystals were used to obtain each set of measurements which included generous overlaps for scaling[20] and separate measurement of the reflexions in the Bijvoet pairs *hkl* and *khl*. Semi-empirical corrections were made for absorption[21]. The data-processing system was designed at every stage to detect errors and to avoid their introduction and to eliminate any need for comprehensive manual checking.

The complete set of 2 Å data comprises some 9,040 reflexions of which 1,640 are centric, a property of this space group which greatly facilitates the determination and refinement of the heavy-atom parameters. The main heavy-atom positions were all found from $(\Delta F)^2$ maps in projection, and subsidiary sites were found from $\Delta F$ maps in two and three dimensions when preliminary signs and phases had been found. The best parameters were determined by the use of the quasi-three-dimensional set of centric reflexions, *hk*0, *k*0*l*, *hhl*, 0*kl*, in an adaptation of Hart's method[22]. The final parameters are shown in Table 1. The large proportion of centric reflexions also made possible an examination of the variation in apparent occupancy with scattering angle, thus revealing the shape of the appropriate scattering factor curve for each heavy-atom. These curves agree well with those expected for a heavy central atom surrounded by a shell of light atoms replacing a sphere of uniform electron density.

The temperature factors shown in Table 1 most deserve comment. Those obtained for the most important sites (of which there are no more than two for any derivative)

Table 1. HEAVY-ATOM PARAMETERS

| | Site | $X$ | $Y$ | $Z$ | $O$ | $B$ | $E$ | $N$ | $R$ |
|---|---|---|---|---|---|---|---|---|---|
| MHTS | I | 0·2068 | 0·6138 | 0·0507 | 39·2 | 17·8 | 58 | 1247 | 0·60 |
| | II | 0·2415 | 0·6393 | 0·9326 | 8·8 | 14·9 | | | |
| $UO_2F_5$ | I | 0·1783 | 0·5849 | 0·7204 | 55·5 | 21·0 | 74 | 1277 | 0·52 |
| | II | 0·0974 | 0·8976 | 0·4650 | 29·3 | 24·2 | | | |
| $UO_2(OH)_n$ | I | 0·0961 | 0·8938 | 0·4664 | 47·1 | 19·2 | 80 | 1140 | 0·57 |
| | II | 0·1898 | 0·5901 | 0·7168 | 42·1 | 124·8 | | | |
| | III | 0·0446 | 0·7266 | 0·5515 | 9·0 | 190·2 | | | |
| | IV | 0·0869 | 0·8976 | 0·4866 | 11·4 | 68·4 | | | |
| | V | 0·2024 | 0·6388 | 0·6781 | 28·6 | 42·8 | | | |

*O*, Occupancy of heavy-atom site—electrons; *B*, isotropic temperature factor constant—$Å^2$; *E*, root mean square difference between observed and calculated heavy-atom changes for centric reflexions—electrons; *N*, number of centric reflexions in the range $0·01 < \sin^2\theta < 0·15$ used in refinement; *R*, reliability index for observed and calculated heavy-atom changes of centric reflexions.

are all comparable with the overall value for the protein crystals themselves. The very large values obtained for other sites of (5) show that these sites are of little importance at high angles and may not represent true sites of heavy-atom attachment. The minor site of (3) MHTS is clearly the $SO_3^-$ group of this molecule: it occupies the same position as the equivalent group in the not strictly isomorphous PCMBS derivative[23].

The refinement programme gave the root mean square 'lack of closure' errors required in the isomorphous phase determination[17] that are shown in Table 1. Analysis of the agreement obtained between measurements of symmetry equivalent reflexions gave the corresponding values required for the anomalous-scattering phase determination[18]. These were found to vary as a function of sin θ having the values $19 + 455 \sin^2\theta/\lambda^2$ for derivatives (3) and (4) and $26·5 + 485 \sin^2\theta/\lambda^2$ for derivative (5).

The phases of the 9,040 reflexions were calculated on the Elliott 803 *B* computer and had a mean figure of merit of 0·60. The variation with angle was very similar to that obtained in the comparable analysis of sperm-whale myoglobin[24]; the mean value fell to 0·38 at the 2 Å limit. The 'best'[17] electron-density distribution, unsharpened, was calculated from these phases and 'figures-of-merit' on the University of London *Atlas* computer. The density was calculated at 1/120ths of the cell edge along *a* and *b* and 1/60ths along *c*, and was tabulated in a form suitable for immediate contouring to a scale of 0·75 in. ≡ 1 Å. The Elliott 803 *B* computer was used to generate that part of the map which was expected, from the 6 Å resolution work, to include one molecule, from the arbitrary asymmetric unit calculated by *Atlas*. The absolute scale of the measurements is known only roughly (from statistical arguments and reasonable occupancies for the heavy-atoms) and the map was contoured at estimated 0·25 $eÅ^{-3}$ intervals.

## The Fourier Map

The electron-density corresponding to a single molecule extends over the full length of the *c*-axis of the unit cell and in the other axial directions lies within the ranges $-\frac{1}{4} < x < +\frac{1}{4}$, $0 < Y < \frac{1}{2}$. The contour map was plotted on 60 sections of constant $Z$ so as to extend safely beyond these ranges of $X$ and $Y$. Two separate blocks of ten sections each, which are representative of the whole map, are shown in Figs. 3 and 4. No $F(000)$ term was included in the Fourier synthesis, so that density values are all relative to the average crystal density. Only contours of higher than average density are shown in the maps, but there are no patches of high negative density even at the heavy atom positions, and it is immediately apparent that the general background is satisfactorily uniform. In regions which must be occupied by water and salt[9], such as those near the two-fold rotation axes, there are many small peaks, which are consistent with some degree of order in the liquid of crystallization. Within the molecular boundary the density indicates clearly the course of the polypeptide chain and the nature of many side-chains.

At this resolution atoms are not expected to appear in the image as separate peaks of electron density, but groups of atoms connected only by ionic or van der Waals' interactions or by hydrogen bonds are expected to be resolved. It is satisfactory therefore to find a continuous ribbon of high density with characteristic features at regular intervals to represent the main polypeptide chain with its carbonyl groups and with side-chains protruding

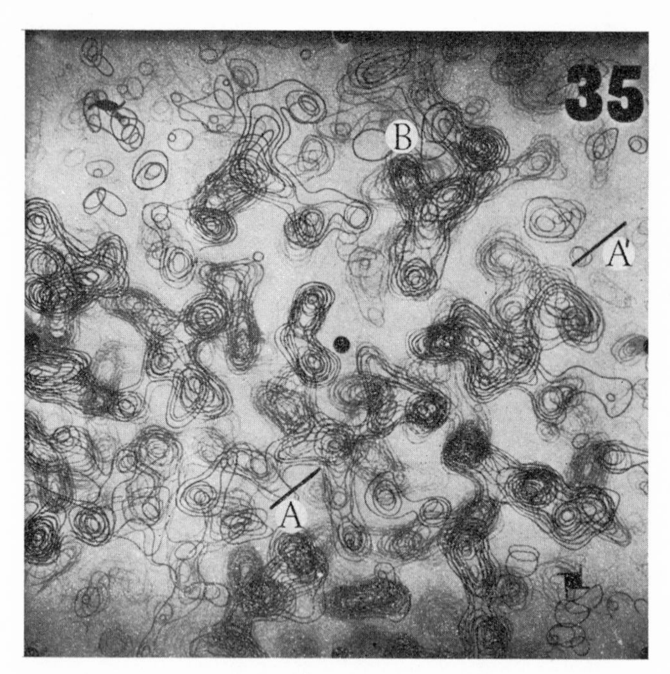


Fig. 3. Sections $Z = 35$—44/60 of the 2 Å resolution electron density map. Contours at intervals of 0·25 $eÅ^{-3}$. *AA'* shows the course of a length of α-helix lying in the plane of the sections. *B* indicates a length of helix more nearly normal to the sections

from it. In some regions the conformation of the main chain is clearly α-helical. For example, Fig. 3 includes the density corresponding to two lengths of α-helix, one of which runs approximately parallel to the sections (*AA'*) while the other is more nearly perpendicular to the sections (*B*). Inspection of the density in comparison with models shows immediately that the helices are right-handed running towards the terminal amino end of the polypeptide chain in the direction *AA'* and at *B* down into the map.

It is in such helical regions that the side-chains are most easily identified. Thus the peak of high density a little below the helix *AA'* and connected back to it clearly represents a disulphide bridge between the helix and more extended chain near the bottom of the picture. To the left of the disulphide bridge, density characteristic of another side-chain, a phenylalanine, can be seen emerging from the helix at an α-carbon position four residues removed from the cystine towards the terminal carboxyl end of the chain. This observation, when combined with the sequence of Fig. 2, immediately identifies the disulphide bridge as that between residues 30 and 115 and was, in fact, the starting-point of a detailed correlation between map and sequence. The helix at *B* corresponds to a part of the chain nearer the terminal amino end and has features clearly consistent with the sequence Arg-Lys-Met-Ala-Ala-Ala of residues 14–9.

In the non-helical regions, the courses of the main chain and the side-chains are rather less easily recognized by mere inspection of the map. Nevertheless, no real difficulty was encountered in following the main chain and nowhere did the density corresponding to it fall to the general background-level. As in the corresponding study of myoglobin[25] however, it was often simpler to recogniz the density corresponding to particula well-marked side-chains than to follow th main-chain density directly. Some of thes clearly identifiable side-chains in a non helical region are shown in Fig. 4. They include most prominently the indole ring of tryptophan residues 28, 108 and 111 and the sulphur atom of methionine 10 which come together with tyrosine 23 in this part of the structure to form curiously regular hydrophobic box.

Of course, not all the residues are a clearly recognizable as this; but using th criteria formulated in the examination o myoglobin[25] it is possible to identify many of them with considerable confidence Nevertheless, the existence of almost complete knowledge of the primary structure has been of enormous value, and no attempt was made to analyse the X-ray image without recourse to it. No disagreements with the chemical information were obvious at this stage of analysi except in those few small regions in which the chemical studies[2,4] do not agree These are discussed here.

## General Description of the Molecule

The molecule is roughly ellipsoidal with dimensions about 45 × 30 × 30 Å. Th general arrangement of the polypeptide chain is indicated in the schematic diagram of Fig. 5. It is clearly even mor complicated than that of myoglobin, and a comprehensive description cannot yet be given. Some importan features, however, are immediately apparent. First the α-helical content is relatively low. Detailed in

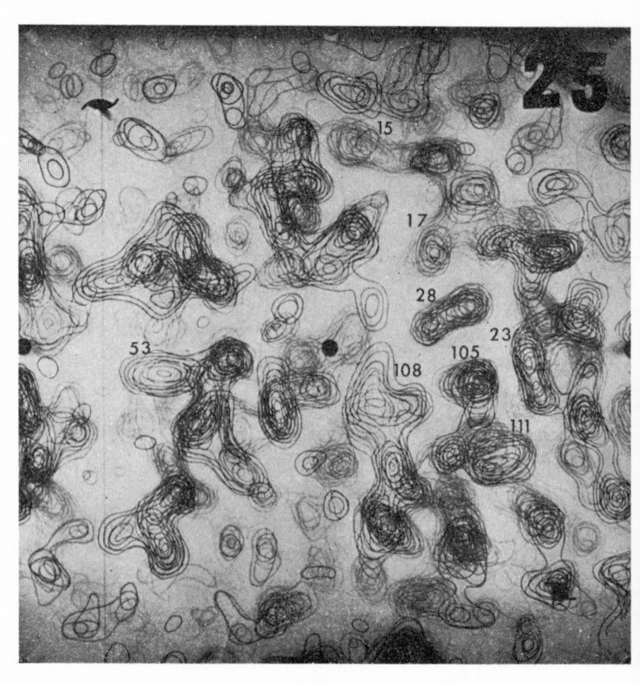


Fig. 4. Sections $Z = 25$—34/60 of the 2 Å resolution electron density map. The densities corresponding to a number of amino-acid side-chains are numbered in accordance with Fig. 2

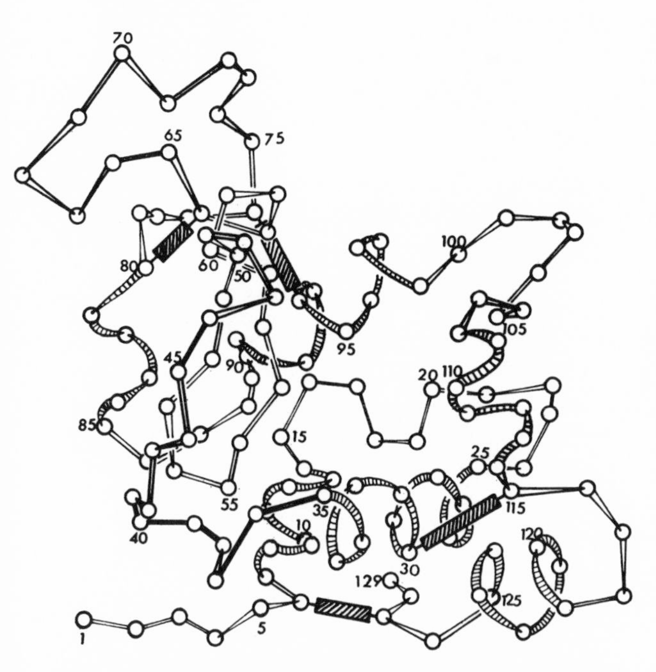


Fig. 5. Schematic drawing of the main chain conformation (by W. L. Bragg)

vestigations of the helical parameters have not yet been made, but about six lengths of helix have been recognized, some of them very short, as indicated in Figs. 5 and 6. If a residue is regarded as being in a helix if one or other of its possible hydrogen bonds is properly made, then about 55 of the 129 residues appear at this stage to be included in helices. This suggests a maximum helix content of about 42 per cent, in fair agreement with the predictions of optical rotatory dispersion[26,27].

The remainder of the molecule is less easily described, the folding of residues 35–80 being particularly complex. In this region three lengths of chain are roughly anti-parallel, two of them, residues 41–54, forming a nearly closed loop. Nearby are the two closely related disulphide bridges connected by short lengths of chain which include the proline residues (70 and 79).

The disulphide bridges each have a helix on at least one side, although the half-cystine is generally the last helical residue or at least in the last turn of its helix. Only one bridge, 30–115, has a helix on both sides and 115 is a terminal residue.

The inside and outside of the molecule are less easily defined than they are for myoglobin, in which the interior of the molecule consists almost exclusively of hydrophobic side-chains[28]. The lysozyme molecule appears to have a hydrophobic spine consisting mainly of the six tryptophan side-chains and including the hydrophobic box already described (Fig. 4). Three of these tryptophan side-chains, 62, 63 and 123, protrude, however, beyond the molecular boundary and there are in addition a number of strongly hydrophobic side-chains clearly on the molecular surface (for example, Val 2, Phe 3, Phe 34, Leu 17).

On the other hand, the parts of the molecule most shielded from contact with surrounding liquid appear to include Ser(91) and Gln(57). In addition, residue 58, which may be asparagine[2], can be regarded as internal. All the lysine and arginine side-chains are external.

## Comparison with Low Resolution Results

It is interesting to compare this molecule with the model obtained at 6 Å resolution. The boundary of the molecule was correctly chosen in the revised 6 Å work (Fig. 1) and was incorrect only in minor details in the earlier investigation[11]. Furthermore, the very general interpretation was correct in that the complex folding indicated by the right-hand part of Fig. 1 does indeed include the two disulphide bridges 64–80 and 76–94. On the other hand, it is somewhat surprising to discover that the 6 Å maps contain no significant indications of the positions of three of the disulphide bridges, possibly because they lie in diffraction minima associated with neighbouring helices. Clearly it is very unlikely indeed that the 6 Å model could have been interpreted correctly in detail, even in the light of the primary structure, and this is true also of a model at 5 Å resolution which is not described here.

## Comparison with Chemical Data

The independent investigations of the primary structure[2,4,5] now agree in all but a few details on the sequence of amino-acid residues in the lysozyme molecule. Two of the remaining doubts[5] are concerned with the distinction between aspartic acid and asparagine, which are not readily distinguishable by the X-ray method. The other three difficulties, which are more serious, can be resolved to some extent by examination of the present map.

Thus, residues 92 and 93 are included in a length of helix and their side-chains have strongly contrasted appearances. That of 92 is clearly forked close to the helix and is surrounded by a clear region of low electron density, while the side-chain of residue 93 is longer and approaches much more nearly to neighbouring density. It is probable, therefore, that the sequence 92–93 is valine–asparagine[4] rather than asparagine–valine[2]. In the second case, the side-chains associated with residues 40–42, which are in a non-helical region, appear to be much more consistent

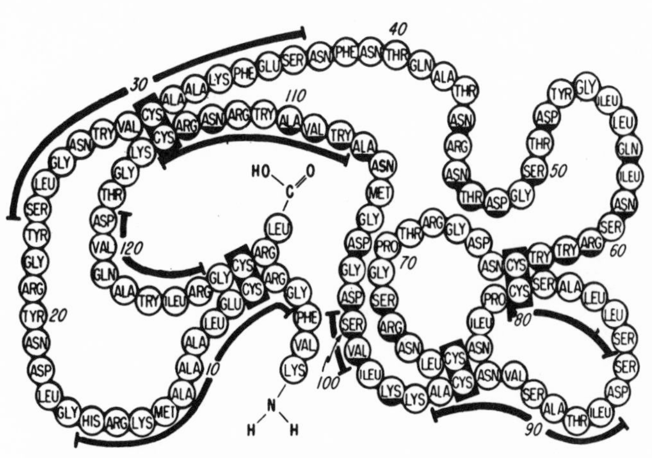


Fig. 6. Primary structure (ref. 5) showing (continuous line) the residues found in α-helical conformation and those in the apparent binding site (underlined)

(by similar arguments) with the sequence Thr-Gln-Ala[4] than with Gln-Ala-Thr[2]. The third ambiguity, between isoleucine and asparagine in positions 58 and 59, has already been referred to. In this case the differences between the two densities are less clear, but on balance that associated with residue 58 seems more consistent with asparagine[2] than with isoleucine[4] even though this residue can be regarded as being inside the molecule. Its environs include the disulphide bridge 76–94 and Leu 83, but also serine 91, and there is an unexplained neighbouring patch of density which could represent an included water molecule. It is hoped that the detailed analysis of side-chain interactions which is now in progress will resolve this uncertainty.

In addition to the primary structure there are, of course, very many chemical data on the physical and chemical properties of lysozyme which cannot be discussed here, but which are expected to be interpretable eventually in terms of the detailed structure. Something must now be said, however, about amino-acid residues probably concerned in the enzymatic activity, a subject on which there is so far rather little definite chemical evidence.

## Inhibitor Binding Site

The following article describes the location at low resolution of the site of attachment of *N*-acetyl-glucosamine and its dimer, chitobiose, both of which molecules act as competitive inhibitors of lysozyme. They attach themselves, at least in this crystal structure, in a cleft in the surface of the molecule which runs nearly vertically down the middle of Fig. 5 and which can be seen clearly in the low-resolution model (Fig. 1, ref. 7). The amino-acid residues bounding this portion of the molecular surface come from various parts of the polypeptide chain and are indicated in Fig. 6. Clearly, further examination is required to establish whether this site represents the binding site of substrate molecules, but for the present it seems a reasonable hypothesis that it does so and that the properties of these residues may well repay detailed investigation. In accordance with chemical evidence[29–32], they include three tryptophans, now shown to be 62, 63 and 108, but not the single histidine which is located on the opposite side of the molecule.

We thank Sir Lawrence Bragg, who drew the diagram of Fig. 5, and Prof. R. King for their advice, and the Managers of the Royal Institution for their support; the Director of the London University Computer Unit for facilities on *Atlas*; and Mrs. R. Arthanari, Mrs. W. Browne, Mrs. S. J. Cole, Mrs. J. A. Conisbee, Miss M. Hibbs, Mrs. K. Sarma and Messrs. A. R. Knott, S. B. Morris and J. E. T. Thirkell for their assistance. We also thank the Medical Research Council and the U.S. National Institutes of Health for financial support.

[1] Fleming, A., *Proc. Roy. Soc.*, B, **93**, 306 (1922).
[2] Jollès, J., Jauregui-Adell, J., and Jollès, P., *Biochim. Biophys. Acta*, **78**, 68 (1963).
[3] Jollès, P., Jauregui-Adell, J., and Jollès, J., *C.R. Acad. Sci., Paris*, **258**, 3926 (1964).
[4] Canfield, R. E., *J. Biol. Chem.*, **238**, 2698 (1963).
[5] Canfield, R. E., and Liu, A. K., *J. Biol. Chem.* (in the press).
[6] Brown, J. R., *Biochem. J.*, **92**, 13P (1964).
[7] Johnson, L. N., and Phillips, D. C. (following article).
[8] Alderton, G., and Fevold, J., *J. Biol. Chem.*, **164**, 1 (1946).
[9] Palmer, K. J., Ballantyre, M., and Galvin, J. A., *J. Amer. Chem. Soc.*, **70**, 906 (1948).
[10] Steinrauf, L. K., *Acta Cryst.*, **12**, 77 (1959).
[11] Blake, C. C. F., Fenn, R. H., North, A. C. T., Phillips, D. C., and Poljak, R. J., *Nature*, **196**, 1173 (1962).
[12] Holmes, K. C., and Leberman, R., *J. Mol. Biol.*, **6**, 439 (1963).
[13] Poljak, R. J., *J. Mol. Biol.*, **6**, 244 (1963).
[14] Arndt, U. W., and Phillips, D. C., *Acta Cryst.*, **14**, 807 (1961).
[15] Phillips, D. C., *J. Sci. Instrum.*, **41**, 123 (1964).
[16] Arndt, U. W., North, A. C. T., and Phillips, D. C., *J. Sci. Instrum.*, **41**, 421 (1964).
[17] Blow, D. M., and Crick, F. H. C., *Acta Cryst.*, **12**, 794 (1959).
[18] North, A. C. T., *Acta Cryst.*, **18**, 212 (1965).
[19] Crick, F. H. C., and Magdoff, B. S., *Acta Cryst.*, **9**, 901 (1956).
[20] Hamilton, W. C., Rollett, J. S., and Sparks, R. A., *Acta Cryst.*, **18**, 129 (1965).
[21] North, A. C. T., Phillips, D. C., and Mathews, F. S. (in preparation).
[22] Hart, R. G., *Acta Cryst.*, **14**, 1188 (1961).
[23] Blake, C. C. F., Fenn, R. H., and Phillips, D. C. (in preparation).
[24] Kendrew, J. C., Dickerson, R. E., Strandberg, B. E., Hart, R. G., Davies, D. R., Phillips, D. C., and Shore, V. C., *Nature*, **185**, 422 (1960).
[25] Kendrew, J. C., Watson, H. C., Strandberg, B. E., Dickerson, R. E., Phillips, D. C., and Shore, V. C., *Nature*, **190**, 663 (1961).
[26] Urnes, P., and Doty, P., *Adv. Prot. Chem.*, **16**, 401 (1961).
[27] Hamaguchi, K., and Imahori, K., *J. Biochem.* (Japan), **55**, 388 (1964).
[28] Kendrew, J. C., *Brookhaven Symp. Quant. Biol.*, **15**, 216 (1962).
[29] Kravchenko, N. A., Kleopina, G. V., and Kaverzneva, E. D., *Biochim. Biophys. Acta*, **92**, 412 (1964).
[30] Bernier, I., and Jollès, P., *C.R. Acad. Sci., Paris*, **253**, 745 (1961).
[31] Ramachandran, L. K., and Rao, G. J. S., *Biochim. Biophys. Acta*, **59**, 507 (1962).
[32] Hartdegen, F. J., and Rupley, J. A., *Biochim. Biophys. Acta*, **92**, 625 (1964).

Reprinted from *Nature* **206**:761–763 (1965)

# STRUCTURE OF SOME CRYSTALLINE LYSOZYME-INHIBITOR COMPLEXES DETERMINED BY X-RAY ANALYSIS AT 6 Å RESOLUTION

By MISS LOUISE N. JOHNSON and DR. D. C. PHILLIPS*

Davy Faraday Research Laboratory, Royal Institution, 21 Albemarle Street, London, W.1

A RECENT investigation of the azide derivative of sperm-whale myoglobin[1] has shown that the interactions of proteins with small molecules can be examined in crystals when the phases of reflexions from isomorphous unsubstituted-protein crystals are known. The difference Fourier method is used in which the changes in structure amplitude caused by the introduction of the small molecule are combined with the known phases to give a map showing the change in electron density. The work on myoglobin[1] was done at 2 Å resolution; but we have now shown that interesting results can be obtained by this method at 6 Å resolution even when the substituted molecules comprise only light atoms. The method depends on the existence of precise phase information for the protein crystals, information which is now available for lysozyme[2], and by its means we have been able to investigate the structural relationship between lysozyme and a number of compounds which are related to the substrate of the enzyme and which act as competitive inhibitors of its action. In this way we have been able to locate a part of the molecule which may be responsible for its enzymatic activity.

Lysozyme has a β(1–4) glucosaminidase activity with the ability to hydrolyse a mucopolysaccharide component of some bacterial cell walls releasing *N*-acetyl amino sugars derived from glucosamine and muramic acid[3]. The proposed structure of a tetrasaccharide[4] isolated from the cell wall of *Micrococcus lysodeikticus*, which is made up of alternate *N*-acetylglycosamine and *N*-acetyl muramic acid units, is shown in Fig. 1 in which the bond probably hydrolysed by lysozyme is indicated by an arrow. Lysozyme will also hydrolyse chitin[5], the (1–4) linked linear chain polymer of *N*-acetylglucosamine, and Wenzel *et al.*[6] have shown that lysozyme will promote the cleavage of the trimer, tri-*N*-acetyl chitotriose,

* Medical Research Council External Staff.

Fig. 1. Tetrasaccharide from *Micrococcus lysodeikticus* with the bond hydrolysed by lysozyme ref. 4) shown by an arrow

releasing dimer and monomer molecules. They reported that *N*-acetylglucosamine inhibits the activity of lysozyme but that glucosamine does not, and recently Rupley[7] has reported that the dimer, di-*N*-acetylchitobiose, is also an inhibitor and that it is not hydrolysed.

With this background we have examined the binding to lysozyme in the crystal structure[2] of five compounds: *N*-acetylglucosamine, di-*N*-acetylchitobiose, 6-iodo α-methyl *N*-acetylglucosaminide, glucosamine (hydrochloride) and muramic acid. The first three of these, which are competitive inhibitors of lysozyme, bind specifically to one and the same site on the enzyme whereas the last two which do not inhibit do not bind specifically to this site either.

## Crystallographic Methods

Crystals of lysozyme chloride were grown in the usual way at *p*H 4·7 (ref. 2) and then soaked for 24–48 h in a solution of the inhibitor with concentration about 0·6 M. A precession photograph was then taken of the *hk*0 reflexions to 3 Å resolution and the intensities were compared with those of the native protein crystals. If there were changes complete three-dimensional data were collected to 6 Å resolution, processed in the usual way and scaled to the native data. (An outline of the experimental methods is included in the previous article[2].) Typically the average $F$ value was about 255 (arbitrary units) while the mean absolute difference in amplitude between the derivative and the native protein was about 30 with some differences as large as 86. About 400 $\Delta F$ values were used with associated phases to calculate each three-dimensional difference map at 6 Å resolution. The calculations were done on the Elliott 803 *B* computer using a programme written by Mr. R. M. Simmons.

## Results

(*a*) *N-acetylglucosamine.* The three-dimensional difference synthesis gave a remarkably clear map with only one peak which could be interpreted as representing the *N*-acetylglucosamine (NAG) molecule. The peak height was 0·16 e/Å$^3$ and the peak contained 48 electrons distributed throughout a volume of 460 Å$^3$. Luzzatti[8] has shown that in an electron-density synthesis the height of peaks representing atoms not taken into account in the phase determination approaches 50 per cent of their true height when the ratio of known to unknown atoms is large. Furthermore, in the present situation a reduction in peak heights is expected due to the replacement of water molecules. The peak height calculated for a molecule of NAG placed in the position of the peak was found to be 0·5 e/Å$^3$ so that the observed value is in satisfactory agreement with the value expected for a molecule of NAG in this position displacing some water.

(*b*) 6-*Iodo α-methyl N-acetylglucosaminide.* This derivative was synthesized for us in this laboratory by Dr. J. W. H. Oldham. It gave rise to a single peak in the same position as that occupied by NAG itself and with a peak height of 0·22 e/Å$^3$. The peak contained 64 electrons distributed throughout a volume of 530 Å$^3$. The peak was significantly higher than that obtained for the uniodinated compound, but the iodine atom was not resolved.

(*c*) *Di-N-acetylchitobiose.* This dimer, in which two NAG molecules are joined by a β(1–4) linkage, was kindly provided by Dr. J. A. Rupley. The difference synthesis showed that the dimer binds to the same place as the monomer, giving rise to a peak of height 0·2 e/Å$^3$ with 100 electrons distributed throughout 960 Å$^3$.

The position occupied by chitobiose in relation to the lysozyme molecule is shown in Fig. 2. The model of the lysozyme molecule is based on the structure analysis at 6 Å resolution[2], while the shaded peak representing chitobiose was constructed from the difference synthesis. The peak due to the single NAG molecule is shown as the dark region in the peak due to the dimer.

(*d*) *Glucosamine hydrochloride.* The addition of glucosamine gave rise to significant changes in the reflexion intensities, but difference maps showed no single peak where the NAG had bound but rather a large number of smaller peaks scattered over the unit cell. From this result it appears that glucosamine will bind to lysozyme perhaps through its amino-group, but that it does not bind specifically in one place.

Fig. 2. Photograph of the model of a lysozyme molecule obtained by X-ray analysis at 6 Å resolution together with the increase in electron density observed in the presence of di-*N*-acetylchitobiose (hatched). The increase in electron density due to *N*-acetylglucosamine is shown as the darker part of the chitobiose

(*e*) *Muramic acid.* Muramic acid, kindly supplied to s by Dr. R. H. Gigg, likewise showed no signs of specific inding to lysozyme.

*Inhibition experiments.* The effects of NAG, 6-iodo -methyl NAG and glucosamine on the activity of ysozyme were investigated by means of the usual method f assay[9]. The results confirmed the earlier conclusion[6] hat NAG inhibits competitively while glucosamine does ot, and showed further that 6-iodo α-methyl NAG has very similar effect to that of NAG.

## Discussion

Although it is only when this work is repeated at higher esolution that detailed information concerning the nteractions between the protein and inhibitor can be btained, these results at 6 Å resolution are highly ncouraging and the following conclusions can already be lrawn.

There is good evidence that the site indicated by the nhibitors is indeed the region of the enzyme responsible or its activity. The inhibitors are structurally analogous o the substrate and inhibit competitively. The inhibitors ind specifically to this one site whereas closely similar nolecules which are not inhibitors do not. The site is lifferent from those to which the heavy atoms bind.

It is possible to make preliminary statements concern- ng the nature of the groups on the substrate which re involved in the binding. It appears the *N*-acetyl group is essential. Molecules such as glucosamine and nuramic acid which do not contain this group do not ind and do not inhibit. The fact that 6-iodo α-methyl V-acetylglucosaminide binds and inhibits suggests that either the reducing group nor the oxygen in the six osition is essential for binding. Studies of other deriva- ives of NAG, in which different groups have been locked[10], are now in progress and it is hoped that they vill show, even in advance of high-resolution analysis, which groups are involved in the binding.

The results described so far do not reveal the site at which the *N*-acetylmuramic acid (NAM) moiety of the ell-wall substrate is bound. Unfortunately, this com- ound, though synthesized many times, is not readily available, and we have not yet obtained any for examina- ion. However, a preliminary investigation of the inter- ction of lysozyme with penicillin V (ref. 11), probably a tructural analogue of *N*-acetylmuramic acid[12], suggests hat it is bound at a site adjacent to that occupied by NAG and above it in Fig. 2. Experiments with *N*-acetyl- nuramic acid itself and the NAG–NAM dimer[13] are now eing planned.

From Fig. 2 it can be seen that the inhibitor molecules lie well embedded in the enzyme molecule in a crevice in its surface. This is interesting in view of the many theories of enzyme action which envisage just such a situation. At no point does the density due to the inhibi- tor molecules penetrate the density representing the enzyme, but in two regions they are very close. The amino-acid residues in these regions can now be identified in the image of the enzyme at 2 Å resolution and they are shown in the previous article[2]. It is encouraging that they do not include the single histidine, which is not now believed to play a direct part in the enzyme-substrate interaction[14], and that they do include three of the trypto- phans, since it has been reported[15,16] that the maintenance, intact, of four of the six indole residues is essential for enzymatic activity. Clearly the high-resolution investiga- tions of enzyme-inhibitor complexes which are now being started are needed to show in detail which amino-acid residues are involved in the binding. Similar experiments with substrate analogues resistant to hydrolysis and with true substrates such as tri-*N*-acetylchitotriose[7], using methods pioneered by Doscher and Richards[17], may well reveal any structural changes in the enzyme that take place during the reaction that it controls.

We thank our colleagues in this laboratory for their help in these experiments; Sir Lawrence Bragg and Prof. R. King for advice; the Managers of the Royal Institution for the provision of facilities; and the Medical Research Council and the U.S. National Institutes of Health for financial support. One of us (L. N. J.) thanks the Department of Scientific and Industrial Research for a research studentship.


1 Stryer, L., Kendrew, J. C., and Watson, H. C., *J. Mol. Biol.*, **8**, 96 (1964).
2 Blake, C. C. F., Koenig, D. F., Mair, G. A., North, A. C. T., Phillips, D. C., and Sarma, V. R., *Nature* (preceding article).
3 Salton, M. R. J., *Biochim. Biophys. Acta*, **22**, 495 (1956).
4 Jeanloz, R. W., Sharon, N., and Flowers, H. M., *Biochim. Biophys. Res. Comm.*, **13**, 20 (1963).
5 Berger, L. R., and Weiser, R. S., *Biochim. Biophys. Acta*, **26**, 517 (1957).
6 Wenzel, M., Lenk, H. P., and Schutte, E., *Z. Physiol. Chem.*, **327**, 13 (1962).
7 Rupley, J. A., *Biochim. Biophys. Acta*, **83**, 245 (1964).
8 Luzzatti, V., *Acta Cryst.*, **6**, 142 (1953).
9 Shugar, D., *Biochim. Biophys. Acta*, **8**, 302 (1952).
10 Oldham, J. W. H. (to be published).
11 Johnson, L. N. (to be published).
12 Collins, J. F., and Richmond, M., *Nature*, **195**, 145 (1962).
13 Sharon, N., paper presented at the Third Intern. Symp. on Fleming's Lysozyme, Milan, April 3–5 (1964).
14 Kravchenko, N. A., Kleopina, G. V., and Kaverzneva, E. D., *Biochim. Biophys. Acta*, **92**, 412 (1964).
15 Bernier, I., and Jollès, P., *C.R. Acad. Sci., Paris*, **253**, 745 (1961).
16 Rao, G. J. S., and Ramachandran, L. K., *Biochim. Biophys. Acta*, **59**, 507 (1962).
17 Doscher, M. S., and Richards, F. M., *J. Biol. Chem.*, **238**, 2399 (1963).

Reprinted from *Natl. Acad. Sci. (USA) Proc.* **44**:162–166 (1958)

# *ON THE ENZYMIC ACTIVITY OF SUBTILISIN-MODIFIED RIBONUCLEASE**

BY FREDERIC M. RICHARDS

DEPARTMENT OF BIOCHEMISTRY, YALE UNIVERSITY, NEW HAVEN, CONNECTICUT

*Communicated by Joseph S. Fruton, December 19, 1957*

The initial stage of the proteolysis of bovine pancreatic ribonuclease (RNase) by subtilisin results in the production of a modified RNase molecule whose enzymic activity is very similar to that of the native material.[1, 2] The presence of this altered form of the enzyme (RNase-S) is easily demonstrated by its chromatographic behavior on the Amberlite resin IRC-50, on which it moves as a discrete peak clearly separated from and following the major component (RNase-A) of the native enzyme. The finding of two *N*-terminal amino acid residues and the separation of two peptides from the performic acid-oxidized material indicate that the single peptide chain of RNase-A has been cleaved at one bond and that the two parts remain associated in RNase-S. These original observations have now been confirmed and extended.

Subtilisin appears to attack first the alanyl-seryl bond between residues Nos. 20 and 21, counting from the *N*-terminal end of the RNase-A chain. By careful treatment with trichloroacetic acid it has been possible to separate this *N*-terminal peptide (RNase-S-Pep) from the rest of the molecule (RNase-S-Prot). These fractions have only traces of residual enzymic activity. On mixing a solution of the trichloroacetic acid precipitate with an equivalent amount of the supernatant fluid, the full enzymic activity of the unfractionated material is regenerated.

## EXPERIMENTAL

*Preparation of RNase-S.*—RNase-A was isolated from crystalline ribonuclease, Armour Lot 381-059, by chromatography[3] on IRC-50, dialyzed to remove the phosphate buffer, and lyophilized. This material was treated with subtilisin at pH 8, as described previously,[2] except that the temperature was lowered to 7° C. At this temperature, it was possible to obtain 80–90 per cent conversion of RNase-A to RNase-S without significant further proteolysis. The uptake of alkali, as measured on the pH-Stat,[4] was about 0.9 moles $OH^-$/mole of protein. An end-group determination on the total digest by the fluorodinitrobenzene procedure revealed bis-DNP-lysine and DNP-serine; other DNP-amino acids were present in insignificant amount. The ninhydrin color of the mixture increased about 10 per cent during the digestion, while no change occurred in the enzymic activity. The total digest was chromatographed on IRC-50 under conditions identical to those used above in the preparation of RNase-A. A trace of material showing enzymic activity was found at the RNase-A position. The RNase-S was eluted more slowly and was separated from the faster-moving material. The bulk of the salt was removed by dialysis for 2 days against distilled water. For final desalting, the solution was run over a mixed-bed resin of IR-120 and IRA-400 in the $H^+$ and $OH^-$ forms, respectively. The eluate was lyophilized and used for the experiments described below.

*Site of Subtilisin Attack.*—RNase-S was oxidized with performic acid, and the products were separated by electrophoresis on glass paper at pH 6 in pyridine-acetate buffer. The eluate of the faster-moving material was hydrolyzed for 24 hours in constant boiling HCl and analyzed by column chromatography,[5] using the sulfonic acid resin XE-69. The ratios of the amino acids obtained closely approximated the following: $Asp_1(MetSO_2)_1Thr_2Ser_3Glu_3Ala_5Phe_1Lys_2His_1Arg_1$. This peptide is characterized by a high content of alanine and the absence of cysteic acid. The composition corresponds exactly[6, 7, 8] to the *N*-terminal 20 amino acids of RNase-A. The presence of a new *N*-terminal serine residue in the protein part of RNase-S also agrees with the known sequence of RNase-A. Thus RNase-S-Pep is composed of the *N*-terminal 20 amino acids of the original RNase-A, whereas RNase-S-Prot accounts for the other 104 amino acids, within the accuracy of the available data.

*Enzyme Assay Procedures.*—Ribonuclease activity employing ribonucleic acid as a substrate was measured either by the Kunitz spectrophotometric procedure[9] or by the acid-soluble nucleotide method, as described by Anfinsen *et al.*[10] A few measurements were made using uridine-2′,3′-phosphate as substrate, also as a spectrophotometric procedure.[11] The accuracy of any of these methods as employed here is no better than ±10 per cent of the measured activity.

*Trichloroacetic Acid Fractionation of RNase-S.*—A solution was prepared containing 1 mg/ml of RNase-S and 4 per cent *w/v* of trichloroacetic acid at 0° C. No precipitate formed at this temperature. A precipitate appeared when the solution was allowed to warm gradually to 30° C. After an additional hour, the precipitate was removed by centrifugation. The supernatant fluid, after continuous extraction with ether to remove the trichloroacetic acid, contained principally RNase-S-Pep. The precipitated protein (RNase-S-Prot) was dissolved, reprecipitated with trichloroacetic acid as before, redissolved, and made up to the original solution volume.

*Regeneration of Enzymic Activity.*—When assayed separately, both the RNase-S-Pep and the RNase-S-Prot solutions showed less than 5 per cent of the activity of unfractionated RNase-S. When equal volumes of the two were mixed, 100 per cent of the original activity was found. No lag was observed in the recovery of this activity, preincubation of the mixture being unnecessary. A lag period of 1 minute could easily have been observed when the Kunitz assay procedure was used. The recovery of activity was essentially instantaneous and complete even in the very dilute conditions of the assay procedures, where both the protein and the peptide components were present at concentrations of less than $1 \times 10^{-6}$ *M*. The very strong interaction between the peptide and protein is emphasized by the data (cf.

Fig. 1) obtained upon the addition of increasing amounts of RNase-S-Pep to a constant amount of RNase-S-Prot. The shape of the curve resembles that obtained in studies[12] on the recombination of the flavin and protein parts of the old yellow enzyme.

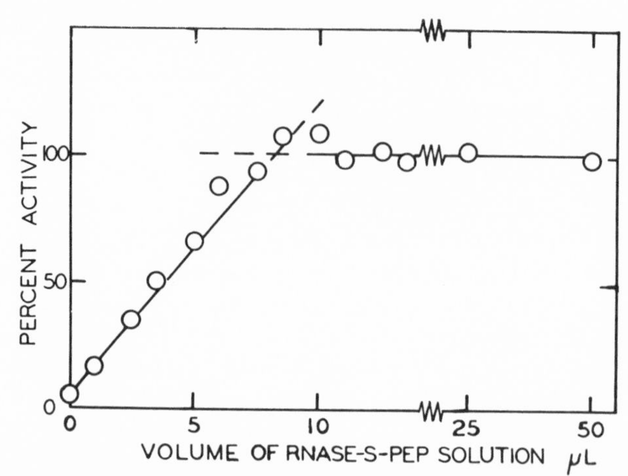


FIG. 1.—Regeneration of ribonuclease activity. The activity of 10-μl. aliquots of RNase-S-Prot solution is shown as a function of the volume of added RNase-S-Pep, both solutions at a concentration equivalent to RNase-S at 1 mg/ml. The maximum regenerated activity thus corresponds to that of 10 μg. of RNase-S. The assays were carried out by the acid-soluble nucleotide method, using 1 ml. of 0.5 per cent ribonucleic acid in 0.1 *M* acetate buffer pH 5.0 as substrate. The solutions to be assayed were added in the volumes indicated. The mixtures were incubated at 25° C. for 15 minutes. The measured activities are expressed as a percentage of that obtained with 10 μg. of RNase-S.

Not only was the peptide obtained by trichloroacetic acid fractionation effective in the regeneration of enzymic activity, but the peptide separated electrophoretically from performic acid-oxidized RNase-S also produced 100 per cent of the expected activity when recombined with unoxidized RNase-S-Prot. The only difference between the latter peptide and the RNase-S-Pep is the conversion of the single methionine residue to the corresponding sulfone. It may be concluded, therefore, that the oxidation state of the sulfur of this methionine residue is not important for the enzymic activity, as estimated by the assay procedures used.

*Effect of Trypsin on the Protein and Peptide Components.*—If a solution of RNase-S-Prot (equivalent to a concentration of RNase-S of 1 mg/ml) is treated at pH 7.5 and 27° C. with crystalline trypsin (10 μg/ml), the ability to regenerate ribonuclease activity, upon mixing with RNase-S-Pep, is completely lost within a few hours. Treatment of RNase-S-Pep with trypsin (under similar experimental conditions) abolishes the regeneration of activity on mixing with RNase-S-Prot. The activity of native RNase is unaffected even after several days of incubation with trypsin under the same conditions. When examined by electrophoresis on paper, a trypsin digest of RNase-S-Pep shows three components, as would be expected from the presence of two bonds known to be sensitive to trypsin in this peptide.[13] The extent of the action of trypsin on RNase-S-Prot is unknown at this time. Neither of these inactive trypsin-digested samples caused any detectable inhibition when added to an assay mixture of undigested RNase-S-Pep and RNase-S-Prot.

*Effect of Carboxypeptidase on the Peptide Component.*—A solution of RNase-S-Pep (derived from RNase-S at 1 mg/ml) was incubated for several days at pH 8 and 27° C. with carboxypeptidase (40 μg/ml). No significant change was observed in the ability of this solution to regenerate ribonuclease activity upon the addition of the protein component. During carboxypeptidase treatment, however, the ninhydrin color did increase, and the presence of free amino acids was demonstrated by reaction with fluorodinitrobenzene.

*Effect of Periodate on the Protein and Peptide Components.*—A solution of RNase-S-Prot (derived from RNase-S at 1 mg/ml), adjusted to pH 8, and made 0.01 *M* with respect to sodium periodate, was slowly inactivated over a period of 16 hours at 27° C. Native RNase and RNase-S-Pep were largely unaffected by this treatment. An interpretation of this result must await a clear demonstration of the nature of the periodate action. Because of the long time interval involved, residues other than the *N*-terminal serine may be involved.[14]

*Effect of Photo-oxidation.*—Solutions of native RNase, RNase-S-Pep, and RNase-S-Prot in phosphate buffer (pH 8) containing 0.005 per cent methylene blue were shaken in a 37° C. bath and illuminated for 1 hour with a 100-watt bulb at a distance of 1 foot, conditions approximating those described by Weil and Seibles.[15] All three solutions were rendered inactive by this treatment. Furthermore, the inactive RNase-S fractions caused no inhibition of the activity of the untreated control solutions. To the extent that the data of Weil and Seibles may be applied to this experiment, it is tempting to speculate that at least two histidine residues may be required for ribonuclease activity, one of which is the histidine residue occurring in the RNase-S-Pep preparation. The only other residue in this peptide presumably susceptible to photo-oxidation is methionine, and the oxidation of this residue has already been shown to be without effect on the potential activity of the peptide.

*Effect of Concentrated Urea.*—RNase-S was dialyzed for 4 days at 25° C. against 8 *M* urea at neutral pH. The dialyzate showed no enzyme activity by itself, but, upon addition to an RNase-S-Prot solution, ribonuclease activity was regenerated. The peptide thus appeared to be dissociated from the protein by 8 *M* urea. The residual protein, freed from urea by further dialysis against water, showed very little activity, and none appeared upon the addition of RNase-S-Pep. Considerable denaturation of the RNase-S-Prot had evidently occurred. The activity of native RNase is unaffected by similar treatment. If assays are carried out in the presence of 8 *M* urea, no regeneration of activity is seen on mixing RNase-S-Pep and RNase-S-Prot. In agreement with earlier data,[16–18] control experiments showed RNase to be active under these conditions, suggesting that the presence of peptide bond No. 20 in the native enzyme prevents the disrupting effect of urea.

*Residual Activity of the Protein and Peptide Fractions.*—The slight enzymic activity of the RNase-S-Pep fraction may be attributed to the presence of a small amount of RNase-S-Prot, which is slightly soluble in 4 per cent trichloroacetic acid. Successive precipitations of the protein fraction leave similar amounts of activity in the supernatant fluid each time (when assayed in the presence of added RNase-S-Pep). The residual activity (about 5 per cent) of the protein precipitate, however, is not affected by reprecipitation, even though no appreciable amounts of peptide can be demonstrated in supernatant fluids other than the first. This residual activity survives the trypsin digestion described above and is present when assays are performed in 8 *M* urea. It is uncertain whether this activity is caused by a small contamination with RNase-A, which was not separated during the chromatography on IRC-50, or whether it is an intrinsic property of the RNase-S-Prot molecule.

### DISCUSSION

A distinction is usually made between enzymes composed solely of amino acid residues and those that contain a non-amino acid prosthetic group. The separation of the RNase-S system into an inactive protein and a dialyzable, highly specific "cofactor" suggests that this distinction may not be so sharp as supposed. This finding makes it possible to study the enzymic activity of ribonuclease in a manner similar to that successfully applied in the case of the conjugated catalytic proteins. In particular, it will be of interest to examine the possibility that the peptide fragment of ribonuclease contains the active site and that the remainder of the protein serves the function of an "apoenzyme."

### SUMMARY

As a result of proteolysis by subtilisin, it has been possible to remove a peptide from the *N*-terminal end of the ribonuclease molecule. This peptide and the residual protein separately have less than 5 per cent of the original activity of the enzyme. Under certain conditions recombination of the two parts results in full restoration of the activity. The effect of various enzymic and chemical modifications of the two parts on this restoration of activity is described.

* This study was aided by grants from the National Science Foundation, from the National Institutes of Health, from the James Hudson Brown Memorial Fund of the Yale University School of Medicine, and from the American Cancer Society.

[1] S. M. Kalman, K. Linderstrøm-Lang, M. Ottesen, and F. M. Richards, *Biochim. et Biophys. Acta*, **16**, 297, 1955.

[2] F. M. Richards, *Compt. rend. Lab. Carlsberg, Ser. chim.*, **29**, 329, 1955.

[3] C. H. W. Hirs, S. Moore, and W. H. Stein, *J. Biol. Chem.*, **200**, 493, 1953.

[4] C. F. Jacobsen, J. Leonis, K. Linderstrøm-Lang, and M. Ottesen in D. Glick *Methods of Biochemical Analysis* (New York: Interscience Publishers, 1957) **4**, 171.

[5] D. H. Spackman, W. H. Stein, and S. Moore, *Federation Proc.*, **15**, 358, 1956.

[6] C. H. W. Hirs, W. H. Stein, and S. Moore, *J. Biol. Chem.*, **221**, 151, 1956.

[7] C. B. Anfinsen, *Federation Proc.*,**16**, 783, 1957.

[8] C. H. W. Hirs, W. H. Stein, and S. Moore, *I.U.P.A.C. Symposium on Protein Structure, Paris, July 1957* (New York: John W. Wiley)(in press).

[9] M. Kunitz, *J. Biol. Chem.*, **164**, 563, 1946.

[10] C. B. Anfinsen, R. R. Redfield, W. L. Choate, J. Page, and W. R. Carroll, *J. Biol. Chem.*, **207**, 201, 1954.

[11] F. M. Richards, *Compt. rend. Lab. Carlsberg, Ser. Chim.*, **29**, 315, 1955.

[12] H. Theorell, *Biochem. Z.*, **278**, 263, 1935.

[13] C. H. W. Hirs, S. Moore, and W. H. Stein, *J. Biol. Chem.*, **219**, 623, 1956.

[14] S. Fujii, K. Arakawa, and N. Aoyagi, *J. Biochem. (Japan)*, **44**, 471, 1957.

[15] L. Weil and T. S. Seibles, *Arch. Biochem.*, **54**, 368, 1955.

[16] C. B. Anfinsen, W. F. Harrington, A. Hvidt, K. Linderstrøm-Lang, M. Ottesen, and J. Schellman, *Biochim. et Biophys. Acta*, **17**, 141, 1955.

[17] C. B. Anfinsen, *Compt. rend Lab. Carlsberg, Ser. chim.*, **30**, 13, 1956.

[18] M. Sela and C. B. Anfinsen, *Biochim. et Biophys. Acta*, **24**, 229, 1957.

# 69


Reprinted from *Nature* **213**:862–865 (1967)

# TERTIARY STRUCTURE OF RIBONUCLEASE

**G. Kartha, J. Bello, and D. Harker**

*Center for Crystallographic Research, Buffalo, New York*

We started our investigations of the tertiary structure of the enzyme ribonuclease in about 1950 at the Protein Structure Project at the Polytechnic Institute of Brooklyn and have continued them in the Biophysics Department of the Roswell Park Memorial Institute since 1959. During the past four years, they have resulted in three dimensional electron density distributions of the protein molecule in the crystalline state; these maps progressively showed more detail as more X-ray diffraction data at higher resolution were used. Recently we have calculated a map at 2 Å resolution, using the data from the free protein crystal and from seven heavy atom derivative crystals. We believe that this map clearly shows the structure of the molecule. Comparison of this map with the primary structure of the protein, as elucidated by biochemical methods during the past few years, makes it possible to locate the amino-acid residues of which the molecule is made. Many of these residues, including four cystine disulphide bridges, can be located unambiguously. In contrast to myoglobin, the absence of any appreciable amount of α helix makes a complete description of this molecule difficult; it must await a more detailed map. The purpose of this article is to show the general course of the polypeptide chain in the ribonuclease molecule, and to describe some features of biochemical relevance. We have also reason to believe that we have located the active site of this enzyme molecule.

## Primary Structure

All our X-ray diffraction investigations have been carried out on bovine pancreatic ribonuclease. The primary structure of this protein has been determined[1], so that we know the number, nature and sequence of the amino-acids with a high degree of certainty. This covalent structure is shown in Fig. 1 and it can be seen that the protein contains a single polypeptide chain of 124 residues in a specific sequence. This chain is internally cross-linked by four disulphide bridges. The molecule does not possess any free SH or similar groups which facilitate specific heavy atom tagging for X-ray studies. Much information[2] from chemical investigations suggests probable non-covalent interactions between the side groups, when the main chain is folded up into its native configuration; some of these interactions seem essential for the integrity of the active centre.

## X-ray Data

Crystals of ribonuclease used in these investigations were grown from 55 per cent 2-methyl-2,4-pentanediol (MPD) at *p*H 5·0 (usually in the presence of phosphate buffer) and were monoclinic in space group $P2_1$ with lattice constants shown in Table 1. There are two ribonuclease molecules per unit cell and the unit cell contains about 40 per cent by weight of the solvent. The molecular weight based on the covalent structure is 13,683.

The diffraction data were collected using either the Eulerian cradle[3] on General Electric diffractometers

Table 1. MODIFICATION II SPACE GROUP $P2_1$ $Z=2$

| | | *a* (Å) | *b* (Å) | *c* (Å) | β | Intensity measurements completed 1/19/67 (Å) |
|---|---|---|---|---|---|---|
| Free protein (STD) | (3*P*65) | 30·13 | 38·11 | 53·29 | 105·75 | 1·8 |
| *Cis*-diglycine pt. | (9*P*12) | 29·87 | 38·39 | 53·19 | 105·95 | 2·0 |
| $K_2PtCl_4$ + D-serine | (9*P*22) | 30·08 | 38·28 | 53·23 | 105·56 | 2·0 |
| $UO_2(NO_3)_2$ + L-valine | (9*P*24) | 30·13 | 38·18 | 52·77 | 105·53 | 2·4 |
| $UO_2(NO_3)_2$ + arsenazo | (9*P*28) | 30·09 | 38·12 | 52·81 | 105·55 | 2 |
| Sodium arsenate | (*R*2–*F*9) | 30·11 | 38·01 | 52·95 | 105·70 | 2 |

(*XRD*-3, *XRD*-6) or, later, the G.E. goniostat. Copper $K\alpha$ radiation was used ($\lambda = 1·5418$ Å). Earlier, the arcs were manually set from precomputed tables using the stationary crystal, stationary counter method, counting the X-ray quanta at the peak position, with intervening balanced pairs of nickel or cobalt filters (Ross filters); each counting time was 10 sec. Approximate absorption corrections were applied to the intensities. During the past year some of the high resolution data used in these investigations were measured using similar techniques but by setting the arcs by an automated *XRD*-5 with prepunched cards containing arc settings for reflexions sorted in an order suitable for efficient collection of data. Statistical studies of data collected from the same or similar crystals indicated an accuracy of 4–5 per cent in |F| for reflexions within the 3 Å resolution sphere. In the 3–2 Å range, however, the overall reproducibility fell to about 10 per cent, mainly as a result of the comparative weakness of many of the reflexions. A copper target was operated at 20 m.amp and 40 kVp. was used as a source of radiation. Wherever possible, measurements were extended to Bijvoet reflexion pairs to take account of anomalous dispersion.

## Structure Determination

Attempts were made to determine the structure of ribonuclease by means of the method of multiple isomorphous series[4]—a technique which had led to the solution of the myoglobin[5] and lysozyme[6] structures. A search for suitable heavy atom derivatives, conducted over the past few years, revealed a few likely ones, some of which, though satisfactory for low resolution work, were either not easily reproduced or not suitable at 2 Å resolution. Thus not all the derivatives used in computing the 4 Å map except the *cis*-diglycine platinum (CDG) were used in the later investigations. This (CDG) derivative contained two main heavy atom sites per molecule; anomalous scattering measurements were made in this case. The 2 Å map also included isomorphous derivative data from five other crystals and limited data to 3 Å from *tris*-(ethylenediamine) platinum (IV) chloride (TEP) for which the anomalous scattering data had been measured earlier.

During the course of the past 3 years, electron density maps were prepared at four resolutions and a summary of these maps is given in Table 2. It is emphasized that some of these derivatives had heavy atoms in similar positions, and the similarity of the derivatives in some

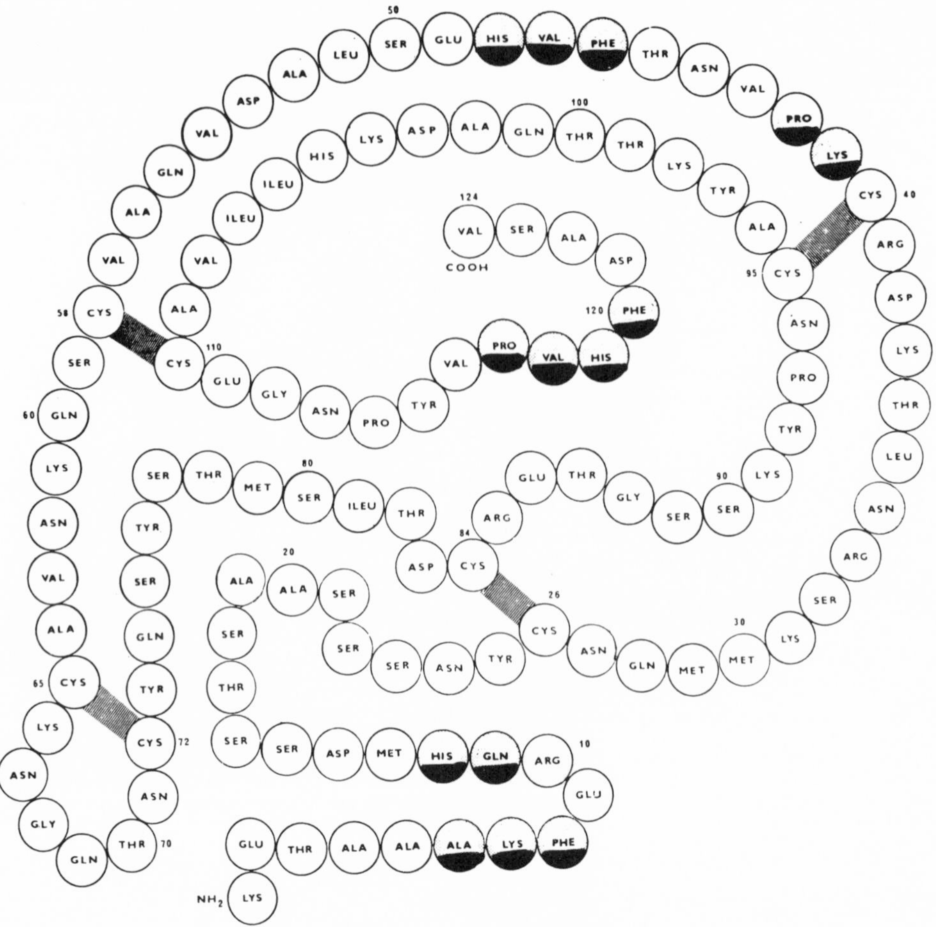


Fig. 1. Covalent structure of bovine pancreatic ribonuclease. (After ref. 1.)

pairs is so great that one could even have replaced the two independent sets of results by a single one. In fact this was done for the two sets of data for CDG derivatives (soaked in different concentration) in the 2 Å map, even though they were treated as independent derivatives in the 2·4 Å map. Table 3 gives a list of the derivatives, the heavy atom sites and occupancies used in the phase evaluation for the 2 Å map. These heavy atom positions were determined and refined by Patterson, difference Fourier and least squares methods[8-10]. The protein phase angles were evaluated by a combination of isomorphous series and anomalous dispersion data[11,12] from the different derivatives, taking care to give proper weights to the phases obtained using any given derivative, and to th reliability of the data from that derivative. The Fouri maps were computed in *xy* or *yz* sections and the resul printed out in coded form suitable for direct contourin on the output sheets from the IBM 7044 computer.

Many features of the molecule are visible in the 3 model computed with 2,340 protein reflexions: (*a*) th position of the amino end which sticks out quite ind pendently of the rest of the molecule; (*b*) the depressio on the surface of the molecule where the phosphate i is located (as will be shown later); (*c*) the position three S–S bridges of high density near which the ma chain density shows the topology expected in the

Table 2. PROGRESS IN FOURIER RESOLUTION OF ELECTRON DENSITY MAPS ACHIEVED IN THIS INVESTIGATION

| Date | Resolution | No. of "reflexions"* used in map | No. of derivative crystals from which data were collected and used in phase evaluation. At higher resolution only some had their anomalous scattering effects measured | No. for which anomalous scattering was also used | Comments |
|---|---|---|---|---|---|
| July 1963 | 4 Å | 1,020 | 7 | 7 | Showed general shape of molecule and also two regio of high density which looked like S-S bridges |
| June 1964 | 3 Å | 2,340 | In addition to above data from seven derivatives, isomorphous series and anomalous scattering data from CDG and TEP | 2 | Showed molecular boundary, three S-S bridges, ami end of chain and, later, we could also infer the po tion of the active site from arsenated RNase |
| Aug. 1966 | 2·4 Å | 4,895 | Data from eight derivative crystals including two separate sets of measurements for CDG | 1 | All S-S bridges now unambiguous. Whole of ma chain could be traced with reasonable certain except for a few regions of ambiguity |
| Dec. 1966 | 2 Å | 7,294 | Data for seven crystals; the two separate sets of CDG data were combined into one | 1 | Ambiguities in tracing the main chain are removed a we can locate all residues with a reasonable degr of certainty |

* Each "reflexion" combines information from several derivatives.

**Table 3.** HEAVY ATOM POSITIONS AND OCCUPANCIES USED IN EVALUATING THE 2 Å MAP

| | Site | $x$ | $y$ | $z$ | Occupancy electrons |
|---|---|---|---|---|---|
| (1) *cis*-Diglycine platinum (II) (CDG) | | | | | |
| | 1 | 0·207 | 0·028 | 0·936 | 53·7 |
| | 2 | 0·410 | 0·500 | 0·425 | 54·9 |
| | 3 | 0·285 | 0·905 | 0·871 | 17·9 |
| | 4 | 0·252 | 0·711 | 0·008 | 11·2 |
| | 5 | 0·483 | 0·676 | 0·495 | 7·3 |
| (2) *tris*-(Ethylenediamine) Pt (IV) (TEP) chloride | | | | | |
| | 1 | 0·126 | 0·654 | 0·975 | 27·4 |
| | 2 | 0·216 | 0·850 | 0·990 | 19·0 |
| (3) D-Serine-Pt (II) complex and Pt($en_3$) $Cl_4$ (PTD) | | | | | |
| Two main sites same as | 1 | 0·193 | 0·036 | 0·940 | 39·8 |
| two main sites of CDG | 2 | 0·416 | 0·500 | 0·429 | 38·1 |
| | 3 | 0·286 | 0·905 | 0·875 | 39·7 |
| | 4 | 0·250 | 0·715 | 0·008 | 11·2 |
| | 5 | 0·467 | 0·500 | 0·353 | 9·0 |
| | 6 | 0·490 | 0·681 | 0·495 | 6·2 |
| (4) Complex of L-valine with uranyl ion | | | | | |
| | 1 | 0·255 | 0·710 | 0·008 | 28·0 |
| | 2 | 0·486 | 0·682 | 0·506 | 21·3 |
| | 3 | 0·253 | 0·818 | 0·495 | 9·5 |
| | 4 | 0·802 | 0·523 | 0·061 | 6·2 |
| | 5 | 0·418 | 0·495 | 0·430 | 9·5 |
| (5) Uranyl complex of 1,8-dihydroxy-2,7-*bis* (*o*-arsonaphenylazo)-naphthalene-3,6-disulphonic acid UAZ | | | | | |
| | 1 | 0·264 | 0·710 | 0·009 | 46·5 |
| | 2 | 0·484 | 0·678 | 0·504 | 39·8 |
| | 3 | 0·250 | 0·808 | 0·495 | 19·0 |
| | 4 | 0·797 | 0·534 | 0·062 | 7·3 |
| | 5 | 0·418 | 0·500 | 0·430 | 15·7 |
| | 6 | 0·418 | 0·558 | 0·590 | 11·2 |
| (6) Arsenated RNase NA1 | | | | | |
| | 1 | 0·443 | 0·461 | 0·388 | 23·5 |
| (7) Arsenated RNase NA2. Basically same as NA1 | | | | | |
| | 1 | 0·451 | 0·462 | 0·386 | 24·6 |

regions. It was also seen that the region near the amino end is partly helical and that the molecules are well separated by regions of very low density.

The 2 Å map involving 7,294 reflexions and data from seven derivatives was computed in sections of constant $z$ at intervals of $x/60$, $y/88$ and $z/112$—these intervals were so chosen that the printed output had a scale of 1 cm to an angstrom. No $F(000)$ term was added to the series and the contours were drawn on an arbitrary scale of 3 units, with the first contour corresponding to 3. In this scale the disulphide bridges had a density in the range 22–17—the only other peak which had a value of more than 17 in the present map was near the main chain at a region close to the residue 46, and this region had a peak value of 19. The region between the molecules rarely rises to more than contour level of 3 and there are no regions of large negative density.

A schematic drawing of the main chain as deduced from the 2 Å contour map appears in Fig. 2. Starting with the amino end, which was quite easily seen even in the 3 Å map, it is possible to proceed along a ribbon of high density corresponding to the main chain by counting along it at appropriate residue distances and using the positions of bulky side groups as a check. In this way it was possible to proceed up to the carboxyl end of the chain, and the four disulphide bridges and their known positions in the amino-acid sequence in Fig. 1 acted as a good check in case any adjustments were needed. At the scale (1 cm = 1 Å) to which the map was drawn it was not easy to recognize all the side groups purely from the shape of the contours, but most of the bigger side chains could be identified by their bulk. It is hoped that the larger scale model now under construction will make it possible to recognize independently many of the side groups by their shapes and bulk. The present map has only been used to trace the course of the α-carbon atoms of the main chain assuming the correctness of the covalent structure of Fig. 1. No attempt has been made at present to check the correctness of the chemical sequence by identifying the side groups from X-ray data alone.

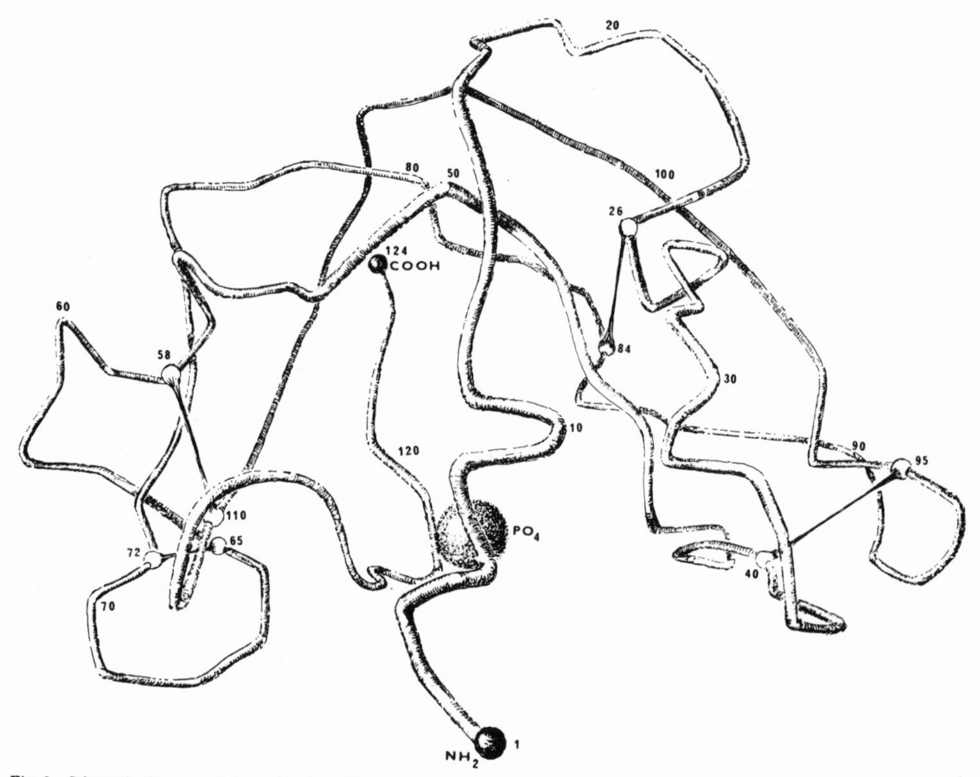


Fig. 2. Schematic diagram of the main chain folding in the ribonuclease molecule. (Thanks are due to Mr. John C. Wallace, who drew from the model.)

## Conformation of the Main Chain

The molecule is roughly kidney shaped with approximate dimensions of 38 × 28 × 22 Å with a deep depression in the middle of one side. Two of the disulphide bridges (26–84) and (40–95) are on one side of this constriction and the other two (58–110) and (65–72) are close together and on the opposite end. Between these two ends the main chains run in three roughly antiparallel sections; the sections between residues 40–58 and 98–110 running in one direction and region 75–90 running in the opposite direction. Quite clearly the molecule has comparatively low helical content. The only obvious helical segment is about two turns in the region of 5–12 near the amino end and possibly two turns each in the regions 28–35 and 51–58, even though the latter are not quite obvious. There are a couple of other regions where there is a suggestion of helical conformation.

The molecule clearly has a much more exposed structure than myoglobin; the smallest of the dimensions is about 20 Å, and no part of the molecule is shielded from the surrounding medium by more than one layer of main chain. This makes it necessary for some of the hydrophobic side chains which would otherwise be shielded from the solvent to lie near the surface. As a general rule, however, most of the segments which consist predominantly of polar side groups—as, for example, residues 66–71 and 85–91—do show up clearly at the outside of the molecule with their side groups pointing outwards into the solution.

## Location of the Active Site

From an examination of the position of the phosphate ion in crystalline ribonuclease, we now have indirect information about the location of the site of enzymatic activity of the molecule. The position of the phosphate ion was first established by crystallizing ribonuclease from a solution containing, not the usual phosphate group, but the electrostatically very similar arsenate group. The resulting crystal was very closely isomorphous with the phosphate crystal. An electron density difference map computed between the phosphate and arsenate crystals at a resolution of 4 Å, and using the known protein phases, resulted in a single peak per asymmetric unit on a very clear background. The position of the arsenate group was also established independently of any knowledge of the protein phases from a three dimensional difference Patterson map using data from the phosphate and arsenate crystals. The location of this region with respect to the protein molecule showed that the phosphate is embedded in the depression of the kidney-shaped surface of the molecule.

In fact, the isomorphism between the phosphate and the arsenate crystals was so good that, despite the fact that the actual electron density difference between an arsenate and phosphate group is not very large, complete three dimensional X-ray data were collected for the arsenate crystals, and these data were used as heavy atom derivative crystal data for protein phase angle evaluation.

## Comparison with Chemical Evidence

Assuming that the active site is indeed near the location of the phosphate ion—an assumption for which there is much chemical evidence and which we can use as a plausible hypothesis—we can locate the regions of the main chain and the amino-acid residues surrounding it. These residues are shaded in Fig. 1. These regions occur at different parts of the main chain, and detailed examination of the characteristic arrangement creating the required charge distribution, and other side chain interactions in this region, is likely to throw some light on the nature of the active site. Even though in our map we are dealing with an arrangement which results when a phosphate ion is bound at the active site, it is possible that the mechanism of attachment and the conformation around this region will not be very different when othe[r] substrate analogues are bound there.

It is seen that the residues closest to the phosphat[e] are residue 119 near the carboxyl end and residue 12 nea[r] the amino end. Both of these are histidines, and muc[h] chemical evidence[13,14] indicates their close relationshi[p] with the activity of the molecule[15]. Other residues furthe[r] out, but which might be of importance, are lysine 7, lysine 41, histidine 48, all of which are reasonably clos[e] to the phosphate site. It is also seen from the presen[t] map that the amino end residues up to 21 or 22 stan[d] out clearly apart from the rest of the molecule, excep[t] for the region near about histidine 12 which comes clos[e] to the active site forming its third side. This agree[s] well with the possibility of cleaving off of this part of th[e] chain by the enzyme subtilisin, leaving the rest of th[e] molecule basically undisturbed.

A fuller description of the ribonuclease molecule and comparison of the intramolecular relationships of th[e] functional group with chemical evidence for active sit[e] groups, intermolecular contacts, and solvent interactio[n] are at present being prepared. Furthermore, we are als[o] preparing a fuller description of the heavy atom "dyes" their preparation and mode of use as well as X-ray dif[f]fraction techniques used in the solution of the structure.

We thank the Dean Langmuir Foundation, the Rocke[-]feller Foundation, and the Damon Runyon Foundatio[n] for providing the initial support of this project, and th[e] National Science Foundation and the National Institute[s] of Health for continuing support during recent year[s]. The Roswell Park Memorial Institute and the New Yor[k] State Department of Health, as well as the Roswell Par[k] Division of Health Research Incorporated, all contribute[d] to the success of this venture by providing space an[d] computing facilities.

In the early days of the work on the structure [of] ribonuclease, important contributions were made b[y] several scientists no longer connected with this project among them the following deserve especial mention an[d] thanks: Dr. B. Magdoff, Dr. V. Luzzati, Dr. M. V. King, Dr. A. Tulinsky, Dr. E. von Sydow, Dr. F. H. C. Crick, Dr. T. C. Furnas, jun., Dr. R. Worthington, Dr. A. DeVries, Dr. D. Harris, Dr. H. H. Mills, Dr. P. Parthasarathy and Dr. R. Davis.

We also thank the following assistants: Miss F. Elain[e] DeJarnette for mounting protein crystals and for collectin[g] most of the X-ray diffraction data during the past si[x] years; Mrs. C. Vincent and Miss K. Go for assisting i[n] data handling and in drawing electron density maps an[d] constructing models; Mrs. Theresa Falzone for excellent assistance in the preparation and for handling the crysta[ls] and heavy atom dyes; Misses Elsa Swyers and Sylvi[a] Scapa for preparing some heavy atom dyes; Mrs. Edit[h] Pignataro for collecting data while the project was a[t] the Polytechnic Institute of Brooklyn; Mr. W. G. Web[er] for constructing, and to some extent designing, th[e] mechanical parts of the single crystal counter X-ra[y] diffractometer.

Received February 20, 1967.

[1] Smyth, D. G., Stein, W. H., and Moore, S., *J. Biol. Chem.*, **238** (1963).
[2] Anfinsen, C. B., *Brookhaven Symp. Enzyme Models and Enzyme Structu[re]* 184 (1962).
[3] Furnas, T. C., and Harker, D., *Rev. Sci. Instrum.*, **26**, 449 (1955).
[4] Harker, D., *Acta Cryst.*, **9** (1956).
[5] Kendrew, J. C., Dickerson, R. E., Strandberg, B. E., Hart, R. G., Davi[es,] D. R., Phillips, D. C., and Shore, V. C., *Nature*, **185**, 422 (1960).
[6] Blake, C. C. F., Koenig, D. F., Mair, G. A., North, A. C. T., Phillips, D. C. and Sarma, V. R., *Nature*, **206**, 757 (1965).
[7] Kartha, G., Bello, J., Harker, D., and DeJarnette, F. E., *Aspects of Prote[in] Structure*, 13 (Academic Press, New York, 1963).
[8] Kartha, G., and Parthasarathy, R., *Acta Cryst.*, **18**, 745, 749 (1965).
[9] Kartha, G., *Acta Cryst.*, **19**, 883 (1965).
[10] Mathews, B. W., *Acta Cryst.*, **20**, 82, 230 (1965).
[11] Kartha, G., Amer. Cryst. Assoc. Meet., Boseman, July 1964.
[12] North, A. C. T., *Acta Cryst.*, **18**, 212 (1965).
[13] Barnard, E. A., and Stein, W. D., *Biochim. Biophys. Acta*, **37**, 371 (1960).
[14] Stark, G. R., Stein, W. H., and Moore, S., *J. Biol. Chem.*, **236**, 436 (1961).
[15] Richards, F. M., and Vithayathil, P. J., *Brookhaven Symp. in Molecul[ar] Biology*, **13**, 115 (1960).

# 70


Reprinted from *J. Biol. Chem.* **242**:3984–3988 (1967)

# The Structure of Ribonuclease-S at 3.5 A Resolution*

(Received for publication, May 31, 1967)

H. W. Wyckoff, Karl D. Hardman,‡ N. M. Allewell, Tadashi Inagami,§ L. N. Johnson,¶ and Frederic M. Richards

*From the Department of Molecular Biophysics, Yale University, New Haven, Connecticut 06520*

SUMMARY

**The electron density map of ribonuclease-S calculated from x-ray diffraction data on the protein and three heavy atom derivatives at 3.5 A resolution is interpretable in terms of main chain and side chain conformation with the aid of pre-existing chemical sequence data and general stereochemical knowledge. Stereoscopic pictures of part of the map and a skeletal model are presented. Features of the structure include 15% helix, 15% hydrophobic core, and appreciable antiparallel-$\beta$ chain pairing. The configura-**

* This work was supported by grants from the National Institutes of Health, the National Science Foundation, and Yale University.

‡ United States Public Health Service Postdoctoral Fellow.

§ Present address, Department of Biochemistry, Vanderbilt University School of Medicine, Nashville, Tennessee 37203.

¶ Present address, Laboratory of Molecular Biophysics, Zoology Department Oxford University, Oxford, England.

**tion of the main chain and assignment of —S—S— bridges closely resembles the structure of RNase-A of Kartha, Bello, and Harker (3) except where there is a chemical difference. The structure is also compatible with much of the relevant chemical literature.**

This communication reports the initial interpretation of an electron density map of ribonuclease-S calculated at 3.5 A resolution from x-ray diffraction data for the protein and three heavy atom derivatives. A resume of the steps leading to the determination of the structure of RNase-S at 6 A resolution was presented in the previous communication (1). These steps included location within the unit cell of the competitive inhibitors uridine 2′,(3′)-phosphate and 5-iodouridine 2′,(3′)-phosphate and the heavy atom ligands: (*a*) uranyl ion, $UO_2^{++}$, at 2 major sites and 5 minor sites, (*b*) a reaction product of dichloroethylenediamine platinum (II), $Pt(en)Cl_2$, at 1 site, and (*c*) tetracyanoplatinate (II) anion, $Pt(CN)_4^{--}$, at 5 sites.

Isomorphous and anomalous dispersion data collection for each of the three heavy atom ligands was extended to include the 469 centric reflections and 1208 Friedel pairs required for a 3.5 A map. Procedures for intensity measurements and calculation of the "best" phase angles and "best" weighting factors were the same as those used at 6 A (1). The new data were collected and scaled in shells to minimize the effects of disorder caused by the ligands. A separate crystal was used for each shell for each ligand because of radiation damage. The radial distribution of the least squares residuals and the average weighting factors (known as the figure of merit) are given in Table I. The uranyl derivative is excellent, the others are adequate, and the phase angle determinations are good according to this analysis.

A three dimensional electron density map, a portion of which

[1] Stereoviewers may be obtained from various suppliers, for example: Abrams Instrument Company, Lansing, Michigan, model CF8; Wards Natural Science Establishment, Inc., Rochester, New York, model 25 W 2951.

TABLE I
*Measures of agreement*

| d spacing range | R factors (two dimensional)[a] | | | Figure[b] of merit (three dimensional) |
|---|---|---|---|---|
| | Uranyl ion | Dichloroethylenediamine platinum (II) | Tetracyanoplatinate (II) anion | |
| *A* | | | | |
| ∞–6.0 | 0.16 | 0.36 | 0.32 | 0.96 |
| 6.0–4.6 | 0.27 | 0.42 | 0.35 | 0.93 |
| 4.6–3.9 | 0.38 | 0.54 | 0.46 | 0.86 |
| 3.9–3.5 | 0.47 | 0.62 | 0.47 | 0.87 |
| 3.5–3.2 | 0.53 | 0.67 | 0.55 | 0.86 |
| 3.2–3.0 | 0.54 | 0.58 | | |

[a] $R = \sum||\Delta F_{obs}| - |\Delta F_{calc}||/\sum|\Delta F_{obs}|$, summed over all centric reflections.
[b] Figure of merit = $\sum$ (weighting factor)/$N$, summed over all reflections.

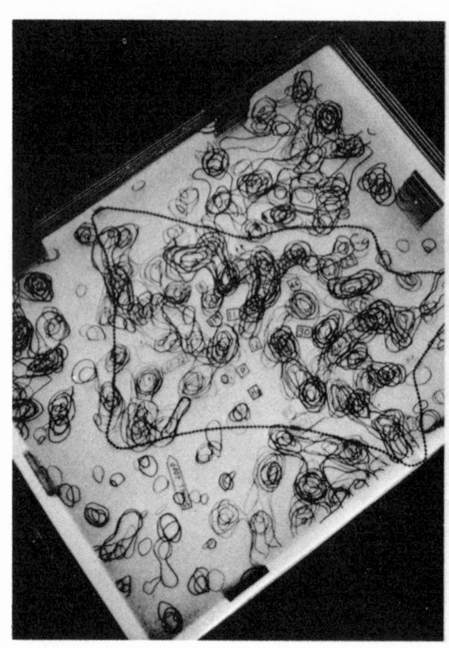
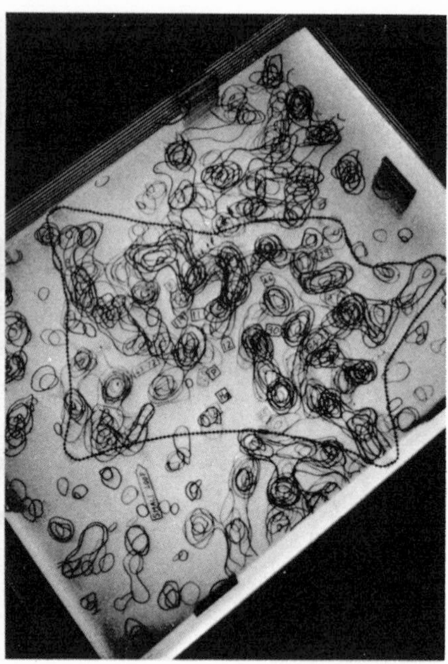

FIG. 1. Stereoscopic[1] view of portion of three dimensional 3.5 A resolution electron density map of RNase-S. Eleven horizontal sections are shown corresponding to one-fourth of the depth of the unit cell and located in the middle of the molecule which is outlined by the beaded chain. The lowest contour is at one-fifth of the maximum. *P* marks the center of gravity of bound uridylic acid. *I* marks the iodine position in 5-iodouridine 2′,(3′)-phosphate. *Hg* locates the mercury atom in 5′-methylmercurithiouridylic acid (not discussed in text). Disulfides 65 to 72 and 26 to 84 are marked as are residues 8, 11, 12, 29, 30, 46, 85, 119, and 120. Residues 113 and 114 are in the *upper left corner*. The *upper right corner* is the residue 20 terminus. The isolated peak near the mercury label may be a sulfate position.

is shown in Fig. 1, was calculated at 3.5 A resolution (scale, 1 cm per A). Almost immediately it was clear that the map was interpretable in terms of the peptide backbone and many large side chains. No useful interpretation could have been made in the absence of the chemical information of the sequence (2). It cannot be known, of course, how long the interpretation would have taken in the absence of the chain model of RNase-A of Kartha, Harker, and Bello (3) and the comparison of that model with the low resolution RNase-S map which had already been made (1).

Construction of a Kendrew (Cambridge Repetition) skeletal wire model,[2] scale 2 cm per A, was started immediately. Close attention was paid to fitting the electron density as well as possible while avoiding packing problems and unlikely chemical conformations. Confidence in the map and its interpretation developed as it became clear that many large side chains known from the sequence did in fact fall into appropriate regions of high density as one proceeded from one disulfide cross-link to another or as one worked in from the ends. The electron density in the main chain was continuously above the lowest level contoured in the map (one-fifth of the maximum level) except for five places in addition to the chemical discontinuity between residues 20 and 21. All of the —S—S— bridges were seen in the 3.5 A map in contradistinction to the 6 A map. All of the methionine residues were seen. All of the ring structures were visible as branches or partially isolated peaks whose center of gravity could be easily located but in most cases no flattening was apparent to reveal the orientation of the plane of the ring. Three of the four arginine residues were seen and about half of the glutamic and aspartic residues were visible with the amides frequently being more clearly defined than the free acids. No lysine residues were visible. In all, about 70 to 80% of all side chains were visible at least as far as the $\beta$ carbon atom.

Pictures of the model are shown in Fig. 2. A list of prominent features includes: (*a*) three sections of helix (2 to 12, 26 to 33, and 50 to 58) totaling approximately 15% of the molecule are all in the first half, and all on the "back" surface; (*b*) the charged residues are on the surface or "lost" in the liquid with the possible exceptions of Asp 14 and His 48; (*c*) a hydrophobic core involving approximately 15% of the molecule includes 16 methyl groups, 8 methylene groups, 2 phenylalanine and 2 methionine residues, 1 disulfide bond, 1 proline and part of 1 histidine residue; (*d*) the S-Peptide (1 to 20) occupies one-third of the surface of the hydrophobic core; (*e*) terminal residues 20 and 21 are widely separated and solvated; (*f*) the S-Protein (21 to 124) consists of two wings joined by three extended chains bordering the hydrophobic core; (*g*) a long twisted and bent antiparallel-$\beta$ pair of chains form two parallel V's consisting of residues 71 through 92 and 94 through 110 with 77 and 104 at the apices; (*h*) three hairpin turns (two of which involve proline and tyrosine residues) occur at residues 38, 93, and 114; (*i*) of the four disulfide bridges none is completely buried; and (*j*) completely buried residues include Tyr 97, Phe 8, Phe 46, and Met 30 and largely buried are His 48, Tyr 25, Met 13, and Met 79 (48 and 25 would likely be completely buried in RNase-A).

The location of the active center is implied by the position of the binding site of uridylic acid and further information is provided by the location of the iodine component of 5-iodouridylic, reported previously. Histidine residues 12 and 119 are near the nucleotide binding site as predicted by Crestfield, Stein, and Moore (4). It does not appear possible for the rings to be coplanar, but they are close enough so that it is unlikely that a water molecule could be placed between them. Although the orientation of the rings with respect to rotation about the $\beta$ bond cannot be determined from the map, it can be stated that position 1 of histidine 12 is not accessible and this explains the alkylation of position 3 instead of position 1 (4). In the orientation of Fig. 2 the uracil moiety is below the histidines and in a groove bounded in part by the hydrophobic residues Val 43 and Phe 120 and the hydroxyl-bearing residues Thr 45 and Ser 123. Position 2 of the uracil ring must be in this groove near the histidines since position 5 is in the liquid near Ala 122 and the ribose is above the uracil. Gln 11 is above His 12 and Asn 44 is to the right of 11. Asp 121 is in a groove to the left of His 119. Lys 7 and Lys 41 are not clearly defined in the map, but the $\epsilon$-amino groups could move close to the active site as predicted by Hirs (5) and Marfey, Uziel, and Little (6). Arg 39 and Lys 66 are slightly further away. No experiments to date have indicated any substantial movement of parts of the enzyme on binding of small molecules.

The S-Peptide appears to be held in place largely by hydrophobic interactions of Phe 8, Met 13, His 12, and Ala 4. Backbone-to-backbone bonding might occur through the carbonyl function of His 12. Arg 33 might hydrogen bond to the backbone near Asp 14 and a charge-charge interaction between these residues exists. Asp 14 is also near His 48, Tyr 25, and Phe 46. Specific data concerning the importance of Met 13 in the S-Peptide-S-Protein binding interaction (7) show that oxidation to the sulfone reduces the binding constant by more than 300-fold without reducing the enzymatic activity of the complex. Met 13 is buried in the hydrophobic core but can be swung out to accommodate the extra oxygen atoms without directly disturbing the orientation of the surrounding residues but clearly at the expense of considerable hydrophobic bonding. Hofmann *et al.* (8) have shown that Asp 14 is important in the binding of peptide to the protein.

The accessibility of some specific residues can be ranked as follows. His 105 and His 119 are freely accessible, His 12 is partly buried, and His 48 is rather inaccessible. This is consistent with the photooxidation results of Kenkare and Richards (9) and the alkylation results cited above (4). Of the methionines, 29 is most accessible and is the $PtCl_4^{--}$ reactive site with the platinum bonding at a position 2.0 to 2.1 A from the unperturbed sulfur position. Met 79 is partially exposed, Met 13 is largely buried, and Met 30 is completely buried. Stark[3] has reported selective methylation of Met 29. Phe 120 is partly exposed in the pyrimidine recognition site, while Phe 8 and Phe 46 are completely buried.

The tyrosines are all partly constrained. Tyr 73, 76, and 115 have their hydroxyl ends exposed, consistent with the normal titration of only 3 tyrosine residues reported by Shugar (10) and identified by various authors (11–13). Tyr 92 may be hydrogen bonded to the backbone at Asp 38. The hydroxyl of Tyr 25 is near the carboxyl of Asp 14 and this residue would be buried if residues 20 and 21 were connected in the most direct way. Tyr 97 is completely buried and probably hydrogen bonded to the backbone at Lys 41. Specific predictions of pairing of 25 to 14, 38 to 92, and 97 to 83 were made by Li, Riehm, and Scheràga (14) on the basis of physical and chemical evidence. Partial analysis of a two dimensional electron density difference map

[2] Ealing Corporation, 2225 Massachusetts Avenue, Cambridge, Massachusetts 02140.

[3] G. R. Stark, private communication.

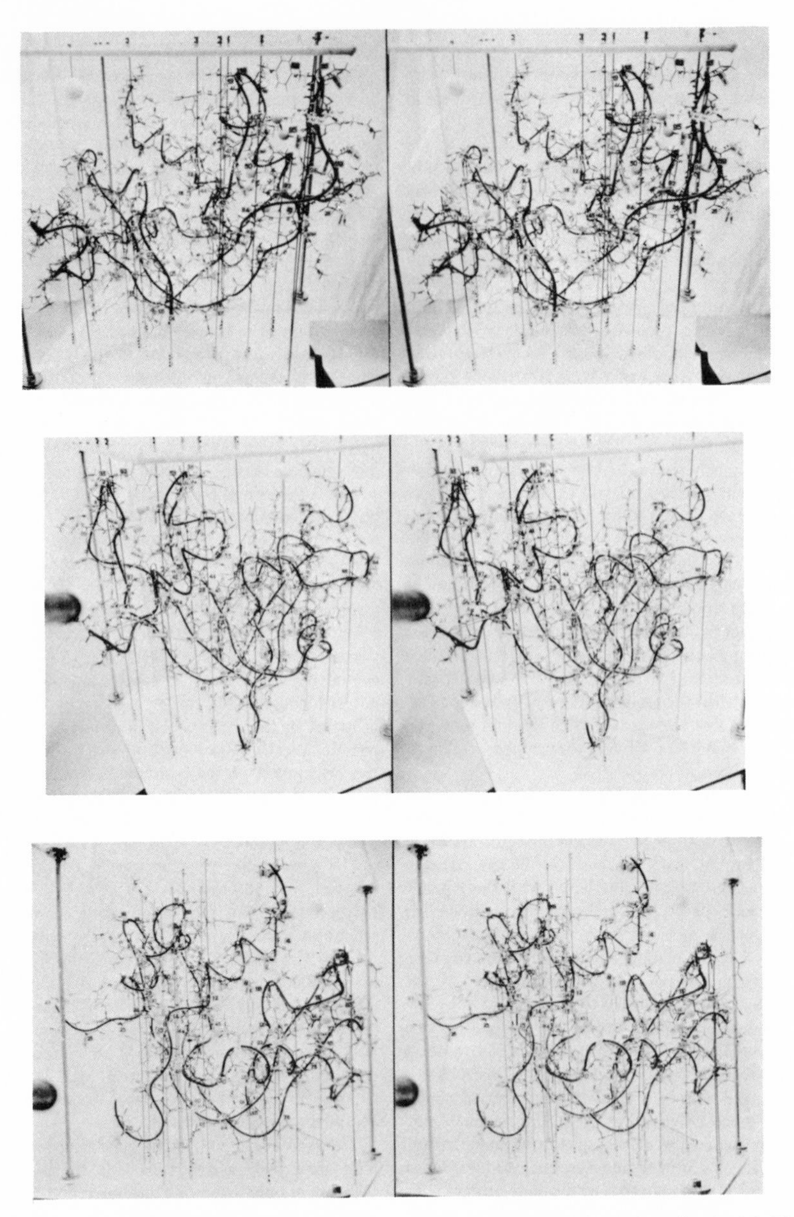

FIG. 2. Stereoscopic view of a skeletal model of RNase-S deduced from the 3.5 A resolution map and chemical sequence data. *Upper* is a view looking somewhat down from the "front" with the active center in the middle. *Middle* is looking down from the "back right" corner. *Lower* is looking slightly up from the "back left" corner. The sma balls locate sulfur atoms. The large ball hanging from the top suppo plate shows the van der Waals size of a paraffinic hydrogen atom. Th numbers are most legible at very low magnification.

following mild iodination in the crystal revealed two sites which were substantiated by least squares refinements. They were at Tyr 73 and Tyr 115. Woody, Friedman, and Scheraga (13) have shown that Tyr 73, Tyr 76, and Tyr 115 are selectively iodinated in solution.

Remaining comments include the fact that the electron density in the region near Pro 114 has proved difficult to match unle this residue is in the *cis* configuration. The other three prolin are *trans*. There is a clustering of positive charge on the surfa as predicted by Loeb and Saroff (15) while carboxyl groups a more disperse. Lys 7 and Lys 41 could be cross-linked with small reagent as found by Marfey *et al.* (6) even though the carbon atoms are far apart. Finally, it would appear that residues 1, 15 through 20, 21 through 23, and 124 could be removed without serious consequence to the structure or its activity. Deslysyl RNase (Lys 1) is known to be active (16). Potts, Young, and Anfinsen (17) removed residues 15 through 20 from the S-Peptide and found it still active. Val 124 can also be removed without deactivating RNase-A (18). Specific predictions of helicity of segments 2 to 11, 43 to 61, and 105 to 113 were made by Prothero (19) and segments 24 to 31, 43 to 49, 50 to 58, and 104 to 112 by Schiffer and Edmundson (20) and a total helix content of 17% has been deduced from optical measurements (21). Our findings noted above are that residues 2 to 12, 26 to 33, and 50 to 58 are helical with a total content approximately 15%.

As noted, many of these steric features are germane to and compatible with the extensive chemical literature on RNase-A and RNase-S. The only marked difference in the chain between this structure and that proposed for RNase-A by Kartha *et al.* (3) is in the region of peptide bond 20. Following subtilisin proteolysis the two new ends move a considerable distance apart, 10 to 15 A. Apparent differences regarding the proximity of residues 46 and 48 to the active site and the kink in the backbone near residue 97 in RNase-A may be matters of interpretation of the electron density rather than true differences in the structure. The positions of bound phosphate and arsenate in RNase-A crystals appear to be consistent with our inhibitor location and orientation.

*Acknowledgments*—We wish to extend special thanks to F. Raymond Salemme and J. David Weinland for technical assistance during this project.

REFERENCES

1. Wyckoff, H. W., Hardman, K. D., Allewell, N. M., Inagami, T., Tsernoglou, D., Johnson, L. N., and Richards, F. M., *J. Biol. Chem.*, **242**, 3749 (1967).
2. Smyth, D. G., Stein, W. H., and Moore, S., *J. Biol. Chem.*, **238**, 227 (1963).
3. Kartha, G., Bello, J., and Harker, D., *Nature*, **213**, 862 (1967).
4. Crestfield, A. M., Stein, W. H., and Moore, S., *J. Biol. Chem.*, **238**, 2413, 2421 (1963).
5. Hirs, C. H. W., *Brookhaven Symp. Biol.*, **15**, 154 (1962).
6. Marfey, P. S., Uziel, M., and Little, J., *J. Biol. Chem.*, **240**, 3270 (1965).
7. Vithayathil, P. J., and Richards, F. M., *J. Biol. Chem.*, **235**, 2343 (1960).
8. Hofmann, K., Finn, F. M., Limetti, M., Montibeller, J., and Zanetti, G., *J. Amer. Chem. Soc.*, **88**, 3633 (1966).
9. Kenkare, U. W., and Richards, F. M., *J. Biol. Chem.*, **241**, 3197 (1966).
10. Shugar, D., *Biochem. J.*, **52**, 142 (1952).
11. Cha, C. Y., and Scheraga, H. A., *J. Biol. Chem.*, **238**, 2958, 2965 (1963).
12. Donovan, L. G., *Biochim. Biophys. Acta*, **78**, 474 (1963).
13. Woody, R. W., Friedman, M. E., and Scheraga, H. A., *Biochemistry*, **5**, 2034 (1966).
14. Li, L., Riehm, J. P., and Scheraga, H. A., *Biochemistry*, **5**, 2043 (1966).
15. Loeb, G. I., and Saroff, H. A., *Biochemistry*, **3**, 1819 (1964).
16. Eaker, D., *J. Polymer Sci.*, **49**, 45 (1961).
17. Potts, J. T., Young, D. M., and Anfinsen, C. B., *J. Biol. Chem.*, **238**, 2593 (1963).
18. Sela, M., Anfinsen, C. B., and Harrington, W. F., *Biochem. Biophys. Acta*, **26**, 502 (1957).
19. Prothero, J. W., *Biophys. J.*, **6**, 367 (1966).
20. Schiffer, M., and Edmundson, A. B., *Biophys. J.*, **7**, 121 (1967).
21. Urnes, P., and Doty, P., *Advance. Protein Chem.*, **16**, 401 (1961).

## Part XIII

# ENZYME SYNTHESIS: RIBONUCLEASE

# Editor's Comments on Paper 71

**71 GUTTE and MERRIFIELD**
*The Total Synthesis of an Enzyme with Ribonuclease A Activity*

In the year 1898, a mere three years after the death of Pasteur, Émile Roux, Pasteur's eminent and long-time associate, gave a most notable address at the University of Lille, which we have already encountered in our comments on Papers 13 and 21. While discussing Buchner's demonstration of cell-free fermentation the year before, Roux stressed the chemical nature of yeast "alcoholase." He continued: "It is not unreasonable to believe that the day will come when we will be as capable as yeast to make intact alcoholases. To think that it will be possible to obtain an enzyme by synthesis! This means that we should not be put off by all their qualities, their complex composition, and their fragility which reminds one of the fragility of living beings!" (Roux 1898, p. 839, editor's translation). This statement is remarkable not only because the chemical nature of enzymes was, as we saw, not clearly recognized by 1898, but because it reveals such a clear and rapid recognition of some of the revolutionary implications of Buchner's work. More than eighty years have passed since this address was given. Roux's dream has not been realized for the enzymes of alcoholic fermentation, but conceptually his ideas are now far less remote than they must have appeared to his audience. At the present time one enzyme, ribonuclease, has been obtained by chemical synthesis.

We have seen in the last section that ribonuclease has been a favorite for various structural and mechanistic studies. It was, with lysozyme, the first enzyme whose primary sequence was determined, and one of the first for which the three-dimensional structure was elucidated. This enzyme was chosen for organic synthesis not only because it is a relatively small protein molecule but because of the ready renaturation of the denatured enzyme. Two synthetic approaches were used for building the ribonuclease molecule. The one by Gutte and Merrifield at the Rockefeller University, included here, used the solid-phase method, invented by Merrifield, to synthesize the complete molecule (ribonuclease-A). The other approach by a large group of workers at the Merck Sharp and Dohme Research Laboratories (see Denkewalter et al. 1969; Hirschmann et al. 1969) used the more conventional approach of stepwise synthesis of ribonuclease-S from 19 separately synthesized fragments. A detailed report on the solid-phase synthesis of ribonuclease-A was published later (Gutte and Merrifield 1971).

Reprinted from *Am. Chem. Soc. J.* **91**:501–502 (1969)

# THE TOTAL SYNTHESIS OF AN ENZYME WITH RIBONUCLEASE A ACTIVITY[1]

**B. Gutte and R. B. Merrifield**

We wish to report our preliminary experiments on the chemical synthesis of a protein with high and specific ribonuclease activity. A protected linear polypeptide of 124 amino acid residues with the sequence of bovine pancreatic ribonuclease A (RNase A)[2] was synthesized by the solid-phase method.[3] After the peptide was removed from the resin support it was purified, and the disulfide bonds were formed by air oxidation to produce the synthetic enzyme. The product was active toward ribonucleic acid and 2′,3′-cyclic cytidine phosphate and was completely inactive toward deoxyribonucleic acid, 2′,3′-cyclic guanosine phosphate, and certain purine-containing dinucleotides, demonstrating its selective requirement for typical RNase A substrates. The conclusion that a protein with the composition and structure of ribonuclease A has been synthesized is further supported by the close agreement between the synthetic protein and natural RNase A with respect to the Michaelis constant, amino acid analyses, electrophoretic and chromatographic behavior, and peptide maps of tryptic digests.

The synthesis was carried out in a stepwise manner starting from the C terminus of RNase A with 0.42 mmol of *t*-butyloxycarbonyl-L-valine esterified to 2 g of 1% cross-linked polystyrene resin support. The general procedures[3–6] of the automated solid-phase method were followed. The following $N^{\alpha}$-*t*-butyloxycarbonyl (Boc)[7] derivatives were used: Asp($\beta$-OBzl), Glu($\gamma$-OBzl), Cys(Bzl), Ser(Bzl), Thr(Bzl), Tyr(Bzl), Lys(Z), Arg($NO_2$), Met(O). Histidine was unprotected. Coupling was usually mediated by N,N′-dicyclohexylcarbodiimide (DCCI) (threefold excess, 5 hr). Boc groups were removed with 50% (v/v) trifluoroacetic acid (TFA) in methylene chloride. Amino acid analysis of the fully protected 124-residue peptide–resin indicated a yield of 17% based on the amount of valine originally esterified to the resin, which means that an average of about 1.4% of the peptide chain was lost from the resin at each cycle of the synthesis.

Cleavage of the polypeptide from its solid support together with removal of all protecting groups was achieved in a single step by treatment with anhydrous HF[8] in the presence of anisole and TFA for 90 min at 0–15°. The yield of this cleavage step was 41%. The product was converted to the S-sulfonate [RNase($SSO_3^-$)$_8$] with $Na_2SO_3$ and $Na_2S_4O_6$ in 8 *M* urea at pH 7.5.[9] Paper electrophoresis at pH 2.25 showed a major Pauly-positive spot, which moved toward the cathode with the same mobility ($R_{His}$ 0.28) as that of the S-sulfonate of natural RNase, and two minor components. The synthetic RNase($SSO_3^-$)$_8$ was purified first on Dowex 1-X2 in 2.4 *M* formic acid–4 *M* urea and then by gel filtration on Sephadex G-50 in 5 *M* urea–0.1 *M* $NH_4OAc$, pH 5.2. The faster moving Sephadex fraction eluted at the same volume as natural RNase($SSO_3^-$)$_8$ and was homogeneous in the paper electrophoresis system; yield of purified RNase($SSO_3^-$)$_8$, 216 mg; amino acid analyses are given in Table I.

**Table I.** Amino Acid Analyses[a]

| Amino acid | Natural RNase Lit.[2] | Natural RNase Found | Synthetic RNase-($SSO_3^-$)$_8$ | Synthetic RNase | Natural RNase HF–TFA treated |
|---|---|---|---|---|---|
| Lys | 10 | 10.0 | 10.3 | 9.1 | 9.2 |
| His | 4 | 3.8 | 3.3 | 3.3 | 3.5 |
| Arg | 4 | 4.1 | 4.6 | 4.2 | 3.9 |
| Asp | 15 | 15.7 | 15.9 | 16.0 | 16.2 |
| Thr | 10 | 9.8 | 9.4 | 10.5 | 10.4 |
| Ser | 15 | 14.0 | 15.3 | 15.5 | 15.5 |
| Glu | 12 | 12.1 | 12.4 | 12.1 | 12.8 |
| Pro | 4 | 3.8 | 3.9 | 3.9 | 3.6 |
| Gly | 3 | 3.1 | 3.1 | 3.3 | 3.4 |
| Ala | 12 | 12.0 | 12.0 | 12.0 | 12.0 |
| Cys | 8 | 7.8 | 6.7 | 7.2 | 7.4 |
| Val | 9 | 8.6 | 9.4 | 8.9 | 8.8 |
| Met | 4 | 4.1 | 3.9 | 3.8 | 3.4 |
| Ile | 3 | 2.3 | 2.1 | 2.1[b] | 2.2 |
| Leu | 2 | 2.2 | 2.9 | 2.6 | 2.3 |
| Tyr | 6 | 5.6 | 5.1 | 5.2 | 5.9 |
| Phe | 3 | 3.3 | 3.2 | 3.5 | 3.2 |

[a] Samples were hydrolyzed in 6 *N* HCl in sealed, evacuated tubes, 24 hr, 110°, and analyzed on a Spinco 120B amino acid analyzer. Cystine was determined on performic acid oxidized samples. Values are expressed as moles per mole of RNase with alanine set at 12.0. All values are uncorrected. [b] A 48-hr hydrolysis time increased the value of isoleucine to 2.87.

(1) Supported by Grant A-1260 from the U. S. Public Health Service.

(2) (a) C. H. W. Hirs, S. Moore, and W. H. Stein, *J. Biol. Chem.*, **235**, 633 (1960); (b) D. G. Smyth, W. H. Stein, and S. Moore, *ibid.*, **237**, 1845 (1962); (c) J. T. Potts, A. Berger, J. Cooke, and C. B. Anfinsen, *ibid.*, **237**, 1851 (1962); (d) E. Gross and B. Witkop, *ibid.*, **237**, 1856 (1962); (e) D. G. Smyth, W. H. Stein, and S. Moore, *ibid.*, **238**, 227 (1963).
(3) (a) R. B. Merrifield, *J. Am. Chem. Soc.*, **85**, 2149 (1963); (b) *Biochemistry*, **3**, 1385 (1964); (c) *Science*, **150**, 178 (1965); (d) *Recent Progr. Hormone Res.*, **23**, 451 (1967); (e) *Advan. Enzymol.*, **32**, in press.
(4) G. R. Marshall and R. B. Merrifield, *Biochemistry*, **4**, 2394 (1965).
(5) A. Marglin and R. B. Merrifield, *J. Am. Chem. Soc.*, **88**, 5051 (1966).
(6) R. B. Merrifield, J. M. Stewart, and N. Jernberg, *Anal. Chem.*, **38**, 1905 (1966).

(7) Nomenclature and abbreviations follow the tentative rules of the IUPAC–IUB Commission on Biochemical Nomenclature, *J. Biol. Chem.*, **241**, 2491 (1966); **242**, 555 (1967).

(8) (a) S. Sakakibara, Y. Shimonishi, Y. Kishida, M. Okada, and H. Sugihara, *Bull. Chem. Soc. Japan*, **40**, 2164 (1967); (b) J. Lenard and A. B. Robinson, *J. Am. Chem. Soc.*, **89**, 181 (1967).
(9) (a) J. L. Bailey, *Biochem. J.*, **67**, 21P (1957); (b) J. L. Bailey and R. D. Cole, *J. Biol. Chem.*, **234**, 1733 (1959).

The RNase($SSO_3^-$)$_8$ was reduced to RNase(SH)$_8$ with mercaptoethanol in 8 *M* urea, pH 8.5, 20 hr. The mixture was passed through Sephadex G-25 in 0.1 *M* HOAc to remove urea and excess mercaptoethanol. The protein-containing fraction was diluted to 0.02 mg/ml in 0.1 *M* Tris, pH 8.3, and oxidized in air at 25° for 20 hr to form the four disulfide bridges of RNase.[10] The crude product was acidified, lyophilized, and desalted on Sephadex G-25 in 0.1 *M* acetic acid. It was then fractionated on IRC-50 in 0.2 *M* sodium phosphate buffer, pH 6.47.[11] Five fractions were detected at 280 mμ. The largest one, fraction I, eluted at the same volume as reduced and reoxidized natural RNase. It was desalted on Sephadex G-25 in 0.1 *M* acetic acid and was lyophilized (yield, 85 mg). This synthetic ribonuclease was indistinguishable from the native enzyme by paper electrophoresis at pH 2.5 ($R_{His}$ 0.58). Peptide maps from tryptic digests[12] of performic acid oxidized samples showed the 14 expected[12] ninhydrin-positive spots in the same relative positions as the natural ribonuclease control. There was one small additional spot in the synthetic preparation near the position of free lysine. Amino acid analyses of acid hydrolysates compared well with those of natural RNase which had been treated with HF and TFA (Table I). Enzymic digestion (papain followed by aminopeptidase M)[13] was complete. It also showed that 79% of the methionine sulfoxide residues had been converted to methionine during the mercaptoethanol reduction of the RNase($SSO_3^-$)$_8$. The remainder was recovered as methionine sulfone.

The synthetic enzyme (fraction I) showed a specific activity of 13% by two methods[14,15] with yeast RNA as substrate, and 24% with 2′,3′-cyclic cytidine phosphate[16] as substrate when compared with pure natural bovine pancreatic ribonuclease A. Fraction II contained some activity, but the other fractions were inactive. The synthetic enzyme was completely inert toward DNA, 2′,3′-cyclic guanosine phosphate, or 5′-(3′-guanylyl)cytidylic acid (GpCp)[17] under conditions where DNase[18] or RNase $T_1$[19] were active and also was without effect on 5′-(3′-adenylyl)adenylic acid (ApAp), demonstrating the high substrate specificity to be expected of RNase A. A further indication that the synthetic material contained the same active enzyme species as natural RNase was obtained from the Michaelis constant. Initial velocities were measured spectrophotometrically[14] and the $K_m$ values, calculated from Eadie plots, were found to be 2.4 mg/ml for natural RNase and 2.5 mg/ml for the synthetic product. These results provide direct evidence for the hypothesis[10a,20] that the linear amino acid sequence of a protein contains all the information necessary to direct the formation of an active enzyme.

Although the physical and chemical methods so far applied to the synthetic protein indicate a good degree of homogeneity, the presence of very closely related molecules cannot be excluded, and the failure to obtain a fully active enzyme means that our product is not yet pure. Assembly of the 124 amino acid residues into the protected, resin-bound straight-chain precursor of RNase required 369 chemical reactions and 11,931 steps of the automated peptide synthesis machine without any intermediate isolation steps. Deficiencies in these peptide-forming reactions, in the work-up conditions, and in the final oxidation and refolding of the protein probably all contributed to the decreased activity, but an assessment of the relative effects of each step much await the completion of further work.

These experiments demonstrate for the first time that a protein molecule with true enzymic activity toward its natural substrate can be totally synthesized from the component amino acids.

(10) (a) F. H. White, Jr., *ibid.*, **236**, 1353 (1961); (b) C. B. Anfinsen and E. Haber, *ibid.*, **236**, 1361 (1961); (c) C. B. Anfinsen, E. Haber, M. Sela, and F. H. White, Jr., *Proc. Natl. Acad. Sci. U. S.*, **47**, 1309 (1961).

(11) A. M. Crestfield, W. H. Stein, and S. Moore, *J. Biol. Chem.*, **238**, 618 (1963).

(12) C. H. W. Hirs, S. Moore, and W. H. Stein, *ibid.*, **219**, 623 (1956).

(13) H. Keutmann and J. T. Potts, Jr., personal communication.

(14) M. Kunitz, *J. Biol. Chem.*, **164**, 563 (1946).

(15) (a) C. B. Anfinsen, R. R. Redfield, W. L. Choate, J. Page, and W. R. Carroll, *ibid.*, **207**, 201 (1954); (b) G. Kalnitsky, J. P. Hummel, and C. Dierks, *ibid.*, **234**, 1512 (1959).

(16) We are indebted to Dr. M. C. Lin for performing this assay. See R. G. Fruchter and A. M. Crestfield, *ibid.*, **240**, 3868 (1965).

(17) R. B. Merrifield and D. W. Woolley, *ibid.*, **197**, 521 (1952).

(18) M. Kunitz, *J. Gen. Physiol.*, **33**, 349 (1950).

(19) K. Takahashi, W. H. Stein, and S. Moore, *J. Biol. Chem.*, **242**, 4682 (1967).

(20) (a) C. B. Anfinsen, *Brookhaven Symp. Biol.* **15**, 194 (1962); (b) C. J. Epstein, R. F. Goldberger, D. M. Young, and C. B. Anfinsen, *Arch. Biochem. Biophys. Suppl.*, **1**, 223 (1962); (c) D. Givol, F. De-Lorenzo, R. F. Goldberger, and C. B. Anfinsen, *Proc. Natl. Acad. Sci. U. S.*, **53**, 676 (1965).

# CONCLUSION

> To think that physics or chemistry ought to be defined in terms of matter or physiology in terms of life is more than an egregious blunder; it is a threat to the existence of science. It implies that people know what matter is without studying physics or chemistry, and what life is without studying physiology.
>
> [*The New Leviathan,* R. G. Collingwood, 1942]

With the application of the two classical requirements of organic chemistry for structure determination, namely, the total synthesis of an enzyme and the determination of its three-dimensional structure, the field of enzymology has reached a peak. We have traced two long and tortuous paths toward this peak. One of these paths starts with the study of active principles of unknown structure by such pioneers as Reaumur, Spallanzani, Planche, Kirchhoff, Thénard, Payen and Persoz, Beaumont, and Schwann; it leads to the inklings of enzymes as distinct molecules, probably of proteinous nature, in the writings of Berthelot, Traube, and Fischer, and goes via Willstätter's claim that enzymes are not proteins to the isolation by Sumner and by Northrop and Kunitz of crystalline proteins with high enzymatic activity. This path culminated in the total synthesis of an enzyme as a pure protein. The other path led to enzymes as eminently knowable, three-dimensional structures with whose help many aspects of the interaction between enzymes and substrates can at long last be studied in molecular detail. This path intertwines with the preceding one in many places since historical analysis is necessarily arbitrary. It includes landmark studies on enzyme-substrate complexes, on enzyme specificity, kinetics, active sites, on control mechanisms, on the dissection of catalytic behavior. Long gone are the days when the secret of life was regarded to be enshrouded in a multifunctional mysterious something called protoplasm. With the vigorous application of the synthetic and analytical criteria of the chemist, the enzyme has at last become a full member, rather than remaining a jealous observer, of the pantheon of "real molecules."

However, in addition to chemical tales, our papers have also told another story. This story, resulting inevitably from the chemical tales, is of even greater conceptual import: As investigations into the nature, occurrence, and activity of the agents of chemical change in biological materials proceeded, vitalistic interpretations of the particular biological phenomena under scrutiny were gradually whittled away and replaced in wider and wider areas by chemical and physicochemical interpretations. I say "whittled away" advisedly since one only rarely finds thinkers who have the far-sightedness to generalize from present facts that force the abolition of vitalistic theories in circumscribed areas, to future facts that will challenge the broadest possible assumptions of these theories. Our story shows such far-sightedness in just a few cases: Thénard and Berzelius exemplify this kind of vision in their emphasis on a prevalent "force" that acts both in living and in non-living systems. Berthelot and Traube quite unambiguously manifest this vision in their emphasis on chemical explanations of biological phenomena, and it is voiced again by Emil Fischer. The clearest and most emphatic statement of this view is given by Berthelot (1860, quoted in part by Fruton 1972, p. 53): "To banish life from all explanations relative to organic chemistry, that is the aim of our studies... . It is necessary...to regard it [fermentation] in as abstract a manner as possible by studying it exclusively in terms of its most general traits."[1] It is most fascinating to find that closely related ideas had been expressed already more than two hundred years earlier by iatrochemists of the seventeenth century. Thus Joseph Duchesne (latinized to Quercetanus) (1544–1609) felt that "chemistry was to serve as a key to all nature" (Debus 1971), while François de le Boë (latinized to Franciscus Sylvius) (1614–1672), professor of medicine at the University of Leiden and founder of the first chemical laboratory at any university (Underwood 1972), "believed that all action in the body could be explained by chemical reactions. He rejected the idea of a vital force which could distinguish bodily reactions from inorganic reactions in the laboratory" (Leicester 1974, p. 102).

When new facts that challenge accepted theories are discovered, one can observe three main types of response in addition to the rare visionary induction: (1) The facts are accepted but their wider implications are not recognized; (2) the facts are challenged, most frequently to retain the accepted theory or theories; and (3) the facts are accepted, but an attempt is made to save as much as possible of old theory by a redefinition of the shrinking boundaries of its application. The history of the development of enzymology has shown us examples of each of these kinds of response: (1) This represents the careful, strictly empirical approach. We find it in Reaumur's paper, in Spallanzani's treatise, in Kirchhoff's paper, in Beaumont's book. In each case the chemical analogy to the phenomenon being studied (meat or starch breakdown) and, in some of Spallanzani's experiments, the persistence of a phenomenon in the dead animal, is clearly stated, but no extrapolations are made; (2) Buchner's discovery of cell-free fermentation was initially challenged by many[2] and its acceptance was due more to the rise of a new generation of biochemists than to the conversion of skeptics (see Kohler 1972); (3) this is a variant of the first approach but it is more combative and more instructive in a study of the development of concepts: it tries to salvage the most deeply entrenched ideas. We have seen examples of the third type of approach in the work of Pasteur, who regarded Berthelot's demonstration that yeast invertase could be solubilized as a mere exception to the rule that fermentative processes cannot be dissociated from the life of the cell. The same attitude was also responsible for Pasteur's vigorous denial of Traube's and particularly of Liebig's chemical approach to fermentative and related processes. As is frequently the case, one here finds vision to be put down as foolhardiness, as unwarranted extrapolation from factual observation, or simply as misinterpretation of the facts. The attitude that informed Pasteur's views was also expressed in the widely prevalent distinction between unorganized or soluble ferments and organized or insoluble ferments. As we saw, this distinction was mistakenly codified by Kühne with the invention of the word *enzyme* for the former group. In the same vein, a discovery is often recognized to go beyond the topic studied, but limits are imposed on interpretation because of the tacit acceptance of the prevailing climate of opinion. We saw a most interesting example of this attitude in Buchner's unwillingness to admit that his soluble yeast preparation, or zymase, was an enzyme because some of its properties were different from those exhibited by the substances defined as enzymes at that time.

A concise and elegant account of the development of enzymology (and also of virology) as a begrudging retreat of complex vitalistic explanations in the face of simpler physicochemical explanations is given by J. H. Northrop (1961) in an essay ("Biochemists, Biologists, and William of Occam") which "should be read by all biochemists" (J. T. Edsall 1962). A few sentences from this essay must suffice here: "The history of biochemistry is a chronicle of a series of controversies, in several of which I have been more or less engaged. These controversies exhibit a common pattern. There is a complicated hypothesis, which usually entails an element of mystery and several unnecessary assumptions. This is opposed by a more simple explanation, which contains no unnecessary assumptions. The complicated one is always the popular one at first, but the simpler one, as a rule, eventually is found to be correct. This process frequently requires 10 to 20 years. The reason for this long time lag was explained by Max Planck. He remarked that scientists never change their minds, but eventually they die."[3]

At the beginning of this work, it was stated that enzymology is a scientific discipline that necessarily abstracts from the totality of living processes. The results arising from this detachment or abstraction do, however, have quite inescapable implications, as we have seen. Thus Felix B. Ahrens (1902), stimulated by Buchner's still recent results, concluded a detailed historical review of the problem of fermentation with the words: "The ancient conflict over the question 'What is fermentation?' has ended: fermentation is a chemical process" (quoted by Fruton 1972, p. 22). Again, in the incisive words of Jacques Loeb (1906, p. 22): "Through the discovery of Buchner, Biology was relieved of another fragment of mysticism. The splitting of sugar into $CO_2$ and alcohol is no more the effect of a 'vital principle' than the splitting up of cane sugar by invertase." An earlier expression of the same sentiment is given by Alfred Jaquet in his paper describing the discovery of an enzyme-dependent oxidative activity in tissue extract. This paper is from a time (1892) just before Buchner's work (1897) when essentially only hydrolytic enzymes had been described, and other chemical activities in living tissues had not been purged of vitalistic associations. Jaquet writes as follows: "[T]he realization that oxidation in the animal organism occurs under the influence of an enzyme, does away with a bit of vital force [ein Stück Lebenskraft] which has again surfaced in recent times." It is most remarkable how strongly the antivitalistic implications of enzymological discoveries around the turn of the century were stressed in these writings.

From the vantage point of Buchner's discovery of cell-free fermentation our story is seen to trace a circuitous return, via disputatious arguments about the nature and activity of yeast cells, to a viewpoint of ferments or enzymes that a more innocent earlier age had taken for granted. Thénard, Kirchhoff, Schwann, and Berzelius emphasized the analogies between what we now call enzymatic reactions and the corresponding nonenzymatic ones.

Their statements exhibit the enthusiasm of discovery, not the zeal of contention. The temper of their times mirrored the temper of the earlier times of Reaumur, Stevens, Spallanzani, Irvine and Fabbroni, who considered it perfectly natural to approach certain biological questions experimentally and to answer them in chemical terms. The importance of Buchner's results lies perhaps less in the facts that he discovered than in the attitude that the interpretation of these facts forced upon those who believed that chemical and vital processes cannot be dissociated from each other. Facts are powerful substrata for the engendering of emotions. Buchner's facts directed contemporary emotions into the attitude expressed by Irvine in 1785: "Certain mixtures of farinaceous matter with sweet matter may be wholly changed into saccharine matter, and the sweet matter with which it is mixed might be considered *in the same light with yest in the vinous fermentation*" (emphasis added).

From one point of view, hence, the contentious period dominated by Liebig, Berthelot, Traube and, later, by Hoppe-Seyler on the one hand, and by Pasteur on the other, and the categorization of their attitudes by Kühne with his term *enzyme,* may be regarded as a mere interlude between the age of Reaumur to Berzelius and the age initiated by Buchner. Yet there is a subtle difference between the earlier and the later attitudes. Irvine compared the hydrolysis of starch with vinous fermentation precisely because he did not know that his "yest" was alive. Similarly, in 1787 when Fabbroni (see Lechevalier and Solotorovsky 1965) says that the "sugar-decomposing substance in vinous effervescence is of vegeto-animal nature" he does not recognize that he is dealing with a living organism. While the earlier attitude simply stated yeast to be a sugar-decomposing *substance* without knowing what it was, the later attitude was based on the experimentally established recognition that the chemical approach, only assumed before, was justified *in spite of the living nature of yeast.*

There are probably two reasons for the choice of yeast rather than of some more tractable biological material as the testing ground between biological and chemical attitudes: (1) Yeast was readily available and intimately connected with the age-old process of alcoholic fermentation, a process at once mysterious and appealing and (2) the demonstration by Schwann and others in 1836 that yeast was alive appeared to many to be irreconcilable with the commonly held view of fermentation as a chemical process. Organic chemistry was beginning to escape the shackles of association with assumed distinct biological forces, and the assignment of yeast to biology may have been viewed as a reversal of this trend.[4] In the words of Eduard Buchner's entertaining Nobel Lecture: "No sooner had it been realized, said Liebig, that all life processes in plants, just as in animals, must be conceived as physical and chemical processes, than along came unscientific people and tried to make acts of life out of simple chemical processes" (Buchner 1907).

The highly influential and aggressive Liebig took up the cudgels on behalf of the chemical approach. He did not have the vision of the brilliant but much more timid Theodor Schwann in whose mind chemical events could be readily reconciled with living processes (Watermann 1960). Schwann, in exile in Belgium, turning more and more to religious meditations and subject to episodes of depression and anxiety (Florkin 1975, p. 244), remained silent on the matter. The circumstance that, as we now know, yeast happens to have a notoriously tough cell wall, and the accident that Pasteur and Berthelot used a strain of yeast that gave negative results, enabled the various promulgators of the biological and of the chemical viewpoints to contend for supremacy for many years. Our story shows that the development of science is more than a recitation of an apparently inevitable unidirectional dictatorship of facts over the prejudices of fashion. A powerful additional determinant of this development is provided by the interplay of personalities and by the manner in which personalities weave fashions into the selection and interpretation of their facts.

The experimental results that follow from the physicochemical approach to living processes are due not only to properties intrinsic to the phenomena selected for study but also to the operation of the selection process itself. The physicochemical approach has been singularly fruitful because, in the words of Victor Henri (1903), it can "be analyzed experimentally and hence can permit a deeper understanding of the mechanism of the phenomena studied." This deeper understanding is made possible not only in terms of the physicochemical results by themselves but in terms of their wider implications, of which we considered an example above. The physicochemical approach, which has been so productive in the inanimate world, has now become a stock-in-trade not only of the biochemist's but also of the biologist's study of living systems. This approach, which has been developed in the present work from the viewpoint of enzymology, continues to encroach on domains of biological thinking from which it had originally been deliberately excluded.

The physicochemical approach is not at all new in Western thinking. At the very beginning of Greek philosophy, we find

Thales' notion that one substance, water, is the substratum of nature. In contrast to such analytical ideas, we find the idea of a vital principle such as Aristotle's entelechy. The conflict between these ideas is just one manifestation of the more general conflict between matter and form (see J. D. Bernal 1954, Ch. 11, Sec. 7). Thus the "sharp division. . .between those who stressed the uniqueness of living matter and those who believed the body to be a mechanical engine" (H. M. Leicester 1974, p. 111) has a venerable tradition. It is possible that nonvitalistic iatrochemical ideas could be formulated in the seventeenth century with somewhat greater ease than we might at first expect since they could be fitted into a framework of still prevalent alchemical attitudes. In later centuries, however, as exemplified by the papers included in the present work, we see the development of a new attitude: There was no pervasive pattern of authority for the physicochemical view of living processes to fall back on; in order to be accepted, this view depended on the generalizing potential of "irreducible and stubborn facts" (William James, quoted by A. N. Whitehead 1925) rather than on a pre-existing framework of assumed but unproven ideas.

In other biological fields antivitalistic ideas were also advanced around the middle of the nineteenth century. Thus "intimidated by vitalists who smugly declared it was all in vain" (D. Fleming 1964, p. viii) "a quadrumvirate of rising physiologists"—Emil du Bois-Reymond, Ernst Brücke, Hermann von Helmholtz, and Karl Ludwig—decided around 1847 to reject "any explanation of life processes which appealed to nonphysical vital properties or forces" (Turner 1972). In the words of du Bois-Reymond, "Brücke and I have sworn to each other to validate the basic truth that in an organism no other forces have any effect than the common physicochemical ones. . ." (quoted by Lesky 1973). Fleming (1964) makes the important point that this decision "was intended as a program for research rather than the enunciation of a *Weltanschauung*. . . . Mechanism was coextensive with scientific knowledge, but not with the range of legitimate curiosity."

More than a hundred years later one can see a fascinating reversal: In the research laboratory a rejection of the purely physicochemical approach may be found useful on occasion, but the adoption of a corresponding philosophical approach to nature does not at all follow. Thus, for example, K. F. Schaffner (1967) states:

> . . .given the current state of biological science, there may be good heuristic reasons for not attempting in all possible areas to develop physico-chemical explanations of biological phenomena, and good reasons for attempting to formulate spe-

> cifically biological theories. This, however, is an argument which supports an irreducibility thesis for *methodological* reasons. Any attempt to twist this into a claim of *real* irreducibility for all time is, in the light of recent work in molecular biology, logically untenable, empirically unwarranted, and heuristically useless. (See also endnote 3, following.)

There is no doubt that the advances in enzymology in the nineteenth century, particularly near the end of that century, contributed massively toward a swaying of scientific attitudes away from vitalistic approaches. However, the results obtained by enzymological and other studies did not suffice to limit the vagaries of biological thinking. Thus it was the embryologist-turned-philosopher Hans Driesch (1867–1941), who at the beginning of the twentieth century powerfully resuscitated vitalistic ideas (Driesch 1908) based, characteristically, on his own—and as it turns out, erroneously interpreted[5]—important research results. Such ideas did not hold sway for long. In 1911 Jacques Loeb gave a celebrated address whose object was "to discuss the question whether our present knowledge gives us any hope that ultimately life, i.e. the sum of all life phenomena, can be unequivocally explained in physico-chemical terms" (Loeb 1912). These words, although advanced as a question, are strongly reminiscent of the "1847 School" of physiologists and those of Berthelot in 1860. There is, however, a fundamental distinction: While earlier workers' views were dictated by faith based on induction, the latter's deductions more than fifty years later were fashioned by conviction based on further evidence. This change in emphasis, subtle but insistent, represents what is usually known as progress.

Ernest Nagel stated in a classic book (1961):

> Vitalism of the substantive type advocated by Driesch and other biologists during the preceding century and the earlier decades of the present one is now almost entirely a dead issue in the philosophy of biology. The issue has ceased to be focal, perhaps less as a consequence of the methodological and philosophical criticisms to which vitalism has been subjected, than because of the sterility of vitalism as a guide in biological research. . .[6]

One cannot of course pin down a certain instant when a given idea formally disappears from the stage of accepted opinion. C. H. Waddington (1905–1975) has described the process well: Around the time of his student days "the whole controversy evaporated" (Waddington 1961). However, even later the vitalistic attitude had not disappeared completely. It had, as it were, to be teased out of recondite regions by the shovel of experimental evidence. Thus

Francis Crick is quoted to have changed from physics to biology because "he was impatient to throw light into the remaining shadowy sanctuaries of vitalistic illusions" (Judson 1979). In a series of lectures delivered in 1966 Crick examined various examples of vitalistic writing from the preceding few years and concluded, reluctantly, that we are far from having seen the last of such ideas (Crick 1966). Indeed, Michael Polanyi subsequent to these lectures forcefully argued for a vitalistic type of approach to an understanding of biological phenomena (see, for example, Polanyi 1967, 1968). Jacques Monod (1970), again, finds vitalism not dead at all: "Certain schools of thought. . .challenge the value of the *analytical* approach to systems as complex as living beings. According to the holist schools which, phoenix-like, spring up anew with every generation, only failure awaits attempts to reduce the properties of a very complex organization to the 'sum' of the properties of its parts." He adds, not particularly impressed by ancient tradition or by the arguments of many vitalistically inclined scholars in the past and present: "A most foolish and wrongheaded quarrel it is. . . ."[7] In a highly perceptive article on the mechanism-vitalism controversy, Hein (1972) also concludes that this controversy is unending, but she refrains from taking sides. In her opinion the point of view taken by any one individual on this issue—and she quotes eminent contemporary biologists on both sides of the fence—is indeed determined not by the examination of scientific evidence, but by "attitudes and prejudices prior to inquiry." She contends that "the mechanism-vitalism dispute is but one of a number of. . . fundamental disagreements which will be perpetuated as long as people ask questions and seek rational answers" (Hein 1972, p. 188).[8]

With the rejection of a vitalistic approach to a study of living phenomena, one does two things: one expresses a particular mental attitude, and one embarks on an experimentally feasible course of investigation. As a result of these two contributing factors, the phenomena studied assume a new, broad and, as pointed out, a self-consistent and cross-fertilizing meaning that could not have been predicted in every instance and in every detail. The physicochemical approach inevitably merges with the biological approach. A fresh and at times unexpected view of biological processes is obtained. In this volume we have seen that the study of enzymes—as an important exercise in this approach—has provided potent examples not just of the discovery of facts but of the accompanying change of attitudes. Moreover, we also saw that the empirical and conceptual abstractions required of such a study led to the coalescence of various fields of investigation. We thus obtain a pragmatic justification for the ultimate validity of these abstractions. This approach has worked in the past and can confidently be expected to work in the future as well.

Our study has taken us across many years and into many countries. It has surveyed agreement and contention. One controversy stands out. It concerns the nature of alcoholic fermentation. This controversy, as so many others in our story, has passed into history, but a little word, *enzyme,* born out of that controversy, remains. This word reminds us that yeast, *zyme,* was the central actor in a long, drawn-out drama that engaged emotions, stirred prejudices, and stimulated experiments. By now *Escherichia coli,* an agent of putrefaction and decay isolated from the intestine, has become the most popular of microorganisms, but yeast, still an object of fundamental and varied research, remains, in the wineries of France where Pasteur encountered it and elsewhere, silent witness to man's progress and to man's age-old fascination with the change of sugar into alcohol. But this simple word, *enzyme,* is much more than a memorial to the faded glories of obliterated questions. It was invented to stand at the crossroads of a conflict between misinterpreted fact and a favored theory of vital processes. Thus this word persists as a symbol of the ever-present interplay between yesterday's fruitful distortion and tomorrow's unanticipated observation. In this sense, then, the word *enzyme,* by forging a bond of continuity between the yesterdays and the tomorrows of our science, beckons us on to new discovery and new reversals until our work and our times are yet another page in an endlessly evolving, endlessly fascinating book of inquiry into the workings of undeterrable nature.

## NOTES

1. "Bannir la vie de toutes les explications relatives à la chimie organique, tel est le but de nos études. . . . Aussi paraît-it nécessaire de. . .concevoir [la fermentation] d'une manière aussi abstraite que possible, en le déterminant d'une manière exclusive par ses caractères les plus généraux."
2. For instance, it was challenged by J. R. Green, Sc.D., F.R.S. (1897): "For the present. . .I must contend, in opposition to Buchner, that at any rate our English Yeasts do not contain any alcohol-producing enzyme." It must be stressed that the factual com-

ponent of Green's statement was at least as strong as the patriotic. Thus Harden (1914, p. 25), following his discussion of Pasteur's negative results states: "The English top yeasts as a rule give poor results (see Dixon and Atkins, 1913) and sometimes yield totally inactive maceration extract." Green later confirmed Buchner's results and became a strong supporter. Kohler (1972, p. 350) cites the background of Green as an example indicating, again, that previous bias rather than the insistence of new observations fashioned his acceptance of Buchner's views (see also the Editor's Comments on Paper 21).

3. Some scientists have actually rejected the pragmatic advantage of Occam's razor and have felt that "Closer examination of biological problems never corroborates this principle" (that Nature is always simple) (Willstätter 1949, p. 366). This view is echoed in A. N. Whitehead's exhortation "Seek simplicity and distrust it," but much more beautifully by Albert Einstein: "Everything should be made as simple as possible but not simpler" (quoted by E. Racker 1976). See Mitchell (1979) for the source of the quotation from Planck.
4. When considered in this light, Liebig was really a vitalist! In the words of Url Lanham: "Liebig's insistent denial that fermentation could be going on *inside* the yeast cell was really a kind of vitalism, for it seemed to say that the process inside a living cell could not be chemical in nature" (Lanham 1968).
5. See D. Fleming, 1964, pp. xxv–xxvi.
6. An earlier version of topics treated in this book was published in which elegant and clear arguments are presented for the rejection of the organismic approach, which to a large extent has replaced the vitalistic approach as an "alternative to physicochemical theories of living processes" (Nagel 1950–1951).
7. "C'est là une très mauvaise et très stupide querelle. . ." (Monod 1970).
8. The Berthelot-Pasteur dispute about the nature of fermentation (Part II) is a case in point on the interpretation of available facts to foster personal prejudice. The conclusion reached by Hein is elegantly illustrated by Duclaux's comment (p. **111**) on the futility of this kind of dispute.

# REFERENCES

Ahrens, F. B. 1902. Das Gärungsproblem [The fermentation problem]. *Samml. Chem. chem.-tech. Vorträge* **7**:445–494.

Alberty, R. A. 1953. The Relationship Between Michaelis Constants, Maximum Velocities and the Equilibrium Constant for an Enzyme-catalyzed Reaction. *Am. Chem. Soc. J.* **75**:1928–1932.

Alberty, R. A. 1958. On the Determination of Rate Constants for Coenzyme Mechanisms. *Am. Chem. Soc. J.* **80**:1777–1782.

Alberty, R. A. 1959. The Rate Equation for an Enzyme Reaction. In *The Enzymes*, 2nd ed., vol. 1, P. D. Boyer, H. Lardy, and K. Myrbäck, eds. New York: Academic, pp. 143–155.

Alworth, W. L. 1972. *Stereochemistry and Its Application in Biochemistry*. New York, Wiley, p. 111.

Anfinsen, C. B. 1973. Principles That Govern the Folding of Protein Chains. *Science* **181**:223–230.

Anson, M. L. 1935. Crystalline Carboxypolypeptidase. *Science* **81**:467–468.

Anson, M. L. 1937. Carboxypeptidase. I. The Preparation of Crystalline Carboxypeptidase. *J. Gen. Physiol.* **20**:663–669.

Aristotle. *History of Animals*. See Peck, 1965.

Aristotle. *Parts of Animals*. See Peck, 1945.

Arnold, L. J., Jr., K. You, W. S. Allison, and N. O. Kaplan. 1976. Determination of the Hydride Transfer Stereospecificity of Nicotinamide Adenine Dinucleotide Linked Oxidoreductases by Proton Magnetic Resonance. *Biochemistry* **15**:4844–4849.

Atkinson, D. E. 1970. Enzymes as Control Elements in Metabolic Regulation. In *The Enzymes*, 3rd ed., vol. 1, P. D. Boyer, ed. New York: Academic, pp. 465–466.

Balls, A. K. 1942. Liebig and the Chemistry of Enzymes and Fermentation. In *Liebig and after Liebig, a Century of Progress in Agricultural Chemistry*. F. R. Moulton, ed. Washington, D.C.: Am. Assoc. Adv. Sci., pp. 30–39.

Barker, H. A. 1978. Explorations of Bacterial Metabolism. *Ann. Rev. Biochem.* **47**:1–33.

Barth, J. 1972. *Chimera*. New York: Random House, p. 7.

Beaglehole, J. C., ed. 1962. *The Endeavour Journal of Joseph Banks, 1768–1771*. Sydney, Aust.: The Trustees of the Public Library of New South Wales in association with Angus and Robertson, Ltd., footnote on p. 360.

Bernal, J. D. 1954. *Science in History.* London: Watts & Co. and New York: Cameron Associates (Also New York: Hawthorn Books, 1965 and Cambridge, Mass.: The MIT Press, 1971.)

Bernard, C. 1849. Recherches sur les usages du suc pancréatique dans la digestion. *C. R.* **28**:249–253.

Berthelot, M. 1860. *Chimie Organique Fondée sur la Synthèse,* vol. II. Paris: Mallet-Bachelier, pp. 655–656.

Berthelot, M. 1876. Observations sur la Communication de M. Pasteur, et sur la théorie des fermentations. *C. R.* **83**:8–9.

Bertrand, G. 1895. Sur la laccase et sur le pouvoir oxydant ce cette diastase. *C. R.* **120**:266–269.

Berzelius, J. J. 1835. On Dry Distilled Racemic Acid. *Swedish Academy of Sciences, Communication,* p. 142. (English translation in *Acta Chem. Scand.* **14**:1677–1680, 1960).

Berzelius, J. J. 1840. *Lehrbuch der Chemie* [Textbook of chemistry], 3rd ed., vol. 9, F. Wöhler, trans. (German), Section: Der Magensaft [Gastric juice]. Dresden: Arnoldische Buchhandlung, pp. 209–210.

Björk, W., and J. Porath. 1959. Fractionation of Snake Venom by the Gel-Filtration Method. *Acta Chem. Scand.* **13**:1256–1259.

Bloomfield, V., L. Peller, and R. A. Alberty. 1962. Multiple Intermediates in Steady-State Enzyme Kinetics. II. Systems Involving Two Reactants and Two Products. *Am. Chem. Soc. J.* **84**:4367–4374.

Bodenstein, M. 1913. Eine Theorie der photochemischen Reaktionsgeschwindigkeiten [A theory of photochemical reaction velocities]. *A. Physik. Chem.* **85**:329–397.

Bodman, G. 1952. Jöns Jacob Berzelius. In *Swedish Men of Science, 1650–1950,* S. Lindroth, ed. Stockholm: Almqvist & Wiksell, pp. 160–171.

Browne, P. 1789. *Civil and Natural History of Jamaica,* 2nd ed. London: B. White and Son, p. 360.

Brücke, E. 1874. *Vorlesungen über Physiologie,* vol. I. Vienna: Wilhelm Braumüller, p. 293.

Bruice, T. C. 1970. Proximity Effects and Enzyme Catalysis. In *The Enzymes,* 3rd ed., vol. 2, P. D. Boyer, ed. New York: Academic, pp. 217–279.

Buchner, E. 1897. Alkoholische Gährung ohne Hefezellen (Zweite Mittheilung) [Alcoholic fermentation without yeast cells, second communication]. *Dtsch. Chem. Ges. Ber.* **30**:1110–1113.

Buchner, E. 1898. Ueber zellenfreie Gährung [On cell-free fermentation]. *Dtsch. Chem. Ges. Ber.* **31**:568–574.

Buchner, E. 1907. Cell-free Fermentation, Nobel Lecture, December 11, 1907. In *Nobel Lectures, Chemistry 1901–1921.* Amsterdam: Elsevier, 1966, pp. 103–120.

Buchner, E., and R. Rapp. 1897. Alkoholische Gährung ohne Hefezellen. *Dtsch. Chem. Ges. Ber.* **30**:2668–2678.

Buchner, E., and R. Rapp. 1898. Alkoholische Gährung ohne Hefezellen. *Dtsch. Chem. Ges. Ber.* **31**:209–217.

Buchner, E., H. Buchner, and M. Hahn. 1903. *Die Zymasegärung, Untersuchungen über den Inhalt der Hefezellen und die biologische Seite des Gärungsproblems* [Fermentation by zymase, investigations on the content of yeast cells and the biological aspect of the problem of fermentation]. Munich and Berlin: R. Oldenbourg.

Bulloch, W. 1938. *The History of Bacteriology.* Oxford: Oxford University Press (also New York: Dover Publications, 1979).

Buss, W. C., and K. Stalter. 1978. Stimulation of Eukaryotic Transcription by Glycerol and Polyhydroxylic Compounds. *Biochemistry* **17**:4825–4832.

Bussemaker, U. C., and C. Daremberg. 1851. Oeuvres d'Oribase, vol. 1. Paris: L'Imprimerie Nationale, pp. 272, 617.

Butler, S. 1898. *The Iliad of Homer, Rendered Into English Prose for the Use of Those Who Cannot Read the Original*, Book V. London: Longmans, lines 901–903.

Bylankin, I. N. 1950. Iz Istorii Otechestvennoi Biokhimii [From the history of the beginning of biochemistry]: A. J. Danilevsky. *Biokhimiya* **15**:97–104.

Canfield, R. E. 1963. The Amino Acid Sequence of Egg White Lysozyme. *J. Biol. Chem.* **238**:2698–2707.

Chance, B. 1947. An Intermediate Compound in the Catalase-Hydrogen Peroxide Reaction. *Acta Chem. Scand.* **1**:236–267.

"Chemicus." 1974. Pages d'histoire: À propos de Pasteur. *L'Actualité Chimique No. 5 (May)*, Soc. Chim. France, pp. 27–32.

Chittenden, R. H. 1930. *The Development of Physiological Chemistry in the United States*. Am. Chem. Soc. Monogr. Ser. No. 54. New York: The Chemical Catalog Co., pp. 15–39.

Citri, N. 1973. Conformational Adaptability of Enzymes. *Adv. Enzymol.* **27**:397–648.

Clark, R. W. 1980. *Freud, The Man and the Cause*. New York: Random House, p. 43

Cleland, W. W. 1977. Citation Classics. *Current Contents (Life Sci.)* **20** (28):8.

Cleland, W. W. 1979a. Statistical Analysis of Enzyme Kinetic Data. *Methods Enzymol.* **63**:103–138.

Cleland, W. W. 1979*b*. Substrate Inhibition. *Methods Enzymol.* **63**:500–513.

Clement, P. A., and H. B. Hoffleit, trans. 1969. Table-Talk VI, 10. In Plutarch's *Moralia*, vol. VIII. Cambridge, Mass.: Harvard Univ. Press, and London: William Heinemann Ltd., p. 511.

Cohn, W. E. 1949. Separation of Mononucleotides by Anion-exchange Chromatography. *Am. Chem. Soc. J.* **71**:2275–2276.

Cohn, W. E. 1957. Methods of Isolation and Characterization of Mono- and Polynucleotides by Ion Exchange Chromatography. *Methods Enzymol.* **3**:724–743.

Cohnheim, J. 1863. *Zur Kenntniss der zuckerbildenden* Fermente [On knowledge of the sugar-forming ferments]. *Virchow's Arch. pathol. Anat. Physiol. klin. Med.* **28**:241–253 (vol. VIII of series 2).

Coley, N. G. 1973. *From Animal Chemistry to Biochemistry*. Amersham, Bucks: Hulton Educational Publications.

Collingwood, R. G. 1942. *The New Leviathan, or Man, Society, Civilization and Barbarism*. Oxford: Clarendon Press, p. 4.

Columella, L. J. M. See Forster and Heffner, 1954.

*Commémoration de Victor Henri*. 1953. J. Chim. Phys. (Paris), vol. 50, pp. 601–616.

Cornish-Bowden, A. 1979. *Fundamentals of Enzyme Kinetics*. London: Butterworths, p. 130.

Corvisart, L. 1857. Sur une fonction peu connue du pancréas, la digestion des aliments azotés. *C. R.* **44**:720.

Corvisart, L. 1859. Sur Le rôle du pancréas dans la digestion. *C. R.* **49**:43–45.
Crick, F. 1966. *Of Molecules and Men.* Seattle : Univ. Washington Press.
Crosland, M. P. 1962. The Symbols of Berzelius. In *Historical Studies in the Language of Chemistry,* Part 4. Cambridge, Mass.: Harvard Univ. Press, pp. 265–281. (Also New York: Dover Publications, 1978, pp. 265–281.)
Dalziel, K. 1957. Initial Steady State Velocities in the Evaluation of Enzyme-Coenzyme-Substrate Reaction Mechanisms. *Acta Chem. Scand.* **11**:1706–1723.
Davis, J. G. 1965. *Cheese,* vol. 1. New York: American Elsevier, p. 4.
Debus, A. G. 1971. Joseph Duchesne. In *Dictionary of Scientific Biography,* vol. 4, C. C. Gillispie, ed. New York : C. Scribner's Sons, p. 209.
Delbrück, M., and A. Schrohe. 1904. *Hefe, Gärung und Fäulnis* [Yeast, fermentation and putrefaction]. Berlin: Paul Parey, p. 40.
Denkewalter, R. G., D. F. Veber, F. W. Holly, and R. Hirschmann. 1969. Studies on the Total Synthesis of an Enzyme. I. Objective and Strategy. *Am. Chem. Soc. J.* **91**:502–503.
Désormes, C. B., and N. Clément. 1806. Théorie de la fabrication de l'acide sulfurique. *Ann. Chim. Phys. (Paris)* **59**:329–339.
Dixon, H. H., and W. R. G. Atkins. 1913. The Extraction of Zymase by Means of Liquid Air (Preliminary Note). *R. Dublin Soc. Sci. Proc.* **14**:1–8. (Not seen in the original, quoted by Harden 1914, pp. 25, 26, 140.)
Dixon, M. 1971. The History of Enzymes and of Biological Oxidations. In *The Chemistry of Life, Eight Lectures on the History of Biochemistry,* J. Needham, ed. Cambridge: Cambridge Univ. Press, pp. 15–37.
Dixon, M., and E. C. Webb. 1964. *Enzymes,* 2nd ed. New York: Academic.
Dixon, M., and E. C. Webb (assisted by C. J. R. Thorne and K. F. Tipton). 1979. *Enzymes,* 3rd ed. New York: Academic.
Dixon, N. E., C. Gazzola, C. J. Asher, D. S. W. Lee, R. L. Blakeley, and B. Zerner. 1980. Jack Bean Urease (EC 3.5.1.5). II. The Relationship Between Nickel, Enzymatic Activity, and the "Abnormal" Ultraviolet Spectrum. The Nickel Content of Jack Beans. *Can. J. Biochem.* **58**: 474–480.
Driesch, H. 1908. *The Science and Philosophy of the Organism,* Gifford Lectures, University of Aberdeen, 1907–1908. London: A. and C. Black.
Duclaux, P. É. 1883. Chimie Biologique. In *Encyclopédie Chimique,* vol. 9, sec. 1, M. Fremy, ed. Paris: Dunod, Editeur.
Duclaux, P. É. 1896. *Pasteur, Histoire d'un Esprit.* Sceaux: Charaire. (Quoted from *Pasteur, The History of a Mind,* E. F. Smith and F. Hedges, trans. Philadelphia, Pa.: W. B. Saunders Co., 1920, p. 212.)
Duclaux, P. É. 1897. Concerning an Article by W. L. Hiepe. *Inst. Pasteur Ann.* **11**:348–351.
Edsall, J. T. 1962. Proteins as Macromolecules: An Essay on the Development of the Macromolecule Concept and Some of Its Vicissitudes. In *Perspectives in the Biochemistry of Large Molecules,* M. Heidelberger, C. H. Li, K. O. Pedersen, and J. Porath, eds. Arch. Biochem. Biophys. Suppl. 1, pp. 12–20.
*Enzyme Nomenclature 1978.* Recommendations of the Nomenclature Committee of the International Union of Biochemistry on the Nomenclature and Classification of Enzymes. New York: Academic, 1979.
Färber, E. 1937. Vorgeschichte der Enzymologie [Prehistory of enzymology]. *Enzymologia* **3**:xiii–xiv.

Farber, E. 1972. Emil Hermann Fischer. In *Dictionary of Scientific Biography*, vol. 5, C. C. Gillispie, ed. New York: C. Scribner's Sons, p. 5.

Feigl, F. 1960. *Spot Tests in Organic Analysis*, 6th ed., R. E. Oesper, trans. Amsterdam: Elsevier.

Fernbach, A., and L. Hubert. 1900. De l'influence des phosphates et de quelques autres matières sur la diastase protéolitique du malt. *C. R.* **131**:293–295.

Fischer, E. 1894. Einfluss der Configuration auf die Wirkung der Enzyme. II. *Dtsch. Chem. Ges. Ber.* **27**:3479–3483.

Fischer, E. 1895. Einfluss der Configuration auf die Wirkung der Enzyme. III. *Dtsch. Chem. Ges. Ber.* **28**:1429–1438.

Fischer, E. 1907. Synthetical Chemistry in Its Relation to Biology (Faraday Lecture). *J. Chem. Soc. Trans.* **91**:1749–1765.

Fischer, E. 1922. *Aus meinem Leben* [From My Life]. Berlin: J. Springer.

Fleming, D., ed. 1964. Introduction. In *The Mechanistic Conception of Life* by J. Loeb. Cambridge, Mass.: The Belknap Press, Harvard Univ. Press, p. viii.

Florkin, M. 1972. *A History of Biochemistry*. Comprehensive Biochemistry, vol. 30, M. Florkin and E. H. Stotz, eds. Amsterdam: Elsevier, pp. 240–241.

Florkin, M. 1975. Theodor Ambrose Hubert Schwann. In *Dictionary of Scientific Biography*, vol. 12, C. C. Gillispie, ed. New York: C. Scribner's Sons, pp. 240–245.

Forster, E. S., and E. H. Heffner, trans. 1954. Book VII, viii, 1–4, In Lucius Junius Moderatus Columella's *On Agriculture*, vol. II. Cambridge, Mass.: Harvard Univ. Press, and London: William Heinemann Ltd., pp. 285–287.

Fourcroy, A. F. 1799. D'un mémoire du cit. Fabroni, sur les fermentations vineuse, putride, acéteuse, et sur l'éthérification; lu à la société philomatique le 3 fructidor an 7; et Réflexions sur la nature et les produits de ces phénomènes. *Ann. chim.* **31**:299–327.

Fowler, A. V., and I. Zabin. 1977. The Amino Acid Sequence of $\beta$-galactosidase of *Escherichia coli*. *Natl. Acad. Sci. (USA) Proc.* **74**:1507–1510.

Fowler, A. V., and I. Zabin. 1978. Amino Acid Sequence of $\beta$-galactosidase. XI. Peptide Ordering Procedure and the Complete Sequence. *J. Biol. Chem.* **253**:5521–5525.

Freud, S. 1955. *The Interpretation of Dreams*, J. Strachey, trans. New York: Basic Books, Inc., p. 422. (Also *Die Traumdeutung*, 1st ed. in German, 1900, Leipzig and Vienna: Deuticke.)

Frieden, C. 1957. The Calculation of an Enzyme-Substrate Dissociation Constant from the Over-all Initial Velocity for Reactions Involving Two Substrates. *Am. Chem. Soc. J.* **79**:1894–1896.

Frieden, C. 1959. Glutamic Dehydrogenase. III. The Order of Substrate Addition in the Enzymatic Reaction. *J. Biol. Chem.* **234**:2891–2896.

Fruton, J. S. 1970. The Specificity and Mechanism of Pepsin Action. *Adv. Enzymol.* **33**:401–443.

Fruton, J. S. 1972. *Molecules and Life: Historical Essays on the Interplay of Chemistry and Biology*. New York: Wiley-Interscience.

Fruton, J. S. 1978. Enzymes in the Middle Ages? *Trends Biochem. Sci.* **3**:N281.

Garavito, R. M., M. G. Rossmann, P. Argos, and W. Eventoff. 1977. Convergence of Active Center Geometries. *Biochemistry* **16**:5065–5071.

Geison, G. L. 1974. Louis Pasteur. In *Dictionary of Scientific Biography*, vol. 10, C. C. Gillispie, ed. New York: C. Scribner's Sons, p. 414.

Gelotte, B. 1960. Studies on Gel-Filtration: Sorption Properties of the Bed Material Sephadex. *J. Chromatogr.* **3**:330–342.

Gelotte, B., and A.-B. Krantz. 1959. Purification of Pepsin by Gel Filtration. *Acta Chem. Scand.* **13**:2127.

Gerhardt, C. 1856. *Traité de Chimie Organique*, vol. IV. Paris: Firmin Didot, p. 538.

Gosse, P. H. 1851. *A Naturalist's Sojourn in Jamaica*. London: Longman, Brown, Green, and Longmans, p. 257.

Gottschalk, A. 1950. Principles Underlying Enzyme Specificity in the Domain of Carbohydrates. *Adv. Carbohydr. Chem.* **5**:49–78.

Gow, A. S. F., trans. 1950. Section 16. In *Theocritus*. Cambridge: Cambridge Univ. Press, p. 136.

Green, J. R. 1897. The Supposed Alcoholic Enzyme in Yeast. *Ann. Bot.* **11**: 555–562.

Gutte, B., and R. B. Merrifield. 1971. The Synthesis of Ribonuclease A. *J. Biol. Chem.* **246**:1922–1941.

Halban, H. v. 1941. Victor Henri, 1872–1940. *Naturforsch. Ges. Zürich Vierteljahrsschr.* **86**:307–320.

Haldane, J. B. S. 1930. *Enzymes*. London: Longmans, Green & Co. (Facsimile reprint in paperback in 1965, Cambridge, Mass.: The MIT Press.)

Harden, A. 1914. *Alcoholic Fermentation*, 2nd ed. London: Longmans, Green & Co., p. 25.

Harries, C. 1917. Eduard Buchner. *Dtsch. Chem. Ges. Ber.* **50**:1843–1876.

Hein, H. 1972. The Endurance of the Mechanism-Vitalism Controversy. *J. Hist. Biol.* **5**:159–188.

Henri, V. 1903. *Lois Générales de l'Action des Diastases*. Paris: Librairie Scientifique A. Hermann, Preface.

Henri, Mme. V. 1914*a*. Etude de l'action metabiotique des rayons ultraviolets. Production des formes de mutation de la bactéridie charbonneuse. *C. R.* **158**:1032–1035.

Henri, Mme. V. 1914*b*. Étude de l'action metabiotique des rays ultraviolets. Modification des caractères morphologiques et biochimiques de la bactéridie charbonneuse. Hérédité des caractères acquis. *C. R.* **159**:340–343.

Henri, V., and Mme. V. Henri. 1914. Étude de l'action métabiotique des rayons ultraviolets. Théorie de la production de formes microbiennes par l'action sur les différentes fonctions nutritives. *C. R.* **159**:413–415.

Henri, V., and M. Landau. 1914. Sur l'application de la spectroscopie à l'étude des équilibres chimiques. Les systèmes formés par l'acide oxalique et les sels d'uranyle. *C. R.* **158**:181–183.

Herriott, R. M. 1938. Isolation, Crystallization, and Properties of Swine Pepsinogen. *J. Gen. Physiol.* **21**:501–540.

Herriott, R. M., and J. H. Northrop. 1936. Isolation of Crystalline Pepsinogen from Swine Gastric Mucosae and Its Autocatalytic Conversion into Pepsin. *Science* **83**:469–470.

Heynsius, A. 1884. Ueber das Verhalten der Eiweissstoffe zu Salzen von Alkalien und von alkalischen Erden [On the behavior of the proteins toward the salts of alkalis and of alkaline earths]. *Pfluger's Arch. Physiol.* **34**:330–334.

Hirs, C. H. W. 1955. Chromatography of Enzymes on Ion Exchange Resins. *Methods Enzymol.* **1**:113–125.

Hirschmann, R., R. F. Nutt, D. F. Veber, R. A. Vitall, S. L. Varga, T. A. Jacob, F. W. Holly, and R. G. Denkewalter. 1969. Studies on the Total Synthesis of an Enzyme. V. The Preparation of Enzymatically Active Material. *Am. Chem. Soc. J.* **91**:507–508.

Hoffmann-Ostenhof, O. 1978. The Origin of the Term Enzyme. *Trends Biochem. Sci.* **3**:186–188.

Hofmeister, F. 1900. Willy Kühne. *Dtsch. Chem. Ges. Ber.* **33**:3875–3880.

Hofmeister, F. 1901. *Die Chemische Organisation der Zelle*. Braunschweig: F. Vieweg und Sohn.

Hooper, W. D., and H. B. Ash, trans. 1934. Marcus Terentius Varro On Agriculture, Book II, xi, 3–5. In Marcus Porcius Cato's *On Agriculture* and Marcus Terentius Varro's *On Agriculture*. Cambridge, Mass.: Harvard Univ. Press, and London: William Heinemann Ltd., p. 415.

Hoppe-Seyler, F. 1878. Ueber Gährungsprozesse [On processes of fermentation]. *Hoppe-Seyler's Z. Physiol. Chem.* **2**:1–28.

Hubbard, R. 1976. 100 Years of Rhodopsin. *Trends Biochem. Sci.* **1**:154–158.

Hughes, G. 1750. *The Natural History of Barbados*. London, pp. 181–182.

Huxley, L. 1901. *Life and Letters of Thomas Henry Huxley*, vol. I. New York: D. Appleton & Co., p. 515.

Hwang, K., and A. C. Ivy. 1951. A Review of the Literature on the Potential Therapeutic Significance of Papain. *N. Y. Acad. Sci. Ann.* **54**:161–207.

*Iliad*, Homer's. See Leaf, 1960.

Irvine, W. 1805. *Essays, Chiefly on Chemical Subjects* by the late William Irvine, M.D., F.R.S. Ed. and by his son William Irvine, M.D. London: Printed for J. Mawman, 22 in the Poultry, pp. 317–319.

Jakoby, W. B. 1971. Crystallization as a Purification Technique. *Methods Enzymol.* **22**:248–252.

Jameson, D. L., ed. 1977. *Evolutionary Genetics*, Benchmark Papers in Genetics, vol. 8, Stroudsburg, Pa.: Dowden, Hutchinson & Ross.

Jaquet, A. 1892. Ueber die Bedingungen der Oxydationsvorgänge in den Geweben [On the conditions for oxidative processes in the tissues]. *Naunyn-Schmiedeberg's Arch. exp. Pathol. Pharmakol.* **29**:386–396.

Jencks, W. P. 1969. *Catalysis in Chemistry and Enzymology*. New York: McGraw-Hill, pp. 1–2.

Jencks, W. P. 1975. Binding Energy, Specificity, and Enzymatic Catalysis: The Circe Effect. *Adv. Enzymol.* **43**:219–410.

Johnson, W. C., L. L. Quill, and F. Daniels. 1947. Rare Earths Separation Developed on Manhattan Project. *Chem. Eng. News* **25**:2494.

Jones, D. B., and C. O. Johns. 1916–1917. Some Proteins from the Jack Bean, *Canavalia Ensiformis. J. Biol. Chem.* **28**:67–75.

Jones, W. H. S., trans. 1951. Book XXIII, lxiv, 127–128. In Pliny's *Natural History*, vol. VI. Cambridge, Mass.: Harvard Univ. Press, and London: William Heinemann Ltd., p. 499. (See also Book XXIII, lxiii, 117–118 on p. 493.)

Jorpes, J. E. 1966. *Jac. Berzelius, His Life and Work*. Stockholm: Almqvist & Wiksell.

Judson, H. F. 1979. *The Eighth Day of Creation: The Makers of the Revolution in Biology*. New York: Simon and Schuster, p. 109.

Kent. A. 1950 William Irvine, M.D. In *An Eighteenth Century Lectureship in Chemistry: Essays and Bicentenary Addresses Relating to the Chemistry Department (1747) of Glasgow University (1451)*. A. Kent, ed., Glasgow University Publications No. 82. Glasgow: Jackson, Son & Co., pp. 140–141. (See p. 141 especially.)

Kirkegaard, S. 1843. *Either/Or*. D. F. Swenson and L. M. Swenson, trans., H. A. Johnson, ed., rev. ed., 1959, vol. 1. Garden City: Doubleday, p. 47.

Kohler, R. 1971. The Background to Eduard Buchner's Discovery of Cell-Free Fermentation. *J. Hist. Biol.* **4**:35–61.

Kohler, R. E. 1972. The Reception of Eduard Buchner's Discovery of Cell-Free Fermentation. *J. Hist. Biol.* **5**:327–353.

Kohler, R. E. 1975. The History of Biochemistry: A Survey. *J. Hist. Biol.* **8**: 275–318.

Kohler, R. E., Jr. 1973. The Enzyme Theory and the Origin of Biochemistry. *Isis* **64**:181–196.

Kornberg, A. 1980. *DNA Replication*. San Francisco: W. H. Freeman, pp. 333–334.

Koshland, D. E., Jr. 1953. Stereochemistry and the Mechanisms of Enzymatic Reactions. *Biol. Rev.* **28**:416–436.

Koshland, D. E., Jr. 1954. *Group Transfer as an Enzymatic Substitution Mechanism*. A Symposium on the Mechanism of Enzyme Action, W. D. McElroy and B. Glass, eds. Baltimore, Md.: The Johns Hopkins Press, pp. 608–641.

Koshland, D. E., Jr. 1973. Protein Shape and Biological Control. *Sci. Am.* **229**(4):52–64 (Oct.).

Koshland, D. E., Jr., G. Nemethy, and D. Filmer. 1966. Comparison of Experimental Binding Data and Theoretical Models in Proteins Containing Subunits. *Biochemistry* **5**:365–385.

Koshland, D. E., Jr., D. H. Strumeyer, and W. J. Ray, Jr. 1962. Amino Acids Involved in the Action of Chymotrypsin. In *Enzyme Models and Enzyme Structure*. Brookhaven Symposia in Biology Number 5, pp. 101–133.

Kossel, A. 1901. Ueber den gegenwärtigen Stand der Eiweisschemie [On the present status of protein chemistry]. *Dtsch. Chem. Ges. Ber.* **34**: 3214–3245. (See p. 3229 especially.)

Kraut, H., and E. Tria. 1937. Über kristallisiertes Pepsin nach Northrop und eiweissfreies Pepsin nach Brücke. *Biochem. Z.* **290**:277–288.

Krebs, H. A., and W. A. Johnson. 1937. The Role of Citric Acid in Intermediate Metabolism in Animal Tissues. *Enzymologia* **4**:148–156.

Kühne, W., and R. H. Chittenden. 1883. Ueber die nächsten Spaltungsproducte der Eiweisskörper [On the first products of hydrolysis of proteins]. *Z. Biol.* **19**:159–208.

Kühne, W., and R. H. Chittenden. 1886*a*. Globulin and Globulosen. *Z. Biol.* **22**:409–422.

Kühne, W., and R. H. Chittenden. 1886*b*. Bemerkungen zur Chemie der Albumosen und Peptone [Remarks concerning the chemistry of the albumoses and peptones]. *Z. Biol.* **22**:423–458.

Kunitz, M. 1939. Isolation from Beef Pancreas of a Crystalline Protein Possessing Ribonuclease Activity. *Science* **90**:112–113.

Kunitz, M. 1940. Crystalline Ribonuclease. *J. Gen. Physiol.* **24**:15–32.

Kuznetsov, V. I. 1966. The Development of Basic Ideas in the Field of Catalysis. *Chymia* **11**:179–204.

Lanham, U. 1968. *Origins of Modern Biology.* New York: Columbia Univ. Press, p. 236.

Leaf, W., ed. 1960. *The Iliad* (reprint of the 1900-1902 London edition). Amsterdam: A. M. Hakkert, p. 255.

Lebedeff, A. 1911. Extraction de la zymase par simple macération. C. R. **152**:49–51.

Lebedew, A. v. 1911. Darstellung des aktiven Hefensaftes durch Maceration [Perparation of active yeast extract by maceration]. *Hoppe-Seyler's Z. Physiol. Chem.* **73**:447–452.

Lechevalier, H. A., and M. Solotorovsky. 1965. *Three Centuries of Microbiology.* New York: Dover Publications, 1974 ed., pp. 20–21.

Leicester, H. M. 1974. *Development of Biochemical Concepts from Ancient to Modern Times.* Cambridge, Mass.: Harvard Univ. Press.

Lemay, P. 1949. Désormes et Clément découvrent et expliquent la catalyse. *Chymia* **2**:45–49.

Lesky, E. 1973. Ernst Wilhelm von Brücke. In *Dictionary of Scientific Biography,* vol. 2, C. C. Gillispie, ed. New York: C. Scribner's Sons, p. 531.

Levy, H. R., P. Talalay, and B. Vennesland. 1962. The Steric Course of Enzymatic Reactions at Meso Carbon Atoms: Application of Hydrogen Isotopes. In *Progress in Stereochemistry,* vol. 3, P. B. D. De la Mare and W. Klyne, eds. London: Butterworths, pp. 299–349.

Lichine, A. 1971. History of Wine. Chapter 1 in *Alexis Lichine's Encyclopedia of Wines and Spirits,* in collaboration with W. Fifield and with the assistance of J. Bartless and J. Stockwood. New York: A. A. Knopf, pp. 1–2.

Liebig, J. 1840. *Die organische Chemie in ihrer Anwendung auf Agricultur und Physiologie* [The application of organic chemistry to agriculture and physiology]. Braunschweig: Verlag von Friedrich Vieweg und Sohn, pp. 204–206.

Lienhard, G. E. 1973. Enzymatic Catalysis and Transition-State Theory. *Science* **180**:149–154.

Lindner, E. B., A. Elmqvist, and J. Porath. 1959. Gel Filtration as a Method for Purification of Protein-bound Peptides Exemplified by Oxytocin and Vasopressin. *Nature* **184**:1565–1566.

Lineweaver, H., and D. Burk. 1934. The Determination of Enzyme Dissociation Constants. *Am. Chem. Soc. J.* **56**:658–666.

Lipmann, F. 1938. Gärversuche mit Mazerations-Extrakten aus Bäckerhefe [Fermentation experiments with extracts of macerated bakers' yeast]. *C. R. Trav. Lab. Carlsberg, Ser. Chim.* **22**:317–320.

Lipscomb, W. N., J. C. Coppola, J. A. Hartsuck, M. L. Ludwig, H. Muirhead, J. Searl, and T. A. Steitz. 1966. The Structure of Carboxypeptidase A. III. Molecular Structure at 6Å Resolution. *J. Mol. Biol.* **19**: 423–441.

Loeb, J. 1906. *The Dynamics of Living Matter.* New York: Columbia Univ. Press, p. 22.

Loeb, J. 1912. *The Mechanistic Conception of Life: Biological Essays.* Chicago, Ill.: Univ. Chicago Press, p. 3. (Reprinted in 1964 with Foreword by D. Fleming, ed. Cambridge, Mass.: The Belknap Press, Harvard Univ. Press, p. 5.)

Loewus, F. A., P. Ofner, H. F. Fisher, F. H. Westheimer, and B. Vennesland. 1953. The Enzymatic Transfer of Hydrogen. II. The Reaction Catalyzed by Lactic Dehydrogenase. *J. Biol. Chem.* **202**:699–704.

Ludwig, M. L., J. A. Hartsuck, T. A. Steitz, H. Muirhead, J. C. Coppola, G. N. Reeke, and W. N. Lipscomb. 1967. The Structure of Carboxypeptidase A. IV. Preliminary Results at 6Å Resolution. *Natl. Acad. Sci. (USA) Proc.* **57**:511–514.

Luisi, P. L. 1979. Why Are Enzymes Macromolecules? *Naturwissenschaften* **66**:498–504.

Mathews, B. W., P. B. Sigler, R. Anderson, and D. M. Blow. 1967. Three-dimensional Structure of Tosyl-α-chymotrypsin. *Nature* **214**:652–656.

Méhu, C. J. M. 1878. Méthode d'extraction des pigments d'origine animale; applications diverses du sulfate d'ammoniaque. *J. Pharm. Chim.* ser. 4, **28**:159–165.

Michaelis, L. 1914. *Die Wasserstoffionenkonzentration.* Berlin: J. Springer, p. 180.

Michaelis, L. 1958. Autobiography, with additions by D. A. MacInnes and S. Granick. *Biograph. Mem. Natl. Acad. Sci. (USA).* **31**:282–321 (esp. p. 286).

Michaelis, L., and H. Davidsohn (1911a). Die Wirkung der Wasserstoffionen auf das Invertin [The action of hydrogen ions on invertin]. *Biochem. Z.* **35**:386–412.

Michaelis, L., and H. Davidsohn 1911*b*. Die Abhängigkeit der Trypsinwirkung von der Wasserstoffionenkonzentration [The dependence of trypsin activity on hydrogen ion concentration]. *Biochem. Z.* **36**: 280–290.

Michaelis, L., and M. L. Menten. 1913. Die Kinetik der Invertinwirkung [The kinetics of invertin activity]. *Biochem. Z.* **49**:333–369.

Michaelis, L., and M. Rothstein. 1920. Zur Theorie der Invertasewirkung [On the theory of invertase activity]. *Biochem. Z.* **110**:217–233.

Mitchell, P. 1979. Keilin's Respiratory Chain Concept and Its Chemiosmotic Consequences. *Science* **206**:1148–1159.

Mitscherlich, E. 1834. Ueber die Aetherbildung [On the formation of ether]. *Ann. Phys.* **31**:273–282.

Mittasch, A. 1938. *Katalyse und Determinismus, ein Beitrag zur Philosophie der Chemie* [Catalysis and determinism, a contribution to the philosophy of chemistry]. Berlin: J. Springer, pp. 1–2.

Mittasch, A., and E. Theis. 1932. *Von Davy und Dobereiner bis Deacon, ein halbes Jahrhundert Grenzflächenkatalyse* [From Davy and Döbereiner to Deacon, a half century of contact catalysis]. Berlin: Verlag Chemie.

Monod, J. 1970. *Le Hasard et la Nécessité: Essai sur la Philosophie Naturelle de la Biologie Moderne.* Paris: Editions du Seuil, p. 93. (Also as *Chance and Necessity: An Essay on the Natural Philosophy of Modern Biology*, A. Wainhouse, trans. New York: A. A. Knopf, 1971, and New York: Vintage Books, 1972, p. 79.)

Moore, S., and W. H. Stein. 1952. Chromatography. *Ann. Rev. Biochem.* **21**:521–546.

Moore, S., and W. H. Stein. 1963. Chromatographic Determination of Amino Acids by the Use of Automatic Recording Equipment. *Methods Enzymol.* **6**:819–831.

Multhauf, R. P. 1966. *The Origins of Chemistry.* London: Oldbourne Book Co. Ltd., p. 13, and New York: Franklin Watts, Inc., 1967.

Myrbäck, K. 1960. Invertase. In *The Enzymes*, 2nd ed., vol. 4, P. D. Boyer, H. Lardy, and K. Myrbäck, eds. New York: Academic, pp. 379–396.

Nachmansohn, D. 1979. *German-Jewish Pioneers in Science 1900–1933.* Berlin: Springer-Verlag, pp. 195–231.

Nagel, E. 1950–1951. Mechanismic Explanation and Organismic Biology: *Philos. Phenomenol. Res.* **11**:327–338.

Nagel, E. 1961. The Standpoint of Organismic Biology. In *The Structure of Science.* New York: Harcourt, Brace & World, pp. 428–429.

Needham, J. 1965. Preface. In *Science and Civilisation in China,* vol. 1: *Introductory Orientations.* Cambridge: Cambridge Univ. Press, p. 7.

Negelein, E., and H.-J. Wulff. 1937*a*. Kristallisation des Proteins der Acetaldehydreduktase. *Biochem. Z.* **289**:436–437.

Negelein, E., and H.-J. Wulff. 1937*b*. Diphosphopyridinproteid, Alkohol, Acetaldehyd. *Biochem. Z.* **293**:351–389.

Neumeister, R. 1888. Bemerkungen zur Chemie der Albumosen und Peptone [Remarks concerning the chemistry of the albumoses and peptones]. *Z. Biol.* **24**:267–271.

Nietzsche, F. W. 1878. *Human, All-too-Human, A Book for Free Spirits,* Ninth Division, Aphorism 483, from the Complete Works of *F. Nietzsche,* vol. 6, H. Zimmern, trans., 1909. Edinburgh: T. N. Foulis, p. 355.

*Nobel Lectures.* 1964. Chemistry, 1942–1962, Biography of J. B. Sumner. Amsterdam: Elsevier, p. 123.

Nord, F. F. 1940. Facts and Interpretations in the Mechanism of Alcoholic Fermentation. *Chem. Rev.* **26**:423–477.

Northrop, J. H. 1939. *Crystalline Enzymes: The Chemistry of Pepsin, Trypsin and Bacteriophage.* New York: Columbia Univ. Press.

Northrop, J. H. 1946. The Preparation of Pure Enzymes and Virus Proteins, Nobel Lecture December 12, 1946. In *Nobel Lectures, Chemistry, 1942–1962.* Amsterdam: Elsevier, 1964, pp. 124–134.

Northrop, J. H. 1961. Biochemists, Biologists, and William of Occam. *Ann. Rev. Biochem.* **30**:1–10.

Northrop, J. H., M. Kunitz, and R. M. Herriott. 1948. *Crystalline Enzymes,* 2nd ed. New York: Columbia Univ. Press.

*Notes and Queries.* 1889. Vegetable Rennet. ser. 7, **8**:231–232.

Numbers, R. L. 1979. William Beaumont and the Ethics of Human Experimentation. *J. Hist. Biol.* **12**:113–135.

Ostwald, W. 1894. Review of a paper by F. Stohmann dealing with the caloric value of constituents of foodstuffs, and postulating a catalytic force. *Z. Physik. Chem.* **15**:705–706.

Pagel, W. 1956. Van Helmont's Ideas on Gastric Digestion and the Gastric Juice. *Hist. Med. Bull.* **30**:524–536.

Partington, J. R. 1962. *A History of Chemistry,* vol. 3. London: Macmillan & Co., pp. 60, 62.

Partington, J. R. 1964. *A History of Chemistry,* vol. 4. London: Macmillan & Co.

Partington, J. R. 1970. *A History of Chemistry,* vol 1, part 1. London: Macmillan & Co., pp. 98, 118.

Pasteur, L. 1858. Mémoire sur la fermentation de l'acide tartrique. *C. R.* **46**:615–618.

Pasteur, L. 1860. Mémoire sur la fermentation alcoolique. *Ann. chim. phys.* ser. 3, **58**:323–426.

Pasteur, L. 1878. Réponse à M. Berthelot. *C. R.* **87**:1053–1059.

Pasteur, L. 1879. *Examen Critique d'un Écrit Posthume de Claude Bernard sur la Fermentation.* Paris. Gauthier-Villars.

Pauling, L. 1946. Molecular Architecture and Biological Reactions. *Chem. Eng. News* **24**:1375–1377.
Peck, A. L., trans. 1945. Book III, xv, 576a. In Aristotle's *Parts of Animals.* Cambridge, Mass.: Harvard Univ. Press and London: William Heinemann Ltd., p. 301.
Peck, A. L., trans. 1965. Book III, xx–xxi, 522b. In Aristotle's *History of Animals,* vol. I. Cambridge, Mass.: Harvard Univ. Press and London: William Heinemann Ltd., p. 229.
Pekelharing, C. A. 1896/97. Ueber eine neue Bereitungsweise des Pepsins [On a new method to prepare pepsin]. *Hoppe-Seyler's Z. Physiol. Chem.* **22**:233–244.
Pekelharing, C. A. 1902. Mittheilungen über Pepsin [Reports on pepsin]. *Hoppe-Seyler's Z. Physiol. Chem.* **35**:8–30.
Peterson, E. A., and H. A. Sober. 1956. Chromatography of Proteins. I. Cellulose Ion-exchange Adsorbents. *Am. Chem. Soc. J.* **78**:751–755.
Phillips, D. C. 1966. The Three-dimensional Structure of an Enzyme Molecule. *Sci. Am.* **215**(5):78–90 (Nov.).
Pigman, W. W. 1944. Specificity, Classification, and Mechanism of Action of Glycosidases. *Adv. Enzymol.* **4**:41–74.
Plantefol, L. 1968. Le genre du mot enzyme. *C. R.* **266**:40–45.
Pliny. See Jones, 1951 and Rackham, 1945.
Plutarch. See Clement and Hoffleit, 1969.
Polanyi, M. 1967. Life Transcending Physics and Chemistry. *Chem. Eng. News* **45**:54–66 (Aug. 21).
Polanyi, M. 1968. Life's Irreducible Structure. *Science* **160**:1308–1312.
Popják, G. 1970. Stereospecificity of Enzymic Reactions. In *The Enzymes,* 3rd ed., vol. 2, P. D. Boyer, ed. New York: Academic, pp. 115–215.
Porath, J. 1960. Gel Filtration of Proteins, Peptides and Amino Acids. *Biochim. Biophys. Acta* **39**:193–207.
Porter, R. R. 1955. The Partition Chromatography of Enzymes. *Methods Enzymol.* **1**:98–112.
Prout, W. 1824. On the Nature of the Acid and Saline Matters Usually Existing in the Stomachs of Animals. *Philos. Trans.* **114**:45–49.
Purich, D. L., ed. 1979. Enzyme Kinetics and Mechanism. Part A. Initial Rate and Inhibitor Methods. *Methods Enzymol.* **63**:1–547.
Quastel, J. H. 1926. A Theory of the Mechanism of Oxidations and Reductions *in Vivo. Biochem. J.* **20**:166–194.
Quastel, J. H. 1980. Beginnings of Modern Neurochemistry. *Trends Biochem. Sci.* **5**:199–200.
Quastel, J. H., and W. R. Wooldridge. 1928. Some Properties of the Dehydrogenating Enzymes of Bacteria. *Biochem. J.* **22**:689–702.
Racker, E. 1976. *A New Look at Mechanisms in Bioenergetics.* New York: Academic, p. 175.
Rackham, H., trans. 1945. Book XVI, lxxii, 181–182. In Pliny's *Natural History,* vol. IV. Cambridge, Mass.: Harvard Univ. Press and London: William Heinemann Ltd., p. 505.
Richards, F. M., and P. J. Vithayathil. 1959. The Preparation of Subtilisin-modified Ribonuclease and the Separation of the Peptide and Protein Components. *J. Biol. Chem.* **234**:1459–1465.
Richards, F. M., and H. W. Wyckoff. 1971. Bovine Pancreatic Ribonuclease. In *The Enzymes,* 3rd ed., vol. 4, P. D. Boyer, ed. New York: Academic, pp. 647–806.

Ringer, W. E. 1915. Weitere Studien am Pekelharingschen Pepsin [Further studies with Pekelharing's pepsin].*Hoppe-Seyler's Z. Physiol. Chem.* **95**:195–258.

Robbins, B. H. 1930. A Proteolytic Enzyme in Ficin, the Anthelmintic Principle of *Leche de Higueron. J. Biol. Chem.* **87**:251–257.

Robinson, T. 1960. Michael Tswett. *Chymia* **6**:146–161.

Rose, W. C. See Young, 1803.

Roux, É. 1898. La fermentation alcoolique et l'évolution de la microbie. *Rev. Sci. (Rev. Rose)* ser. 4, **10**:833–840.

Samuelson, O. 1953. *Ion Exchangers in Analytical Chemistry*. New York: Wiley, pp. 238–240.

Sarton, G. 1952. *A History of Science: Ancient Science Through the Golden Age of Greece*. Cambridge, Mass.: Harvard Univ. Press, preface. (Reprinted in 1970, New York: W. W. Norton & Co.)

Schaffner, K. F. 1967. Antireductionism and Molecular Biology. *Science* **157**:644–647.

Schwimmer, S. 1954. Industrial Production and Utilization of Enzymes from Flowering Plants. *Econ. Bot.* **8**:99–113.

Segal, H. L. 1959. The Development of Enzyme Kinetics. In *The Enzymes*, 2nd ed., vol. 1, P. D. Boyer, H. L. Lardy, and K. Myrbäck, eds. New York: Academic, pp. 1–48.

Segal, H. L. 1973. Enzymatic Interconversion of Active and Inactive Forms of Enzymes. *Science* **180**:25–32.

Sigler, P. B., D. M. Blow, B. W. Mathews, and R. Henderson. 1968. Structure of Crystalline a-Chymotrypsin. II. A Preliminary Report Including a Hypothesis for the Activation Mechanism. *J. Mol. Biol.* **35**:143–164.

Smyth, D. G., W. H. Stein, and S. Moore. 1963. The Sequence of Amino Acid Residues in Bovine Pancreatic Ribonuclease: Revisions and Confirmations. *J. Biol. Chem.* **238**:227–234.

Sober, H. A., F. J. Gutter, M. M. Wyckoff, and E. A. Peterson. 1956. Chromatography of Proteins. II. Fractionation of Serum Protein on Anion-exchange Cellulose. *Am. Chem. Soc. J.* **78**:756–763.

Söderbaum, H. G., ed. 1916. *Jac. Berzelius Bref. V. Correspondance entre Berzelius et G. J. Mulder (1834–1847)*. Uppsala: Almqvist & Wiksells.

Sørensen, S. P. L. 1912. Über die Messung und Bedeutung der Wasserstoffionenkonzentration bei biologischen Prozessen [On the measurement and significance of the hydrogen ion concentration in biological processes] *Ergebn. Physiol.* **12**:393–532.

Stadtman, E. R. 1970. Mechanisms of Enzyme Regulation. In *The Enzymes*, 3rd. ed., vol. 1, P. D. Boyer, ed. New York: Academic, pp. 397–459.

Stent, G. S., and R. Calendar. 1978. *Molecular Genetics, An Introductory Narrative*. San Francisco: W. H. Freeman, p. 661.

Stern, K. G. 1945. Constitution of the Prosthetic Group of Catalase. *Nature* **136**:302.

Stephenson, M. 1949. *Bacterial Metabolism*. London: Longmans, Green and Co., p. 6. (Facsimile reprint in paperback in 1966, Cambridge, Mass.: The M. I. T. Press.)

Sumner, J. B. 1919. The Globulins of the Jackbean, *Canavalia Ensiformis. J. Biol. Chem.* **37**:137–142.

Sumner, J. B., and A. L. Dounce. 1937. Crystalline Catalase. *J. Biol. Chem.* **121**:417–424.

Sung, S. C. 1979. J. H. Quastel. *Trends Biochem. Sci.* **4**:N101–N102.

Synge, R. L. M. 1962. Tsvett, Willstätter, and the Use of Adsorption for Purification of Proteins. In *Perspectives in the Biochemistry of Large Molecules*, M. Heidelberger, C. H. Li, K. O. Pedersen, and J. Porath, eds. Arch. Biochem. Biophys. Suppl. 1, pp. 1–6.
Thénard, L. J. 1803. Mémoire sur la fermentation vineuse. *Ann. Chim. Phys.* **46**:294–320.
Thénard, L. J. 1820. Mémoire sur la combinaison de l'oxigène avec l'eau, et sur les propriétés extraordinaires que possède l'eau oxigénée. *Acad. R. Sci. Inst. France, Mém. Annee 1818*, ser. 2, **3**:385–488.
Theorell, H. 1934. Reindarstellung (Kristallisation) des gelben Atmungsfermentes und die reversible Spaltung desselben. *Biochem. Z.* **272**: 155–156.
Theorell, H. 1935. Das gelbe Oxydationsferment. *Biochem. Z.* **278**:263–290.
Tiselius, A., S. Hjertén, and O. Levin. 1956. Protein Chromatography on Calcium Phosphate Columns. *Arch Biochem. Biophys.* **65**:132–155.
Tompkins, E. R., J. X. Khym, and W. E. Cohn. 1947. Ion-exchange as a Separations Method. I. The Separation of Fission-produced Radioisotopes, Including Individual Rare Earths, by Complexing Elution from Amberlite Resin. *Am. Chem. Soc. J.* **69**:2769–2777.
Turner, R. S. 1972. Hermann von Helmholtz. In *Dictionary of Scientific Biography*, vol. 6, C. C. Gillispie, ed. New York: C. Scribner's Sons, p. 242.
Underwood, E. A. 1972. Franciscus Sylvius and His Iatrochemical School. *Endeavour* **31**:73–76.
Varro, M. T. See Hooper and Ash, 1934.
Velluz, L. 1964. *Vie de Berthelot*. Paris: Librairie Plon.
Vennesland, B. 1955. Some Applications of Deuterium to the Study of Enzyme Mechanisms. *Faraday Soc. Discuss.* **20**:240–248.
Vennesland, B. 1956. Steric Specificity of Hydrogen Transfer in Pyridine Nucleotide Dehydrogenase Reactions. *J. Cell. Comp. Physiol.* **47** (suppl. 1):201–216.
Vennesland, B. 1958. Stereospecificity of Hydrogen Transfer in Pyridine Nucleotide Dehydrogenase Reactions. *Fed. Proc.* **17**:1150–1157.
Vennesland, B., and F. H. Westheimer. 1954. Hydrogen Transport and Steric Specificity in Reactions Catalyzed by Pyridine Nucleotide Dehydrogenases. In *The Mechanism of Enzyme Action*, W. D. McElroy and B. Glass, eds. Baltimore, Md.: Johns Hopkins Press, pp. 357–379.
Vickery, H. B. 1971. Russell Henry Chittenden. In *Dictionary of Scientific Biography*, vol. 3, C. C. Gillispie, ed. New York: C. Scribner's Sons, pp. 256–258.
Voet, J. G., and R. H. Abeles. 1970. The Mechanism of Action of Sucrose Phosphorylase: Isolation and Properties of a $\beta$-linked Covalent Glucose-Enzyme Complex. *J. Biol. Chem.* **245**:1020–1031.
Volhard, J. 1909. *Justus von Liebig*, vol. II. Leipzig: J. A. Barth, p. 74.
Waddington, C. H. 1961. *The Nature of Life: The Main Problems and Trends of Thought in Modern Biology*. New York: Harper and Row, p. 19.
Walden, P. 1949. Aus der Entwicklungsgeschichte der Enzymologie von ihren Anfängen bis zum Anbruch des zwanzigsten Jahrhunderts [From the history of the development of enzymology from its beginnings until the start of the twentieth century]. In *Ergebnisse der Enzym-*

*forschung*, vol. 10, R. Weidenhagen, ed. Leipzig: Akademische Verlagsgesellschaft Geest & Portig K.-G., pp. 1–64.

Warburg, O. 1938. Chemische Konstitution von Fermenten. In *Ergebnisse der Enzymforschung*, vol. 7, F. F. Nord and R. Weidenhagen, eds. Leipzig: Akademische Verlagsgesellschaft m.b.H., pp. 210–245.

Warburg, O. 1946. *Schwermetalle als Wirkungsgruppen von Fermenten*. Berlin: Verlag Dr. Werner Saenger. (Also 2nd ed., 1949, Freiburg i. Br.: Editio Cantor, and as *Heavy Metal Prosthetic Groups and Enzyme Action*, A. Lawson, trans. Oxford: Clarendon Press.)

Warburg, O. 1949. *Wasserstoffübertragende Fermente* [Hydrogen-transferring ferments]. Freiburg i. Br.: Editio Cantor.

Watermann, R. 1960. *Theodor Schwann, Leben und Werk*. Düsseldorf: Verlag L. Schwann, pp. 88–89, 120–128.

Webb, B. H., and A. H. Johnson. 1965. *Fundamentals of Dairy Chemistry*. Westport, Conn.: The Avi Publishing Co., pp. 732–733.

Whitehead, A. N. 1925. *Science and the Modern World*. New York: The Free Press, 1953, p. 3.

Willstätter, R. 1927. *Problems and Methods in Enzyme Research*, George Fisher Baker Non-Resident Lectureship in Chemistry at Cornell University. Ithaca, N.Y.: Cornell Univ. Press.

Willstätter, R. 1933. Problems of Modern Enzyme Chemistry, Willard Gibbs Medal Address. *Chem. Rev.* **13**:501–512.

Willstätter, R. 1949. *Aus meinem Leben*, A. Stoll, ed. Weinheim: Verlag Chemie. (Also as *From My Life*, L. S. Hornig, trans. New York: W. A. Benjamin, 1965.)

Willstätter, R., and M. Rohdewald. 1940. Über enzymatische Systeme der Zucker-Umwandlung im Muskel [On enzyme systems of sugar metabolism in muscle]. *Enzymologia* **8**:1–63.

Wöhler, F., and J. Liebig. 1837. Ueber die Bildung des Bittermandelöls [On the formation of oil of bitter almonds]. *Ann. Pharm.* **22**:1–24.

Wöhler, F., and J. Liebig (published anonymously). 1839. Das enträthselte Geheimniss der geistigen Gährung [The deciphered secret of spirituous fermentation]. *Ann. Pharm.* **29**:100–104.

Wolfenden, R. 1972. Analog Approaches to the Structure of the Transition State in Enzyme Reactions. *Acc. Chem. Res.* **5**:10–18.

Wong, J. T., and C. S. Hanes. 1962. Kinetic Formulations for Enzymic Reactions Involving Two Substrates. *Can. J. Biochem. Physiol.* **40**:763–804.

Wurtz, A. 1880. Sur la papaïne. Nouvelle contribution à l'histoire des ferments solubles. *C. R.* **91**:787–791.

Wurtz, A., and E. Bouchut. 1879. Sur le ferment digestif du *Carica papaya*. *C. R.* **89**:425–429.

Yoshida, H. 1883. Chemistry of Lacquer (Urushi), Part I. *Chem. Soc. J., Trans.* **43**:472–486.

Young, J. R. 1803. *An Experimental Inquiry into the Principles of Nutrition and the Digestive Process*. Philadelphia, Pa.: Eaken & Mecum. (Facsimile reprint with an introductory essay by W. C. Rose published in 1959 by the University of Illinois Press.)

Zechmeister, L. 1950. *Progress in Chromatography, 1938–1947*. London: Chapman & Hall Ltd., pp. 254–260.

Zechmeister, L., and L. Cholnoky. 1938. *Die chromatographische Adsorptionsmethode*. Vienna: J. Springer, pp. 264–265. (Also as *Principles and Practice of Chromatography,* A. L. Bacharach and F. A. Robinson, transl., pp. 283–284.)

Zittle, C. A. 1953. Adsorption Studies of Enzymes and Other Proteins. *Adv. Enzymol.* **14**:319–374.

[*Editor's Note:* The following references were inadvertantly omitted from the preceding list.]

Newberg, C., and I. Mandl. 1950. Invertase. In *The Enzymes,* vol. 1, J. B. Sumner and K. Myrback, eds. New York: Academic Press, pp. 527–550.

Westheimer, F. H., H. F. Fisher, E. E. Conn, and B. Vennesland. 1951. The Enzymatic Transfer of Hydrogen from Alcohol to DPN. *Am. Chem. Soc. J.* **73**:2403.

# BIBLIOGRAPHY

Coley, N. G. 1973. *From Animal Chemistry to Biochemistry.* Amersham, Bucks: Hulton Educational Publications.

Dixon, M. 1971. The History of Enzymes and of Biological Oxidations. In *The Chemistry of Life, Eight Lectures on the History of Biochemistry,* J. Needham, ed. Cambridge: Cambridge Univ. Press, pp. 15–37.

Florkin, M. 1972–. A History of Biochemistry, Section VI (vols. 30–34) of *Comprehensive Biochemistry,* M. Florkin and E. H. Stotz, eds. Amsterdam: Elsevier. (By early 1980 only volume 34 had not yet been published.)

Fruton, J. S. 1972. *Molecules and Life: Historical Essays on the Interplay of Chemistry and Biology.* New York: Wiley-Interscience.

Gillispie, C. C., ed. 1970–1978. *Dictionary of Scientific Biography,* 15 vols. New York: C. Scribner's Sons.

Harden, A. 1911. Historical Introduction. In *Alcoholic Fermentation.* London: Longmans, Green and Co., pp. 1–17.

Hoffmann-Ostenhof, O. 1954. Historischer Überblick über die Entwicklung der Enzymologie [Historical survey on the development of enzymology]. In *Enzymologie.* Vienna: Springer-Verlag, pp. 4–13.

Kohler, R. 1971. The Background to Eduard Buchner's Discovery of Cell-free Fermentation. *J. Hist. Biol.* **4**:35–61.

Kohler, R. E. 1972. The Reception of Eduard Buchner's Discovery of Cell-free Fermentation. *J. Hist. Biol.* **5**:327–353.

Kohler, R. E. 1975. The History of Biochemistry: A Survey. *J. Hist. Biol.* **8**: 275–318.

Kohler, R. E., Jr. 1973. The Enzyme Theory and the Origin of Biochemistry. *Isis* **64**:181–196.

Leicester, H. M. 1974. *Development of Biochemical Concepts from Ancient to Modern Times.* Cambridge, Mass.: Harvard Univ. Press.

Lieben, F. 1935. *Geschichte der Physiologischen Chemie* [History of physiological chemistry]. Leipzig: F. Deuticke.

Needham, J., ed. 1971. *The Chemistry of Life, Eight Lectures on the History of Biochemistry.* Cambridge: Cambridge Univ. Press.

Walden, P. 1949. Aus der Entwicklungsgeschichte der Enzymologie von ihren Anfängen bis zum Anbruch des zwanzigsten Jahrhunderts [From the history of the development of enzymology from its beginnings until the start of the twentieth century]. In *Ergebnisse der Enzymforschung,* vol. 10, R. Weidenhagen, ed. Leipzig: Akademische Verlagsgesellschaft Geest & Portig K.-G., pp. 1–64.

*Bibliography*

To locate references the following were indispensable: for journals between 1800 and 1900 the *Catalogue of Scientific Papers Compiled by the Royal Society of London* (four series: 1800–1863, 1863–1873, 1874–1883, 1884–1900, and one supplementary volume for 1800–1883; Cambridge Univ. Press), and for journals and books from c. 1500 to the middle of the nineteenth century the *Chemisch-Pharmaceutisches Bio- und Bibliographikon,* edited by Fritz Ferchl, published in 1938 by Arthur Nemayer, Mittenwald (Bavaria) and reprinted in 1971 by Martin Sändig oHG, 6229 Niederwalluf bei Wiesbaden, West Germany. Biographical data could be retrieved with the help of Joseph S. Fruton's impressive *Selective Bibliography of Biographical Data for the History of Biochemistry Since 1800,* second edition (Am. Philos. Soc. Library Publ. No. 7, Phila.).

For a discussion of the historical approach to biochemistry see "Dialogue: A Discussion among Historians of Science and Scientists," F. L. Holmes, chairman, in *The Origins of Modern Biochemistry: A Retrospect on Proteins,* P. R. Srinivasan, J. S. Fruton, and J. T. Edsall, eds. N.Y. Acad. Sci. Annals vol. 325, 1979, pp. 149–169.

# AUTHOR CITATION INDEX

# SUBJECT INDEX

# About the Editor

HERBERT C. FRIEDMANN is an associate professor of biochemistry at the University of Chicago. He is a member of the American Society of Biological Chemists, of the American Chemical Society, and of the American Society for Microbiology. His research deals with biosynthetic processes in bacteria and, on occasion, in plants. He is particularly interested in aspects of amino acid, nucleotide, porphyrin, and vitamin $B_{12}$ metabolism. Investigations in his laboratory have led to the discovery and study of a variety of new enzymes. In 1978 he received an award from the University of Chicago for excellence in undergraduate teaching.